Introduction to Linear Algebra

Introduction to Linear Algebra

DANIEL NORMAN
Queen's University

PUBLISHER: Ron Doleman
MANAGING EDITOR: Linda Scott
EDITOR: Santo D'Agostino
COPY EDITOR: Debbie Davies
COVER AND TEXT DESIGN: Artplus Limited
PAGE LAYOUT: Nelson Gonzalez

Canadian Cataloguing in Publication Data

Norman, Daniel, date
Introduction to linear algebra

Includes index.
ISBN 0–201–60210–5

1. Algebras, Linear. I. Title.

QA184.N67 1994 512′.5 C94–932007–2

ISBN 0–201–60210–5

Printed in Canada

12 WC 08

TABLE OF CONTENTS

A NOTE TO INSTRUCTORS

This book was written after several years' experience teaching a compulsory course in linear algebra to first-year engineering students at Queen's during their second term (or semester). The ideal audience would therefore consist of students who have had some exposure to calculus, to vectors in two and three dimensions, and to physics. However, the book could be used with students with a less extensive background. Earlier drafts of the book have been used by several instructors for several years with engineering students, and on two occasions, instructors have used a draft version with a general audience of Arts and Science students.

In writing the book, I was particularly conscious of the need for students to relate linear algebra to their earlier mathematics, to develop geometrical understanding, and to see clear indications of applications.

Mathematical Content

My views on content are formed partly from teaching advanced calculus and differential equations to these same students in later years, and partly from informal discussions with some engineering faculty. This book is quite traditional in aiming at eigenvectors and diagonalization; I believe that in order to explain these topics coherently, some discussion of linear transformations and change of basis is required. Some of the mathematical features of the presentation are as follows.

- A major concern is to introduce the idea of linear transformation *early*, in a geometrical context, and to use it to explain aspects of matrix multiplication, matrix inversion, and features of solutions of systems of linear equations. This occurs naturally in $\mathbb{R}^2$, $\mathbb{R}^3$, and $\mathbb{R}^n$, and does not require the introduction of abstract vector spaces. Geometrical transformations provide intuitively satisfying illustrations of the important concepts.
- In order to be able to use linear transformations in Chapter 3, we first need to remind students about vectors in two and three dimensional spaces, and to show that most of the relevant ideas extend easily to n dimensions. By using the dot product to define projections onto and perpendicular to a one-dimensional subspace, we give early examples of linear mappings. The introduction to these ideas should encourage students to relate the vectors of linear algebra with the vectors they encountered in earlier mathematics courses or in physics courses.
- Chapter 2 provides a standard treatment of the solution of systems of linear equations by elimination, and the corresponding discussion of the row reduction of augmented or coefficient matrices.
- Because of the students' interests and because of time constraints, I have

worked almost exclusively in $\mathbb{R}^n$ and its subspaces. General vector spaces are defined, and illustrated briefly by spaces such as spaces of polynomials or of functions, but I believe that a good grasp of the theory in $\mathbb{R}^n$ is a sufficient goal in a first course. For similar reasons, "one-to-one", "onto", and "isomorphism" occur only in one section (which I omit with my engineering class); special cases of these ideas are used (without the names) in the discussion of invertibility (Section 3–5).

- Topics have been ordered to give students a chance to work with an idea before it is used in a much more involved or specialized setting. For example, spanning sets appear in Section 3–4, but linear independence and bases are delayed until Section 4–2. Coordinates with respect to an arbitrary basis appear in Section 4–3, change of basis in Section 4–6, and similarity and diagonalization of matrices in Section 6–1. The diagonalization of symmetric matrices (Section 8–1) is separated from the more general case (Chapter 6) by the discussion of orthonormal bases and orthogonal matrices.
- Definitions, theorems, and proofs are important and I have tried to state them carefully; I believe that most proofs are important in showing the connections between ideas, so that students should see them. A few proofs, however, do not do much for developing a feeling for the subject, and these are postponed to the end of the relevant section (or in a very few cases, omitted).
- Determinants are postponed as long as possible (Chapter 5, just before the discussion of eigenvalues). I believe that it is positively unhelpful to students to have determinants available too soon. Some students are far too eager to latch onto mindless algorithms involving determinants (for example, for checking linear independence of three vectors in three-dimensional space), rather than actually coming to terms with the defining ideas.
- Because of suggestions from reviewers, I have added an appendix on complex numbers, and included sections on systems, vector spaces, and diagonalization with complex numbers. These sections (2–4, 4–8, 6–6, 7–5, and 8–6) could be presented in the order of the text, or collected and dealt with as a separate "chapter" at the end of the course (thus reviewing many of the essential concepts). Alternatively, an instructor could easily integrate this material into the appropriate chapters.

Physical Applications

One of the difficulties in any linear algebra course is that the applications of linear algebra are not so immediate or so intuitively appealing as those of elementary calculus. Most convincing applications of linear algebra require a fairly lengthy buildup of background that would be inappropriate in a linear algebra text. However, without some of these applications, many students would find it difficult to remain motivated to learn linear algebra.

I have tried to respond to this concern in the following ways.

- There is an introductory Note to Students suggesting some of the application areas.
- There is a "half-time pep-talk" consisting of an Essay on Linearity and Superposition in Physics, just before the introduction of the abstract material students find most difficult in Chapter 4.
- Applications are discussed in Sections 2–3 (electrical circuits, trusses, linear programming), 2–4 (more circuits), 4–5 (note on controllability), 5–2 and 5–4 (notes on the Jacobian determinant), 6–4 (diagonalization of a system of linear differential equations arising from a mixing problem), 6–5 (Markov processes, the power method for determining eigenvalues), 7–3 (method of least squares), 8–3 (conics and quadric surfaces), 8–4 (the Hessian matrix and extreme values), and 8–5 (strain, the inertia tensor).

A Course Outline

The following table indicates the sections in each chapter that I consider to be "central material". Material in later chapters depends on some familiarity with the material of these central sections. In using this book, I would expect students to read Section 2–3 and the Essay for themselves. In our one-term course (35 lectures), we treat the first two chapters quite briskly, since most of our students have had some experience with some of this material; we omit all of the "optional" sections except for 6–4 (or 6–5) and 8–3, as well as the sections using complex numbers; I would also de-emphasize some of the more technical proofs (for example, in the sections on dimension and rank, or symmetric matrices).

Chapters	Central Material	Optional	Complex Numbers
1	1,2,3,4		
2	1,2	3	4
3	1,2,3,4,5	6	
4	1,2,3,4,5,6	7	8
Essay		Essay	
5	1,2,3,4		
6	1,2,3	4;5	6
7	1,2	3;4	5
8	1,2	3;4;5	6
Appendix			Complex Numbers

Computers

As explained in the Note on the Exercises, there are some exercises in the book that require access to a microcomputer with suitable software. Students should realize that the theory does not apply only to matrices of small size with integer entries. However, there are many ideas to be learned in linear algebra so I have not discussed numerical methods. Some numerical issues (accuracy and efficiency) are discussed in notes and exercises.

Notation

My chief deviation from standard notation is the use of brackets to denote a column vector. Thus [**x**] denotes the n by 1 matrix whose entries are the coordinates of the vector **x**. I know that some instructors find this unnecessary and distracting, but my experience has been that students are initially confused when we use **x** to denote both $(x_1, x_2, \ldots, x_n)$ and the corresponding column, so I prefer to have a notation that clearly denotes the column. Moreover, it is sometimes useful to have a notation that indicates explicitly that in some equation or calculation, vectors must be treated as columns. Later in the book, as students gain experience, I treat these brackets as optional; this seems to me a natural evolution: as we become more comfortable with a subject, we become more willing to use more compact notations.

I have also introduced the notation $a_{j\rightarrow}$ to denote the jth row of matrix A, and the notation $a_{\downarrow k}$ to denote the kth column of A.

Acknowledgements

I can no longer remember what I learned from whom, but I know I have learned things about linear algebra and how to teach it from many colleagues, including Leo Jonker, Ole Nielsen, Tony Geramita, Bruce Kirby, Bill Woodside, Jim Verner, Ron Hirschorn, Cedric Schubert, Eddy Campbell, and Norman Rice; Norman has been particularly patient and helpful with my questions and grumbling. After teaching from earlier drafts of the book, David Pollack, Bob Erdahl, Leslie Roberts, Balwant Singh, Ian Hughes, Norm Pullman, Ernst Kani, and Naomi Shaked-Monderer suggested improvements and additional exercises. In particular, the exercises on magic squares and Fibonacci sequences were suggested to me by David Pollack, and Morris Orzech and Bob Erdahl suggested two better ways of showing that a "right inverse" is a "left inverse". Conversations with Marc Maes, Malcolm Griffin, and Bill Ross were helpful in presenting some applications, as were notes from a seminar by Jim Allen (Physics) and conversations with Bill Kamphuis (Civil Engineering) and Jim McLellan (Chemical Engineering). It is a pleasure to thank them all.

I am also grateful to colleagues at other universities who reviewed the book to assist me in revising the book. Detailed comments were provided by the

following people:
H. R. Atkinson (Windsor),
N. A. Beirnes (Regina),
G. Claessens (Industriele Hogeschool Leuven),
D. Dubrovsky (Dawson College),
J. Gelfgren (Umea, Sweden),
T. G. Kucera (Manitoba),
J. Labute (McGill),
J. Macki (Alberta),
R. Riefer (Memorial),
R. Stanczak (Toronto),
P. Stewart (Dalhousie),
B. Textorius (Linkoping, Sweden)

Earlier reviews were provided by:
K. L. Chowdhury (Calgary),
G. Ord (Western Ontario),
C. Roth (McGill)
J. B. Sabat (New Brunswick)

I also benefited from comments from some reviewers whose names are not known to me. Many of these comments resulted in improvements; all of them caused me to rethink the presentation. Santo D'Agostino (Ryerson) was commissioned to act as a developmental editor, and his detailed questions, criticisms, and suggestions were extremely helpful.

All of these colleagues, at Queen's and elsewhere, have helped to improve the book, but any errors or faults of the book remain the responsibility of the author.

I am very grateful to my publisher, Ron Doleman of Addison-Wesley (Canada). He has been encouraging and helpful throughout. I also thank Linda Scott, Debbie Davies, and Abdulqafar Abdullahi of Addison-Wesley for their help, and Nelson Gonzalez for efficiently converting manuscript to readable text.

Finally, I express my thanks to my wife Marilyn, for her patience with my preoccupation with this book over several years — and for much else.

A NOTE ON THE EXERCISES

In each section, the exercises are divided into A, B, C, and D Exercises.

A Exercises are intended to provide a sufficient variety and number of standard computational problems for students to master the techniques of the course; answers are provided at the back of the text. The author's view is that an *average* student may find that there are more exercises than she or he requires, but students should at least read all of these to make sure they understand what they are supposed to be able to do. Complete solutions are provided in a Student Solutions Manual.

B Exercises are essentially duplicates of the A Exercises with no answers provided, for instructors who want such exercises for homework. In a few cases, the B Exercises are not exactly parallel to the A Exercises.

C Exercises require the use of a microcomputer and suitable software. I do not know whether all of these exercises can easily be done in all packages students may use with linear algebra. The number of computer-related exercises is limited, but students should do enough to feel comfortable using a computer to do matrix calculations. Another important feature of these exercises is that it reminds students that this theory is a theory using real numbers and not only integers or simple fractions.

D Exercises generally require students to work with general cases, and to write simple arguments, or to invent examples. These are important aspects of mastering mathematical ideas, and all students should attempt at least some of these — and not get discouraged if they make slow progress with an unfamiliar kind of activity. However, many of the D Exercises are not particularly difficult. In a short course, instructors will have to be rather selective in assigning D Exercises. Answers are not provided, but for D Exercises marked with an asterisk, solutions may be found in the Student Solutions Manual. Answers to Quizzes are included at the back of the text.

In addition to the exercises at the end of each section, there is a sample Chapter Quiz in the Chapter Review at the end of each chapter. Students should be aware that their instructors may have a different idea of what constitutes an appropriate test on this material.

At the end of each chapter, there are some Further Exercises; some of these are similar to the D Exercises, and some provide an extended investigation of some idea or application of linear algebra.

The answers at the back of this book were provided by the author with the help of Peter Chamberlain (an undergraduate at Queen's University) and Jennifer MacKenzie. Peter provided solutions to the A Exercises in

Chapters 2–8 and all the Chapter Quizzes, and Jennifer provided solutions for Chapter 1. This solutions were checked and revised by the author, and compared with solutions prepared earlier where these existed. Most of the answers have been checked by computer. The answers at the back are excerpted from these solutions.

The author is grateful to Peter and Jennifer for their assistance.

A NOTE TO STUDENTS

LINEAR ALGEBRA — WHO NEEDS IT?

Engineers

Suppose you become a control engineer and have to design or upgrade an automatic control system. The system may be controlling a manufacturing process, or perhaps an airplane landing system. You will probably start with a linear model of the system, requiring linear algebra for its solution. To include feedback control, your system must take account of many measurements (for the example of the airplane, position, velocity, pitch, etc.) and it will have to assess this information very rapidly in order to determine the correct control responses. A standard part of such control systems is a Kalman-Bucy filter, which is not so much a piece of hardware as a piece of mathematical machinery for doing the required calculations. Linear algebra is an essential component of the Kalman-Bucy filter.

If you become a structural engineer or a mechanical engineer, you may be concerned with the problem of vibrations in structures or machinery. To understand the problem, you will have to know about eigenvalues and eigenvectors and how they determine the normal modes of oscillation. Eigenvalues and eigenvectors are some of the central topics in linear algebra.

An electrical engineer will need linear algebra to analyze circuits and systems; a civil engineer will need linear algebra to determine the internal forces in static structures, and to understand principal axes of strain.

In addition to these fairly specific uses, engineers will also find that they need to understand linear algebra to understand systems of differential equations, and some aspects of the calculus of functions of two or more variables. Moreover, the ideas and techniques of linear algebra are central to numerical techniques for solving problems of heat and fluid flow, which are a major concern in mechanical engineering. And the ideas of linear algebra underly advanced techniques such as Laplace transforms and Fourier analysis.

Physicists

Linear algebra is important in physics partly for the reasons described above. In addition, it is essential in applications such as the inertia tensor in general rotating motion. Linear algebra is an absolutely essential tool in quantum physics (where, for example, energy levels may be determined as eigenvalues of linear operators) and relativity (where understanding change of coordinates is one of the central issues).

Life and social scientists

Input-output models, described by matrices, are often used in economics, and similar ideas can be used in modelling populations where one needs to keep track of sub-populations (generations, for example, or genotypes). In all sciences, statistical analysis of data is of great importance, and much of this analysis uses linear algebra; for example, the method of least squares (for regression) can be understood in terms of projections in linear algebra.

Managers

A manager or decision maker in industry will have to make decisions about the best allocation of resources: enormous amounts of computer time around the world are devoted to linear programming algorithms that solve such allocation problems. The same sorts of techniques play a role in some areas of mine management. Linear algebra is essential here as well.

Mathematicians

Linear algebra with its applications is a subject of continuing research. Moreover, linear algebra is vital to all mathematicians, because it provides essential ideas and tools in areas as diverse as abstract algebra, differential equations, advanced calculus of functions of several variables, differential geometry, functional analysis, and numerical analysis.

So who needs linear algebra? Almost every kind of engineer or scientist will find linear algebra an important tool.

Will these applications be explained in this book?

Unfortunately, most of these applications require too much specialized background to include in a first year linear algebra book. Since I wrote the book with an audience of engineering students in mind, I have tried to give introductions to some engineering and science applications in Sections 2–3, 2–4, 4–5, 6–4, 6–5, and 8–5, but these have to be just introductions. In addition, an Essay on Linearity and Superposition in Physics is intended to motivate you to take seriously the topics in Chapter 4.

Students not in engineering or physical science do not need to understand the physical applications in order to follow the rest of the book. However, most of the applications do not require much prior knowledge of physics, and I encourage you to try to read them (at least superficially).

Is linear algebra hard?

Most people who have learned linear algebra and calculus believe that the ideas of elementary calculus (such as limit and integral) are more difficult than those of introductory linear algebra, and that most calculus courses include harder problems than most linear algebra courses. So, at least by this comparison, linear algebra is not hard. Still, some students find learning linear algebra difficult. I think two factors contribute to the difficulty students have.

First, students do not see what linear algebra is good for. That is why it is important to read the applications in the text; even if you do not understand them completely, they will give you some sense of where linear algebra fits into a broader picture.

Second, some students mistakenly see mathematics as a collection of recipes for solving standard problems and are uncomfortable with the fact that linear algebra is "abstract" and includes a lot of "theory". Students should realize that computers carry out all of these recipes faster and more accurately than humans can — there will be no long-term payoff for merely learning the procedures. However, practicing the procedures on specific examples is often an important step towards a much more important goal: understanding the *concepts* used in linear algebra to formulate and solve problems, and learning to interpret the results of calculations. Such understanding requires us to come to terms with some theory.

The reason mathematics is widely useful is that it is *abstract*: the same good idea can unlock the problems of control engineers, civil engineers, physicists, managers, and social scientists only because the idea has been abstracted from a particular setting. One technique solves many problems only because someone has established a *theory* of how to deal with these kinds of problems. Definitions are the way we try to capture the important ideas that are discovered to be widely useful, and theorems are ways we summarize the useful general facts about the kind of problems we are studying. Proofs are important because they help convince us that a statement is true, but they are also important because they give us practice in using the important ideas; they also make it easier to learn the subject because they show us how ideas hang together so we do not have to memorize too many disconnected facts.

Many of the concepts introduced in linear algebra are natural and easy but some seem unnatural and "technical" to beginners. Don't avoid these apparently more difficult ideas: use examples and theorems to see how these ideas are an essential part of the story of linear algebra. By learning the "vocabulary" and "grammar" of linear algebra, you will be equipping yourself with concepts and techniques that engineers, scientists, and mathematicians find invaluable for tackling an extraordinarily rich variety of problems.

Earlier versions of this book were used as notes or in a preliminary edition from 1991 to 1994. I am grateful to several students, particularly Bruce Storms, who pointed out errors and suggested improvements.

CHAPTER

Some Geometry in Euclidean Spaces

Some of the material of this chapter will be familiar to many students, but some ideas are introduced that will be new to most. Although this material is not strictly LINEAR ALGEBRA, it is important for its own sake, and it provides the context and the vocabulary for many applications and examples in linear algebra.

SECTION

1–1 Vectors in $\mathbb{R}^2$, $\mathbb{R}^3$, $\mathbb{R}^n$; The Vector Equation of a Line

We begin by considering the two-dimensional plane in Cartesian coordinates. Choose an origin O, and two mutually perpendicular axes, called the x_1-axis and the x_2-axis, as shown in Figure 1–1–1. Then the magnitude of the coordinate a_1 of a point A is the distance from A to the x_2-axis, with a_1 positive if A is to the right of this axis and negative if A is to the left. Similarly, the magnitude of the coordinate a_2 is the distance from A to the x_1-axis, with a_2 positive for A above the x_1-axis and negative for A below. We will sometimes speak of the point A(a_1, a_2), or simply (a_1, a_2). You have already learned how to plot graphs of equations in this plane.

For applications in many areas of mathematics, and in many subjects such as physics and economics, it is useful to view "ordered pairs" such as (a_1, a_2) more abstractly, and provide rules for adding such pairs and for multiplying them by constants.

DEFINITION

$\mathbb{R}^2$ is the set of all ordered pairs of real numbers (x_1, x_2) with **addition** defined by $(x_1, x_2)+(y_1, y_2) = (x_1+y_1, x_2+y_2)$, and with **multiplication by real numbers** defined by $k(x_1, x_2) = (kx_1, kx_2)$.

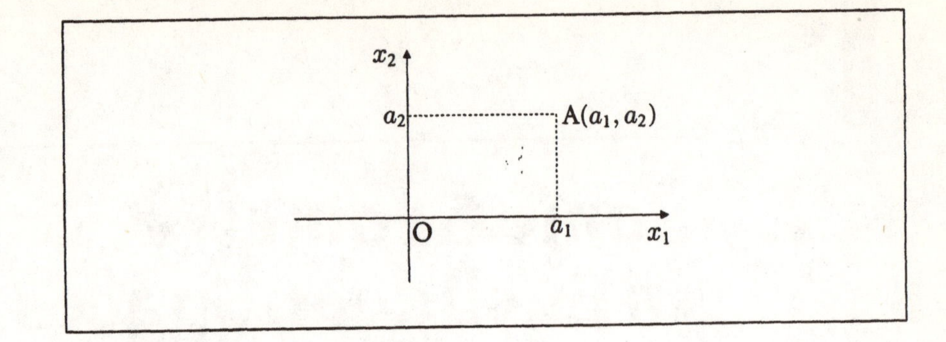

FIGURE 1-1-1. *Coordinates in the plane.*

(x_1, x_2) will be called a **vector** in $\mathbb{R}^2$, and we shall use the notation $\mathbf{x} = (x_1, x_2)$ to denote such vectors. When (a_1, a_2) is considered as a point in the plane, a_1 and a_2 are called **coordinates;** when (a_1, a_2) is thought of as a vector in $\mathbb{R}^2$, a_1 and a_2 are called the **components** of **a**. Real numbers are sometimes referred to as **scalars** to distinguish them from vectors. We can use familiar pictures in the plane to illustrate vectors in $\mathbb{R}^2$.

(Some students are uncomfortable with the fact that we may consider (x_1, x_2) as either a "vector" or a "point", but it is sometimes helpful to have two ways of looking at the same object.)

It is quite common to draw an "arrow" from $(0, 0)$ to (x_1, x_2) when we wish to emphasize that **x** is a vector. Note, however, that the points between $(0, 0)$ and (x_1, x_2) should *not* be thought of as points "on the vector". The representation of a vector as an arrow is particularly common in physics, where "force" and "acceleration" are vector quantities, which can conveniently be represented by an arrow of suitable magnitude and direction. (The magnitude, or length, of a vector is discussed in Section 1–2.)

The addition of two vectors is illustrated in Figure 1–1–2: construct a parallelogram with vectors **a** and **b** as adjacent sides; then **a** + **b** is the vector corresponding to the vertex of the parallelogram opposite to the origin. Observe that the components really are added according to the definition. People sometimes speak of the "parallelogram rule for addition" (or sometimes, they leave out half of the picture and speak of a "triangle rule"). Note that the construction can also be described in terms of points: let A be the point corresponding to the vector **a**, and let B be the point corresponding to the vector **b**; let C be the point such that the figure OACB is a parallelogram, and let **c** be the vector corresponding to C. Then $\mathbf{c} = \mathbf{a} + \mathbf{b}$.

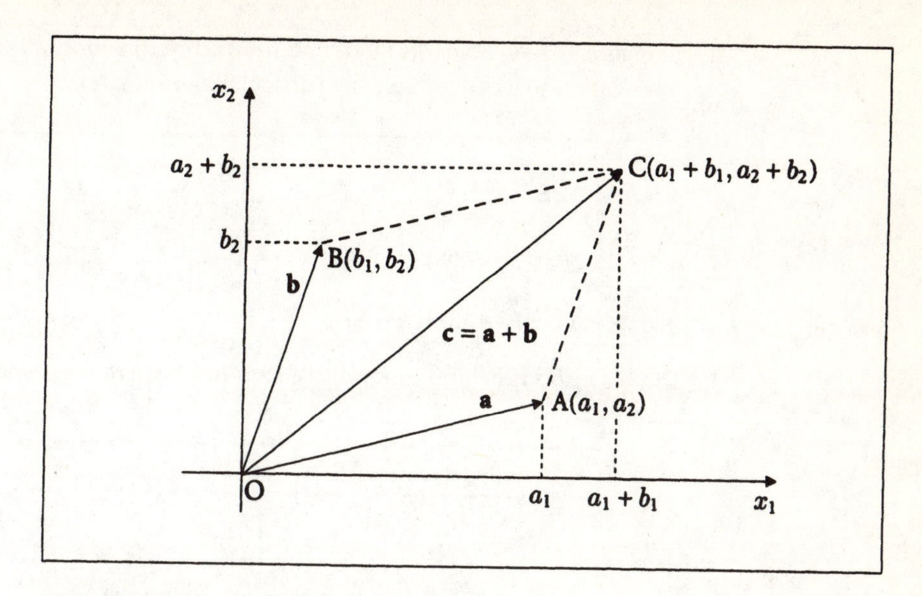

FIGURE 1-1-2. *Addition of the vectors* **a** *and* **b**.

EXAMPLE 1

$(2, 4) + (5, 1) = (7, 5)$; $(1, 2) + (3, -7) = (4, -5)$.

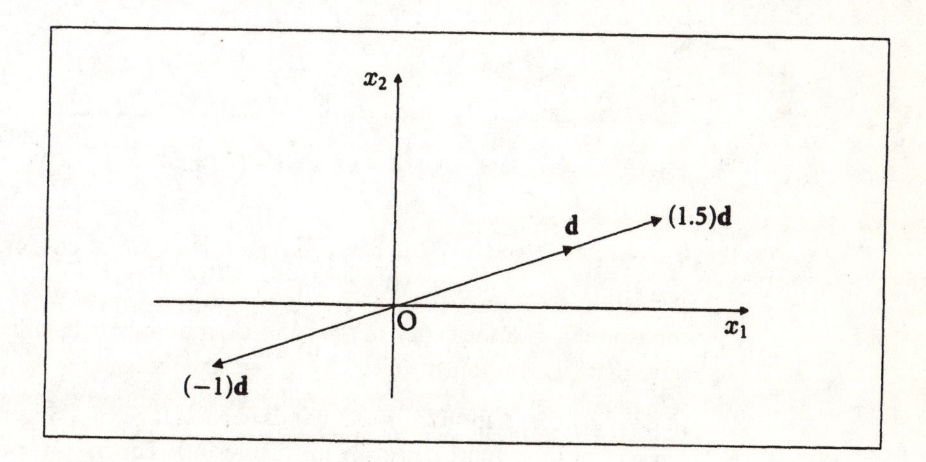

FIGURE 1-1-3. *Scalar multiples of the vector* **d**.

Similarly, the multiplication of a vector by a real number is illustrated in Figure 1–1–3. Note that multiplication by a negative number reverses the direction of the vector. It is important to note that $\mathbf{a} - \mathbf{b}$ is given by $\mathbf{a} + (-1)\mathbf{b}$, as illustrated in Figure 1–1–4. Students often find subtraction the most difficult operation to visualize, so it is worth paying particular attention to it.

EXAMPLE 2

$3(2, -3.02) = (6, -9.06);$ $\qquad -2(3, 4) = (-6, -8);$

$(4, -3) - (2, -5) = (2, 2).$

You may find it helpful to make sketches to illustrate these and other simple examples.

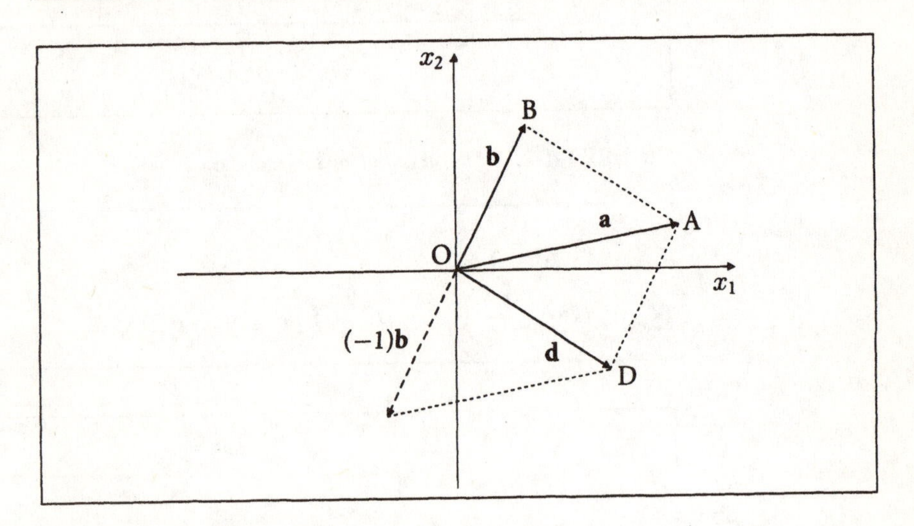

FIGURE 1-1-4. *Subtraction:* $\mathbf{d} = \mathbf{a} - \mathbf{b} = \mathbf{a} + (-1)\mathbf{b}$.

The vectors $(1, 0)$ and $(0, 1)$ play a special role in our discussion of $\mathbb{R}^2$, so we give them special names. We write

$$\mathbf{i} = (1, 0) \text{ and } \mathbf{j} = (0, 1)$$

and we say that **i** and **j** form the **standard basis** for $\mathbb{R}^2$. (We shall discuss the concept of basis more generally in Chapter 4.) The basis vectors **i** and **j** are important because any vector $\mathbf{a} = (a_1, a_2)$ in $\mathbb{R}^2$ can be written as a sum of scalar multiples of **i** and **j**:

$$\mathbf{a} = (a_1, a_2) = a_1(1, 0) + a_2(0, 1) = a_1\mathbf{i} + a_2\mathbf{j}.$$

It is usual to use the phrase **linear combination** to mean "sum of scalar multiples", so any vector in $\mathbb{R}^2$ can be written as a linear combination of the standard basis vectors.

One other vector in $\mathbb{R}^2$ deserves special mention: the **zero vector,**

$$\mathbf{0} = (0, 0).$$

Some important properties of the zero vector are that for any $\mathbf{a}$ in $\mathbb{R}^2$: (1) $\mathbf{0} + \mathbf{a} = \mathbf{a}$; (2) $\mathbf{a} + (-\mathbf{a}) = \mathbf{0}$; (3) $0\mathbf{a} = \mathbf{0}$.

The Vector Equation of a Line in $\mathbb{R}^2$

In Figure 1–1–3, it is apparent that the set of all multiples of the vector $\mathbf{d}$ makes up a line through the origin. We make this our definition of a line in $\mathbb{R}^2$: **a line through the origin in $\mathbb{R}^2$ is a set of the form**

$$\{\mathbf{x} \mid \mathbf{x} = t\mathbf{d}, \text{ where } \mathbf{d} \text{ is a fixed vector in } \mathbb{R}^2,\ \ t \in \mathbb{R}\}.$$

If you are unfamiliar with the set notation, the braces {} denote "the set", and the bar | is read "such that", so that the whole expression reads "the set of all $\mathbf{x}$ such that $\mathbf{x}$ is a scalar multiple of a fixed vector $\mathbf{d}$ in $\mathbb{R}^2$, where the scalar multiplier t is an arbitrary real number". If it is clear from the context that $\mathbf{d}$ is a fixed vector in $\mathbb{R}^2$, the set might be written $\{t\mathbf{d} \mid t \in \mathbb{R}\}$. Often we do not use the formal set notation, but simply write the **vector equation of the line:** $\mathbf{x} = t\mathbf{d}$, $t \in \mathbb{R}$. The vector $\mathbf{d}$ is called the **direction vector** of the line.

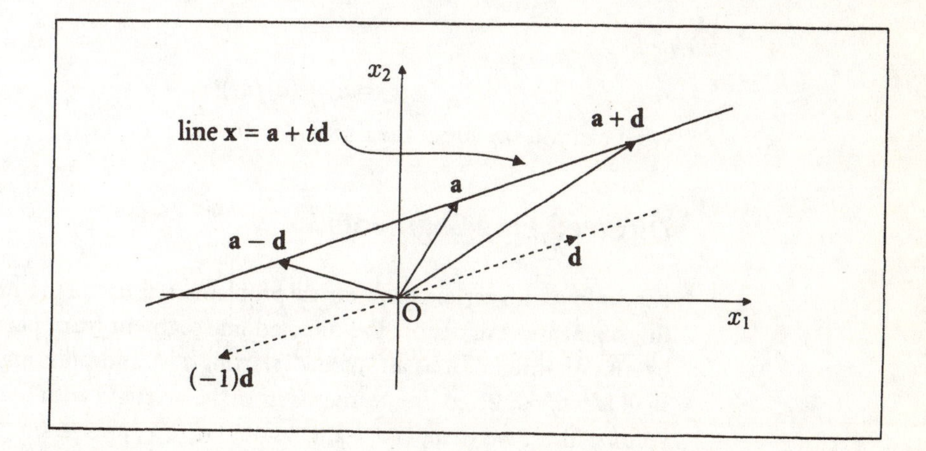

FIGURE 1-1-5. *The line with equation* $\mathbf{x} = \mathbf{a} + t\mathbf{d}$ *is parallel to the line with equation* $\mathbf{x} = t\mathbf{d}$.

We define the **line through point a with direction vector d** to be the set

$\{\mathbf{x} \mid \mathbf{x} = \mathbf{a} + t\mathbf{d},\ \mathbf{a}, \mathbf{d}$ fixed vectors in $\mathbb{R}^2$, $t \in \mathbb{R}\}$. This line is parallel to the line with equation $\mathbf{x} = t\mathbf{d}$, $t \in \mathbb{R}$, because of the parallelogram rule for addition. As shown in Figure 1–1–5, each point on the line through **a** can be obtained from a corresponding point on the line $\mathbf{x} = t\mathbf{d}$ by adding the vector **a**. We say that the line through **a** is obtained from the line through the origin by "translation by **a**". More generally, two lines are parallel if the direction vector of one line is a non-zero multiple of the direction vector of the other line.

EXAMPLE 3

A vector equation of the line through the point $(2, -3)$ with direction vector $(-4, 5)$ is $\mathbf{x} = (2, -3) + t(-4, 5)$, $t \in \mathbb{R}$.

Sometimes the components of a vector equation are written separately:

$$\mathbf{x} = \mathbf{a} + t\mathbf{d} \text{ becomes } \begin{cases} x_1 = a_1 + td_1 \\ x_2 = a_2 + td_2 \end{cases} \qquad t \in \mathbb{R}$$

These are referred to as the **parametric** equations of the line. The familiar scalar form of the equation of the line is obtained by eliminating the parameter t: provided that $d_1 \neq 0$, $d_2 \neq 0$,

$$\frac{x_1 - a_1}{d_1} = t = \frac{x_2 - a_2}{d_2}$$

or

$$x_2 = a_2 + (d_2/d_1)(x_1 - a_1).$$

What can you say about the line if $d_1 = 0$ or $d_2 = 0$?

Directed Line Segments

For dealing with certain geometrical problems, it is useful to introduce **directed line segments.** We denote the directed line segment from point A to point B by $\overrightarrow{AB}$. We think of it as an "arrow" starting at A and pointing towards B. We shall *identify* directed line segments from the origin O with the corresponding vectors: thus, we write $\overrightarrow{OA} = \mathbf{a}$, $\overrightarrow{OB} = \mathbf{b}$, $\overrightarrow{OC} = \mathbf{c}$, and so on. A directed line segment that starts at the origin (such as $\overrightarrow{OA}$) is sometimes called the "position vector" (of the point A).

For many problems, we are interested only in the direction and length of the directed line segment and not in the point where it is located. For example, in

Figure 1–1–2, we may wish to treat the line segment $\overrightarrow{BC}$ as if it were the same as $\overrightarrow{OA}$. Taking our cue from this example, for arbitrary points A, B, C in $\mathbb{R}^2$, we define $\overrightarrow{BC}$ to be **equivalent** to $\overrightarrow{OA}$ if $\mathbf{c} - \mathbf{b} = \mathbf{a}$. In this case, we have used one directed line segment $\overrightarrow{OA}$ starting from the origin in our definition.

More generally, for arbitrary points B, C, D, E in $\mathbb{R}^2$, we define $\overrightarrow{BC}$ to be **equivalent** to $\overrightarrow{DE}$ if they are both equivalent to the same $\overrightarrow{OA}$ (for some A), that is, if

$$\mathbf{c} - \mathbf{b} = \mathbf{a} \text{ and } \mathbf{e} - \mathbf{d} = \mathbf{a} \text{ for the same } \mathbf{a}.$$

We can abbreviate this by simply requiring that

$$\mathbf{c} - \mathbf{b} = \mathbf{e} - \mathbf{d}.$$

For example, if B is point (4, 7), C is (9, 3), D is (−2, 6), and E is (3, 2), then $\overrightarrow{BC}$ is equivalent to $\overrightarrow{DE}$ because $\mathbf{c} - \mathbf{b} = (5, -4)$ and $\mathbf{e} - \mathbf{d} = (5, -4)$.

In some problems where it is not necessary to distinguish between equivalent directed line segments, we "identify" them (that is, we treat them as the *same* object) and write statements such as $\overrightarrow{AB} = \overrightarrow{CD}$. Indeed, we identify them with the corresponding line segment starting at the origin, so in the example in the previous paragraph, we write $\overrightarrow{BC} = \overrightarrow{DE} = (5, -4)$.

(Writing $\overrightarrow{BC} = \overrightarrow{DE}$ is a bit sloppy — an "abuse of notation" — because $\overrightarrow{BC}$ is not really the same object as $\overrightarrow{DE}$. However, introducing the precise language of "equivalence classes" and more careful notation to deal with directed line segments does not seem to be helpful at this stage. By introducing directed line segments, we are encouraged to think about "vectors" that are located at arbitrary points in space, and this is helpful in solving some geometrical problems as we shall see below. If you pay attention to what you are doing, you will quickly learn when it is useful and acceptable to identify equivalent vectors. If you know elementary physics, you may know that identifying equivalent "force vectors" is appropriate when you are finding the total force on a body, but when you are concerned with the moments of forces (torques), you must not identify force vectors applied at different points of the body.)

To show how we use this notation, let us review vector addition. The calculation illustrated in Figure 1–1–2 is:

$$\begin{aligned}\mathbf{a} + \mathbf{b} &= \overrightarrow{OA} + \overrightarrow{OB} = \overrightarrow{OA} + \overrightarrow{AC} \quad \text{(since } \overrightarrow{OB} \text{ and } \overrightarrow{AC} \text{ are equivalent)}\\ &= \overrightarrow{OC} = \mathbf{c} \quad \text{(by the definition of vector addition)}\end{aligned}$$

Notice that thinking of the sum as $\overrightarrow{OA} + \overrightarrow{AC}$ gives a helpful picture of vector addition: to get the sum **a** and **b** we move from O to A (since $\mathbf{a} = \overrightarrow{OA}$) and then from A to C (since $\mathbf{b} = \overrightarrow{OB}$ and $\overrightarrow{OB}$ is equivalent to $\overrightarrow{AC}$).

It is also worth reviewing subtraction of vectors. In Figure 1–1–4,

$$\begin{aligned}\mathbf{a} - \mathbf{b} &= \overrightarrow{OA} - \overrightarrow{OB}\\ &= \overrightarrow{OA} + \overrightarrow{BO}\\ &= \overrightarrow{OA} + \overrightarrow{AD}\\ &= \overrightarrow{OD} = \mathbf{d}.\end{aligned}$$

Directed line segments can be used to prove simple geometrical facts.

EXAMPLE 4

Prove that the two diagonals of a parallelogram bisect each other.

SOLUTION: We may use the parallelogram OACB in Figure 1–1–2. First notice that it would be just as good to prove that the midpoints of the two diagonals coincide. The midpoint of the diagonal $\overrightarrow{OC}$ is given by

$$\frac{1}{2}\overrightarrow{OC} = \frac{1}{2}\mathbf{c} = \frac{1}{2}(\mathbf{a} + \mathbf{b}).$$

To get to the midpoint of $\overrightarrow{AB}$, start from A and go halfway towards B: thus, the midpoint is given by

$$\overrightarrow{OA} + \frac{1}{2}\overrightarrow{AB} = \mathbf{a} + \frac{1}{2}(\mathbf{b} - \mathbf{a}) = \frac{1}{2}(\mathbf{a} + \mathbf{b}),$$

and the midpoint of the second diagonal is indeed identical with the midpoint of the first diagonal.

These ideas can also be used to find points in the plane, as in the following example.

EXAMPLE 5

(a) Find the coordinates of the point 1/3 of the way from the point A to the point B. (See Figure 1–1–6.)

(b) Find the coordinates of this point if A = (1, 2) and B = (3, −2).

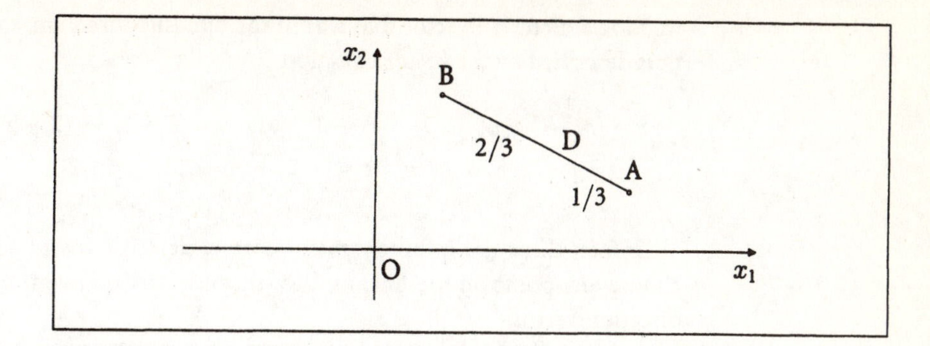

FIGURE 1-1-6. D *is the point 1/3 of the way from* A *to* B.

SOLUTION: (a) Let $\mathbf{a}$ be the vector corresponding to the point A, and let $\mathbf{b}$ correspond to B. Then the required point D corresponds to the vector

$$\overrightarrow{OD} = \overrightarrow{OA} + \frac{1}{3}\overrightarrow{AB} = \mathbf{a} + \frac{1}{3}(\mathbf{b} - \mathbf{a}) = \frac{2}{3}\mathbf{a} + \frac{1}{3}\mathbf{b}.$$

(b) When A = (1, 2), B = (3, −2),

$$\begin{aligned}\overrightarrow{OD} &= (1,2) + \frac{1}{3}\Big((3,-2) - (1,2)\Big) \\ &= (1,2) + \frac{1}{3}(2,-4) = \left(\frac{5}{3}, \frac{2}{3}\right),\end{aligned}$$

so D is $\left(\frac{5}{3}, \frac{2}{3}\right)$.

We can also solve Example 5 by using the vector equation of the line that contains the points A and B. The direction vector of this line is $\overrightarrow{AB}$, so the vector equation of the line is $\mathbf{x} = \mathbf{x}(t) = \overrightarrow{OX} = \overrightarrow{OA} + t\overrightarrow{AB}$, $t \in \mathbb{R}$. The desired point D is then given by $\mathbf{x}(1/3) = \overrightarrow{OA} + (1/3)\overrightarrow{AB}$, as before.

EXAMPLE 6

Find a vector equation of the line through (1, 2) and (3, −1).

SOLUTION: A vector equation of this line is

$$\mathbf{x} = (1,2) + t\Big((3,-1) - (1,2)\Big) = (1,2) + t(2,-3), \quad t \in \mathbb{R}.$$

However, a little thought indicates that we would have the same line if we started at the second point and "moved" towards the first point — or even if we took a direction vector that was in the opposite direction; the same line is thus described by the vector equations:

$$\mathbf{x} = (3,-1) + u(-2,3), \quad u \in \mathbb{R} \quad \text{or} \quad \mathbf{x} = (3,-1) + p(2,-3), \quad p \in \mathbb{R},$$
$$\text{or even} \quad \mathbf{x} = (1,2) + r(-2,3), \quad r \in \mathbb{R}.$$

In fact, there are infinitely many correct descriptions of a line: we may choose any point on the line, and we may use any non-zero multiple of the direction vector.

Vectors and Lines in $\mathbb{R}^3$

Almost everything we have done so far works perfectly well in three dimensions. We have to learn to draw and interpret two-dimensional pictures of $\mathbb{R}^3$. Figure 1-1-7 shows how this is done: the first coordinate axis (x_1) is usually pictured coming out of the page (or blackboard), the second (x_2) to the right, and the third (x_3) towards the top of the picture. To give some sense of perspective, the first axis is usually shown slightly below the horizontal and rotated somewhat to the left or to the right. It is often very difficult to make precise and helpful drawings of objects in $\mathbb{R}^3$.

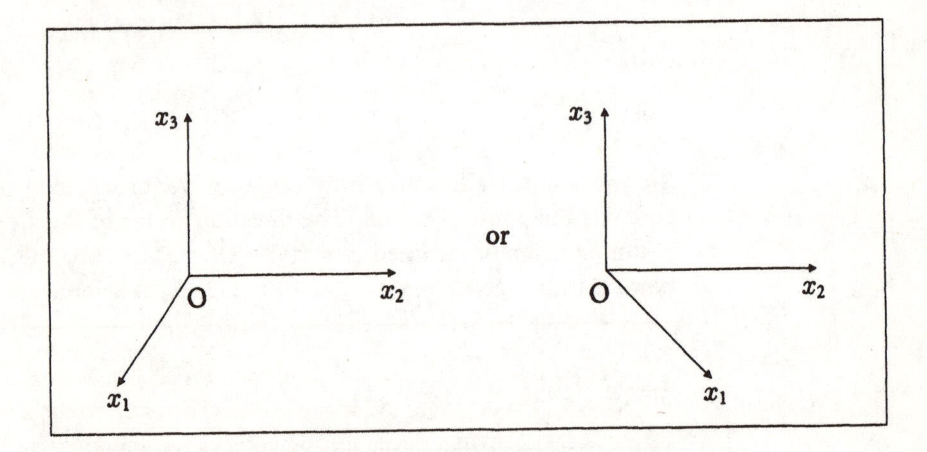

FIGURE 1-1-7. *Two pictures of the positive coordinate axes in* $\mathbb{R}^3$.

It should be noted that we are adopting the *convention* that the coordinate axes form a "right-handed system". Since the idea of right-handed systems will be important later in the discussion of the cross-product, we discuss it at some length. One way to visualize a right-handed system is to spread out the thumb, index finger, and middle finger of your right hand. The thumb is the first (x_1)

axis, the index finger is the second (x_2), and the middle finger is the third (x_3). (See Figure 1–1–8.) A second way to recognize a right-handed system is to hold your right hand with the thumb pointing in the direction of the *third* axis and your fingers curled; then the fingers point from the first axis to the second.

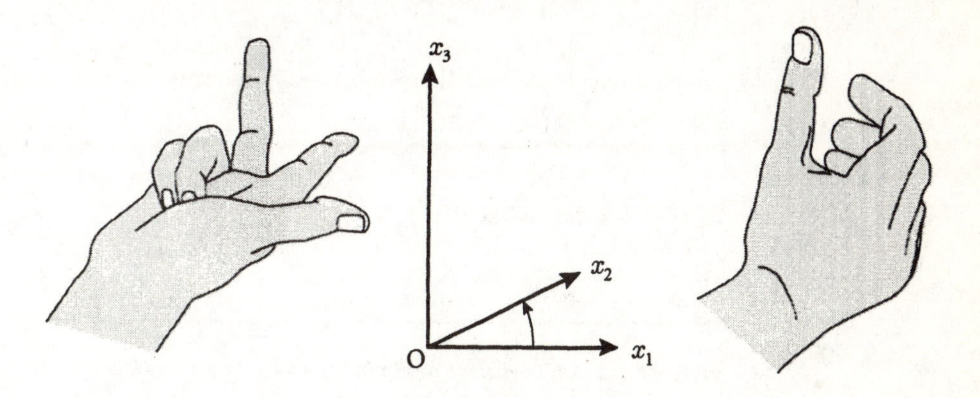

FIGURE 1–1–8. *Two ways to identify right-handed systems.*

We now define $\mathbb{R}^3$ to be the three-dimensional analog of $\mathbb{R}^2$.

$\mathbb{R}^3$ is the set of all triples (x_1, x_2, x_3) of real numbers with **addition** defined by $(x_1, x_2, x_3) + (y_1, y_2, y_3) = (x_1 + y_1, x_2 + y_2, x_3 + y_3)$, and **multiplication** by a scalar k defined by $k(x_1, x_2, x_3) = (kx_1, kx_2, kx_3)$.

x_1, x_2, and x_3 are the **components** of the **vector x** in $\mathbb{R}^3$.

Addition still follows the parallelogram rule. It may help you to visualize this if you realize that two vectors in $\mathbb{R}^3$ must lie within a plane in $\mathbb{R}^3$ so that the two-dimensional picture is still valid. See Figure 1–1–9.

It is useful to introduce a standard basis for $\mathbb{R}^3$ just as we did for $\mathbb{R}^2$. Define $\mathbf{i} = (1, 0, 0)$, $\mathbf{j} = (0, 1, 0)$, and $\mathbf{k} = (0, 0, 1)$. Then any vector $\mathbf{x} = (x_1, x_2, x_3)$ in $\mathbb{R}^3$ can be written as a linear combination of **i**, **j**, and **k**:

$$\mathbf{x} = x_1(1, 0, 0) + x_2(0, 1, 0) + x_3(0, 0, 1) = x_1\mathbf{i} + x_2\mathbf{j} + x_3\mathbf{k}.$$

As before, a **linear combination** means a sum of scalar multiples.

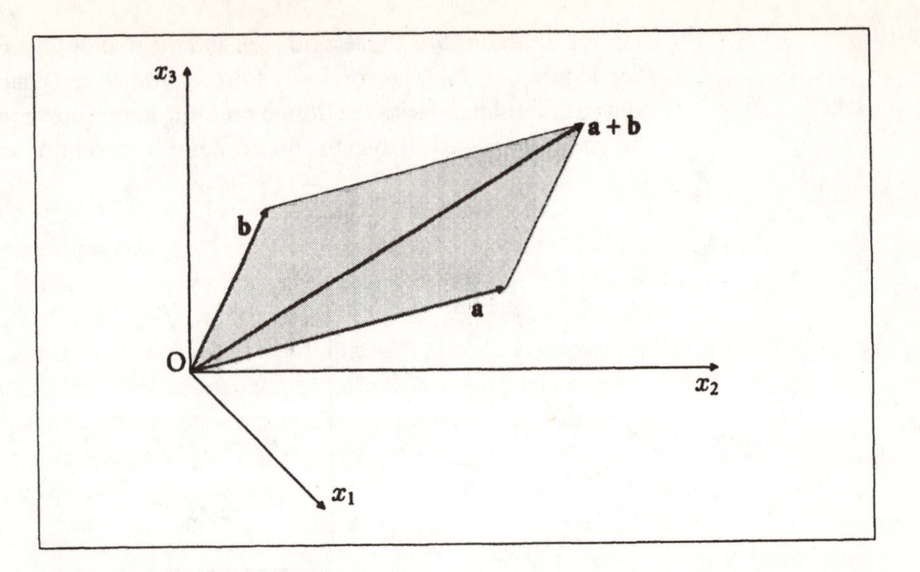

FIGURE 1-1-9. *In the plane determined by* **a** *and* **b**, **a** + **b** *is found by the parallelogram rule.*

Just as in $\mathbb{R}^2$, we introduce the zero vector $\mathbf{0} = (0,0,0)$ in $\mathbb{R}^3$, with the same properties as in $\mathbb{R}^2$.

Directed line segments are introduced in the three-dimensional case just as they were in the two-dimensional case.

A line through the point A in $\mathbb{R}^3$ (corresponding to vector **a**) with direction vector **d** can be described by a vector equation:

$$\mathbf{x} = \mathbf{a} + t\mathbf{d}, \quad t \in \mathbb{R}.$$

It is very important to realize that a line in $\mathbb{R}^3$ cannot be described by a single scalar linear equation as was the case in $\mathbb{R}^2$. We shall see in Section 1–2 that such an equation describes a plane in $\mathbb{R}^3$.

EXAMPLE 7

Find a vector equation of the line that passes through the points $(1,5,-2)$ and $(4,-1,3)$.

SOLUTION:

$$\mathbf{x} = (1,5,-2) + t(4-1,-1-5,3-(-2)) = (1,5,-2) + t(3,-6,5), \quad t \in \mathbb{R}.$$

Note that the corresponding parametric equations are $x_1 = 1 + 3t$, $x_2 = 5 - 6t$, $x_3 = -2 + 5t$.

In $\mathbb{R}^2$, two lines fail to have a point of intersection only if they are parallel. In $\mathbb{R}^3$, a pair of lines can fail to have a point of intersection if the lines are parallel (which means that the direction vector of one line is a non-zero multiple of the direction vector of the second line) or if they are **skew**. An easy way to visualize one pair of skew lines is as follows: let the first line correspond to a rope stretched along the ceiling of a room from one corner to the opposite corner and let the second line correspond to a rope stretched along the floor and joining the two corners that are not connected by the first rope; the direction vector of the second line is not a multiple of the direction vector of the first, but the lines do not have a point of intersection. To determine if two lines do have a point of intersection, we must solve a system of equations.

EXAMPLE 8

Do the lines

$$\begin{aligned} \ell_1 : \mathbf{x}^{(1)}(t) &= (1,0,1) + t(1,1,-1), \quad t \in \mathbb{R}, \\ \ell_2 : \mathbf{x}^{(2)}(s) &= (2,3,4) + s(0,-1,2), \quad s \in \mathbb{R}, \end{aligned}$$

have a point of intersection? If so, find it. (When we are considering two lines, it is advisable to use different letters to denote the parameters on the two lines; note what confusion would arise if we used the same parameters for both lines in the following solution.)

SOLUTION: Are there values of t and s such that $\mathbf{x}^{(1)}(t) = \mathbf{x}^{(2)}(s)$?

We compare the coordinates in parametric form:

$$\begin{aligned} x_1 &: \quad 1 + t = 2 + 0s \\ x_2 &: \quad 0 + t = 3 + (-1)s \\ x_3 &: \quad 1 - t = 4 + 2s. \end{aligned}$$

From the first of these equations, $t = 1$, then from the second, $s = 2$; when these values are substituted into the third equation, it is not satisfied. The system of equations has no solution, so there is no point of intersection. (We shall be considering systematic procedures for examining systems of equations in Chapter 2. For simple problems such as this, just solve the first pair of equations and check the third, as we have done.) Of course, if there were a solution for all three equations simultaneously, we could use the value of t in ℓ_1 (or s in ℓ_2) to find the common point **x**.

Vectors in $\mathbb{R}^n$

It is a simple matter to extend these ideas to n dimensions. The objects are called n-tuples or n-vectors and denoted $(x_1, x_2, \ldots, x_n)$. Addition and multiplication by scalars are defined as one might expect:

$$(x_1, x_2, \ldots, x_n) + (y_1, y_2, \ldots, y_n) = (x_1 + y_1, x_2 + y_2, \ldots, x_n + y_n),$$
$$k(x_1, x_2, \ldots, x_n) = (kx_1, kx_2, \ldots, kx_n).$$

We introduce the standard basis for $\mathbb{R}^n$: define

$$\mathbf{e}_1 = (1, 0, 0, \ldots, 0), \quad \mathbf{e}_2 = (0, 1, 0, 0, \ldots, 0), \ldots, \quad \mathbf{e}_n = (0, \ldots, 0, 1).$$

Then any vector $\mathbf{x} = (x_1, x_2, \ldots, x_n)$ can be written as a linear combination of the basis vectors:

$$\mathbf{x} = x_1\mathbf{e}_1 + x_2\mathbf{e}_2 + \cdots + x_n\mathbf{e}_n.$$

As in $\mathbb{R}^2$ and $\mathbb{R}^3$, we have the zero vector $\mathbf{0} = (0, 0, \ldots, 0)$.

The vector equation of a line in $\mathbb{R}^n$ is analogous to the vector equation of lines in $\mathbb{R}^2$ and $\mathbb{R}^3$.

Since $\mathbb{R}^n$ is not a space we can visualize for $n > 3$, we have to rely on pictures from $\mathbb{R}^2$ and $\mathbb{R}^3$ to suggest what is going on in $\mathbb{R}^n$.

Students sometimes do not see any point in discussing n-dimensional space because it does not seem to correspond to any physically realistic geometry. We indicate briefly some places where more than three dimensions are important. To discuss the motion of a particle, a physicist needs to specify its position (three variables), and its velocity (three more variables); altogether she has six variables, and some of the analysis of the motion will be easiest to understand if it is carried out in a six-dimensional space. An engineer wanting to analyze the control of a wheeled vehicle may begin with many variables: the position and velocity of the centre of mass of the vehicle itself, as well as of each wheel, and various angle variables; while she will not try to visualize n-space, she will carry out the analysis in n-space, for some $n > 3$. An economist seeking to model the Canadian economy will use many variables: one standard model has well over 1500 variables. Of course, calculations in such huge models are carried out by computer, but *the ideas of vector geometry and linear algebra are necessary in order to decide which calculations are required and what the results mean.*

EXERCISES 1–1

A Exercises have answers at the back. See the introductory Note on the Exercises.

A1 For each of the following, compute the sum, difference, or product, and illustrate with a sketch.

(a) $(1,3)+(2,-4)$ (b) $(3,2)-(4,1)$
(c) $(4,-2)+(-1,3)$ (d) $(-3,-4)-(-2,5)$
(e) $3(-1,4)$ (f) $-2(3,-2)$

A2 Determine the following:

(a) $(2,3,-4)-(5,1,-2)$ (b) $(2,1,-6,-7)+(-3,1,-4,5)$
(c) $-6(4,-5,-6)$ (d) $-2(-5,1,1,-2)$

A3 Let $\mathbf{x}=(1,2,-2)$ and $\mathbf{y}=(2,-1,3)$. Determine

(a) $2\mathbf{x}-3\mathbf{y}$; (b) $-3(\mathbf{x}+2\mathbf{y})+5\mathbf{x}$;
(c) $\mathbf{z}$ such that $\mathbf{y}-2\mathbf{z}=3\mathbf{x}$; (d) $\mathbf{z}$ such that $\mathbf{z}-3\mathbf{x}=2\mathbf{z}$.

A4 Consider the points $P(2,3,1)$, $Q(3,1,-2)$, $R(1,4,0)$, $S(-5,1,5)$. Determine $\vec{PQ}$, $\vec{PR}$, $\vec{PS}$, $\vec{QR}$, $\vec{SR}$, and verify that $\vec{PQ}+\vec{QR}=\vec{PR}=\vec{PS}+\vec{SR}$.

A5 Write the vector equation of the line passing through the given point with the given direction vector.

(a) point $(3,4)$, direction vector $(-5,1)$
(b) point $(2,0,5)$, direction vector $(4,-2,-11)$
(c) point $(4,0,1,5,-3)$, direction vector $(-2,0,1,2,-1)$

A6 Write a vector equation for the line that passes through the given points.

(a) $(-1,2)$ and $(2,-3)$ (b) $(4,1)$ and $(-2,-1)$
(c) $(1,3,-5)$ and $(-2,-1,0)$ (d) $\left(\frac{1}{2},\frac{1}{4},1\right)$ and $\left(-1,1,\frac{1}{3}\right)$
(e) $(1,0,-2,-5)$ and $(-3,2,-1,2)$

A7 For each of the following lines in $\mathbb{R}^2$, determine parametric equations and a vector equation.

(a) $x_2=3x_1+2$ (b) $2x_1+3x_2=5$
(c) the line that passes through $(2,-1)$ and has slope 3

A8 Find the midpoint of the line segment joining the given points.

(a) $(2,1,1)$ and $(-3,1,-4)$ (b) $(2,-1,0,3)$ and $(-3,2,1,-1)$

A9 Find points that divide the line segment joining the given points into three equal parts.
(a) $(2,4,1)$ and $(-1,1,7)$
(b) $(-1,1,5)$ and $(4,2,1)$

A10 Given the points P and Q, and the real number r, determine the point R such that $\overrightarrow{PR} = r\overrightarrow{PQ}$. Make a rough sketch to illustrate the idea.
(a) $P(1,4,-5)$ and $Q(-3,1,4)$; $r = 1/4$.
(b) $P(2,1,1,6)$ and $Q(8,7,6,0)$; $r = -1/3$.
(c) $P(2,1,-2)$ and $Q(-3,1,4)$; $r = 4/3$.

A11 Determine the points of intersection (if any) of the pairs of lines.
(a) $\mathbf{x} = (1,2) + t(3,5),\ t \in \mathbb{R}$, and $\mathbf{x} = (3,-1) + s(4,1),\ s \in \mathbb{R}$.
(b) $\mathbf{x} = (2,3,4) + t(1,1,1),\ t \in \mathbb{R}$, and $\mathbf{x} = (3,2,1) + s(3,1,-1),\ s \in \mathbb{R}$.
(c) $\mathbf{x} = (3,4,5) + t(1,1,1),\ t \in \mathbb{R}$, and $\mathbf{x} = (2,4,1) + s(2,3,-2),\ s \in \mathbb{R}$.
(d) $\mathbf{x} = (1,0,1) + t(3,-1,2),\ t \in \mathbb{R}$, and $\mathbf{x} = (5,0,7) + s(-2,2,2),\ s \in \mathbb{R}$.

A12 (a) A set of points in $\mathbb{R}^n$ is **collinear** if they all lie on one line. By considering directed line segments, give a general method for determining whether a given set of three points is collinear.
(b) Determine whether the points $P(1,2,2,1)$, $Q(4,1,4,2)$, and $R(-5,4,-2,-1)$ are collinear. Show how you decide.
(c) Determine whether the points $S(1,0,1,2)$, $T(3,-2,3,1)$, and $U(-3,4,-1,5)$ are collinear. Show how you decide.

B Exercises are similar to A but no answers are provided.

B1 For each of the following, compute the sum, difference, or product, and illustrate with a sketch.
(a) $(1,-2) + (-1,3)$
(b) $(4,-3) - (2,3)$
(c) $2(-2,-1)$
(d) $-3(1,-2)$

B2 Determine the following:
(a) $(5,-4,-3) - (2,-1,3)$
(b) $(4,1,-2,1) - (2,1,4,-3)$
(c) $4(2,-5,-1)$
(d) $-6(-1,0,-1,2)$

B3 Let $\mathbf{x} = (3,5,-1)$ and $y = (-5,2,1)$. Determine
(a) $2\mathbf{x} - 3\mathbf{y}$;
(b) $-2(\mathbf{x} - \mathbf{y}) - 3\mathbf{y}$;
(c) $\mathbf{z}$ such that $\mathbf{y} - 2\mathbf{z} = 3\mathbf{x}$.

B4 (a) Consider the points $P(1,4,1)$, $Q(4,3,-1)$, $R(-1,4,2)$, $S(8,6,-5)$. Determine $\overrightarrow{PQ}$, $\overrightarrow{PR}$, $\overrightarrow{PS}$, $\overrightarrow{QR}$, $\overrightarrow{SR}$, and verify that $\overrightarrow{PQ} + \overrightarrow{QR} = \overrightarrow{PR} = \overrightarrow{PS} + \overrightarrow{SR}$.
(b) Consider the points $P(3,-2,1)$, $Q(2,7,-3)$, $R(3,1,5)$, $S(-2,4,-1)$. Determine $\overrightarrow{PQ}$, $\overrightarrow{PR}$, $\overrightarrow{PS}$, $\overrightarrow{QR}$, $\overrightarrow{SR}$, and verify that $\overrightarrow{PQ} + \overrightarrow{QR} = \overrightarrow{PR} = \overrightarrow{PS} + \overrightarrow{SR}$.

B5 Write the vector equation of the line passing through the given point with the given direction vector.
(a) point $(-3,4)$, direction vector $(4,-3)$
(b) point $(2,3,-1)$, direction vector $(2,-4,8)$
(c) point $(3,1,-2,7,-1)$, direction vector $(2,-3,4,2,0.33)$

B6 Write a vector equation for the line that passes through the given points.

(a) $(2, -6, 3)$ and $(-1, 5, 2)$ (b) $\left(1, -1, \frac{1}{2}\right)$ and $\left(\frac{1}{2}, \frac{1}{3}, 1\right)$

(c) $(3, 1, -4, 0)$ and $(1, 2, 1, -3)$

B7 For each of the following lines in $\mathbb{R}^2$, determine parametric equations and a vector equation.

(a) $x_2 = -2x_1 + 3$ (b) $x_1 + 2x_2 = 3$

(c) the line that passes through $(-3, 2)$ and has slope -2

B8 Find the midpoint of the line segment joining the given points.

(a) $(1, -2, 3)$ and $(4, 1, 3)$ (b) $(6, -4, -1, 2)$ and $(0, 1, 2, 5)$

B9 Find points that divide the line segment joining the given points into three equal parts.

(a) $(4, -1, 2)$ and $(-2, 5, -1)$ (b) $(1, 1, 2, -2)$ and $(3, -1, 1, 4)$

B10 Given the points P and Q, and the real number r, determine the point R such that $\overrightarrow{PR} = r\overrightarrow{PQ}$. Make a rough sketch to illustrate the idea.

(a) $P(4, -2, 3)$ and $Q(1, 4, -2)$; $r = 3/5$.

(b) $P(2, -1, -2)$ and $Q(5, -2, 3)$; $r = 1/3$.

(c) $P(3, -1, 3, -2)$ and $Q(-4, 4, -1, 3)$; $r = -1/5$.

(d) $P(2, 4, -2)$ and $Q(3, -2, -3)$; $r = 1.7$.

B11 Determine the points of intersection (if any) of the pairs of lines.

(a) $\mathbf{x} = (-1, 2) + t(2, -3)$, $t \in \mathbb{R}$, and $\mathbf{x} = (7, -1) + s(1, 6)$, $s \in \mathbb{R}$.

(b) $\mathbf{x} = (1, 4, 0) + t(2, 3, -1)$, $t \in \mathbb{R}$, and $\mathbf{x} = (2, 7, -3) + s(1, 0, 1)$, $s \in \mathbb{R}$.

(c) $\mathbf{x} = (2, -1, 2) + t(1, -1, -1)$, $t \in \mathbb{R}$, and $\mathbf{x} = (5, -1, 5) + s(1, 2, 5)$, $s \in \mathbb{R}$.

(d) $\mathbf{x} = (3, 1, -1) + t(2, 1, 1)$, $t \in \mathbb{R}$, and $\mathbf{x} = (3, 8, 1) + s(1, 4, 2)$, $s \in \mathbb{R}$.

(e) $\mathbf{x} = (1, 3, 2) + t(4, -1, -1)$, $t \in \mathbb{R}$, and $\mathbf{x} = (5, -3, 2) + s(-2, 1, 1)$, $s \in \mathbb{R}$.

B12 (You will need the answer to A12 (a) to answer this.)

(a) Determine whether the points $P(2, 1, 1)$, $Q(1, 2, 3)$, and $R(4, -1, -3)$ are collinear. Show how you decide.

(b) Determine whether the points $S(1, 1, 0, -1)$, $T(6, 2, 1, 3)$, and $U(-4, 0, -1, -5)$ are collinear. Show how you decide.

COMPUTER RELATED EXERCISES

C Exercises require a microcomputer with a suitable software package. See the Note on the Exercises.

C1 Find out how to enter vectors and take their sum or difference. Try to take the sum of a 3-vector and a 4-vector to see how your software responds. Can you multiply a vector by a scalar?

CONCEPTUAL EXERCISES

D Exercises are not necessarily difficult, but they are not routine calculations. It is important that students do some of these.

***D1** Let A, B, C be points in $\mathbb{R}^2$ corresponding to vectors **a**, **b**, **c** respectively.
(a) Explain in terms of directed line segments why

$$\overrightarrow{AB} + \overrightarrow{BC} + \overrightarrow{CA} = \mathbf{0}.$$

(b) Verify the equation of part (a) by expressing $\overrightarrow{AB}$, $\overrightarrow{BC}$, and $\overrightarrow{CA}$ in terms of **a**, **b**, and **c**.

***D2** Let A, B, C, D be points in the plane such that ABCD is a quadrilateral. Show that if the diagonals AC and BD bisect each other, then ABCD is a parallelogram.

D3 Let A, B, C be points in $\mathbb{R}^2$, and let Z be the midpoint of the (undirected) line segment AB, X be the midpoint of BC, and Y be the midpoint of AC. Show that

$$\overrightarrow{ZX} = \frac{1}{2}\overrightarrow{AC},\ \overrightarrow{XY} = \frac{1}{2}\overrightarrow{BA},\ \overrightarrow{YZ} = \frac{1}{2}\overrightarrow{CB}.$$

D4 Let A, B, C, D be points in $\mathbb{R}^2$ and let W, X, Y, Z be the midpoints of AB, BC, CD, DA respectively. Show that WXYZ is a parallelogram. [Hint: Use D5 to obtain an expression for $\overrightarrow{WX}$.]

D5 Show that the three medians of a triangle meet in a common point, and that this point occurs 2/3 of the way along the median. (Recall that a median joins a vertex to the midpoint of the opposite side.) Hint: One possible approach is as follows: by using rotations and translations, we may assume that the vertices of the triangle are $O(0,0)$, $A(a_1,0)$, and $B(b_1,b_2)$. Determine the vector equation of the line through O in the direction of the midpoint of AB and the vector equation of the line from A toward the midpoint of OB. For what values of the parameters do these intersect?

D6 (a) Show that if two parallel lines in $\mathbb{R}^2$ have a point of intersection, then the lines coincide.
(b) Show that two lines in $\mathbb{R}^2$ that are not parallel to each other must have a point of intersection. [Hint: In $\mathbb{R}^2$, $\mathbf{d} = k\mathbf{b}$ if and only if $b_1d_2 = b_2d_1$.]
(c) Explain why parts (a) and (b) justify the following statement: In $\mathbb{R}^2$ two distinct lines fail to have a point of intersection if and only if they are parallel.

SECTION
1–2

Length; Dot Product; Equation of a Plane

Length in $\mathbb{R}^2$, $\mathbb{R}^3$, $\mathbb{R}^n$

The length of a vector in $\mathbb{R}^2$ is defined by the usual distance formula.

DEFINITION

If $\mathbf{x}$ is a vector in $\mathbb{R}^2$, its length is defined to be

$$\|\mathbf{x}\| = \sqrt{x_1^2 + x_2^2}.$$

EXAMPLE 9

$\|(2,3)\| = \sqrt{4+9} = \sqrt{13}$. (Note that for simple calculations such as this, used to illustrate basic ideas, it is more informative to leave the result in simple algebraic form, rather than to give a decimal approximation. Although a decimal approximation may be essential to interpreting the result practically, it is not an *exact* answer to the problem.)

There is a natural way to define the length of a vector in $\mathbb{R}^3$ or $\mathbb{R}^n$. The word **norm** is often used as a synonym for length when we are speaking of vectors. The appropriate formula in $\mathbb{R}^3$ can be obtained from a two-step calculation using the formula for $\mathbb{R}^2$, as shown in Figure 1–2–1. Consider the point X corresponding to $\mathbf{x} = (x_1, x_2, x_3)$; let P be the point $(x_1, x_2, 0)$; observe that OPX is a right angled triangle, so that

$$\|\mathbf{x}\|^2 = \|\overrightarrow{\mathrm{OP}}\|^2 + \|\overrightarrow{\mathrm{PX}}\|^2 = (x_1^2 + x_2^2) + x_3^2.$$

This suggests the following definition.

DEFINITION

If $\mathbf{x} \in \mathbb{R}^3$, the **norm** of $\mathbf{x}$ is $\|\mathbf{x}\| = \sqrt{x_1^2 + x_2^2 + x_3^2}$.

If $\mathbf{x} \in \mathbb{R}^n$: $\|\mathbf{x}\| = \sqrt{x_1^2 + x_2^2 + \cdots + x_n^2}$.

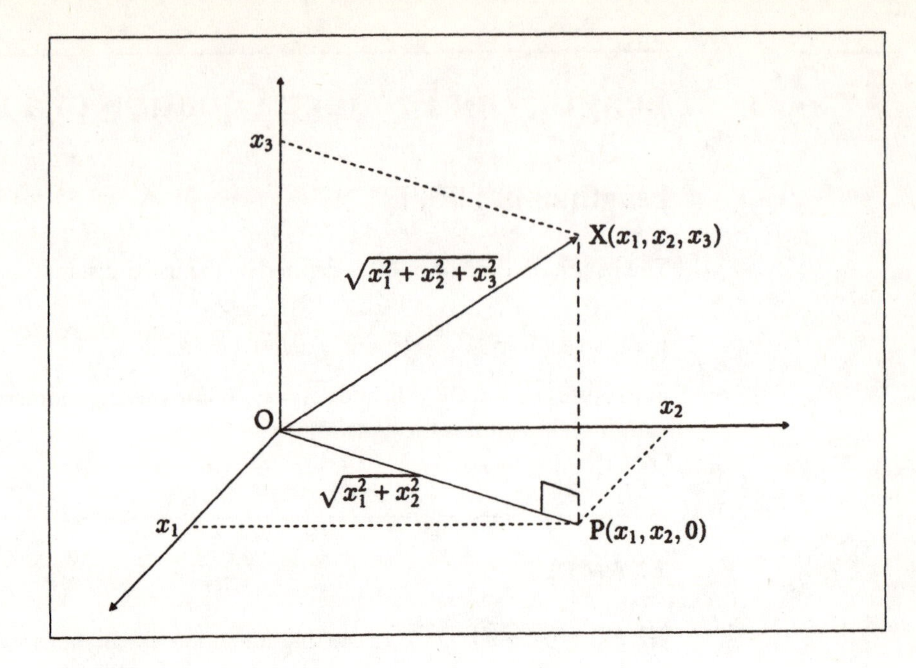

FIGURE 1-2-1. *Length in* $\mathbb{R}^3$.

Some important properties of the length function in $\mathbb{R}^n$.

(1) $\|\mathbf{x}\| \geq 0$ for all $\mathbf{x}$ in $\mathbb{R}^n$, and $\|\mathbf{x}\| = 0$ if and only if $\mathbf{x} = \mathbf{0} = (0, 0, \ldots, 0)$ (the zero vector in $\mathbb{R}^n$).

(2) $\|k\mathbf{x}\| = |k|\,\|\mathbf{x}\|$ for all real numbers k, and for all $\mathbf{x}$ in $\mathbb{R}^n$.

(3) (The Triangle Inequality): $\|\mathbf{x} + \mathbf{y}\| \leq \|\mathbf{x}\| + \|\mathbf{y}\|$, for all $\mathbf{x}$ and $\mathbf{y}$ in $\mathbb{R}^n$.

Properties (1) and (2) are easily checked from the definition of $\|\mathbf{x}\|$. Property (3) reflects the "triangle rule for addition of vectors:" $\mathbf{x}$, $\mathbf{y}$, and $\mathbf{x} + \mathbf{y}$ may be taken as the three sides of a triangle (see Figure 1–1–2 with $\mathbf{x} = \mathbf{a}$, $\mathbf{y} = \mathbf{b}$). The Triangle Inequality states that the length of one side is less than or equal to the sum of the lengths of the other two sides. An algebraic proof of this inequality will be given later in this section in terms of the dot product.

The distance between two points P and Q is calculated as $\|\overrightarrow{PQ}\|$.

EXAMPLE 10

Find the distance between the points $(-1, 3, 4)$ and $(2, -5, 1)$ in $\mathbb{R}^3$.

SOLUTION: The distance is $\sqrt{(2+1)^2 + (-5-3)^2 + (1-4)^2} = \sqrt{82}$.

Angles and the Dot Product

The problem of determining the angle between two vectors in $\mathbb{R}^2$ leads to the important idea of the **dot product** of two vectors. Consider Figure 1–2–2: a standard formula from trigonometry (the Law of Cosines) says that the length of the side $\|\overrightarrow{AB}\|$ is given in terms of the lengths of the other sides and the angle θ between them by

$$\|\overrightarrow{AB}\|^2 = \|\overrightarrow{OA}\|^2 + \|\overrightarrow{OB}\|^2 - 2\,\|\overrightarrow{OA}\|\,\|\overrightarrow{OB}\|\cos\theta.$$

Substitute $\overrightarrow{OA} = \mathbf{a}$, $\overrightarrow{OB} = \mathbf{b}$, and $\overrightarrow{AB} = \mathbf{a} - \mathbf{b}$ into the trigonometric formula and simplify; the result is the equation:

$$2(a_1b_1 + a_2b_2) = 2\,\|\mathbf{a}\|\,\|\mathbf{b}\|\cos\theta.$$

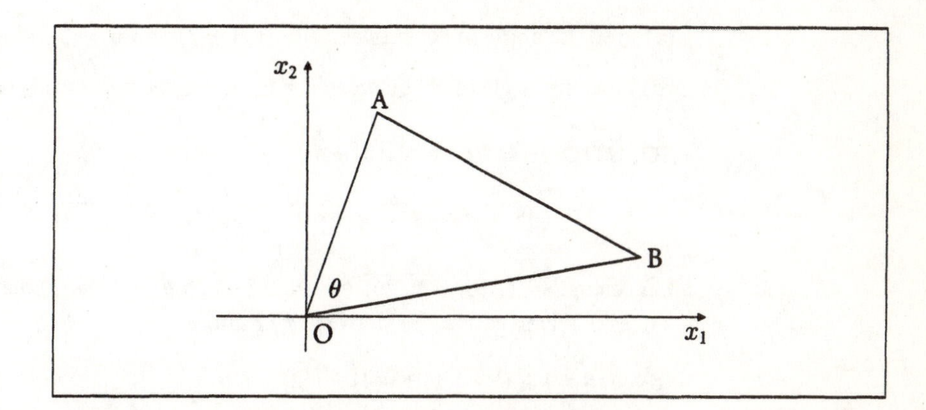

FIGURE 1-2-2. $\|\overrightarrow{AB}\|^2 = \|\overrightarrow{OA}\|^2 + \|\overrightarrow{OB}\|^2 - 2\|\overrightarrow{OA}\|\|\overrightarrow{OB}\|\cos\theta$.

For vectors in $\mathbb{R}^3$, a similar calculation gives

$$a_1b_1 + a_2b_2 + a_3b_3 = \|\mathbf{a}\|\,\|\mathbf{b}\|\cos\theta.$$

Because the same sort of expression appears both times, we make the following definition.

DEFINITION

The **dot product** of two vectors **a** and **b** in $\mathbb{R}^n$ is

$$\mathbf{a}\cdot\mathbf{b} = a_1b_1 + a_2b_2 + a_3b_3 + \cdots + a_nb_n.$$

The dot product is also sometimes called the **scalar product** or **inner product;** note that the value of a dot product is a real number (that is, a scalar).

Observe that in $\mathbb{R}^2$ and $\mathbb{R}^3$ the cosine of the angle between non-zero vectors can be calculated by means of the dot product: if $\mathbf{a} \neq \mathbf{0}$ and $\mathbf{b} \neq \mathbf{0}$, the angle θ between them is given by the equation

$$\cos\theta = \frac{\mathbf{a} \cdot \mathbf{b}}{\|\mathbf{a}\|\,\|\mathbf{b}\|},$$

where θ is always chosen to satisfy $0 \leq \theta \leq \pi$. We make this the *definition* of the angle between vectors in $\mathbb{R}^n$.

EXAMPLE 11

(a) Find the angle in $\mathbb{R}^3$ between $\mathbf{a} = (1, 4, -2)$ and $\mathbf{b} = (3, -1, 4)$.

(b) Find the angle in $\mathbb{R}^5$ between $\mathbf{c} = (2, 4, 0, 1, 3)$ and $\mathbf{d} = (1, 1, 4, 2, 0)$.

SOLUTION: (a) $\mathbf{a} \cdot \mathbf{b} = 1(3) + 4(-1) + (-2)(4) = -9$;

$$\|\mathbf{a}\| = \sqrt{1 + 16 + 4} = \sqrt{21}; \quad \|\mathbf{b}\| = \sqrt{9 + 1 + 16} = \sqrt{26};$$

hence $\cos\theta = (-9)/\sqrt{21(26)} \approx -0.38516$, so $\theta \approx 1.966$ radians. (Note that since $\cos\theta$ is negative, θ is between $\pi/2$ and π.)

(b) $\cos\theta = \big(2(1) + 4(1) + 0(4) + 1(2) + 3(0)\big)/(\sqrt{30}\sqrt{22}) \approx 0.31140$, so $\theta \approx 1.254$ radians.

Notice that two vectors $\mathbf{a}$ and $\mathbf{b}$ in $\mathbb{R}^2$ or $\mathbb{R}^3$ are perpendicular to each other if the angle θ between them is $\pi/2$, and then $\cos\theta = 0$; but this means that $\mathbf{a} \cdot \mathbf{b} = 0$. Conversely, if $\mathbf{a} \cdot \mathbf{b} = 0$, either the angle between $\mathbf{a}$ and $\mathbf{b}$ must be $\pi/2$ and the vectors are mutually perpendicular, or one (or both) of the vectors must be the zero vector. It is usual to introduce a new word to describe this property in $\mathbb{R}^n$ (and we shall use this word in $\mathbb{R}^2$ and $\mathbb{R}^3$ as well).

DEFINITION

Two vectors $\mathbf{a}$ and $\mathbf{b}$ in $\mathbb{R}^n$ are **orthogonal** to each other if $\mathbf{a} \cdot \mathbf{b} = 0$.

Notice that the definition includes the possibility that one or both of $\mathbf{a}$ and $\mathbf{b}$ is the zero vector; the zero vector is orthogonal to every vector in $\mathbb{R}^n$.

EXAMPLE 12

$(1, 0, 3, -2)$ is orthogonal to $(2, 3, 0, 1)$ since their dot product is zero.

A very useful relation between the dot product in $\mathbb{R}^n$ and the norm in $\mathbb{R}^n$ is:

$$\mathbf{a} \cdot \mathbf{a} = a_1a_1 + a_2a_2 + \cdots + a_na_n = \|\mathbf{a}\|^2, \quad \text{or} \quad \|\mathbf{a}\| = \sqrt{\mathbf{a} \cdot \mathbf{a}}.$$

A Fresh Start and Some Important Properties of the Dot Product

Sometimes it is helpful to realize that *all the formulas and properties for lengths and angles follow from the definition of dot product.* Pretend that we know nothing about lengths, angles, and dot products. We begin with the formal **definition of scalar product** in $\mathbb{R}^n$:

$$\mathbf{a} \cdot \mathbf{b} = a_1b_1 + a_2b_2 + a_3b_3 + \cdots + a_nb_n.$$

From this definition some **important properties** follow.

(i) $\mathbf{a} \cdot \mathbf{a} \geq 0$, for all $\mathbf{a}$ in $\mathbb{R}^n$, and $\mathbf{a} \cdot \mathbf{a} = 0$ if and only if $\mathbf{a} = \mathbf{0}$.

(ii) $\mathbf{a} \cdot \mathbf{b} = \mathbf{b} \cdot \mathbf{a}$, for all $\mathbf{a}$ and $\mathbf{b}$ in $\mathbb{R}^n$.

(iii) $\mathbf{a} \cdot (\mathbf{b} + \mathbf{c}) = \mathbf{a} \cdot \mathbf{b} + \mathbf{a} \cdot \mathbf{c}$, for all $\mathbf{a}$, $\mathbf{b}$, and $\mathbf{c}$ in $\mathbb{R}^n$.

(iv) $(k\mathbf{a}) \cdot \mathbf{b} = k(\mathbf{a} \cdot \mathbf{b})$, for all $\mathbf{a}$ and $\mathbf{b}$ in $\mathbb{R}^n$, for all k in $\mathbb{R}$.

You should verify all of these properties, using the definition. Notice that from (ii) and (iii) we also deduce that $(\mathbf{b} + \mathbf{c}) \cdot \mathbf{a} = \mathbf{b} \cdot \mathbf{a} + \mathbf{c} \cdot \mathbf{a}$, and similarly from (ii) and (iv), we have $\mathbf{a} \cdot (k\mathbf{b}) = k(\mathbf{a} \cdot \mathbf{b})$ as well.

Because of property (i) we can now **define the norm** of a vector in $\mathbb{R}^n$ (remember that we have temporary amnesia and have forgotten what we knew about the norm). We **define** the norm by

$$\|\mathbf{a}\| = \sqrt{\mathbf{a} \cdot \mathbf{a}} = \sqrt{a_1^2 + a_2^2 + \cdots + a_n^2}.$$

This norm still satisfies properties (1) $\|\mathbf{x}\| \geq 0$ for all $\mathbf{x}$ in $\mathbb{R}^n$, with $\|\mathbf{x}\| = 0$ if and only if $\mathbf{x} = \mathbf{0}$; and (2) $\|k\mathbf{x}\| = |k|\,\|\mathbf{x}\|$ for all k in $\mathbb{R}$ and all $\mathbf{x}$ in $\mathbb{R}^n$. The Triangle Inequality will be proved as a consequence of another important property of the scalar product, called the **Cauchy-Schwarz (or Cauchy-Schwarz-Buniakowski) inequality**:

(v) $|\mathbf{a} \cdot \mathbf{b}| \leq \|\mathbf{a}\| \, \|\mathbf{b}\|$, for all **a** and **b** in $\mathbb{R}^n$, with equality if and only if one of **a** or **b** is a scalar multiple of the other.

If we were allowed to use what we know about dot products and angles, this would be easy to prove, because $|\cos\theta| \leq 1$ for all θ, but we are supposed to use only the definition of scalar product. We consider two proofs: one is a straightforward computation, while the second is a "rabbit-out-of-the-hat" proof, which has the advantage of working for general real "inner products" (introduced in Section 7–4).

Proof of the Cauchy-Schwarz-Buniakowski Inequality

First, notice that the inequality is trivially true if **a** or **b** is the zero vector, so we suppose that neither is zero. Next, observe that what we must prove is equivalent to $(\|\mathbf{a}\| \, \|\mathbf{b}\|)^2 - |\mathbf{a} \cdot \mathbf{b}|^2 \geq 0$.

(Computational proof) We carry out the computation in $\mathbb{R}^2$; a similar argument is valid in $\mathbb{R}^n$ for larger n.

$$\begin{aligned}(\|\mathbf{a}\|\|\mathbf{b}\|)^2 - \|\mathbf{a} \cdot \mathbf{b}\|^2 &= (a_1^2 + a_2^2)(b_1^2 + b_2^2) - (a_1 b_1 + a_2 b_2)^2 \\ &= a_1^2 b_1^2 + a_1^2 b_2^2 + a_2^2 b_1^2 + a_2^2 b_2^2 - (a_1^2 b_1^2 + 2a_1 a_2 b_1 b_2 + a_2^2 b_2^2) \\ &= a_1^2 b_2^2 + a_2^2 b_1^2 - 2a_1 a_2 b_1 b_2 = (a_1 b_2 - a_2 b_1)^2 \geq 0,\end{aligned}$$

and the required inequality is established. Moreover, $|\mathbf{a} \cdot \mathbf{b}| = \|\mathbf{a}\| \, \|\mathbf{b}\|$ if and only if $a_1 b_2 - a_2 b_1 = 0$, or (*temporarily ignoring cases where components may be zero*) $a_1/b_1 = a_2/b_2$, which means that $\mathbf{a} = k\mathbf{b}$. *Now, as an exercise, examine the cases where* b_1 *or* b_2 *may be zero.*

■

(General proof) By property (i) we have $(t\mathbf{a}+\mathbf{b}) \cdot (t\mathbf{a}+\mathbf{b}) \geq 0$, for any **a** and **b**, and for any real t. Use (iii) to expand, and obtain:

$$(\mathbf{a} \cdot \mathbf{a})t^2 + (2\mathbf{a} \cdot \mathbf{b})t + (\mathbf{b} \cdot \mathbf{b}) \geq 0, \quad \text{for all real } t. \tag{\#}$$

Note that $\mathbf{a} \cdot \mathbf{a} > 0$, because we assumed that $\mathbf{a} \neq 0$. Now a quadratic expression $At^2 + Bt + C$ (with $A > 0$) is never negative if and only if the corresponding equation $At^2 + Bt + C = 0$ has complex roots or coincident real roots, and from the formula for solving quadratic equations, this is true if and only if $B^2 - 4AC \leq 0$. Thus inequality (#) implies that $(2\mathbf{a} \cdot \mathbf{b})^2 - 4(\mathbf{a} \cdot \mathbf{a})(\mathbf{b} \cdot \mathbf{b}) \leq 0$, which is the required inequality. Moreover, $|\mathbf{a} \cdot \mathbf{b}|^2 - \|\mathbf{a}\|^2\|\mathbf{b}\|^2 = 0$ if and only if $(\mathbf{a} \cdot \mathbf{a})t^2 + (2\mathbf{a} \cdot \mathbf{b})t + (\mathbf{b} \cdot \mathbf{b}) = 0$ for some real t, and then $t\mathbf{a} + \mathbf{b} = \mathbf{0}$, so **b** is a multiple of **a**.

■

Proof of the Triangle Inequality

The required statement $\|\mathbf{x}+\mathbf{y}\| \le \|\mathbf{x}\| + \|\mathbf{y}\|$ is eqivalent to

$$\|\mathbf{x}+\mathbf{y}\|^2 \le (\|\mathbf{x}\| + \|\mathbf{y}\|)^2.$$

This squared form will allow us to use the dot product conveniently. Thus, we consider

$$\begin{aligned}\|\mathbf{x}+\mathbf{y}\|^2 - (\|\mathbf{x}\| + \|\mathbf{y}\|)^2 &= (\mathbf{x}+\mathbf{y})\cdot(\mathbf{x}+\mathbf{y}) - (\|\mathbf{x}\|^2 + 2\|\mathbf{x}\|\,\|\mathbf{y}\| + \|\mathbf{y}\|^2)\\ &= \mathbf{x}\cdot\mathbf{x}+\mathbf{x}\cdot\mathbf{y}+\mathbf{y}\cdot\mathbf{x}+\mathbf{y}\cdot\mathbf{y} - (\mathbf{x}\cdot\mathbf{x} + 2\|\mathbf{x}\|\,\|\mathbf{y}\| + \mathbf{y}\cdot\mathbf{y})\\ &= 2\mathbf{x}\cdot\mathbf{y} - 2\|\mathbf{x}\|\,\|\mathbf{y}\|.\end{aligned}$$

The right hand side is less than or equal to zero by the Cauchy-Schwarz inequality. Therefore $\|\mathbf{x}+\mathbf{y}\| \le (\|\mathbf{x}\| + \|\mathbf{y}\|)$.

The angle θ between $\mathbf{a}$ and $\mathbf{b}$ can now be *defined* in terms of the scalar product by the formula:

$$\cos\theta = \frac{\mathbf{a}\cdot\mathbf{b}}{\|\mathbf{a}\|\,\|\mathbf{b}\|}, \qquad 0 \le \theta \le \pi.$$

From (i) and (v) it follows that $-1 \le \cos\theta \le 1$ for any pair of non-zero vectors, and that the angle between $\mathbf{a}$ and $\mathbf{b}$ is the same as the angle between $\mathbf{b}$ and $\mathbf{a}$, so that this definition of angle is plausible.

From now on, we shall use these properties of scalar (dot) products, norms, and angles, without worrying about the order in which they were obtained.

The (Scalar) Equation of a Plane in $\mathbb{R}^3$

The easiest plane in $\mathbb{R}^3$ to visualize is the x_1x_2-plane:

$$\{\mathbf{x} \mid \mathbf{x} \in \mathbb{R}^3 \quad \text{and} \quad x_3 = 0\}.$$

We shall usually not use the full set notation, and simply describe this as the plane with equation $x_3 = 0$.

This simple example illustrates one fact that often gives beginners difficulty: in $\mathbb{R}^3$, a single scalar linear equation describes a plane and not a line. (A **linear** equation in $\mathbb{R}^3$ is an equation of the form $a_1x_1 + a_2x_2 + a_3x_3 = d$.)

How do we find the equation of a general plane in $\mathbb{R}^3$? Suppose that we want to find the equation of the plane through the point $A(a_1, a_2, a_3)$, with **normal vector** $\mathbf{n} = (n_1, n_2, n_3)$; we *define* $\mathbf{n}$ to be a normal to the plane if $\mathbf{n}$ is a non-zero vector orthogonal (perpendicular) to any directed line segment $\overrightarrow{PQ}$ lying in the plane. (That is, **n** is orthogonal to $\overrightarrow{PQ}$ for any points P and Q in the plane; see Figure 1–2–3.) To find the equation of this plane, let X be any other point in the plane. Then **n** is orthogonal to $\overrightarrow{AX}$ so

$$0 = \mathbf{n} \cdot \overrightarrow{AX} = \mathbf{n} \cdot (\mathbf{x} - \mathbf{a}) = n_1(x_1 - a_1) + n_2(x_2 - a_2) + n_3(x_3 - a_3).$$

This equation, which must be satisfied by the coordinates of a point X in the plane, can be rewritten

$$n_1x_1 + n_2x_2 + n_3x_3 = d, \quad \text{where} \quad d = n_1a_1 + n_2a_2 + n_3a_3.$$

This is the standard equation of this plane. For computational purposes the form $\mathbf{n} \cdot (\mathbf{x} - \mathbf{a}) = 0$ is often the easiest to use.

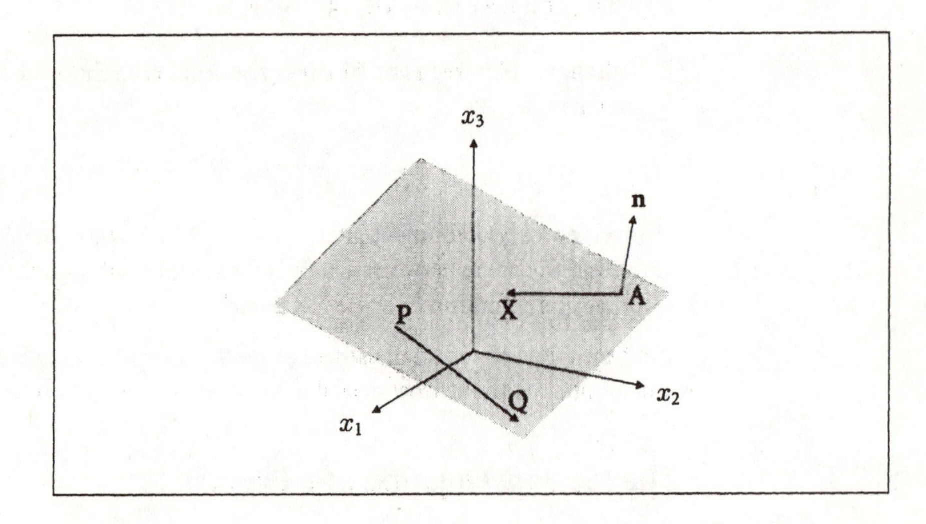

FIGURE 1-2-3. *The normal* **n** *is orthogonal to every directed line segment lying in the plane, so* **n** *is orthogonal to* $\overrightarrow{AX}$ *and* $\overrightarrow{PQ}$.

EXAMPLE 13

Find the equation of the plane that passes through the point $(2, 3, -1)$ and has normal vector $(1, -4, 1)$.

SOLUTION: The equation is

$$(1, -4, 1) \cdot \big(x_1 - 2, x_2 - 3, x_3 - (-1)\big) = 0,$$

or

$$x_1 - 2 - 4(x_2 - 3) + (x_3 + 1) = 0,$$

or

$$x_1 - 4x_2 + x_3 = -11.$$

It is important to note that the reasoning that leads to the equation of a plane can be reversed in order to identify the set of points satisfying an equation of the form $c_1x_1 + c_2x_2 + c_3x_3 = k$. If $c_1 \neq 0$, this can be rewritten as

$$c_1x_1 - k + c_2x_2 + c_3x_3 = 0, \text{ or } \mathbf{c} \cdot \left(x_1 - (k/c_1), x_2, x_3\right) = 0,$$

where $\mathbf{c} = (c_1, c_2, c_3)$. This equation describes a plane through the point $(k/c_1, 0, 0)$, with normal vector **c**. If $c_2 \neq 0$ we could have combined the k with the x_2 term and found that the plane passed through the point $(0, k/c_2, 0)$. In fact we could find a point in the plane in many ways, but the normal vector will always be **c** or a non-zero multiple of **c**.

EXAMPLE 14

Describe the set of points in $\mathbb{R}^3$ satisfying the linear equation

$$5x_1 - 6x_2 + 7x_3 = 11.$$

SOLUTION: We wish to rewrite the equation in the form $\mathbf{n} \cdot (\mathbf{x} - \mathbf{a}) = 0$. Thus, we rewrite it as

$$(5x_1 - 11) - 6x_2 + 7x_3 = 0,$$

or

$$5(x_1 - \frac{11}{5}) - 6(x_2 - 0) + 7(x_3 - 0) = 0.$$

Thus, we identify the set as a plane with normal vector $(5, -6, 7)$, passing through the point $(\frac{11}{5}, 0, 0)$. Alternatively, if we put $x_1 = x_3 = 0$, we find that $x_2 = -\frac{11}{6}$, so the plane passes through $(0, -\frac{11}{6}, 0)$. Or we could take the point $(0, 0, \frac{11}{7})$.

Two planes are defined to be **parallel** if the normal vector to the first plane is a non-zero multiple of the normal vector to the second. Thus, for example, the plane $x + 2y - z = 1$ is parallel to the plane $2x + 4y - 2z = 7$.

Two planes are **orthogonal** to each other if their normal vectors are orthogonal. For example, the plane $x_1 + x_2 + x_3 = 0$ is orthogonal to the plane $x_1 + x_2 - 2x_3 = 3$ because $(1, 1, 1) \cdot (1, 1, -2) = 0$.

A line is **orthogonal** to a plane if the direction vector of the line is a non-zero multiple of the normal vector to the plane. For example, the line $\mathbf{x} = t(1, 1, 2)$ is orthogonal to the plane $2x_1 + 2x_2 + 4x_3 = 5$.

A line is **parallel** to a plane if the line is parallel to a line lying in the plane. This means that a direction vector of the line is orthogonal to the normal vector to the plane. Thus the line $\mathbf{x} = t(1, 1, 1)$ is parallel to the plane $x_1 - x_3 = 5$ because $(1, 1, 1) \cdot (1, 0, -1) = 0$.

EXAMPLE 15

Find the equation of the plane that contains the point $(2, 4, -1)$ and is parallel to the plane $2x + 3y - 5z = 6$.

SOLUTION: The equation must be of the form $2x + 3y - 5z = d$. Since $(2, 4, -1)$ must satisfy the equation,

$$d = 2(2) + 3(4) - 5(-1) = 21,$$

and the required equation is

$$2x + 3y - 5z = 21.$$

We shall be able to look at more interesting examples of the equation of a plane after we have introduced the cross-product in Section 1–4.

The Vector Equation of a Plane

Our first example of a plane was the plane with (scalar) equation $x_3 = 0$. In set notation, it could also be described in the form

$$\{(x_1, x_2, 0) \mid x_1 \in \mathbb{R}, x_2 \in \mathbb{R}\} = \{x_1\mathbf{i} + x_2\mathbf{j} \mid x_1 \in \mathbb{R}, x_2 \in \mathbb{R}\}$$

where $\mathbf{i} = (1, 0, 0)$ and $\mathbf{j} = (0, 1, 0)$, as in Section 1–1. Sometimes it is preferable to use other letters instead of x_1 and x_2 in this expression, say r and s. Then we say that we are considering the plane with vector equation $\mathbf{x} = r\mathbf{i} + s\mathbf{j}$ (with $r, s \in \mathbb{R}$ understood).

Next, consider the plane through the origin and containing the points A(1, 1, 1) and B(1, −2, 3). If X is any other point in this plane,

$$\overrightarrow{OX} = r\overrightarrow{OA} + s\overrightarrow{OB}, \quad r \in \mathbb{R},\ s \in \mathbb{R}.$$

(We can show that this description of a plane is consistent with the point-normal description when we have introduced the cross-product in Section 1–4.) In terms of the corresponding vectors **x**, **a**, and **b**, the plane has vector equation

$$\mathbf{x} = r\mathbf{a} + s\mathbf{b} = r(1, 1, 1) + s(1, -2, 3).$$

Make a sketch to illustrate this plane.

Now suppose that this plane is *translated* by adding the vector $\mathbf{p} = (0, 0, 4)$. (It is easy to check that (0, 0, 4) is not a point in the original plane.) The equation of the translated plane is

$$\mathbf{x} = \mathbf{p} + r\mathbf{a} + s\mathbf{b}, \quad r \in \mathbb{R},\ s \in \mathbb{R}.$$

Note that the point (0, 0, 4) is a point in this new plane (when $r = s = 0$). Note that the point (1, 1, 1) is not a point in the new plane: instead, the point D(1, 1, 5) corresponding to the vector $\mathbf{p} + \mathbf{a}$ is a point in the plane; similarly the point E corresponding to $\mathbf{p} + \mathbf{b}$ is a point in the plane. It seems natural to say that the directed line segments $\overrightarrow{PD}$ and $\overrightarrow{PE}$ lie in the plane, but we have seen that we must not say that **a** and **b** lie in the plane, even though we (sloppily) write $\overrightarrow{PD} = \mathbf{a}$ and $\overrightarrow{PE} = \mathbf{b}$. We may say that **a** and **b** are parallel to this plane. (*When directed line segments were introduced, we warned that it was an abuse of notation to write statements such as* $\overrightarrow{PD} = \mathbf{a}$. *We have just encountered one place where this can cause confusion unless we are careful.*)

Note that just as in the case of vector equations of lines, for planes, the vector equation is not unique: for one plane, there are many choices for the point in the plane, and many choices for the two line segments lying in the plane.

EXAMPLE 16

Write a vector equation for the plane passing through the points P(1, 0, 2), Q(2, 3, 1), and R(−2, 1, 3).

SOLUTION: A vector equation for this plane is

$$\mathbf{x} = \overrightarrow{OP} + r\overrightarrow{PQ} + s\overrightarrow{PR} = (1,0,2) + r(1,3,-1) + s(-3,1,1), \quad r,s \in \mathbb{R}.$$

Usually when someone asks you for "the equation of a plane", it is the *scalar* equation that is required. We shall find out how to obtain the scalar equation of a plane from its vector equation in Section 1–4.

Analogs to the Plane in Higher Dimensions

Suppose that $\mathbf{n}$ and $\mathbf{a}$ are vectors in $\mathbb{R}^n$ for some $n > 3$; suppose that $\mathbf{n} \neq \mathbf{0}$. We may again consider the set with (scalar) equation $\mathbf{n} \cdot (\mathbf{x} - \mathbf{a}) = 0$, or

$$n_1x_1 + n_2x_2 + n_3x_3 + \cdots + n_nx_n = d,$$

where

$$d = n_1a_1 + n_2a_2 + n_3a_3 + \cdots + n_na_n.$$

This set is called a **hyperplane**; it contains the point $\mathbf{a}$ and has normal vector $\mathbf{n}$. When we consider the idea of "dimension" carefully, we shall see that the hyperplane is of dimension $n-1$. For now, note that in n-dimensional space a point may move with n "degrees of freedom"; imposing one condition (the equation above) reduces this by 1 to $(n-1)$.

We may also consider sets of the form $\mathbf{x} = r\mathbf{u} + s\mathbf{v}$, $r, s \in \mathbb{R}$, where $\mathbf{u}$ and $\mathbf{v}$ are vectors in $\mathbb{R}^n$. In such a set, the point $\mathbf{x}$ has only two "degrees of freedom" (choose r, choose s), so such a set is a 2-dimensional subset of $\mathbb{R}^n$. It is called a 2-plane.

EXERCISES 1–2

A1 Calculate the lengths of the given vectors.

(a) $(2,-5)$

(b) $(2,3,-2)$

(c) $(1,\frac{1}{5},-3)$

(d) $(1,-1,0,2)$

A2 Determine the distance from P to Q if

(a) P is $(2,3)$ and Q is $(-4,1)$;

(b) P is $(1,1,-2)$ and Q is $(-3,1,1)$;

(c) P is $(4,-6,1)$ and Q is $(-3,5,1)$;

(d) P is $(2,1,1,5)$ and Q is $(4,6,-2,1)$.

A3 Verify the triangle inequality and the Cauchy-Schwarz inequality if
(a) $\mathbf{x} = (4, 3, 1)$ and $\mathbf{y} = (2, 1, 5)$;
(b) $\mathbf{x} = (1, -1, 2)$ and $\mathbf{y} = (-3, 2, 4)$.

A4 Determine the angle (in radians) between the vectors $\mathbf{a}$ and $\mathbf{b}$ if
(a) $\mathbf{a} = (2, 1, 4)$ and $\mathbf{b} = (4, -2, 1)$;
(b) $\mathbf{a} = (1, -2, 1)$ and $\mathbf{b} = (3, 1, 0)$;
(c) $\mathbf{a} = (5, 1, 1, -2)$ and $\mathbf{b} = (2, 3, -2, 1)$.

A5 Determine whether the given pair of vectors is orthogonal.
(a) $(1, 3, 2), (2, -2, 2)$
(b) $(-3, 1, 7), (2, -1, 1)$
(c) $(2, 1, 1), (-1, 4, 2)$
(d) $(4, 1, 0, -2), (-1, 4, 3, 0)$

A6 Determine all values of k for which the vectors are orthogonal.
(a) $(3, -1), (2, k)$
(b) $(3, -1), (k, k^2)$
(c) $(1, 2, 3), (3, -k, k)$
(d) $(1, 2, 3), (k, k, -k)$

A7 Find the scalar equation of the plane containing the given point with the given normal.
(a) point $(-1, 2, -3)$, normal $(2, 4, -1)$
(b) point $(2, 5, 4)$, normal $(3, 0, 5)$
(c) point $(1, -1, 1)$, normal $(3, -4, 1)$

A8 Determine the scalar equation of the hyperplane passing through the given point with the given normal.
(a) point $(1, 1, -1, -2)$, normal $(3, 1, 4, 1)$
(b) point $(2, -2, 0, 1)$, normal $(0, 1, 3, 3)$

A9 Determine a normal vector for the plane or hyperplane.
(a) $3x_1 - 2x_2 + x_3 = 7$ in $\mathbb{R}^3$
(b) $-4x_1 + 3x_2 - 5x_3 - 6 = 0$ in $\mathbb{R}^3$
(c) $x_1 - x_2 + 2x_3 - 3x_4 = 5$ in $\mathbb{R}^4$

A10 Find an equation for the plane through the given point and parallel to the given plane.
(a) point $(1, -3, -1)$, plane $2x_1 - 3x_2 + 5x_3 = 17$
(b) point $(0, -2, 4)$, plane $x_2 = 0$

A11 Determine the point of intersection of the given line and plane.
(a) $\mathbf{x} = (2, 3, 1) + t(1, -2, -4)$, $t \in \mathbb{R}$, and $3x_1 - 2x_2 + 5x_3 = 11$.
(b) $\mathbf{x} = (1, 1, 2) + t(1, -1, -2)$, $t \in \mathbb{R}$, and $2x_1 + x_2 - x_3 = 5$.

A12 Find a vector equation for the plane that passes through the given points.
(a) $(3, 2, 1), (-4, 1, 7), (2, 0, 0)$
(b) $(-1, -4, 3), (-2, 4, 6), (3, 1, -4)$
(c) $(1, 0, 0), (0, 1, 0), (0, 0, 1)$

A13 Given the plane $2x_1 - x_2 + 3x_3 = 5$, for each of the following lines, determine if the line is parallel to the plane, orthogonal to the plane, or neither parallel nor orthogonal. If the answer is "neither", determine the angle between the direction vector of the line and the normal vector to the plane.

(a) $\mathbf{x} = (3, 0, 4) + t(-1, 1, 1)$, $t \in \mathbb{R}$
(b) $\mathbf{x} = (1, 1, 2) + t(-2, 1, -3)$, $t \in \mathbb{R}$
(c) $\mathbf{x} = (3, 0, 0) + t(1, 1, 2)$, $t \in \mathbb{R}$
(d) $\mathbf{x} = (-1, -1, 2) + t(4, -2, 6)$, $t \in \mathbb{R}$
(e) $\mathbf{x} = (0, 0, 0) + t(0, 3, 1)$, $t \in \mathbb{R}$

B1 Calculate the lengths of the given vectors.
(a) $(2, -6, -3)$ (b) $(-5, 1, 2)$ (c) $(3, 1, -3, -1)$

B2 Determine the distance from P to Q if
(a) P is $(-2, -2, 5)$ and Q is $(-4, 1, 4)$;
(b) P is $(3, 1, -3)$ and Q is $(-1, 4, 5)$;
(c) P is $(5, -2, -3, 6)$ and Q is $(2, 5, -4, 3)$.

B3 Verify the triangle inequality and the Cauchy-Schwarz inequality if
(a) $\mathbf{x} = (2, -6, -3)$ and $\mathbf{y} = (-3, 4, 5)$;
(b) $\mathbf{x} = (4, 1, -2)$ and $\mathbf{y} = (3, 5, 1)$.

B4 Determine the angle (in radians) between the vectors **a** and **b** if
(a) $\mathbf{a} = (2, 3, -5)$ and $\mathbf{b} = (-2, 5, -1)$;
(b) $\mathbf{a} = (2, 1, 0, 7)$ and $\mathbf{b} = (-3, 0, 5, 1)$.

B5 Determine whether the given pair of vectors is orthogonal.
(a) $(1, 4, 1)$, $(-4, 1, -4)$ (b) $(1, 3, 1)$, $(3, -1, 0)$
(c) $(1, 2, 1, 2)$, $(3, 1, -3, -1)$

B6 Determine all values of k for which the vectors are orthogonal.
(a) $(1, 2, 1)$, $(k, 2k, 4)$ (b) $(1, -1, 1)$, $(k, 3k, k^2)$

B7 Find the scalar equation of the plane containing the given point with the given normal vector.
(a) point $(-3, -3, 1)$, normal vector $(-1, 4, 7)$
(b) point $(6, -2, 5)$, normal vector $(4, -2, 1)$

B8 Find the scalar equation of the hyperplane containing the given point with the given normal vector.
(a) point $(2, 1, 1, 5)$, normal vector $(3, -2, -5, 1)$
(b) point $(3, 1, 0, 7)$, normal vector $(2, -4, 1, -3)$

B9 Determine a normal vector for the plane or hyperplane.
(a) $-x_1 - 2x_2 + 5x_3 = 7$ in $\mathbb{R}^3$
(b) $x_1 + 4x_2 - x_4 = 2$ in $\mathbb{R}^4$

B10 Find an equation for the plane through the given point and parallel to the given plane.
(a) point $(3, -1, 7)$, plane $5x_1 - x_2 - 2x_3 = 6$
(b) point $(-1, 2, -5)$, plane $2x_2 + 3x_3 = 7$

B11 Find the point of intersection of the line with the plane.
(a) line $\mathbf{x} = (1, -5, 3) + t(3, 2, -1)$, $t \in \mathbb{R}$, plane $2x_1 + 3x_2 - 7x_3 = 11$
(b) line $\mathbf{x} = (2, 1, 5) + t(2, -1, 4)$, $t \in \mathbb{R}$, plane $x_1 - 3x_2 + 4x_3 = 15$

B12 Find a vector equation for the plane that passes through the given points.
(a) $(1, 1, 4)$, $(2, 1, -4)$, $(0, 1, 1)$
(b) $(2, -1, -1)$, $(5, 1, 2)$, $(-3, 1, 0)$

B13 Given the plane $3x_1 + x_2 - x_3 = 0$, for each of the following lines, determine if the line is parallel to the plane, orthogonal to the plane, or neither parallel nor orthogonal. If the answer is "neither", determine the angle between the direction vector of the line and the normal vector to the plane.
(a) $\mathbf{x} = (0, 0, 1) + t(1, 2, 2)$, $t \in \mathbb{R}$
(b) $\mathbf{x} = (0, 1, 2) + t(6, 2, -2)$, $t \in \mathbb{R}$
(c) $\mathbf{x} = (2, 1, 0) + t(-1, -1, -4)$, $t \in \mathbb{R}$

CONCEPTUAL EXERCISES

D1 (a) Using intuitive geometrical arguments, what can you say about the vectors $\mathbf{a}$, $\mathbf{n}$, and $\mathbf{d}$ if the line with vector equation $\mathbf{x} = \mathbf{a} + t\mathbf{d}$ and the plane with scalar equation $\mathbf{n} \cdot \mathbf{x} = k$ have no point of intersection?
(b) Confirm your answer to part (a) by determining when it is possible to find a value of the parameter t that gives a point of intersection.

***D2** Prove, as a consequence of the triangle inequality, that
$\left| \|\mathbf{x}\| - \|\mathbf{y}\| \right| \leq \|\mathbf{x} - \mathbf{y}\|$. (Hint: $\|\mathbf{x}\| = \|\mathbf{x} - \mathbf{y} + \mathbf{y}\|$)

D3 Determine the equation of the set of points in $\mathbb{R}^3$ that are equidistant from points $\mathbf{a}$ and $\mathbf{b}$. Explain why the set is a plane, and determine its normal.

***D4** Consider the following statement: "If $\mathbf{a} \cdot \mathbf{b} = \mathbf{a} \cdot \mathbf{c}$ then $\mathbf{b} = \mathbf{c}$."
(a) If the statement is true, prove it. If the statement is false, provide a counterexample. (A counterexample is an example, meaning specific vectors $\mathbf{a}$, $\mathbf{b}$, $\mathbf{c}$, for which the statement is false.)
(b) If we specify $\mathbf{a} \neq \mathbf{0}$, does that change the result?

D5 (a) Let $\mathbf{n}$ be a unit vector in $\mathbb{R}^3$. Let α be the angle between $\mathbf{n}$ and the x-axis, let β be the angle between $\mathbf{n}$ and the y-axis, and let γ be the angle between $\mathbf{n}$ and the z-axis. Explain why

$$\mathbf{n} = (\cos\alpha, \cos\beta, \cos\gamma).$$

[Hint: What is $\mathbf{n} \cdot \mathbf{i}$?]
Because of this equation, the components n_1, n_2, n_3 are sometimes called **direction cosines.**

(b) Explain why

$$\cos^2 \alpha + \cos^2 \beta + \cos^2 \gamma = 1.$$

(c) Give a 2-dimensional version of direction cosines, and explain the connection to the identity

$$\cos^2 \theta + \sin^2 \theta = 1.$$

D6 Consider a cube such that each edge has length s. Let the four vertices on one face of the cube be A, B, C, D (taken in order) and let A′, B′, C′, D′ be the corresponding vertices on the opposite face.

(a) A solid with four vertices (not in a common plane) is called a tetrahedron, and it is *regular* if all edges are of equal length. Show that A′C′BD is a regular tetrahedron.

(b) Let P be the centre of the cube. Determine the angle A′PC′. [Hint: Locate A at the origin in $\mathbb{R}^3$, let B, D, A′ be points on the coordinate axes, and use vector methods.]

Remark: Calculations of angles in figures such as this are useful in studying molecules and crystals.

SECTION

1–3 Projection and Minimum Distance

The idea of projection is one of the important applications of the dot product. Suppose that we want to know "how much of a given vector **b** points along some other given vector **a**" (see Figure 1–3–1). In elementary physics, this is exactly what is required when a force is "resolved" into its components along certain directions (for example, into its vertical and horizontal components). When we define projections it is helpful to think of examples in two or three dimensions, but the ideas do not really depend on whether the vectors are in $\mathbb{R}^2$, $\mathbb{R}^3$, or $\mathbb{R}^n$. We begin with all vectors located at (or "starting at") the origin.

First consider the case where the vector $\mathbf{a} = \mathbf{i}$ in $\mathbb{R}^2$; how much of an arbitrary vector $\mathbf{b} = (b_1, b_2)$ points along $\mathbf{i}$? If we want simply the scalar magnitude, the answer is just the first component b_1 of $\mathbf{b}$. For later comparison, note that this component can be computed using the dot product: $b_1 = \mathbf{b} \cdot \mathbf{i}$. Note that b_1 can be positive or negative. We shall speak of the **component** of **b** along **i**, and write $cpt_{\mathbf{i}}(\mathbf{b}) = \mathbf{b} \cdot \mathbf{i} = b_1$.

In some problems, what is required is the **vector** in the direction of **i** with length equal to the component of **b** along **i**; this is easily seen to be $b_1\mathbf{i}$. This will be called the **projection of b onto i**, and is denoted $\mathbf{proj}_{\mathbf{i}}(\mathbf{b})$. For comparison with the general case below, note that $\mathbf{proj}_{\mathbf{i}}(\mathbf{b}) = b_1\mathbf{i} = (\mathbf{b} \cdot \mathbf{i})\mathbf{i}$.

Similarly, $cpt_{\mathbf{j}}(\mathbf{b}) = b_2$ and $\mathbf{proj}_{\mathbf{j}}(\mathbf{b}) = b_2\mathbf{j} = (\mathbf{b} \cdot \mathbf{j})\mathbf{j}$.

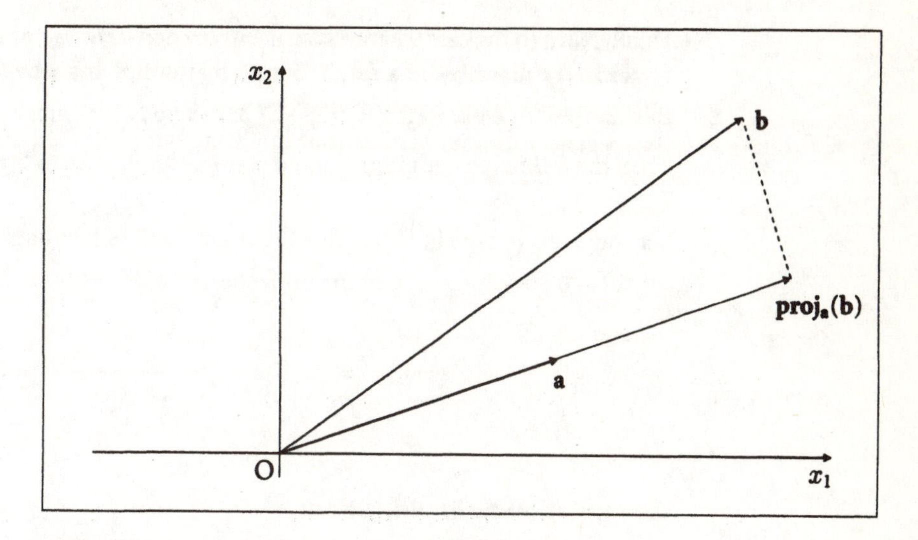

FIGURE 1-3-1. $\mathbf{proj}_{\mathbf{a}}(\mathbf{b})$ *is a vector in the direction of* **a**.

Next consider the case where **a** has arbitrary direction, but is a **unit vector**: a unit vector is a vector of length 1, so that $\|\mathbf{a}\| = 1$. Draw a perpendicular from (the point) **b** to the line through the origin with direction vector **a**. (Figure 1–3–1 with $\|\mathbf{a}\| = 1$.) The appropriate definitions are as follows:

if $\|\mathbf{a}\| = 1$, $cpt_{\mathbf{a}}(\mathbf{b}) = \|\mathbf{b}\| \cos\theta = \|\mathbf{b}\|\,\|\mathbf{a}\| \cos\theta = \mathbf{b} \cdot \mathbf{a}$ and

$$\mathbf{proj}_{\mathbf{a}}(\mathbf{b}) = (\mathbf{b} \cdot \mathbf{a})\mathbf{a}.$$

Note that the formulas above for component along **i** or projection onto **i** fit this pattern.

EXAMPLE 17

Find the projection of $\mathbf{b} = (5, 3, -2)$ onto the unit vector $\mathbf{a} = (1/\sqrt{3}, 1/\sqrt{3}, 1/\sqrt{3})$.

SOLUTION:

$$\mathbf{proj_a(b)} = (\mathbf{b} \cdot \mathbf{a})\mathbf{a} = \left(5/\sqrt{3} + 3/\sqrt{3} - 2/\sqrt{3}\right)\left(1/\sqrt{3}, 1/\sqrt{3}, 1/\sqrt{3}\right)$$
$$= 6/\sqrt{3}\left(1/\sqrt{3}, 1/\sqrt{3}, 1/\sqrt{3}\right) = (2, 2, 2).$$

Finally, turn to the case where $\mathbf{a}$ is an arbitrary non-zero vector in $\mathbb{R}^n$. **The unit vector in the direction of a** can be found by multiplying $\mathbf{a}$ by $(1/\|\mathbf{a}\|)$; denote this unit vector by $\hat{\mathbf{a}}$, so that $\hat{\mathbf{a}} = \dfrac{\mathbf{a}}{\|\mathbf{a}\|}$. (*You should check that* $\hat{\mathbf{a}}$ *is a unit vector.*) Then the following definition should seem natural; refer to Figure 1–3–1.

DEFINITION

If $\mathbf{a}$ and $\mathbf{b}$ are vectors in $\mathbb{R}^n$, define the **component** of $\mathbf{b}$ along $\mathbf{a}$ to be $cpt_{\mathbf{a}}(\mathbf{b}) = \mathbf{b} \cdot \hat{\mathbf{a}} = \mathbf{b} \cdot \dfrac{\mathbf{a}}{\|\mathbf{a}\|}$, and the **projection** of $\mathbf{b}$ onto $\mathbf{a}$ to be

$$\mathrm{proj}_{\mathbf{a}}(\mathbf{b}) = (\mathbf{b} \cdot \hat{\mathbf{a}})\hat{\mathbf{a}} = \left(\mathbf{b} \cdot \frac{\mathbf{a}}{\|\mathbf{a}\|}\right)\frac{\mathbf{a}}{\|\mathbf{a}\|}.$$

Computational simplification:

To avoid square roots, rewrite the formula for projection in the form

$$\mathbf{proj_a(b)} = \frac{\mathbf{b} \cdot \mathbf{a}}{\|\mathbf{a}\|^2}\,\mathbf{a}.$$

Notice that the formulas in this definition simplify to the earlier formulas if $\mathbf{a}$ is already a unit vector.

EXAMPLE 18

Let $\mathbf{a} = (4, 3, -1)$ and $\mathbf{b} = (-2, 5, 3)$. Determine $\mathbf{proj_a(b)}$ and $\mathbf{proj_b(a)}$.

SOLUTION:

$$\|\mathbf{a}\|^2 = 4^2 + 3^2 + (-1)^2 = 26$$

$$\begin{aligned}\mathbf{proj_a(b)} &= \frac{\mathbf{b}\cdot\mathbf{a}}{\|\mathbf{a}\|^2}\,\mathbf{a}\\ &= \frac{1}{26}\Big((-2,5,3)\cdot(4,3,-1)\Big)(4,3,-1)\\ &= \frac{4}{26}(4,3,-1) = \left(\frac{8}{13},\frac{6}{13},\frac{-2}{13}\right)\end{aligned}$$

$$\begin{aligned}\mathbf{proj_b(a)} &= \Big((4,3,-1)\cdot(-2,5,3)/(4+25+9)\Big)(-2,5,3)\\ &= \frac{4}{38}(-2,5,3) = \left(\frac{-4}{19},\frac{10}{19},\frac{6}{19}\right)\end{aligned}$$

REMARKS

(1) This example illustrates that in general, $\mathbf{proj_a(b)} \neq \mathbf{proj_b(a)}$; of course, we should not expect equality because $\mathbf{proj_a(b)}$ is in the direction of **a**, whereas $\mathbf{proj_b(a)}$ is in the direction of **b**.

(2) Our notation has emphasized the fact that $cpt_{\mathbf{a}}$ and $\mathbf{proj_a}$ are functions. Both functions have the appropriate $\mathbb{R}^n$ as "domain of definition", but there is a crucial distinction: $cpt_{\mathbf{a}}$ is **real** (scalar)-**valued,** whereas $\mathbf{proj_a}$ is **vector-valued** (that is, the result is a vector in the appropriate $\mathbb{R}^n$). Now that we have emphasized that they are functions, we shall often abbreviate by omitting the parentheses and write $cpt_{\mathbf{a}}\mathbf{b}$ and $\mathbf{proj_a b}$.

(3) Notice that $|cpt_{\mathbf{a}}\mathbf{b}| = \|\mathbf{proj_a b}\|$. The absolute value is necessary because $cpt_{\mathbf{a}}\mathbf{b}$ may be negative.

The Perpendicular Part

When you resolve forces in physics, you often want the component of a force **F** *perpendicular* to a given direction **d**. This "perpendicular part" of **F** has important geometrical applications. In $\mathbb{R}^2$ we could find the perpendicular part of **F** by finding a vector in the unique direction perpendicular to **d** (an easy problem in $\mathbb{R}^2$): call this vector $\mathbf{d}^{\perp}$. Then calculate $cpt_{\mathbf{d}^{\perp}}\mathbf{F}$ to get the required component (or $\mathbf{proj_{d^{\perp}} F}$ if the vector part of **F** perpendicular to **d** is required). To follow a similar strategy in $\mathbb{R}^n$ for $n > 2$ would require a lot more mathematical machinery. We now describe a much simpler approach that works for many problems of this sort.

We begin by restating the problem: in $\mathbb{R}^n$, given a fixed vector **a**, and any other vector **x**, express **x** as the sum of a vector parallel (or anti-parallel) to **a** and a vector orthogonal to **a**. (*Draw yourself a picture.*) That is,

write $\mathbf{x} = \mathbf{w} + \mathbf{z}$, *where* $\mathbf{w} = k\,\mathbf{a}$ *for some real number k, and* $\mathbf{z} \cdot \mathbf{a} = 0$.

If this is possible, what can we say about **w** and **z**? To find out, we use what turns out to be a very useful and common trick: take the dot product of **x** with **a**.

$$\mathbf{a} \cdot \mathbf{x} = \mathbf{a} \cdot (\mathbf{w} + \mathbf{z}) = \mathbf{a} \cdot (k\mathbf{a} + \mathbf{z}) = k(\mathbf{a} \cdot \mathbf{a}) + \mathbf{a} \cdot \mathbf{z} = k\|\mathbf{a}\|^2 + 0$$

Therefore, k must be defined to be $\dfrac{\mathbf{x} \cdot \mathbf{a}}{\|\mathbf{a}\|^2}$, so that in fact,

$$\mathbf{w} = k\mathbf{a} = \mathbf{proj}_{\mathbf{a}}\mathbf{x},$$

as we might have expected. One bonus of approaching the general problem this way is that it is now clear that this is the only way to choose **w** to satisfy the problem. Next, since $\mathbf{x} = \mathbf{proj}_{\mathbf{a}}\mathbf{x} + \mathbf{z}$, it follows that

$$\mathbf{z} = \mathbf{x} - \mathbf{proj}_{\mathbf{a}}\mathbf{x}.$$

Is this **z** really orthogonal to **a**? To check, calculate

$$\begin{aligned}\mathbf{z} \cdot \mathbf{a} = (\mathbf{x} - \mathbf{proj}_{\mathbf{a}}\mathbf{x}) \cdot \mathbf{a} &= \mathbf{x} \cdot \mathbf{a} - \left(\left(\frac{\mathbf{x} \cdot \mathbf{a}}{\|\mathbf{a}\|^2}\right)\mathbf{a}\right) \cdot \mathbf{a} \\ &= \mathbf{x} \cdot \mathbf{a} - \left(\frac{\mathbf{x} \cdot \mathbf{a}}{\|\mathbf{a}\|^2}\right)\|\mathbf{a}\|^2 \\ &= \mathbf{x} \cdot \mathbf{a} - \mathbf{x} \cdot \mathbf{a} = 0,\end{aligned}$$

so **z** is orthogonal to **a** as required. Since it is often useful to construct a vector **z** in this way, we introduce a name for it.

If **a** and **x** are vectors in $\mathbb{R}^n$, define the **projection of x perpendicular to a to be**

$$\mathbf{perp}_{\mathbf{a}}(\mathbf{x}) = \mathbf{x} - \mathbf{proj}_{\mathbf{a}}(\mathbf{x}).$$

Notice that $\mathbf{perp}_{\mathbf{a}}$ is again a **vector-valued** function on $\mathbb{R}^n$. Notice also that $\mathbf{x} = \mathbf{proj}_{\mathbf{a}}\mathbf{x} + \mathbf{perp}_{\mathbf{a}}\mathbf{x}$. ($\mathbf{perp}_{\mathbf{a}}\mathbf{x}$ is not called the "component" perpendicular to **a** because components are scalar-valued.) See Figure 1–3–2.

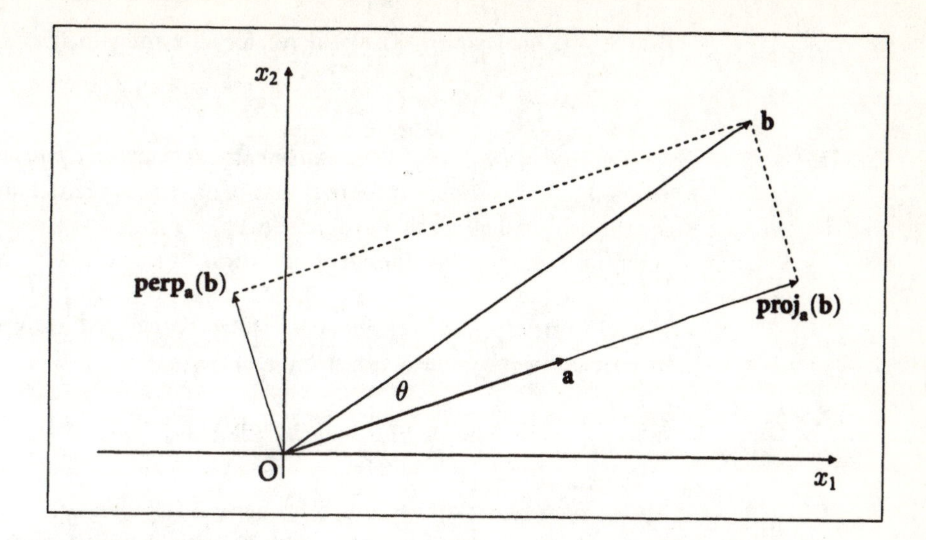

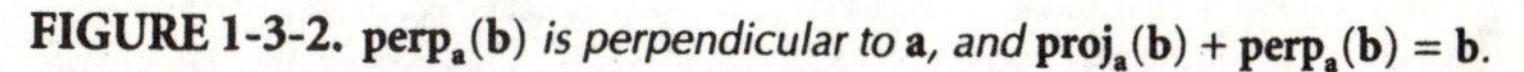
FIGURE 1-3-2. $\mathbf{perp_a(b)}$ *is perpendicular to* $\mathbf{a}$, *and* $\mathbf{proj_a(b)} + \mathbf{perp_a(b)} = \mathbf{b}$.

EXAMPLE 19

Determine $\mathbf{proj}_{(2,1,1)}(1,5,1)$ and $\mathbf{perp}_{(2,1,1)}(1,5,1)$.

SOLUTION:

$$\begin{aligned}\mathbf{proj}_{(2,1,1)}(1,5,1) &= \frac{(1,5,1)\cdot(2,1,1)}{\|(2,1,1)\|^2}(2,1,1)\\ &= \left(\frac{2+5+1}{4+1+1}\right)(2,1,1)\\ &= (8/6)(2,1,1) = (8/3,4/3,4/3);\end{aligned}$$

$$\mathbf{perp}_{(2,1,1)}(1,5,1) = (1,5,1) - (8/3,4/3,4/3) = (-5/3,11/3,-1/3).$$

There is a convenient check available: $(2,1,1)$ should be orthogonal to $(-5/3,11/3,-1/3)$; is it?

Some Properties of Projections

Projections will appear several times in this book, and some of their special properties are very important. Two of these are called the **linearity** properties:

(L1) $\mathbf{proj_a}(\mathbf{x}+\mathbf{y}) = \mathbf{proj_a}(\mathbf{x}) + \mathbf{proj_a}(\mathbf{y})$, for all $\mathbf{x}$ and $\mathbf{y}$ in $\mathbb{R}^n$;

(L2) $\mathbf{proj_a}(k\mathbf{x}) = k\,\mathbf{proj_a}(\mathbf{x})$, for all $\mathbf{x}$ in $\mathbb{R}^n$, and all k in $\mathbb{R}$.

(*You should check that these statements are true; they follow easily from the definition of projection.*) It follows that $\mathbf{perp_a}$ also satisfies the corresponding equations. We shall see that $\mathbf{proj_a}$ and $\mathbf{perp_a}$ are just two cases among the many functions satisfying the linearity properties.

$\mathbf{proj_a}$ and $\mathbf{perp_a}$ also have one special property called the **projection** property; we write it for $\mathbf{proj_a}$, but it is also true for $\mathbf{perp_a}$:

$$\mathbf{proj_a}(\mathbf{proj_a}(\mathbf{x})) = \mathbf{proj_a}(\mathbf{x}) \text{ for all } \mathbf{x} \text{ in } \mathbb{R}^n.$$

One final remark may be helpful here. In $\mathbb{R}^2$, $\mathbf{perp_j}(\mathbf{x}) = \mathbf{proj_i}(\mathbf{x})$ for all $\mathbf{x}$; more generally $\mathbf{perp_d}(\mathbf{x}) = \mathbf{proj_{d^\perp}}(\mathbf{x})$. It is generally true that the problem of finding the perpendicular part of a vector can be converted to a projection problem, but to do this we need the concepts of subspace (Chapter 3) and projection onto a subspace (Chapter 7).

Minimum Distance

A Distance Problem

What is the distance from the point B(4, 3) to the line with vector equation $\mathbf{x} = (1,2) + t(-1,1)$? In this and similar problems, distance always means the minimum distance; geometrically, we see that the minimum distance is found along a line segment from B perpendicular to the given line. A formal proof that minimum distance requires perpendicularity can be given using Pythagoras's theorem. (See Exercise D4. A proof is given in a more general setting in Section 7–2.)

To answer the particular question, take *any* point on the line $\mathbf{x} = (1,2) + t(-1,1)$: the obvious choice is $(1,2)$, which we call A. From Figure 1–3–3, we see that the required distance is the length of $\mathbf{perp}_{(-1,1)}\overrightarrow{AB}$. Since $\overrightarrow{AB} = (4,3) - (1,2) = (3,1)$, the distance is

$$\begin{aligned}\|\mathbf{perp}_{(-1,1)}(3,1)\| &= \|(3,1) - \mathbf{proj}_{(-1,1)}(3,1)\| \\ &= \left\|(3,1) - \left(\frac{-3+1}{1+1}\right)(-1,1)\right\| \\ &= \|(3,1) + 1(-1,1)\| \\ &= \|(2,2)\| = 2\sqrt{2}.\end{aligned}$$

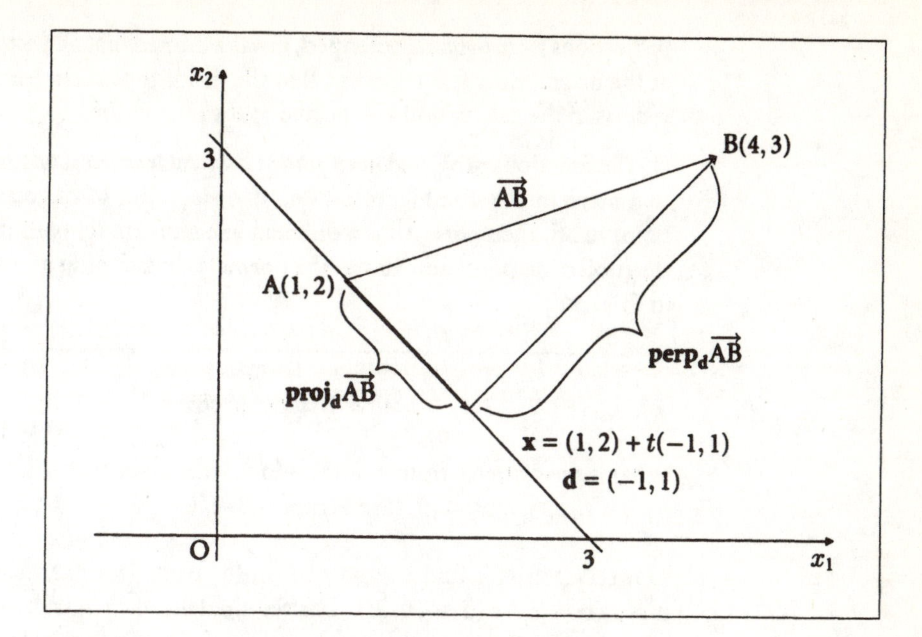

FIGURE 1-3-3. *The distance from the point* B *to the line* $\mathbf{x} = \mathbf{a} + t\mathbf{d}$ *is* $\|\mathbf{perp_d}(\mathbf{b} - \mathbf{a})\| = \|\mathbf{perp_d}\,\overrightarrow{AB}\|$, *where* $\mathbf{d} = (-1, 1)$.

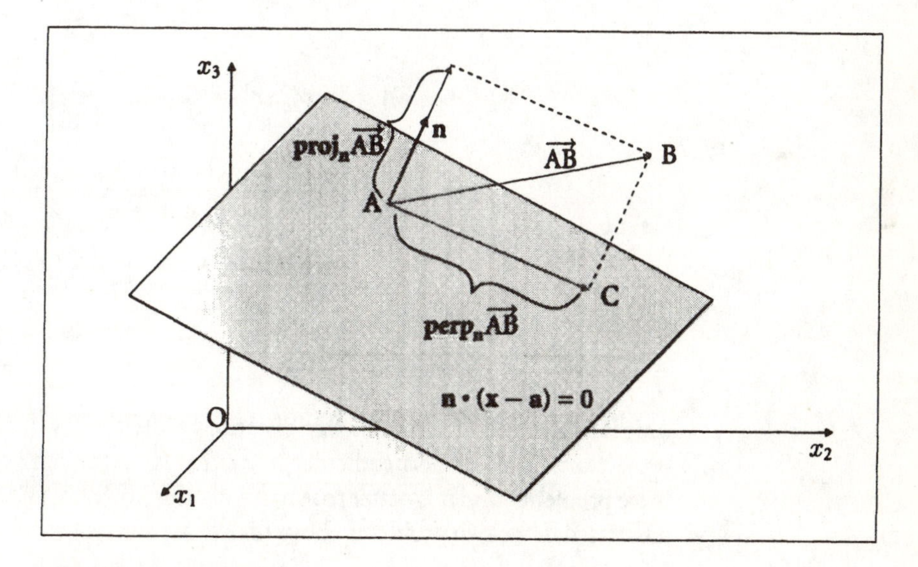

FIGURE 1-3-4. $\mathbf{proj_n}\overrightarrow{AB}$ *and* $\mathbf{perp_n}\overrightarrow{AB}$, *where* $\mathbf{n}$ *is normal to the plane.*

Notice that in this problem and in similar problems, we take advantage of the fact that the direction vector **d** can be thought of as "starting" at any point; when $\mathbf{perp}_{(-1,1)}\overrightarrow{AB}$ is calculated, both vectors are located at the point A. When projections were originally defined, it was assumed that all vectors were located at the origin; now it is apparent that the definitions make sense as long as all vectors in the calculation are located at the same point.

The functions $\mathbf{proj_n}$ and $\mathbf{perp_n}$ are of particular interest when $\mathbf{n}$ is the normal to a plane in $\mathbb{R}^3$. See Figure 1–3–4. If A is a point in the plane and B is any point in $\mathbb{R}^3$, then $\mathbf{perp_n}\overrightarrow{AB}$ is a directed line segment lying in the plane. Thus, the projection perpendicular to the normal is in fact projection onto the plane in this case.

EXAMPLE 20 Another distance problem

What is the distance from a point B in $\mathbb{R}^3$ to a plane with equation $n_1x_1 + n_2x_2 + n_3x_3 = d$? (See Figure 1–3–4, with $\mathbf{n} \cdot \mathbf{a} = d$.)

SOLUTION: We find any point A in the plane (for example, if $n_1 \neq 0$, we may take A to be $(d/n_1, 0, 0)$). The required distance must be measured along a line perpendicular to the plane, that is, along a line with direction vector $\mathbf{n}$. Thus from Figure 1–3–4, the distance is the absolute value of the component of $\overrightarrow{AB} = (\mathbf{b} - \mathbf{a})$ along the normal vector $\mathbf{n}$:

$$
\begin{aligned}
|cpt_{\mathbf{n}}(\mathbf{b}-\mathbf{a})| = \|\mathbf{proj_n}\overrightarrow{AB}\| &= \left|(\mathbf{b}-\mathbf{a}) \cdot \frac{\mathbf{n}}{\|\mathbf{n}\|}\right| \\
&= \left|\frac{(b_1 - d/n_1, b_2, b_3) \cdot (n_1, n_2, n_3)}{\sqrt{n_1^2 + n_2^2 + n_3^2}}\right| \\
&= \left|\frac{n_1b_1 + n_2b_2 + n_3b_3 - d}{\sqrt{n_1^2 + n_2^2 + n_3^2}}\right|.
\end{aligned}
$$

This is a standard formula for this distance problem. However, the lengths of projections along or perpendicular to a suitable vector can be used for all of these problems, and it is better to learn to use this powerful and versatile idea as illustrated in the two problems above than to memorize complicated formulas.

EXAMPLE 21 Finding the nearest point

In some applications we need to determine the point in the plane that is nearest to the point B. Let us call this point C, as in Figure 1–3–4. Then we can determine

C by observing that

$$\overrightarrow{OC} = \overrightarrow{OA} + \overrightarrow{AC} = \overrightarrow{OA} + \textbf{perp}_{\mathbf{n}}\overrightarrow{AB}.$$

However, we get an easier calculation if we observe from the figure that

$$\overrightarrow{OC} = \overrightarrow{OB} + \overrightarrow{BC} = \mathbf{b} + \textbf{proj}_{\mathbf{n}}\overrightarrow{BA};$$

notice that we need $\overrightarrow{BA}$ here instead of $\overrightarrow{AB}$. Exercise D5 asks you to check that these two calculations of $\overrightarrow{OC}$ are consistent.

If the plane in this problem happens to pass through the origin, then we may take A=O, and the point in the plane that is closest to B is given by $\textbf{perp}_{\mathbf{n}}\overrightarrow{OB}$.

ALTERNATIVE SOLUTION: Write the (vector) equation of the line through **b** with direction vector **n**. Determine the point where the line meets the plane by substituting the parametric equations for x_1, x_2, x_3 into the equation of the plane and solving for the parameter; then use the parameter to determine the point **x**. You can also use this alternative method to find the distance from this point to **b**.

EXAMPLE 22 An application in the analysis of data

An investigator carries out 10 measurements on each person in his class who can be persuaded to have the measurements taken. Thus for each person, he has a "data point" $(d_1, d_2, \ldots, d_{10})$. His theoretical model about these particular measurements says that the data points should satisfy the equation $n_1d_1 + n_2d_2 + n_3d_3 + \cdots + n_{10}d_{10} = 0$, for some given coefficients $n_1, n_2, \ldots, n_{10}$. That is, **d** should be a point in the hyperplane $\mathbf{n} \cdot \mathbf{x} = 0$, for some given 10-vector **n**. Call the points in this hyperplane "model points". Now any measurement is subject to error, so the data points do not lie exactly in the hyperplane. It seems reasonable to assign to any data point the closest model point. How does he find the model point corresponding to a data point **d**?

SOLUTION: This is simply the n-dimensional version of finding the nearest point to a plane, in the special case where the plane passes through the origin. The point on the hyperplane that is at the minimum distance from **d** is the point obtained by perpendicular projection onto the hyperplane, that is, the model point corresponding to the data point **d** is $\textbf{perp}_{\mathbf{n}}\mathbf{d}$.

Many questions of finding a best approximation, or of finding the best way to "fit" data to a model, boil down to problems of finding minimum *distances in some $\mathbb{R}^n$. They are often described as "least squares fits", and they can often best be solved with the aid of projection techniques. This idea is discussed in Section 7–3.*

EXERCISES 1–3

A1 For each of the given pairs of vectors $\mathbf{a}, \mathbf{b}$, check that $\mathbf{a}$ is a unit vector, determine $\mathbf{proj}_{\mathbf{a}}(\mathbf{b})$ and $\mathbf{perp}_{\mathbf{a}}(\mathbf{b})$, and check your results by verifying that $\mathbf{proj}_{\mathbf{a}}(\mathbf{b}) + \mathbf{perp}_{\mathbf{a}}(\mathbf{b}) = \mathbf{b}$ and $\mathbf{a} \cdot \mathbf{perp}_{\mathbf{a}}(\mathbf{b}) = 0$ in each case.

(a) $\mathbf{a} = (0, 1)$ and $\mathbf{b} = (3, -5)$
(b) $\mathbf{a} = (3/5, 4/5)$ and $\mathbf{b} = (-4, 6)$
(c) $\mathbf{a} = (0, 1, 0)$, $\mathbf{b} = (-3, 5, 2)$
(d) $\mathbf{a} = (1/3, -2/3, 2/3)$, $\mathbf{b} = (4, 1, -3)$

A2 Consider the force represented by the vector $\mathbf{F} = (10, 18, -6)$, and let $\mathbf{u} = (2, 6, 3)$.

(a) Determine a unit vector in the direction of $\mathbf{u}$.
(b) Find the component of $\mathbf{F}$ in the direction of $\mathbf{u}$.
(c) Determine the projection of $\mathbf{F}$ onto $\mathbf{u}$.
(d) Determine the projection of $\mathbf{F}$ perpendicular to $\mathbf{u}$.

A3 The same instructions as A2 with $\mathbf{F} = (3, 11, 2)$ and $\mathbf{u} = (3, 1, -2)$.

A4 Determine

(a) $\mathbf{proj}_{(2,3,-2)}(4, -1, 3)$ and $\mathbf{perp}_{(2,3,-2)}(4, -1, 3)$;
(b) $\mathbf{proj}_{(1,1,-2)}(4, 1, -2)$ and $\mathbf{perp}_{(1,1,-2)}(4, 1, -2)$;
(c) $\mathbf{proj}_{(-2,1,-1)}(5, -1, 3)$ and $\mathbf{perp}_{(-2,1,-1)}(5, -1, 3)$;
(d) $\mathbf{proj}_{(-1,2,1,-3)}(2, -1, 2, 1)$ and $\mathbf{perp}_{(-1,2,1,-3)}(2, -1, 2, 1)$.

A5 For the given point and line, find by projection the point on the line that is closest to the given point, and use **perp** to find the distance from the point to the line.

(a) point $(0, 0)$, line $\mathbf{x} = (1, 4) + t(-2, 2)$, $t \in \mathbb{R}$
(b) point $(2, 5)$, line $\mathbf{x} = (3, 7) + t(1, -4)$, $t \in \mathbb{R}$
(c) point $(1, 0, 1)$, line $\mathbf{x} = (2, 2, -1) + t(1, -2, 1)$, $t \in \mathbb{R}$
(d) point $(2, 3, 2)$, line $\mathbf{x} = (1, 1, -1) + t(1, 4, 1)$, $t \in \mathbb{R}$

A6 Use a projection (onto or perpendicular to) to find the distance from the point to the plane.

(a) point $(2, 3, 1)$, plane $3x_1 - x_2 + 4x_3 = 5$
(b) point $(-2, 3, -1)$, plane $2x_1 - 3x_2 - 5x_3 = 5$
(c) point $(0, 2, -1)$, plane $2x_1 - x_3 = 5$
(d) point $(-1, -1, 1)$, plane $2x_1 - x_2 - x_3 = 4$

A7 For the given point and hyperplane in $\mathbb{R}^4$, determine by a projection the point in the hyperplane that is closest to the given point.

(a) point $(2, 4, 3, 4)$, hyperplane $3x_1 - x_2 + 4x_3 + x_4 = 0$
(b) point $(-1, 3, 2, -2)$, hyperplane $x_1 + 2x_2 + x_3 - x_4 = 4$

B1 For each of the given pairs of vectors $\mathbf{a}, \mathbf{b}$, check that $\mathbf{a}$ is a unit vector, determine $\mathbf{proj_a}(\mathbf{b})$ and $\mathbf{perp_a}(\mathbf{b})$, and check your results by verifying that $\mathbf{proj_a}(\mathbf{b}) + \mathbf{perp_a}(\mathbf{b}) = \mathbf{b}$ and $\mathbf{a} \cdot \mathbf{perp_a}(\mathbf{b}) = 0$ in each case.

(a) $\mathbf{a} = (0, 1)$ and $\mathbf{b} = (4, 3)$
(b) $\mathbf{a} = (4/5, -3/5)$ and $\mathbf{b} = (-2, 5)$
(c) $\mathbf{a} = (0, 0, 1)$, $\mathbf{b} = (2, -4, 7)$
(d) $\mathbf{a} = (2/3, 2/3, -1/3)$, $\mathbf{b} = (-2, 3, -2)$

B2 Consider the force represented by the vector $\mathbf{F} = (12, -15, 4)$, and let $\mathbf{u} = (-1, 8, 2)$.

(a) Determine a unit vector in the direction of $\mathbf{u}$.
(b) Find the component of $\mathbf{F}$ in the direction of $\mathbf{u}$.
(c) Determine the projection of $\mathbf{F}$ onto $\mathbf{u}$.
(d) Determine the projection of $\mathbf{F}$ perpendicular to $\mathbf{u}$.

B3 The same instructions as B2 with $\mathbf{F} = (5, 13, 3)$ and $\mathbf{u} = (2, -4, 3)$.

B4 Determine

(a) $\mathbf{proj}_{(1,-2,1)}(5, 1, -2)$ and $\mathbf{perp}_{(1,-2,1)}(5, 1, -2)$;
(b) $\mathbf{proj}_{(0,1,-1)}(4, 4, 2)$ and $\mathbf{perp}_{(0,1,-1)}(4, 4, 2)$;
(c) $\mathbf{proj}_{(1,0,-1)}(6, 2, 6)$ and $\mathbf{perp}_{(1,0,-1)}(6, 2, 6)$;
(d) $\mathbf{proj}_{(2,0,1,1)}(-1, 2, -1, 2)$ and $\mathbf{perp}_{(2,0,1,1)}(-1, 2, -1, 2)$.

B5 For the given point and line, find by projection the point on the line that is closest to the given point, and use **perp** to find the distance from the point to the line.

(a) point $(3, -5)$, line $\mathbf{x} = (2, -4) + t(3, -4)$, $t \in \mathbb{R}$
(b) point $(0, 0, 2)$, line $\mathbf{x} = (2, 1, 0) + t(-1, 2, 2)$, $t \in \mathbb{R}$
(c) point $(1, -3, 0)$, line $\mathbf{x} = (1, -1, -1) + t(1, 1, 1)$, $t \in \mathbb{R}$
(d) point $(-1, 2, -3)$, line $\mathbf{x} = (4, 1, -2) + t(2, 1, 4)$, $t \in \mathbb{R}$

B6 Use a projection (onto or perpendicular to) to find the distance from the point to the plane.

(a) point $(2, -3, -1)$, plane $2x_1 - 3x_2 - 5x_3 = 7$
(b) point $(2, 2, 3)$, plane $2x_1 + x_2 - 4x_3 = 5$
(c) point $(2, -1, 2)$, plane $x_1 - x_2 - x_3 = 6$

B7 For the given point and hyperplane in $\mathbb{R}^4$, determine by a projection the point in the hyperplane that is closest to the given point.

(a) point $(1, 3, 0, -1)$, hyperplane $2x_1 - 2x_2 + x_3 + 3x_4 = 0$
(b) point $(5, 2, 3, 7)$, hyperplane $2x_1 + x_2 + 4x_3 + 3x_4 = 40$

B8 Use a projection to determine the distance between the two parallel lines $\mathbf{x} = (3, -1, 2, 5) + s(1, 2, 0, 1)$, $s \in \mathbb{R}$, and $\mathbf{x} = (2, 7, 1, 0) + t(1, 2, 0, 1)$, $t \in \mathbb{R}$.

COMPUTER RELATED EXERCISES

C1 Determine if your computer software handles projections and dot products. If it does, let $\mathbf{a} = (1.12, 2.10, 7.03, 4.15, 6.13)$ and $\mathbf{b} = (1.00, -1.01, 1.02, -1.03, 1.04)$, and determine $\mathbf{proj_a}(\mathbf{b})$ and $\mathbf{perp_a}(\mathbf{b})$. (You may have to do a subtraction to calculate $\mathbf{perp_a}(\mathbf{b})$.) Also determine $\mathbf{proj_b}(\mathbf{a})$.

CONCEPTUAL EXERCISES

***D1** Verify that the functions $\mathbf{proj_a}$ and $\mathbf{perp_a}$ defined on $\mathbb{R}^n$ for an arbitrary vector $\mathbf{a}$ do have the linearity properties.

D2 (a) Given $\mathbf{a}$ and $\mathbf{b}$ in $\mathbb{R}^3$, verify that the composite map $C : \mathbb{R}^3 \to \mathbb{R}^3$ defined by $C(\mathbf{x}) = \mathbf{proj_a}(\mathbf{proj_b}\mathbf{x})$ also has the linearity properties.
(b) Suppose that $C(\mathbf{x}) = \mathbf{0}$ for all $\mathbf{x}$ in $\mathbb{R}^3$, where C is defined as in part (a) of this question; what can you say about $\mathbf{a}$ and $\mathbf{b}$? Explain.

D3 By one of the linearity properties, we know that $\mathbf{proj_a}(-\mathbf{x}) = -\mathbf{proj_a}(\mathbf{x})$. Check, and explain geometrically, that $\mathbf{proj_{-a}}(\mathbf{x}) = \mathbf{proj_a}(\mathbf{x})$. It is sufficient to do this in $\mathbb{R}^2$.

***D4** (a) (Pythagoras's theorem) Use the fact that $\|\mathbf{x}\|^2 = \mathbf{x} \cdot \mathbf{x}$ to prove that $\|\mathbf{u} + \mathbf{v}\|^2 = \|\mathbf{u}\|^2 + \|\mathbf{v}\|^2$ if and only if $\mathbf{u} \cdot \mathbf{v} = 0$.
(b) Let ℓ be the line in $\mathbb{R}^n$ with equation $\mathbf{x} = t\mathbf{d}$, and let $\mathbf{a}$ be any point that is *not* on ℓ. Let $\mathbf{x}$ be a point on ℓ, and prove that the smallest value of $\|\mathbf{a} - \mathbf{x}\|^2$ is obtained when $\mathbf{x} = \mathbf{proj_d}(\mathbf{a})$ (that is, when $\mathbf{a} - \mathbf{x}$ is perpendicular to $\mathbf{d}$). Hint: Consider $\|\mathbf{a} - \mathbf{x}\| = \|\mathbf{a} - \mathbf{proj_d}(\mathbf{a}) + \mathbf{proj_d}(\mathbf{a}) - \mathbf{x}\|$.

D5 By using the definition of $\mathbf{perp_n}$, and the fact that $\vec{AB} = -\vec{BA}$, show that, as discussed in Example 21,

$$\vec{OA} + \mathbf{perp_n}\vec{AB} = \vec{OB} + \mathbf{proj_n}\vec{BA}.$$

D6 Find the scalar equation of the plane such that each point of the plane is equidistant from the points $A(2, 2, 5)$ and $B(-3, 4, 1)$ in two ways:
(a) write and simplify the equation $\|\vec{AX}\| = \|\vec{BX}\|$;
(b) determine a point on the plane and the normal vector by geometrical arguments.

D7 (a) Let $\mathbf{a} = (1, 1, -1)$ and $\mathbf{x} = (2, 5, 3)$. Show that $\mathbf{proj_a}(\mathbf{perp_a}(\mathbf{x})) = \mathbf{0}$.
(b) For general $\mathbf{a}$ in $\mathbb{R}^3$, prove that for any $\mathbf{x}$ in $\mathbb{R}^3$, $\mathbf{proj_a}(\mathbf{perp_a}(\mathbf{x})) = \mathbf{0}$ (an algebraic proof!).
(c) Explain geometrically why $\mathbf{proj_a}(\mathbf{perp_a}(\mathbf{x})) = \mathbf{0}$ for every $\mathbf{x}$. (A picture in $\mathbb{R}^2$ might help.)

SECTION

1–4 The Vector Cross-Product in $\mathbb{R}^3$; Volume

Given a pair of vectors $\mathbf{a}$ and $\mathbf{b}$ in $\mathbb{R}^3$, how can one find a third vector $\mathbf{c}$ that is orthogonal to both $\mathbf{a}$ and $\mathbf{b}$? This problem arises in many natural ways; for example, to find the scalar equation of a plane whose vector equation is $\mathbf{x} = r\mathbf{u} + s\mathbf{v}$, we must find the normal vector $\mathbf{n}$ orthogonal to $\mathbf{u}$ and $\mathbf{v}$. In physics, it is observed that the force on an electrically charged particle moving in a magnetic field is in the direction orthogonal to the velocity of the particle and to the vector describing the magnetic field.

If $\mathbf{c}$ is orthogonal to both $\mathbf{a}$ and $\mathbf{b}$, it must satisfy the equations

$$\mathbf{a} \cdot \mathbf{c} = a_1c_1 + a_2c_2 + a_3c_3 = 0$$

and

$$\mathbf{b} \cdot \mathbf{c} = b_1c_1 + b_2c_2 + b_3c_3 = 0.$$

In Chapter 2, we shall develop systematic methods for solving such equations for (c_1, c_2, c_3); for the present, we simply give a solution:

$$\mathbf{c} = (a_2b_3 - a_3b_2, a_3b_1 - a_1b_3, a_1b_2 - a_2b_1).$$

You should verify that this $\mathbf{c}$ does satisfy $\mathbf{a} \cdot \mathbf{c} = 0$ and $\mathbf{b} \cdot \mathbf{c} = 0$. Also notice from the form of the equations $\mathbf{a} \cdot \mathbf{c} = 0$ and $\mathbf{b} \cdot \mathbf{c} = 0$ that any multiple of this vector would also satisfy the equations.

DEFINITION

The **vector cross-product** of vectors $\mathbf{a} = (a_1, a_2, a_3)$ and $\mathbf{b} = (b_1, b_2, b_3)$ in $\mathbb{R}^3$ is denoted by $\mathbf{a} \times \mathbf{b}$ and defined by

$$\begin{aligned}\mathbf{a} \times \mathbf{b} &= (a_2b_3 - a_3b_2, a_3b_1 - a_1b_3, a_1b_2 - a_2b_1)\\ &= (a_2b_3 - a_3b_2)\mathbf{i} + (a_3b_1 - a_1b_3)\mathbf{j} + (a_1b_2 - a_2b_1)\mathbf{k}.\end{aligned}$$

EXAMPLE 23

Calculate the cross-product of $(2, 3, 5)$ and $(-1, 1, 2)$.

SOLUTION: $(2, 3, 5) \times (-1, 1, 2) = (6 - 5, -5 - 4, 2 - [-3]) = (1, -9, 5)$.

Important notes:

(1) The cross-product of **a** and **b** is itself a new *vector.*

(2) The cross-product is a construction that is defined only in $\mathbb{R}^3$. (There is a generalization to higher dimensions, but it is considerably more complicated and it will not be considered in this book.)

The formula for the cross-product is a little awkward to remember and it is common to use a "determinant" to help remember it. We shall study the determinant in detail later and only introduce now what is needed for the cross-product. First *define* a 2 by 2 determinant: $\begin{vmatrix} a & b \\ c & d \end{vmatrix} = ad - bc.$

Next, write $\mathbf{a} \times \mathbf{b}$ as a "3 by 3 determinant" and "expand"; the following equalities are *true by definition*:

$$\mathbf{a} \times \mathbf{b} = \begin{vmatrix} \mathbf{i} & \mathbf{j} & \mathbf{k} \\ a_1 & a_2 & a_3 \\ b_1 & b_2 & b_3 \end{vmatrix} = \mathbf{i}\begin{vmatrix} a_2 & a_3 \\ b_2 & b_3 \end{vmatrix} - \mathbf{j}\begin{vmatrix} a_1 & a_3 \\ b_1 & b_3 \end{vmatrix} + \mathbf{k}\begin{vmatrix} a_1 & a_2 \\ b_1 & b_2 \end{vmatrix}$$
$$= \mathbf{i}(a_2b_3 - a_3b_2) - \mathbf{j}(a_1b_3 - a_3b_1) + \mathbf{k}(a_1b_2 - a_2b_1)$$

Notice that in the first expression (the 3 by 3 determinant) for $\mathbf{a} \times \mathbf{b}$, we write the standard basis vectors of $\mathbb{R}^3$ in the first row, the components of **a** in the second row, and the components of **b** in the third row. When we "expand", we pick out the basis vectors with a special rule for the signs: *note carefully that* **j** *must be given a minus sign in order for this procedure to provide the correct answer.* (Notice that in the second component of the answer, the terms are rearranged from the terms in the original definition of $\mathbf{a} \times \mathbf{b}$.) In the expansion, each basis vector is multiplied by a 2 by 2 determinant; the entries in this 2 by 2 determinant are obtained by *deleting* the first row of the 3 by 3 determinant and the column containing the corresponding basis vector. That is, delete the first column for **i**, the second column for **j**, and the third for **k**.

EXAMPLE 24

$$(3, -2, 1) \times (2, 3, 7) = \begin{vmatrix} \mathbf{i} & \mathbf{j} & \mathbf{k} \\ 3 & -2 & 1 \\ 2 & 3 & 7 \end{vmatrix} = \mathbf{i}\begin{vmatrix} -2 & 1 \\ 3 & 7 \end{vmatrix} - \mathbf{j}\begin{vmatrix} 3 & 1 \\ 2 & 7 \end{vmatrix} + \mathbf{k}\begin{vmatrix} 3 & -2 \\ 2 & 3 \end{vmatrix}$$
$$= (-14 - 3)\mathbf{i} - (21 - 2)\mathbf{j} + (9 - [-4])\mathbf{k}$$
$$= -17\mathbf{i} - 19\mathbf{j} + 13\mathbf{k} = (-17, -19, 13)$$

By construction, $\mathbf{a} \times \mathbf{b}$ is orthogonal to $\mathbf{a}$ and $\mathbf{b}$, so the direction of $\mathbf{a} \times \mathbf{b}$ is known except for sign: does it point "up" or "down"? **The general rule is as follows:** the three vectors $\mathbf{a}$, $\mathbf{b}$, and $\mathbf{a} \times \mathbf{b}$, taken in this order, form a right-handed system. (If necessary, review the idea of right-handedness from Section 1–1.) Let us see how this works for simple cases where we can easily draw pictures.

EXAMPLE 25

Verify that

$$\mathbf{i} \times \mathbf{j} = \mathbf{k}, \quad \mathbf{j} \times \mathbf{k} = \mathbf{i}, \quad \mathbf{k} \times \mathbf{i} = \mathbf{j},$$

but

$$\mathbf{j} \times \mathbf{i} = -\mathbf{k}, \quad \mathbf{k} \times \mathbf{j} = -\mathbf{i}, \quad \mathbf{i} \times \mathbf{k} = -\mathbf{j},$$

and check that in every case, the three vectors taken in order form a right-handed system. These simple examples also suggest some of the general properties of the cross-product.

Some Properties of the Cross-Product

(i) $\mathbf{a} \times \mathbf{b} = -\mathbf{b} \times \mathbf{a}$, for any $\mathbf{a}$ and $\mathbf{b}$ in $\mathbb{R}^3$.

(ii) $\mathbf{a} \times \mathbf{a} = \mathbf{0}$ (the zero *vector*).

(iii) If $\mathbf{a} \times \mathbf{b} = \mathbf{0}$, then either [at least one of $\mathbf{a}$ or $\mathbf{b}$ is the zero vector] or [there is some non-zero real number k such that $\mathbf{b} = k\mathbf{a}$].

The first two properties are straightforward consequences of the definition. The third can be proved from the definition with a little care: if the first component of $\mathbf{a} \times \mathbf{b}$ is zero, and if a_2 and a_3 are not zero, then $b_2/a_2 = b_3/a_3$; similarly, from the second component, with a_1 also not zero, $b_1/a_1 = b_3/a_3$; thus there is a common ratio, call it k, and $\mathbf{b} = k\mathbf{a}$, as claimed. The cases where one of the components is zero require (easy) special arguments. However, the general result also follows from a formula for $\|\mathbf{a} \times \mathbf{b}\|$ that is proved below.

Two other properties that can be checked by straightforward calculation are:

(iv) $\mathbf{a} \times (\mathbf{b} + \mathbf{c}) = \mathbf{a} \times \mathbf{b} + \mathbf{a} \times \mathbf{c}$, and

(v) $(k\mathbf{a}) \times \mathbf{b} = k(\mathbf{a} \times \mathbf{b})$, for any real number k.

It is also important to note that one rule we might expect does not in fact hold: $\mathbf{a} \times (\mathbf{b} \times \mathbf{c}) \neq (\mathbf{a} \times \mathbf{b}) \times \mathbf{c}$, in general. Note that this means that *the parentheses cannot be omitted* in such a product. (There are formulas available for these triple vector products but we shall not need them: see Exercise F3 in Further Exercises at the end of this chapter.)

The Length of the Cross-Product

Given **a** and **b**, the direction of their cross-product is known. What is the *length* of the cross-product of **a** and **b**? The answer follows from a straightforward calculation; we only give an outline and leave you to check the details.

$$\|\mathbf{a} \times \mathbf{b}\|^2 = (a_2b_3 - a_3b_2)^2 + (a_3b_1 - a_1b_3)^2 + (a_1b_2 - a_2b_1)^2.$$

Expand by the binomial theorem, and add and subtract the term $(a_1^2b_1^2 + a_2^2b_2^2 + a_3^2b_3^2)$. The resulting terms can be rearranged so as to be seen to be equal to

$$(a_1^2 + a_2^2 + a_3^2)(b_1^2 + b_2^2 + b_3^2) - (a_1b_1 + a_2b_2 + a_3b_3)^2.$$

Thus,

$$\begin{aligned}\|\mathbf{a} \times \mathbf{b}\|^2 &= \|\mathbf{a}\|^2\|\mathbf{b}\|^2 - (\mathbf{a} \cdot \mathbf{b})^2 = \|\mathbf{a}\|^2\|\mathbf{b}\|^2 - \|\mathbf{a}\|^2\|\mathbf{b}\|^2 \cos^2\theta \\ &= \|\mathbf{a}\|^2\|\mathbf{b}\|^2(1 - \cos^2\theta) = \|\mathbf{a}\|^2\|\mathbf{b}\|^2 \sin^2\theta.\end{aligned}$$

It follows that the length of the cross-product is

$$\|\mathbf{a} \times \mathbf{b}\| = \|\mathbf{a}\|\|\mathbf{b}\| \sin\theta.$$

(Recall that $0 \leq \theta \leq \pi$, so that $\sin\theta \geq 0$ and the length of the cross-product is non-negative as required.)

To interpret this formula, consider Figure 1–4–1. The two vectors **a** and **b** determine a parallelogram. Take the length of **a** to be the base of the parallelogram; then the altitude is the length of $\mathbf{perp_a b}$. From standard trigonometry, this length is $\|\mathbf{b}\| \sin\theta$, so that the area of the parallelogram is

$$(\text{base}) \times (\text{altitude}) = \|\mathbf{a}\|\|\mathbf{b}\| \sin\theta,$$

and by the calculation above, this is equal to $\|\mathbf{a} \times \mathbf{b}\|$.

SUMMARY

$\mathbf{a} \times \mathbf{b}$ is a vector orthogonal to **a** and **b**, such that **a**, **b**, and $\mathbf{a} \times \mathbf{b}$ form a right-handed system; the length of $\mathbf{a} \times \mathbf{b}$ is the area of the parallelogram determined by **a** and **b**.

(A little nit-picking: if $\mathbf{a} \times \mathbf{b} = \mathbf{0}$, *then* $\mathbf{a}$, $\mathbf{b}$ *and* $\mathbf{a} \times \mathbf{b}$, *cannot form a right-handed system, and the Summary is not quite right. Write a correct version for yourself.)*

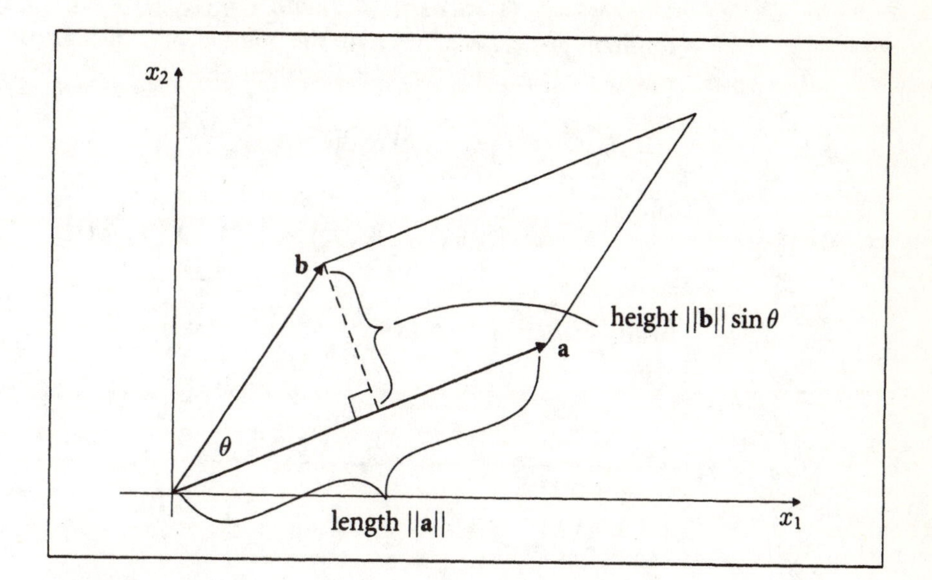

FIGURE 1-4-1. *The area of the parallelogram is* $||\mathbf{a}||\,||\mathbf{b}||\sin\theta$.

EXAMPLE 26

In Example 24, it was shown that $(3, -2, 1) \times (2, 3, 7) = (-17, -19, 13)$. Therefore, the area of the parallelogram determined by the vectors $(3, -2, 1)$ and $(2, 3, 7)$ is $||(-17, -19, 13)|| = \sqrt{819}$.

Some Problems on Lines, Planes, and Distances

The cross-product allows us to answer many questions about lines, planes, and distances in $\mathbb{R}^3$.

Finding the Normal to a Plane and the Distance Between Skew Lines

In Section 1–2, the vector equation of a plane was given in the form $\mathbf{x} = \mathbf{a} + r\mathbf{u} + s\mathbf{v}$. By definition, the normal vector $\mathbf{n}$ must be perpendicular to $\mathbf{u}$ and $\mathbf{v}$, and therefore will be given by $\mathbf{n} = \mathbf{u} \times \mathbf{v}$, provided that $\mathbf{u} \times \mathbf{v} \neq \mathbf{0}$. (If $\mathbf{u} \times \mathbf{v} = \mathbf{0}$, we do not really have a plane: *why?*)

EXAMPLE 27

The lines $\mathbf{x} = (1,3,2) + t(1,0,2)$ and $\mathbf{x} = (1,3,2) + s(-1,2,1)$ must lie in a common plane, since they have the point $(1,3,2)$ in common. Find a scalar equation of the plane that contains these lines.

SOLUTION: The normal to the plane is

$$\mathbf{n} = (1,0,2) \times (-1,2,1) = \begin{vmatrix} \mathbf{i} & \mathbf{j} & \mathbf{k} \\ 1 & 0 & 2 \\ -1 & 2 & 1 \end{vmatrix} = (-4,-3,2),$$

so an equation of the plane is $(-4,-3,2) \cdot (x_1 - 1, x_2 - 3, x_3 - 2) = 0$, or

$$-4x_1 - 3x_2 + 2x_3 = -9, \quad \text{or} \quad 4x_1 + 3x_2 - 2x_3 = 9.$$

EXAMPLE 28

Find a scalar equation of the plane that contains the three points $P(1,-2,1)$, $Q(2,-2,-1)$, and $R(4,1,1)$.

SOLUTION: The normal to the plane is given by

$$\mathbf{n} = \overrightarrow{PQ} \times \overrightarrow{PR} = (1,0,-2) \times (3,3,0) = \begin{vmatrix} \mathbf{i} & \mathbf{j} & \mathbf{k} \\ 1 & 0 & -2 \\ 3 & 3 & 0 \end{vmatrix} = (6,-6,3),$$

and the equation of the plane is $(6,-6,3) \cdot \big((x_1,x_2,x_3) - (1,-2,1)\big) = 0$, or

$$2x_1 - 2x_2 + x_3 = 7.$$

If two lines are skew, there is of course no plane that contains them both. However, they do lie in parallel planes; this can be seen as follows. Suppose that the skew lines are $\mathbf{x} = \mathbf{a} + s\mathbf{c}$ and $\mathbf{x} = \mathbf{b} + t\mathbf{d}$. Then the cross-product of the two direction vectors $\mathbf{n} = \mathbf{c} \times \mathbf{d}$ is perpendicular to both lines, so the plane through $\mathbf{a}$ with normal $\mathbf{n}$ contains the first line, and the plane through $\mathbf{b}$ with normal $\mathbf{n}$ contains the second line. Since the two planes have the same normal vector, they are parallel planes. These ideas together with the idea of projection provide a method for finding the distance between two skew lines.

EXAMPLE 29

Find the distance between the skew lines

$$\mathbf{x} = (1, 4, 2) + s(2, 0, 1),\ s \in \mathbb{R}, \text{ and } \mathbf{x} = (2, -3, 1) + t(1, 1, 3),\ t \in \mathbb{R}.$$

SOLUTION: The lines are certainly not parallel because the direction vector $(2, 0, 1)$ is not a multiple of the second direction vector $(1, 1, 3)$. It might be wise to first check that the lines are skew, but if they are not, the distance will turn out to be zero, so we omit that step. By the discussion in the preceding paragraph, the lines lie in parallel planes, with normal $\mathbf{n} = (2, 0, 1) \times (1, 1, 3) = (-1, -5, 2)$. Then the distance between them is the absolute value of the component along $\mathbf{n}$ of *any* directed line segment joining a point on the first plane (for example, the point $(1, 4, 2)$ on the first line) to a point on the second plane (for example, the point $(2, -3, 1)$ on the second line). Thus the distance is

$$\left|cpt_{\mathbf{n}}\Big((2, -3, 1) - (1, 4, 2)\Big)\right| = \Big|(1, -7, -1) \cdot (-1, -5, 2)/\|(-1, -5, 2)\|\Big|$$
$$= 32/\sqrt{30}.$$

ALTERNATIVE SOLUTION: First calculate the cross-product $(-1, -5, 2)$ as above. Then write the equation of the plane with this normal vector, passing through the first line: $(-1, -5, 2) \cdot \big(\mathbf{x} - (1, 4, 2)\big) = 0$. Now find the distance to this plane from the point $(2, -3, 1)$ on the second line.

If you know about partial derivatives, you could also solve this problem by minimizing a function of the two parameters s and t.

Finding the Line of Intersection of Two Planes

Unless two planes in $\mathbb{R}^3$ are parallel, their set of intersection will be a line. The direction vector of this line lies in both planes, so it is perpendicular to both of the normals; it can therefore be obtained as the cross-product of the two normals. The vector equation of the line can then be written, once we find one point that lies on both planes.

EXAMPLE 30

Find a vector equation of the line of intersection of the two planes $x_1 + x_2 - 2x_3 = 3$ and $2x_1 - x_2 + 3x_3 = 6$.

SOLUTION: The direction vector is $d = \begin{vmatrix} \mathbf{i} & \mathbf{j} & \mathbf{k} \\ 1 & 1 & -2 \\ 2 & -1 & 3 \end{vmatrix} = (1, -7, -3)$. One easy way to find a point on the line is to put $x_3 = 0$, then solve the remaining equations: $x_1 + x_2 = 3$ and $2x_1 - x_2 = 6$. The solution of these two equations is $x_1 = 3$, $x_2 = 0$. Hence a vector equation of the line of intersection is

$$\mathbf{x} = (3, 0, 0) + t(1, -7, -3), \quad t \in \mathbb{R}.$$

ALTERNATIVE SOLUTION: A vector equation of the line can also be obtained by "solving" the system of equations:

$$\begin{aligned} x_1 + x_2 - 2x_3 &= 3 \\ 2x_1 - x_2 + 3x_3 &= 6. \end{aligned}$$

by elimination. Multiply the first equation by (-2) and add it to the second. The system becomes:

$$\begin{aligned} x_1 + \ x_2 - 2x_3 &= 3 \\ -3x_2 + 7x_3 &= 0. \end{aligned}$$

x_3 can be chosen arbitrarily, say $x_3 = s$ (any real number); then $x_2 = (7/3)s$, and $x_1 = 3 - (7/3)s + 2s = 3 - (1/3)s$, and the points that satisfy the equations of both planes are of the form

$$x = (3 - (1/3)s, (7/3)s, s) = (3, 0, 0) + (1/3)s(-1, 7, 3), \quad s \in \mathbb{R}.$$

This is exactly the same line as we obtained by the first method since multiplying the direction vector by a non-zero constant does not change the line.

The Triple Scalar Product and Volumes in $\mathbb{R}^3$

Three vectors **a**, **b**, and **c** in $\mathbb{R}^3$ may be taken to be the three adjacent edges of a parallelepiped; a parallelepiped is a solid that has six faces, such that opposite faces are parallel to each other (see Figure 1–4–2). Is there an expression for the volume of the parallelepiped in terms of the three vectors? To obtain such an expression, observe that the parallelogram determined by **a** and **b** can be regarded as the base of the solid parallelepiped; this base has area $\|\mathbf{a} \times \mathbf{b}\|$. With respect to this base, the altitude of the solid is the absolute value of the component of **c** along the normal vector to the base, that is, $\dfrac{|\mathbf{c} \cdot (\mathbf{a} \times \mathbf{b})|}{\|\mathbf{a} \times \mathbf{b}\|}$. To get the volume of the parallelepiped, multiply this altitude by the area of the base ($\|\mathbf{a} \times \mathbf{b}\|$):

the volume of the parallelepiped is $|\mathbf{c} \cdot (\mathbf{a} \times \mathbf{b})|$.

The product $\mathbf{c} \cdot (\mathbf{a} \times \mathbf{b})$ is called the **scalar triple product** of **c**, **a**, and **b**. Notice that the result is a real number (that is, a scalar).

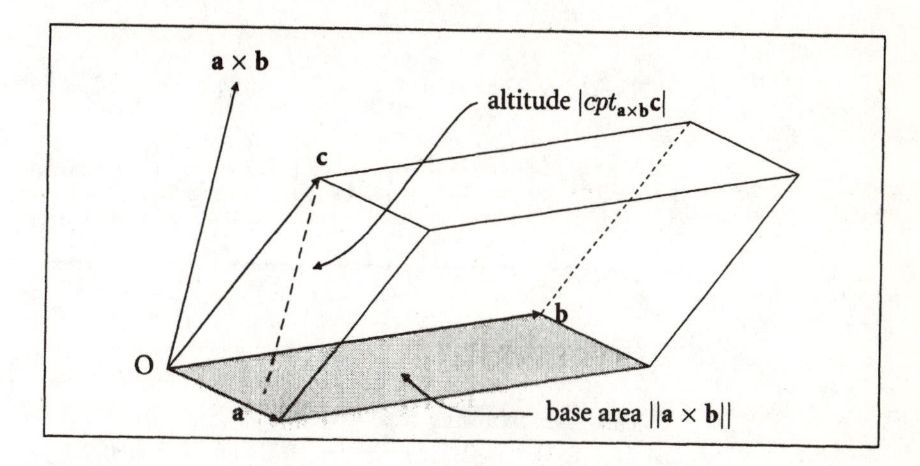

FIGURE 1-4-2. *The parallelepiped with adjacent edges* **a**, **b**, **c** *has volume given by* $|\mathbf{c} \cdot (\mathbf{a} \times \mathbf{b})|$.

The sign of the scalar triple product also has an interpretation. Recall that the ordered triple of vectors $\{\mathbf{a}, \mathbf{b}, \mathbf{a} \times \mathbf{b}\}$ is right-handed; we can think of $\mathbf{a} \times \mathbf{b}$ as the "upwards" normal vector to the plane with vector equation $\mathbf{x} = s\mathbf{a} + t\mathbf{b}$. Some other vector **c** is then "upwards", and $\{\mathbf{a}, \mathbf{b}, \mathbf{c}\}$ (in that order) is right-handed, if and only if $cpt_{(\mathbf{a} \times \mathbf{b})}\mathbf{c}$ is positive, and this component is positive if and only if $\mathbf{c} \cdot (\mathbf{a} \times \mathbf{b})$ is positive. If $\mathbf{c} \cdot (\mathbf{a} \times \mathbf{b})$ is negative, then $\{\mathbf{a}, \mathbf{b}, \mathbf{c}\}$ is a left-handed system.

It is worth noting that the scalar triple product is also given by determinant

$$\begin{vmatrix} c_1 & c_2 & c_3 \\ a_1 & a_2 & a_3 \\ b_1 & b_2 & b_3 \end{vmatrix} = c_1 \begin{vmatrix} a_2 & a_3 \\ b_2 & b_3 \end{vmatrix} - c_2 \begin{vmatrix} a_1 & a_3 \\ b_1 & b_3 \end{vmatrix} + c_3 \begin{vmatrix} a_1 & a_2 \\ b_1 & b_2 \end{vmatrix} = \mathbf{c} \cdot (\mathbf{a} \times \mathbf{b}).$$

It is often useful to note that $\mathbf{c} \cdot (\mathbf{a} \times \mathbf{b}) = \mathbf{a} \cdot (\mathbf{b} \times \mathbf{c}) = \mathbf{b} \cdot (\mathbf{c} \times \mathbf{a})$. This is straightforward but tedious to verify. It will be easy once the properties of the determinant are established in Chapter 5.

EXAMPLE 31

Find the volume of the parallelepiped determined by the vectors $(1,0,2)$, $(2,1,1)$, and $(-1,-1,3)$.

SOLUTION: The volume is $(1,0,2)\cdot\Big((2,1,1)\times(-1,-1,3)\Big)$

$$=(1,0,2)\cdot\left(\begin{vmatrix}1 & 1\\ -1 & 3\end{vmatrix}, -\begin{vmatrix}2 & 1\\ -1 & 3\end{vmatrix}, \begin{vmatrix}2 & 1\\ -1 & -1\end{vmatrix}\right)=1(4)+0(-7)+2(-1)=2.$$

EXERCISES 1–4

A1 Calculate the following cross-products.
(a) $(1,-5,2)\times(-2,1,5)$
(b) $(2,-3,-5)\times(4,-2,7)$
(c) $(-1,0,1)\times(0,4,5)$

A2 Let $\mathbf{p}=(-1,4,2)$, $\mathbf{q}=(3,1,-1)$, $\mathbf{r}=(2,-3,-1)$. Check by calculation that the following general properties hold.
(a) $\mathbf{p}\times\mathbf{p}=\mathbf{0}$
(b) $\mathbf{p}\times\mathbf{q}=-\mathbf{q}\times\mathbf{p}$
(c) $\mathbf{p}\times 3\mathbf{r}=3(\mathbf{p}\times\mathbf{r})$
(d) $\mathbf{p}\times(\mathbf{q}+\mathbf{r})=\mathbf{p}\times\mathbf{q}+\mathbf{p}\times\mathbf{r}$
(e) $\mathbf{p}\times(\mathbf{q}\times\mathbf{r})\neq(\mathbf{p}\times\mathbf{q})\times\mathbf{r}$

A3 Calculate the area of the parallelogram determined by the following vectors.
(a) $(1,2,1)$ and $(2,3,-1)$
(b) $(1,0,1)$ and $(1,1,4)$
(c) $(1,2)$ and $(-2,5)$ (Hint: Think of these as vectors $(1,2,0)$ and $(-2,5,0)$ in $\mathbb{R}^3$.)
(d) $(-3,1)$ and $(4,3)$

A4 In each case, determine whether the given pair of lines has a point of intersection; if so, determine the scalar equation of the plane containing the lines, and if not, determine the distance between the lines.
(a) $\mathbf{x}=(1,3,1)+s(-2,-1,1)$ and $\mathbf{x}=(0,1,4)+t(3,0,1)$, $s,t\in\mathbb{R}$
(b) $\mathbf{x}=(1,3,1)+s(-2,-1,1)$ and $\mathbf{x}=(0,1,7)+t(3,0,1)$, $s,t\in\mathbb{R}$

A5 Same instructions as in A4.
(a) $\mathbf{x}=(2,1,4)+s(2,1,-2)$ and $\mathbf{x}=(-2,1,5)+t(1,3,1)$, $s,t\in\mathbb{R}$
(b) $\mathbf{x}=(0,1,3)+s(1,-1,4)$ and $\mathbf{x}=(0,-1,5)+t(1,1,2)$, $s,t\in\mathbb{R}$

A6 Determine the scalar equation of the plane with vector equation
(a) $\mathbf{x}=(1,4,7)+s(2,3,-1)+t(4,1,0)$, $\quad s,t\in\mathbb{R}$;
(b) $\mathbf{x}=(2,3,-1)+s(1,1,0)+t(-2,1,2)$, $\quad s,t\in\mathbb{R}$;
(c) $\mathbf{x}=(1,-1,3)+s(2,-2,1)+t(0,3,1)$, $\quad s,t\in\mathbb{R}$.

A7 Determine the scalar equation of the plane that contains the following points.
(a) $(2, 1, 5)$, $(4, -3, 2)$, $(2, 6, -1)$
(b) $(3, 1, 4)$, $(-2, 0, 2)$, $(1, 4, -1)$
(c) $(-1, 4, 2)$, $(3, 1, -1)$, $(2, -3, -1)$

A8 Determine a vector equation of the line of intersection of the given planes by both of the methods described in Section 1-4.
(a) $x_1 + 3x_2 - x_3 = 5$ and $2x_1 - 5x_2 + x_3 = 7$
(b) $2x_1 - 3x_3 = 7$ and $x_2 + 2x_3 = 4$

A9 Find the volume of the parallelepiped determined by the following vectors.
(a) $(4, 1, -1)$, $(-1, 5, 2)$, and $(1, 1, 6)$
(b) $(-2, 1, 2)$, $(3, 1, 2)$, and $(0, 2, 5)$

B1 Calculate the following cross-products.
(a) $(2, -1, 6) \times (4, 4, 5)$
(b) $(-3, 1, 0) \times (1, 5, 2)$
(c) $(2, -4, \frac{1}{2}) \times (-1, 1, 3)$

B2 Let $\mathbf{p} = (4, 1, 1)$, $\mathbf{q} = (0, 4, -2)$, $\mathbf{r} = (1, -2, 3)$. Check by calculation that the following general properties hold.
(a) $\mathbf{p} \times \mathbf{p} = \mathbf{0}$
(b) $\mathbf{p} \times \mathbf{q} = -\mathbf{q} \times \mathbf{p}$
(c) $\mathbf{p} \times 3\mathbf{r} = 3(\mathbf{p} \times \mathbf{r})$
(d) $\mathbf{p} \times (\mathbf{q} + \mathbf{r}) = \mathbf{p} \times \mathbf{q} + \mathbf{p} \times \mathbf{r}$
(e) $\mathbf{p} \times (\mathbf{q} \times \mathbf{r}) = (\mathbf{p} \times \mathbf{q}) \times \mathbf{r}$

B3 Calculate the area of the parallelogram determined by the following vectors.
(a) $(2, 1, 3)$ and $(3, -1, 5)$
(b) $(2, 1, 4)$ and $(5, -1, -1)$
(c) $(2, 3)$ and $(-5, 2)$ (Hint: Think of these as vectors $(2, 3, 0)$ and $(-5, 2, 0)$ in $\mathbb{R}^3$.)
(d) $(1, 0, -5)$ and $(-3, 1, 1)$

B4 In each case, determine whether the given pair of lines has a point of intersection; if so, determine the scalar equation of the plane containing the lines, and if not, determine the distance between the lines.
(a) $\mathbf{x} = (-1, -1, 0) + s(2, 2, 1)$ and $\mathbf{x} = (5, -1, 0) + t(1, -2, -1)$, $s, t \in \mathbb{R}$
(b) $\mathbf{x} = (-1, -2, 4) + s(-2, 1, 2)$ and $\mathbf{x} = (0, 2, 1) + t(1, 1, 1)$, $s, t \in \mathbb{R}$

B5 Same instructions as in B4.
(a) $\mathbf{x} = (0, 2, 1) + s(1, 1, 1)$ and $\mathbf{x} = (-1, 4, 4) + t(-2, 1, 2)$, $s, t \in \mathbb{R}$
(b) $\mathbf{x} = (-3, -3, 0) + s(2, 2, 1)$ and $\mathbf{x} = (-5, 1, 0) + t(1, -2, -1)$, $s, t \in \mathbb{R}$

B6 Determine the scalar equation of the plane with vector equation
(a) $\mathbf{x} = (4, -2, -5) + s(-1, 3, -2) + t(1, 4, 5), \quad s, t \in \mathbb{R}$;
(b) $\mathbf{x} = (3, 4, 5) + s(-2, 1, 1) + t(-1, 0, 2), \quad s, t \in \mathbb{R}$;
(c) $\mathbf{x} = (-1, 0, 1) + s(3, 1, 1) + t(-1, -1, 3), \quad s, t \in \mathbb{R}$.

B7 Determine the scalar equation of the plane that contains the following points.
(a) $(5, 2, 1)$, $(-3, 2, 4)$, $(8, 1, 6)$
(b) $(5, -1, 2)$, $(-1, 3, 4)$, $(3, 1, 1)$
(c) $(0, 2, 1)$, $(3, -1, 1)$, $(1, 3, 0)$

B8 (a) Determine a vector equation of the line of intersection of the planes $x_1 + 4x_2 + x_3 = 5$ and $3x_1 - 7x_2 - x_3 = 6$ by both of the methods described in Section 1–4.

(b) Same as (a) with planes $x_1 - 3x_2 - 2x_3 = 4$ and $3x_1 + 2x_2 + x_3 = 2$.

B9 Determine the volume of the parallelepiped determined by the following vectors.

(a) $(5, -1, 1), (1, 4, -1), (1, 2, -6)$ (b) $(1, 1, 4), (1, 3, 4), (-2, 1, -5)$

CONCEPTUAL EXERCISES

***D1** Show that $(\mathbf{a} - \mathbf{b}) \times (\mathbf{a} + \mathbf{b}) = 2\mathbf{a} \times \mathbf{b}$.

D2 Show that if $\mathbf{x}$ is a point on the line through $\mathbf{a}$ and $\mathbf{b}$, then

$$\mathbf{x} \times (\mathbf{b} - \mathbf{a}) = \mathbf{a} \times \mathbf{b}.$$

D3 Consider the following statement: "If $\mathbf{a} \neq \mathbf{0}$, and if $\mathbf{a} \times \mathbf{b} = \mathbf{a} \times \mathbf{c}$, then $\mathbf{b} = \mathbf{c}$." If the statement is true, prove it. If it is false, give a counterexample.

D4 Explain why $\mathbf{a} \times (\mathbf{b} \times \mathbf{c})$ must be a vector in the plane with vector equation $\mathbf{x} = s\mathbf{b} + t\mathbf{c}$, $s, t \in \mathbb{R}$. (That is, explain why this product must be a linear combination of the vectors $\mathbf{b}$ and $\mathbf{c}$.)

***D5** What does it mean geometrically if $\mathbf{a} \cdot (\mathbf{b} \times \mathbf{c}) = 0$?

Chapter Review

Suggestions for Student Review

Organizing your own review is an important step towards mastering new material; it is more valuable than memorizing someone else's list of "key ideas". To retain new concepts as useful tools, you must be able to state definitions and be able to make connections between various ideas and techniques; you should also be able to give (even better, create) instructive examples. The suggestions below are not intended to be an exhaustive checklist; instead, they suggest the kinds *of activity and questioning that will help you gain a confident grasp of the material.*

1 Find some person or persons to *talk* with about mathematics—there's lots of evidence this is the best way to learn. Be sure you do your share of *asking* and *answering* —a little bit of embarrassment is a small price for real learning. Also, be sure to get lots of practice in writing answers independently.

2 Draw pictures to illustrate addition of vectors, subtraction of vectors, and multiplication of a vector by a scalar (general case). (Section 1–1)

3 Explain how you find a vector equation for a line, and make up examples to show why the vector equation for a given line is not unique. (Someone once said "If you can't explain it, you don't understand it.") (Section 1–1)

4 State the relation (or relations) between length in $\mathbb{R}^3$ and the dot product in $\mathbb{R}^3$. Make up examples to illustrate. (Section 1–2)

5 Explain how projection onto a vector **a** is defined in terms of the dot product, with a picture to illustrate. Define the part of a vector **x** perpendicular to **a**, and verify (in the general case) that it is perpendicular to **a**. (Section 1–3)

6 Explain with a picture how projection helps us to solve minimum distance problems (point to line in $\mathbb{R}^2$, point to plane in $\mathbb{R}^3$). (Section 1–3)

7 Discuss the role of the normal vector to a plane in determining the scalar equation of the plane. Explain how you can get from a scalar equation of a plane to a vector equation for the plane, or from a vector equation to the scalar equation. (Sections 1–2 and 1–4)

8 State the important algebraic properties and geometric properties of the cross-product. (Section 1–4)

Chapter Quiz

Your instructor may have different ideas of an appropriate level of difficulty for a test on this material.

1 Determine a vector equation of the line passing through points $(-2, 1, -4)$ and $(5, -2, 1)$.

2 Determine the scalar equation of the plane that contains the points $(1, -1, 0)$, $(3, 1, -2)$, and $(-4, 1, 6)$.

3 Determine whether the line $\mathbf{x} = (1, 3, 1) + t(-1, 2, 3)$, $t \in \mathbb{R}$ is parallel to or orthogonal to each of the given planes, or neither parallel nor orthogonal. State the reason for your decision clearly in each case.

(a) $2x_1 + 4x_2 + 6x_3 = 5$
(b) $x_1 - 2x_2 - 3x_3 = 4$
(c) $5x_1 + x_2 + x_3 = 6$

4 Determine the cosine of the angle between the vector $(2, -3, 1)$ and each of the coordinate axes.

5 Find the point on the line $\mathbf{x} = t(3, -2, 3)$, $t \in \mathbb{R}$ that is closest to the point $(2, 3, 4)$. Illustrate your method of calculation with a rough sketch.

6 Find the point on the hyperplane $x_1 + x_2 + x_3 + x_4 = 1$ that is closest to the point $(3, -2, 0, 2)$, and determine the distance from the point to the plane.

7 Determine a non-zero vector that is orthogonal to both $(1, 2, 0)$ and $(-3, 1, 1)$.

8 Verify that the four points $(1, 1, 1)$, $(2, 1, 2)$, $(0, 2, 0)$, and $(-1, 2, -1)$ are the vertices of a parallelogram, and determine the area of the parallelogram.

9 Prove that the solid parallelepiped determined by vectors **a**, **b**, and **c** has the same volume as the parallelepiped determined by $\mathbf{a} + k\mathbf{b}$, **b**, and **c**.

10 State whether each of the following statements is true or false; each statement is to be interpreted in $\mathbb{R}^3$. If the statement is true, explain briefly; if the statement is false, give a counterexample (a specific example in which we can see that the general statement is not true.)

(i) If two lines are not parallel, they must intersect.

(ii) If line ℓ_1 is parallel to line ℓ_2, and line ℓ_2 is parallel to line ℓ_3, then ℓ_1 is parallel to ℓ_3.

(iii) Any three points lie in exactly one plane.

(iv) The dot product of a vector with itself cannot be zero.

(v) For any vectors $\mathbf{a}$ and $\mathbf{b}$, $\mathbf{proj}_{\mathbf{a}}\mathbf{b} = \mathbf{proj}_{\mathbf{b}}\mathbf{a}$.

(vi) The area of the parallelogram determined by vectors $\mathbf{a}$ and $\mathbf{b}$ is the same as the area of the parallelogram determined by $\mathbf{a}$ and $(\mathbf{b} + 3\mathbf{a})$.

Further Exercises

These exercises are intended to be a little more challenging than the exercises at the end of each section. Some explore topics beyond the material discussed in the text.

F1 Consider the statement: "If $\mathbf{a} \neq \mathbf{0}$, and both $\mathbf{a} \cdot \mathbf{b} = \mathbf{a} \cdot \mathbf{c}$ and $\mathbf{a} \times \mathbf{b} = \mathbf{a} \times \mathbf{c}$, then $\mathbf{b} = \mathbf{c}$." Either prove the statement or give a counterexample.

F2 Suppose that $\mathbf{a}$ and $\mathbf{b}$ are mutually orthogonal unit vectors in $\mathbb{R}^3$. Prove that for every $\mathbf{x}$ in $\mathbb{R}^3$,

$$\mathbf{perp}_{\mathbf{a}\times\mathbf{b}}(\mathbf{x}) = \mathbf{proj}_{\mathbf{a}}(\mathbf{x}) + \mathbf{proj}_{\mathbf{b}}(\mathbf{x}).$$

F3 In Exercise D4 of Section 1–4, you were asked to show that

$$\mathbf{a} \times (\mathbf{b} \times \mathbf{c}) = s\mathbf{b} + t\mathbf{c} \text{ for some } s, t \text{ in } \mathbb{R}.$$

(If you haven't done this, do it now.)

(a) By direct calculation, prove that $\mathbf{a} \times (\mathbf{b} \times \mathbf{c}) = (\mathbf{a} \cdot \mathbf{c})\mathbf{b} - (\mathbf{a} \cdot \mathbf{b})\mathbf{c}$.

(b) Prove that $\mathbf{a} \times (\mathbf{b} \times \mathbf{c}) + \mathbf{b} \times (\mathbf{c} \times \mathbf{a}) + \mathbf{c} \times (\mathbf{a} \times \mathbf{b}) = \mathbf{0}$.

F4 Show that if A, B, and C are collinear points, and $\overrightarrow{OA} = \mathbf{a}$, $\overrightarrow{OB} = \mathbf{b}$, $\overrightarrow{OC} = \mathbf{c}$, then

$$(\mathbf{a} \times \mathbf{b}) + (\mathbf{b} \times \mathbf{c}) + (\mathbf{c} \times \mathbf{a}) = \mathbf{0}.$$

F5 Prove that

(a) $\mathbf{a} \cdot \mathbf{b} = \frac{1}{4}\|\mathbf{a} + \mathbf{b}\|^2 - \frac{1}{4}\|\mathbf{a} - \mathbf{b}\|^2$

(b) $\|\mathbf{a} + \mathbf{b}\|^2 + \|\mathbf{a} - \mathbf{b}\|^2 = 2\|\mathbf{a}\|^2 + 2\|\mathbf{b}\|^2$

(c) Interpret (a) and (b) in terms of a parallelogram determined by vectors $\mathbf{a}$ and $\mathbf{b}$.

F6 (a) Let $\mathcal{T}$ denote the triangle with vertices $(0,1,2)$, $(1,2,3)$ and $(1,0,1)$. Show that the point $P(0,3/2,5/2)$ lies in the same plane as $\mathcal{T}$.

(b) Determine whether P is inside $\mathcal{T}$.

F7 Four points A, B, C, D in $\mathbb{R}^3$ may be collinear; or they may be coplanar (in a common plane) but not collinear, in which case they are the vertices of a plane quadrilateral; or they may not lie in any common plane. In the last case, the points taken in the given order may be considered as the vertices of a *folded quadrilateral* formed by the two triangles ABC and ADC joined along the "fold line" AC.

(a) For each of the sets given below, determine whether the points are collinear, form a plane quadrilateral, of form a folded quadrilateral.

(i) $(1,2,3)$, $(2,4,0)$, $(3,6,-3)$, $(4,8,-6)$

(ii) $(2,4,1)$, $(0,6,2)$, $(-1,4,4)$, $(1,2,3)$

(iii) $(2,1,1)$, $(2,5,-1)$, $(3,4,-1)$, $(3,0,1)$

(iv) $(0,0,0)$, $(2,2,1)$, $(5,0,0)$, $(3,2,-1)$

(b) Review how you decide whether a plane quadrilateral is a parallelogram, rectangle, or square, and give criteria for deciding whether a folded quadrilateral is a folded parallelogram, rectangle, or square. Decide whether the quadrilaterals of part (a) have these special properties.

(c) Construct a folded square such that the angle between the two triangles is some specified angle between 0 and π. [Hint: Try $A = (-1,0,0)$, $C = (1,0,0)$.]

CHAPTER

Systems of Linear Equations and Row Echelon Form

The standard method of solving systems of linear equations is elimination. It can be represented by row reduction of a matrix to its "reduced row echelon form". This is the most fundamental procedure in linear algebra. Obtaining and interpreting the reduced row echelon form of a matrix will play an important role in almost everything we do in the rest of this book.

SECTION

2–1 Systems of Linear Equations; Elimination and Back-Substitution; Matrices and Elementary Row Operations

A linear equation in n variables $(x_1, x_2, x_3, \ldots, x_n)$ is an equation that can be written in the form:

$$a_1x_1 + a_2x_2 + a_3x_3 + \cdots + a_nx_n = b;$$

here the real numbers $a_1, a_2, \ldots, a_n$ are called the **coefficients** of the equation, and b is also a real number (usually referred to as "the right-hand side" or "constant term"). The x_j $(j = 1, 2, \ldots, n)$ are often thought of as "unknowns" or "variables" to be solved for. A vector $(s_1, s_2, \ldots, s_n)$ in $\mathbb{R}^n$ is a **solution of** the equation if the equation is satisfied when the numbers $s_1, s_2, \ldots, s_n$ are substituted for $x_1, x_2, \ldots, x_n$, in order.

As a simple example, consider the equation $x_1 + 2x_2 = 4$: $(2, 1)$, $(3, 0.5)$, and $(6, -1)$ are all solutions of this equation. We want to develop a systematic procedure for finding solutions.

In many problems, it is necessary to find a vector $(x_1, x_2, \ldots, x_n)$ in $\mathbb{R}^n$ that simultaneously satisfies several linear equations. For example, in Chapter 1, the common solutions of the linear equations of two planes determined the vector equation of the line of intersection of the planes. The problem of whether two lines in $\mathbb{R}^3$ had a point of intersection turned into a problem of trying to satisfy three linear equations in the two parameters.

When we are dealing with several linear equations in the same variables at the same time, we speak of a **system of linear equations.** Such systems arise very frequently in almost every conceivable area where mathematics is applied: in analyzing stresses in complicated structures; as a fundamental tool in the numerical analysis of the flow of fluids or heat; in questions of allocating resources or managing inventory; in questions of determining appropriate controls to guide aircraft or robots. Some examples appear in Section 2–3.

The general system of m linear equations in n variables is written in the form

$$\begin{aligned}
a_{11}x_1 + a_{12}x_2 + a_{13}x_3 + \cdots + a_{1n}x_n &= b_1 \\
a_{21}x_1 + a_{22}x_2 + a_{23}x_3 + \cdots + a_{2n}x_n &= b_2 \\
&\cdots \\
a_{m1}x_1 + a_{m2}x_2 + a_{m3}x_3 + \cdots + a_{mn}x_n &= b_m.
\end{aligned}$$

Note that for each coefficient, the first index indicates which equation the coefficient appears in, and the second index indicates which variable the coefficient multiplies. Note also the indices on the right-hand sides indicating which equation the constant appears in.

We want to establish a standard procedure for determining all solutions of such a system — *if there are any solutions!* Remember that there may be no solutions for the parameter values that give the point of intersection of two lines in $\mathbb{R}^3$; and there will generally be many solutions of the system of two equations describing the intersection of two planes in $\mathbb{R}^3$. It will be convenient to speak of the **solution set** of a system, meaning the set of all solutions of the system.

The standard procedure is **elimination:** by multiplying and adding some of the original equations, *some* of the variables are eliminated from *some* of the equations. If the process is successful, we will end up with a system where the solutions are easily determined almost by inspection. We begin with some examples and explain the general rules as we proceed.

EXAMPLE 1

Find all solutions of the system of linear equations

$$\begin{aligned} x_1 + x_2 - 2x_3 &= 4 \\ x_1 + 3x_2 - x_3 &= 7 \\ 2x_1 + x_2 - 5x_3 &= 7. \end{aligned}$$

SOLUTION: To solve this system by elimination, begin by eliminating x_1 from all equations except the first.

Add (-1) times the first equation to the second equation. The first and third equations are unchanged, so the system is now:

$$\begin{aligned} x_1 + x_2 - 2x_3 &= 4 \\ 2x_2 + x_3 &= 3 \\ 2x_1 + x_2 - 5x_3 &= 7. \end{aligned}$$

Note two important things about this step. First, if the numbers x_1, x_2, x_3 satisfy the original system before the step, then they certainly satisfy the revised system after the step. (This follows from the rule of arithmetic that if $P = Q$ and $R = S$, then $P + R = Q + S$, so when we add two equations, both of which are satisfied, the resulting sum equation is satisfied.)

Second, the step is reversible: to get back the original system just add (1) times the first equation to the revised second equation. It follows that if the numbers x_1, x_2, x_3 satisfy the revised system, they also satisfy the original system. Thus the original system and the revised system have exactly the same solutions.

We say that two systems of equations are **equivalent** if they have the same set of solutions. *It is absolutely fundamental to the method of elimination that elimination steps leave the solution set unchanged and that each step is reversible. Every system produced during an elimination procedure is equivalent to the original system: every system produced during an elimination procedure has the same solution set.*

Add (-2) times the first equation to the third equation.

$$\begin{aligned} x_1 + x_2 - 2x_3 &= 4 \\ 2x_2 + x_3 &= 3 \\ -1x_2 - 1x_3 &= -1 \end{aligned}$$

Again, note that this is a reversible step that does not change the solution set. Note also that x_1 has been eliminated from all except the first equation, so now we leave the first equation and turn our attention to x_2.

(Although we will not modify or use the first equation in the next several steps, keep writing the entire system after each step. This is important because it leads to a good general procedure for dealing with large systems.)

It is convenient to work with an equation (other than the first) in which x_2 has coefficient +1. We could multiply the second equation by 1/2, but to avoid fractions if possible, follow these steps:

Interchange the second and third equations. (Another reversible step that does not change the solution set.)

$$\begin{aligned} x_1 + x_2 - 2x_3 &= 4 \\ -1x_2 - 1x_3 &= -1 \\ 2x_2 + x_3 &= 3 \end{aligned}$$

Multiply the second equation by (-1). (Reversible, does not change the solution set.)

$$\begin{aligned} x_1 + x_2 - 2x_3 &= 4 \\ x_2 + x_3 &= 1 \\ 2x_2 + x_3 &= 3 \end{aligned}$$

Add (-2) **times the second equation to the third equation.**

$$\begin{aligned} x_1 + x_2 - 2x_3 &= 4 \\ x_2 + x_3 &= 1 \\ -1x_3 &= 1 \end{aligned}$$

Multiply the third equation by (-1).

$$\begin{aligned} x_1 + x_2 - 2x_3 &= 4 \\ x_2 + x_3 &= 1 \\ x_3 &= -1 \end{aligned}$$

In the third equation, all the variables except x_3 have been eliminated; by elimination, we have solved for x_3. We could continue and eliminate x_3 from the second and first equations, and x_2 from the first equation, by similar steps, but it is usual to complete the solution process by **back-substitution.**

For this purpose, it is important to observe the form of the equations in our final system. Each equation has a **leading variable**: that is, a first variable with non-zero coefficient; in fact, we have arranged that the leading coefficients are all 1's, and we shall always assume that this is the case when we speak of leading variables. The leading variable in the second equation is to the right of the leading variable in the first, and the leading variable in the third is to the right of the leading variable in the second.

To proceed by back-substitution, first observe that $x_3 = -1$. Substitute this value back into the second equation and find that

$$x_2 = 1 - x_3 = 1 - (-1) = 2.$$

Next, substitute these values back into the first equation to obtain

$$x_1 = 4 - x_2 + 2x_3 = 4 - 2 + 2(-1) = 0.$$

Thus, the only solution of this system is $(x_1, x_2, x_3) = (0, 2, -1)$. Since the final system is equivalent to the original system, this solution is the unique solution of the problem.

Check that $(0, 2, -1)$ does satisfy the original system of equations. You should take advantage of the fact that this check of the correctness of your elimination procedure is always available to you.

The system just solved is a particularly simple one. However, the solution procedure introduced all the kinds of steps that are ever needed in the process of elimination. It is worth reviewing them.

Steps in Elimination

(1) Multiply one equation by a non-zero constant.

(2) Interchange two equations.

(3) Add a multiple of one equation to another equation.

A Warning. Do *not* combine steps of type (1) and type (3) into one step of the form "Add a multiple of one equation to a *multiple* of another equation". Although such a combination would not lead to errors in this chapter, it would lead to errors when we apply these ideas in Chapter 5.

Note that in our elimination we systematically worked from left to right: we first eliminated x_1 from all equations below the first, then x_2 from all equations below the second, leaving only x_3 in the third. For reasons discussed after Example 2, you should always work from left to right.

EXAMPLE 2

Determine x_1, x_2, x_3, and x_4 satisfying the linear equations $x_1 + 2x_3 + x_4 = 14$ and $x_1 + 3x_3 + 3x_4 = 19$.

> *REMARK*
>
> Notice that neither equation contains x_2. This may seem peculiar, but it happens in some applications that one of the variables of interest does not appear in the linear equations. If it truly is one of the variables of the problem, it is wrong to ignore it. Rewrite the equations to make it explicit.
>
> $$\begin{aligned} x_1 + 0x_2 + 2x_3 + x_4 &= 14 \\ x_1 + 0x_2 + 3x_3 + 3x_4 &= 19 \end{aligned}$$

SOLUTION: Add (-1) times the first equation to the second equation.

$$\begin{aligned} x_1 + 0x_2 + 2x_3 + x_4 &= 14 \\ x_3 + 2x_4 &= 5 \end{aligned}$$

x_2 is not shown in the second equation because the leading variable must have a non-zero coefficient. We cannot carry elimination any further: since there are only two equations, there can be at most two leading variables (in this case, x_1 and x_3), and the solution procedure must now be completed by back-substitution.

The equations do not completely determine both x_3 and x_4: one of them can be chosen arbitrarily, and the equations can still be satisfied. As part of our systematic solution procedure, **we always start from the right.** x_4 is chosen arbitrarily, say $x_4 = t$, $t \in \mathbb{R}$ (this means that t can be any real number). Then the second equation can be solved for the leading variable x_3 :

$$x_3 = 5 - 2x_4 = 5 - 2t.$$

Next look at x_2: the equations can be satisfied for any value of x_2, so write $x_2 = s$, $s \in \mathbb{R}$. Finally solve the first equation for its leading variable x_1:

$$x_1 = 14 - 2x_3 - x_4 = 14 - 2(5 - 2t) - t = 4 + 3t.$$

The conclusion is that any vector of the form

$$(x_1, x_2, x_3, x_4) = (4 + 3t, s, 5 - 2t, t), \quad s \in \mathbb{R}, \quad t \in \mathbb{R}$$

is a solution of the system of equations, and that any solution is of this form. We say that this is the **general solution** of the system. For many purposes, it is useful to recognize that this solution can be "split up into a constant part, a part in t, and a part in s". More properly, we say that the general solution can be written as a sum of a constant vector, a vector multiplied by s, and a vector multiplied by t:

$$(x_1, x_2, x_3, x_4) = (4, 0, 5, 0) + s(0, 1, 0, 0) + t(3, 0, -2, 1), \quad s \in \mathbb{R}, \quad t \in \mathbb{R}.$$

This will be the standard format for displaying general solutions. It is acceptable to leave x_2 in place of s and x_4 in place of t in this format (and then you must say $x_2 \in \mathbb{R}$, $x_4 \in \mathbb{R}$). Note that there are infinitely many solutions for the system of this example. In this solution, s and t are called **parameters**: there are two parameters in this general solution. It is also worth noting in this case that the general solution is the vector equation of a 2-plane in $\mathbb{R}^4$. Again, we can check this solution by substitution into the original equations.

We could have solved for x_1 and x_4 in terms of x_3. In fact, in both of the examples we could have proceeded in slightly different ways. However, there are big advantages to having one systematic procedure that can be applied to all systems, and one standard way of displaying general solutions. One advantage arises when we summarize important facts about solving systems as Theorems; general statements of this kind make sense only if we all use the same systematic solution procedure, and the same format for displaying solutions. A second advantage arises when we use computers to solve systems; computers require systematic procedures, and computer outputs can be interpreted only if we understand the format for displaying solutions. Always eliminate working from left to right, then back-substitute beginning at the variable on the right and working back to the left.

REMARK

The solution procedure we have introduced is known as **Gaussian elimination with back-substitition.** A slight variation of this procedure is introduced in the next section.

EXAMPLE 3

Find the general solution to the system of linear equations:

$$\begin{aligned} x_1 + x_2 \qquad &= 1 \\ x_2 + x_3 &= 2 \\ x_1 + 2x_2 + x_3 &= -2 \end{aligned}$$

SOLUTION: Add (-1) times the first equation to the third.

$$\begin{aligned} x_1 + x_2 \qquad &= 1 \\ x_2 + x_3 &= 2 \\ x_2 + x_3 &= -3 \end{aligned}$$

Notice that it is impossible to satisfy the last two equations simultaneously, so we can already conclude that this system has no solution, but let us see how the standard procedure deals with this system.

Add (-1) times the second equation to the third.

$$\begin{aligned} x_1 + x_2 \qquad &= 1 \\ x_2 + x_3 &= 2 \\ 0x_1 + 0x_2 + 0x_3 &= -5 \end{aligned}$$

It is impossible to find (x_1, x_2, x_3) to satisfy the last equation, so it is impossible to find (x_1, x_2, x_3) to satisfy this system of equations. **The system has no solution,** and we say that the system of linear equations is **inconsistent.** (Notice that the last equation of this system could be written in the form $0 = -5$; however, it might be very confusing to leave the left side in this way with no variables!)

The Matrix Representation of a System of Linear Equations

After you have solved a few systems of equations, you may realize that you could write the solutions out faster if you could omit the letters x_1, x_2, etc., as long as you could keep the coefficients lined up properly. We shall agree to simply write out the coefficients and the right-hand sides in a rectangular array, called a **matrix.** Thus, the general linear system of equations at the beginning of this section is represented by the matrix

$$\left[\begin{array}{cccccccc|c} a_{11} & a_{12} & \cdots & a_{1k} & \cdots & a_{1n} & b_1 \\ a_{21} & a_{22} & \cdots & a_{2k} & \cdots & a_{2n} & b_2 \\ \vdots & \vdots & \ddots & \vdots & \ddots & \vdots & \vdots \\ a_{j1} & a_{j2} & \cdots & a_{jk} & \cdots & a_{jn} & b_j \\ \vdots & \vdots & \ddots & \vdots & \ddots & \vdots & \vdots \\ a_{m1} & a_{m2} & \cdots & a_{mk} & \cdots & a_{mn} & b_m \end{array}\right].$$

This is called the **augmented matrix** of the system; it is "augmented" because it includes as its last column the right-hand sides of the equations of the system. The matrix without this last column is called the **coefficient matrix** of the system:

$$\begin{bmatrix} a_{11} & a_{12} & \cdots & a_{1k} & \cdots & a_{1n} \\ a_{21} & a_{22} & \cdots & a_{2k} & \cdots & a_{2n} \\ \vdots & \vdots & \ddots & \vdots & \ddots & \vdots \\ a_{j1} & a_{j2} & \cdots & a_{jk} & \cdots & a_{jn} \\ \vdots & \vdots & \ddots & \vdots & \ddots & \vdots \\ a_{m1} & a_{m2} & \cdots & a_{mk} & \cdots & a_{mn} \end{bmatrix}.$$

This coefficient matrix with typical entry a_{jk} in the jth **row** and the kth **column** is often denoted by A. For reasons that will become clear in Chapter 3, we shall speak of the system of linear equations $A\mathbf{x} = \mathbf{b}$, and we use $\mathbf{b}$ to denote the column $\begin{bmatrix} b_1 \\ b_2 \\ \vdots \\ b_m \end{bmatrix}$ (the same notation as we use for the vector $(b_1, b_2, \ldots, b_m)$).

Thus, we write $\left[A \mid \mathbf{b}\right]$ as the augmented matrix of the system $A\mathbf{x} = \mathbf{b}$.

In the augmented matrix the vertical line separating A from $\mathbf{b}$ is optional. With care and experience it is not really necessary to have this notational crutch, and sometimes hand-written calculations are easier to read if the line is omitted. Computer displays generally omit the line. However, the coefficient matrix of a system and the augmented matrix of a system play different roles, and until you feel sure of what you are doing you may wish to use the vertical line. It will often be omitted in this book.

Each row in the augmented matrix corresponds to an equation in the system of linear equations. Notice that multiplying the jth equation by a constant corresponds to multiplying the jth row by the same constant; interchanging the ith and j^{th} equations corresponds exactly to interchanging the ith and j^{th}

rows; adding a multiple of the ith equation to the jth equation corresponds to adding the same multiple of the ith row to the jth row. Thus the **Steps in Elimination** correspond to the following **elementary row operations.**

Elementary Row Operations:

(1) **Multiply one row by a non-zero constant.**

(2) **Interchange two rows.**

(3) **Add a multiple of one row to another row.**

As with Steps in Elimination, do not combine operations of types (1) and (3) into one operation. It may be helpful to restate (3):

(3) (restated) **Replace a row by (itself plus a multiple of another).**

The process of performing elementary row operations on a matrix to bring it into some simpler form is called **row reduction.**

Recall that if a system of equations was obtained from another system by one of the allowed elimination steps, the systems were said to be equivalent. For matrices, if matrix M is transformed into matrix N by an elementary row operation (or by a sequence of elementary row operations), we say that M is **row equivalent** to N. (To say that matrix M is "equivalent" to matrix N means something different.) Just as elimination steps are reversible, so are elementary row operations; it follows that if M is row equivalent to N, then N is row equivalent to M, so we may say that "M and N are row equivalent". It also follows that if A is row equivalent to B, and B is row equivalent to C, then A is row equivalent to C.

EXAMPLE 4

Let us see how the elimination of Example 1 appears in matrix notation. The augmented matrix for the system, with the optional vertical line omitted, is

$$\begin{bmatrix} 1 & 1 & -2 & 4 \\ 1 & 3 & -1 & 7 \\ 2 & 1 & -5 & 7 \end{bmatrix}.$$

The first step in the elimination was to add (-1) times the first equation to the second; here we add (-1) times the first row to the second.

$$\begin{bmatrix} 1 & 1 & -2 & 4 \\ 1 & 3 & -1 & 7 \\ 2 & 1 & -5 & 7 \end{bmatrix} R_2 + (-1)R_1 \rightarrow \begin{bmatrix} 1 & 1 & -2 & 4 \\ 0 & 2 & 1 & 3 \\ 2 & 1 & -5 & 7 \end{bmatrix}.$$

Observe that we have introduced a notation to indicate this row operation. R_2 denotes the second row, and $(-1)R_1$ denotes the first row multiplied by (-1), so $R_2 + (-1)R_1$ means add (-1) times row 1 to row 2. Note that we indicate first the row R_2 to which a multiple of another row is added. We shall denote interchange of rows 2 and 3 (for example) by $R_2 \updownarrow R_3$. We use an arrow $(\rightarrow)$ to indicate that we obtain the new matrix by this step. Some people use $\sim$ ("equivalent to") where we use the arrow. However, it would be wrong to connect the two matrices by an equal sign: the matrices are *not* equal. You should also avoid connecting them by an implication arrow $(\Rightarrow)$, which means something else.

When you become confident with elementary row operations, you may omit these indicators of which row operation you are using. However, they make it easier for you to check your own work, and instructors may require them in work submitted for grading. Now we return to Example 4.

To each step in the elimination of Example 1, there corresponds an elementary row operation; these elementary row operations make up the following row reduction, as indicated by the notation:

$$\begin{bmatrix} 1 & 1 & -2 & 4 \\ 0 & 2 & 1 & 3 \\ 2 & 1 & -5 & 7 \end{bmatrix} R_3 - 2R_1 \rightarrow \begin{bmatrix} 1 & 1 & -2 & 4 \\ 0 & 2 & 1 & 3 \\ 0 & -1 & -1 & -1 \end{bmatrix} R_2 \updownarrow R_3 \rightarrow$$

$$\begin{bmatrix} 1 & 1 & -2 & 4 \\ 0 & -1 & -1 & -1 \\ 0 & 2 & 1 & 3 \end{bmatrix} -1R_2 \rightarrow \begin{bmatrix} 1 & 1 & -2 & 4 \\ 0 & 1 & 1 & 1 \\ 0 & 2 & 1 & 3 \end{bmatrix} R_3 - 2R_2 \rightarrow$$

$$\begin{bmatrix} 1 & 1 & -2 & 4 \\ 0 & 1 & 1 & 1 \\ 0 & 0 & -1 & 1 \end{bmatrix} -1R_3 \rightarrow \begin{bmatrix} 1 & 1 & -2 & 4 \\ 0 & 1 & 1 & 1 \\ 0 & 0 & 1 & -1 \end{bmatrix}.$$

All the row operations corresponding to the elimination in Example 1 have now been performed. We must now **interpret** this final matrix as the **augmented** matrix of a system of linear equations **equivalent** to the original system of linear equations. (Of course, all of the matrices obtained during the process are also the matrices of equivalent systems; that is, all of these matrices are **row equivalent.**) The last row of the final matrix represents the equation $x_3 = -1$; the second row corresponds to $x_2 + x_3 = 1$, so $x_2 = 2$; the first row gives $x_1 + x_2 - 2x_3 = 4$, so $x_1 = 0$. We have exactly the same solution as in Example 1.

For later reference, the matrices corresponding to Examples 2 and 3 are shown below, without all the details of the row operations or the back-substitution.

EXAMPLE 5

Row reduction that corresponds to the elimination in Example 2.

$$\begin{bmatrix} 1 & 0 & 2 & 1 & 14 \\ 1 & 0 & 3 & 3 & 19 \end{bmatrix} R_2 - R_1 \rightarrow \begin{bmatrix} 1 & 0 & 2 & 1 & 14 \\ 0 & 0 & 1 & 2 & 5 \end{bmatrix}$$

EXAMPLE 6

Row reduction that corresponds to the elimination in Example 3.

$$\begin{bmatrix} 1 & 1 & 0 & 1 \\ 0 & 1 & 1 & 2 \\ 1 & 2 & 1 & 2 \end{bmatrix} \rightarrow \cdots \rightarrow \begin{bmatrix} 1 & 1 & 0 & 1 \\ 0 & 1 & 1 & 2 \\ 0 & 0 & 0 & -5 \end{bmatrix}$$

EXAMPLE 7

Find the general solution of the system

$$\begin{aligned} 3x_1 + 8x_2 - 18x_3 + x_4 &= 35 \\ x_1 + 2x_2 - 4x_3 \phantom{{}+x_4} &= 11 \\ x_1 + 3x_2 - 7x_3 + x_4 &= 10 \end{aligned}$$

SOLUTION: Write the augmented matrix of the system and row reduce.

$$\begin{bmatrix} 3 & 8 & -18 & 1 & 35 \\ 1 & 2 & -4 & 0 & 11 \\ 1 & 3 & -7 & 1 & 10 \end{bmatrix} R_1 \updownarrow R_2 \rightarrow \begin{bmatrix} 1 & 2 & -4 & 0 & 11 \\ 3 & 8 & -18 & 1 & 35 \\ 1 & 3 & -7 & 1 & 10 \end{bmatrix} R_2 - 3R_1 \rightarrow$$

$$\begin{bmatrix} 1 & 2 & -4 & 0 & 11 \\ 0 & 2 & -6 & 1 & 2 \\ 1 & 3 & -7 & 1 & 10 \end{bmatrix} R_3 - R_1 \rightarrow \begin{bmatrix} 1 & 2 & -4 & 0 & 11 \\ 0 & 2 & -6 & 1 & 2 \\ 0 & 1 & -3 & 1 & -1 \end{bmatrix} R_2 \updownarrow R_3 \rightarrow$$

$$\begin{bmatrix} 1 & 2 & -4 & 0 & 11 \\ 0 & 1 & -3 & 1 & -1 \\ 0 & 2 & -6 & 1 & 2 \end{bmatrix} R_3 - 2R_2 \rightarrow \begin{bmatrix} 1 & 2 & -4 & 0 & 11 \\ 0 & 1 & -3 & 1 & -1 \\ 0 & 0 & 0 & -1 & 4 \end{bmatrix} \rightarrow$$

$$\begin{bmatrix} 1 & 2 & -4 & 0 & 11 \\ 0 & 1 & -3 & 1 & -1 \\ 0 & 0 & 0 & 1 & -4 \end{bmatrix}$$

Now interpret the final matrix as the augmented matrix of a system of equations in four variables. Start from the right: from the last equation, $x_4 = -4$. The equations do not specify both x_2 and x_3 uniquely; choose x_3 arbitrarily, say $x_3 = t$, where t is any real number ($t \in \mathbb{R}$). Then the second equation gives $x_2 = -1 - x_4 + 3x_3 = 3 + 3t$. Finally, from the first equation, $x_1 = 5 - 2x_3 = 5 - 2t$. Thus, the general solution is

$$\mathbf{x} = (5 - 2t, 3 + 3t, t, -4) = (5, 3, 0, -4) + t(-2, 3, 1, 0), \quad t \in \mathbb{R}.$$

This is the vector equation of a line in $\mathbb{R}^4$. Check this solution by substituting these values for x_1, x_2, x_3, x_4 into the original equations.

Row Echelon Form

How do we know which row operations to perform? And how do we know when we have done enough and should start back-substitution? To answer these questions, recall that the aim is to get as many leading variables in the system as possible, and to have each leading variable eliminated from all the equations after the equation in which it first appears. Moreover, any two equations should be arranged so that the leading variable in the upper equation is to the left of the leading variable in the lower equation.

Thus, in order to be able to solve the system by back-substitution without any further elimination, we require that the elementary row operations be carried out on the *augmented* matrix until the *coefficient* matrix is in a special form called **row echelon form.**

A matrix is in row echelon form if:

(1) In each row the first coefficient that is not zero is a 1; it is called a "leading 1".

(2) When two successive rows are compared, if both have leading 1's, the leading 1 in the upper row is to the left of the leading 1 in the lower row.

(3) When all the entries in a row are zeros, this row appears below all rows that contain a leading 1.

> *REMARK*
>
> It follows from these properties that all the entries in a column below a leading 1 must be zero (for otherwise (1) or (2) would be violated).

Let us review our Examples to see how they fit this row echelon form.

EXAMPLE 8

(a) The matrix $\begin{bmatrix} 1 & 1 & -2 & 4 \\ 0 & 1 & 1 & 1 \\ 0 & 0 & 1 & -1 \end{bmatrix}$ of Example 4 is in row echelon form.

(b) The matrix $\begin{bmatrix} 1 & 0 & 2 & 1 & 14 \\ 0 & 0 & 1 & 2 & 5 \end{bmatrix}$ of Example 5 is in row echelon form.

(c) The *augmented* matrix $\left[\begin{array}{ccc|c} 1 & 1 & -2 & 1 \\ 0 & 1 & 1 & 2 \\ 0 & 0 & 0 & -5 \end{array}\right]$ of Example 6 is *not* in row echelon form; it could be brought into row echelon form by multiplying the last row by $(-1/5)$. Note, however, that in this matrix, the *coefficient* matrix is in row echelon form.

(d) The matrix $\begin{bmatrix} 1 & 2 & -4 & 0 & 11 \\ 0 & 1 & -3 & 1 & -1 \\ 0 & 0 & 0 & 1 & -4 \end{bmatrix}$ of Example 7 is in row echelon form.

> *REMARK*
>
> Note that in the examples there are many 1's that are not leading 1's.

Any matrix can be row reduced to row echelon form by the following steps.

First, consider the first column of the matrix; if it consists entirely of zero entries, move to the next column. If it contains some non-zero entry, interchange rows (if necessary) so the top entry in the *column* is non-zero and multiply by a constant so that this entry becomes 1 — a leading 1 for its row. By elementary row operations of the third kind, make all entries below this 1 into zeros. Next, consider the submatrix consisting of all columns to the right of the column we have just worked on, and all rows below the row with the most recent leading 1. Repeat the procedure just described for this submatrix to obtain the next leading 1, with zeros below it. Keep repeating this whole procedure until we have "used up" all rows and columns of the original matrix. The resulting matrix is in row echelon form.

Important Remark. For a given matrix A, the row echelon form is not unique, that is, you and I might correctly determine *different* row echelon forms for the same matrix A; however, it can be shown that any two row echelon forms for the same matrix A must agree on the position of the leading 1's. (This fact may seem obvious but is not easy to prove; it follows from Exercise F6 in Chapter 4 Review.) Moreover, the set of solutions obtained from one row echelon form for A must be the same as the solution set obtained from any other row echelon form for A, because the two row echelon forms for A are both row equivalent to A and therefore to each other.

Consistent Systems and Unique Solutions

We shall see that it is often important to be able to recognize whether a given system is **consistent**: it is consistent if it has a solution or solutions. It is also important in many applications to know whether the system has a **unique solution** (that is, exactly one solution).

To illustrate the possibilities, consider a system of three linear equations in three variables; each equation can be considered as the equation of a plane in $\mathbb{R}^3$. A solution of the system determines a point of intersection of the three planes that we call $\mathcal{P}_1$, $\mathcal{P}_2$, $\mathcal{P}_3$. Figure 2–1–1 illustrates an inconsistent system: there is no point common to all three planes if the line of intersection of $\mathcal{P}_1$ and $\mathcal{P}_3$ is parallel to the line of intersection of $\mathcal{P}_2$ and $\mathcal{P}_3$. On the other hand, there is a unique solution if the three planes intersect in a single point as in Figure 2–1–2; there are infinitely many solutions if the three planes contain a common line as in Figure 2–1–3.

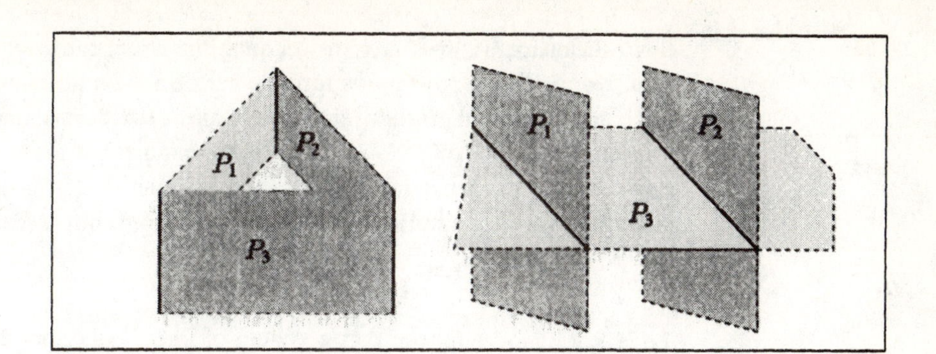

FIGURE 2-1-1. *Two cases where three planes have no common point of intersection.*

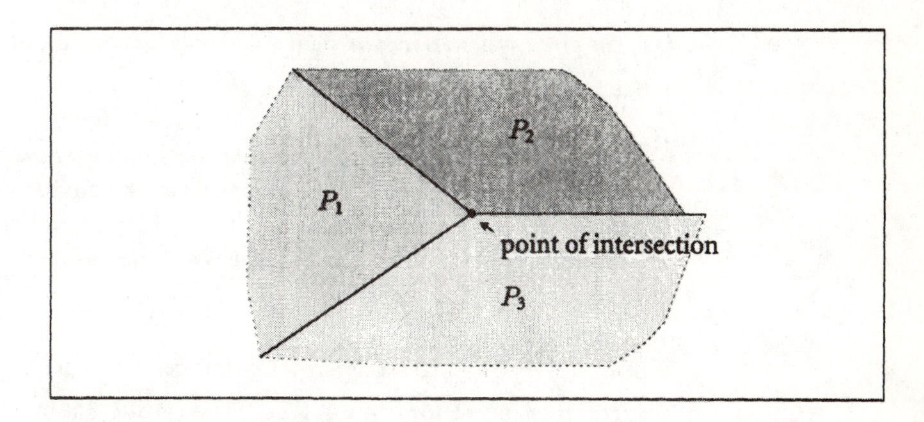

FIGURE 2-1-2. *Three planes with one point in common: the corresponding system of equations has a unique solution.*

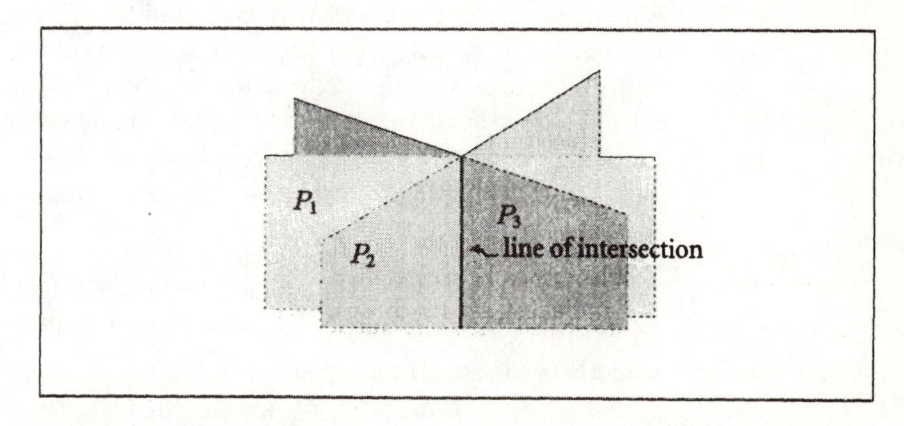

FIGURE 2-1-3. *Three planes that meet in a common line: the corresponding system is consistent, with infinitely many solutions.*

Row echelon form allows us to answer questions about consistency and uniqueness. Notice that the row operations are performed on the **augmented** matrix, but that the solution procedure requires that the **coefficient** matrix be brought into row echelon form. (You may bring the augmented matrix into row echelon form if you wish, and interpret that. It sometimes involves an extra row operation, which is a nuisance for hand calculations, but just as easy for most computer programs.)

THEOREM 1. Suppose that a system of linear equations has augmented matrix $\left[A \mid \mathbf{b}\right]$, and that $\left[A \mid \mathbf{b}\right]$ is row equivalent to $\left[S \mid \mathbf{c}\right]$, where the coefficient matrix S of the second system is in row echelon form. Then:

(1) The given system is **inconsistent** if and only if some row of $\left[S \mid \mathbf{c}\right]$ is of the form $\left[0\,0\ldots 0 \mid c\right]$ with $c \neq 0$;

(2) If the system is **consistent**, there are two possibilities. *Either* the number of leading 1's in S is equal to the number of variables in the system, and then the system has a **unique solution**, *or* the number of leading 1's is less than the number of variables, and then the system has infinitely many solutions (and not a unique solution).

Proof. (1) If $\left[S \mid \mathbf{c}\right]$ contains a row of the form $\left[0\,0\ldots 0 \mid c\right]$ with $c \neq 0$, then clearly there is no vector $(x_1, x_2, \ldots, x_n)$ that satisfies the system of equations (recall Example 3) and the system is inconsistent. On the other hand, if it contains no such row, then each row must *either* be of the form $\left[0\,0\ldots 0 \mid 0\right]$, which corresponds to an equation satisfied by *any* $(x_1, x_2, \ldots, x_n)$, or *else* contain a leading 1. We may ignore the rows that consist entirely of zeros, leaving only rows with leading 1's. In the latter case, the corresponding system can clearly be solved by assigning arbitrary values to the non-leading variables, and then determining the leading variables by back-substitution. Thus, if there is no row of the form $\left[0\,0\ldots 0 \mid \mathbf{c}\right]$, the system cannot be inconsistent.

(2) Now consider the case of a consistent system. The number of leading variables cannot be greater than the number of columns in the coefficient matrix; if it is equal, then each variable is a leading variable, and thus is determined uniquely by the system corresponding to $\left[S \mid \mathbf{c}\right]$. If some variables are not leading variables, they may be chosen arbitrarily—they are parameters in the general solution, and there are infinitely many solutions.

■

EXAMPLE 9

The following examples illustrate the possibilities discussed in the theorem.

(a) If $\left[A \mid \mathbf{b}\right]$ is row equivalent to $\begin{bmatrix} 1 & 3 & -1 & 4 \\ 0 & 1 & 2 & -3 \\ 0 & 0 & 0 & 4 \end{bmatrix}$, the system $A\mathbf{x} = \mathbf{b}$ is inconsistent.

(b) If $\left[A \mid \mathbf{b}\right]$ is row equivalent to $\begin{bmatrix} 1 & -2 & 3 & 1 \\ 0 & 1 & 3 & -5 \\ 0 & 0 & 2 & 8 \end{bmatrix}$, the system $A\mathbf{x} = \mathbf{b}$ is consistent, with a unique solution. Notice that the coefficient matrix is not quite in row echelon form, because the leading non-zero coefficient in the third row is 2 instead of 1. However, we can easily determine consistency of the system from the matrix in this form.

(c) If $\left[A \mid \mathbf{b}\right]$ is row equivalent to $\begin{bmatrix} 1 & -2 & 3 & 1 \\ 0 & 1 & 3 & -5 \\ 0 & 0 & 0 & 0 \end{bmatrix}$, the system $A\mathbf{x} = \mathbf{b}$ is consistent with infinitely many solutions, because x_3 is not a leading variable, so it may be chosen arbitrarily.

Remarks on Choosing a Pivot

In carrying out the row reduction to row echelon form, in (almost) each column you must choose the entry that becomes (by interchange and multiplication) the leading 1 in that column. This is called choosing a pivot. In principle, it does not matter which non-zero entry you choose from a column to use as a pivot in the procedure just described. In practice, it can have considerable impact on the amount of work required, and on the accuracy of the result.

For hand calculations on simple integer examples, it is sensible to go to some trouble to postpone fractions, because avoiding fractions may reduce both the effort required, and the chances of making errors.

EXAMPLE 10

This example illustrates alternative row reductions.

$$\begin{bmatrix} 3 & 7 & 5 \\ 2 & -5 & 11 \end{bmatrix} \xrightarrow{R_1 - R_2} \begin{bmatrix} 1 & 12 & -6 \\ 2 & -5 & 11 \end{bmatrix} \rightarrow \begin{bmatrix} 1 & 12 & -6 \\ 0 & -29 & 23 \end{bmatrix} \rightarrow \cdots$$

may be preferable to

$$\begin{bmatrix} 3 & 7 & 5 \\ 2 & -5 & 11 \end{bmatrix} \xrightarrow{(1/3)R_1} \begin{bmatrix} 1 & 7/3 & 5/3 \\ 2 & -5 & 11 \end{bmatrix} \rightarrow \cdots$$

Or you might prefer

$$\begin{bmatrix} 3 & 7 & 5 \\ 2 & -5 & 11 \end{bmatrix} \rightarrow \begin{bmatrix} 6 & 14 & 10 \\ 2 & -5 & 11 \end{bmatrix} \rightarrow \begin{bmatrix} 6 & 14 & 10 \\ 6 & -15 & 33 \end{bmatrix} \rightarrow \begin{bmatrix} 6 & 14 & 10 \\ 0 & -29 & 23 \end{bmatrix} \rightarrow \cdots$$

Notice that it is not required that you convert the pivot element to 1 before obtaining zeros in other entries in that column.

These alternatives may lead to different row echelon forms, but all of these forms are row equivalent and lead to the same solution set for the system corresponding to the original matrix.

In computer calculations, the choice of the pivot may affect accuracy. The problem is that real numbers are represented to only a finite number of digits in a computer, so inevitably some round-off or truncation error occurs. When you are doing a large number of arithmetic operations, these errors can accumulate, and they can be particularly serious if at some stage you subtract two **nearly** equal numbers. The following artificial example gives some idea of the difficulties that might be encountered.

The system

$$\begin{aligned} 0.1000x_1 + 0.9990x_2 &= 1.000 \\ 0.1000x_1 + 1.000\ x_2 &= 1.006 \end{aligned}$$

is easily found to have solution $x_2 = 6.000$, $x_1 = -49.94$. Notice that all coefficients were given to four digits. Suppose all the entries are rounded to three digits: the system becomes

$$\begin{aligned} 0.100x_1 + 0.999x_2 &= 1.00 \\ 0.100x_1 + 1.00\ x_2 &= 1.01 \end{aligned}$$

and the solution is now $x_2 = 10$, $x_1 = -89.9$. Notice that in spite of the fact that there was only a small change in one term on the right-hand side, the resulting solution is not close to the solution of the original problem. Geometrically, this can be understood by observing that the solution is the intersection point of two nearly parallel lines, so that a small displacement of one line causes a major shift of the intersection point. Difficulties of this kind may arise in higher dimensional systems of equations that arise in real applications.

Careful choice of pivots in computer programs can reduce the error caused by these sorts of problems. However, some matrices are "ill conditioned" — even with high precision calculations, the solutions produced by computers for systems with such matrices may be unreliable. And in applications, the entries in the matrices may be experimentally determined, and small errors in the entries may result in large errors in calculated solutions, no matter how much precision is used in computation. To understand this problem better, you need to know something about sources of error in numerical computation, and more linear algebra. We shall not discuss it further in this book, but you should be aware of the difficulty if you use computers to solve systems of equations.

A Word Problem

To illustrate the application of systems of equations to "word problems", we give a simple example. More interesting applications usually require a fair bit of background; two applications from physics/engineering are discussed in Section 2–3.

EXAMPLE II

A boy has a jar full of coins. Altogether there are 180 nickels, dimes, and quarters. The number of dimes is one half of the total number of nickels and quarters. The value of the coins is $16.00. How many of each kind of coin does he have?

SOLUTION: Let n be the number of nickels, d the number of dimes, and q the number of quarters. *Don't overlook this important step of explicitly naming the variables of the problem.* Then,

$$n + d + q = 180.$$

The second piece of information we are given is that

$$d = \frac{1}{2}(n + q).$$

We rewrite this in a form suitable for inclusion in a system of equations:

$$n - 2d + q = 0.$$

Finally, we have the value of the coins (in cents!):

$$5n + 10d + 25q = 1600.$$

Thus n, d, and q satisfy the system of linear equations

$$\begin{aligned} n + d + q &= 180 \\ n - 2d + q &= 0 \\ 5n + 10d + 25q &= 1600. \end{aligned}$$

Write the augmented matrix and row reduce:

$$\begin{bmatrix} 1 & 1 & 1 & 180 \\ 1 & -2 & 1 & 0 \\ 5 & 10 & 25 & 1600 \end{bmatrix} \to \cdots \to \begin{bmatrix} 1 & 1 & 1 & 180 \\ 0 & 1 & 0 & 60 \\ 0 & 0 & 1 & 20 \end{bmatrix}.$$

Solve the final system by back-substitution and interpret: the boy has 100 nickels, 60 dimes, and 20 quarters.

EXERCISES 2–1

A1 Solve each of the following systems by back-substitution (no elimination required), and write the general solution in the standard form.

(a) $\begin{aligned} x_1 - 3x_2 &= 5 \\ x_2 &= 4 \end{aligned}$

(b) $\begin{aligned} x_1 + 2x_2 - x_3 &= 7 \\ x_3 &= 6 \end{aligned}$

(c) $\begin{aligned} x_1 + 3x_2 - 2x_3 &= 4 \\ x_2 + 5x_3 &= 2 \\ x_3 &= 2 \end{aligned}$

(d) $\begin{aligned} x_1 - 2x_2 + x_3 + 4x_4 &= 7 \\ x_2 - x_4 &= -3 \\ x_3 + x_4 &= 2 \end{aligned}$

(e) $\begin{aligned} x_1 + 2x_2 - 3x_3 + x_4 + 2x_5 &= 2 \\ x_2 + x_3 - 2x_4 - 7x_5 &= -3 \\ x_4 - 4x_5 &= 1 \end{aligned}$

A2 For each of the following matrices, carry out the indicated sequence of elementary row operations. Show steps clearly.

(a) $\begin{bmatrix} 7 & 3 & 5 & 4 \\ 5 & 2 & 1 & -2 \\ 1 & 2 & 5 & -2 \end{bmatrix}$: (i) interchange rows 1 and 3; (ii) to row 2, add (-5) times (new) row 1; (iii) to row 3, add (-7) times row 1; (iv) multiply row 2 by $(-1/8)$; (v) add 11 times row 2 to row 3.

(b) Same matrix as in part (a): (i) to row 2, add $(-5/7)$ times row 1; (ii) to row 3, add $(-1/7)$ times row 1; (iii) multiply row 2 by (-7); (iv) to row 3, add $(-11/7)$ times row 2.

(c) $\begin{bmatrix} 1 & 13 & -9 & 8 \\ 1 & 1 & -1 & 2 \\ 5 & 1 & 3 & 3 \\ -6 & -8 & 3 & 5 \end{bmatrix}$: (i) add (-1) times row 1 to row 2; (ii) add (-5) times row 1 to row 3; (iii) add 6 times row 1 to row 4.

(d) Same matrix as in (c): (i) interchange rows 1 and 2; (ii) add (-1) times row 1 to row 2; (iii) add (-5) times row 1 to row 3; (iv) add 6 times row 2 to row 4.

A3 (a) Which of the matrices A, B, C, D given below is in row echelon form? For each matrix that is not in row echelon form, explain why it is not.
(b) Now consider each matrix to be the augmented matrix of a system of equations. Decide in each case whether the *coefficient* matrix of the system is in row echelon form; if it is not, say why it is not. Decide whether each system is consistent.

$$A = \begin{bmatrix} 1 & 2 & 3 & 4 \\ 0 & 1 & -2 & -3 \\ 0 & 0 & 0 & 3 \end{bmatrix} \quad B = \begin{bmatrix} 0 & 1 & 2 & 3 \\ 0 & 0 & 1 & 1 \\ 0 & 0 & 0 & 0 \end{bmatrix}$$

$$C = \begin{bmatrix} 1 & -1 & -2 & -3 \\ 0 & 1 & 2 & 0 \\ 0 & 1 & 0 & 3 \end{bmatrix} \quad D = \begin{bmatrix} 1 & 0 & 2 & 1 \\ 0 & 2 & 1 & 0 \\ 0 & 0 & 1 & 1 \end{bmatrix}$$

A4 Row reduce the following matrices to obtain a row equivalent matrix in row echelon form. (Note that different row echelon forms are possible for one given matrix, but that all correct answers will agree on the position of the leading 1's.) Show your row operations.

(a) $\begin{bmatrix} 4 & 1 & 1 \\ 1 & -3 & 2 \end{bmatrix}$ (b) $\begin{bmatrix} 2 & -2 & 5 & 8 \\ 1 & -1 & 2 & 3 \\ -1 & 1 & 0 & 2 \end{bmatrix}$

(c) $\begin{bmatrix} 1 & -1 & -1 \\ 2 & -1 & -2 \\ 5 & 0 & 0 \\ 3 & 4 & 5 \end{bmatrix}$ (d) $\begin{bmatrix} 2 & 0 & 2 & 0 \\ 1 & 2 & 3 & 4 \\ 1 & 4 & 9 & 16 \\ 3 & 6 & 13 & 20 \end{bmatrix}$

(e) $\begin{bmatrix} 0 & 1 & 2 & 1 \\ 1 & 2 & 1 & 1 \\ 3 & -1 & -4 & 1 \\ 2 & 1 & 3 & 6 \end{bmatrix}$ (f) $\begin{bmatrix} 3 & 1 & 8 & 2 & 4 \\ 1 & 0 & 3 & 0 & 1 \\ 0 & 2 & -2 & 4 & 3 \\ -4 & 1 & 11 & 3 & 8 \end{bmatrix}$

A5 Each of the following matrices is the augmented matrix of a system of linear equations. In each case, the *coefficient* matrix is in row echelon form, or *nearly* in row echelon form. Determine whether the system is consistent, and if so, determine the general solution.

(a) $A = \begin{bmatrix} 1 & 2 & -1 & 2 \\ 0 & 1 & 3 & 4 \\ 0 & 0 & 0 & -5 \end{bmatrix}$ (b) $B = \begin{bmatrix} 1 & 0 & 1 & 0 & 1 \\ 0 & 1 & 1 & 1 & 2 \\ 0 & 0 & 0 & 1 & 3 \end{bmatrix}$

(c) $C = \begin{bmatrix} 1 & 1 & -1 & 3 & 1 \\ 0 & 0 & 2 & 1 & 3 \\ 0 & 0 & 0 & 1 & -2 \end{bmatrix}$

A6 For each of the following systems of linear equations,

(i) write the augmented matrix;

(ii) obtain a row equivalent matrix such that the coefficient matrix is in row echelon form;

(iii) comment on whether the system is consistent or inconsistent, and if consistent, on whether the solution is unique;

(iv) if the system is consistent, write its general solution in the standard form, and interpret the solution as a point, or a line, or a plane, etc.

(a) $$\begin{aligned} 3x_1 - 5x_2 &= 2 \\ x_1 + 2x_2 &= 4 \end{aligned}$$

(b) $$\begin{aligned} x_1 + 2x_2 + x_3 &= 5 \\ 2x_1 - 3x_2 + 2x_3 &= 6 \end{aligned}$$

(c) $$\begin{aligned} x_1 + 2x_2 - 3x_3 &= 8 \\ x_1 + 3x_2 - 5x_3 &= 11 \\ 2x_1 + 5x_2 - 8x_3 &= 19 \end{aligned}$$

(d) $$\begin{aligned} -3x_1 + 6x_2 + 16x_3 &= 36 \\ x_1 - 2x_2 - 5x_3 &= -11 \\ 2x_1 - 3x_2 - 8x_3 &= -17 \end{aligned}$$

(e) $$\begin{aligned} x_1 + 2x_2 - x_3 &= 4 \\ 2x_1 + 5x_2 + x_3 &= 10 \\ 4x_1 + 9x_2 - x_3 &= 19 \end{aligned}$$

(f) $$\begin{aligned} x_1 + 2x_2 - 3x_3 \phantom{{}+4x_4} &= -5 \\ 2x_1 + 4x_2 - 6x_3 + x_4 &= -8 \\ 6x_1 + 13x_2 - 17x_3 + 4x_4 &= -21 \end{aligned}$$

(g) $$\begin{aligned} 2x_2 - 2x_3 \phantom{{}+3x_4} + 2x_5 &= 2 \\ x_1 + 2x_2 - 3x_3 + x_4 + 4x_5 &= 1 \\ 2x_1 + 4x_2 - 5x_3 + 3x_4 + 8x_5 &= 3 \\ 2x_1 + 5x_2 - 7x_3 + 3x_4 + 10x_5 &= 5 \end{aligned}$$

A7 The following matrices are the augmented matrices of systems of linear equations. Determine the values of *a*, *b*, *c*, *d* for which the systems are consistent. In cases where the system is consistent, determine whether the system has a unique solution.

(a) $$\begin{bmatrix} 2 & 4 & -3 & 6 \\ 0 & b & 7 & 2 \\ 0 & 0 & a & a \end{bmatrix}$$

(b) $$\begin{bmatrix} 1 & -1 & 4 & -2 & 5 \\ 0 & 1 & 2 & 3 & 4 \\ 0 & 0 & d & 5 & 7 \\ 0 & 0 & 0 & cd & c \end{bmatrix}$$

A8 A fruit-seller has apples, bananas, and oranges. Altogether he has 1500 pieces of fruit. On average, each apple weighs 120 grams, each banana weighs 140 grams, and each orange weighs 160 grams. He can sell apples for 25 cents each, bananas for 20 cents each, and oranges for 30 cents each. If the fruit weighs 208 kilograms, and the total selling price is $380.00, how many of each kind of fruit does he have?

A9 A student is taking courses in Algebra, Calculus, and Physics at a college where grades are given in percentages. To determine her standing for a Physics prize, a weighted average is calculated based on 50% of her Physics grade, 30% of her Calculus grade, and 20% of her Algebra grade; the weighted average is 84. For an Applied Mathematics prize, a weighted average based on 1/3 of each of the three grades is calculated to be 83. For a Pure Mathematics prize, her average based on 50% of her Calculus grade and 50% of her Algebra grade is 82.5. What are her grades in the individual courses?

B1 Solve each of the following systems by back-substitution (no elimination required), and write the general solution in the standard form.

(a) $$\begin{aligned} x_1 - 2x_2 - x_3 &= 5 \\ x_2 + 3x_3 &= 4 \\ x_3 &= -2 \end{aligned}$$

(b) $$\begin{aligned} x_1 - 3x_2 + x_3 &= 1 \\ x_2 + 2x_3 &= -1 \end{aligned}$$

(c) $$\begin{aligned} x_1 + 3x_2 - x_3 + 2x_4 &= -2 \\ x_2 - x_3 + 2x_4 &= -1 \\ x_3 + 3x_4 &= 3 \end{aligned}$$

(d) $$\begin{aligned} x_1 + 3x_2 - 2x_3 - 2x_4 + 2x_5 &= -2 \\ x_2 - 2x_3 - 2x_4 + 3x_5 &= 4 \\ x_3 + x_4 - 2x_5 &= -3 \end{aligned}$$

B2 For each of the following matrices, carry out the indicated sequence of elementary row operations. Show steps clearly.

(a) $\begin{bmatrix} 2 & -4 & 6 & 1 \\ 2 & 1 & 10 & -4 \\ 1 & 1 & 7 & -1 \end{bmatrix}$: (i) interchange rows 1 and 3; (ii) add (-2) times row 1 to row 2; (iii) add (-2) times row 1 to row 3; (iv) add (-6) times row 2 to row 3.

(b) $\begin{bmatrix} 0 & 1 & 0 & 1 \\ 1 & 2 & 3 & 4 \\ 1 & 4 & 9 & 16 \\ 2 & 0 & 2 & 0 \end{bmatrix}$: (i) interchange rows 1 and 4; (ii) interchange rows 2 and 4; (iii) multiply row 1 by 1/2; (iv) add (-1) times row 1 to row 3; (v) add (-1) times row 1 to row 4; (vi) multiply row 3 by 1/4.

B3 (a) Which of the matrices A, B, C, D given below is in row echelon form? For each matrix that is not in row echelon form, explain why it is not.
(b) Now consider each matrix to be the augmented matrix of a system of equations. Decide in each case whether the *coefficient* matrix of the system is in row echelon form; if it is not, say why it is not. Decide whether each system is consistent.

$$A = \begin{bmatrix} 0 & 1 & -2 & 5 \\ 0 & 0 & 1 & 2 \\ 0 & 0 & 0 & 1 \end{bmatrix} \qquad B = \begin{bmatrix} 1 & 0 & 1 & 1 \\ 0 & 3 & 0 & 6 \\ 0 & 0 & 1 & 2 \end{bmatrix}$$

$$C = \begin{bmatrix} 1 & 2 & 2 & 4 \\ 1 & 0 & 0 & 3 \\ 0 & 0 & 0 & 1 \end{bmatrix} \qquad D = \begin{bmatrix} 1 & 0 & 2 & 1 \\ 0 & 1 & 2 & 1 \\ 0 & 0 & 1 & -3 \end{bmatrix}$$

B4 Row reduce the following matrices to obtain a row equivalent matrix in row echelon form. (Note that different row echelon forms are possible for one given matrix, but that all correct answers will agree on the position of the leading 1's.)

(a) $\begin{bmatrix} 2 & 4 & 2 & -1 \\ 1 & 3 & 3 & 0 \\ 2 & 5 & 5 & -4 \end{bmatrix}$ (b) $\begin{bmatrix} 5 & 7 & -2 & 7 & 4 \\ 1 & 1 & 0 & 1 & 0 \\ 3 & 0 & 3 & -6 & 5 \end{bmatrix}$

(c) $\begin{bmatrix} 2 & 1 & 0 & 7 \\ 1 & 1 & -1 & 2 \\ 3 & 2 & 0 & 12 \\ 6 & 4 & -1 & 25 \end{bmatrix}$ (d) $\begin{bmatrix} 1 & 2 & 1 & 3 & 0 \\ 2 & 5 & 2 & 6 & 1 \\ 3 & 7 & 4 & 9 & 3 \\ 2 & 6 & 2 & 6 & 5 \end{bmatrix}$

B5 Each of the following matrices is the augmented matrix of a system of linear equations. In each case, the *coefficient* matrix is in row echelon form, or *nearly* in row echelon form. Determine whether the system is consistent, and if so, determine the general solution.

(a) $A = \begin{bmatrix} 1 & -1 & 1 & 0 & 1 \\ 0 & 1 & 0 & -1 & 2 \\ 0 & 0 & 2 & 4 & -3 \end{bmatrix}$ (b) $B = \begin{bmatrix} 1 & -2 & 1 & 0 & 1 \\ 0 & 0 & 1 & 2 & 3 \\ 0 & 0 & 0 & 0 & 1 \end{bmatrix}$

(c) $C = \begin{bmatrix} 2 & 1 & 0 & 1 & 2 \\ 0 & 1 & 0 & 1 & 1 \\ 0 & 0 & 1 & 1 & 2 \end{bmatrix}$

B6 For each of the following systems of linear equations,

(i) write the augmented matrix;

(ii) obtain a row equivalent matrix such that the coefficient matrix is in row echelon form;

(iii) comment on whether the system is consistent or inconsistent, and if consistent, on whether the solution is unique;

(iv) if the system is consistent, write its general solution in the standard form.

(a) $\begin{aligned} 2x_1 + x_2 + 5x_3 &= -4 \\ x_1 + x_2 + x_3 &= -2 \end{aligned}$

(b) $\begin{aligned} 2x_1 + x_2 - x_3 &= 6 \\ x_1 - 2x_2 - 2x_3 &= 1 \\ -x_1 + 12x_2 + 8x_3 &= 7 \end{aligned}$

(c) $\begin{aligned} x_2 + x_3 &= 2 \\ x_1 + x_2 + x_3 &= 3 \\ 2x_1 + 3x_2 + 3x_3 &= 9 \end{aligned}$

(d) $\begin{aligned} x_1 + x_2 &= -7 \\ 2x_1 + 4x_2 + x_3 &= -16 \\ x_1 + 2x_2 + x_3 &= 9 \end{aligned}$

(e) $\begin{aligned} x_1 + x_2 + 2x_3 + x_4 &= 3 \\ x_1 + 2x_2 + 4x_3 + x_4 &= 7 \\ x_1 + x_4 &= -21 \end{aligned}$

B7 The following matrices are the augmented matrices of systems of linear equations. Determine the values of a, b, c, d for which the systems are consistent. In cases where the system is consistent, determine whether the system has a unique solution.

(a) $\begin{bmatrix} 1 & -2 & 4 & 7 \\ 0 & a^2-1 & a & 3 \\ 0 & 0 & b & -3 \end{bmatrix}$ (b) $\begin{bmatrix} 1 & 0 & 2 & 5 & 2 \\ 0 & c & c & 0 & 1 \\ 0 & 0 & c & 0 & c \\ 0 & 0 & 0 & cd & c+d \end{bmatrix}$

B8 A bookkeeper is trying to determine the prices that a manufacturer was charging by examining old sales slips, which show the number of various items shipped and the total price. He finds that 20 armchairs, 10 sofa beds, and 8 double beds cost \$15,200; that 15 armchairs, 12 sofa beds, and 10 double beds cost \$15,700; and that 12 armchairs, 20 sofa beds, and 10 double beds cost \$19,600. Determine the cost for each item (or explain why the sales slips must be in error).

B9 (Requires knowledge of forces and moments.)
A rod 10 m long is pivoted at its centre; it swings in the horizontal plane. Forces of magnitude F_1, F_2, F_3 are applied perpendicular to the rod in the directions indicated by the arrows in Figure 2-1-4; F_1 is applied at the left end of the rod, F_2 is applied at a point 2 m to the right of centre, and F_3 at a point 4 m to the right. The total force on the pivot is zero, the moment about the centre is zero, and the sum of the magnitudes of the forces is 80 newtons. Write a system of three equations for F_1, F_2, F_3; write the corresponding augmented matrix; and use the standard procedure to find F_1, F_2, F_3.

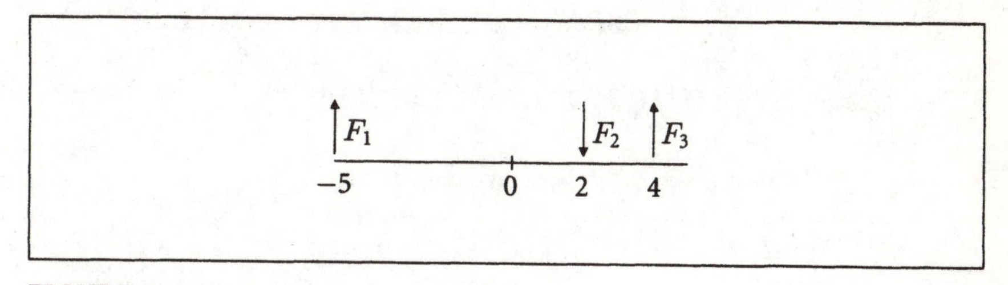

FIGURE 2-1-4. *The forces of Exercise B9.*

B10 Students at Linear University write an examination in linear algebra. An average mark is computed for 100 students in Business, an average is computed for 300 students in Liberal Arts, and an average is computed for 200 students in Science. The average of these three averages is 85%. However, the overall average for the 600 students is 86%. Also, the average for the 300 students in Business and Science is 4 marks higher than the average for the students in Liberal Arts. Determine the average for each group of students by solving a system of linear equations.

COMPUTER RELATED EXERCISES

C1 If your computer software allows you to perform row operations one at a time,

$$\text{let } A = \begin{bmatrix} 35 & 45 & 18 & 13 \\ 17 & 65 & -61 & 7 \\ 23 & 19 & 6 & 41 \end{bmatrix}.$$

(a) What is the matrix that is obtained from A at the end of the following sequence of row operations?
To row 1, add (-2) times row 2.
To row 3, add (-23) times row 1.

Interchange rows 2 and 3.

(b) What is the matrix obtained from A by the same row operations performed in reverse order?

C2 You may wish to redo problems A2, A4, A6, B2, B4, and B6 using a computer.

C3 Suppose that a system of linear equations has the augmented matrix

$$\begin{bmatrix} 1.121 & -2.015 & 2.131 & 4.612 \\ 2.501 & 3.214 & 4.130 & 3.115 \\ -1.639 & -12.473 & -1.827 & 8.430 \end{bmatrix}.$$

Determine the solution of this system. (Answer: $(-33.77, 0.740, 20.63)$) Now change the entry in the second row and third column from 4.130 to 4.080 and find the solution. (Answer: $(26.37, -2.131, -13.72)$) The answer changes greatly with one small change in one coefficient.

CONCEPTUAL EXERCISES

***D1** Consider the linear system in x, y, z, w:

$$\begin{aligned} x + y + w &= b \\ 2x + 3y + z + 5w &= 6 \\ z + w &= 4 \\ 2y + 2z + aw &= 1. \end{aligned}$$

For what values of the constants a and b is the system (i) inconsistent? (ii) consistent with a unique solution? (iii) consistent with a non-unique solution?

D2 Recall that in $\mathbb{R}^3$ two planes $\mathbf{n} \cdot \mathbf{x} = c$ and $\mathbf{m} \cdot \mathbf{x} = d$ are parallel if and only if the normal vector $\mathbf{m}$ is a non-zero multiple of the normal vector $\mathbf{n}$. Row reduce a suitable augmented matrix to explain why two parallel planes must either coincide or else have no points in common.

SECTION

2–2 Reduced Row Echelon Form; Rank; Homogeneous Systems

For the purpose of determining the solution to a particular system of linear equations, elimination with back-substitution as described in Section 2–1 is the standard basic procedure. There are, however, situations and applications where it is advantageous to carry the elimination steps (elementary row operations) "as far as possible" so as to avoid the need for back-substitution. For example, such **complete elimination** will be used in determining the inverse of a matrix in Chapter 3.

To see what further elimination steps (elementary row operations) might be worthwhile, recall that elimination proceeds by establishing certain leading variables, and eliminating a leading variable from the equations below the equation where it is the leading variable. The only further elimination steps that simplify the system are steps that eliminate a leading variable from equations *above* the equation in which it is the leading variable.

EXAMPLE 12

In Section 2–1, Example 1, elimination was performed until the system was found to be equivalent to the system:

$$\begin{aligned} x_1 + x_2 - 2x_3 &= 4 \\ x_2 + x_3 &= 1 \\ x_3 &= -1. \end{aligned}$$

We continue to solve this system by complete elimination. Add (-1) times the third equation to the second. Then add (2) times the third to the first. The system takes the form

$$\begin{aligned} x_1 + x_2 \qquad &= 2 \\ x_2 \qquad &= 2 \\ x_3 &= -1. \end{aligned}$$

The leading variable x_3 has now been eliminated from the first and second equations. Now add (-1) times the second equation to the first to obtain

$$\begin{aligned} x_1 \qquad &= 0 \\ x_2 \qquad &= 2 \\ x_3 &= -1. \end{aligned}$$

The system has been solved by **complete elimination**. Each leading variable has been eliminated from every equation except the one in which it is the leading variable. (This procedure is often called **Gauss-Jordan elimination** to distinguish it from the Gaussian elimination with back-substitution of Section 2–1.)

A matrix corresponding to a system on which complete elimination has been carried out is in a special kind of row echelon form.

DEFINITION

A matrix R is said to be in **reduced row echelon form** if

(1) it is in row echelon form, and

(2) in a column with a leading 1, all the entries except the leading 1 are zeros.

Notice that for row echelon form we required only that the entries below the leading 1 were 0, but now we are also requiring that the entries above the leading 1 are 0. As in the case of row echelon form, it is easy to see that every matrix is **row equivalent** to a matrix in reduced row echelon form. However, in this case we get a stronger result. To state the theorem, it is useful to describe a matrix with m rows and n columns as an m by n matrix.

THEOREM 1. For any given m by n matrix A there is a *unique* m by n matrix R in reduced row echelon form that is row equivalent to A.

Discussion. That A can be row reduced to *some* matrix in reduced row echelon form can be proved by arguments like those in Section 2–1. The proof that there is *only one* such reduced row echelon form that is row equivalent to A requires more ideas; you are asked to give the proof as Exercise F6 of Chapter 4 Review.

EXAMPLE 13

Obtain reduced row echelon forms that are row equivalent to the matrices of Examples 4 and 5 in Section 2–1.

SOLUTION: (a) The original augmented matrix for Example 4 in Section 2–1 was $\begin{bmatrix} 1 & 1 & -2 & 4 \\ 1 & 3 & -1 & 7 \\ 2 & 1 & -5 & 7 \end{bmatrix}$; call this matrix B. We found that B is row equivalent to the matrix $S = \begin{bmatrix} 1 & 1 & -2 & 4 \\ 0 & 1 & 1 & 1 \\ 0 & 0 & 1 & -1 \end{bmatrix}$, which is in row echelon form. By the complete elimination above, we see that another matrix row equivalent to B is $R = \begin{bmatrix} 1 & 0 & 0 & 0 \\ 0 & 1 & 0 & 2 \\ 0 & 0 & 1 & -1 \end{bmatrix}$. This matrix is also in row echelon form, but it is also in the more special **reduced** row echelon form. Thus, for the matrix B, we have given two matrices S and R that are row equivalent to B and in row echelon form. However, there is only one matrix row equivalent to B and in **reduced** row echelon form, and it is R.

(b) For the matrix of Example 5, the row reduction is as follows:

$$\begin{bmatrix} 1 & 0 & 2 & 1 & 14 \\ 1 & 0 & 3 & 3 & 19 \end{bmatrix} \to \begin{bmatrix} 1 & 0 & 2 & 1 & 14 \\ 0 & 0 & 1 & 2 & 5 \end{bmatrix} \quad \text{(row echelon form)}$$

$$\to \begin{bmatrix} 1 & 0 & 0 & -3 & 4 \\ 0 & 0 & 1 & 2 & 5 \end{bmatrix} \quad \text{(reduced row echelon form).}$$

Because of the uniqueness of the reduced row echelon form, we may speak of *the* reduced row echelon form that is row equivalent to a given matrix B. Often, we are a little sloppy and speak of the reduced row echelon form *of* B, or of reducing B to its reduced row echelon form. Reduced row echelon form is not more efficient than the methods of Section 2–1 for solving systems of equations, but the uniqueness of the reduced row echelon form is sometimes useful, as we shall see. For small systems with integer coefficients, you may also find it just as easy to solve by using reduced row echelon form.

To reduce a matrix to its reduced row echelon form by hand calculations, it is usually easiest to first proceed as in Section 2–1 to obtain row echelon form. Then begin with the leading 1 that is furthest to the *right* and perform row operations to make the other entries in its column zero. Move back to the left, getting zero in all entries above each leading 1 (the row echelon form already took care of the entries below each leading 1).

If you were writing a computer program to obtain reduced row echelon form, it might seem more natural not to obtain a row echelon form first. Instead, you might find the leftmost leading 1 and obtain zeros in *all* other entries in its column. Move to the next leading 1 and obtain zeros in all other entries in its column. That is, obtain zeros below *and* above the leading 1 before moving on to the next leading 1. However, this is a poor strategy because it requires more multiplications and additions than the previous strategy. Not only does this require more time, but it also allows more opportunities for loss of accuracy. This is discussed in Exercise F2 at the end of this chapter.

Some Shortcuts and Some Bad Moves

When carrying out elementary row operations, you may get weary of rewriting the matrix every time. It is natural to think of combining a couple of elementary row operations in one rewriting. For example,

$$\begin{bmatrix} 1 & 1 & -2 & 4 \\ 1 & 3 & -1 & 7 \\ 2 & 1 & -5 & 7 \end{bmatrix} \begin{matrix} \\ R_2 - R_1 \\ R_3 - 2R_1 \end{matrix} \rightarrow \begin{bmatrix} 1 & 1 & -2 & 4 \\ 0 & 2 & 1 & 3 \\ 0 & -1 & -1 & -1 \end{bmatrix}.$$

Choosing one particular row (in this case, the first) and adding multiples of it to several other rows is perfectly acceptable, and should lead to no errors. It is usually safe to combine interchanging two rows and multiplying one row by a constant, or in larger matrices, doing one step that involves the first and second rows and another that involves the third and fourth.

There is one combination of steps that does cause errors (there was a warning about this in Section 2–1). The following example illustrates the problem:

$$\begin{bmatrix} 2 & \dots \\ 3 & \dots \end{bmatrix} \begin{matrix} \\ 2R_2 - 3R_1 \end{matrix} \rightarrow \begin{bmatrix} 2 & \dots \\ 0 & \dots \end{bmatrix}.$$

This is really **two** elementary row operations: multiply a row (the second) by a non-zero constant, and then add to a row a multiple of another row. These two operations should always be shown separately;

$$\begin{bmatrix} 2 & \dots \\ 3 & \dots \end{bmatrix} \begin{matrix} \\ 2R_2 \end{matrix} \rightarrow \begin{bmatrix} 2 & \dots \\ 6 & \dots \end{bmatrix} \begin{matrix} \\ R_2 - 3R_1 \end{matrix} \rightarrow \begin{bmatrix} 2 & \dots \\ 0 & \dots \end{bmatrix}.$$

In fact, no problem would arise with such combinations of row operations if we used elementary row operations only for solving equations. However, we shall also use elementary row operations in evaluating the determinant in Chapter 5, and in that setting, ignoring the fact that you have multiplied the second row by a constant will certainly lead to error.

Avoid the following serious error:

$$\begin{bmatrix} 1 & 1 & 3 \\ 1 & 2 & 4 \end{bmatrix} \begin{matrix} R_1 - R_2 \\ R_2 - R_1 \end{matrix} \rightarrow \begin{bmatrix} 0 & -1 & -1 \\ 0 & 1 & 1 \end{bmatrix}. \qquad \text{(WRONG!)}$$

This is clearly nonsense, because the final matrix should have a leading 1 in the first column. By performing one row operation, we **change** one row, and thereafter we must use that row in its new changed form. This may be clearer if we write the steps one at a time:

$$\begin{bmatrix} 1 & 1 & 3 \\ 1 & 2 & 4 \end{bmatrix} \begin{matrix} \\ R_2 - R_1 \end{matrix} \rightarrow \begin{bmatrix} 1 & 1 & 3 \\ 0 & 1 & 1 \end{bmatrix}.$$

Now it is clearly impossible to make both entries in the first column zero.

Rank of a Matrix

The number of leading 1's in the reduced row echelon form of a matrix is clearly an important piece of information in determining whether the system has solutions, and if so, how many.

DEFINITION

The **rank of a matrix** M is the number of leading 1's in the reduced row echelon form that is row equivalent to M.

The rank of M is also equal to the number of leading 1's in any row echelon form that is row equivalent to M, since that number must be equal to the number in the reduced row echelon form. Since the row echelon form is not unique, however, it is more tiresome to try to give clear arguments in terms of row echelon form. It is sometimes useful to say that the rank of M is equal to the number of **non-zero** rows in the reduced row echelon form of M. Later (in Section 4-5), we shall meet more conceptual ways of describing rank.

Theorem 1 of Section 2–1, concerning the existence of solutions for a system of linear equations, can conveniently be restated in terms of rank.

THEOREM 2. Suppose that $A\mathbf{x} = \mathbf{b}$ is a system of m linear equations in n variables.

(a) The system is consistent if and only if the rank of the coefficient matrix A is equal to the rank of the augmented matrix $\left[A \mid \mathbf{b}\right]$.

(b) If a system is consistent and the rank of the coefficient matrix A is equal to the number of variables, then there is a unique solution; otherwise, a consistent system has infinitely many solutions.

Proof. Notice first that the reduced row echelon form of A simply consists of the first n columns of the reduced row echelon form of $\left[A \mid \mathbf{b}\right]$. Now, the system is inconsistent if and only if the reduced row echelon form of $\left[A \mid \mathbf{b}\right]$ contains a row of the form $\left[0\ 0 \ldots 0 \mid 1\right]$. But if there are rows like this, the rank of $\left[A \mid \mathbf{b}\right]$ is certainly greater than the rank of A; of course, the rank of $\left[A \mid \mathbf{b}\right]$ cannot be less than the rank of A, so the system is *inconsistent* exactly when the two ranks are not equal and consistent when the two ranks are equal. The rest of the theorem follows just as it did in Theorem 1 of Section 2–1.

■

COROLLARY (*An important special case in Section 3-5*). **If the rank of the coefficient matrix A is equal to the number of *equations* in the system $A\mathbf{x} = \mathbf{b}$ (or equivalently, equal to the number of rows in A), then the system is consistent for *every* right-hand side $\mathbf{b}$.**

Proof. If the rank of A is equal to the number of equations, then the rank of A must be the rank of $\left[A \mid \mathbf{b}\right]$ no matter what $\mathbf{b}$ is, so the system is consistent for every $\mathbf{b}$.

A more direct way to see why the result is true is to observe that if the rank of A is equal to the number of equations, then, the last row of the reduced row echelon form of A must contain a leading 1, so the system must be consistent. ■

When the solution of a consistent system of linear equations is not unique, it is sometimes important to determine how many parameters there are in the standard general solution of the system. The parameters of the solution are exactly the *non-leading* variables, which can be assigned arbitrary real values, so if the system has m equations in n variables, **the number of parameters is (n – rank of the coefficient matrix).**

EXAMPLE 14

In Example 13 (b) we saw that the augmented matrix $\begin{bmatrix} 1 & 0 & 2 & 1 & 14 \\ 1 & 0 & 3 & 3 & 19 \end{bmatrix}$ had reduced echelon form $\begin{bmatrix} 1 & 0 & 0 & -3 & 4 \\ 0 & 0 & 1 & 2 & 5 \end{bmatrix}$. The rank of the coefficient matrix is two, the number of variables is four, and the number of parameters in the general solution is two (4 – 2). This agrees with what we found when we determined the general solution in Example 2 (Section 2–1).

In Example 7 (Section 2–1), we saw that

$$\begin{bmatrix} 3 & 8 & -18 & 1 & 35 \\ 1 & 2 & -4 & 0 & 11 \\ 1 & 3 & -7 & 1 & 10 \end{bmatrix} \to \begin{bmatrix} 1 & 2 & -4 & 0 & 11 \\ 0 & 1 & -3 & 1 & -1 \\ 0 & 0 & 0 & 1 & -4 \end{bmatrix}.$$

The corresponding system has 4 variables; the rank of the coefficient matrix is 3 (the number of leading 1's in the final form of the coefficient matrix); the number of parameters in the general solution is $4 - 3 = 1$. This agrees with the solution in Example 7.

Notice that, according to the Corollary to Theorem 2, in both of these examples the system would be consistent no matter how we changed the right-hand side **b** of the system, because in each case the rank of the coefficient matrix is equal to the number of equations.

Contrast this with Example 3; as we saw in Example 6, the augmented matrix in this case reduced to $\begin{bmatrix} 1 & 1 & 0 & 1 \\ 0 & 1 & 1 & 2 \\ 0 & 0 & 0 & -5 \end{bmatrix}$. The rank of the coefficient matrix is two, but the rank of the augmented matix is three, so the system is inconsistent. Even if we changed the original right-hand side **b** in Example 3, the system would be inconsistent unless we chose a **b** such that after row reduction we got a 0 in the bottom right corner.

We shall introduce vocabulary for talking more precisely about the rank of a matrix and the number of solutions of the corresponding system in Chapter 4.

Homogeneous Linear Equations

Frequently systems of linear equations appear where all of the terms on the right-hand side are zero. For example, in $\mathbb{R}^3$ a vector **x** that is orthogonal to two given vectors **a** and **b** must satisfy the equations $a_1x_1 + a_2x_2 + a_3x_3 = 0$ and $b_1x_1 + b_2x_2 + b_3x_3 = 0$ (see Section 1–4). Later, we shall be concerned with eigenvectors, which can be found as the solutions of equations with right-hand side zeros.

A linear equation is **homogeneous** if the right-hand side is zero. A system of linear equations is **homogeneous** if *all* of the equations of the system are homogeneous.

Since a homogeneous system is a special case of the systems already discussed, no new tools or techniques are needed to solve them. However, it is usual to work with the coefficient matrix, and not the augmented matrix, since the last column of the augmented matrix consists of zeros, and nothing of interest happens to these zeros when you carry out elementary row operations. The rank of the augmented matrix is necessarily equal to the rank of the coefficient matrix for a homogeneous system.

A homogeneous system is always satisfied by the zero vector. That is, $\mathbf{x} = \mathbf{0} \in \mathbb{R}^n$ is always a solution of a homogeneous system in n variables. It is called the **trivial solution.** The following theorem is thus a special case of the previous theorem.

THEOREM 3. Consider a homogeneous system of m linear equations in n variables, with coefficient matrix A. Then,

(i) the system is always consistent, since it has the trivial solution;

(ii) the system has the trivial solution as its unique solution if and only if the rank of A is n; otherwise, there are infinitely many solutions with ($n-$ rank of A) parameters in the standard general solution.

No new proof is required, since this is a special case of the previous theorem, and the special features are explained in the remarks preceding the theorem. *The special case of this theorem where $m < n$ is very important.*

THEOREM 4. If $m < n$, a homogeneous system of m linear equations in n variables always has non-trivial (that is, non-zero) solutions.

Proof. The rank of the coefficient matrix is less than or equal to m, and $m < n$, so part (ii) of Theorem 3 applies. ■

As a final remark on homogeneous systems, note that in the standard form of the general solution of such a system, $\mathbf{x} = \mathbf{a} + s\mathbf{b} + t\mathbf{c} + \cdots$, the constant vector $\mathbf{a}$ is always the zero vector; this is apparent if you consider the reduced row echelon form of the augmented matrix of the system.

EXAMPLE 15

Determine whether the given system is consistent, and if so, determine the number of parameters in its general solution.

$$\begin{aligned} x_1 + 2x_2 + 2x_3 + x_4 + 4x_5 &= 0 \\ 3x_1 + 7x_2 + 7x_3 + 3x_4 + 13x_5 &= 0 \\ 2x_1 + 5x_2 + 5x_3 + 2x_4 + 9x_5 &= 0. \end{aligned}$$

SOLUTION: The system is homogeneous, so it is certainly consistent, with at least the trivial solution. We row reduce the coefficient matrix:

$$\begin{bmatrix} 1 & 2 & 2 & 1 & 4 \\ 3 & 7 & 7 & 3 & 13 \\ 2 & 5 & 5 & 2 & 9 \end{bmatrix} \to \begin{bmatrix} 1 & 0 & 0 & 1 & 2 \\ 0 & 1 & 1 & 0 & 1 \\ 0 & 0 & 0 & 0 & 0 \end{bmatrix}.$$

The rank is 2, the number of variables is 5, so there are 3 parameters in the general solution. *Check by writing out the general solution.*

EXERCISES 2–2

A1 Determine the reduced row echelon form for each of the following matrices, and state the rank of the matrix.

(a) $A = \begin{bmatrix} 2 & 0 & 1 \\ 0 & 1 & 2 \\ 1 & 1 & 1 \end{bmatrix}$ (b) $B = \begin{bmatrix} 1 & 2 & 3 \\ 2 & 1 & 2 \\ 2 & 3 & 4 \end{bmatrix}$

(c) $C = \begin{bmatrix} 1 & 1 & 1 & 1 \\ 1 & 1 & 1 & 0 \\ 1 & 1 & 0 & 0 \end{bmatrix}$ (d) $D = \begin{bmatrix} 2 & -1 & 2 & 8 \\ 1 & -1 & 0 & 2 \\ 3 & -2 & 3 & 13 \end{bmatrix}$

(e) $E = \begin{bmatrix} 1 & 1 & 0 & 1 \\ 0 & 1 & 1 & 2 \\ 2 & 3 & 1 & 4 \\ 1 & 2 & 3 & 4 \end{bmatrix}$ (f) $F = \begin{bmatrix} 0 & 1 & 0 & 2 & 5 \\ 3 & 1 & 8 & 5 & 3 \\ 1 & 0 & 3 & 2 & 1 \\ 2 & 1 & 6 & 7 & 1 \end{bmatrix}$

A2 Consider each of the following matrices, already in reduced row echelon form.
(i) Suppose that it is the augmented matrix of a non-homogeneous system. Comment on the consistency of the system; if it is consistent, is the solution unique? If there are parameters in the standard form of the general solution, how many? Finally, if it is consistent, write the general solution in standard form.
(ii) Suppose that it is the coefficient matrix of a homogeneous system. Comment on the consistency of the system. Is there a unique solution? If not, how many parameters are there in the general solution? Finally, write the general solution in standard form.

(a) $A = \begin{bmatrix} 1 & 0 & 2 & 0 \\ 0 & 1 & -1 & 0 \\ 0 & 0 & 0 & 1 \end{bmatrix}$ (b) $B = \begin{bmatrix} 1 & 0 & 2 & 0 & 0 \\ 0 & 1 & -1 & 0 & -2 \\ 0 & 0 & 0 & 1 & 1 \end{bmatrix}$

(c) $C = \begin{bmatrix} 1 & 0 & 0 & 0 & 0 \\ 0 & 1 & 1 & 0 & 0 \\ 0 & 0 & 0 & 1 & 0 \\ 0 & 0 & 0 & 0 & 1 \end{bmatrix}$

A3 For each homogeneous system, write the coefficient matrix, and determine the rank and the number of parameters in the general solution. Then determine the general solution.

(a)
$$\begin{aligned} 2x_2 - 5x_3 &= 0 \\ x_1 + 2x_2 + 3x_3 &= 0 \\ x_1 + 4x_2 - 3x_3 &= 0 \end{aligned}$$

(b)
$$\begin{aligned} 3x_1 + x_2 - 9x_3 &= 0 \\ x_1 + x_2 - 5x_3 &= 0 \\ 2x_1 + x_2 - 7x_3 &= 0 \end{aligned}$$

(c)
$$\begin{aligned} x_1 - x_2 + 2x_3 - 3x_4 &= 0 \\ 3x_1 - 3x_2 + 8x_3 - 5x_4 &= 0 \\ 2x_1 - 2x_2 + 5x_3 - 4x_4 &= 0 \\ 3x_1 - 3x_2 + 7x_3 - 7x_4 &= 0 \end{aligned}$$

(d)
$$\begin{aligned} x_2 + 2x_3 + 2x_4 \qquad &= 0 \\ x_1 + 2x_2 + 5x_3 + 3x_4 - x_5 &= 0 \\ 2x_1 + x_2 + 5x_3 + x_4 - 3x_5 &= 0 \\ x_1 + x_2 + 4x_3 + 2x_4 - 2x_5 &= 0 \end{aligned}$$

B1 Determine the reduced row echelon form for each of the following matrices, and state the rank of the matrix.

(a) $A = \begin{bmatrix} 2 & 5 & 3 \\ 1 & 2 & 2 \\ 1 & 3 & 2 \end{bmatrix}$ (b) $B = \begin{bmatrix} 2 & 4 & 8 \\ 1 & 1 & 3 \\ 1 & -1 & 1 \end{bmatrix}$

(c) $C = \begin{bmatrix} 1 & 1 & 2 & 1 \\ 2 & 1 & 4 & 3 \\ 0 & 3 & 2 & 1 \end{bmatrix}$ (d) $D = \begin{bmatrix} 2 & 1 & 1 & 1 \\ 3 & 2 & 1 & 1 \\ 4 & 3 & 2 & 1 \\ 1 & 0 & -1 & 0 \end{bmatrix}$

(e) $E = \begin{bmatrix} 1 & 1 & 2 & 1 & -2 \\ 2 & 2 & 4 & 3 & -6 \\ 0 & 1 & 2 & 2 & -4 \\ 3 & 2 & 4 & 2 & -4 \end{bmatrix}$

B2 Consider each of the following matrices, already in reduced row echelon form.
(i) Suppose that it is the augmented matrix of a non-homogeneous system. Comment on the consistency of the system; if it is consistent, is the solution unique? If there are parameters in the standard form of the general solution, how many? Finally, if it is consistent, write the general solution in standard form.
(ii) Suppose that it is the coefficient matrix of a homogeneous system. Comment on the consistency of the system. Is there a unique solution? If not, how many parameters are there in the general solution? Finally, write the general solution in standard form.

(a) $A = \begin{bmatrix} 1 & 3 & 0 & -1 \\ 0 & 0 & 1 & 2 \\ 0 & 0 & 0 & 0 \end{bmatrix}$ (b) $B = \begin{bmatrix} 1 & 1 & 0 & -2 & 0 \\ 0 & 0 & 1 & 1 & 0 \\ 0 & 0 & 0 & 0 & 1 \end{bmatrix}$

(c) $C = \begin{bmatrix} 1 & 2 & 0 & 0 & 0 & -3 \\ 0 & 0 & 1 & -5 & 0 & 4 \\ 0 & 0 & 0 & 0 & 1 & 1 \\ 0 & 0 & 0 & 0 & 0 & 0 \end{bmatrix}$

B3 For each homogeneous system, write the coefficient matrix, and determine the rank and the number of parameters in the general solution. Then determine the general solution.

(a)
$$\begin{aligned} x_1 + 5x_2 - 3x_3 &= 0 \\ 3x_1 + 5x_2 - 9x_3 &= 0 \\ x_1 + x_2 - 3x_3 &= 0 \end{aligned}$$

(b)
$$\begin{aligned} x_1 + 4x_2 - 2x_3 &= 0 \\ 2x_1 \qquad - 3x_3 &= 0 \\ 4x_1 + 8x_2 - 7x_3 &= 0 \end{aligned}$$

(c)
$$\begin{aligned} x_1 + x_2 + x_3 - 2x_4 &= 0 \\ 2x_1 + 7x_2 \qquad - 14x_4 &= 0 \\ x_1 + 3x_2 \qquad - 6x_4 &= 0 \\ x_1 + 4x_2 \qquad - 8x_4 &= 0 \end{aligned}$$

(d)
$$\begin{aligned} x_1 + 3x_2 + x_3 + x_4 + 2x_5 &= 0 \\ 2x_2 + x_3 \qquad - x_5 &= 0 \\ x_1 + 2x_2 + 2x_3 + x_4 \qquad &= 0 \\ x_1 + 2x_2 + x_3 + x_4 + x_5 &= 0 \end{aligned}$$

COMPUTER RELATED EXERCISES

C1 Determine the reduced row echelon form of the matrix of problem C1 of Section 2–1.

C2 Determine the reduced row echelon form of the augmented matrix of the system of linear equations

$$\begin{aligned} 2.01x + 3.45y + 2.23z &= 4.13 \\ 1.57x + 2.03y - 3.11z &= 6.11 \\ 2.23x + 7.10y - 4.28z &= 0.47. \end{aligned}$$

If the system is consistent, determine the general solution.

CONCEPTUAL EXERCISES

***D1** It is desired to find a vector $\mathbf{x} \neq \mathbf{0}$ in $\mathbb{R}^3$ that is simultaneously orthogonal to given vectors **a**, **b**, and **c** in $\mathbb{R}^3$.

(a) Write equations that must be satisfied by **x**.

(b) What condition must be satisfied by the rank of the matrix

$A = \begin{bmatrix} a_1 & a_2 & a_3 \\ b_1 & b_2 & b_3 \\ c_1 & c_2 & c_3 \end{bmatrix}$ if there are to be non-trivial solutions? Explain.

D2 (a) Suppose that $\begin{bmatrix} 1 & 0 & 2 \\ 0 & 1 & -1 \end{bmatrix}$ is the coefficient matrix of a homogenous system of linear equations. Find the general solution of the system and indicate why it describes a line through the origin.

(b) Suppose that a 2 by 3 matrix A (not the matrix of part (a)) is the coefficient matrix of a homogeneous system and that the rank of A is 2. Explain why the solution set is a line through the origin in $\mathbb{R}^3$. What could you say if the rank was 1?

(c) Let **a**, **b**, and **c** be three vectors in $\mathbb{R}^4$. Write conditions on a vector **x** such that **x** is orthogonal to **a**, **b**, and **c**. (This should lead to a homogenous system with coefficient matrix C whose rows are **a**, **b**, and **c**.) What does the rank of C tell us about the set of vectors **x** in $\mathbb{R}^4$ that are orthogonal to **a**, **b**, and **c**?

D3 What can you say about the consistency of a system of m linear equations in n variables and the number of parameters in the general solution if:

(a) $m = 5$, $n = 7$, the rank of the coefficient matrix is 4?

(b) $m = 3$, $n = 6$, the rank of the coefficient matrix is 3?

(c) $m = 5$, $n = 4$, the rank of the augmented matrix is 4?

***D4** A system of linear equations has augmented matrix $\begin{bmatrix} 1 & a & b & 1 \\ 1 & 1 & 0 & a \\ 1 & 0 & 1 & b \end{bmatrix}$. For which values of a and b is the system consistent? Are there values for which there is a unique solution? Determine the general solution.

SECTION

2–3 Some Applications of Systems of Linear Equations

Resistor Circuits in Electricity

The flow of electrical current in simple electrical circuits is described by simple linear laws. In an electrical circuit, the **current** has a direction, and therefore, has a sign attached to it; **voltage** is also a signed quantity; **resistance** is a positive scalar. The laws for electrical circuits are the following.

Ohm's Law

If an electrical current of magnitude I amperes is flowing through a resistor with resistance R ohms, then the drop in the voltage across the resistor is $V = IR$, measured in volts. The filament in a light bulb and the heating element of an electrical heater are familiar examples of electrical resistors. (See Figure 2–3–1.)

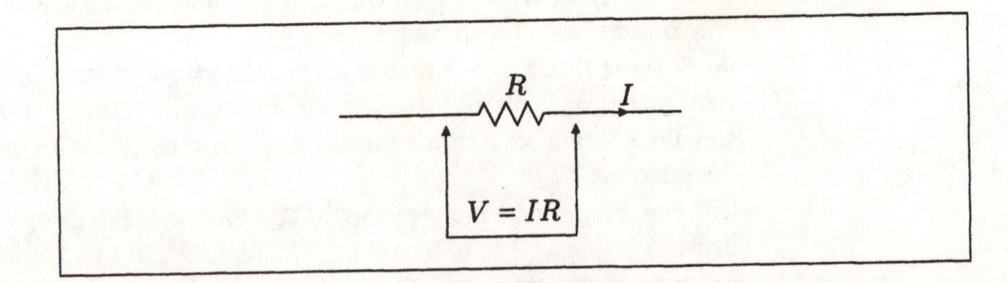

FIGURE 2-3-1. *Ohm's law: the voltage across the resistor is* $V = IR$.

Kirchhoff's Laws

(1) At a node or junction where several currents enter, the signed sum of the currents entering the node is zero. (See Figure 2–3–2.)

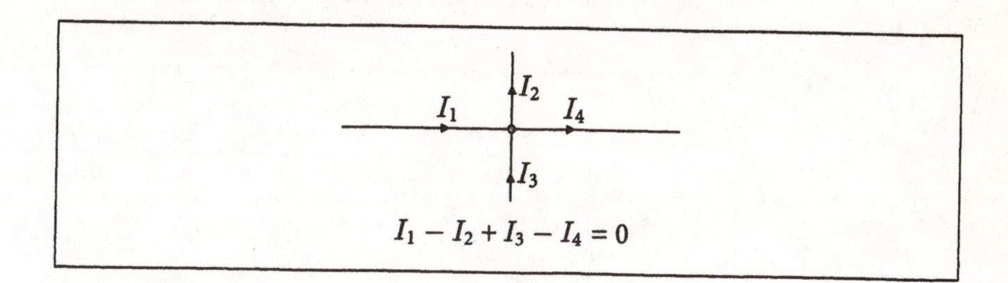

FIGURE 2-3-2. *One of Kirchhoff's laws:* $I_1 - I_2 + I_3 - I_4 = 0$.

(2) In a closed loop consisting only of resistors and an **electromotive force** E (for example, E might be due to a battery), the sum of the voltage drops across resistors is equal to E. (Figure 2–3–3.)

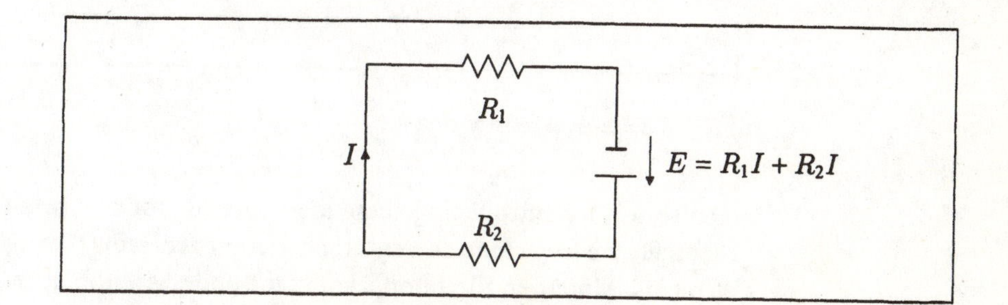

FIGURE 2-3-3. *The other Kirchhoff law:* $E = R_1I + R_2I$.

Note that we adopt the convention of drawing an arrow to show the direction of I or of E. These can be assigned arbitrarily, and then the circuit laws will determine whether the quantity has a positive or negative sign. It is important to be consistent in using these assigned directions when you write down Kirchhoff's law for loops.

Sometimes it is necessary to determine the current flowing in each of the loops of a network of loops as shown in Figure 2–3–4. (If the sources of electromotive force are distributed in various places, it will not be sufficient to deal with the problems as a collection of resistors "in parallel and/or in series".) In such problems, it is convenient to introduce the idea of the "current in the loop", which will be denoted i. The true current across any circuit element is given as the algebraic (signed) sum of the "loop currents" flowing through that circuit element. For example, in Figure 2–3–4, the circuit consists of four loops, and a loop current has been indicated in each loop. Across the resistor R_1 in the figure, the true current is simply the loop current i_1 ; however, across the resistor R_2, the true current (directed from top to bottom) is $i_1 - i_2$. Similarly, across R_4, the true current (from right to left) is $i_1 - i_3$.

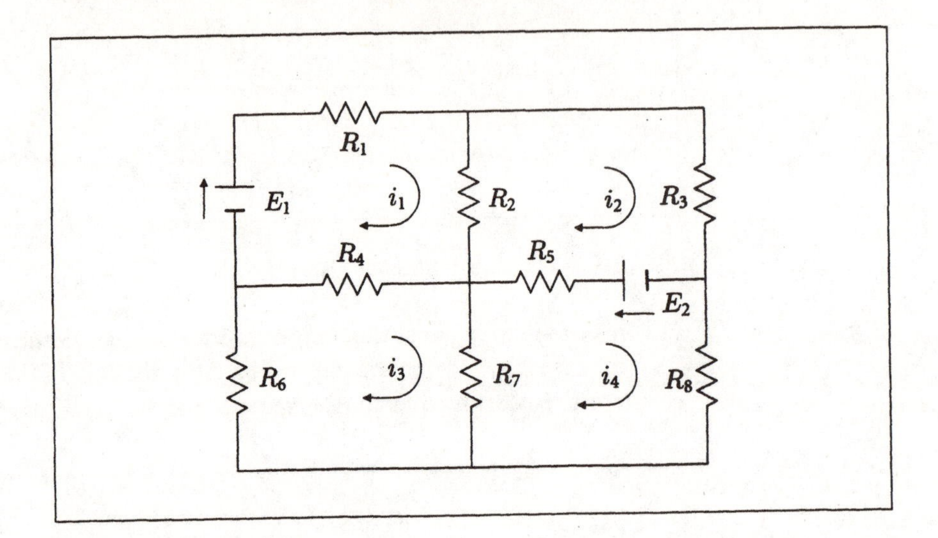

FIGURE 2-3-4. *A resistor circuit.*

The reason for introducing these loop currents for our present problem is that there are fewer loop currents than there are currents through individidual elements. Moreover, Kirchhoff's law at the nodes is automatically satisfied, so we do not have to write nearly so many equations.

To determine the currents in the loops, it is necessary to write down Kirchhoff's second law with Ohm's law describing the voltage drops across the resistors. For Figure 2–3–4, the resulting equation for the top left loop is:

$$R_1 i_1 + R_2(i_1 - i_2) + R_4(i_1 - i_3) = E_1.$$

For the top right loop: $R_3 i_2 + R_5(i_2 - i_4) + R_2(i_2 - i_1) = E_2.$

For the bottom left loop: $R_6 i_3 + R_4(i_3 - i_1) + R_7(i_3 - i_4) = 0.$

For the bottom right loop: $R_8 i_4 + R_7(i_4 - i_3) + R_5(i_4 - i_2) = -E_2.$

Multiply out and collect terms to display the equations as a system in the variables i_1, i_2, i_3, i_4. The augmented matrix of the system is

$$\left[\begin{array}{cccc|c} (R_1 + R_2 + R_4) & -R_2 & -R_4 & 0 & E_1 \\ -R_2 & (R_2 + R_3 + R_5) & 0 & -R_5 & E_2 \\ -R_4 & 0 & (R_4 + R_6 + R_7) & -R_7 & 0 \\ 0 & -R_5 & -R_7 & (R_5 + R_7 + R_8) & -E_2 \end{array}\right].$$

To determine the loop currents, this augmented matrix must be reduced to row echelon form. There is no particular purpose in finding an explicit solution for this general problem, and in a linear algebra course, there is no particular value in plugging in particular values for E_1 and E_2 and the seven resistors. Instead, the point of this example is to show that even for a fairly simple electrical circuit with the most basic elements (resistors), the analysis requires you to be competent in dealing with systems of linear equations that are large enough that systematic, efficient methods of solution are essential.

Obviously, as the number of loops in the network grows, so does the number of variables and the number of equations. For larger systems, it is important to know whether you have the correct number of equations to determine the unknowns. The theorems of Section 2–1, the idea of rank, and the idea of linear independence (defined in Section 4–2) are all important.

Resistor circuits are not the most interesting or important electrical circuits. In Section 2–4, we shall see how to use systems of linear equations with complex coefficients to determine currents in simple alternating current circuits with inductors and capacitors as well as resistors. More complicated circuits include non-linear elements, and more complicated analysis is required, but linear algebra will continue to be an important tool.

The moral of this example

Linear algebra is an essential tool for dealing with the large systems of linear equations that may arise in dealing with circuits; really interesting examples cannot be given without a short course on electrical circuits and their components.

Planar Trusses

It is common to use trusses, such as the one shown in Figure 2–3–5, in construction. For example, many bridges employ some variation of this design. As part of the problem of designing such structures, it is necessary to determine the **axial forces** in each **member** of the structure (that is, the force along the long axis of the member). To keep this simple, only two-dimensional trusses with hinged joints will be considered; it will be assumed that any displacements of the joints under loading are small enough to be negligible.

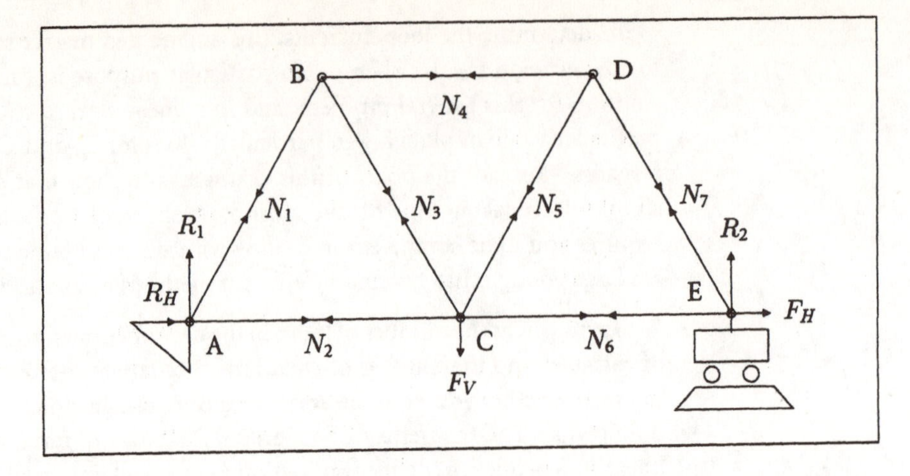

FIGURE 2-3-5. *A simple planar truss. All triangles are equilateral, with sides of length s.*

The external loads (such as vehicles on a bridge, or wind or waves) are assumed to be given. The reaction forces at the supports (shown as $R_1\mathbf{j}$, $R_2\mathbf{j}$, and $R_H\mathbf{i}$ in the figure) are also external forces; these forces must have values such that the total external force on the structure is zero. To get enough information to design a truss for a particular application, you must really determine the forces in the members under various loadings. To illustrate the kinds of equations that arise, we shall consider only the very simple case of a vertical force $-F_V\mathbf{j}$ acting at C and a horizontal force $F_H\mathbf{i}$ acting at E. Notice that in this figure, the right-hand end of the truss is allowed to undergo small horizontal displacements; it turns out that if a reaction force were applied here as well, the equations would not uniquely determine all the unknown forces (the structure would be "statically indeterminate") and other considerations would have to be introduced.

The geometry of the truss is assumed given: here it will be assumed that the triangles are equilateral with all sides equal to s metres.

First consider the equations that indicate that the total force on the structure is zero, and that the total moment about some convenient point due to those forces is zero. Note that the axial forces along the members do not appear in this first set of equations.

Total Horizontal Force: $\quad R_H + F_H = 0.$

Total Vertical Force: $\quad R_1 + R_2 - F_V = 0.$

Moment about A: $\quad -F_V(s) + R_2(2s) = 0, \quad \text{so} \quad R_2 = (1/2)F_V = R_1.$

Next, we consider the system of equations obtained from the fact that the sum of the forces at each joint must be zero (the moments are automatically zero because the forces along the members act through the joint).

At a joint, each member at that joint exerts a force in the direction of the axis of the member. It will be assumed that each member is in tension, so it is "pulling" away from the joint; if it were compressed, it would be "pushing" at the joint. As indicated in the figure, the force exerted on the joint A by the upper left-hand member has magnitude N_1; with the conventions that forces to the right are positive and forces up are positive, the force vector exerted by this member on the joint A is $\left((1/2)N_1, (\sqrt{3}/2)N_1\right)$. On the joint B, the same member will exert a force $\left(-(1/2)N_1, -(\sqrt{3}/2)N_1\right)$. If N_1 is positive, the force is a tension force; if N_1 is negative, there is compression.

For each of the joints A, B, C, D, E, there are two equations, the first for the sum of the horizontal forces, the second for the sum of the vertical forces.

$$
\begin{array}{lcccccccccl}
\text{A1} & (1/2)N_1 & +N_2 & & & & & & +R_H & = & 0 \\
\text{A2} & (\sqrt{3}/2)N_1 & & & & & & & +R_1 & = & 0 \\
\text{B1} & (-1/2)N_1 & & +(1/2)N_3 & +N_4 & & & & & = & 0 \\
\text{B2} & (-\sqrt{3}/2)N_1 & & -(\sqrt{3}/2)N_3 & & & & & & = & 0 \\
\text{C1} & & -N_2 & -(1/2)N_3 & & +(1/2)N_5 & +N_6 & & & = & 0 \\
\text{C2} & & & (\sqrt{3}/2)N_3 & & +(\sqrt{3}/2)N_5 & & & & = & F_V \\
\text{D1} & & & & -N_4 & -(1/2)N_5 & & +(1/2)N_7 & & = & 0 \\
\text{D2} & & & & & -(\sqrt{3}/2)N_5 & & -(\sqrt{3}/2)N_7 & & = & 0 \\
\text{E1} & & & & & & -N_6 & -(1/2)N_7 & & = & -F_H \\
\text{E2} & & & & & & & (\sqrt{3}/2)N_7 & +R_2 & = & 0
\end{array}
$$

Notice that if the reaction forces are treated as unknowns, this is a system of 10 equations in 10 unknowns. The geometry of the truss and its supports determines the coefficient matrix of this system, and it could be shown that the system is necessarily consistent, and in fact has a unique solution in this case. Notice also that if the horizontal force equations (A1, B1, C1, D1, E1) are added together, the sum is the Total Horizontal Force equation, and similarly the sum of the vertical force equations is the Total Vertical Force equation. A suitable combination of the equations would also produce the moment equation, so if those three equations are solved as above, then the 10 joint equations will still be a consistent system for the remaining 7 axial force variables.

For this particular truss, the system of equations is quite easy to solve, since some of the variables are already "leading variables". For example, if $F_H = 0$, from A2 and E2 it follows that $N_1 = N_7 = (-1/\sqrt{3})F_V$ and then B2, C2, and D2 give $N_3 = N_5 = (1/\sqrt{3})F_V$; then A1 and E1 imply that $N_2 = N_6 = (1/2\sqrt{3})F_V$, and C2 implies that $N_4 = -(1/\sqrt{3})F_V$. Note that the members AC, BC, CD, and CE are under tension, and AB, BD, and DE experience compression, which makes intuitive sense.

This system would still be easy to solve even if other external loads were considered (always assumed to be applied at the joints). However, since the system must be solved under various loadings, it is essential to have a good systematic general procedure for solving such systems. There are computer packages for carrying out this sort of analysis.

This is a particularly simple truss. In the real world, trusses often involve many more members, and use more complicated geometry; trusses may also be three-dimensional. Therefore, the systems of equations that arise may be considerably larger and more complicated. It is also sometimes essential to introduce considerations other than the equations of equilibrium of forces in statics. To study these questions, you need to know the basic facts of linear algebra.

It is worth noting that in the system of equations above, each of the quantities $N_1, N_2, \dots, N_7$ appears with a non-zero coefficient in only some of the equations. Since each member touches only two joints, this sort of special structure will often occur in the equations that arise in the analysis of trusses. A deeper knowledge of linear algebra is important in understanding how such special features of linear equations may be exploited to produce efficient solution methods.

Linear Programming

Linear programming is a procedure for deciding the "best" way to allocate resources. "Best" may mean fastest, or most profitable, or cheapest, or best by whatever criterion is appropriate. For linear programming to be applicable, the problem must have some special features. These will be illustrated by an example (a rather artificial one).

In a somewhat primitive economy, a man decides to earn a living by making hinges and gate latches. He is able to obtain a supply of 25 kilograms a week of suitable metal at a price of 2 cowrie shells per kilogram. His design requires 500 grams to make a hinge and 250 grams to make a gate latch. With his primitive tools, he finds that he can make a hinge in one hour, and it takes 3/4 of an hour to make a gate latch. He is willing to work 60 hours a week. The going price is 3 cowrie shells for a hinge and 2 cowrie shells for a gate latch. How many hinges and how many gate latches should he produce each week in order to maximize his net income?

To analyze the problem, let x be the number of hinges produced per week, and let y be the number of gate latches. Then the amount of metal used is $(0.5x+0.25y)$ kilograms. Clearly this must be less than or equal to 25 kilograms:

$$0.5x + 0.25y \leq 25,$$

or

$$2x + y \leq 100.$$

Such an inequality is called a **constraint** on x and y; it is a linear constraint because the corresponding equation is linear.

Our producer also has a time constraint: the time taken making hinges plus the time taken making gate latches cannot exceed 60 hours; therefore,

$$1x + 0.75y \leq 60,$$

or

$$4x + 3y \leq 240.$$

Obviously, also $x \geq 0$, and $y \geq 0$.

The producer's net revenue for selling x hinges and y gate latches is $R(x, y) = 3x + 2y - 2(25)$ cowrie shells. This is called the **objective** function for the problem. The mathematical problem can now be stated as follows:

Find the point (x, y) that maximizes the objective function

$$R(x, y) = 3x + 2y - 50,$$

subject to the linear constraints $x \geq 0$, $y \geq 0$, $2x + y \leq 100$, $4x + 3y \leq 240$.

This is a **linear programming problem** because it asks for the maximum (or minimum) of a **linear** objective function, subject to **linear** constraints. It is useful to introduce one bit of special vocabulary: the **feasible** set for the problem is the set of (x, y) satisfying all of the constraints. The solution procedure relies on the fact that the feasible set for a linear programming problem has a special kind of shape (see Figure 2–3–6 for the feasible set for this particular problem). *Any line that meets the feasible set either meets the set in a single line segment or only touches the set on its boundary.* In particular, because of the way the feasible set is defined in terms of linear inequalities, it turns out that it is impossible for one line to meet the feasible set in two separate pieces.

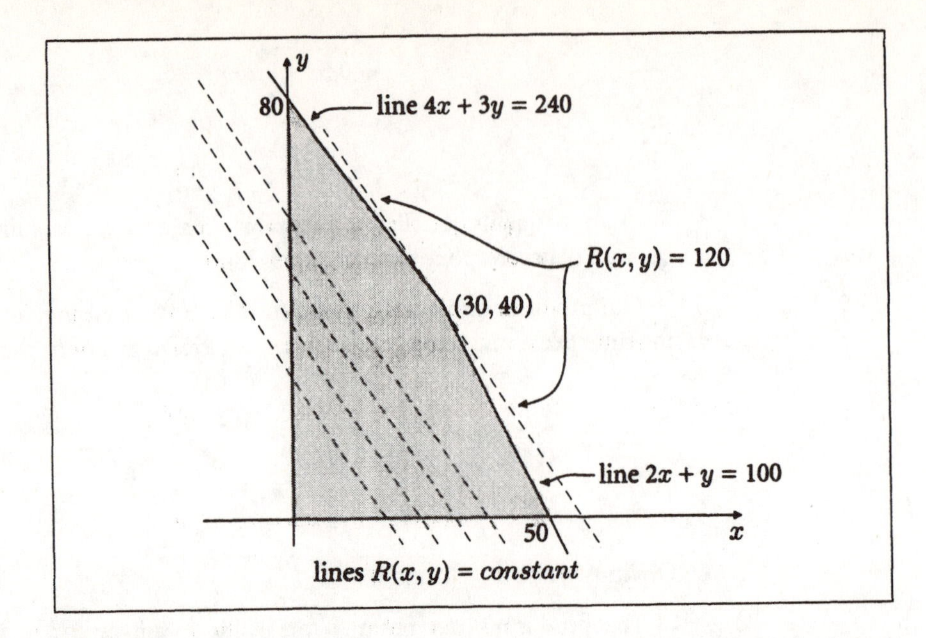

FIGURE 2-3-6. *The feasible region for the linear programming example. The dashed lines are level sets of the objective function R.*

For example, the shaded region in Figure 2–3–7 cannot possibly be the feasible set for a linear programming problem, because some lines meet the region in *two* line segments. (This property of feasible sets is not very hard to prove, but since this is only a brief illustration, the proof is omitted.)

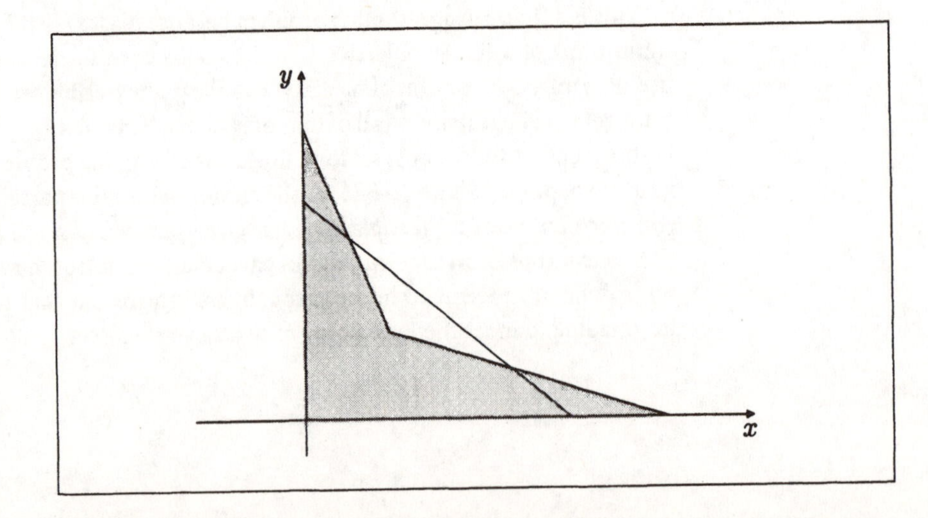

FIGURE 2-3-7. *The shaded region cannot be the feasible region for a linear programming problem because it meets a line in two segments.*

Now consider sets of the form $R(x, y)$ = constant; these are called *level sets* of R. These sets obviously form a family of parallel lines, and some of them are shown in Figure 2–3–6. Choose some point in the feasible set: check that $(20, 20)$ is such a point. Then the line

$$R(x, y) = R(20, 20) = 50$$

meets the feasible set in a line segment. $(30, 30)$ is also a feasible point (*check*), and

$$R(x, y) = R(30, 30) = 100$$

also meets the feasible set in a line segment. You can tell that $(30, 30)$ is not a boundary point of the feasible set because it satisfies all the constraints with **strict** inequality; boundary points must satisfy one of the constraints with equality.

As we move further from the origin into the first quadrant, $R(x, y)$ increases. The biggest possible value for $R(x, y)$ will occur at a point where the set $R(x, y)$ = constant (for some constant to be determined) just touches the feasible set. For larger values of $R(x, y)$, the set $R(x, y)$ = constant does not meet the feasible set at all, so there are no feasible points that give such bigger values of R. The touching must occur at a **vertex**, that is, at an intersection point of two of the boundary lines. (In general, the line $R(x, y)$ = constant for the largest possible constant could touch the feasible set along a line segment that makes up part of the boundary, but such a line segment has two vertices as endpoints, so it is correct to say that the touching occurs at a vertex.)

For this particular problem, the vertices of the feasible set are easily found to be $(0, 0)$, $(50, 0)$, $(0, 80)$, and the solution of the system of equations:

$$\begin{aligned} 2x + y &= 100, \\ 4x + 3y &= 240. \end{aligned}$$

The solution of this system is $(30, 40)$. Now compare the value of $R(x, y)$ at all of these vertices. $R(0, 0) = -50$; $R(50, 0) = 100$; $R(0, 80) = 110$; $R(30, 40) = 120$. Clearly the vertex $(30, 40)$ gives the best net revenue, so the producer should make 30 hinges and 40 gate latches each week.

General Remarks

Such problems of allocating resources are extremely common in business and in government. Problems such as scheduling ship transits through the Welland Canal between Lake Ontario and Lake Erie can be analyzed in this way. Oil companies must make choices about the grades of crude oil to use in their refineries, and about the amounts of various refined products to produce. Such problems often involve tens or even hundreds of variables, and similar numbers of constraints. The boundaries of the feasible set are hyperplanes in some $\mathbb{R}^n$, where n is large. Although the basic principles of the solution method remain the same as in this example (look for the best vertex), the problem is much more complicated because there are so many vertices. In fact, it is not trivial just to find vertices; simply solving all possible combinations of systems of boundary equations is not good enough. Note in the simple two-dimensional example that the point $(60, 0)$ is the intersection point of two of the lines ($y = 0$ and $4x + 3y = 240$) that make up the boundary, but it is not a vertex of the feasible region because it fails to satisfy the constraint $2x + y \leq 100$. For higher dimensional problems, drawing pictures is not good enough, and an organized approach is called for.

The standard method for solving linear programming problems has been the **simplex method**, which finds an initial vertex and then prescribes a method for moving to another vertex, improving the value of the objective function with each step. Many people have had linear programming problems to solve, with large sums of money depending on good answers, so such problems have been among the biggest users of computer time. It is still a matter of interest to improve solution methods. In the 1980s and 1990s, new methods have been proposed.

Again, it has been possible to hint at a major important application area for linear algebra, but to pursue it would require the development of specialized mathematical tools. Moreover, interesting applications require detailed information from some specialized area such as resource constraints in manufacturing, or in government, so we cannot pursue them here.

EXERCISES 2–3

A1 Determine the augmented matrix of the system of linear equations determining the loop currents indicated in Figure 2–3–8.

A2 Determine the system of equations for the reaction forces and axial forces in members for the truss shown in Figure 2–3–9.

A3 Find the maximum value of the objective function $x + y$ subject to the constraints $0 \leq x \leq 100$, $0 \leq y \leq 80$, $4x + 5y \leq 600$. Sketch the feasible region.

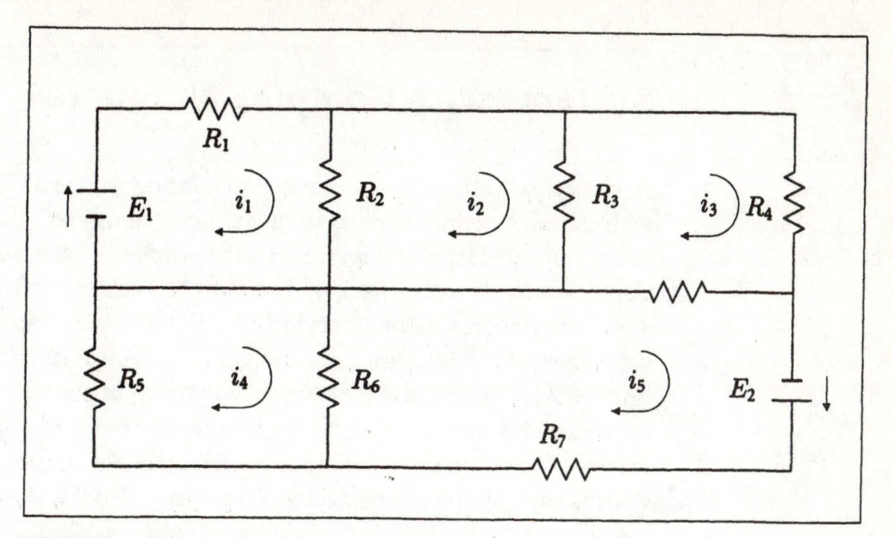

FIGURE 2-3-8. *The circuit for Exercise A1.*

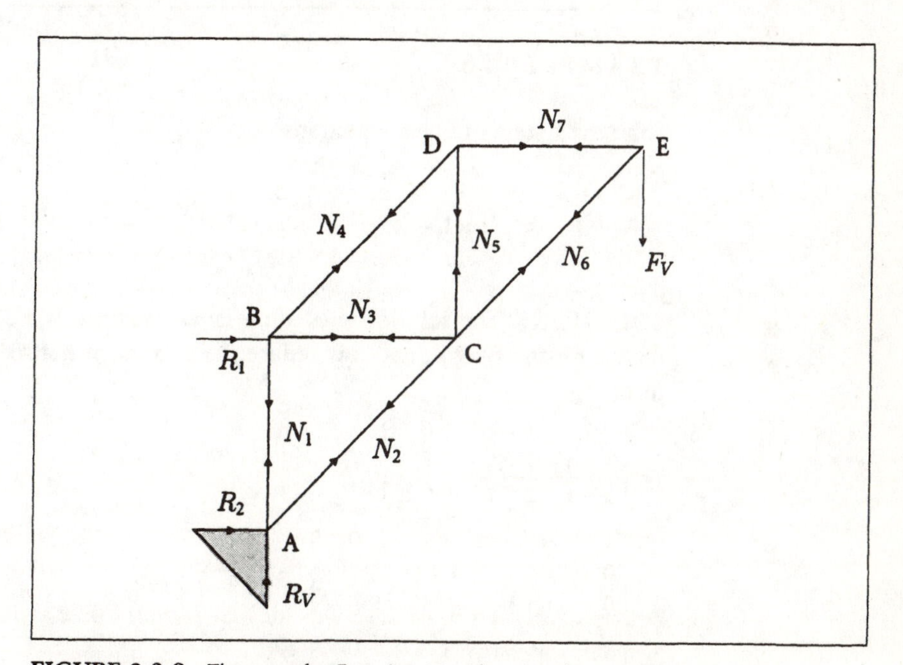

FIGURE 2-3-9. *The truss for Exercise A2. The triangles are right-angled and isosceles, with sides of length s, s, and $\sqrt{2}s$.*

SECTION

2–4 Systems with Complex Numbers

In some applications, it is necessary to consider systems of linear equations with complex coefficients and complex right-hand sides. One physical application discussed later in this section is the problem of determining currents in electrical circuits with capacitors and inductive coils as well as resistance. The arithmetic of complex numbers includes all the usual operations of addition, subtraction, multiplication, and division (complex arithmetic is reviewed in Appendix 1), so we can solve systems with complex coefficients by exactly the same elimination/row reduction procedures we use for systems with real coefficients. We should expect, however, that we will obtain complex solutions for systems with complex coefficients. Recall that i denotes the complex number such that $i^2 = -1$.

EXAMPLE 16

Solve the system of linear equations:

$$\begin{array}{rcrcrcl} z_1 & + & z_2 & + & z_3 & = & 0 \\ (1-i)z_1 & + & z_2 & & & = & i \\ (3-i)z_1 & + & 2z_2 & + & z_3 & = & 1+2i. \end{array}$$

SOLUTION: The solution procedure is, as usual, to write the augmented matrix for the system, and row reduce the coefficient matrix to row echelon form:

$$\begin{bmatrix} 1 & 1 & 1 & 0 \\ 1-i & 1 & 0 & i \\ 3-i & 2 & 1 & 1+2i \end{bmatrix} \begin{array}{l} \\ R_2-(1-i)R_1 \rightarrow \\ R_3-(3-i)R_1 \end{array}$$

$$\begin{bmatrix} 1 & 1 & 1 & 0 \\ 0 & i & -1+i & i \\ 0 & -1+i & -2+i & 1+2i \end{bmatrix} -iR_2 \rightarrow$$

$$\begin{bmatrix} 1 & 1 & 1 & 0 \\ 0 & 1 & 1+i & 1 \\ 0 & -1+i & -2+i & 1+2i \end{bmatrix} \begin{array}{l} \\ \rightarrow \\ R_3+(1-i)R_2 \end{array}$$

$$\begin{bmatrix} 1 & 1 & 1 & 0 \\ 0 & 1 & 1+i & 1 \\ 0 & 0 & i & 2+i \end{bmatrix} \begin{array}{l} \\ \\ -iR_3 \end{array} \rightarrow \begin{bmatrix} 1 & 1 & 1 & 0 \\ 0 & 1 & 1+i & 1 \\ 0 & 0 & 1 & 1-2i \end{bmatrix}.$$

We now can write the solution by back-substitution, as usual:

$$z_3 = 1 - 2i,\ z_2 = 1 - (1+i)z_3 = 1 - (1+i)(1-2i) = -2+i,$$
$$z_1 = -z_2 - z_3 = 1+i,$$

so $\mathbf{z} = (1+i, -2+i, 1-2i)$.

Complex Numbers in Electrical Circuit Equations

This application requires some knowledge of calculus and physics; it may be omitted with no loss of continuity.

For purposes of the following discussion only, we switch to a notation commonly used by engineers and physicists and denote by j the complex number such that $j^2 = -1$, so that we can use i to denote current.

In Section 2–3, we discussed electrical circuits with resistors. We now also consider capacitors and inductors, and alternating current. A simple capacitor can be thought of as two conducting plates separated by vacuum or some dielectric. Charge can be stored on these plates, and it is found that the voltage across a capacitor at time t is proportional to the charge stored at that time:

$$V(t) = \frac{Q(t)}{C},$$

where Q is the charge and the constant C is called the *capacitance* of the capacitor.

The usual model of an inductor is a coil; because of magnetic effects, it is found that with time-varying current $i(t)$ the voltage across an inductor is proportional to the rate of change of current:

$$V(t) = L\frac{di(t)}{dt},$$

where the constant of proportionality L is called the *inductance.*

As in the case of resistor circuits, Kirchhoff's Law applies: the sum of the voltage drops across the circuit elements must be equal to the applied electromotive force (voltage). Thus, for a simple loop with inductance L, capacitance C, resistance R, and applied electromotive force $E(t)$ (Figure 2-4-1), the circuit equation is

$$L\frac{di(t)}{dt} + Ri(t) + \frac{1}{C}Q(t) = E(t).$$

For our purposes, it is easier to work with the derivative of this equation and use the fact that $\frac{dQ}{dt} = i$:

$$L\frac{d^2i(t)}{dt^2} + R\frac{di(t)}{dt} + \frac{1}{C}i(t) = \frac{dE(t)}{dt}.$$

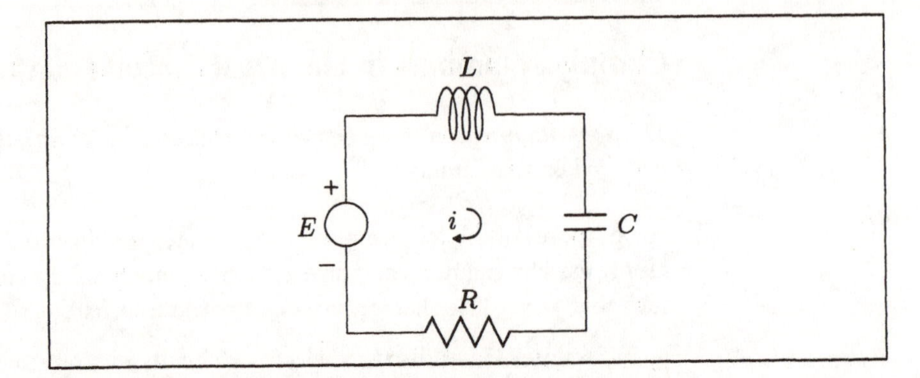

FIGURE 2-4-1. *Kirchhoff's voltage law applies to an alternating current circuit.*

In general, the solution to such an equation will involve the *superposition* (sum) of a *steady state* solution and a *transient* solution. Here we will be looking only for the steady state solution, in the special case where the applied electromotive force, and hence any current, is a single-frequency sinusoidal function. Thus we can assume that

$$E(t) = Be^{j\omega t} \quad \text{and} \quad i(t) = Ae^{j\omega t},$$

where A and B are complex numbers that determine the amplitudes and phases of voltage and current, and ω is 2π multiplied by the frequency. Then

$$\frac{di}{dt} = j\omega\, Ae^{j\omega t} = j\omega\, i \qquad \text{and} \qquad \frac{d^2i}{dt^2} = (j\omega)^2\, i = -\,\omega^2\, i,$$

and the circuit equation can be rewritten

$$-\,\omega^2\, Li + j\omega\, Ri + \frac{1}{C}i = \frac{dE}{dt}.$$

Now consider a network of circuits with resistors, capacitors, inductors, electromotive force, and currents as shown in Figure 2-4-2. As in Section 2-3, the currents are loop currents, so that the actual current across some circuit elements is the difference of two loop currents. (For example, across R_1, the actual current is $i_1 - i_2$.) From our assumption that we have only one single-frequency source, we may conclude that the steady state loop currents must be of the form

$$i_1(t) = A_1 e^{j\omega t}, \; i_2(t) = A_2 e^{j\omega t}, \; i_3(t) = A_3 e^{j\omega t}.$$

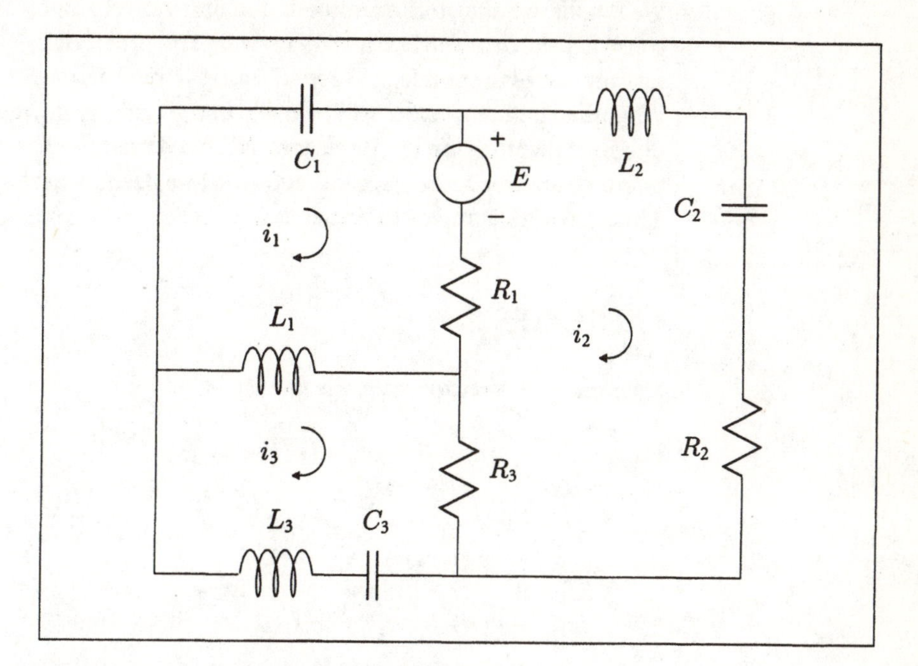

FIGURE 2-4-2. *An alternating current network.*

By applying the Kirchhoff laws to the top left loop, we find that

$$\left[-\omega^2 L_1 (A_1 - A_3) + j\omega R_1 (A_1 - A_2) + \frac{1}{C} A_1\right] e^{j\omega t} = -j\omega B e^{j\omega t}.$$

If we write the corresponding equations for the other two loops, reorganize each equation, and divide out the non-zero common factor $e^{j\omega t}$, we obtain the following system of linear equations for the three variables A_1, A_2, A_3:

$$\left[\left(-\omega^2 L_1 + \frac{1}{C_1}\right) + j\omega R_1\right] A_1 - j\omega R_1 A_2 + \omega^2 L_1 A_3 = -j\omega B$$

$$-j\omega R_1 A_1 + \left[\left(-\omega^2 L_2 + \frac{1}{C_2}\right) + j\omega (R_1 + R_2 + R_3)\right] A_2 - j\omega R_3 A_3 = j\omega B$$

$$\omega^2 L_1 A_1 - j\omega R_3 A_2 + \left[-\omega^2 (L_1 + L_3) + \frac{1}{C_3} + j\omega R_3 \right] A_3 = 0$$

Thus, we have a system of three linear equations with complex coefficients for the three variables A_1, A_2, A_3, which we can solve by standard elimination. We emphasize that this example is for illustrative purposes only: we have constructed a completely arbitrary network and provided the solution method for only part of the problem, in a special case. A much more extensive discussion is required before a reader will be ready to start examining realistic circuits to discover what they can do. But even this limited example illustrates the general point: to analyze some electrical networks, we need to solve systems of linear equations with complex coefficients.

EXERCISES 2–4

A1 Determine whether the system is consistent, and if so, determine the general solution.

(a)
$$\begin{array}{rcrcrcl} z_1 & + & iz_2 & + & (1+i)z_3 & = & 1-il \\ -2z_1 & + & (1-2i)z_2 & - & 2z_3 & = & 2i \\ 2iz_1 & - & 2z_2 & - & (2+3i)z_3 & = & -1+3i \end{array}$$

(b)
$$\begin{array}{rcrcrcrcl} z_1 & + & (1+i)z_2 & + & 2z_3 & + & z_4 & = & 1-i \\ 2z_1 & + & (2+i)z_2 & + & 5z_3 & + & (2+i)z_4 & = & 4-i \\ iz_1 & + & (-1+i)z_2 & + & (1+2i)z_3 & + & 2iz_4 & = & 1 \end{array}$$

B1 Determine whether the system is consistent, and if so, determine the general solution.

(a)
$$\begin{array}{rcrcrcl} & & z_2 & - & iz_3 & = & 1+3i \\ iz_1 & - & z_2 & + & (-1+i)z_3 & = & 1+2i \\ 2z_1 & + & 2iz_2 & + & (3+2i)z_3 & = & 4 \end{array}$$

(b)
$$\begin{array}{rcrcrcrcl} z_1 & + & (2+i)z_2 & + & iz_3 & & & = & 1+i \\ iz_1 & + & (-1+2i)z_2 & & & + & 2iz_4 & = & -i \\ z_1 & + & (2+i)z_2 & + & (1+i)z_3 & + & 2iz_4 & = & 2-i \end{array}$$

Chapter Review

Suggestions for Student Review

Try to do these before looking up answers in the text.

1 Explain why elimination works as a method for solving systems of linear equations. (Section 2–1, Example 1)

2 When you row reduce an augmented matrix $\left[A \mid \mathbf{b} \right]$ to solve a system of linear equations, why can you stop when the coefficient matrix is in row echelon

form? How do you use this form to decide if the system is consistent and if it has a unique solution? (Section 2–1)

3 How is reduced row echelon form different from row echelon form? (Section 2–2)

4 (a) Write the augmented matrix of a consistent non-homogeneous system of 3 linear equations in 4 variables, such that the coefficient matrix is in row echelon form (but not reduced row echelon form) and of rank 3.
(b) Determine the general solution of your system.
(c) Perform the following sequence of elementary row operations on your augmented matrix:
  (i) interchange the first and second rows;
  (ii) add the (new) second row to the first row;
  (iii) add twice the second row to the third row;
  (iv) add the third row to the second.
(d) Regard the result of (c) as the augmented matrix of a system, and solve that system directly (don't just use the reverse operations in (c)). Check that your general solution agrees with (b).

5 How can you use the rank of A and the rank of $\left[A \mid \mathbf{b}\right]$ to decide whether the system with augmented matrix $\left[A \mid \mathbf{b}\right]$ is consistent? (Section 2–2, Theorem 2)

6 For homogeneous systems, how can you use the row echelon form to determine whether there are non-trivial solutions, and if there are, how many parameters there are in the general solution? Is there any case where we know (by inspection) that a homogeneous system has non-trivial solutions? (Section 2–2, Theorems 3, 4)

Chapter Quiz

1 Determine whether the system is consistent by row reducing its augmented matrix. If it is consistent, determine the general solution.

$$\begin{aligned} x_2 - 2x_3 + x_4 &= 2 \\ 2x_1 - 2x_2 + 4x_3 - x_4 &= 10 \\ x_1 - x_2 + x_3 \phantom{{}- x_4} &= 2 \\ x_1 \phantom{{}- x_2} + x_3 \phantom{{}- x_4} &= 9 \end{aligned}$$

Show your row operations clearly.

2 Find a matrix in reduced row echelon form that is row equivalent to

$$A = \begin{bmatrix} 0 & 3 & 3 & 0 & -1 \\ 1 & 1 & 3 & 3 & 1 \\ 2 & 4 & 9 & 6 & 1 \\ -2 & -4 & -6 & -3 & -1 \end{bmatrix}.$$

Show your row operations clearly.

3 The matrix

$$A = \begin{bmatrix} 1 & 2 & 3 & a & 2 \\ 0 & 2 & 1 & 0 & -3 \\ 0 & 0 & (b+2) & 0 & b \\ 0 & 0 & 0 & (c^2-1) & (c+1) \end{bmatrix}$$

is the augmented matrix of a system of linear equations.

(a) Determine all values of (a, b, c) such that the system is consistent and all values of (a, b, c) such that the system is inconsistent.

(b) Determine all values of (a, b, c) such that the system has a unique solution.

4 (a) Determine the set of all vectors in $\mathbb{R}^5$ that are orthogonal to all of the vectors $(1, 1, 3, 1, 4)$, $(2, 1, 5, 0, 0)$, and $(3, 2, 8, 5, 9)$.

(b) Let **a**, **b**, and **c** be three vectors in $\mathbb{R}^5$. Explain why there must be non-zero vectors orthogonal to all of **a**, **b**, and **c**.

5 Indicate whether the following statements are TRUE or FALSE, and in each case, justify your answer by a counterexample or a brief explanation.

(a) A consistent system must have a unique solution.

(b) If there are more equations than variables in a non-homogeneous system of linear equations, then the system must be inconsistent.

(c) Some homogeneous systems of linear equations have unique solutions.

Further Exercises

These exercises are intended to be challenging. Not all will be of interest to all students.

F1 The purpose of this exercise is to explore the relation between the general solution of the system $A\mathbf{x} = \mathbf{b}$ and the general solution of the **corresponding homogeneous** system $A\mathbf{x} = \mathbf{0}$. This relation will be studied with different tools in Section 3–4. We begin by considering some special examples where the coefficient matrix is in reduced row echelon form.

(a) Let $R = \begin{bmatrix} 1 & 0 & r_{13} \\ 0 & 1 & r_{23} \end{bmatrix}$. Show that the general solution of the homogeneous system $R\mathbf{x} = \mathbf{0}$ is

$$\mathbf{x}_H = t\mathbf{v}, \quad t \in \mathbb{R},$$

where $\mathbf{v}$ is expressed in terms of r_{13} and r_{23}. Show that the general solution of the non-homogeneous system $R\mathbf{x} = \mathbf{c}$ is

$$\mathbf{x}_N = \mathbf{p} + \mathbf{x}_H,$$

where $\mathbf{p}$ is expressed in terms of $\mathbf{c}$, and $\mathbf{x}_H$ is as above.

(b) Let $R = \begin{bmatrix} 1 & r_{12} & 0 & 0 & r_{15} \\ 0 & 0 & 1 & 0 & r_{25} \\ 0 & 0 & 0 & 1 & r_{35} \end{bmatrix}$. Show that the general solution of the homogeneous system $R\mathbf{x} = \mathbf{0}$ is

$$\mathbf{x}_H = t_1\mathbf{v}_1 + t_2\mathbf{v}_2, \quad t_i \in \mathbb{R},$$

where each of $\mathbf{v}_1$ and $\mathbf{v}_2$ can be expressed in terms of the entries r_{ij}. Express each $\mathbf{v}_i$ explicitly. Then show that the general solution of $R\mathbf{x} = \mathbf{c}$ can be written

$$\mathbf{x}_N = \mathbf{p} + \mathbf{x}_H,$$

where $\mathbf{p}$ is expressed in terms of the components of $\mathbf{c}$, and $\mathbf{x}_H$ is the solution of the corresponding homogeneous solution.

The pattern should now be apparent; if not, try again with another special case of R. The hard thing in the next part of this exercise is to create a good labelling system so that you can say what you want to say.

(c) Let R be a matrix in reduced row echelon form, with m rows, n columns, and rank k. Show that the general solution of the homogeneous system $R\mathbf{x} = \mathbf{0}$ is

$$\mathbf{x}_H = t_1\mathbf{v}_1 + \cdots + t_{n-k}\mathbf{v}_{n-k}, \quad t_i \in \mathbb{R},$$

where each $\mathbf{v}_i$ is expressed in terms of the entries in R. Suppose that the system $R\mathbf{x} = \mathbf{c}$ is consistent, and show that its general solution is

$$\mathbf{x}_N = \mathbf{p} + \mathbf{x}_H,$$

where $\mathbf{p}$ is expressed in terms of the components of $\mathbf{c}$, and $\mathbf{x}_H$ is the solution of the corresponding homogeneous solution.

(d) Use the result of (c) to discuss the relation between the general solution of the consistent system $A\mathbf{x} = \mathbf{b}$ and the corresponding homogeneous system $A\mathbf{x} = \mathbf{0}$.

F2 *Comparing the efficiency of row reduction procedures*

When we use a computer to solve large systems of linear equations, we want to keep the number of arithmetic operations as small as possible. This reduces the time taken for calculation, which is important in many industrial and commmercial applications. It also tends to improve accuracy: every arithmetic operation is an opportunity to *lose* accuracy through truncation or round-off, or subtraction of two nearly equal numbers, etc.

We want to count the number of multiplications and/or divisions in solving a system by elimination. We focus on these operations because they are more time-consuming than addition or subtraction, and the number of additions is approximately the same as the number of multiplications. We make certain assumptions: the system $A\mathbf{x} = \mathbf{b}$ has n equations and n variables, and it is consistent with a unique solution. (Equivalently, A has n rows, n columns, and rank n.) We assume for simplicity that no row interchanges are required. (If row interchanges are required, they can be handled by renaming "addresses" in the computer.) Also, we do not insist that leading coefficients be 1's until the very end of the calculation (that means, we do not immediately divide all entries in the first row by a_{11}.)

(a) How many multiplications and divisions are required to reduce $\left[A \mid \mathbf{b}\right]$ to a form $\left[C \mid \mathbf{d}\right]$ such that all entries below the diagonal are zero: $c_{jk} = 0$ if $j > k$? (We know that we can obtain such a form with all diagonal entries non-zero because the rank k is n.)

Hints

(1) To carry out the obvious first row operation, compute $\frac{a_{21}}{a_{11}}$ —one division. Since we know what will happen in the first column, we do not multiply a_{11} by $\frac{a_{21}}{a_{11}}$, but we must multiply every other element of the first row of $\left[A \mid \mathbf{b}\right]$ by this factor and subtract the product from the corresponding element of the second row—n multiplications.

(2) Obtain zeros in the remaining entries in the first column, then move to the $(n-1)$ by n block consisting of the reduced version of $\left[A \mid \mathbf{b}\right]$ with the first row and the first column deleted.

(3) Note that $1 + 2 + \cdots + n = \frac{n(n+1)}{2}$ and

$$1^2 + 2^2 + \cdots + n^2 = \frac{n(n+1)(2n+1)}{6}.$$

(4) The biggest term in your answer should be $n^3/3$. Note that n^3 is much greater than n^2 when n is large; for example, compare n^3 with n^2 when $n = 20$.

(a) Determine how many multiplications and divisions are required to solve the system with augmented matrix $\left[C \mid \mathbf{d}\right]$ of part (a) by back-substitution. [Answer: $n(n+1)/2$.]

Show that the number of multiplications and divisions required to row reduce $\left[C \mid \mathbf{d}\right]$ to the form $\left[I \mid \mathbf{s}\right]$ (where $\mathbf{s}$ is the solution) is the same as the number used in solving the system by back-substitution. Conclude that the Gauss-Jordan procedure is as efficient as Gaussian elimination with back-substitution. For large n, the number of multiplications or divisions is roughly $\frac{n^3}{3}$.

(b) Suppose that we do a "clumsy" Gauss-Jordan procedure: we do not first obtain near-row echelon form ("near" because we have not divided each row by its leading coefficient); instead, we obtain zeros in all entries of the first column except the first, then zeros in all entries of the second column except the second, and so on. Show that the number of multiplications and divisions required in this procedure is roughly $\frac{n^3}{2}$, so that this procedure requires approximately 50% more operations than the more efficient procedures.

CHAPTER

Matrix Multiplication, Transformations, Inverses

Matrices were used essentially as bookkeeping devices in the previous chapter. They also possess interesting algebraic properties, which means that they have wider and more powerful applications than is suggested by their use in solving systems of equations.

SECTION

3–1 Matrix Addition and Multiplication; Block Multiplication

Equality of Matrices; Addition; Multiplication by Scalars

We have seen that the idea of a matrix is useful in solving systems of linear equations. However, we shall see that matrices show up in different kinds of problems, and *it is important to be able to think of matrices as "things" that are worth studying and playing with — things that may have no connection with a system of equations.* A **matrix is a rectangular array of numbers.** A typical matrix is

$$A = \begin{bmatrix} a_{11} & a_{12} & \cdots & a_{1k} & \cdots & a_{1n} \\ a_{21} & a_{22} & \cdots & a_{2k} & \cdots & a_{2n} \\ \vdots & \vdots & \ddots & \vdots & \ddots & \vdots \\ a_{j1} & a_{j2} & \cdots & a_{jk} & \cdots & a_{jn} \\ \vdots & \vdots & \ddots & \vdots & \ddots & \vdots \\ a_{m1} & a_{m2} & \cdots & a_{mk} & \cdots & a_{mn} \end{bmatrix}.$$

A has m rows and n columns, and we say that its **size** is m **by** n. Two matrices A and B are **equal** if and only if they have the same size and if each entry of A is equal to the corresponding entry of B, that is, if $a_{jk} = b_{jk}$ for each j from 1 to m and for each k from 1 to n. (Write: $a_{jk} = b_{jk}, j = 1, \dots, m; k = 1, \dots, n$ or $a_{jk} = b_{jk}$ for all j and k [range of j and k understood].)

Two matrices A and B can be **added** only if they are of the same size, and then their **sum** $A + B$ is defined to be the matrix whose jk^{th} entry is $a_{jk} + b_{jk}$, for all j and k. Similarly, the jk^{th} entry of $A - B$ is $a_{jk} - b_{jk}$.

EXAMPLE 1

(a) $\begin{bmatrix} 2 & 3 \\ 4 & 1 \end{bmatrix} + \begin{bmatrix} 5 & 1 \\ -2 & 7 \end{bmatrix} = \begin{bmatrix} 7 & 4 \\ 2 & 8 \end{bmatrix}$

(b) $\begin{bmatrix} 3 & 1 \\ 0 & -5 \end{bmatrix} - \begin{bmatrix} 4 & 1 \\ -2 & 5 \end{bmatrix} = \begin{bmatrix} -1 & 0 \\ 2 & -10 \end{bmatrix}$

(c) $\begin{bmatrix} 2 & 3 \\ 6 & -2 \end{bmatrix} + \begin{bmatrix} 5 & 3 & 0 \\ 2 & 3 & -2 \end{bmatrix}$ is not defined because the matrices are not of the same size.

A matrix A can also be **multiplied** by a scalar c: the matrix cA is defined to be the matrix whose typical entry is ca_{jk}. Thus, each entry has been multiplied by c. We can also combine multiplication by scalars with addition to form linear combinations: the typical entry in the linear combination $cA + dB$ is $ca_{jk} + db_{jk}$.

EXAMPLE 2

(a) $5\begin{bmatrix} 2 & 3 \\ 4 & 1 \end{bmatrix} = \begin{bmatrix} 10 & 15 \\ 20 & 5 \end{bmatrix}$

(b) $2\begin{bmatrix} 1 & 3 \\ 0 & -1 \end{bmatrix} + 3\begin{bmatrix} 4 & 0 \\ 1 & 2 \end{bmatrix} = \begin{bmatrix} 2+12 & 6+0 \\ 0+3 & -2+6 \end{bmatrix} = \begin{bmatrix} 14 & 6 \\ 3 & 4 \end{bmatrix}$

An Introduction to Matrix Multiplication

The operations so far are very natural and easy. Multiplication of two matrices is more complicated, but a simple example illustrates that there is one useful natural definition. Suppose that we wish to "change variables": we wish to work with new variables y_1 and y_2 that are defined in terms of the original variables x_1 and x_2 by the equations

$$y_1 = a_{11}x_1 + a_{12}x_2$$
$$y_2 = a_{21}x_1 + a_{22}x_2.$$

This is a system of linear equations like those in Chapter 2.

It is convenient to write these equations in matrix form. Let [**x**] denote the **column matrix** $\begin{bmatrix} x_1 \\ x_2 \end{bmatrix}$, [**y**] the column matrix $\begin{bmatrix} y_1 \\ y_2 \end{bmatrix}$, and A the coefficient matrix $\begin{bmatrix} a_{11} & a_{12} \\ a_{21} & a_{22} \end{bmatrix}$. Then the change of variables equations can be written in the matrix form [**y**] = A[**x**], provided that we **define the product** of A and [**x**] according to the following rule:

$$A[\mathbf{x}] = \begin{bmatrix} a_{11} & a_{12} \\ a_{21} & a_{22} \end{bmatrix} \begin{bmatrix} x_1 \\ x_2 \end{bmatrix} = \begin{bmatrix} a_{11}x_1 + a_{12}x_2 \\ a_{21}x_1 + a_{22}x_2 \end{bmatrix}.$$

It is instructive to rewrite the entries in the right-hand matrix as dot products, so that the equation becomes:

$$\begin{bmatrix} a_{11} & a_{12} \\ a_{21} & a_{22} \end{bmatrix} \begin{bmatrix} x_1 \\ x_2 \end{bmatrix} = \begin{bmatrix} a_{11}x_1 + a_{12}x_2 \\ a_{21}x_1 + a_{22}x_2 \end{bmatrix} = \begin{bmatrix} (a_{11}, a_{12}) \cdot (x_1, x_2) \\ (a_{21}, a_{22}) \cdot (x_1, x_2) \end{bmatrix}.$$

Thus, in order for the right-hand side of the original equations to be represented correctly by the matrix product A[**x**], **the entry in the first row of A[x] must be the dot product of the first row of A (regarded as a vector) with the vector x** (we regard the entries in the matrix [**x**] as the components of a vector so that the dot product is defined); **the entry in the second row must be the dot product of the second row of A with x.**

Suppose there is a second change of variables from [**y**] to [**z**]:

$$z_1 = b_{11}y_1 + b_{12}y_2$$
$$z_2 = b_{21}y_1 + b_{22}y_2.$$

In matrix form, this is written [**z**] = B[**y**]. Now suppose that these changes are performed one after the other, so that the values for y_1 and y_2 from the first change of variables are substituted into the second pair of equations:

$$z_1 = b_{11}(a_{11}x_1 + a_{12}x_2) + b_{12}(a_{21}x_1 + a_{22}x_2)$$
$$z_2 = b_{21}(a_{11}x_1 + a_{12}x_2) + b_{22}(a_{21}x_1 + a_{22}x_2).$$

After simplification (which you should check) this can be written

$$\begin{bmatrix} z_1 \\ z_2 \end{bmatrix} = \begin{bmatrix} (b_{11}a_{11} + b_{12}a_{21}) & (b_{11}a_{12} + b_{12}a_{22}) \\ (b_{21}a_{11} + b_{22}a_{21}) & (b_{21}a_{12} + b_{22}a_{22}) \end{bmatrix} \begin{bmatrix} x_1 \\ x_2 \end{bmatrix}.$$

We want this to be equivalent to $[\mathbf{z}] = B[\mathbf{y}] = BA[\mathbf{x}]$, so that the product BA must be

$$BA = \begin{bmatrix} (b_{11}a_{11} + b_{12}a_{21}) & (b_{11}a_{12} + b_{12}a_{22}) \\ (b_{21}a_{11} + b_{22}a_{21}) & (b_{21}a_{12} + b_{22}a_{22}) \end{bmatrix}.$$

Thus, the product matrix BA must be defined by the following rule *(check that the dot products do give the correct entries in the explicit form above)*:

the 11 (row 1, column 1) entry of BA is $(b_{11}, b_{12}) \cdot (a_{11}, a_{21})$;

the 12 (row 1, column 2) entry of BA is $(b_{11}, b_{12}) \cdot (a_{12}, a_{22})$;

the 21 (row 2, column 1) entry of BA is $(b_{21}, b_{22}) \cdot (a_{11}, a_{21})$;

the 22 (row 2, column 2) entry of BA is $(b_{21}, b_{22}) \cdot (a_{12}, a_{22})$.

EXAMPLE 3

$$\begin{bmatrix} 2 & 3 \\ 4 & 1 \end{bmatrix} \begin{bmatrix} 5 & 1 \\ -2 & 7 \end{bmatrix} = \begin{bmatrix} 2(5) + 3(-2) & 2(1) + 3(7) \\ 4(5) + 1(-2) & 4(1) + 1(7) \end{bmatrix} = \begin{bmatrix} 4 & 23 \\ 18 & 11 \end{bmatrix}$$

SUMMARY

The entry in the jth row and k^{th} column of the product BA is the dot product of the jth row of B with the kth column of A.

Note that the definition uses rows from the first matrix B and columns from the second matrix A, so the order of the matrices matters. This rule is consistent with the rule described for the product $A[\mathbf{x}]$ above. These special cases lead to the following general rule.

The General Rule for Matrix Multiplication

It will be convenient to introduce a notation for rows and columns (unfortunately, there is no standard notation). Let A be an m by n matrix. Denote the jth row of A by $a_{j\rightarrow}$; strictly speaking, we should distinguish between the

1 by n **row matrix** $[\, a_{j1} \quad a_{j2} \quad a_{j3} \quad \ldots \quad a_{jn} \,]$ and the corresponding **n-vector** $(a_{j1}, a_{j2}, a_{j3}, \ldots, a_{jn})$, but no confusion should arise from using $a_{j\rightarrow}$ to denote both. Where it is essential to emphasize either the matrix or vector character of $a_{j\rightarrow}$, write out the full unambiguous form. It is sometimes helpful to distinguish between a column and its corresponding vector, so let $[a_{\downarrow k}]$ denote the m by 1 **column matrix** consisting of the kth column of A, and let $a_{\downarrow k}$ denote the corresponding m-vector $(a_{1k}, a_{2k}, a_{3k}, \ldots, a_{mk})$. Note that the dot product is defined for vectors.

DEFINITION

The product BA of two matrices B and A is defined to be the matrix whose jk^{th} entry is $b_{j\rightarrow} \cdot a_{\downarrow k}$. This product is defined *only if the number of entries in a row of B is equal to the number of entries in a column of A* (otherwise, the dot product is not defined.)

General Case

If B is m by n and A is n by q, then

$$BA = \begin{bmatrix} b_{1\rightarrow} \\ b_{2\rightarrow} \\ \vdots \\ b_{m\rightarrow} \end{bmatrix} \begin{bmatrix} [a_{\downarrow 1}] & [a_{\downarrow 2}] & \ldots & [a_{\downarrow q}] \end{bmatrix}$$

$$= \begin{bmatrix} b_{1\rightarrow} \cdot a_{\downarrow 1} & b_{1\rightarrow} \cdot a_{\downarrow 2} & \ldots & b_{1\rightarrow} \cdot a_{\downarrow q} \\ b_{2\rightarrow} \cdot a_{\downarrow 1} & b_{2\rightarrow} \cdot a_{\downarrow 2} & \ldots & b_{2\rightarrow} \cdot a_{\downarrow q} \\ \vdots & \vdots & \ddots & \vdots \\ b_{m\rightarrow} \cdot a_{\downarrow 1} & b_{m\rightarrow} \cdot a_{\downarrow 2} & \ldots & b_{m\rightarrow} \cdot a_{\downarrow q} \end{bmatrix}$$

Note that in this case the product BA is m by q.

EXAMPLE 4

Check the following products.

(a) $$\begin{bmatrix} 2 & 3 & 0 & 1 \\ 4 & -1 & 2 & -1 \end{bmatrix} \begin{bmatrix} 3 & 1 \\ 1 & 2 \\ 2 & 3 \\ 0 & 5 \end{bmatrix} = \begin{bmatrix} 9 & 13 \\ 15 & 3 \end{bmatrix}$$

(b) $$\begin{bmatrix} 1 & 1 & 2 \\ -2 & -1 & 3 \\ 0 & 0 & 1 \end{bmatrix} \begin{bmatrix} 5 & 6 \\ 4 & 7 \\ 2 & 5 \end{bmatrix} = \begin{bmatrix} 13 & 23 \\ -8 & -4 \\ 2 & 5 \end{bmatrix}$$

(c) $\begin{bmatrix} 2 & 3 \\ 1 & -3 \end{bmatrix} \begin{bmatrix} 2 & -3 \\ 4 & 1 \\ 5 & 7 \end{bmatrix}$ is not defined because there are 2 entries in each row of the first matrix but 3 entries in each column of the second.

Note that matrices do not need to be **square** in order for products to be defined. (A matrix is square if the number of rows equals the number of columns.) Note in Example 4 (a) that the product BA of a 2 by 4 matrix B and a 4 by 2 matrix A is defined and the product has size 2 by 2; in Example 4 (b), the product of a 3 by 3 matrix and a 3 by 2 has size 3 by 2. Generally, because we take dot products of rows of the left-hand matrix with columns of the right-hand matrix, the product of an m by n matrix (on the left) and a p by q matrix (on the right) is defined only if $n = p$ and then the product has size m by q.

Some consequences of the definition

If the matrices A, B, and C are of the correct size so that the required products are defined, and c is a real number, then:

(1) $A(B + C) = AB + AC$ (since $a_{j\rightarrow} \cdot (b_{\downarrow k} + c_{\downarrow k}) = a_{j\rightarrow} \cdot b_{\downarrow k} + a_{j\rightarrow} \cdot c_{\downarrow k}$, by the properties of dot products);

(2) $A(cB) = (cA)B = c(AB)$ (since $a_{j\rightarrow} \cdot (c b_{\downarrow k}) = (c a_{j\rightarrow}) \cdot b_{\downarrow k} = c(a_{j\rightarrow} \cdot b_{\downarrow k})$).

You should check these for some simple examples. See Exercises A3, B3.

Important fact

The matrix product is not commutative: usually $AB \neq BA$. In fact, if BA is defined, it is not necessarily true that AB is even defined. For example, if B is 2 by 2, and A is 2 by 3, then BA is defined but AB is not. However, even if both AB and BA are defined, they are usually not equal. $AB = BA$ is true only in very special circumstances.

EXAMPLE 5

Show that if $A = \begin{bmatrix} 2 & 3 \\ 4 & -1 \end{bmatrix}$ and $B = \begin{bmatrix} 5 & 1 \\ -2 & 7 \end{bmatrix}$, then $AB \neq BA$.

SOLUTION:

$$AB = \begin{bmatrix} 2 & 3 \\ 4 & -1 \end{bmatrix} \begin{bmatrix} 5 & 1 \\ -2 & 7 \end{bmatrix} = \begin{bmatrix} 4 & 23 \\ 22 & -3 \end{bmatrix}$$

but

$$BA = \begin{bmatrix} 5 & 1 \\ -2 & 7 \end{bmatrix} \begin{bmatrix} 2 & 3 \\ 4 & -1 \end{bmatrix} = \begin{bmatrix} 14 & 14 \\ 24 & -13 \end{bmatrix}.$$

Therefore $AB \neq BA$.

The Summation Notation and Matrix Multiplication

Some calculations with matrix products are most easily described in terms of the standard summation notation. The Greek letter $\sum$ (sigma) is used to denote summation. We shall be summing over a subscript; the subscript and its range are indicated below and above the $\sum$. Thus, by definition

$$\sum_{j=1}^{n} a_j = a_1 + a_2 + \cdots + a_n.$$

In connection with this summation notation we sometimes denote the jk^{th} entry of a matrix A by

$$a_{jk} = (A)_{jk}.$$

This may seem pointless for a single matrix, but it is useful for representing the product. If A is m by n and B is n by p, then the jk^{th} entry of the product AB is

$$(AB)_{jk} = a_{j\rightarrow} \cdot b_{\downarrow k} = a_{j1}b_{1k} + a_{j2}b_{2k} + \cdots + a_{jn}b_{nk}$$
$$= \sum_{f=1}^{n} a_{jf}b_{fk} = \sum_{f=1}^{n} (A)_{jf}(B)_{fk}.$$

We shall use this notation in the proof of Theorem 1 below.

More Properties of Matrix Multiplication

Since the commutative rule does not hold for matrix products, we might wonder about other familiar rules of arithmetic. For example, is it true for matrix multiplication that $A(BC) = (AB)C$?

EXAMPLE 6

Let $A = \begin{bmatrix} 1 & 2 \\ -3 & -2 \end{bmatrix}$, $B = \begin{bmatrix} 2 & 1 \\ 4 & 1 \end{bmatrix}$, and $C = \begin{bmatrix} -1 & 2 \\ 5 & 3 \end{bmatrix}$. Then,

$$(AB)C = \begin{bmatrix} 10 & 3 \\ -14 & -5 \end{bmatrix} \begin{bmatrix} -1 & 2 \\ 5 & 3 \end{bmatrix} = \begin{bmatrix} 5 & 29 \\ -11 & -43 \end{bmatrix}$$

and

$$A(BC) = \begin{bmatrix} 1 & 2 \\ -3 & -2 \end{bmatrix} \begin{bmatrix} 3 & 7 \\ 1 & 11 \end{bmatrix} = \begin{bmatrix} 5 & 29 \\ -11 & -43 \end{bmatrix}.$$

In this case, at least, $(AB)C = A(BC)$.

THEOREM 1. The matrix product is **associative:** if AB and BC are both defined, then so are $A(BC)$ and $(AB)C$, and then $A(BC) = (AB)C$.

Proof. We must show that $A(BC)$ and $(AB)C$ have the same size, and that

$$(A(BC))_{jk} = ((AB)C)_{jk} \quad \text{for all } j, k.$$

Since AB and BC are both defined, we may assume that A is m by n, B is n by p, and C is p by q. It is straightforward to check that $(AB)C$ and $A(BC)$ are both m by q, so the sizes agree. Next consider

$$((AB)C)_{jk} = \sum_{f=1}^{p} (AB)_{jf} c_{fk} = \sum_{f=1}^{p} \left(\sum_{h=1}^{n} a_{jh} b_{hf} \right) c_{fk}.$$

But a finite sum can be rearranged, so the last expression is equal to

$$\sum_{h=1}^{n} a_{jh} \left(\sum_{f=1}^{p} b_{hf} c_{fk} \right) = (A(BC))_{jk}.$$

Thus, we have shown that

$$((AB)C)_{jk} = (A(BC))_{jk}, \quad j = 1, \ldots, m; \quad k = 1, \ldots, q.$$

Hence, the associativity of matrix multiplication is proved.

■

> *REMARK*
>
> Because of the equality, we may write the product in the form ABC—it will not matter whether we calculate $A(BC)$ or $(AB)C$.

In ordinary multiplication, we expect to be able to use the **Cancellation Law:** if $ab = ac$ and $a \neq 0$, then $b = c$. *However, cancellation is almost never valid in the case of matrix multiplication.*

EXAMPLE 7

Let $A = \begin{bmatrix} 0 & 0 \\ 0 & 1 \end{bmatrix}$, $B = \begin{bmatrix} 5 & 6 \\ 7 & 8 \end{bmatrix}$, and $C = \begin{bmatrix} 2 & 3 \\ 7 & 8 \end{bmatrix}$. Then,

$$AB = \begin{bmatrix} 0 & 0 \\ 0 & 1 \end{bmatrix} \begin{bmatrix} 5 & 6 \\ 7 & 8 \end{bmatrix} = \begin{bmatrix} 0 & 0 \\ 7 & 8 \end{bmatrix},$$
$$AC = \begin{bmatrix} 0 & 0 \\ 0 & 1 \end{bmatrix} \begin{bmatrix} 2 & 3 \\ 7 & 8 \end{bmatrix} = \begin{bmatrix} 0 & 0 \\ 7 & 8 \end{bmatrix},$$

so $AB = AC$. However, $B \neq C$.

We must distinguish carefully between a general cancellation law such as this and the following theorem, which we will use many times.

THEOREM 2. Suppose that A and B are m by n matrices such that $A[\mathbf{x}] = B[\mathbf{x}]$ for every $\mathbf{x}$ in $\mathbb{R}^n$. Then $A = B$.

You are asked to prove this, with hints, in Exercise 3–1–D1.

> *REMARK*
>
> It is the assumption that equality holds *for every* $\mathbf{x}$ that distinguishes this from a cancellation law.

The Zero Matrix and the Identity Matrix

The m by n **zero matrix** is the m by n matrix such that all entries are zero. It is denoted by 0 or, if it is important to emphasize the size, by $0_{m,n}$. Thus, for example,

$$0_{2,3} = \begin{bmatrix} 0 & 0 & 0 \\ 0 & 0 & 0 \end{bmatrix}.$$

There is a zero matrix for each size m by n. Zero matrices are chiefly useful for writing equations such as $A + X = 0$. Note that if A is an m by n matrix, then $0_{m,m}A = 0_{m,n}$ and $A0_{n,q} = 0_{m,q}$.

An even more important special matrix is the n by n **identity matrix**, which is denoted by I or, if it is necessary to indicate the size, by I_n. It is always a square matrix. Each entry on the **main diagonal** (that is, each entry whose row index is the same as its column index) is 1. All the off-diagonal entries are 0. Thus, for example,

$$I_3 = \begin{bmatrix} 1 & 0 & 0 \\ 0 & 1 & 0 \\ 0 & 0 & 1 \end{bmatrix}.$$

A particularly important property of identity matrices is that for any m by n matrix A, $I_mA = A = AI_n$. *You should check this for a simple case.* (Note that the identity matrices must always be chosen to be of the correct size.)

The Transpose of a Matrix

The transpose of a matrix is important later in this book. Since it is a simple operation, we introduce it now. To *transpose a matrix, turn its rows into columns and its columns into rows.* The transpose of A is denoted by A^T.

EXAMPLE 8

Determine the transpose of $A = \begin{bmatrix} -1 & 6 & -4 \\ 3 & 5 & 2 \end{bmatrix}$.

SOLUTION: $A^T = \begin{bmatrix} -1 & 6 & -4 \\ 3 & 5 & 2 \end{bmatrix}^T = \begin{bmatrix} -1 & 3 \\ 6 & 5 \\ -4 & 2 \end{bmatrix}$

For the general matrix A at the beginning of this section,

$$A = \begin{bmatrix} a_{11} & a_{12} & \cdots & a_{1k} & \cdots & a_{1n} \\ a_{21} & a_{22} & \cdots & a_{2k} & \cdots & a_{2n} \\ \vdots & \vdots & \ddots & \vdots & \ddots & \vdots \\ a_{j1} & a_{j2} & \cdots & a_{jk} & \cdots & a_{jn} \\ \vdots & \vdots & \ddots & \vdots & \ddots & \vdots \\ a_{m1} & a_{m2} & \cdots & a_{mk} & \cdots & a_{mn} \end{bmatrix}$$

$$A^T = \begin{bmatrix} a_{11} & a_{21} & \cdots & a_{j1} & \cdots & a_{m1} \\ a_{12} & a_{22} & \cdots & a_{j2} & \cdots & a_{m2} \\ \vdots & \vdots & \ddots & \vdots & \ddots & \vdots \\ a_{1k} & a_{2k} & \cdots & a_{jk} & \cdots & a_{mk} \\ \vdots & \vdots & \ddots & \vdots & \ddots & \vdots \\ a_{1n} & a_{2n} & \cdots & a_{jn} & \cdots & a_{mn} \end{bmatrix}.$$

Notice that the entries of the first row of A have become the entries in the first column of its transpose A^T, the second row of A turns into the second column of A^T, and so on. Similarly, the entries in the first column of A appear in the first row of A^T, and so on. However, you must read with care: in A^T the number a_{jk} appears in the kth row and j^{th} column: in general $(A^T)_{kj} = (A)_{jk}$. Note also that a_{mn} appears in the nth row and m^{th} column of A^T; if A is m by n, A^T is n by m.

Some Properties of the Transpose

How does the operation of transposition combine with addition, scalar multiplication, or matrix multiplication? In most of these cases, the answer is easy, but the answer given for matrix multiplication in the following theorem is not obvious.

THEOREM 3. (1) For any matrix A, $(A^T)^T = A$.

(2) For any matrices A and B such that $A + B$ is defined, $(A+B)^T = A^T + B^T$.

(3) For any matrix A and any scalar k, $(kA)^T = kA^T$.

(4) If the product AB is defined, then $(AB)^T = B^T A^T$.

Proof. The proof of (1), (2), and (3) consists of writing the left and right sides of each equation and checking that they are equal. (If you have difficulty seeing this, write general 2 by 2 matrices A and B and write the left and right sides of each of (1), (2), and (3).)

The proof of (4) is not straightforward. Note that by definition of the transpose, $((AB)^T)_{jk} = (AB)_{kj}$; but,

$$(AB)_{kj} = (\text{the } k^{\text{th}} \text{ row of } A) \cdot (\text{the } j^{\text{th}} \text{ column of } B);$$

by transposition, the kth row of A is the kth column of A^T and the jth column of B is the jth row of B^T so that

$$((AB)^T)_{jk} = (AB)_{kj} = (k^{\text{th}} \text{ column of } A^T) \cdot (j^{\text{th}} \text{ row of } B^T).$$

Since the dot product is commutative ($\mathbf{a} \cdot \mathbf{b} = \mathbf{b} \cdot \mathbf{a}$), this becomes

$$((AB)^T)_{jk} = (j^{\text{th}} \text{ row of } B^T) \cdot (k^{\text{th}} \text{ column of } A^T) = (B^T A^T)_{jk},$$

and this is true for all j and k. Hence, $(AB)^T = B^T A^T$ as claimed.

■

EXAMPLE 9

If $A = \begin{bmatrix} 2 & 3 \\ 4 & 1 \end{bmatrix}$ and $B = \begin{bmatrix} 5 & 1 \\ -2 & 7 \end{bmatrix}$, verify that $(AB)^T = B^T A^T$.

SOLUTION: $AB = \begin{bmatrix} 2 & 3 \\ 4 & 1 \end{bmatrix} \begin{bmatrix} 5 & 1 \\ -2 & 7 \end{bmatrix} = \begin{bmatrix} 4 & 23 \\ 18 & 11 \end{bmatrix}$, so

$$(AB)^T = \begin{bmatrix} 4 & 18 \\ 23 & 11 \end{bmatrix}.$$

On the other hand,

$$B^T A^T = \begin{bmatrix} 5 & 1 \\ -2 & 7 \end{bmatrix}^T \begin{bmatrix} 2 & 3 \\ 4 & 1 \end{bmatrix}^T = \begin{bmatrix} 5 & -2 \\ 1 & 7 \end{bmatrix} \begin{bmatrix} 2 & 4 \\ 3 & 1 \end{bmatrix} = \begin{bmatrix} 4 & 18 \\ 23 & 11 \end{bmatrix}.$$

Thus, $(AB)^T = B^T A^T$, as expected.

Block Multiplication

Consider the product of two 3 by 3 matrices:

$$AB = \begin{bmatrix} a_{1\to} \cdot b_{\downarrow 1} & a_{1\to} \cdot b_{\downarrow 2} & a_{1\to} \cdot b_{\downarrow 3} \\ a_{2\to} \cdot b_{\downarrow 1} & a_{2\to} \cdot b_{\downarrow 2} & a_{2\to} \cdot b_{\downarrow 3} \\ a_{3\to} \cdot b_{\downarrow 1} & a_{3\to} \cdot b_{\downarrow 2} & a_{3\to} \cdot b_{\downarrow 3} \end{bmatrix}.$$

Inspect the first column of this product: it involves only the first column of B. In fact, the first column of this product is $A \begin{bmatrix} b_{11} \\ b_{21} \\ b_{31} \end{bmatrix} = A[b_{\downarrow 1}]$. If the second and third columns are treated in a similar fashion, the product can be written in the form:

$$AB = A\Big[[b_{\downarrow 1}] \quad [b_{\downarrow 2}] \quad [b_{\downarrow 3}]\Big] = \Big[A[b_{\downarrow 1}] \quad A[b_{\downarrow 2}] \quad A[b_{\downarrow 3}]\Big].$$

There is no saving in computation in writing the product of two matrices in this way, but for some applications we can see better what the computation means or represents. For example, we shall use this rule in determining how the volume of a parallelepiped changes under a matrix linear transformation (what this means will be explained in Chapter 5).

This is a very simple example of a general way of looking at matrix multiplication in terms of **blocks**: in this example, the blocks were the columns of B. We could also regard the rows of A as blocks and write

$$AB = \begin{bmatrix} a_{1\to} B \\ a_{2\to} B \\ a_{3\to} B \end{bmatrix}.$$

In the next section, the following version of block multiplication will play a particularly important role:

$$A[\mathbf{x}] = \begin{bmatrix} a_{11} & a_{12} & a_{13} \\ a_{21} & a_{22} & a_{23} \\ a_{31} & a_{32} & a_{33} \end{bmatrix} \begin{bmatrix} x_1 \\ x_2 \\ x_3 \end{bmatrix} = \begin{bmatrix} a_{11}x_1 + a_{12}x_2 + a_{13}x_3 \\ a_{21}x_1 + a_{22}x_2 + a_{23}x_3 \\ a_{31}x_1 + a_{32}x_2 + a_{33}x_3 \end{bmatrix}$$

$$= \begin{bmatrix} a_{11}x_1 \\ a_{21}x_1 \\ a_{31}x_1 \end{bmatrix} + \begin{bmatrix} a_{12}x_2 \\ a_{22}x_2 \\ a_{32}x_2 \end{bmatrix} + \begin{bmatrix} a_{13}x_3 \\ a_{23}x_3 \\ a_{33}x_3 \end{bmatrix} = x_1[a_{\downarrow 1}] + x_2[a_{\downarrow 2}] + x_3[a_{\downarrow 3}]$$

$$= \Big[[a_{\downarrow 1}] \quad [a_{\downarrow 2}] \quad [a_{\downarrow 3}]\Big] \begin{bmatrix} x_1 \\ x_2 \\ x_3 \end{bmatrix}.$$

By considering the columns of A as blocks, we see that $A[\mathbf{x}]$ may be written as a linear combination of the columns of A, with the components of $\mathbf{x}$ as the coefficients in the linear combination $x_1[a_{\downarrow 1}] + x_2[a_{\downarrow 2}] + x_3[a_{\downarrow 3}]$. This has been written out for the particular case of a 3 by 3 matrix and a 3-vector $\mathbf{x}$, but it is easy to see that the same result is true for $A[\mathbf{x}]$ with A of any size.

There are more general statements about the products of two matrices, each of which has been partitioned into blocks. In addition to clarifying the meaning of some calculations, block multiplication is used in organizing calculations with very large matrices.

We shall state just one more version here. Roughly speaking, as long as the sizes of the blocks are chosen so that the products of the blocks are defined and fit together as required, block multiplication works, and is defined by an extension of the usual rules of matrix multiplication to the blocks.

Suppose that A and B are any two matrices such that AB is defined and that A and B are **partitioned** into blocks as indicated:

$$A = \begin{bmatrix} A_1 \\ A_2 \end{bmatrix}, \qquad B = [\, B_1 \quad B_2 \,].$$

For example, A_1 might consist of 2 rows and A_2 of 3 rows while B_1 has 3 columns and B_2 has 4 columns. (The number of columns in A must equal the number of rows in B since AB is defined; it does not matter what that common number is.) Now the product of a 2 by 1 matrix and a 1 by 2 matrix is given by

$$\begin{bmatrix} a_1 \\ a_2 \end{bmatrix} [\, b_1 \quad b_2 \,] = \begin{bmatrix} a_1 b_1 & a_1 b_2 \\ a_2 b_1 & a_2 b_2 \end{bmatrix}.$$

Similarly, for the partitioned block matrices

$$\begin{bmatrix} A_1 \\ A_2 \end{bmatrix} [\, B_1 \quad B_2 \,] = \begin{bmatrix} A_1 B_1 & A_1 B_2 \\ A_2 B_1 & A_2 B_2 \end{bmatrix}.$$

If these blocks are of the size suggested, then the top left block A_1B_1 has size 2 by 3, the top right block A_1B_2 is 2 by 4, the block A_2B_1 is 3 by 3, and the block A_2B_2 is 3 by 4; notice that these product blocks fit together properly to give a matrix with $(2+3) = 5$ rows and $(3+4) = 7$ columns.

EXERCISES 3–1

A1 Calculate the following:

(a) $\begin{bmatrix} 2 & -2 & 3 \\ 4 & 1 & -1 \end{bmatrix} + \begin{bmatrix} -3 & -4 & 1 \\ 2 & -5 & 3 \end{bmatrix}$ (b) $(-3)\begin{bmatrix} 1 & -2 \\ 2 & 1 \\ 4 & -2 \end{bmatrix}$

(c) $\begin{bmatrix} 2 & 3 \\ 1 & -2 \end{bmatrix} - 3\begin{bmatrix} 1 & -2 \\ 4 & 5 \end{bmatrix}$

A2 Calculate the following products or explain why the product is not defined.

(a) $\begin{bmatrix} -2 & 3 \\ 3 & 4 \end{bmatrix}\begin{bmatrix} 4 & 5 & -3 \\ -1 & 3 & 2 \end{bmatrix}$ (b) $\begin{bmatrix} 2 & 0 & 3 \\ 1 & 1 & 1 \\ -1 & 3 & 2 \end{bmatrix}\begin{bmatrix} 6 & -2 \\ 3 & 1 \\ 0 & 5 \end{bmatrix}$

(c) $\begin{bmatrix} 2 & 3 \\ -1 & -1 \\ 5 & 3 \end{bmatrix}\begin{bmatrix} 4 & 3 & 2 & 1 \\ -4 & 0 & 3 & -2 \end{bmatrix}$ (d) $\begin{bmatrix} 2 & 3 \\ -4 & 2 \end{bmatrix}\begin{bmatrix} 1 & 2 \\ 1 & 0 \\ 0 & -3 \end{bmatrix}$

A3 Check that $A(B + C) = AB + AC$ and that $A(3B) = 3(AB)$ for the given matrices.

(a) $A = \begin{bmatrix} 2 & 3 \\ -3 & 1 \end{bmatrix}$, $B = \begin{bmatrix} -3 & 2 \\ 1 & 0 \end{bmatrix}$, $C = \begin{bmatrix} -2 & -3 \\ -2 & 4 \end{bmatrix}$

(b) $A = \begin{bmatrix} 2 & -1 & -2 \\ 3 & 2 & 1 \end{bmatrix}$, $B = \begin{bmatrix} -1 & 2 \\ -3 & 4 \\ 0 & 3 \end{bmatrix}$, $C = \begin{bmatrix} 5 & -2 \\ -3 & 0 \\ 4 & 3 \end{bmatrix}$

A4 For the given matrices A, B, check whether $A + B$ and AB are defined, and if so, check that $(A + B)^T = A^T + B^T$ and $(AB)^T = B^T A^T$.

(a) $A = \begin{bmatrix} 1 & 2 \\ 1 & 3 \\ -2 & 1 \end{bmatrix}$, $B = \begin{bmatrix} -4 & -3 \\ 1 & -1 \\ 3 & 2 \end{bmatrix}$

(b) $A = \begin{bmatrix} 2 & -4 & 5 \\ 4 & 1 & -3 \end{bmatrix}$, $B = \begin{bmatrix} -3 & -4 \\ 5 & -2 \\ 1 & 3 \end{bmatrix}$

A5 Let $A = \begin{bmatrix} 2 & 5 \\ -1 & 3 \end{bmatrix}$, $B = \begin{bmatrix} -1 & 3 & -4 \\ 3 & 5 & 2 \end{bmatrix}$, $C = \begin{bmatrix} 1 & 4 \\ 1 & 3 \\ 4 & -3 \end{bmatrix}$,

$D = \begin{bmatrix} 4 & 3 & 2 & 1 \\ -1 & 0 & 1 & 2 \\ 2 & 1 & 0 & 3 \end{bmatrix}$.

Determine the following products or state why they do not exist.

(a) AB (b) BA (c) AC (d) DC (e) CD (f) $C^T D$

(g) Verify that $A(BC) = (AB)C$.

(h) Verify that $(AB)^T = B^T A^T$.

(i) Without doing any arithmetic, determine $D^T C$.

A6 Let $A = \begin{bmatrix} 2 & 3 & 1 \\ 3 & -1 & 4 \\ -1 & 0 & 1 \end{bmatrix}$, and let $[\mathbf{x}] = \begin{bmatrix} 1 \\ 2 \\ 4 \end{bmatrix}$, $[\mathbf{y}] = \begin{bmatrix} 3 \\ 1 \\ -1 \end{bmatrix}$, $[\mathbf{z}] = \begin{bmatrix} 0 \\ -1 \\ 1 \end{bmatrix}$.

(a) Determine $A[\mathbf{x}]$, $A[\mathbf{y}]$, and $A[\mathbf{z}]$.

(b) Use the result of (a) to determine $A \begin{bmatrix} 1 & 3 & 0 \\ 2 & 1 & -1 \\ 4 & -1 & 1 \end{bmatrix}$.

A7 Consider $A[\mathbf{x}]$ as a linear combination of columns of A to determine $\mathbf{x}$ if $\begin{bmatrix} 2 & 3 & 1 \\ 3 & -1 & 4 \\ -1 & 0 & 1 \end{bmatrix}[\mathbf{x}]$ is equal to (a) $\begin{bmatrix} 2 \\ 3 \\ -1 \end{bmatrix}$; (b) $\begin{bmatrix} 4 \\ 6 \\ -2 \end{bmatrix}$; (c) $\begin{bmatrix} 3 \\ 12 \\ 3 \end{bmatrix}$.

A8 Calculate the following products.

(a) $\begin{bmatrix} -3 \\ 1 \end{bmatrix} \begin{bmatrix} 2 & 6 \end{bmatrix}$ (b) $\begin{bmatrix} 2 & 6 \end{bmatrix} \begin{bmatrix} -3 \\ 1 \end{bmatrix}$

(c) $\begin{bmatrix} 2 \\ -1 \\ 3 \end{bmatrix} \begin{bmatrix} 5 & 4 & -3 \end{bmatrix}$ (d) $\begin{bmatrix} 5 & 4 & -3 \end{bmatrix} \begin{bmatrix} 2 \\ -1 \\ 3 \end{bmatrix}$

A9 Verify the following case of block multiplication by calculating both sides of the equation and comparing.

$$\left[\begin{array}{cc|cc} 2 & 3 & -4 & 5 \\ -4 & 1 & 2 & 1 \end{array}\right] \left[\begin{array}{cc} 6 & 3 \\ -2 & 4 \\ \hline 1 & 3 \\ -3 & 2 \end{array}\right] = \begin{bmatrix} 2 & 3 \\ -4 & 1 \end{bmatrix} \begin{bmatrix} 6 & 3 \\ -2 & 4 \end{bmatrix} + \begin{bmatrix} -4 & 5 \\ 2 & 1 \end{bmatrix} \begin{bmatrix} 1 & 3 \\ -3 & 2 \end{bmatrix}$$

B1 Calculate the following:

(a) $\begin{bmatrix} 3 & -2 \\ -4 & 1 \\ 3 & 7 \end{bmatrix} + \begin{bmatrix} 5 & 4 \\ 1 & 4 \\ -6 & -9 \end{bmatrix}$ (b) $(-5) \begin{bmatrix} 2 & 3 & -6 & -2 \\ -7 & 1 & 0 & 5 \end{bmatrix}$

(c) $\begin{bmatrix} 4 & 2 & 3 \\ -2 & 1 & 5 \end{bmatrix} - 4 \begin{bmatrix} -2 & -1 & 5 \\ 6 & 7 & 1 \end{bmatrix}$

B2 Calculate the following products or explain why the product is not defined.

(a) $\begin{bmatrix} -3 & 2 \\ 5 & -1 \end{bmatrix} \begin{bmatrix} 3 & 1 & -2 \\ 2 & -3 & -1 \end{bmatrix}$ (b) $\begin{bmatrix} 0 & 3 & -1 \\ -1 & 2 & -1 \\ 1 & 1 & 3 \end{bmatrix} \begin{bmatrix} 7 & -3 \\ 2 & -1 \\ 5 & 0 \end{bmatrix}$

(c) $\begin{bmatrix} 3 & -1 \\ 2 & 4 \\ 2 & 7 \end{bmatrix} \begin{bmatrix} 3 & 1 & -2 & 2 \\ -2 & 1 & -2 & 3 \end{bmatrix}$ (d) $\begin{bmatrix} 1 & 2 & -1 \\ 2 & 3 & 1 \end{bmatrix} \begin{bmatrix} -4 & 1 & 5 \\ 6 & 3 & -1 \end{bmatrix}$

B3 Check that $A(B + C) = AB + AC$ and that $A(3B) = 3(AB)$ for the given matrices.

(a) $A = \begin{bmatrix} 3 & 5 \\ -2 & -1 \end{bmatrix}$, $B = \begin{bmatrix} 4 & 2 & -1 \\ 1 & -6 & 8 \end{bmatrix}$, $C = \begin{bmatrix} -2 & 1 & 4 \\ 2 & 2 & -3 \end{bmatrix}$

(b) $A = \begin{bmatrix} 3 & 4 \\ 1 & 0 \\ -3 & 5 \end{bmatrix}$, $B = \begin{bmatrix} 2 & -3 \\ 1 & 5 \end{bmatrix}$, $C = \begin{bmatrix} -4 & -6 \\ 3 & 1 \end{bmatrix}$

B4 For the given matrices A, B, check whether $A + B$ and AB are defined, and if so, check that $(A + B)^T = A^T + B^T$ and $(AB)^T = B^T A^T$.

(a) $A = \begin{bmatrix} 1 & 1 & 4 \\ -2 & 1 & 6 \end{bmatrix}$, $B = \begin{bmatrix} 5 & 4 \\ -2 & 3 \\ 0 & -1 \end{bmatrix}$

(b) $A = \begin{bmatrix} 1 & 3 & 1 \\ -2 & 1 & 4 \\ 1 & 0 & -3 \end{bmatrix}$, $B = \begin{bmatrix} 6 & -2 & 2 \\ 1 & 1 & 3 \\ 2 & -3 & -4 \end{bmatrix}$

B5 Let $A = \begin{bmatrix} 2 & 1 \\ -1 & -2 \\ 4 & -3 \end{bmatrix}$, $B = \begin{bmatrix} 3 & -4 \\ 4 & 5 \end{bmatrix}$, $C = \begin{bmatrix} -2 & 5 & -2 \\ -2 & 1 & 0 \end{bmatrix}$,

$D = \begin{bmatrix} 1 & 3 & -5 \\ 0 & 2 & 1 \\ -3 & 2 & 1 \\ 1 & 1 & -1 \end{bmatrix}$.

Determine the following products or state why they do not exist.

(a) AB (b) BA (c) AC (d) DC (e) DA (f) CD^T

(g) Verify that $B(CA) = (BC)A$.

(h) Verify that $(AB)^T = B^T A^T$.

(i) Without doing any arithmetic, determine DC^T.

B6 Let $A = \begin{bmatrix} 1 & 0 & 2 \\ -1 & 2 & 3 \\ -1 & 1 & 0 \end{bmatrix}$, and let $[\mathbf{x}] = \begin{bmatrix} 4 \\ 0 \\ -2 \end{bmatrix}$, $[\mathbf{y}] = \begin{bmatrix} -1 \\ 1 \\ 2 \end{bmatrix}$, $[\mathbf{z}] = \begin{bmatrix} 5 \\ 1 \\ 3 \end{bmatrix}$.

(a) Determine $A[\mathbf{x}]$, $A[\mathbf{y}]$, and $A[\mathbf{z}]$.

(b) Use the result of (a) to determine $A \begin{bmatrix} 4 & 1 & -5 \\ 0 & -1 & -1 \\ -2 & -2 & -3 \end{bmatrix}$. (Watch the signs!)

B7 Consider $A[\mathbf{x}]$ as a linear combination of columns of A to determine $\mathbf{x}$ if

$\begin{bmatrix} 3 & 3 & -5 \\ 0 & -1 & -1 \\ -2 & -4 & -3 \end{bmatrix} [\mathbf{x}]$ is equal to (a) $\begin{bmatrix} 3 \\ -1 \\ -4 \end{bmatrix}$; (b) $\begin{bmatrix} -6 \\ 0 \\ 4 \end{bmatrix}$; (c) $\begin{bmatrix} 15 \\ 3 \\ 9 \end{bmatrix}$.

B8 Calculate the following products.

(a) $\begin{bmatrix} 2 \\ 1 \\ 5 \end{bmatrix} \begin{bmatrix} -3 & 1 & 2 \end{bmatrix}$ (b) $\begin{bmatrix} -3 & 1 & 2 \end{bmatrix} \begin{bmatrix} 2 \\ 1 \\ 5 \end{bmatrix}$

B9 Verify the following case of block multiplication by calculating both sides of the equation and comparing.

$$\left[\begin{array}{cc|cc} 1 & -1 & -2 & 4 \\ 3 & 2 & 1 & 3 \end{array}\right]\left[\begin{array}{cc} 4 & 3 \\ 2 & 5 \\ \hline 6 & 1 \\ -1 & 3 \end{array}\right]$$

$$= \begin{bmatrix} 1 & -1 \\ 3 & 2 \end{bmatrix}\begin{bmatrix} 4 & 3 \\ 2 & 5 \end{bmatrix} + \begin{bmatrix} -2 & 4 \\ 1 & 3 \end{bmatrix}\begin{bmatrix} 6 & 1 \\ -1 & 3 \end{bmatrix}$$

COMPUTER RELATED EXERCISES

C1 Use a computer to check your results for Exercises A2 and A5.

C2 Use a computer to determine the following matrix products and the transpose of the products. Note that one of the matrices appears twice; if you are careful you should not have to enter it twice.

(a) $\begin{bmatrix} 2.12 & 5.35 \\ -1.97 & 3.56 \end{bmatrix}\begin{bmatrix} -1.02 & 3.47 & -4.94 \\ 3.33 & 5.83 & 2.29 \end{bmatrix}$

(b) $\begin{bmatrix} -1.02 & 3.47 & -4.94 \\ 3.33 & 5.83 & 2.29 \end{bmatrix}\begin{bmatrix} 1.88 & 4.25 \\ 1.55 & 3.38 \\ 4.67 & -3.73 \end{bmatrix}$

CONCEPTUAL EXERCISES

***D1** Prove Theorem 2 using the following hints.

(a) If $A = B$, obviously $A[\mathbf{x}] = B[\mathbf{x}]$ for every $\mathbf{x}$ in $\mathbb{R}^n$, so we need to prove only the converse.

(b) To prove $A = B$, prove that $A - B = 0$; note that $A[\mathbf{x}] = B[\mathbf{x}]$ for every $\mathbf{x}$ in $\mathbb{R}^n$ if and only if $(A - B)[\mathbf{x}] = [\mathbf{0}]$ for every $\mathbf{x}$ in $\mathbb{R}^n$.

(c) Suppose that $C[\mathbf{x}] = [\mathbf{0}]$ for every $\mathbf{x}$ in $\mathbb{R}^n$; consider the case where $\mathbf{x} = c_{j\rightarrow}$, and conclude that each row of C must be the zero vector.

***D2** Write a proof of the fact that $(AB)^T = B^T A^T$ using the summation notation. (Hint: Write $(A)_{jk}$ instead of a_{jk}.)

D3 Determine the typical (jk^{th}) entry in AA^T and the typical entry in $A^T A$. Explain why it follows that if AA^T or $A^T A$ is the zero matrix, then $A = 0$.

D4 (a) Construct a 2 by 2 matrix A that is not the zero matrix yet satisfies $A^2 = 0$.

(b) Find 2 by 2 matrices A and B with $A \neq B$ and neither $A = 0$ nor $B = 0$, such that $A^2 - AB - BA + B^2 = 0$.

D5 Find as many 2 by 2 matrices as you can that satisfy $A^2 = I$.

SECTION

3–2 Matrix Mappings; Linear Mappings and Transformations

In your earlier mathematics you will have met the idea of a **function.** A function is a rule that assigns to every element of an initial set called the **domain** of the function a value (or "image") in another set called the **codomain** of the function. Some people like to think of elements of the domain as "inputs" of the function with the corresponding values called the "outputs" (so that outputs belong to the codomain). If f is a function with domain U and codomain V, we write $f : U \to V$. If u is an element of U (an "input"), $f(u) = v$ is the corresponding value (or "output"). A simple example of a function is $f : \mathbb{R} \to \mathbb{R}$ defined by $f(x) = 0$ for all x in $\mathbb{R}$; notice that in this example the codomain consists of all real numbers, but that 0 is the only real number that actually occurs as a value of the function f. The codomain may be much larger than the set of actual values of f; we shall discuss the set of values of a function (called the range of f) in Section 3–4.

We shall be particularly interested in functions whose domain is $\mathbb{R}^n$ and whose codomain is $\mathbb{R}^m$. It is common to call such functions **mappings** or **transformations.** One example we have already seen is projection; if $\mathbf{a}$ is a vector in $\mathbb{R}^3$, then we have the mapping $\mathbf{proj_a} : \mathbb{R}^3 \to \mathbb{R}^3$ as defined in Section 1–3.

Matrix Mappings

Let A be an m by n matrix. The rule for multiplication introduced in the preceding section means that for any such matrix, there is a mapping, called the **matrix mapping f_A corresponding to matrix A,** defined as follows. Let $\mathbf{x}$ be a vector in $\mathbb{R}^n$, and let $[\mathbf{x}]$ denote the corresponding column matrix $\begin{bmatrix} x_1 \\ x_2 \\ \vdots \\ x_n \end{bmatrix}$.

Define the image of x under f_A, $f_A(\mathbf{x})$, to be the vector corresponding to the column matrix $A[\mathbf{x}]$. Using the square bracket notation to remind us that we must represent vectors as columns in matrix calculations, we can write the definition in the form:

$$[f_A(\mathbf{x})] = A[\mathbf{x}].$$

EXAMPLE 10

Let $A = \begin{bmatrix} 2 & 3 \\ -1 & 4 \\ 0 & 1 \end{bmatrix}$. Let us calculate the images of some vectors under the map f_A as defined above. To determine $f_A(2,5)$, for example, calculate

$$\begin{bmatrix} 2 & 3 \\ -1 & 4 \\ 0 & 1 \end{bmatrix}\begin{bmatrix} 2 \\ 5 \end{bmatrix} = \begin{bmatrix} 19 \\ 18 \\ 5 \end{bmatrix} \text{ so that } f_A(2,5) = (19,18,5).$$

Similarly, $\begin{bmatrix} 2 & 3 \\ -1 & 4 \\ 0 & 1 \end{bmatrix}\begin{bmatrix} -1 \\ 4 \end{bmatrix} = \begin{bmatrix} 10 \\ 17 \\ 4 \end{bmatrix}$, so $f_A(-1,4) = (10,17,4)$. Note that the domain of f_A is $\mathbb{R}^2$ and the codomain is $\mathbb{R}^3$.

In the general case, since A has m rows, so does $A[\mathbf{x}]$, and hence $f_A(\mathbf{x})$ is a vector in $\mathbb{R}^m$. It is usual to say that the function f_A **maps from** $\mathbb{R}^n$ **to** $\mathbb{R}^m$: for each vector $\mathbf{x}$ in the **domain** $\mathbb{R}^n$, there is a corresponding image vector $f_A(\mathbf{x})$ in the **codomain** $\mathbb{R}^m$. We write $f_A : \mathbb{R}^n \to \mathbb{R}^m$. This mapping may be pictured as in Figure 3–2–1.

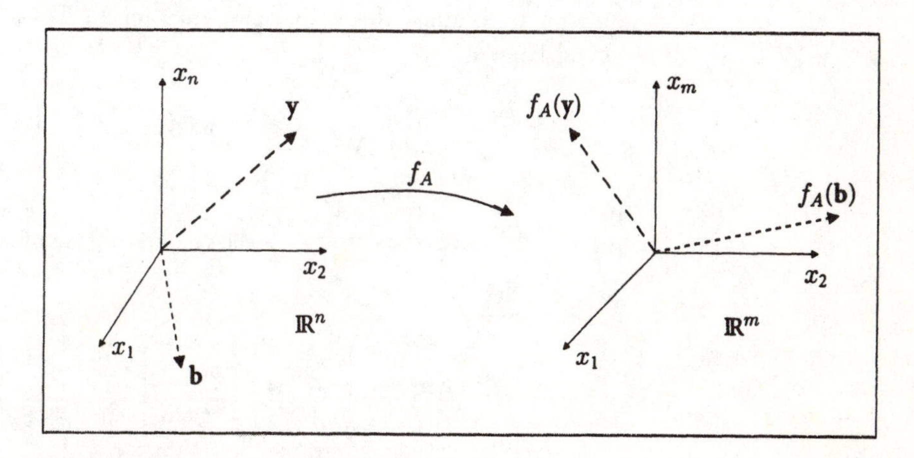

FIGURE 3-2-1. *Picturing the matrix mapping* $f_A : \mathbb{R}^n \to \mathbb{R}^m$.

We shall refer to either the vector $f_A(\mathbf{x})$ or the column matrix $[f_A(\mathbf{x})]$ as the image of $\mathbf{x}$ under f_A. Some people feel that the correspondence between vectors and column matrices is so obvious that the square brackets are unnecessary; they write statements such as $f_A(\mathbf{x}) = A\mathbf{x}$. We shall use the bracket notation in this section and the next as a reminder that vectors are represented as columns in matrix calculations. Later (after due warning), the brackets will sometimes be omitted; however, where we want to emphasize that we must work with columns, we shall continue to use brackets.

EXAMPLE 11

The following calculations with the matrix from Example 10 illustrate some very important general principles.

It is natural to ask what a mapping "does" to a general vector (x_1, x_2) and to the standard basis vectors $(1, 0)$ and $(0, 1)$. We calculate these images.

$$[f_A(1,0)] = \begin{bmatrix} 2 & 3 \\ -1 & 4 \\ 0 & 1 \end{bmatrix} \begin{bmatrix} 1 \\ 0 \end{bmatrix} = \begin{bmatrix} 2 \\ -1 \\ 0 \end{bmatrix}$$

Note that the image of the first basis vector under the map f_A gives exactly the first column of A. Notice that this calculation is an example of block multiplication, as explained in Section 3–1: if we treat the columns of A as the blocks, we saw in Section 3–1 that the product $A \begin{bmatrix} 1 \\ 0 \end{bmatrix}$ can be written as a linear combination of columns of A with coefficient 1 for the first column and 0 for the second. Similarly,

$$[f_A(0,1)] = \begin{bmatrix} 2 & 3 \\ -1 & 4 \\ 0 & 1 \end{bmatrix} \begin{bmatrix} 0 \\ 1 \end{bmatrix} = \begin{bmatrix} 3 \\ 4 \\ 1 \end{bmatrix};$$

the image of the second basis vector is the second column of A.

Next, consider

$$[f_A(x_1,x_2)] = \begin{bmatrix} 2 & 3 \\ -1 & 4 \\ 0 & 1 \end{bmatrix} \begin{bmatrix} x_1 \\ x_2 \end{bmatrix} = \begin{bmatrix} 2x_1 + 3x_2 \\ -1x_1 + 4x_2 \\ 0x_1 + 1x_2 \end{bmatrix}$$
$$= x_1 \begin{bmatrix} 2 \\ -1 \\ 0 \end{bmatrix} + x_2 \begin{bmatrix} 3 \\ 4 \\ 1 \end{bmatrix}.$$

Again notice that, by block multiplication, the image of (x_1, x_2) is a sum of scalar multiples of the columns of A, or a **linear combination of the columns** of A, with x_1 and x_2 as coefficients (and remember that the columns are the images of the standard basis vectors.) We may expect results similar to these for general matrix mappings.

Now rewrite the image column matrix as a vector. Then the equation describing the mapping is

$$f_A(x_1, x_2) = (2x_1 + 3x_2, -1x_1 + 4x_2, 0x_1 + 1x_2).$$

Observe that in the first component $2x_1+3x_2$ of the image vector the coefficients of x_1 and x_2 are given by the first row of A, in the second component $-1x_1+4x_2$ by the second row of A, and in the third component $0x_1 + 1x_2$ by the third row of A. Again, it should be clear from the way that matrix multiplication works that a similar pattern will hold for any matrix mapping.

The next theorem generalizes the patterns we have just seen.

THEOREM 1. Let $\mathbf{e}_1, \mathbf{e}_2, \ldots, \mathbf{e}_n$ be the standard basis vectors of $\mathbb{R}^n$, and let A be an m by n matrix with $f_A : \mathbb{R}^n \to \mathbb{R}^m$, the corresponding matrix mapping. Then,

(i) $[f_A(\mathbf{e}_k)] = [a_{\downarrow k}]$ (the kth column of A is the image of the kth basis vector of $\mathbb{R}^n$ under f_A);

(ii) $[f_A(\mathbf{x})] = A[\mathbf{x}] = x_1[a_{\downarrow 1}] + x_2[a_{\downarrow 2}] + \cdots + x_n[a_{\downarrow n}]$
$= x_1 A[\mathbf{e}_1] + x_2 A[\mathbf{e}_2] + \cdots + x_n A[\mathbf{e}_n]$
(the image of $\mathbf{x}$ is the linear combination of the columns of A with the x_k as coefficients);

(iii) $f_A(\mathbf{x}) = (a_{1\to} \cdot \mathbf{x},\ a_{2\to} \cdot \mathbf{x},\ \ldots,\ a_{m\to} \cdot \mathbf{x})$ (the coefficients of the x_k in the jth component of $f_A(\mathbf{x})$ are given by the jth row of A).

Proof. (i) By block multiplication, since $\mathbf{e}_k$ has a 1 as its kth entry,

$$[f_A(\mathbf{e}_k)] = A[\mathbf{e}_k] = \begin{bmatrix} [a_{\downarrow 1}] & [a_{\downarrow 2}] & \ldots & [a_{\downarrow n}] \end{bmatrix} \begin{bmatrix} 0 \\ \vdots \\ 0 \\ 1 \\ 0 \\ \vdots \\ 0 \end{bmatrix} = [a_{\downarrow k}].$$

(ii) By block multiplication, and by part (i),

$$\begin{aligned}[f_A(\mathbf{x})] &= A[\mathbf{x}] = x_1[a_{\downarrow 1}] + x_2[a_{\downarrow 2}] + \cdots + x_n[a_{\downarrow n}] \\ &= x_1 A[\mathbf{e}_1] + x_2 A[\mathbf{e}_2] + \cdots + x_n A[\mathbf{e}_n] \\ &= x_1[f_A(\mathbf{e}_1)] + x_2[f_A(\mathbf{e}_2)] + \cdots + x_n[f_A(\mathbf{e}_n)].\end{aligned}$$

(iii) Partition A into blocks consisting of rows and treat $[\mathbf{x}]$ as one block:

$$[f_A(\mathbf{x})] = \begin{bmatrix} a_{1\rightarrow} \\ a_{2\rightarrow} \\ \vdots \\ a_{m\rightarrow} \end{bmatrix} [\mathbf{x}] = \begin{bmatrix} a_{1\rightarrow} \cdot \mathbf{x} \\ a_{2\rightarrow} \cdot \mathbf{x} \\ \vdots \\ a_{m\rightarrow} \cdot \mathbf{x} \end{bmatrix}.$$

Write these column matrices as m-vectors and the proof is complete. ■

Linearity of Matrix Mappings

In Section 1–3 we met the linearity properties. A function $L : \mathbb{R}^n \rightarrow \mathbb{R}^m$ satisfies these properties if:

(1) for any $\mathbf{x}$ and $\mathbf{y}$ in $\mathbb{R}^n$, $L(\mathbf{x}+\mathbf{y}) = L(\mathbf{x}) + L(\mathbf{y})$;

(2) for any $\mathbf{x}$ in $\mathbb{R}^n$ and for any k in $\mathbb{R}$, $L(k\mathbf{x}) = kL(\mathbf{x})$.

Property (1) is sometimes described by saying that "L **respects** (or **preserves**) addition"; similarly, "L **respects** (or **preserves**) multiplication by scalars". Notice that the addition $\mathbf{x}+\mathbf{y}$ takes place in $\mathbb{R}^n$, while the addition $L(\mathbf{x}) + L(\mathbf{y})$ takes place in $\mathbb{R}^m$.

Projection mappings were our first examples of mappings with these properties (in Section 1–3). It is an important fact, easy to prove, that **every matrix mapping satisfies the linearity properties.** To see this, suppose that A is a matrix and that f_A is the corresponding matrix mapping. Then for any $\mathbf{x}$ and $\mathbf{y}$ in the domain, and any real number k, by the properties of matrix multiplication:

(1) $[f_A(\mathbf{x}+\mathbf{y})] = A[(\mathbf{x}+\mathbf{y})] = A[\mathbf{x}] + A[\mathbf{y}] = [f_A(\mathbf{x})] + [f_A(\mathbf{y})]$, and clearly this implies the corresponding statement for vectors, so f_A respects addition;

(2) also, $[f_A(k\mathbf{x})] = A[(k\mathbf{x})] = kA[\mathbf{x}] = k[f_A(\mathbf{x})]$, so f_A respects multiplication by scalars, and f_A satisfies both linearity properties.

Linear Mappings (Linear Transformations)

Since it is clumsy to have to say "g satisfies the linearity properties" every time such a map arises, it is convenient to make a definition.

A mapping (or function) $L : \mathbb{R}^n \rightarrow \mathbb{R}^m$ is a **linear mapping** (or a **linear transformation**) if it satisfies the linearity properties:

(1) for any $\mathbf{x}$ and $\mathbf{y}$ in $\mathbb{R}^n$, $L(\mathbf{x}+\mathbf{y}) = L(\mathbf{x}) + L(\mathbf{y})$;

(2) for any $\mathbf{x}$ in $\mathbb{R}^n$ and for any k in $\mathbb{R}$, $L(k\mathbf{x}) = kL(\mathbf{x})$.

The discussion above shows that **every matrix mapping is a linear mapping.**

REMARKS

(1) "Linear transformation" means exactly the same thing as "linear mapping". Some people prefer one or the other, but we shall use both.

(2) It is possible to combine the two linearity properties into one statement: $L : \mathbb{R}^n \rightarrow \mathbb{R}^m$ is a linear mapping if, for every $\mathbf{x}$ and $\mathbf{y}$ in the domain, and for any scalars a and b,

$$L(a\mathbf{x} + b\mathbf{y}) = aL(\mathbf{x}) + bL(\mathbf{y}).$$

You should satisfy yourself that the content of this one statement is equivalent to the content of the two stated above. Since an expression of the form $a\mathbf{x}+b\mathbf{y}$ is called a linear combination of $\mathbf{x}$ and $\mathbf{y}$, we say that a linear transformation respects (or preserves) linear combinations.

(3) Note that statement (ii) in Theorem 1 about matrix mappings is an extended version of the combined linearity property stated in Remark (2).

(4) For the time being, we have defined only linear mappings whose domain is some $\mathbb{R}^n$ and whose codomain is some $\mathbb{R}^m$. In Chapter 4, it will be possible to consider other sets as domain or codomain. In fact, one of the major questions of that chapter is: What sets **can** be the domain or codomain for linear mappings? (You might think about it now—what operations must be possible in the sets?)

(5) So far, our only examples of linear mappings are projections and matrix mappings. We shall see some others in the next section.

(6) There are lots of mappings that are not linear.

EXAMPLE 12

Determine whether the following mappings are linear.

(a) $f : \mathbb{R}^3 \to \mathbb{R}$ defined by $f(\mathbf{x}) = \|\mathbf{x}\|$;

(b) $g : \mathbb{R}^2 \to \mathbb{R}^2$ defined by $g(x_1, x_2) = (x_1^2, x_1x_2)$.

SOLUTION: (a) We must test whether f respects addition and multiplication by scalars for any $\mathbf{x}$ and $\mathbf{y}$ in $\mathbb{R}^3$. So, consider

$$f(\mathbf{x}+\mathbf{y}) = \|\mathbf{x}+\mathbf{y}\| \quad \text{and} \quad f(\mathbf{x}) + f(\mathbf{y}) = \|\mathbf{x}\| + \|\mathbf{y}\|.$$

Are these equal? By the triangle inequality

$$\|\mathbf{x}+\mathbf{y}\| \leq \|\mathbf{x}\| + \|\mathbf{y}\|,$$

and we expect equality only when one of $\mathbf{x}$, $\mathbf{y}$ is a multiple of the other. So we know the answer is "no, these are not equal, and f does not respect addition." To settle the matter, we give a counterexample: if $\mathbf{x} = (1, 0, 0)$ and $\mathbf{y} = (0, 1, 0)$,

$$f(\mathbf{x}+\mathbf{y}) = \|(1, 1, 0)\| = \sqrt{2},$$

but,

$$f(\mathbf{x}) + f(\mathbf{y}) = \|(1, 0, 0)\| + \|(0, 1, 0)\| = 1 + 1 = 2.$$

Thus, it is not true that $f(\mathbf{x}+\mathbf{y}) = f(\mathbf{x}) + f(\mathbf{y})$ for every pair of vectors $\mathbf{x}$, $\mathbf{y}$ in $\mathbb{R}^3$, so f *is not linear.*

(b) We consider multiplication by scalars first for this example:

$$\begin{aligned} g(k\mathbf{x}) &= g(kx_1, kx_2) \\ &= \left((kx_1)^2, (kx_1)(kx_2)\right) \\ &= (k^2x_1^2, k^2x_1x_2). \end{aligned}$$

On the other hand,

$$kg(\mathbf{x}) = k(x_1^2, x_1x_2) = (kx_1^2, kx_1x_2).$$

Clearly, for most values of k (except for $k = 0$ or 1),

$$g(k\mathbf{x}) \neq kg(\mathbf{k}).$$

Thus, g fails to respect multiplication by scalars, so g *is not linear.*

One failure is enough to disqualify a mapping from being linear, but for practice, check that f does not respect multiplication by scalars (try k negative) and that g does not respect addition.

Is Every Linear Mapping a Matrix Mapping?

Every matrix determines a corresponding linear mapping. It is natural to ask whether every linear mapping determines a corresponding matrix, that is, whether every linear mapping can be represented as a matrix mapping. The answer is yes; for each linear mapping $L : \mathbb{R}^n \to \mathbb{R}^m$, there is such a matrix and it will be denoted $[L]$. Since $[L]$ must satisfy the conclusions of Theorem 1, the following theorem should come as no surprise.

THEOREM 2. Let $L : \mathbb{R}^n \to \mathbb{R}^m$ be a linear mapping, and let $\{\mathbf{e}_1, \mathbf{e}_2, \ldots, \mathbf{e}_n\}$ be the standard basis of $\mathbb{R}^n$. Then L can be represented as a matrix mapping with an m by n matrix $[L]$, such that

$$[L(\mathbf{x})] = [L][\mathbf{x}].$$

$[L]$ is the matrix whose k^{th} **column** is $[L(\mathbf{e}_k)]$, $k = 1, 2, \ldots, n$:

$$[L] = \Big[[L(\mathbf{e}_1] \quad [L(\mathbf{e}_2)] \quad \ldots \quad [L(\mathbf{e}_n)] \Big].$$

Proof. We need to determine an expression for $L(\mathbf{x})$. Since L is linear, for any $\mathbf{x}$ in $\mathbb{R}^n$,

$$L(\mathbf{x}) = L(x_1\mathbf{e}_1 + x_2\mathbf{e}_2 + \cdots + x_n\mathbf{e}_n) = x_1L(\mathbf{e}_1) + x_2L(\mathbf{e}_2) + \cdots + x_nL(\mathbf{e}_n).$$

We want this to look like the result of matrix multiplication, so we write the vectors as column matrices; we use brackets to indicate this column form. Thus, for any $\mathbf{x}$ in $\mathbb{R}^n$,

$$[L(\mathbf{x})] = x_1[L(\mathbf{e}_1)] + x_2[L(\mathbf{e}_2)] + \cdots + x_n[L(\mathbf{e}_n)].$$

From experience with block multiplication, we recognize that the right-hand side of this equation can be rewritten, so that

$$[L(\mathbf{x})] = \Big[[L(\mathbf{e}_1)] \quad [L(\mathbf{e}_2)] \quad \ldots \quad [L(\mathbf{e}_n)] \Big] \begin{bmatrix} x_1 \\ x_2 \\ \vdots \\ x_n \end{bmatrix}.$$

Thus, with $[L]$ defined by

$$[L] = \begin{bmatrix} [L(\mathbf{e}_1)] & [L(\mathbf{e}_2)] & \dots & [L(\mathbf{e}_n)] \end{bmatrix},$$

we have

$$[L(\mathbf{x})] = [L][\mathbf{x}].$$

Since the image $[L(\mathbf{x})]$ is determined as the product of matrix $[L]$ with $[\mathbf{x}]$, we have represented L as a matrix mapping, as required.

Note that $[L]$ is the only matrix that gives the desired answer for every $\mathbf{x}$ in $\mathbb{R}^n$ because of Theorem 2 in Section 3–1. Also note that $[L]$ has n columns because there are n basis vectors in $\mathbb{R}^n$, and $[L]$ has m rows because $L(\mathbf{e}_i)$ is a vector in $\mathbb{R}^m$, so a typical column $[L(\mathbf{e}_i)]$ has m entries.

■

REMARK

This theorem may look just like part of Theorem 1. However, here we start with the linear mapping, whereas in Theorem 1 we started with the matrix.

The matrix $[L]$ determined in this theorem depends on the **standard** basis, so the resulting matrix is sometimes called the **standard matrix** of the linear mapping. Later (in Section 4–6), we shall see that we can use other bases ("bases" is the plural of "basis") and get different matrices corresponding to the same linear mapping.

EXAMPLE 13

Consider the mapping $\mathbf{proj}_{(3,4)} : \mathbb{R}^2 \to \mathbb{R}^2$. It was seen in Section 1–3 that such a map satisfies the linearity properties. Determine the matrix $[\mathbf{proj}_{(3,4)}]$.

SOLUTION: By Theorem 2, the first column of the matrix is the image under $\mathbf{proj}_{(3,4)}$ of the first standard basis vector, and this can be computed by the methods of Section 1–2:

$$\begin{aligned} \mathbf{proj}_{(3,4)}(1,0) &= \big((1,0)\cdot(3,4)/25\big)\,(3,4) \\ &= 3/25(3,4) = (9/25, 12/25). \end{aligned}$$

Similarly, the second column is the image of the second basis vector:

$$\mathbf{proj}_{(3,4)}(0,1) = \big((0,1)\cdot(3,4)/25\big)\,(3,4) = (12/25, 16/25).$$

Hence, the standard matrix of this linear mapping is

$$[\mathbf{proj}_{(3,4)}] = \begin{bmatrix} 9/25 & 12/25 \\ 12/25 & 16/25 \end{bmatrix}$$

and

$$[\mathbf{proj}_{(3,4)}(\mathbf{x})] = \begin{bmatrix} 9/25 & 12/25 \\ 12/25 & 16/25 \end{bmatrix} [\mathbf{x}].$$

Theorem 2 is a converse for parts (i) and (ii) of Theorem 1. The next theorem is a converse for part (iii) of Theorem 2.

THEOREM 3. The mapping $L : \mathbb{R}^n \to \mathbb{R}^m$ defined by $L(\mathbf{x}) = (\mathbf{r}_1 \cdot \mathbf{x}, \mathbf{r}_2 \cdot \mathbf{x}, \ldots, \mathbf{r}_m \cdot \mathbf{x})$, for some vectors $\mathbf{r}_1, \mathbf{r}_2, \ldots, \mathbf{r}_m$ in $\mathbb{R}^n$, is linear and it is the matrix mapping of the matrix whose rows are $\mathbf{r}_1, \mathbf{r}_2, \ldots, \mathbf{r}_m$:

$$[L] = \begin{bmatrix} \mathbf{r}_1 \\ \mathbf{r}_2 \\ \vdots \\ \mathbf{r}_m \end{bmatrix}.$$

Proof. Suppose that $L(\mathbf{x}) = (\mathbf{r}_1 \cdot \mathbf{x}, \mathbf{r}_2 \cdot \mathbf{x}, \ldots, \mathbf{r}_m \cdot \mathbf{x})$. Then it is easy to check that L satisfies the linearity properties, and it follows that $[L(\mathbf{x})] = \begin{bmatrix} \mathbf{r}_1 \\ \mathbf{r}_2 \\ \vdots \\ \mathbf{r}_m \end{bmatrix} [\mathbf{x}]$,

so $[L] = \begin{bmatrix} \mathbf{r}_1 \\ \mathbf{r}_2 \\ \vdots \\ \mathbf{r}_m \end{bmatrix}$ (by Theorem 2 of Section 3–1). ■

EXAMPLE 14

Use Theorem 3 to determine the standard matrix of $\mathbf{proj}_{(3,4)}$.

SOLUTION: The rows of the matrix $[\mathbf{proj}_{(3,4)}]$ can be read from the coefficients in the components of $\mathbf{proj}_{(3,4)}(\mathbf{x})$, so we compute this image vector:

$$\begin{aligned}\mathbf{proj}_{(3,4)}(x_1, x_2) &= \Big((x_1, x_2) \cdot (3, 4)/25\Big)(3, 4)\\ &= (1/25)(3x_1 + 4x_2)(3, 4)\\ &= \Big((9/25)x_1 + (12/25)x_2, (12/25)x_1 + (16/25)x_2\Big).\end{aligned}$$

The coefficients in the first component form the first row, and the coefficients in the second component form the second row, so the matrix $[\mathbf{proj}_{(3,4)}]$ must be $\begin{bmatrix} 9/25 & 12/25 \\ 12/25 & 16/25 \end{bmatrix}$, in agreement with the answer in Example 13.

The method of Example 13 (based on Theorem 2) is often a better tool for answering questions about linear mappings than the method of Example 14 (based on Theorem 3).

EXAMPLE 15

(a) The mapping $G : \mathbb{R}^3 \to \mathbb{R}^2$ defined by $G(x_1, x_2, x_3) = (x_1, x_2)$ is linear, by Theorem 3. However, let us check directly whether G respects addition and multiplication by scalars. Consider

$$\begin{aligned}G\Big((x_1, x_2, x_3) + (y_1, y_2, y_3)\Big) &= G(x_1 + y_1, x_2 + y_2, x_3 + y_3)\\ &= (x_1 + y_1, x_2 + y_2)\\ &= G(\mathbf{x}) + G(\mathbf{y})\end{aligned}$$

and

$$G(k\mathbf{x}) = (kx_1, kx_2) = kG(\mathbf{x}),$$

so G is linear.

The standard matrix of the mapping, by Theorem 2, is

$$\begin{aligned}[G] &= \Big[[G(1,0,0)] \quad [G(0,1,0)] \quad [G(0,0,1)]\Big]\\ &= \Big[[(1,0)] \quad [(0,1)] \quad [(0,0)]\Big] = \begin{bmatrix} 1 & 0 & 0 \\ 0 & 1 & 0 \end{bmatrix}.\end{aligned}$$

We could also use Theorem 3 to find $[G]$.

The mapping $H : \mathbb{R}^4 \to \mathbb{R}^2$ defined by $H(x_1, x_2, x_3, x_4) = (x_1, x_3)$ is also easily seen to be linear. The matrix $[H]$ is $\begin{bmatrix} 1 & 0 & 0 & 0 \\ 0 & 0 & 1 & 0 \end{bmatrix}$.

(b) Consider the **injection** mapping $inj : \mathbb{R}^2 \to \mathbb{R}^3$ defined by $inj(x_1, x_2) = (x_1, x_2, 0)$. It is easy to check that this map is linear and that the matrix $[\,inj\,] = \begin{bmatrix} 1 & 0 \\ 0 & 1 \\ 0 & 0 \end{bmatrix}$. Injections can be defined from $\mathbb{R}^n$ into $\mathbb{R}^m$ as long as $n \leq m$.

Linear Transformations from $\mathbb{R}^n$ to Itself

It often happens that the domain and the codomain for a linear mapping are the same. For example, $\mathbf{proj_a}$ and $\mathbf{perp_a}$ in Section 1–2 were maps from $\mathbb{R}^2$ to $\mathbb{R}^2$, or from $\mathbb{R}^3$ to $\mathbb{R}^3$, or from $\mathbb{R}^n$ to $\mathbb{R}^n$. The geometrical linear transformations we shall meet in Section 3–3 all map from $\mathbb{R}^n$ to $\mathbb{R}^n$. In cases such as these where the codomain is the same as the domain, linear transformations are often called **linear operators**. A matrix mapping f_A can be a linear transformation from $\mathbb{R}^n$ to itself only if the matrix A is **square**, so that $\mathbf{x}$ and $f_A(\mathbf{x})$ are vectors in the same $\mathbb{R}^n$.

Figure 3–2–2 illustrates the way in which a linear transformation from $\mathbb{R}^n$ to $\mathbb{R}^n$ is pictured: $\mathbf{x}$ and $L(\mathbf{x})$ are shown as vectors in the same $\mathbb{R}^n$. (However, there are examples with A square where it is helpful to think of the domain and the codomain as distinct copies of the same $\mathbb{R}^n$; in such cases, Figure 3–2–1 is still appropriate.)

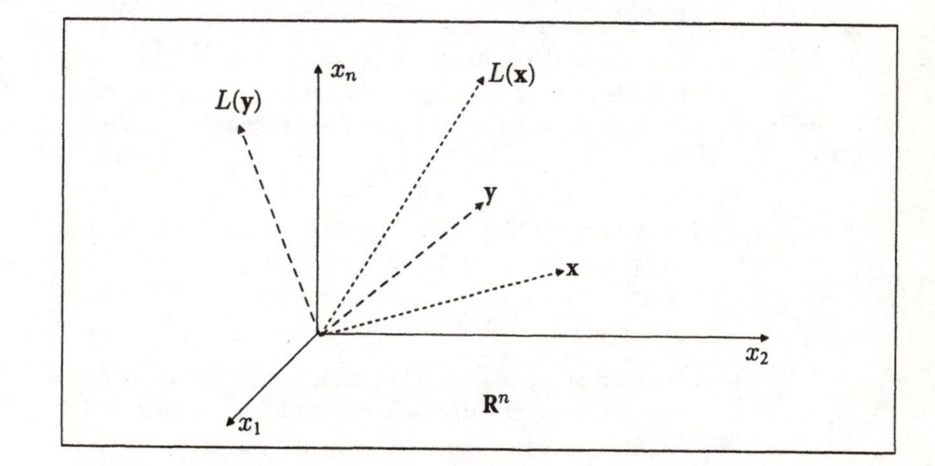

FIGURE 3-2-2. *Picturing the linear transformation $L : \mathbb{R}^n \to \mathbb{R}^n$.*

EXERCISES 3–2

A1 Let $A = \begin{bmatrix} -2 & 3 \\ 3 & 0 \\ 1 & 5 \\ 4 & -6 \end{bmatrix}$, and let f_A be the corresponding matrix mapping.

(a) Determine the domain and codomain of f_A.
(b) Determine $f_A(2,-5)$ and $f_A(-3,4)$.
(c) Write the standard basis vectors in the domain, and determine their images under f_A.
(d) Determine $f_A(\mathbf{x})$.
(e) Check your answers to (c) and (d) by using Theorems 2 and 3 to write the matrix of the linear mapping f_A. (If you do not get A, you have a problem.)

A2 Let $A = \begin{bmatrix} 1 & 2 & -3 & 0 \\ 2 & -1 & 0 & 3 \\ 1 & 0 & 2 & -1 \end{bmatrix}$, and let f_A be the corresponding matrix mapping.

(a) Determine the domain and codomain of f_A.
(b) Determine $f_A(2,-2,3,1)$ and $f_A(-3,1,4,2)$.
(c) Write the standard basis vectors in the domain, and determine their images under f_A.
(d) Determine $f_A(\mathbf{x})$.
(e) Check your answers to (c) and (d) by using Theorems 2 and 3 to write the matrix of the linear mapping f_A. (If you do not get A, you have a problem.)

A3 For each of the following mappings, state the domain and the codomain. Determine whether the mapping is linear by using the definition of linearity: either prove it is linear or give a counterexample to show why it cannot be linear.

(a) $f(x_1, x_2) = (\sin x_1, e^{x_2})$
(b) $g(x_1, x_2) = (2x_1 + 3x_2, x_1 - x_2)$
(c) $h(x_1, x_2) = (2x_1 + 3x_2, x_1 - x_2, (x_1 x_2))$
(d) $k(x_1, x_2, x_3) = (x_1 + x_2, 0, x_2 - x_3)$
(e) $l(x_1, x_2, x_3) = (x_2, |x_1|)$

A4 For each of the following linear mappings, determine the domain, the codomain, and the standard matrix of the mapping.

(a) $L(x_1, x_2, x_3) = (2x_1 - 3x_2 + x_3, x_2 - 5x_3)$
(b) $K(x_1, x_2, x_3, x_4) = (5x_1 + 3x_3 - x_4, x_2 - 7x_3 + 3x_4)$
(c) $M(x_1, x_2, x_3, x_4) = (x_1 - x_3 + x_4, x_1 + 2x_2 - x_3 - 3x_4,$
$x_2 + x_3, x_1 - x_2 + x_3 - x_4)$

A5 Determine the matrix $[\mathbf{proj}_{(-2,1)}]$.

A6 Determine the matrix $[\mathbf{perp}_{(1,4)}]$.

A7 Determine the matrix $[\mathbf{proj}_{(2,2,-1)}]$.

B1 Let $B = \begin{bmatrix} 1 & 4 & -2 \\ -5 & 3 & 1 \end{bmatrix}$, and let f_B be the corresponding matrix mapping.

(a) Determine the domain and codomain of f_B.

(b) Determine $f_B(3, 4, -5)$ and $f_B(-2, 1, -4)$.

(c) Write the standard basis vectors in the domain, and determine their images under f_B.

(d) Determine $f_B(\mathbf{x})$.

B2 Let $B = \begin{bmatrix} 2 & 1 & -0 \\ 0 & 2 & -3 \\ 5 & 7 & 9 \\ 2 & 4 & 8 \end{bmatrix}$, and let f_B be the corresponding matrix mapping.

(a) Determine the domain and codomain of f_B.

(b) Determine $f_B(-4, 2, 1)$ and $f_B(3, -3, 2)$.

(c) Write the standard basis vectors in the domain, and determine their images under f_B.

(d) Determine $f_B(\mathbf{x})$.

B3 For each of the following mappings, state the domain and the codomain. Determine whether the mapping is linear by using the definition of linearity: either prove it is linear or give a counterexample to show why it cannot be linear.

(a) $f(x_1, x_2, x_3) = (2x_2, x_1 - x_3)$

(b) $g(x_1, x_2) = (\cos x_2, x_1 x_2^3)$

(c) $h(x_1, x_2, x_3) = (0, 0, x_1 + x_2 + x_3)$

B4 For each of the following linear mappings, determine the domain, the codomain, and the standard matrix of the mapping.

(a) $L(x_1, x_2, x_3) = (2x_1 - 3x_2, x_2, 4x_1 - 5x_2)$

(b) $M(x_1, x_2, x_3, x_4) = (2x_1 - x_3 + 3x_4, -x_1 - 2x_2 + 2x_3 + x_4, 3x_2 + x_3)$

(c) $N(x_1, x_2, x_3) = (x_3 - x_1, 0, 5x_1 + x_2)$

B5 Determine the matrix $[\mathbf{proj}_{(1,-3)}]$.

B6 Determine the matrix $[\mathbf{perp}_{(2,-1)}]$.

B7 Determine the matrix $[\mathbf{proj}_{(1,-1,4)}]$.

B8 Determine the matrix $[\mathbf{perp}_{(1,-2,2)}]$.

CONCEPTUAL EXERCISES

***D1** Let $\mathbf{a}$ be some fixed vector in $\mathbb{R}^n$, and define a mapping $\mathrm{DOT}_{\mathbf{a}}$ by $\mathrm{DOT}_{\mathbf{a}}(\mathbf{x}) = \mathbf{a} \cdot \mathbf{x}$. Verify that $\mathrm{DOT}_{\mathbf{a}}$ is a linear mapping. What is its codomain? Verify that the matrix of this linear mapping can be written as $[\mathbf{a}]^T$.

D2 If $\mathbf{u}$ is a unit vector, show that $[\mathbf{proj}_{\mathbf{u}}] = [\mathbf{u}][\mathbf{u}]^T$. Note that $[\mathbf{u}][\mathbf{u}]^T$ is an

n by n matrix. (Hint: Theorem 2 of Section 3–1.)

D3 Let $\mathbf{a}$ be some fixed vector in $\mathbb{R}^3$, and define a mapping $\text{CROSS}_{\mathbf{a}}$ by $\text{CROSS}_{\mathbf{a}}(\mathbf{x}) = \mathbf{a} \times \mathbf{x}$. Verify that $\text{CROSS}_{\mathbf{a}}$ is a linear mapping, and determine its codomain and its standard matrix.

SECTION

3–3 Geometrical Transformations; Composition of Linear Mappings

Geometrical transformations have always been of great interest to mathematicians. They also have many important applications. Physicists and engineers often rely on simple geometrical transformations to gain understanding of the properties of the materials or structures they wish to examine. For example, structural engineers use stretches, shears, and rotations to understand the deformation of materials. Material scientists use rotations and reflections to analyze crystals and other fine structures. Many of these simple geometrical transformations in $\mathbb{R}^2$ and $\mathbb{R}^3$ are linear. The following is a brief partial catalog of some of these transformations, and (if they are linear) their matrix representations. ($\mathbf{proj_a}$ and $\mathbf{perp_a}$ belong to the list of geometrical transformations too, but they were discussed in Sections 1–3 and 3–2.)

Translations in the Plane

A typical "translation by $\mathbf{a}$" $T : \mathbb{R}^2 \to \mathbb{R}^2$ is defined by $T(\mathbf{x}) = \mathbf{x} + \mathbf{a}$. (See Figure 3–3-1.) Since $T(k\mathbf{x}) = k\mathbf{x} + \mathbf{a} \neq k(T(\mathbf{x}))$, this is **not a linear transformation** of $\mathbb{R}^2$. Translations in higher dimensional spaces also fail to be linear. Accordingly, there is no matrix associated to these mappings.

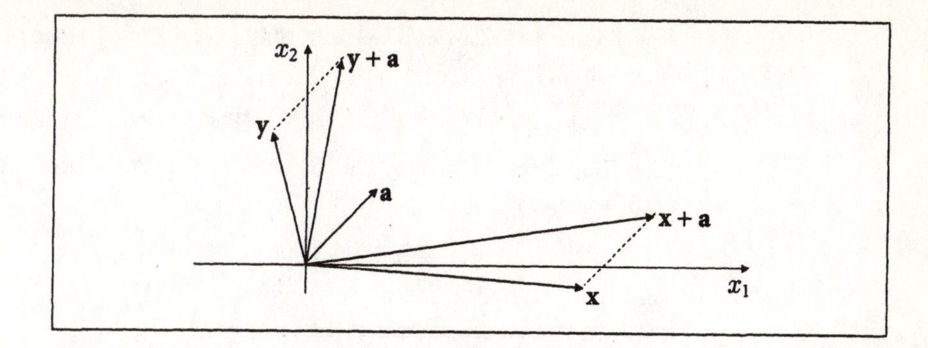

FIGURE 3-3-1. *Translation by* **a** *is not linear.*

Rotations in the Plane

$R_\theta : \mathbb{R}^2 \to \mathbb{R}^2$ is defined to be the transformation that rotates $\mathbf{x}$ counterclockwise through angle θ to the image $R_\theta(\mathbf{x})$ (Figure 3–3–2). Is R_θ linear? Since rotation does not change lengths, we get the same result if we first multiply a vector $\mathbf{x}$ by a scalar k and then rotate through angle θ, or if we first rotate through θ then multiply by k. Thus, $R_\theta(k\mathbf{x}) = kR_\theta(\mathbf{x})$ for any k and any $\mathbf{x}$. Since the **shape** of a triangle is not altered by rotation, the picture for the parallelogram rule of addition should be unchanged under rotation, so $R_\theta(\mathbf{x}+\mathbf{y}) = R_\theta(\mathbf{x}) + R_\theta(\mathbf{y})$, and R_θ is linear. A more formal proof is obtained by showing that rotation can be represented as a matrix linear mapping.

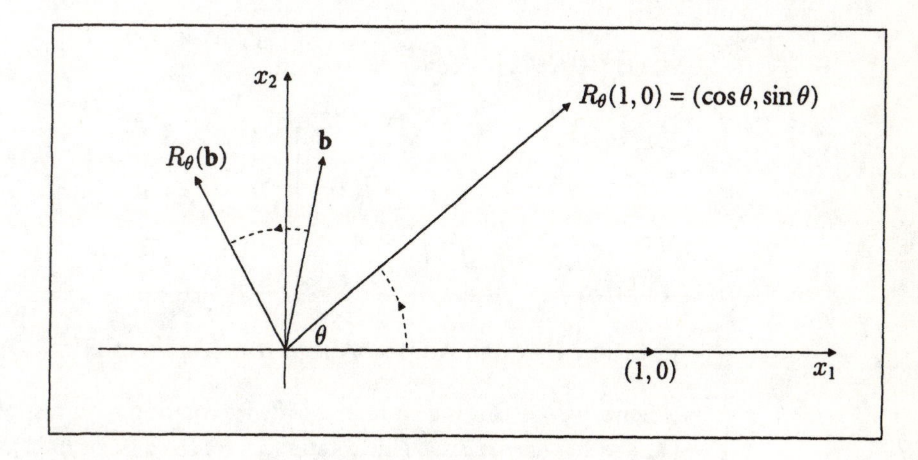

FIGURE 3-3-2. *Counterclockwise rotation through angle θ in the plane.*

From the definition of the cosine and sine of an angle,

$$R_\theta(1,0) = (\cos\theta, \sin\theta).$$

(See Figure 3–3–2, where we note that $R_\theta(1,0)$ is a vector of length 1.) By trigonometry

$$R_\theta(0,1) = (-\sin\theta, \cos\theta).$$

(You should draw a picture and check.) Since these vectors give the columns of the matrix of R_θ,

$$[R_\theta] = \begin{bmatrix} \cos\theta & -\sin\theta \\ \sin\theta & \cos\theta \end{bmatrix}.$$

It follows from a calculation of the product $[R_\theta][\mathbf{x}]$ that

$$R_\theta(\mathbf{x}) = (x_1\cos\theta - x_2\sin\theta, x_1\sin\theta + x_2\cos\theta).$$

It is an exercise in trigonometric identities to check that $[R_\theta][R_\alpha] = [R_{\theta+\alpha}]$, which corresponds to the fact that a rotation through angle α followed by a rotation through angle θ is equal to a single rotation through angle $\theta + \alpha$. Notice that this means that these rotation matrices commute with each other: $[R_\theta][R_\alpha] = [R_\alpha][R_\theta]$. Remember that in general matrix multiplication is not commutative.

EXAMPLE 16

What is the matrix of rotation of $\mathbb{R}^2$ through angle $2\pi/3$?

SOLUTION: Since $\cos\frac{2\pi}{3} = \frac{-1}{2}$, $\sin\frac{2\pi}{3} = \frac{\sqrt{3}}{2}$,

$$\left[R_{\frac{2\pi}{3}}\right] = \begin{bmatrix} \frac{-1}{2} & \frac{-\sqrt{3}}{2} \\ \frac{\sqrt{3}}{2} & \frac{-1}{2} \end{bmatrix}.$$

Rotation Through Angle θ About the x_3-axis in $\mathbb{R}^3$

(Figure 3–3–3: this is a counter clockwise rotation with respect to the right-handed standard basis.) This rotation leaves the x_3-axis unchanged, so that if the transformation is denoted R, then $R(0,0,1) = (0,0,1)$. Together with the previous case, this tells us that the matrix of this rotation must be $\begin{bmatrix} \cos\theta & -\sin\theta & 0 \\ \sin\theta & \cos\theta & 0 \\ 0 & 0 & 1 \end{bmatrix}$. These ideas can be adapted to give rotations about the other coordinate axes. It is natural to wonder whether it is possible to determine the matrix of a rotation about an arbitrary axis in $\mathbb{R}^3$; we shall see how to do this in Chapter 7.

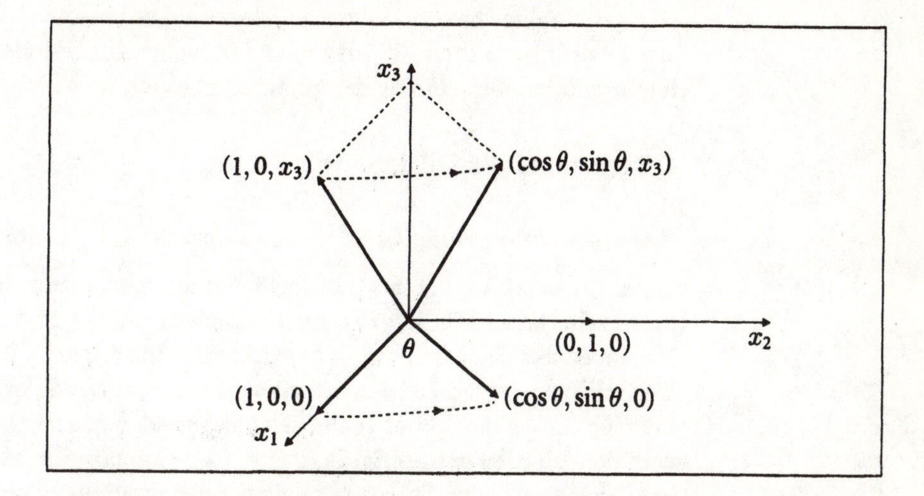

FIGURE 3-3-3. *A right-handed counterclockwise rotation about the x_3-axis in $\mathbb{R}^3$.*

Stretches

Imagine that all lengths in the x_1-direction in the plane are stretched by a scalar factor $k > 0$ while lengths in the x_2-direction are left unchanged (Figure 3–3–4).

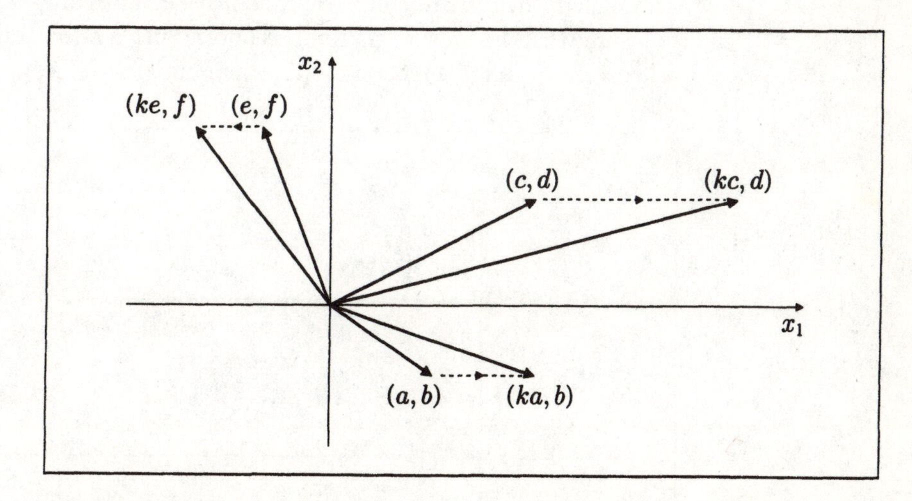

FIGURE 3-3-4. *A stretch by factor k in the x_1-direction.*

This linear transformation, called a "stretch by factor k in the x_1-direction", has matrix $\begin{bmatrix} k & 0 \\ 0 & 1 \end{bmatrix}$. (If $k < 1$, you might prefer to call this a "shrink", but such terminology has not caught on.) It should be obvious that stretches can also be defined in the x_2-direction, and in higher dimensions. Stretches are important in understanding deformations of solids.

Contractions and Dilatations

If a linear transformation $T : \mathbb{R}^2 \to \mathbb{R}^2$ has matrix $\begin{bmatrix} k & 0 \\ 0 & k \end{bmatrix}$ with $k > 0$, then for any $\mathbf{x}$, $T(\mathbf{x}) = k\mathbf{x}$, so that this transformation stretches vectors in all directions by the same factor. Thus, for example, a circle of radius 1 centred at the origin is mapped to a circle of radius k also centred at the origin. If $0 < k < 1$, such a transformation is called a **contraction**; if $k > 1$, it is a **dilatation**. If $k < 0$, the map can be thought of as the **composition** of a contraction or dilatation with the **reflection in the origin** ($R(\mathbf{x}) = -\mathbf{x}$). Composition of mappings will be discussed shortly. There are obvious generalizations of contractions and dilatations to higher dimensions.

Shears

Sometimes a force applied to a rectangle will cause it to deform into a parallelogram as shown in Figure 3–3–5. The change can be described by the transformation $S : \mathbb{R}^2 \to \mathbb{R}^2$ such that $S(1,0) = (1,0)$ and $S(0,1) = (s,1)$. Although the deformation of a real solid may be more complicated, it is usual to assume that the transformation S is linear. Such a linear transformation is called a shear (in the direction of x_1, by amount s).

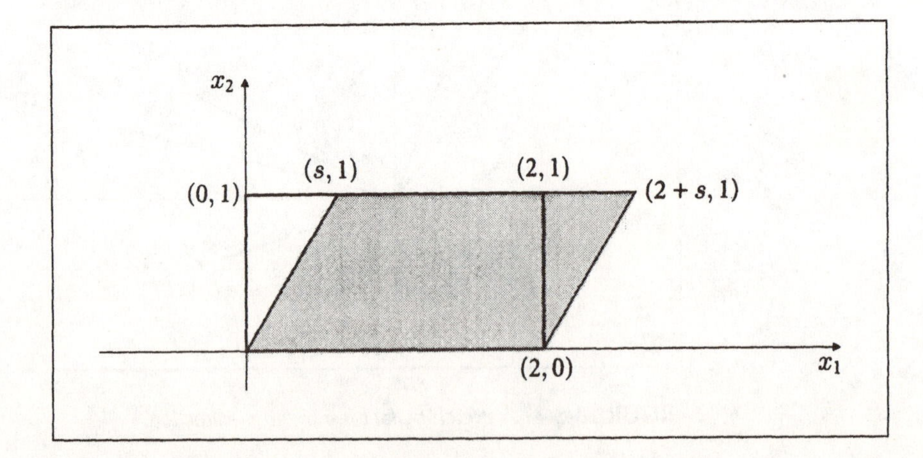

FIGURE 3-3-5. *A shear in the direction of x_1 by amount s.*

Since the action of S on the standard basis vectors is known, its matrix can be written $[S] = \begin{bmatrix} 1 & s \\ 0 & 1 \end{bmatrix}$.

Reflections in Coordinate Axes in $\mathbb{R}^2$ or Coordinate Planes in $\mathbb{R}^3$

Let $R : \mathbb{R}^2 \to \mathbb{R}^2$ be reflection in the x_1-axis (Figure 3–3–6). Then each vector corresponding to a point above the axis is mapped by R to the mirror image vector below. Hence, $R(x_1, x_2) = (x_1, -x_2)$, and it follows that the matrix of this reflection is $\begin{bmatrix} 1 & 0 \\ 0 & -1 \end{bmatrix}$. Similarly, reflection in the x_2-axis has matrix $\begin{bmatrix} -1 & 0 \\ 0 & 1 \end{bmatrix}$.

Next, consider the reflection $T : \mathbb{R}^3 \to \mathbb{R}^3$ that reflects in the x_1x_2-plane (that is, the plane $x_3 = 0$). Points above the plane are reflected to points below the plane. The matrix of this reflection is $\begin{bmatrix} 1 & 0 & 0 \\ 0 & 1 & 0 \\ 0 & 0 & -1 \end{bmatrix}$. You should write the matrices for the reflections in the other two coordinate planes in $\mathbb{R}^3$.

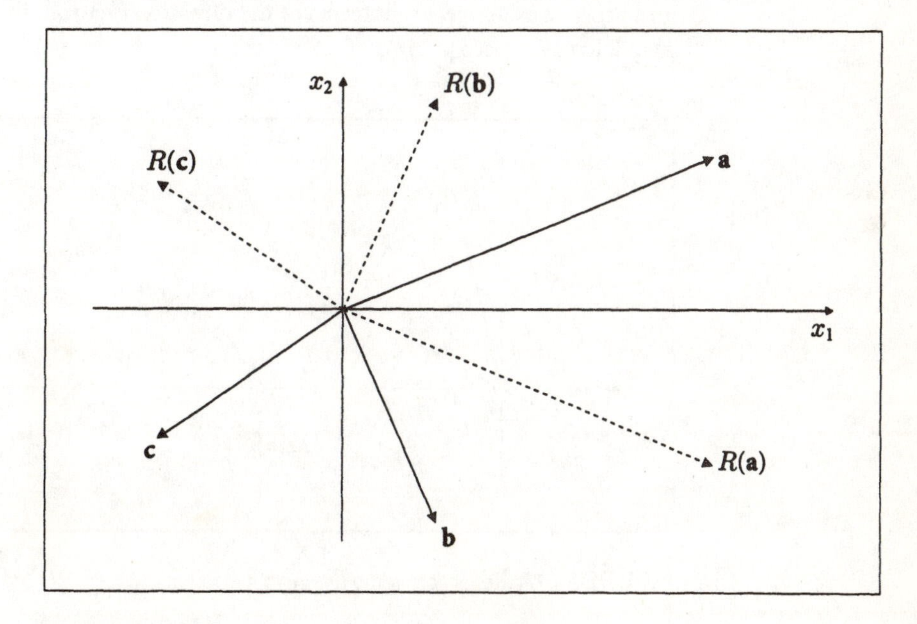

FIGURE 3-3-6. *R is reflection in the x_1-axis (in $\mathbb{R}^2$).*

General Reflections

We consider only reflections in (or "across") lines in $\mathbb{R}^2$ or planes in $\mathbb{R}^3$ that pass through the origin. (Reflections in lines or planes not containing the origin involve translations as well as linear mappings.)

Consider the plane in $\mathbb{R}^3$ with equation $\mathbf{n} \cdot \mathbf{x} = 0$. Since reflection turns out to be related to $\mathbf{proj_n}$, **reflection in the plane with normal vector n** will be denoted $\mathbf{refl_n}$. If a point $\mathbf{a}$ does not lie on the plane, its image under $\mathbf{refl_n}$ is a point on the opposite side of the plane, lying on a line through $\mathbf{a}$ perpendicular to the plane of reflection, at the same distance from the plane as $\mathbf{a}$. (See Figure 3-3-7, which shows reflection in a line.) From the figure, we see that

$$\mathbf{refl_n}(\mathbf{a}) = \mathbf{a} - 2\mathbf{proj_n}(\mathbf{a}).$$

Since $\mathbf{proj_n}$ is linear, it is easy to see that $\mathbf{refl_n}$ is also linear.

It is important to notice that $\mathbf{refl_n}$ is reflection with **normal vector n.** The calculations for reflection in a line in $\mathbb{R}^2$ are similar to those for a plane provided that the equation of the line is given in **scalar** form $\mathbf{n} \cdot \mathbf{x} = 0$. If the **vector** equation of the line is given, $\mathbf{x} = t\mathbf{d}$, then either we must find a normal vector $\mathbf{n}$ and proceed as above, or, in terms of the **direction vector d**, the reflection will map $\mathbf{a}$ to $(\mathbf{a} - 2\mathbf{perp_d a})$.

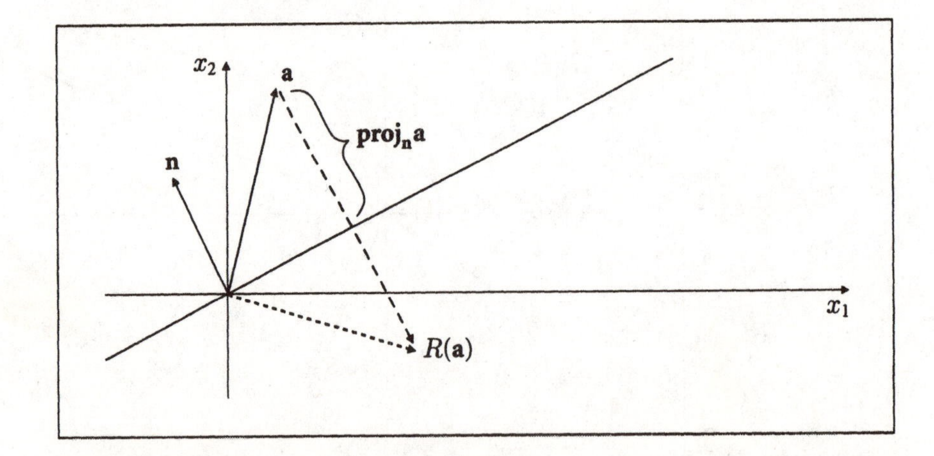

FIGURE 3-3-7. *Reflection in the line with normal* **n**.

Before determining the matrix for a particular choice of $\mathbf{n}$, we pause to remark on the form of the equation for $\mathbf{refl_n}(\mathbf{a})$. Introduce the notation Id to denote the **identity transformation** (in this case, in $\mathbb{R}^3$, but the notation will be used to denote the identity transformation in any space). Then $Id(\mathbf{a}) = \mathbf{a}$ for all $\mathbf{a}$ in

$\mathbb{R}^3$. The equation for $\mathbf{refl_n(a)}$ can then be written

$$\mathbf{refl_n(a)} = Id(\mathbf{a}) - 2\mathbf{proj_n(a)}.$$

In this context, it seems very natural to introduce the ideas of a **sum of linear transformations, and a scalar multiple of a linear transformation,** and write

$$\mathbf{refl_n(a)} = Id(\mathbf{a}) - 2\mathbf{proj_n(a)} = \left(Id + (-2)\mathbf{proj_n}\right)(\mathbf{a}),$$

where we make the last equality true by **definition.** This will be discussed further at the end of the following example.

EXAMPLE 17

Now consider the particular case where $\mathbf{n} = (1,-1,2)$. What is the matrix $[\mathbf{refl}_{(1,-1,2)}]$? As in the projection examples of Section 3–2 (Examples 13 and 14), this question will be answered in two ways.

Method 1. Find the columns of the matrix by calculating the images of the standard basis vectors.

$$\begin{aligned}\mathbf{refl}_{(1,-1,2)}(\mathbf{e}_1) &= \mathbf{e}_1 - 2\frac{(\mathbf{e}_1 \cdot \mathbf{n})}{\|\mathbf{n}\|^2}\mathbf{n} \\ &= (1,0,0) - 2\left(\frac{(1,0,0)\cdot(1,-1,2)}{1+1+4}\right)(1,-1,2) \\ &= (1,0,0) - \frac{2}{6}(1,-1,2) \\ &= (2/3, 1/3, -2/3);\end{aligned}$$

$$\mathbf{refl}_{(1,-1,2)}(0,1,0) = (0,1,0) - \left(\frac{-2}{6}\right)(1,-1,2) = (1/3, 2/3, 2/3);$$

$$\mathbf{refl}_{(1,-1,2)}(0,0,1) = (0,0,1) - \frac{4}{6}(1,-1,2) = (-2/3, 2/3, -1/3).$$

Hence, with these three vectors as **columns,**

$$[\mathbf{refl}_{(1,-1,2)}] = \begin{bmatrix} 2/3 & 1/3 & -2/3 \\ 1/3 & 2/3 & 2/3 \\ -2/3 & 2/3 & -1/3 \end{bmatrix}.$$

Method 2. Calculate $\mathbf{refl}_{(1,-1,2)}(\mathbf{x})$ and read the rows of the matrix from the coefficients of the components.

$$\begin{aligned}
\mathbf{refl}_{(1,-1,2)}(\mathbf{x}) &= \mathbf{x} - 2\left(\frac{(x_1,x_2,x_3)\cdot(1,-1,2)}{6}\right)(1,-1,2)\\
&= (x_1,x_2,x_3) - \left(\frac{1}{3}\right)(x_1 - x_2 + 2x_3)(1,-1,2)\\
&= \Bigg(\left(\frac{2}{3}\right)x_1 + \left(\frac{1}{3}\right)x_2 + \left(\frac{-2}{3}\right)x_3,\\
&\qquad \left(\frac{1}{3}\right)x_1 + \left(\frac{2}{3}\right)x_2 + \left(\frac{2}{3}\right)x_3,\\
&\qquad \left(\frac{-2}{3}\right)x_1 + \left(\frac{2}{3}\right)x_2 + \left(\frac{-1}{3}\right)x_3\Bigg)
\end{aligned}$$

Take the coefficients of the first component as the first row of the matrix, and so on, and find, as above, that

$$[\mathbf{refl}_{(1,-1,2)}] = \begin{bmatrix} 2/3 & 1/3 & -2/3 \\ 1/3 & 2/3 & 2/3 \\ -2/3 & 2/3 & -1/3 \end{bmatrix}.$$

REMARK

Method 2 may appear to be a little shorter, but in fact the calculations are a little more intricate and more likely to lead to error. In general, for finding the matrix of a linear mapping L, Method 1 is to be preferred unless $L(\mathbf{x})$ is given explicitly.

Compositions and Linear Combinations of Linear Mappings

As we considered geometrical transformations, various ways of combining linear transformations appeared. These are now defined more carefully.

DEFINITION

Let L and M be linear mappings from $\mathbb{R}^n$ to $\mathbb{R}^m$, and k any real number. The **sum** $(L+M)$ is defined to be the mapping such that $(L+M)(\mathbf{x}) = L(\mathbf{x}) + M(\mathbf{x})$ for all $\mathbf{x}$ in $\mathbb{R}^n$; **multiplication of L by the scalar** k is defined by $(kL)(\mathbf{x}) = kL(\mathbf{x})$ for all $\mathbf{x}$ in $\mathbb{R}^n$.

It is easy to see that $L+M$ and kL are linear mappings. It is also easy to see that their matrices are given by $[L+M] = [L]+[M]$ and $[kL] = k[L]$ respectively. Thus, in the discussion of the reflection above the matrices could have been written

$$\begin{aligned}[\mathbf{refl_n}] &= [Id + (-2)\mathbf{proj_n}] \\ &= [Id] - 2[\mathbf{proj_n}] \\ &= I - 2[\mathbf{proj_n}].\end{aligned}$$

Here the fact that the matrix of the identity transformation is the identity matrix has been used:

$$[Id] = I.$$

DEFINITION

Consider the linear mappings $L : \mathbb{R}^n \to \mathbb{R}^m$ and $M : \mathbb{R}^m \to \mathbb{R}^p$. The **composition** $M \circ L : \mathbb{R}^n \to \mathbb{R}^p$ is defined by $(M \circ L)(\mathbf{x}) = M(L(\mathbf{x}))$ for all $\mathbf{x}$ in $\mathbb{R}^n$.

The order of the mappings is important in this definition. We say that "M follows L", since L "acts" or "maps" first, and then M acts on the image $L(\mathbf{x})$ of $\mathbf{x}$ under L. Note that the definition only makes sense if the domain of the second map M contains the codomain of the first map L. $M \circ L$ is easily shown to be linear. Figure 3–3–8 is sometimes helpful in thinking about composition.

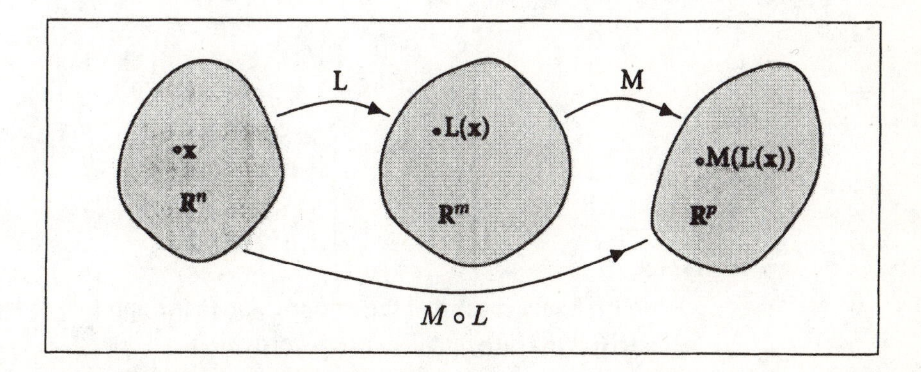

FIGURE 3-3-8. *Composition of mappings.*

It is also easy to see (but a little tedious to write out all the details) that the **matrix of the composition is the product of the matrices of M and L, taken in the correct order:** $[M \circ L] = [M][L]$. In fact the definition of matrix multiplication is chosen to make this equality true — see the introduction to that definition in Section 3–1.

EXAMPLE 18

Recall that composition was mentioned in the discussion of contractions and dilatations. Consider the linear transformation of $\mathbb{R}^2$ defined by $L(\mathbf{x}) = k\mathbf{x}$ where $-1 < k < 0$. This can be rewritten as $L(\mathbf{x}) = -|k|\mathbf{x}$, so that L may be recognized as the composition $(-Id) \circ$ (contraction by factor $|k|$). The mapping $-Id$ is called **reflection in the origin.**

EXAMPLE 19

Determine the matrix of the linear transformation that consists of a shear in the direction of x_2 by amount 0.1 followed by a rotation through angle $\pi/6$.

SOLUTION: Let S denote the shear and R the rotation. The transformation of interest is then $R \circ S$, with matrix $[R][S]$.

$$[S] = \begin{bmatrix} 1 & 0 \\ 0.1 & 1 \end{bmatrix}$$

and

$$[R] = \begin{bmatrix} \cos \pi/6 & -\sin \pi/6 \\ \sin \pi/6 & \cos \pi/6 \end{bmatrix} = \begin{bmatrix} \sqrt{3}/2 & -1/2 \\ 1/2 & \sqrt{3}/2 \end{bmatrix}$$

so

$$\begin{aligned} [R \circ S] &= \begin{bmatrix} \sqrt{3}/2 & -1/2 \\ 1/2 & \sqrt{3}/2 \end{bmatrix} \begin{bmatrix} 1 & 0 \\ 0.1 & 1 \end{bmatrix} \\ &\approx \begin{bmatrix} 0.866 & -0.5 \\ 0.5 & 0.866 \end{bmatrix} \begin{bmatrix} 1 & 0 \\ 0.1 & 1 \end{bmatrix} \\ &\approx \begin{bmatrix} 0.816 & -0.5 \\ 0.587 & 0.866 \end{bmatrix}. \end{aligned}$$

You should check that the composition in the opposite order has a different matrix; it is a different linear transformation.

EXERCISES 3–3

A1 Suppose that S and T are linear mappings with matrices

$$[S] = \begin{bmatrix} 2 & 1 & 3 \\ -1 & 0 & 2 \end{bmatrix} \quad \text{and} \quad [T] = \begin{bmatrix} 1 & 2 \\ -1 & 2 \\ 2 & 3 \end{bmatrix}.$$

(a) Determine the domain and the codomain of each mapping.
(b) Determine the matrices that represent $S \circ T$ and $T \circ S$.

A2 Suppose that S and T are linear mappings with matrices

$$[S] = \begin{bmatrix} -3 & -3 & 0 & 1 \\ 0 & 2 & 4 & 2 \end{bmatrix} \text{ and } [T] = \begin{bmatrix} 1 & 4 \\ -2 & 1 \\ 2 & -1 \\ 3 & -4 \end{bmatrix}.$$

(a) Determine the domain and the codomain of each mapping.
(b) Determine the matrices that represent $S \circ T$ and $T \circ S$.

A3 Determine the matrices of the rotations in the plane through angles
(a) $\pi/2$; (b) π; (c) $-\pi/4$.

A4 (a) In the plane, what is the matrix of a stretch S by factor 5 in the x_2-direction?
(b) Calculate the composition of S **followed by** a rotation through angle θ.
(c) Calculate the composition of S **following** a rotation through angle θ.

A5 Determine the matrices of the following reflections in $\mathbb{R}^2$.
(a) R is reflection in the line $x_1 + 3x_2 = 0$
(b) S is reflection in the line $2x_1 = x_2$

A6 Determine the matrix of the reflections in the following planes in $\mathbb{R}^3$.
(a) $x_1 + x_2 + x_3 = 0$ (b) $2x_1 - 2x_2 - x_3 = 0$

A7 Let L, M, and N be linear mappings with matrices

$$[L] = \begin{bmatrix} 2 & 3 \\ -1 & 4 \\ 0 & 1 \end{bmatrix}, \quad [M] = \begin{bmatrix} 1 & 1 & 2 \\ 3 & -2 & -1 \end{bmatrix}, \quad [N] = \begin{bmatrix} 2 & 1 \\ 3 & 1 \\ -3 & 0 \\ 1 & 4 \end{bmatrix}.$$

Determine which of the following compositions are defined, and determine the domain and codomain for those that are defined.
(a) $L \circ M$ (b) $M \circ L$ (c) $L \circ N$ (d) $N \circ L$ (e) $M \circ N$ (f) $N \circ M$

A8 (a) Let $D : \mathbb{R}^3 \to \mathbb{R}^3$ be the dilatation with factor $k = 5$, and let inj: $\mathbb{R}^3 \to \mathbb{R}^4$ be defined by $inj\,(x_1, x_2, x_3) = (x_1, x_2, 0, x_3)$. Determine the matrix of $inj \circ D$.
(b) Let $P : \mathbb{R}^3 \to \mathbb{R}^2$ be defined by $P(x_1, x_2, x_3) = (x_2, x_3)$, and let S be the shear in $\mathbb{R}^3$ such that $S(x_1, x_2, x_3) = (x_1, x_2, x_3 + 2x_1)$. Determine the matrix of $P \circ S$.
(c) Can you define a shear $T : \mathbb{R}^2 \to \mathbb{R}^2$ such that $T \circ P = P \circ S$ (where P, S are as in part (b))?
(d) Let $Q : \mathbb{R}^3 \to \mathbb{R}^2$ be defined by $Q(x_1, x_2, x_3) = (x_1, x_2)$. Determine the matrix of $Q \circ S$ (where S is the mapping in part (b)).

B1 Suppose that S and T are linear mappings with matrices

$$[S] = \begin{bmatrix} 2 & 1 \\ 1 & -2 \\ 3 & 1 \end{bmatrix} \quad \text{and} \quad [T] = \begin{bmatrix} 3 & -1 & 1 \\ -2 & 1 & 0 \end{bmatrix}.$$

(a) Determine the domain and the codomain of each mapping.
(b) Determine the matrices that represent $S \circ T$ and $T \circ S$.

B2 Suppose that S and T are linear mappings with matrices

$$[S] = \begin{bmatrix} 4 & -3 \\ 1 & 1 \\ 5 & -3 \\ -2 & 0 \end{bmatrix} \quad \text{and } [T] = \begin{bmatrix} 4 & 0 & 2 & 3 \\ -2 & 1 & 3 & 0 \end{bmatrix}.$$

(a) Determine the domain and the codomain of each mapping.
(b) Determine the matrices that represent $S \circ T$ and $T \circ S$.

B3 Determine the matrices of the rotations in the plane through angles
(a) $-\pi/2$; (b) $-\pi$; (c) $3\pi/4$.

B4 (a) In the plane, what is the matrix of a stretch S by factor 0.6 in the x_2 direction?
(b) Calculate the composition of S **followed by** a rotation through angle θ.
(c) Calculate the composition of S **following** a rotation through angle $\pi/4$.

B5 Determine the matrix of the following reflections in $\mathbb{R}^2$.
(a) R is reflection in the line $x_1 - 5x_2 = 0$
(b) S is reflection in the line $3x_1 + 4x_2 = 0$

B6 Determine the matrix of the reflections in the following planes in $\mathbb{R}^3$.
(a) $x_1 - 3x_2 - x_3 = 0$ (b) $2x_1 + x_2 - x_3 = 0$

B7 Let L, M, and N be linear mappings with matrices

$$[L] = \begin{bmatrix} 1 & 4 \\ 3 & 2 \\ -1 & 0 \end{bmatrix}, \quad [M] = \begin{bmatrix} 3 & 2 & 2 & 0 \\ 1 & -1 & -1 & 2 \end{bmatrix}, \quad [N] = \begin{bmatrix} 2 & 1 & 1 \\ 3 & 1 & 4 \\ -3 & 0 & 7 \\ 1 & -4 & 1 \end{bmatrix}.$$

Determine which of the following compositions are defined, and determine the domain and codomain for those that are defined.
(a) $L \circ M$ (b) $M \circ L$ (c) $L \circ N$ (d) $N \circ L$ (e) $M \circ N$ (f) $N \circ M$.

B8 (a) Let $C : \mathbb{R}^3 \to \mathbb{R}^3$ be contraction with factor 1/3, and let *inj*: $\mathbb{R}^3 \to \mathbb{R}^5$ be defined by *inj* $(x_1, x_2, x_3) = (0, x_1, 0, x_2, x_3)$. What is the matrix of *inj* $\circ\, C$?
(b) Let $S : \mathbb{R}^3 \to \mathbb{R}^3$ be the shear defined by $S(x_1, x_2, x_3) = (x_1, x_2 - 2x_3, x_3)$. Determine the matrices of $C \circ S$ and $S \circ C$ where C is the contraction of part (a).
(c) Let $T : \mathbb{R}^3 \to \mathbb{R}^3$ be the shear defined by $T(x_1, x_2, x_3) = (x_1 + 3x_2, x_2, x_3)$.

Determine the matrices of $S \circ T$ and $T \circ S$.

CONCEPTUAL EXERCISES

D1 If L and M are linear mappings from $\mathbb{R}^n$ to $\mathbb{R}^n$, and k is any real number, verify that $L + M$ and kL are linear mappings.

***D2** If $L : \mathbb{R}^n \to \mathbb{R}^m$ and $M : \mathbb{R}^m \to \mathbb{R}^q$ are linear, verify that $M \circ L$ is linear.

D3 Verify that for rotations in the plane,

$$[R_\alpha \circ R_\beta] = [R_\alpha][R_\beta] = [R_{\alpha+\beta}].$$

***D4** In Exercise 3–3–A5, $[R] = \begin{bmatrix} 4/5 & -3/5 \\ -3/5 & -4/5 \end{bmatrix}$ and $[S] = \begin{bmatrix} -3/5 & 4/5 \\ 4/5 & 3/5 \end{bmatrix}$ are reflection matrices. Calculate $[R \circ S]$, and verify that it can be identified as the matrix of a rotation; determine the angle of the rotation. Try to make a picture illustrating how the composition of these reflections is a rotation.

D5 In $\mathbb{R}^3$, calculate the matrix of the composition of reflection in the x_2x_3-plane followed by reflection in the x_1x_2-plane, and identify it as a rotation about some coordinate axis. What is the angle of the rotation?

D6 (a) Construct a 2 by 2 matrix $A \neq I$ such that $A^3 = I$. [Hint: Think geometrically.]

(b) Construct a 2 by 2 matrix $A \neq I$ such that $A^5 = I$.

D7 In Section 1–3, it was stated that $\mathbf{perp_a}$ satisfied the projection property so that $\mathbf{perp_a} \circ \mathbf{perp_a} = \mathbf{perp_a}$. In terms of matrices, this should imply that $[\mathbf{perp_a}][\mathbf{perp_a}] = [\mathbf{perp_a}]$. Verify this for $\mathbf{perp}_{(1,4)}$ (the matrix for this was calculated in Exercise A6 of Section 3–2).

***D8** From geometrical considerations, we know that $\mathbf{refl_n} \circ \mathbf{refl_n} = Id$. Verify the corresponding matrix equation. (Hint: $[\mathbf{refl_n}] = I - 2[\mathbf{proj_n}]$ and $\mathbf{proj_n}$ satisfies the projection property from Section 1–3.)

Special Subsets for Systems and Mappings; Subspaces of $\mathbb{R}^n$

Throughout this section L will denote a linear mapping from $\mathbb{R}^n$ to $\mathbb{R}^m$ and A will denote the m by n standard matrix of L. We shall explore the connection between the properties of L and its matrix $[L] = A$ on the one hand, and the systems of equations $A\mathbf{x} = \mathbf{b}$ and $A\mathbf{x} = \mathbf{0}$ on the other hand.

Notational Change: From now on, we usually omit the brackets [ ] on vectors, so that $\mathbf{x}$ denotes either the vector $(x_1, x_2, \ldots, x_n)$ or the corresponding n by 1 (column) matrix. We shall sometimes continue to use the brackets where it is helpful to emphasize the fact that we are considering columns.

We shall discuss the connections between mappings and systems in terms of certain subsets of the domain $\mathbb{R}^n$ or of the codomain $\mathbb{R}^m$. Recall that a subset of $\mathbb{R}^n$ is any collection of vectors in $\mathbb{R}^n$. For example, the line $\mathbf{x} = (1, 2, 3) + t(4, 5, 6)$, $t \in \mathbb{R}$, is a subset of $\mathbb{R}^3$; the hyperplane consisting of vectors (x_1, x_2, x_3, x_4) satisfying the equation $x_1 - x_2 + x_3 - x_4 = 5$ is a subset of $\mathbb{R}^4$; the set $\{(1, 5), (2, -7)\}$ consisting of just two vectors is a subset of $\mathbb{R}^2$.

We begin by considering special sets associated with systems of equations and with mappings. For reasons explained later in this section, some subsets qualify for the special name of *subspace* of $\mathbb{R}^n$, so we call them *spaces* when we first discuss them. (The general notion of a *vector space* will be introduced in Chapter 4; for now, it is enough to know that each $\mathbb{R}^n$ is a *vector space*.) We then define the concept of subspace of $\mathbb{R}^n$, show how our special sets are related to this concept, and develop the properties of subspaces.

The Solution Space of the Homogeneous System $A\mathbf{x} = \mathbf{0}$; the Nullspace of L

The **solution space** of the homogeneous system $A\mathbf{x} = \mathbf{0}$ is the subset $\{\mathbf{x} \in \mathbb{R}^n \mid A\mathbf{x} = \mathbf{0}\}$. Note that this is a subset of $\mathbb{R}^n$, the *domain* of the linear mapping L with matrix A.

EXAMPLE 20

The solution space of the homogeneous system consisting of one equation

$$x_1 + 2x_2 - 3x_3 = 0$$

is geometrically a plane in $\mathbb{R}^3$ passing through the origin. We can describe this solution space by writing the general solution by back-substitution:

$$\mathbf{x} = s(-2,1,0) + t(3,0,1), \quad s,t \in \mathbb{R}.$$

For later reference, note that the system can be written in the form $A\mathbf{x} = \mathbf{0}$, where $A = [\,1 \quad 2 \quad -3\,]$.

EXAMPLE 21

Let $A = \begin{bmatrix} 1 & 0 & 2 & 0 & 4 \\ 0 & 1 & 3 & 0 & 5 \\ 0 & 0 & 0 & 1 & 6 \end{bmatrix}$, and consider the homogeneous system $A\mathbf{x} = \mathbf{0}$. The solution space is a subset of $\mathbb{R}^5$; it is easily found by back-substitution to be

$$\{s(-2,-3,1,0,0) + t(-4,-5,0,-6,1) \mid s,t \in \mathbb{R}\}.$$

Notice that in both of these examples, the solution space is displayed automatically as the set of all linear combinations of certain vectors.

For the linear mapping L with matrix A, the vectors $\mathbf{x}$ such that $L(\mathbf{x}) = \mathbf{0}$ are exactly the same as the vectors satisfying $A\mathbf{x} = \mathbf{0}$; it is usual to introduce a special name for these vectors. It will be convenient to say that $\mathbf{b}$ is the **image** of $\mathbf{a}$ under L if $L(\mathbf{a}) = \mathbf{b}$; thus, if $L(\mathbf{x}) = \mathbf{0}$, then the image of $\mathbf{x}$ is $\mathbf{0}$.

DEFINITION

The **nullspace** of a linear mapping L is the set of vectors whose image under L is the zero vector $\mathbf{0}$. That is, the nullspace of L is $\{\mathbf{x} \in \mathbb{R}^n \mid L(\mathbf{x}) = \mathbf{0}\}$. *(A mnemonic: the nullspace of L is the set that is nullified (sent to* $\mathbf{0}$*) by L.)*

Note that the nullspace of L is a subset of the domain of L. The word **kernel** is often used in place of nullspace.

THEOREM 1. If $L : \mathbb{R}^n \to \mathbb{R}^m$ is a linear mapping with matrix A, the nullspace of L is the same as the solution space of the homogeneous system $A\mathbf{x} = \mathbf{0}$.

Proof. Immediate consequence of the definitions.

■

Examples 20 and 21 give examples of nullspaces defined in terms of matrices, but in the following example, the nullspace of L can be determined geometrically.

EXAMPLE 22

Let $L : \mathbb{R}^3 \to \mathbb{R}^3$ be $\mathbf{proj}_{(2,-1,3)}$. Since vectors orthogonal to $(2, -1, 3)$ are all mapped to $\mathbf{0}$ by L, the nullspace of L is the set of all vectors orthogonal to $(2, -1, 3)$.

The Solution Set of the Nonhomogeneous System of Linear Equations $A\mathbf{x} = \mathbf{b}$

Next, we want to consider solutions for a nonhomogeneous system $A\mathbf{x} = \mathbf{b}$, and compare this solution set with the solutions for the **corresponding homogeneous** system $A\mathbf{x} = \mathbf{0}$ (that is the system with the same coefficient matrix A). Thus, we consider two nonhomogeneous examples where the coefficient matrices are the same as in Examples 20 and 21.

EXAMPLE 23

The solution set of the nonhomogeneous system

$$x_1 + 2x_2 - 3x_3 = 5$$

is geometrically a plane in $\mathbb{R}^3$ passing through the point $(5, 0, 0)$ (and not through the origin). The coefficient matrix for this system is $A = \begin{bmatrix} 1 & 2 & -3 \end{bmatrix}$, the same as in Example 20. The solution set is a subset of $\mathbb{R}^3$. The solution set can also be described by writing the general solution:

$$\mathbf{x} = (5, 0, 0) + s(-2, 1, 0) + t(3, 0, 1), \quad s, t \in \mathbb{R}.$$

Notice that this solution set is obtained from the solution space of Example 20 by translating the plane by the vector $(5, 0, 0)$.

EXAMPLE 24

Consider the system of equations $A\mathbf{x} = \begin{bmatrix} 7 \\ 8 \\ 9 \end{bmatrix}$, where A is the matrix of Example 21: $A = \begin{bmatrix} 1 & 0 & 2 & 0 & 4 \\ 0 & 1 & 3 & 0 & 5 \\ 0 & 0 & 0 & 1 & 6 \end{bmatrix}$. The general solution in standard form is found by back-substitution to be

$$\mathbf{x} = (7, 8, 0, 9, 0) + s(-2, -3, 1, 0, 0) + t(-4, -5, 0, -6, 1), \quad s, t \in \mathbb{R}.$$

Or, we could say that the solution set is

$$\{(7, 8, 0, 9, 0) + s(-2, -3, 1, 0, 0) + t(-4, -5, 0, -6, 1) \mid \quad s, t \in \mathbb{R}\}.$$

Again, this solution set is obtained from the solution space of the corresponding homogeneous problem (Example 21) by translating by the fixed vector $(7, 8, 0, 9, 0)$.

In both Example 23 and Example 24 the solution set is a translate of the solution space for the corresponding homogeneous problem. These examples illustrate the following theorem.

THEOREM 2. Let $\mathbf{v}$ be a solution of the system of linear nonhomogeneous equations $A\mathbf{x} = \mathbf{b}$.

(i) If $\mathbf{y}$ is any other solution of the same system, then $A(\mathbf{y} - \mathbf{v}) = \mathbf{0}$, so that $(\mathbf{y} - \mathbf{v})$ is a solution of the corresponding homogeneous system $A\mathbf{x} = \mathbf{0}$.

(ii) If $\mathbf{h}$ is any solution of the corresponding homogeneous system $A\mathbf{x} = \mathbf{0}$, then $\mathbf{v} + \mathbf{h}$ is a solution of the nonhomogeneous system $A\mathbf{x} = \mathbf{b}$.

Proof. (i) $A(\mathbf{y} - \mathbf{v}) = A\mathbf{y} - A\mathbf{v} = \mathbf{b} - \mathbf{b} = \mathbf{0}$.

(ii) $A(\mathbf{v} + \mathbf{h}) = A\mathbf{v} + A\mathbf{h} = \mathbf{b} + \mathbf{0} = \mathbf{b}$.

■

The solution $\mathbf{v}$ of the nonhomogeneous solution is sometimes called a **particular solution** of the system $A\mathbf{x} = \mathbf{b}$. Theorem 2 can thus be restated as follows: any solution of the nonhomogeneous system can be obtained by adding a solution of the corresponding homogeneous system to a particular solution. Note that in Example 24, $(7, 8, 0, 9, 0)$ is a particular solution of the nonhomogeneous system, and $(s(-2, -3, 1, 0, 0) + t(-4, -5, 0, -6, 1), s, t \in \mathbb{R})$ is the general solution of the corresponding homogeneous system.

> *REMARK*
>
> Theorem 2 describes a pattern for thinking about solutions to linear problems; this pattern is illustrated for systems of linear algebraic equations in Examples 20–24. This pattern is also the key to important methods for solving linear differential equations.

The Range of *L* and the Columnspace of *A*

Suppose that $\mathbf{y}$ in $\mathbb{R}^m$ is the image of $\mathbf{x}$ under the mapping L (that is, $\mathbf{y} = L(\mathbf{x})$ for some $\mathbf{x}$ in $\mathbb{R}^n$). If you like the language of inputs and outputs, $\mathbf{y}$ is the output corresponding to input $\mathbf{x}$. The set of all images (outputs) under L is called the range of L.

DEFINITION

The **range** of L is defined to be the set

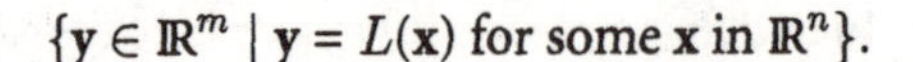

$$\{\mathbf{y} \in \mathbb{R}^m \mid \mathbf{y} = L(\mathbf{x}) \text{ for some } \mathbf{x} \text{ in } \mathbb{R}^n\}.$$

Notice that the range of L is a subset of the *codomain* $\mathbb{R}^m$ of L. For some mappings, the range is easily determined geometrically, as in the next two examples.

EXAMPLE 25

Consider the linear mapping $\mathbf{proj}_{(1,1,3)} : \mathbb{R}^3 \to \mathbb{R}^3$. Every image of this mapping is a multiple of $(1, 1, 3)$, so the range of the mapping is the set of all multiples of the vector $(1, 1, 3)$. On the other hand, the range of $\mathbf{perp}_{(1,1,3)}$ is the set of all vectors orthogonal to $(1, 1, 3)$; that is, the range is the plane through the origin with normal vector $(1, 1, 3)$. In each of these cases, note that the range is a subset of the codomain $\mathbb{R}^3$.

EXAMPLE 26

If L is a rotation, reflection, contraction or dilatation of $\mathbb{R}^3$ to itself, it is easy to see that the range of L is all of $\mathbb{R}^3$.

It is natural to ask whether the range of L can easily be described in terms of the matrix A of L. Since

$$[L(\mathbf{x})] = A[\mathbf{x}] = x_1[a_{\downarrow 1}] + x_2[a_{\downarrow 2}] + \cdots + x_n[a_{\downarrow n}],$$

the image of $\mathbf{x}$ is a linear combination of columns of the matrix A.

The **columnspace** of A is the set of all linear combinations of columns of A (with the columns considered as vectors in $\mathbb{R}^m$, so that the columnspace is a subspace of $\mathbb{R}^m$).

THEOREM 3. The range of L is equal to the columnspace of $A = [L]$.

Proof. See the paragraph before the definition of columnspace. ■

EXAMPLE 27

If L is the mapping with standard matrix $A = \begin{bmatrix} 1 & 1 \\ 2 & 1 \\ 1 & 3 \end{bmatrix}$, then the range of L is the columnspace of A, that is, all vectors of the form $c_1 \begin{bmatrix} 1 \\ 2 \\ 1 \end{bmatrix} + c_2 \begin{bmatrix} 1 \\ 1 \\ 3 \end{bmatrix}$. Writing these vectors in the usual notation for $\mathbb{R}^3$, we see that any vector of the form $c_1(1, 2, 1) + c_2(1, 1, 3)$ is in the range of L.

The range of the mapping L is related to the system of equations $A\mathbf{x} = \mathbf{b}$ as stated in the following theorem.

THEOREM 4. The system of equations $A\mathbf{x} = \mathbf{b}$ is consistent if and only if $\mathbf{b}$ is in the range of the linear mapping L with matrix A (or, equivalently, if and only if $\mathbf{b}$ is in the columnspace of A).

Proof. The system is consistent if and only if there is a vector **x** such that $A\mathbf{x} = \mathbf{b}$; but this is really the definition of the property that **b** is an element of the range of the linear mapping with matrix A. ■

EXAMPLE 28

Suppose that L is a linear mapping with matrix $A = \begin{bmatrix} 1 & 1 \\ 2 & 1 \\ 1 & 3 \end{bmatrix}$. Determine whether $\mathbf{c} = (1, 3, -1)$ and $\mathbf{d} = (2, 1, 9)$ are in the range of L.

SOLUTION: **c** is in the range of L if and only if the system $A\mathbf{x} = \mathbf{c}$ is consistent. Similarly **d** is in the range if and only if $A\mathbf{x} = \mathbf{d}$ is consistent. Since the coefficient matrix is the same for the two systems, we can answer both questions by row reducing the doubly augmented matrix $[A \mid \mathbf{c} \mid \mathbf{d}]$.

$$\left[\begin{array}{cc|c|c} 1 & 1 & 1 & 2 \\ 2 & 1 & 3 & 1 \\ 1 & 3 & -1 & 9 \end{array}\right] \rightarrow \left[\begin{array}{cc|c|c} 1 & 1 & 1 & 2 \\ 0 & -1 & 1 & -3 \\ 0 & 2 & -2 & 7 \end{array}\right] \rightarrow \left[\begin{array}{cc|c|c} 1 & 1 & 1 & 2 \\ 0 & 1 & -1 & 3 \\ 0 & 0 & 0 & 1 \end{array}\right]$$

By considering the reduced matrix corresponding to $[A \mid \mathbf{c}]$ (ignore the last column) we see that the system $A\mathbf{x} = \mathbf{c}$ is consistent, so **c** is in the range of L. However, the system $A\mathbf{x} = \mathbf{d}$ is not consistent, so **d** is not in the range of L.

Subspaces of $\mathbb{R}^n$

For some of the sets considered above we have used the word "space". How do we decide whether a subset of $\mathbb{R}^n$ qualifies for the special title *subspace*, and why do we bother? The short answer is that *subspaces* have special properties with respect to the very important operations of *addition of vectors and multiplication of vectors by scalars.*

EXAMPLE 29

What happens when we take sums or scalar multiples of solutions of Example 20 or of solutions of the corresponding nonhomogeneous system of Example 23?

SOLUTION: The systems are

$$x_1 + 2x_2 - 3x_3 = \begin{cases} 0 & \text{(homogeneous case, Example 20)} \\ 5 & \text{(nonhomogeneous case, Example 23)} \end{cases}$$

Consider two vectors **u** and **v** in the solution space of the homogeneous system. Then

$$[1 \quad 2 \quad -3]\,\mathbf{u} = 0 \quad \text{and} \quad [1 \quad 2 \quad -3]\,\mathbf{v} = 0.$$

Now consider the *sum* $(\mathbf{u}+\mathbf{v})$; is it also in the solution space?

$$[1 \quad 2 \quad -3]\,(\mathbf{u}+\mathbf{v}) = [1 \quad 2 \quad -3]\,\mathbf{u} + [1 \quad 2 \quad -3]\,\mathbf{v} = 0 + 0 = 0,$$

so for arbitrary **u** and **v** in the solution space of $A\mathbf{x} = 0$, the sum $(\mathbf{u}+\mathbf{v})$ is in the solution space. We say that the solution space for the homogeneous problem is *closed under addition.*

Similarly, consider the multiple $k\mathbf{u}$, for any real number k: then

$$[1 \quad 2 \quad -3]\,k\mathbf{u} = k\,[1 \quad 2 \quad -3]\,\mathbf{u} = k0 = 0,$$

so $k\mathbf{u}$ is in the solution space, and we say that the solution space for the homogeneous problem is *closed under multiplication by scalars.*

Now contrast the corresponding nonhomogeneous system $[1 \quad 2 \quad -3]\mathbf{x} = 5$ of Example 23. If **z** and **w** are in the solution set of this problem then,

$$[1 \quad 2 \quad -3]\,(\mathbf{z}+\mathbf{w}) = [1 \quad 2 \quad -3]\,\mathbf{z} + [1 \quad 2 \quad -3]\,\mathbf{w} = 5 + 5 = 10 \neq 5,$$

and $[1 \quad 2 \quad -3]\,k\mathbf{z} = k\,[1 \quad 2 \quad -3]\,\mathbf{z} = k5 \neq 5$, (in general). Thus, the sum $(\mathbf{y}+\mathbf{z})$ is not in the solution set in this case, so this solution set is *not closed under addition*; similarly, the scalar multiple $k\mathbf{z}$ is not in the solution set, so the solution set for the non-homogeneous problem is *not closed under multiplication by scalars.*

EXAMPLE 30

By arguments like those in Example 29, we can show that the solution set of the nonhomogeneous problem in Example 24 is not closed under addition or multiplication by scalars, while the solution *space* of the corresponding homogeneous problem of Example 21 is closed under addition and multiplication by scalars. (Check this yourself.)

We are now ready to introduce the definition of *subspace*, which justifies the distinction we make between solution spaces and solution sets in Examples 29 and 30.

A non-empty subset S of $\mathbb{R}^n$ is called a **subspace** of $\mathbb{R}^n$ if for all vectors $\mathbf{x}$ and $\mathbf{y}$ in S, and every scalar k,

(1) $(\mathbf{x}+\mathbf{y}) \in S$, and

(2) $k\mathbf{x} \in S$.

Condition (1) is described by saying that "S is **closed under addition**"; that is, when addition is performed, the sum lies inside the set S in which we started. Condition (2) is described by saying that "S is **closed under multiplication by scalars**". Thus, a subset is a subspace if and only if it is closed under addition and multiplication by scalars.

WARNING: *"Subspace" is now a technical word, and it must be used in a precise way. You must memorize the definition and learn how to use it to test whether a given subset is a subspace.*

REMARK

Example 29 shows that the solution space of Example 20 is a subspace of $\mathbb{R}^3$, while the solution set of Example 23 is *not* a subspace of $\mathbb{R}^3$. Similarly, Example 30 shows that the solution space of the homogeneous system of Example 21 is a subspace of $\mathbb{R}^5$, while the solution set of the nonhomogeneous system of Example 24 is *not* a subspace of $\mathbb{R}^5$. Before generalizing these results as Theorem 5, let us note some further properties of subspaces.

The definition requires that a subspace be non-empty: a subspace always contains at least one vector. It follows from (2) with $k = 0$ that every subspace of $\mathbb{R}^n$ contains the zero vector ($\mathbf{0}$ in $\mathbb{R}^n$). This fact provides an easy method for disqualifying as *subspaces* any subsets that do not contain the zero vector. For example, in $\mathbb{R}^3$, a plane is a subspace if and only if it contains the origin (see Example 29).

The set $\{\mathbf{0}\}$ consisting of only the zero vector in $\mathbb{R}^n$ is a subspace of $\mathbb{R}^n$. It is easy to see that $\mathbb{R}^n$ is always a subspace of itself. The next two theorems show that other subspaces arise naturally in linear algebra.

THEOREM 5. Let A be an m by n matrix. Then:

(1) the solution space of the homogeneous system $A\mathbf{x} = \mathbf{0}$ is a subspace of $\mathbb{R}^n$;

(2) the solution set of the nonhomogeneous system $A\mathbf{x} = \mathbf{b}$ $(\mathbf{b} \neq \mathbf{0})$ is *not* a subspace of $\mathbb{R}^n$.

Proof. The proof consists of rewriting the arguments of Example 29 for the general case. To prove (1), let $\mathbf{u}$ and $\mathbf{v}$ be any vectors in the solution space of $A\mathbf{x} = \mathbf{0}$, and let k be any scalar. Then,

$$A(\mathbf{u} + \mathbf{v}) = A\mathbf{u} + A\mathbf{v} = \mathbf{0} + \mathbf{0} = \mathbf{0}$$

and

$$A(k\mathbf{u}) = kA\mathbf{u} = k\mathbf{0} = \mathbf{0}.$$

Thus, $\mathbf{u} + \mathbf{v}$ and $k\mathbf{u}$ are also in the solution space, so the solution space is closed under addition and multiplication by scalars. It is therefore a subspace of $\mathbb{R}^n$.

The proof of (2) also follows the pattern in Example 29 and is left to the reader.

■

COROLLARY Let $L : \mathbb{R}^n \to \mathbb{R}^m$ be a linear mapping with matrix A. Then the nullspace of L is a subspace of $\mathbb{R}^n$.

Proof. By Theorem 1, the nullspace of L is the solution space of $A\mathbf{x} = \mathbf{0}$; by Theorem 5, this is a subspace of $\mathbb{R}^n$.

To get practice with the definitions we also give a direct proof. The nullspace is not empty because the nullspace always contains the zero vector. Suppose that $\mathbf{x}$ and $\mathbf{y}$ are vectors in the nullspace of L, and that k is a real number. Then $L(\mathbf{x}) = \mathbf{0}$ and $L(\mathbf{y}) = \mathbf{0}$; by linearity it follows that $L(\mathbf{x} + \mathbf{y}) = L(\mathbf{x}) + L(\mathbf{y}) = \mathbf{0}$, so the nullspace is closed under addition; similarly, $L(k\mathbf{x}) = kL(\mathbf{x}) = k\mathbf{0} = \mathbf{0}$, so the nullspace is also closed under multiplication by scalars. Hence, the nullspace of L is a subspace of $\mathbb{R}^n$.

■

THEOREM 6. If $L : \mathbb{R}^n \to \mathbb{R}^m$ is linear, the range of L is a subspace of $\mathbb{R}^m$.

Proof. *Try to prove this yourself before reading the following proof. How do you express the fact that you need to talk about vectors in the range?*

The range of L is certainly not empty. We need to show that the range of L is closed under addition and multiplication by scalars. Suppose that $\mathbf{y}$ and $\mathbf{z}$ are vectors in the range of L, and that k is a real number. By the definition of range, there must exist vectors $\mathbf{u}$ and $\mathbf{v}$ in the domain $\mathbb{R}^n$ of L such that $L(\mathbf{u}) = \mathbf{y}$ and $L(\mathbf{v}) = \mathbf{z}$. Then by the linearity of L,

$$\mathbf{y} + \mathbf{z} = L(\mathbf{u}) + L(\mathbf{v}) = L(\mathbf{u} + \mathbf{v}),$$

so $\mathbf{y} + \mathbf{z}$ is in the range of L. Similarly, $k\mathbf{y}$ is in the range of L because

$$k\mathbf{y} = kL(\mathbf{u}) = L(k\mathbf{u}).$$

Since the range of L is closed under addition and multiplication by scalars, it is a subspace of $\mathbb{R}^m$.

■

REMARK

It is a consequence of this theorem that the columnspace of an m by n matrix is a subspace of $\mathbb{R}^m$. We shall see that this also follows from the facts about spanning sets introduced below.

We consider two more examples of subsets that are not subspaces.

EXAMPLE 31

Consider the set $U = \{(1,2,3),(2,5,6)\}$ consisting of two vectors in $\mathbb{R}^3$. It cannot be a subspace because $(1,2,3) + (2,5,6) = (3,7,9)$, and $(3,7,9)$ is not a vector in U, so U is not closed under addition. (U also fails to contain $\mathbf{0} = (0,0,0)$, and it also fails to be closed under multiplication by scalars, but one failure is enough to disqualify a subset from being a subspace.)

EXAMPLE 32

Let Q be the set of vectors in $\mathbb{R}^n$ such that $x_1 \geq 0$ and $x_2 \geq 0$. (Q consists of all vectors in the first quadrant together with those on the non-negative coordinate axes.) Q is closed under addition but if $\mathbf{x} \neq \mathbf{0}$ is in Q then obviously $(-1)\mathbf{x}$ is not in Q, so Q is not closed under multiplication by scalars, so Q is not a subspace of $\mathbb{R}^2$.

Spanning Sets

The second main way that subspaces arise is as the set of all linear combinations of some **spanning set**. We have already seen some examples: the columnspace of the matrix A is the set of all linear combinations of $\{[a_{\downarrow 1}], [a_{\downarrow 2}], \ldots, [a_{\downarrow n}]\}$, and the solution space of the homogeneous system of Example 21 is the set of all linear combinations of $\{(2, -3, 1, 0, 0), (4, -5, 0, -6, 1)\}$. We require an easy theorem and a bit more vocabulary.

THEOREM 7. If $S = \{\mathbf{v}_1, \mathbf{v}_2, \ldots, \mathbf{v}_k\}$ is a set of vectors in $\mathbb{R}^n$, and $\mathbb{S}$ is the set of all linear combinations of vectors in S,

$$\mathbb{S} = \{c_1\mathbf{v}_1 + c_2\mathbf{v}_2 + \cdots + c_k\mathbf{v}_k \mid c_j \in \mathbb{R} \text{ for } j = 1, 2, \ldots k\},$$

then $\mathbb{S}$ is a subspace of $\mathbb{R}^n$.

Proof. Let $\mathbf{x}$ and $\mathbf{y}$ be vectors in $\mathbb{S}$. Then for some real numbers c_j and d_j, $(j = 1, 2, \ldots, k)$, $\mathbf{x} = c_1\mathbf{v}_1 + c_2\mathbf{v}_2 + \cdots + c_k\mathbf{v}_k$, and $\mathbf{y} = d_1\mathbf{v}_1 + d_2\mathbf{v}_2 + \cdots + d_k\mathbf{v}_k$, and it follows that

$$\begin{aligned}\mathbf{x} + \mathbf{y} &= c_1\mathbf{v}_1 + c_2\mathbf{v}_2 + \cdots + c_k\mathbf{v}_k + d_1\mathbf{v}_1 + d_2\mathbf{v}_2 + \cdots + d_k\mathbf{v}_k \\ &= (c_1 + d_1)\mathbf{v}_1 + (c_2 + d_2)\mathbf{v}_2 + \cdots + (c_k + d_k)\mathbf{v}_k,\end{aligned}$$

so $\mathbf{x} + \mathbf{y}$ is a linear combination of vectors in S and therefore in $\mathbb{S}$; hence, $\mathbb{S}$ is closed under addition. Similarly,

$$k\mathbf{x} = k(c_1\mathbf{v}_1 + c_2\mathbf{v}_2 + \cdots + c_k\mathbf{v}_k) = kc_1\mathbf{v}_1 + kc_2\mathbf{v}_2 + \cdots + kc_k\mathbf{v}_k,$$

and $k\mathbf{x}$ is also a linear combination of vectors in S, so $\mathbb{S}$ is closed under multiplication by scalars. Therefore $\mathbb{S}$ is a subspace of $\mathbb{R}^n$, as claimed. ■

DEFINITION

If $\mathbb{S}$ is the subspace of $\mathbb{R}^n$ consisting of all linear combinations of the vectors $\mathbf{v}_1, \mathbf{v}_2, \ldots, \mathbf{v}_k$ in $\mathbb{R}^n$, then $\mathbb{S}$ is called the **subspace spanned by** the set $\{\mathbf{v}_1, \mathbf{v}_2, \ldots, \mathbf{v}_k\}$, or the **span** of $\{\mathbf{v}_1, \mathbf{v}_2, \ldots, \mathbf{v}_k\}$. The set $\{\mathbf{v}_1, \mathbf{v}_2, \ldots, \mathbf{v}_k\}$ is called a **spanning set** for the subspace $\mathbb{S}$. The notation used for such cases will be

$$\mathbb{S} = \mathrm{Sp}(\{\mathbf{v}_1, \mathbf{v}_2, \ldots, \mathbf{v}_k\}).$$

EXAMPLE 33

Consider the line $\mathbb{L}$ with vector equation $\mathbf{x} = t\mathbf{a}$, $t \in \mathbb{R}$, in $\mathbb{R}^n$. Then $\mathbb{L}$ is a subspace spanned by $\{\mathbf{a}\}$, and $\{\mathbf{a}\}$ is a spanning set for $\mathbb{L}$: $\mathbb{L} = \mathrm{Sp}(\{\mathbf{a}\})$. Similarly, the 2-plane $\mathbb{M}$ in $\mathbb{R}^n$ with vector equation $\mathbf{x} = s\mathbf{a} + t\mathbf{b}$ is a subspace of $\mathbb{R}^n$ with spanning set $\{\mathbf{a}, \mathbf{b}\}$: $\mathbb{M} = \mathrm{Sp}(\{\mathbf{a}, \mathbf{b}\})$.

A Spanning Set for the Solution Subspace of a Homogeneous System

In Example 20, the solution space is displayed as the subspace spanned by $\{(-2,1,0),(3,0,1)\}$. In Example 21, the solution space is displayed as the subspace spanned by $\{(-2,-3,1,0,0),(-4,-5,0,-6,1)\}$. In fact, the standard way of representing the solution space of a homogeneous system always displays the solution subspace as the set of all linear combinations of a spanning set. Consider the system $A\mathbf{x} = \mathbf{0}$, where A is an m by n matrix. Suppose that the rank of the matrix is r, so there are $n - r$ non-leading variables, each of which is a parameter in the general solution of the homogeneous system; the general solution is then a linear combination of $n - r$ spanning vectors (each spanning vector being multiplied by one of the $n - r$ parameters).

Examples of Spanning Sets

Spanning sets are usually fairly straightforward to deal with, but the next two examples illustrate that some care is needed.

EXAMPLE 34

Consider the line $\mathbb{L}$ in $\mathbb{R}^2$ with vector equation $\mathbf{x}(t) = (2,4) + t(1,2), \quad t \in \mathbb{R}$. At first glance this may appear **not** to be a subspace because it fails to contain $(0,0)$. However, $(0,0)$ is a point on $\mathbb{L}$ because $\mathbf{x}(-2) = (0,0)$. The equation could be rewritten $\mathbf{x}(t) = (t+2)(1,2)$; or introduce a new parameter $s = t+2$, and write $\mathbf{x}(s) = s(1,2)$, $s \in \mathbb{R}$. It follows that $\mathbb{L} = \mathrm{Sp}(\{(1,2)\})$. It is not hard to create higher dimensional examples, which require similar rewriting before it becomes clear that the given subset is a subspace with an obvious spanning set.

The next example is a rather simple illustration of a very important fact: there are many choices of spanning sets for a given subspace.

EXAMPLE 35

$\mathbb{R}^2$ is spanned by the standard basis vectors $\mathbf{e}_1 = (1,0)$ and $\mathbf{e}_2 = (0,1)$, because any $\mathbf{x}$ in $\mathbb{R}^2$ can be written $x_1(1,0) + x_2(0,1)$. What is the subspace of $\mathbb{R}^2$ spanned by $\{(1,0),(1,1)\}$? A typical vector in this subspace is
$c_1(1,0) + c_2(1,1) = (c_1 + c_2, c_2)$. Now in fact, any vector $\mathbf{x}$ in $\mathbb{R}^2$ can be written

in the form $(x_1, x_2) = (c_1 + c_2, c_2)$ by taking $c_1 = x_1 - x_2$ and $c_2 = x_2$, so the subspace spanned by $\{(1,0),(1,1)\}$ is $\mathbb{R}^2$ itself.

It is easy to determine whether a given vector $\mathbf{x}$ lies in the span of a given set $\{\mathbf{v}_1, \mathbf{v}_2, \ldots, \mathbf{v}_k\}$.

EXAMPLE 36

Is the vector $(3,4,2)$ in the subspace of $\mathbb{R}^3$ spanned by $V = \{(1,-1,1),(-2,1,1)\}$?

SOLUTION: Can $(3,4,2)$ be written as a linear combination of $(1,-1,1)$ and $(-2,1,1)$; that is, are there numbers c_1 and c_2 such that $(3,4,2) = c_1(1,-1,1) + c_2(-2,1,1)$? Write this in column form: $\begin{bmatrix} 1 & -2 \\ -1 & 1 \\ 1 & 1 \end{bmatrix} \begin{bmatrix} c_1 \\ c_2 \end{bmatrix} = \begin{bmatrix} 3 \\ 4 \\ 2 \end{bmatrix}$, and observe that $\left[\begin{array}{cc|c} 1 & -2 & 3 \\ -1 & 1 & 4 \\ 1 & 1 & 2 \end{array}\right]$ is the augmented matrix for the system. The reduced row echelon form is $\left[\begin{array}{cc|c} 1 & -2 & 3 \\ 0 & 1 & -7 \\ 0 & 0 & 20 \end{array}\right]$, so the system is inconsistent, and there are no solutions. Therefore, $(3,4,2)$ is not in the subspace spanned by V. (Sometimes it is enough just to check consistency, but sometimes, if the system is consistent, it is of interest to find the coefficients c_1 and c_2.)

Notice that Example 28 illustrated exactly the same solution procedure, which can be summarized as follows.

Procedure for Determining Whether a Vector Lies in Sp(V)

Let V be the set of vectors $\{\mathbf{v}_1, \mathbf{v}_2, \ldots, \mathbf{v}_k\}$ in $\mathbb{R}^n$. Then a vector $\mathbf{x}$ in $\mathbb{R}^n$ is in Sp(V) if and only if there are numbers $c_1, c_2, \ldots, c_k$ such that

$$c_1\mathbf{v}_1 + c_2\mathbf{v}_2 + \ldots + c_k\mathbf{v}_k = \mathbf{x},$$

that is, if the system $c_1[\mathbf{v}_1] + c_2[\mathbf{v}_2] + \ldots + c_k[\mathbf{v}_k] = [\mathbf{x}]$ is consistent.

The Rowspace of a Matrix

The idea of the rowspace of a matrix A is similar to the idea of the columnspace of A. Later (in Section 4–5), we shall need to know how to compare the rowspaces of two row equivalent matrices.

DEFINITION

Given an m by n matrix A, the **rowspace** of A is the subspace of $\mathbb{R}^n$ spanned by the rows of A (regarded as vectors).

EXAMPLE 37

Let $A = \begin{bmatrix} 1 & -2 \\ -1 & 1 \\ 1 & 1 \end{bmatrix}$. The rowspace is the subspace of $\mathbb{R}^2$ spanned by $\{(1, -2), (-1, 1), (1, 1)\}$. You should check that any vector $\mathbf{x}$ in $\mathbb{R}^2$ lies in $\mathrm{Sp}(\{(1, -2), (-1, 1), (1, 1)\})$, so that in this example the rowspace is $\mathbb{R}^2$ itself.

THEOREM 8. If matrix A is row equivalent to matrix B, then the rowspace of A is equal to the rowspace of B. In particular, if B is the reduced row echelon matrix that is row equivalent to matrix A, then the non-zero rows of B form a spanning set for the rowspace of B, and therefore also for the rowspace of A.

Proof. Suppose that matrix B is obtained from A by the elementary row operation of interchanging two rows; then, except for the order, the rows of A are the same as the rows of B, so the rowspace of A is certainly the same as the rowspace of B. Similarly, for the elementary row operation of multiplying a row by a non-zero constant, the rowspace is unchanged. Consider the third kind of elementary row operation, the addition of a multiple of one row to another. To be specific, suppose matrix B is obtained from matrix A by adding k times the first row of A to its second row. Then,

$$b_{1\rightarrow} = a_{1\rightarrow}, \quad b_{2\rightarrow} = a_{2\rightarrow} + ka_{1\rightarrow}, \quad b_{3\rightarrow} = a_{3\rightarrow}, \quad \ldots, \quad b_{m\rightarrow} = a_{m\rightarrow},$$

so,

$$\begin{aligned} &c_1 b_{1\rightarrow} + c_2 b_{2\rightarrow} + \cdots c_m b_{m\rightarrow} \\ &\qquad = c_1 a_{1\rightarrow} + c_2(a_{2\rightarrow} + ka_{1\rightarrow}) + c_3 a_{3\rightarrow} + \cdots + c_m a_{m\rightarrow} \\ &\qquad = (c_1 + c_2 k)a_{1\rightarrow} + c_2 a_{2\rightarrow} + c_3 a_{3\rightarrow} + \cdots + c_m a_{m\rightarrow}. \end{aligned}$$

Thus, a linear combination of the rows of B is seen to be a linear combination of the rows of A, so any vector in the rowspace of B is in the rowspace of A.

Since elementary row operations are reversible, a linear combination of the rows of A must also be a linear combination of the rows of B, so any vector in the rowspace of A is in the rowspace of B. By considering a sequence of row operations, we see that row equivalent matrices must have the same rowspaces. ∎

Subspaces Defined by Equations and Subspaces Defined by Spanning Sets

If a subspace is given as the set of vectors satisfying some homogeneous linear equation, or homogeneous system of linear equations, we have seen that we can write it as the span of a set of vectors simply by writing the general solution in standard form. Suppose a subspace is given as the span of a set of vectors; can we find an equation (or equations) that defines the subspace? A procedure for answering such questions is illustrated by the following example.

EXAMPLE 38

$\text{Sp}(\{(1,2,1,1),(1,1,3,1),(3,5,5,3)\})$ is a subspace of $\mathbb{R}^4$. Determine a homogeneous linear equation (or equations) that defines this set.

SOLUTION: A vector $\mathbf{x}$ is in this set if and only if for some c_1, c_2, c_3, $c_1(1,2,1,1)+c_2(1,1,3,1)+c_3(3,5,5,3) = \mathbf{x}$. Write the vectors as columns, and row reduce the augmented matrix of the resulting nonhomogeneous system.

$$\begin{bmatrix} 1 & 1 & 3 & x_1 \\ 2 & 1 & 5 & x_2 \\ 1 & 3 & 5 & x_3 \\ 1 & 1 & 3 & x_4 \end{bmatrix} \rightarrow \begin{bmatrix} 1 & 1 & 3 & x_1 \\ 0 & -1 & -1 & x_2 - 2x_1 \\ 0 & 2 & 2 & x_3 - x_1 \\ 0 & 0 & 0 & x_4 - x_1 \end{bmatrix} \rightarrow \begin{bmatrix} 1 & 1 & 3 & x_1 \\ 0 & 1 & 1 & 2x_1 - x_2 \\ 0 & 0 & 0 & -5x_1 + 2x_2 + x_3 \\ 0 & 0 & 0 & -x_1 + x_4 \end{bmatrix}$$

The system is consistent if and only if $-5x_1+2x_2+x_3 = 0$ and $-x_1+x_4 = 0$. Thus, $\text{Sp}(\{(1,2,1,1),(1,1,3,1),(3,5,5,3)\})$ is the solution space of the homogeneous system of linear equations $-5x_1 + 2x_2 + x_3 = 0$ and $-x_1 + x_4 = 0$.

EXERCISES 3-4

A1 Let L be the linear mapping with matrix $\begin{bmatrix} 1 & 0 & -1 \\ 0 & 1 & 3 \\ 1 & 2 & 1 \\ 2 & 5 & 1 \end{bmatrix}$.

(a) Is $(3, 1, 6, 1)$ in the range of L? If so, find a vector $\mathbf{x}$ such that $L(\mathbf{x}) = (3, 1, 6, 1)$.
(b) Is $(3, -5, 1, 5)$ in the range of L? If so, find $\mathbf{x}$ such that $L(\mathbf{x}) = (3, -5, 1, 5)$.

A2 For each of the following cases, show that the set is or is not a subspace of the appropriate $\mathbb{R}^n$.
(a) $\{(x_1, x_2, x_3) \mid x_1x_2 = 0\}$
(b) $\{\mathbf{x} \in \mathbb{R}^5 \mid \|\mathbf{x}\|^2 \geq 0\}$
(c) the plane in $\mathbb{R}^3$ with vector equation

$$\mathbf{x} = (2, 3, 4) + s(1, 1, 1) + t(1, 2, 3)$$

(d) $\{\mathbf{x} \in \mathbb{R}^3 \mid \mathbf{proj}_{(1,1,1)}(\mathbf{x}) = (2, 2, 2)\}$
(e) $\{\mathbf{x} \in \mathbb{R}^3 \mid \mathbf{refl}_{(a,b,c)}(\mathbf{x}) = \mathbf{x} \text{ for some fixed } (a, b, c)\}$

A3 Determine whether the following sets are subspaces of $\mathbb{R}^4$, giving clear reasons.
(a) $S = \{\mathbf{x} \in \mathbb{R}^4 \mid 2x_1 = 3x_4 \text{ and } x_2 - 5x_3 = 0\}$
(b) $T = \{\mathbf{x} \in \mathbb{R}^4 \mid x_1 + 2x_3 = 5 \text{ and } x_1 - 3x_4 = 0\}$
(c) $U = \{\mathbf{x} \in \mathbb{R}^4 \mid x_1 = x_3x_4 \text{ and } x_2 - x_4 = 0\}$
(d) $V = \{\mathbf{x} \in \mathbb{R}^4 \mid x_1 + 3x_3 = 0 \text{ and } x_2 + x_3 - x_4 = 0\}$
(e) $W = \{\mathbf{x} \in \mathbb{R}^4 \mid x_1 + x_2 + x_3 = -x_4 \text{ and } x_3 = 2\}$

A4 Let $V = \{(1, 0, 1, 1), (2, 1, 0, 1), (-1, 1, 2, 1)\}$. For each of the following vectors, either express it as a linear combination of the vectors of V or show that it is not a vector in $\text{Sp}(V)$.
(a) $(-3, 2, 8, 4)$ (b) $(5, 4, 6, 7)$

A5 Let $V = \{(1, -1, 1, 0), (-1, 1, 0, 2), (1, 1, -1, -1)\}$. For each of the following vectors, either express it as a linear combination of vectors of V or show that it is not a vector in $\text{Sp}(V)$.
(a) $(3, 2, -1, -1)$ (b) $(-7, 3, 0, 8)$

A6 Determine a spanning set for the nullspace of L if
(a) L is the linear mapping with matrix $\begin{bmatrix} 1 & 0 & 0 & -2 & 5 \\ 0 & 1 & 0 & 3 & 4 \\ 0 & 0 & 1 & 1 & 2 \end{bmatrix}$;
(b) L is the linear mapping with matrix $\begin{bmatrix} 1 & 2 & 0 & 0 & -2 & 0 \\ 0 & 0 & 1 & 0 & -1 & 0 \\ 0 & 0 & 0 & 1 & -5 & 0 \\ 0 & 0 & 0 & 0 & 0 & 1 \end{bmatrix}$.

A7 Is $\{(1, 1, 2), (1, 0, 1)\}$ a spanning set for the plane with equation $x_1 + x_2 - x_3 = 0$? Answer this by the procedure used in Example 38.

A8 By the procedure of Example 38, determine the equation of the subspace of $\mathbb{R}^3$ spanned by $\{(1, 1, 2), (3, -1, 0)\}$.

A9 Discuss the statement "$\{(1,1,2),(1,0,1)\}$ is a spanning set for the plane $x_1 + x_2 - x_3 = 1$".

A10 Determine the matrix of a linear mapping $L : \mathbb{R}^2 \to \mathbb{R}^3$ whose nullspace is $\text{Sp}(\{(1,1)\})$ and whose range is $\text{Sp}(\{(1,2,3)\})$.

A11 Determine the matrix of a linear mapping $L : \mathbb{R}^2 \to \mathbb{R}^3$ whose nullspace is $\text{Sp}(\{(1,-2)\})$ and whose range is $\text{Sp}(\{(1,1,1)\})$.

B1 Let L be the linear mapping with matrix $\begin{bmatrix} 1 & 2 & 1 \\ 1 & 0 & 3 \\ 2 & 1 & -1 \\ 0 & 2 & 2 \end{bmatrix}$.

(a) Is $(5,5,4,4)$ in the range of L? If so, find a vector $\mathbf{x}$ such that $L(\mathbf{x}) = (5,5,4,4)$.

(b) Is $(1,-3,2,1)$ in the range of L? If so, find $\mathbf{x}$ such that $L(\mathbf{x}) = (1,-3,2,1)$.

B2 For each of the following cases, show that the set is or is not a subspace of the appropriate $\mathbb{R}^n$.

(a) $\{(x_1,x_2,x_3) \mid x_1 + x_2 \geq 0\}$

(b) $\{\mathbf{x} \in \mathbb{R}^5 \mid \mathbf{x} \cdot \mathbf{a} = 1, \text{ for some fixed } \mathbf{a}\}$

(c) the plane in $\mathbb{R}^3$ with vector equation

$$\mathbf{x} = (1,0,1) + s(2,1,2) + t(1,1,1)$$

(d) $\{\mathbf{x} \in \mathbb{R}^3 \mid \mathbf{perp}_{(2,3,4)}(\mathbf{x}) = (1,1,1)\}$

(e) $\{\mathbf{x} \in \mathbb{R}^3 \mid \mathbf{proj}_{(1,1,1)}(\mathbf{x}) \in \text{Sp}(\{(1,1,1)\})\}$

B3 Determine whether the following sets are subspaces of $\mathbb{R}^4$, giving clear reasons.

(a) $S = \{\mathbf{x} \in \mathbb{R}^4 \mid 2x_1 - 5x_4 = 7 \text{ and } 3x_2 - 2x_4 = 0\}$

(b) $T = \{\mathbf{x} \in \mathbb{R}^4 \mid x_1 + 2x_3 = 0 \text{ and } x_1 - 3x_4 = 0\}$

(c) $U = \{\mathbf{x} \in \mathbb{R}^4 \mid x_1^2 = x_2^2 = x_3^2 + x_4^2\}$

(d) $V = \{\mathbf{x} \in \mathbb{R}^4 \mid x_1 + x_2 + x_4 = 0 \text{ and } 2x_2 + 3x_3 + x_4 = 0\}$

(e) $W = \{\mathbf{x} \in \mathbb{R}^4 \mid x_1x_2x_4 = 0\}$

B4 Let $V = \{(1,1,0,1),(2,0,0,2),(0,2,-1,1)\}$. For each of the following vectors, either express it as a linear combination of the vectors of V or show that it is not a vector in $\text{Sp}(V)$.

(a) $(-4,-2,2,-6)$ (b) $(6,0,0,3)$

B5 Let $V = \{(1,2,1,0),(1,1,3,1),(1,-1,1,-1)\}$. For each of the following vectors, either express it as a linear combination of the vectors of V or show that it is not a vector in $\text{Sp}(V)$.

(a) $(0,1,4,2)$ (b) $(6,4,10,3)$

B6 Determine a spanning set for the nullspace of L if

(a) L is the linear mapping with matrix $\begin{bmatrix} 1 & 0 & 0 & 2 & 7 \\ 0 & 1 & 0 & 0 & 4 \\ 0 & 0 & 0 & 1 & 4 \\ 0 & 0 & 0 & 0 & 0 \end{bmatrix}$;

(b) L is the linear mapping with matrix $\begin{bmatrix} 1 & -2 & 0 & -4 & 0 & -6 \\ 0 & 0 & 1 & -3 & 0 & -6 \\ 0 & 0 & 0 & 0 & 1 & -5 \end{bmatrix}$.

B7 Is $\{(-1, 1, 1), (1, 1, -3)\}$ a spanning set for the plane with equation $2x_1 + x_2 + x_3 = 0$? Answer this by the procedure used in Example 38.

B8 By the procedure of Example 38, determine a scalar equation of the subspace of $\mathbb{R}^3$ spanned by $\{(1, -1, 2), (1, 2, 5)\}$.

B9 Discuss the statement "$\{(-1, 1, 1), (1, 1, -3)\}$ is a spanning set for the plane $2x_1 + x_2 + x_3 = 20$".

B10 Determine the matrix of a linear mapping $L : \mathbb{R}^2 \to \mathbb{R}^2$ whose nullspace is $\text{Sp}(\{(2, 1)\})$ and whose range is $\text{Sp}(\{(2, 1)\})$.

B11 Determine the matrix of a linear mapping $L : \mathbb{R}^2 \to \mathbb{R}^3$ whose nullspace is $\text{Sp}(\{(1, -1)\})$ and whose range is $\text{Sp}(\{(0, 3, 5)\})$.

CONCEPTUAL EXERCISES

D1 Let $\mathbf{n}$ and $\mathbf{m}$ be distinct non-zero vectors in $\mathbb{R}^3$, and let $\mathbf{b}$ be an arbitrary vector in $\mathbb{R}^2$. Define sets U, V, and W as follows:

$$U = \{\mathbf{x} \in \mathbb{R}^3 \mid x_1 = x_2 - 5x_3\}, \quad V = \{\mathbf{x} \in \mathbb{R}^3 \mid \mathbf{n} \cdot \mathbf{x} = 0 \text{ and } \mathbf{m} \cdot \mathbf{x} = 3\},$$
$$W = \{\mathbf{b} \in \mathbb{R}^2 \mid (\mathbf{n} \cdot \mathbf{x}, \mathbf{m} \cdot \mathbf{x}) = \mathbf{b} \text{ for some } \mathbf{x} \text{ in } \mathbb{R}^3\}.$$

(a) By using either the definition of subspace or the fact that a subspace must contain the zero vector, determine whether each of the sets U, V, W is a subspace (of $\mathbb{R}^2$ or $\mathbb{R}^3$?).
(b) By appealing to a suitable theorem or remark in Section 3–4, determine whether each of U, V, W is a subspace (of $\mathbb{R}^2$ or $\mathbb{R}^3$?)

***D2** (a) Let V be any subset of vectors in $\mathbb{R}^n$. Let $V^\perp$ be the set of vectors that are orthogonal to every vector in V. (That is, $\mathbf{w}$ is in $V^\perp$ if and only if $\mathbf{w} \cdot \mathbf{v} = 0$ for every $\mathbf{v}$ in V.) Show that $V^\perp$ is a subspace of $\mathbb{R}^n$.
(b) Notice that part (a) makes sense even if V is an infinite set of vectors in $\mathbb{R}^n$. If V is a *subspace* of $\mathbb{R}^n$, show that every vector in V is also in $(V^\perp)^\perp$. (It is true that $V = (V^\perp)^\perp$, but ideas from Chapter 4 are needed to prove it.)
(c) If V is a subspace of $\mathbb{R}^n$, $V^\perp$ is called the *orthogonal complement of V in $\mathbb{R}^n$*. For an m by n matrix A, show that the solution space of $A\mathbf{x} = \mathbf{0}$ is the orthogonal complement of the rowspace of A.

D3 (a) Show that the intersection of two subspaces of $\mathbb{R}^n$ is a subspace of $\mathbb{R}^n$. (The intersection of sets U and V is written $U \cap V$ and defined to be the set of elements $\mathbf{x}$ such that $\mathbf{x} \in U$ *and* $\mathbf{x} \in V$.)
(b) Give an example to show that the union of two subspaces of $\mathbb{R}^n$ is not in general a subspace of $\mathbb{R}^n$. (The union $U \cup V$ is defined to be the set of all $\mathbf{x}$ that are in at least one of U or V. An example in $\mathbb{R}^2$ will suffice.)
(c) If U and V are subsets of $\mathbb{R}^n$, $U + V$ is defined to be the set of vectors $\{\mathbf{u} + \mathbf{v} \mid \mathbf{u} \in U, \mathbf{v} \in V\}$. If U and V are subspaces, show that $U + V$ is a subspace of $\mathbb{R}^n$.

D4 Suppose that $L : \mathbb{R}^n \to \mathbb{R}^m$ and $M : \mathbb{R}^m \to \mathbb{R}^p$ are linear mappings.
(a) Show that the range of $M \circ L$ is a subspace of the range of M. Give an example such that the range of $M \circ L$ is not equal to the range of M. (Hint: What if L is the zero mapping? Now give an example where L is not the zero mapping.)
(b) Show that the nullspace of L is a subspace of the nullspace of $M \circ L$.

SECTION

3–5 The Matrix Inverse and Inverse Transformations

The *matrix inverse* is defined in terms of matrix multiplication; the best procedure for finding a matrix inverse is usually row reduction. Since we have the background and the techniques for discussing the matrix inverse, we introduce it at this point. The most important application of the matrix inverse in this book is in the discussion of "change of basis" (Section 4–6, Chapters 6, 7, 8); since that discussion requires several new ideas, it is best to master the matrix inverse now.

DEFINITION

Let A be an n by n matrix. If B is a matrix such that $BA = AB = I$, then B is said to be the **inverse of the matrix** A (and A is the inverse of B). The inverse of A is usually denoted A^{-1}. If A has an inverse, A is said to be **invertible**.

EXAMPLE 39

The matrix $\begin{bmatrix} 2 & -1 \\ -1 & 1 \end{bmatrix}$ is the inverse of the matrix $\begin{bmatrix} 1 & 1 \\ 1 & 2 \end{bmatrix}$, because

$$\begin{bmatrix} 1 & 1 \\ 1 & 2 \end{bmatrix} \begin{bmatrix} 2 & -1 \\ -1 & 1 \end{bmatrix} = \begin{bmatrix} 1 & 0 \\ 0 & 1 \end{bmatrix} = I$$

and

$$\begin{bmatrix} 2 & -1 \\ -1 & 1 \end{bmatrix} \begin{bmatrix} 1 & 1 \\ 1 & 2 \end{bmatrix} = \begin{bmatrix} 1 & 0 \\ 0 & 1 \end{bmatrix} = I.$$

Notice that the definition requires us to calculate both AB and BA.

Notice that in the definition, B is *the* inverse of A. This depends on the easily proved fact that the inverse is unique.

THEOREM 1. *(Uniqueness of the matrix inverse)* If A is a square matrix, and $BA = AB = I$, and $CA = AC = I$, then $B = C$.

Proof. $B = BI = B(AC) = (BA)C = IC = C$. ■

REMARK

Note that the proof uses less than the full assumptions of the theorem: we have proved that if $BA = I = AC$, then $B = C$. Sometimes people say that if $BA = I$, B is a "left inverse" of A. The proof shows that any left inverse must be equal to any right inverse. However, in general settings, a *mapping* may have a right inverse and *not have* a left inverse (see Exercise D4); Theorem 3 will show that for n by n matrices a right inverse is automatically a left inverse.

We can apply the matrix inverse to find the solution of a system of linear equations $A\mathbf{x} = \mathbf{b}$, where the coefficient matrix A is an invertible square matrix with inverse A^{-1}. Multiply both sides of the vector equation $A\mathbf{x} = \mathbf{b}$ on the left by A^{-1}:

$$A^{-1}A\mathbf{x} = A^{-1}\mathbf{b}.$$

Since $A^{-1}A = I$ and $I\mathbf{x} = \mathbf{x}$, we get

$$\mathbf{x} = A^{-1}\mathbf{b},$$

and the solution is displayed as a product of A^{-1} and $\mathbf{b}$.

EXAMPLE 40

Let $A = \begin{bmatrix} 1 & 1 \\ 1 & 2 \end{bmatrix}$.

(a) Find the solution of $A\mathbf{x} = \begin{bmatrix} 2 \\ 4 \end{bmatrix}$.

(b) Find the solution of $A\mathbf{x} = \begin{bmatrix} -3 \\ 1 \end{bmatrix}$.

SOLUTION: (a) By Example 39, $A^{-1} = \begin{bmatrix} 2 & -1 \\ -1 & 1 \end{bmatrix}$. The solution of $A\mathbf{x} = \begin{bmatrix} 2 \\ 4 \end{bmatrix}$ is

$$\mathbf{x} = A^{-1}A\mathbf{x} = A^{-1}\begin{bmatrix} 2 \\ 4 \end{bmatrix} = \begin{bmatrix} 2 & -1 \\ -1 & 1 \end{bmatrix}\begin{bmatrix} 2 \\ 4 \end{bmatrix} = \begin{bmatrix} 0 \\ 2 \end{bmatrix}.$$

(b) Similarly, the solution of $A\mathbf{x} = \begin{bmatrix} -3 \\ 1 \end{bmatrix}$ is

$$\mathbf{x} = \begin{bmatrix} 2 & -1 \\ -1 & 1 \end{bmatrix}\begin{bmatrix} -3 \\ 1 \end{bmatrix} = \begin{bmatrix} -7 \\ 4 \end{bmatrix}.$$

The procedure used in Example 40 is not more efficient for solving systems than solution by row reduction and back-substitution unless A^{-1} is known in advance. (Here efficiency is measured by counting the arithmetic operations used in solving directly and comparing with the number used in determining A^{-1} and the product $A^{-1}\mathbf{b}$). However, the fact that we can write the solution of $A\mathbf{x} = \mathbf{b}$ in the form $\mathbf{x} = A^{-1}\mathbf{b}$ will be very useful when we discuss change of basis in Section 4–6.

Inverse Transformation

It is useful to introduce inverse transformations here because many geometrical transformations provide nice examples of inverses. Note that just as the inverse matrix is defined only for square matrices, the inverse transformation is defined only for linear transformations that map some $\mathbb{R}^n$ to itself.

DEFINITION

If $L : \mathbb{R}^n \to \mathbb{R}^n$ is a linear transformation, and there exists another linear transformation $M : \mathbb{R}^n \to \mathbb{R}^n$ such that $M \circ L = L \circ M = Id : \mathbb{R}^n \to \mathbb{R}^n$, then M is defined to be the **inverse of** L (and L is also the inverse of M). The inverse of L is usually denoted L^{-1}. A linear transformation is **invertible** if it has an inverse.

THEOREM 2. **Suppose that $L : \mathbb{R}^n \to \mathbb{R}^n$ is a linear transformation with standard matrix $[L] = A$, and that $M : \mathbb{R}^n \to \mathbb{R}^n$ is a linear transformation with standard matrix $[M] = B$. Then L is the inverse transformation of M if and only if A is the inverse matrix of B.**

Proof. Composition of mappings is represented by products of matrices, so $L \circ M = Id = M \circ L$ if and only if $AB = I = BA$.

■

For many of the geometrical transformations of Section 3–3, an inverse transformation exists by geometrical arguments, and these provide many examples of inverse matrices. You should determine the matrix for the transformation and its inverse in each of the following cases and check by multiplication that the product is I:

(1) the rotation R_θ of the plane, with $(R_\theta)^{-1} = R_{(-\theta)}$;

(2) in the plane, (stretch by factor k in the x_1-direction)$^{-1}$ = (stretch by factor $1/k$ in the x_1-direction);

(3) (contraction by factor k)$^{-1}$ = (dilatation by factor $(1/k)$);

(4) (shear by amount k)$^{-1}$ = shear by amount $(-k)$ in the same direction;

(5) for any normal $\mathbf{n}$ in $\mathbb{R}^3$, $(\mathbf{refl_n})^{-1} = \mathbf{refl_n}$ (this mapping is its own inverse!).

However, many linear transformations (and therefore matrices) do not have inverses. For example, $\mathbf{proj_n}$ and $\mathbf{perp_n}$ turn out not to have inverses. (This will be explained in Example 42 later in this section, but you may be able to see why it is true if you think about it now.)

Some Properties of the Matrix Inverse

Suppose that A and B are invertible matrices and that k is a non-zero real number. Verify the following statements:

(1) $(kA)^{-1} = (1/k)A^{-1}$;

(2) $(AB)^{-1} = B^{-1}A^{-1}$;

(3) $(A^T)^{-1} = (A^{-1})^T$.

For example, to verify (2), consider

$$B^{-1}A^{-1}(AB) = B^{-1}(A^{-1}A)B = B^{-1}B = I$$

and

$$(AB)(B^{-1}A^{-1}) = AA^{-1} = I.$$

It follows that $B^{-1}A^{-1}$ is the inverse matrix of AB. The proofs of (1) and (3) are similar.

We often use Property (2) in the following form:
the product of invertible matrices is invertible.

A Procedure for Finding a Matrix Inverse

Some students may have learned the cofactor method for finding matrix inverses in a previous course. The cofactor method is useful for some purposes, so it will be discussed in Chapter 5 on determinants. However, it is extremely inefficient for actually calculating inverses for matrices that are 4 by 4 or larger, so it is essential to learn the next procedure.

Suppose that A is a square matrix. Does it have an inverse? And if so, how do we find A^{-1}? Fortunately, one procedure answers both questions. We begin by trying to solve the matrix equation $AX = I$ for the unknown square matrix X. If a solution X can be found, it is a "right inverse" for A (see the Remark following Theorem 1). We can then show (Theorem 3 below) that X must automatically also satisfy $XA = I$ so that X is the required inverse. On the other hand, if the system is inconsistent, there can be no matrix X satisfying the equation, so A is not invertible.

To keep it simple, the procedure will be examined in the case where A is 3 by 3, but it should be clear that it can be applied to any square matrix. Write the matrix equation $AX = I$ in the form:

$$AX = A\Big[[x_{\downarrow 1}] \quad [x_{\downarrow 2}] \quad [x_{\downarrow 3}]\Big] = \begin{bmatrix} 1 & 0 & 0 \\ 0 & 1 & 0 \\ 0 & 0 & 1 \end{bmatrix}$$

$$= \Big[[\mathbf{e}_1] \quad [\mathbf{e}_2] \quad [\mathbf{e}_3]\Big].$$

By block multiplication, this is equivalent to

$$A[x_{\downarrow 1}] = [\mathbf{e}_1], \quad A[x_{\downarrow 2}] = [\mathbf{e}_2], \quad A[x_{\downarrow 3}] = [\mathbf{e}_3],$$

so it is necessary to solve three systems of equations, one for each column of X. Note that each system has a different standard basis vector as its right-hand side, but all have the same coefficient matrix. Since the solution procedure for systems of equations requires that we row reduce the coefficient matrix, we might as well write one "three times augmented matrix" and solve all three systems at once. Therefore, write

$$[A \mid \mathbf{e}_1 \mid \mathbf{e}_2 \mid \mathbf{e}_3] = [\, A \mid I \,]$$

and row reduce to solve. It is best here to obtain the reduced row echelon form of A (that is, solve the systems by complete elimination).

Suppose that A is row equivalent to the identity I, and call the resulting block on the right B so that the reduction gives

$$[\, A \mid I \,] \to [\, I \mid B \,].$$

Now the final matrix must be interpreted by breaking it up into three systems:

$$I[x_{\downarrow 1}] = [b_{\downarrow 1}], \quad I[x_{\downarrow 2}] = [b_{\downarrow 2}], \quad I[x_{\downarrow 3}] = [b_{\downarrow 3}],$$

and it follows that the first column of the desired matrix X is **exactly** the first column of B, the second column of X is the second column of B, and so on. Thus, A is invertible and its inverse has been found.

If the reduced row echelon form of A is not I, it must contain a row of zeros. It follows from Theorem 4 below that in this case the system $AX = I$ can have no solution and A is not invertible. First, we summarize the procedure, and give an example.

Summary of Procedure: To find the inverse of the square matrix A, row reduce the matrix $[\, A \mid I \,]$ until the left block is in reduced row echelon form. If the reduced row echelon form is I so that $[\, A \mid I \,] \to [\, I \mid B \,]$, then $B = A^{-1}$. If the reduced row echelon form of A is not I, then A is not invertible.

EXAMPLE 41

Determine whether $A = \begin{bmatrix} 1 & 1 & 2 \\ 1 & 2 & 2 \\ 2 & 4 & 3 \end{bmatrix}$ is invertible, and if so, determine its inverse.

SOLUTION: Write the matrix $[\, A \mid I \,]$ and row reduce.

$$\left[\begin{array}{ccc|ccc} 1 & 1 & 2 & 1 & 0 & 0 \\ 1 & 2 & 2 & 0 & 1 & 0 \\ 2 & 4 & 3 & 0 & 0 & 1 \end{array}\right] \to \left[\begin{array}{ccc|ccc} 1 & 1 & 2 & 1 & 0 & 0 \\ 0 & 1 & 0 & -1 & 1 & 0 \\ 0 & 2 & -1 & -2 & 0 & 1 \end{array}\right] \to$$

$$\left[\begin{array}{ccc|ccc} 1 & 0 & 2 & 2 & -1 & 0 \\ 0 & 1 & 0 & -1 & 1 & 0 \\ 0 & 0 & -1 & 0 & -2 & 1 \end{array}\right] \to \left[\begin{array}{ccc|ccc} 1 & 0 & 0 & 2 & -5 & 2 \\ 0 & 1 & 0 & -1 & 1 & 0 \\ 0 & 0 & 1 & 0 & 2 & -1 \end{array}\right].$$

Hence, A is invertible and $A^{-1} = \begin{bmatrix} 2 & -5 & 2 \\ -1 & 1 & 0 \\ 0 & 2 & -1 \end{bmatrix}$.

Check. You should check that the inverse has been correctly calculated by verifying that $AA^{-1} = I$. In this case,

$$\begin{bmatrix} 1 & 1 & 2 \\ 1 & 2 & 2 \\ 2 & 4 & 3 \end{bmatrix} \begin{bmatrix} 2 & -5 & 2 \\ -1 & 1 & 0 \\ 0 & 2 & -1 \end{bmatrix} = \begin{bmatrix} 1 & 0 & 0 \\ 0 & 1 & 0 \\ 0 & 0 & 1 \end{bmatrix}.$$

Now we must justify our claim that a "right inverse" is *the* inverse.

THEOREM 3. Suppose that A and B are n by n matrices such that $AB = I$. Then $BA = I$, so that $B = A^{-1}$; moreover, B and A have rank n.

Proof. The procedure says that $AB = I$ if and only if there is a sequence of elementary row operations (EROs) such that

$$[\, A \mid I \,] \to [\, I \mid B \,].$$

This means that the sequence of EROs reduces A to I and I to B. Therefore, we can say that the same sequence of EROs reduces

$$[\, I \mid A \,] \to [\, B \mid I \,].$$

But elementary row operations are reversible, so it follows that there is a sequence of EROs such that

$$[\, B \mid I \,] \to [\, I \mid A \,].$$

From the procedure, we know that this is true if and only if $BA = I$, so that B is the inverse of A, as claimed. A and B have rank n because each is row equivalent to I.

■

We now show that Theorem 3 can be proved without referring to the procedure for finding A^{-1}.

Alternative proof of Theorem 3. We first show that B has rank n, by contradiction. Suppose that B has rank less than n. Then the homogeneous system $B\mathbf{x} = \mathbf{0}$ has non-trivial solutions by Theorem 3 of Section 2–2. But this means that for some non-zero $\mathbf{x}$, $AB\mathbf{x} = \mathbf{0}$, so AB is certainly not equal to I. Since this contradicts our assumption, B must have rank n.

Since B has rank n, the nonhomogeneous system $\mathbf{y} = B\mathbf{x}$ is consistent for every $\mathbf{y}$ in $\mathbb{R}^n$ (Theorem 2, Section 2–2). This means that for every $\mathbf{y}$ there is an $\mathbf{x}$ such that $\mathbf{y} = B\mathbf{x}$. Now consider:

$$BA\mathbf{y} = BA(B\mathbf{x}) = B(AB)\mathbf{x} = BI\mathbf{x} = B\mathbf{x} = \mathbf{y}.$$

Thus, $BA\mathbf{y} = \mathbf{y}$ for every $\mathbf{y}$ in $\mathbb{R}^n$, so $BA = I$ (by Theorem 2 of Section 3–1). Therefore, $AB = I$ and $BA = I$, so that $B = A^{-1}$.

That A must have rank n follows from $BA = I$. It can also be proved directly from $AB = I$.

■

Some Facts About Square Matrices and Solutions of Linear Systems

In Theorem 3 and in the description of the procedure for finding the inverse matrix, some facts were used about systems of equations with square matrices. It is worth stating them clearly as a theorem. Most of the conclusions are simply special cases of results in Sections 2–1, 2–2, and 3–4.

THEOREM 4. Suppose that A is an n by n matrix. Then the following statements are equivalent (that is, one is true if and only if each of the others is true).

(1) The rank of A is n.

(2) The reduced row echelon form of A is I.

(3) The homogeneous system $A\mathbf{x} = \mathbf{0}$ has only the trivial solution.

(4) For every $\mathbf{d}$ in $\mathbb{R}^n$, the nonhomogeneous system $A\mathbf{x} = \mathbf{d}$ is consistent.

(5) The matrix A is invertible.

Proof. (We use the "implication arrow": "P $\Rightarrow$ Q" means "if P then Q" and "P $\Leftarrow$ Q" means "if Q then P"; "P $\Leftrightarrow$ Q" means both. It is common in proving a theorem such as this to prove (1) $\Rightarrow$ (2) $\Rightarrow$ (3) $\Rightarrow$ (4) $\Rightarrow$ (5) $\Rightarrow$ (1), so that any statement implies any other. To emphasize direct connections between some of the statements, we prove more than is required in this case.)

(1) $\Leftrightarrow$ (2): The rank of A is $n \Leftrightarrow$ the reduced row echelon form of A has n leading 1's $\Leftrightarrow A$ has reduced row echelon form equal to I (see Section 2–2).

(1) $\Leftrightarrow$ (3): This follows from the procedure for solving systems; it was stated as part of Theorem 3 in Section 2–2.

(2) $\Rightarrow$ (4): This follows from Theorem 2 in Section 2–2.

(4) $\Rightarrow$ (5): If (4) is true, solve $A\mathbf{x} = \mathbf{e}_i$, $i = 1, 2, \ldots, n$ to obtain the columns of A^{-1}. Hence, (5) is true by Theorem 3.

(5) $\Rightarrow$ (3): If (5) is true, then $A\mathbf{x} = \mathbf{0}$ implies that $\mathbf{x} = A^{-1}A\mathbf{x} = \mathbf{0}$.

It is now possible to construct a chain of implication arrows from any statement to any other (for example, (5) $\Rightarrow$ (3) $\Rightarrow$ (1) $\Rightarrow$ (2) $\Rightarrow$ (4) $\Rightarrow$ (5)), so the proof is complete.

■

More About Inverse Transformations

Sometimes the ideas about inverses are clearer when stated for inverse transformations. The definition of the inverse of a linear transformation was given earlier in this section. Students who have taken calculus may be more familiar with the following definition of the inverse.

Alternative definition: Let $L : \mathbb{R}^n \to \mathbb{R}^n$ be a linear transformation. A linear transformation $M : \mathbb{R}^n \to \mathbb{R}^n$ is defined to be the **inverse of** L if $\mathbf{x} = M(\mathbf{y})$ whenever $\mathbf{y} = L(\mathbf{x})$.

REMARKS

(1) In mathematics, if two definitions are given for the same concept, it is necessary to check that each definition implies the other. Here it is easy to check that the second definition implies the first (since the second definition implies that $L(M(\mathbf{y})) = \mathbf{y}$ for all $\mathbf{y}$ in $\mathbb{R}^n$, and $M(L(\mathbf{x})) = \mathbf{x}$ for all $\mathbf{x}$ in $\mathbb{R}^n$). It is almost as easy to show that the first implies the second, and we leave it to the reader.

(2) If $L : \mathbb{R}^n \to \mathbb{R}^n$ is linear and there is some map M (not **assumed** to be linear) that acts as an inverse, then it can be shown that M is necessarily linear. You are asked to show this in Exercise D3.

There are some direct consequences of the definition (in either form) of the inverse transformation. First, notice that if $\mathbf{y}$ in $\mathbb{R}^n$ is in the domain of the inverse M, then it must be in the range of the original L. Therefore, it follows that if L has an inverse, the range of L must be all of the codomain $\mathbb{R}^n$. This property is the linear transformation version of statement (4) of Theorem 4 about square matrices.

Note also that if $L(\mathbf{x}) = \mathbf{y}$, then an inverse M can be defined by $M(\mathbf{y}) = \mathbf{x}$ only if for each $\mathbf{y}$ in the range of L, there is **exactly** one $\mathbf{x}$ such that $L(\mathbf{x}) = \mathbf{y}$. In particular, if L is invertible, there can be only one vector $\mathbf{x}$ such that $L(\mathbf{x}) = \mathbf{0}$, and it is obviously the zero vector; thus, if L is invertible, the nullspace of L consists of only the zero vector.

We can now give an expanded version of Theorem 4, relating the statements there to statements about the corresponding linear transformation.

THEOREM 5. *(An upgrading of Theorem 4)*

Suppose that $L : \mathbb{R}^n \to \mathbb{R}^n$ is a linear transformation with matrix $[L] = A$. Then the following statements are equivalent to each other and to the statements of Theorem 4.

(6) L is invertible.

(7) The range of L is $\mathbb{R}^n$.

(8) The nullspace of L is $\{\mathbf{0}\}$.

Proof. (5) $\Leftrightarrow$ (6): This was Theorem 2.

(4) $\Leftrightarrow$ (7): The range of L is the columnspace of A by Theorem 3 of Section 3–4, and $A\mathbf{x} = \mathbf{d}$ has a solution if and only if $\mathbf{d}$ is in the columnspace of A by Theorem 4 of Section 3–4.

(3) $\Leftrightarrow$ (8): Theorem 1 of Section 3–4.

Every statement can now be connected to every other statement by a chain of implication arrows. ∎

REMARKS

(3) It is also possible to give alternative proofs — for example, (6) ⇒ (7) and (6) ⇒ (8) by the discussion preceding the theorem; or (8) ⇒ (3) ⇒ (4) ⇒ (7).

(4) It would be possible, following the discussion of linear independence, dimension, and rank in Chapter 4, to give alternative proofs for Theorems 4 and 5.

EXAMPLE 42

Earlier in this section we said that in $\mathbb{R}^n$, for any **n**, $\mathbf{proj_n}$ and $\mathbf{perp_n}$ are not invertible linear transformations. It is now easy to explain why they cannot be invertible. The range of $\mathbf{proj_n}$ consists of multiples of **n**, so $\mathbf{proj_n}$ does not satisfy the condition that its range is all of $\mathbb{R}^3$ and so it cannot be invertible; also, the nullspace of $\mathbf{proj_n}$ consists of all vectors orthogonal to **n**, so the nullspace of $\mathbf{proj_n}$ contains non-zero vectors and $\mathbf{proj_n}$ cannot be invertible. $\mathbf{perp_n}$ fails to be invertible because its range is the set of vectors orthogonal to **n** (not all of $\mathbb{R}^n$), and its nullspace is Sp(**n**) (not just $\{\mathbf{0}\}$).

Finally, recall that the matrix condition $AB = BA = I$ implies that the matrix inverse can be defined only for square matrices. Here is an example that illustrates for linear mappings that the domain and codomain of L must be the same if $L : \mathbb{R}^n \to \mathbb{R}^n$ is to have an inverse.

EXAMPLE 43

Consider the linear mappings $P : \mathbb{R}^4 \to \mathbb{R}^3$ defined by $P(x_1, x_2, x_3, x_4) = (x_1, x_2, x_3)$ and $inj : \mathbb{R}^3 \to \mathbb{R}^4$ defined by $inj(x_1, x_2, x_3) = (x_1, x_2, x_3, 0)$.

It is easy to see that $P \circ inj = Id : \mathbb{R}^3 \to \mathbb{R}^3$, but that $inj \circ P \neq Id : \mathbb{R}^4 \to \mathbb{R}^4$. Thus, P is not an inverse for *inj*.

Notice that P satisfies the condition that its range is all of its codomain, but it fails the condition that its nullspace is trivial. On the other hand, *inj* satisfies the condition that its nullspace is trivial, but it fails the condition that its range is all of its codomain.

Some of the ideas of this section are discussed in a more general setting in Section 4–7.

EXERCISES 3–5

A1 For each of the following matrices, either show that the matrix is not invertible or find its inverse. Do these by hand, using the procedure of this section, and check by multiplication. You may also want to check by computer.

$$A = \begin{bmatrix} 3 & -4 \\ 2 & 5 \end{bmatrix}; \quad B = \begin{bmatrix} 1 & 0 & 1 \\ 2 & 1 & 3 \\ 1 & 0 & 2 \end{bmatrix}; \quad C = \begin{bmatrix} 1 & 0 & 2 \\ 1 & 1 & 3 \\ 3 & 1 & 7 \end{bmatrix}$$

$$D = \begin{bmatrix} 0 & 0 & 1 \\ 0 & 1 & 1 \\ 1 & 1 & 1 \end{bmatrix}; \quad E = \begin{bmatrix} 1 & 1 & 3 & 1 \\ 0 & 2 & 1 & 0 \\ 2 & 2 & 7 & 1 \\ 0 & 6 & 3 & 1 \end{bmatrix}; \quad F = \begin{bmatrix} 1 & 0 & 1 & 0 & 1 \\ 0 & 1 & 0 & 1 & 0 \\ 0 & 0 & 1 & 1 & 1 \\ 0 & 0 & 0 & 1 & 2 \\ 0 & 0 & 0 & 0 & 1 \end{bmatrix}.$$

A2 Let B equal the matrix B from Exercise A1. Use B^{-1} to find the solutions of the following.

(a) $B\mathbf{x} = \begin{bmatrix} 1 \\ 1 \\ 1 \end{bmatrix}$ (b) $B\mathbf{x} = \begin{bmatrix} -1 \\ 0 \\ 1 \end{bmatrix}$

(c) $B\mathbf{x} = \begin{bmatrix} 0 \\ 1 \\ 2 \end{bmatrix} = \left(\begin{bmatrix} 1 \\ 1 \\ 1 \end{bmatrix} + \begin{bmatrix} -1 \\ 0 \\ 1 \end{bmatrix}\right)$

A3 Let $A = \begin{bmatrix} 2 & 1 \\ 3 & 2 \end{bmatrix}$, $B = \begin{bmatrix} 1 & 2 \\ 3 & 5 \end{bmatrix}$.

(a) Find A^{-1} and B^{-1}.

(b) Calculate AB and $(AB)^{-1}$, and check that $(AB)^{-1} = B^{-1}A^{-1}$.

(c) Calculate $(3A)^{-1}$ and check that it is equal to $\frac{1}{3}(A^{-1})$.

(d) Calculate $(A^T)^{-1}$ and check that $(A^T)^{-1} = (A^{-1})^T$.

A4 By geometrical arguments, determine the inverse of each of the following matrices.

(a) the matrix of the rotation $R_{\pi/6}$ in the plane

(b) $\begin{bmatrix} 1 & -3 \\ 0 & 1 \end{bmatrix}$ (c) $\begin{bmatrix} 5 & 0 \\ 0 & 5 \end{bmatrix}$ (d) $\begin{bmatrix} 1 & 0 \\ 0 & -1 \end{bmatrix}$

A5 The mappings in this question are from $\mathbb{R}^2$ to $\mathbb{R}^2$.

(a) Determine the matrix of the shear S by amount 2 in the x_2-direction and the matrix of S^{-1}.

(b) Determine the matrix of the reflection R in the line $x_1 - x_2 = 0$ and the matrix of R^{-1}.

(c) Determine the matrix of $(R \circ S)^{-1}$ and the matrix of $(S \circ R)^{-1}$ (without determining the matrices of $R \circ S$ and $S \circ R$).

B1 For each of the following matrices, either show that the matrix is not invertible or find its inverse. Do these by hand, using the procedure of this section, and check by multiplication. You may also want to check by computer.

$$A = \begin{bmatrix} 1 & -1 & 2 \\ 3 & 1 & 5 \\ 2 & 2 & 3 \end{bmatrix}; \quad C = \begin{bmatrix} 1 & 2 & 0 \\ 2 & 2 & 5 \\ 1 & -1 & 3 \end{bmatrix};$$

$$E = \begin{bmatrix} 1 & -1 & 0 & 2 \\ 0 & 1 & 1 & 0 \\ 2 & -2 & 3 & 5 \\ 1 & 0 & 1 & 3 \end{bmatrix}; \quad F = \begin{bmatrix} 1 & 0 & 0 & 0 & 1 \\ 0 & 0 & 0 & 1 & 0 \\ 0 & 0 & 1 & 0 & 0 \\ 0 & 1 & 0 & 0 & 0 \\ 1 & 0 & 0 & 0 & 0 \end{bmatrix}.$$

B2 Same instructions as B1.

$$G = \begin{bmatrix} 1 & 1 & -2 \\ 2 & 1 & 5 \\ 4 & 3 & 1 \end{bmatrix}; \quad H = \begin{bmatrix} 2 & -1 & 3 \\ 1 & 2 & 2 \\ 1 & 0 & 1 \end{bmatrix};$$

$$J = \begin{bmatrix} 0 & 2 & 2 & 5 \\ 0 & 1 & 0 & 3 \\ 1 & 3 & 1 & 3 \\ 3 & 6 & 0 & 3 \end{bmatrix}; \quad K = \begin{bmatrix} 0 & 0 & 0 & 0 & 1 \\ 0 & 0 & 0 & 1 & 0 \\ 0 & 0 & 1 & 1 & 1 \\ 0 & 1 & 0 & 0 & 0 \\ 1 & 0 & 0 & 0 & 0 \end{bmatrix}.$$

B3 (a) If $B = \begin{bmatrix} 2 & -1 & 1 \\ 0 & 1 & 1 \\ 1 & -1 & -1 \end{bmatrix}$, find B^{-1}.

(b) Use B^{-1} as calculated in part (a) to solve $B\mathbf{x} = \mathbf{d}$ if
(i) $\mathbf{d} = (2, 3, -1)$; (ii) $\mathbf{d} = 3(2, 3, -1)$; (iii) $\mathbf{d} = (4, -2, 3)$.

B4 Let $A = \begin{bmatrix} 2 & 5 \\ 1 & 2 \end{bmatrix}$, $B = \begin{bmatrix} 1 & 2 \\ 3 & 5 \end{bmatrix}$.

(a) Find A^{-1} and B^{-1}.
(b) Calculate AB and $(AB)^{-1}$, and check that $(AB)^{-1} = B^{-1}A^{-1}$.
(c) Calculate $(5A)^{-1}$ and check that it is equal to $\frac{1}{5}(A^{-1})$.
(d) Calculate $(A^T)^{-1}$ and check that $(A^T)^{-1} = (A^{-1})^T$.

B5 By geometrical arguments, determine the inverse of each of the following matrices.

(a) the matrix of the rotation $R_{\pi/4}$ in the plane

(b) $\begin{bmatrix} 1 & 0 \\ 2 & 1 \end{bmatrix}$ (c) $\begin{bmatrix} 5 & 0 \\ 0 & 1 \end{bmatrix}$ (d) $\begin{bmatrix} -1 & 0 \\ 0 & -1 \end{bmatrix}$

B6 The mappings in this question are from $\mathbb{R}^2$ to $\mathbb{R}^2$.
(a) Determine the matrix of the stretch S by factor 3 in the x_2-direction and the matrix of S^{-1}.
(b) Determine the matrix of the reflection R in the line $x_1 + x_2 = 0$ and the matrix of R^{-1}.
(c) Determine the matrix of $(R \circ S)^{-1}$ and the matrix of $(S \circ R)^{-1}$ (without determining the matrices of $R \circ S$ and $S \circ R$).

B7 For each of the following pairs of linear mappings from $\mathbb{R}^3$ to $\mathbb{R}^3$, determine the matrices $[R^{-1}]$, $[S^{-1}]$, and $[(R \circ S)^{-1}]$.
(a) R is the rotation about the x_1-axis through angle $\pi/2$, S is the stretch by factor 0.5 in the x_3-direction.
(b) R is reflection in the plane $x_1 - x_3 = 0$, S is a shear by amount 0.4 in the x_3-direction in the x_1x_3-plane.

COMPUTER RELATED EXERCISES

C1 Use a computer to determine the inverse of

$$A = \begin{bmatrix} 1.23 & 3.11 & 1.01 & 0.00 \\ 2.01 & -2.56 & 3.03 & 0.04 \\ 1.11 & 0.03 & -5.11 & 2.56 \\ 2.14 & -1.90 & 4.05 & 1.88 \end{bmatrix}.$$

(For your own information, invert A^{-1} and consider the question of accuracy.)

CONCEPTUAL EXERCISES

D1 (a) Show that $(A^T)^{-1} = (A^{-1})^T$.
(b) Determine an expression in terms of A^{-1} and B^{-1} for $((AB)^T)^{-1}$.

D2 (a) Suppose that A is an n by n matrix such that $A^3 = I$. Find an expression for A^{-1} in terms of A. [Hint: Find X such that $AX = I$.]
(b) Suppose that B satisfies $B^5 + B^3 + B = I$. Find an expression for B^{-1} in terms of B.

***D3** Suppose that $L : \mathbb{R}^n \to \mathbb{R}^n$ is a linear transformation and that $M : \mathbb{R}^n \to \mathbb{R}^n$ is a function (not assumed to be linear) such that $\mathbf{x} = M(\mathbf{y})$ whenever $\mathbf{y} = L(\mathbf{x})$. Show that M must be linear.

D4 Let S denote the set of all infinite sequences of real numbers. A typical element of S is $\mathbf{x} = (x_1, x_2, x_3, \ldots, x_n, \ldots)$. Define addition $\mathbf{x} + \mathbf{y}$ and scalar multiplication $k\mathbf{x}$ in the obvious way. (In Chapter 4, we shall see that this makes S a "vector space".) Define the left shift $L : S \to S$ by $L(x_1, x_2, x_3, \ldots) = (x_2, x_3, x_4, \ldots)$ and the right shift $R : S \to S$ by

$R(x_1, x_2, x_3, \ldots) = (0, x_1, x_2, x_3, \ldots)$. Check that L and R are linear. Check that $L \circ R(\mathbf{x}) = \mathbf{x}$, but that $R \circ L(\mathbf{x}) \neq \mathbf{x}$. L has a right inverse (R), but it does not have a left inverse. It is important in this example that S is "infinite-dimensional". Dimension is discussed in Chapter 4.

***D5** Prove that if A and B are square matrices such that AB is invertible, then A and B are invertible. (There is a fairly short proof.)

SECTION

3–6 Elementary Matrices and the Inverse of a Matrix

Elementary matrices provide an alternative way of looking at the matrix inverse, but they lead to the same procedure for finding inverses by row reduction. In this book, they appear only in this section and in one proof in Section 5–4. Elementary matrices do have other applications but they will not be discussed in this book.

DEFINITION

A matrix E is an **elementary matrix** if it is obtained from the identity matrix I by a single elementary row operation.

It follows from the definition that an elementary matrix is square. All our examples will be 3 by 3 but n by n elementary matrices can easily be constructed for any n.

EXAMPLE 44

$E_1 = \begin{bmatrix} 1 & 0 & k \\ 0 & 1 & 0 \\ 0 & 0 & 1 \end{bmatrix}$ is elementary for any number $k \neq 0$, since it is obtained from I_3 by adding k times the third row to the first row—a single elementary row operation.

EXAMPLE 45

$E_2 = \begin{bmatrix} 1 & 0 & 0 \\ 0 & 0 & 1 \\ 0 & 1 & 0 \end{bmatrix}$ is elementary since it is obtained from I_3 by interchanging the second and third rows.

EXAMPLE 46

$\begin{bmatrix} 2 & 0 & 3 \\ 0 & 1 & 0 \\ 0 & 0 & 1 \end{bmatrix}$ is **not** elementary since it requires **two** elementary row operations to obtain this matrix from I_3.

The following lemma explains why elementary matrices are introduced: they make it possible to represent elementary row operations by matrix multiplication. (A "lemma" is a "little" theorem that is primarily interesting as a step to a bigger theorem.)

LEMMA 1. If A is an n by n matrix and E is the elementary matrix obtained from I_n by a certain elementary row operation, then the product EA is the matrix obtained from A by the same row operation.

It would be tedious and not very illuminating to write the proof in the general n by n case. Instead, we just illustrate why this works by verifying the conclusion for some simple cases. Let A be a 3 by 3 matrix.

Case 1. Consider the elementary row operation that consists of adding k times the third row to the first. This is the elementary row operation of Example 44, which corresponds to the elementary matrix E_1. Then,

$$\begin{bmatrix} a_{11} & a_{12} & a_{13} \\ a_{21} & a_{22} & a_{23} \\ a_{31} & a_{32} & a_{33} \end{bmatrix} \xrightarrow{R_1 + kR_3} \begin{bmatrix} a_{11} + ka_{31} & a_{12} + ka_{32} & a_{13} + ka_{33} \\ a_{21} & a_{22} & a_{23} \\ a_{31} & a_{32} & a_{33} \end{bmatrix},$$

while

$$E_1A = \begin{bmatrix} 1 & 0 & k \\ 0 & 1 & 0 \\ 0 & 0 & 1 \end{bmatrix} \begin{bmatrix} a_{11} & a_{12} & a_{13} \\ a_{21} & a_{22} & a_{23} \\ a_{31} & a_{32} & a_{33} \end{bmatrix} = \begin{bmatrix} a_{11} + ka_{31} & a_{12} + ka_{32} & a_{13} + ka_{33} \\ a_{21} & a_{22} & a_{23} \\ a_{31} & a_{32} & a_{33} \end{bmatrix},$$

which is the same.

Case 2. Consider the elementary row operation of Example 45, and the matrix E_2.

$$\begin{bmatrix} a_{11} & a_{12} & a_{13} \\ a_{21} & a_{22} & a_{23} \\ a_{31} & a_{32} & a_{33} \end{bmatrix} \; R_2 \updownarrow R_3 \rightarrow \begin{bmatrix} a_{11} & a_{12} & a_{13} \\ a_{31} & a_{32} & a_{33} \\ a_{21} & a_{22} & a_{23} \end{bmatrix},$$

while

$$\begin{aligned} E_2 A &= \begin{bmatrix} 1 & 0 & 0 \\ 0 & 0 & 1 \\ 0 & 1 & 0 \end{bmatrix} \begin{bmatrix} a_{11} & a_{12} & a_{13} \\ a_{21} & a_{22} & a_{23} \\ a_{31} & a_{32} & a_{33} \end{bmatrix} \\ &= \begin{bmatrix} a_{11} & a_{12} & a_{13} \\ a_{31} & a_{32} & a_{33} \\ a_{21} & a_{22} & a_{23} \end{bmatrix}, \end{aligned}$$

and again the conclusion of the theorem is verified.

You might wish to check one case involving the third kind of elementary row operation (multiplication of one row by a non-zero constant).

LEMMA 2. If A is a square matrix, there exists a sequence of elementary matrices, $E_1, E_2, \ldots, E_k$, such that $E_k \ldots E_2 E_1 A$ is equal to the reduced row echelon form of A.

Proof. From Chapter 2, there is a sequence of elementary row operations that brings A into its reduced row echelon form. Call the elementary matrix corresponding to the first operation E_1, the elementary matrix corresponding to the second operation E_2, and so on until the final row operation corresponds to E_k. The conclusion follows immediately from the definition of these elementary matrices and the fact that they correspond to row operations that bring A into reduced row echelon form. Note that the matrix of the first operation must appear immediately to the left of A in the product, then E_2 is immediately to the left of that, and so on.

■

So far, what we have done can be applied to any n by p matrix A. We want to apply these ideas to matrix inversion, so from now on in this section, let A be an n by n matrix. Notice that Lemma 2 implies that if A has rank n, then there is a sequence of elementary row operations such that $E_k \ldots E_2 E_1 A = I$.

To discuss the problem of inverting A, we use only the definition of matrix inverse from Section 3–5 and the fact that the product of invertible matrices is invertible. (Recall that the proof of the latter fact consists of showing that if A and B are invertible, then $B^{-1}A^{-1}$ is the inverse for AB. It is easy to extend the property to more than two factors by a proof by induction.)

LEMMA 3. Elementary matrices are invertible. Moreover, the inverse of an elementary matrix is an elementary matrix.

Discussion. This follows from the fact that elementary row operations are reversible. It is easy to check in simple cases. For example, if E_1 is the matrix of Example 44, then $E_1^{-1} = \begin{bmatrix} 1 & 0 & -k \\ 0 & 1 & 0 \\ 0 & 0 & 1 \end{bmatrix}$, and in Example 45, E_2 is its own inverse.

It is now possible to give a new proof of the essential fact about matrix inversion.

THEOREM 1. If an n by n matrix A has rank n (or equivalently, has reduced row echelon form I), then A is invertible. Moreover, its inverse is the product of the elementary matrices that reduce A to its reduced row echelon form.

Proof. There exists a sequence of elementary matrices, $E_1, E_2, \ldots, E_k$, such that $E_k \ldots E_2E_1A = I$. Since each of the elementary matrices is invertible, we may multiply both sides by E_k^{-1} to get

$$E_k^{-1}(E_kE_{k-1} \ldots E_2E_1A) = E_k^{-1}I \quad \text{or} \quad E_{k-1} \ldots E_2E_1A = E_k^{-1}.$$

Next multiply both sides by E_{k-1}^{-1} to get

$$E_{k-2} \ldots E_2E_1A = (E_{k-1})^{-1}E_k^{-1}.$$

Continue multiplying by the inverse of the elementary matrix on the left until the equation becomes

$$A = E_1^{-1}E_2^{-1} \ldots (E_{k-1})^{-1}E_k^{-1}.$$

Since A is the product of invertible matrices, it is invertible and its inverse is

$$A^{-1} = E_k \ldots E_2E_1.$$

■

To get an effective procedure for finding the inverse of A, note that by block multiplication,

$$E_k \ldots E_2E_1\left[\, A \mid I \,\right] = \left[\text{reduced row echelon form of } A \mid E_k \ldots E_2E_1\right],$$

which corresponds to the fact that under elementary row operations,

$$\left[A \mid I\right] \rightarrow \left[\text{reduced row echelon form of } A \mid E_k \ldots E_2 E_1\right].$$

Thus, as in Section 3–5, the procedure is to row reduce $[A \mid I]$ until the left block is in reduced row echelon form: if that form is I, then the resulting block on the right is the required inverse matrix; if the form is not I, then A is not invertible.

Finally, we restate the result of Theorem 1 and Lemma 3 in a form needed in Section 5–4.

THEOREM 2. If an n by n matrix A has rank n, then it may be represented as the product of elementary matrices: $A = F_1 F_2 \ldots F_k$, where each F_i is elementary.

Proof. By Theorem 1, $A = E_1^{-1} E_2^{-1} \ldots (E_{k-1})^{-1} E_k^{-1}$, and by Lemma 3, each E_i^{-1} is elementary. Let F_i be E_i^{-1}, and then $A = F_1 F_2 \ldots F_k$ as required.

■

A NOTE ON *LU*-DECOMPOSITION

The LU-decomposition is a method for organizing the elimination procedure to solve a system $A\mathbf{x} = \mathbf{b}$; it is commonly used in computer algorithms for solving systems of equations. LU-decomposition relies on some ideas about elementary matrices, so we give a brief outline here.

To simplify, we suppose that A is an n by n invertible square matrix, so that the system $A\mathbf{x} = \mathbf{b}$ has a unique solution for any $\mathbf{b}$. As we shall see, the method proceeds by row reducing the coefficient matrix A and *not* the augmented matrix $\left[A \mid \mathbf{b}\right]$. Also, we shall use *only* row operations of the form: "add a multiple of one row to another."

Omitting row interchanges may seem rather serious: without row interchanges, we cannot bring a matrix such as $\begin{bmatrix} 0 & 1 \\ 1 & 2 \end{bmatrix}$ into row echelon form. However, in a computer, one can keep track of row interchanges without physically moving entries from one location to another by simply keeping a list of numbers $(i_1, i_2, \ldots, i_n)$ to tell us that before beginning elimination in the first column, we interchanged row i_1 with row 1; before beginning elimination in the second column, we interchanged row i_2 with row 2, and so on. To simplify our present outline, we assume that *we can accomplish the desired row reduction without row interchanges.*

Also, we do not use "multiplication of a row by a non-zero constant", so we cannot expect to get "leading 1's": the first non-zero entry in a row (or the leading coefficient in the corresponding equation) may be any non-zero number. Thus, we cannot hope to get row echelon form. However, by row reduction, the matrix A may always be brought into **upper triangular form** U where all the entries *below* the diagonal are zeros: $u_{jk} = 0$ if $j > k$. In our present discussion, the rank of A must be n, so we have extra information: in the upper triangular form for our problem, all diagonal entries must be non-zero. For example, an upper triangular 3 by 3 matrix of rank 3 must be of the form

$$U = \begin{bmatrix} u_{11} & u_{12} & u_{13} \\ 0 & u_{22} & u_{23} \\ 0 & 0 & u_{33} \end{bmatrix},$$

where u_{11}, u_{22}, and u_{33} are all non-zero.

Instead of multiplying the jth row by $(u_{jj})^{-1}$ during the row reduction, we must now multiply by $(u_{jj})^{-1}$ as the final step in determining x_j by back-substitution. For efficiency and numerical accuracy, this revised procedure is actually better: we only multiply one number by $(u_{jj})^{-1}$ instead of multiplying all the other non-zero entries in its row; and there is generally less loss in accuracy in computing $(a+b)/c$ than in computing $(a/c)+(b/c)$.

We are left with row operations of the form "add a non-zero multiple of the kth row to the jth row." Since we are trying to obtain zeros in entries *below* the diagonal, we will use these operations only for $k < j$. For example, for a 3 by 3 matrix A, we will add $\left(-a_{21}/a_{11}\right)$ times the first row to the second row. The corresponding elementary matrix is

$$\begin{bmatrix} 1 & 0 & 0 \\ l_{21} & 1 & 0 \\ 0 & 0 & 1 \end{bmatrix},$$

where $l_{21} = \left(-a_{21}/a_{11}\right)$.

In fact, because we are using only one kind of elementary row operation, each of the corresponding elementary matrices will have 1's on the diagonal, a non-zero entry in *one* below-diagonal position, and zeros elsewhere. The below-diagonal entry will be the factor by which some row is multiplied before being added to another row.

Suppose that the elementary matrices used to reduce A to U are called $E_1, E_2, \ldots, E_k$. By an argument similar to the proof of Lemma 2, we have

$$E_k\, E_{k-1} \ldots E_2\, E_1\, A = U.$$

Because of the special form of the E_j 's described in the previous paragraph, it can be shown that

$$E_k\, E_{k-1} \ldots E_2\, E_1 = L,$$

where L is of the form

$$L = \begin{bmatrix} 1 & 0 & 0 & \ldots & 0 \\ l_{21} & 1 & 0 & \ldots & 0 \\ \vdots & \vdots & \vdots & \ddots & \vdots \\ l_{n1} & l_{n2} & l_{n3} & \ldots & 1 \end{bmatrix}.$$

That is, L is *lower triangular*, with 1's on the diagonal. Moreover, it can be seen that each l_{jk} for $j > k$ is exactly the factor by which the kth row is multiplied before it is added to the jth row during the row reduction.

Now we have $LU = A$ and we have achieved the LU-decomposition. Next we use it to solve the system

$$A\mathbf{x} = \mathbf{b} \qquad \text{or} \qquad LU\mathbf{x} = \mathbf{b}.$$

Introduce a vector $\mathbf{y}$ by $\mathbf{y} = U\mathbf{x}$. Now, if we first solve $L\mathbf{y} = \mathbf{b}$ for $\mathbf{y}$, then solve $U\mathbf{x} = \mathbf{y}$ for $\mathbf{x}$, we have the solution $\mathbf{x}$ of our original system. The system $L\mathbf{y} = \mathbf{b}$ is solved by *forward*-substitution (that is, find y_1 first, then $y_2, \ldots$, and finally y_n), and then $U\mathbf{x} = \mathbf{y}$ is solved by back-substitution (that is, x_n first,$\ldots$, x_1 last).

If we keep a record of L and U, we can use this decomposition to solve $A\mathbf{x} = \mathbf{b}$, $A\mathbf{x} = \mathbf{c}, \ldots, A\mathbf{x} = \mathbf{p}$ for different right-hand sides $\mathbf{b}, \mathbf{c}, \ldots, \mathbf{p}$. It can be shown that this procedure is better (for accuracy and efficiency) than computing A^{-1} and using it to solve these systems.

For some computer applications, the available memory may be limited (particularly in older computers). It is useful to note that L and U can be stored *together* in the space originally occupied by A (that is in n^2 locations, rather than in $2n^2$). To see how to do this, note that the diagonal entries of L are all 1's, so there is no need to store them; the other entries of L are all below the diagonal, and the entries of U are all on the diagonal or above. If A is 3 by 3, we store the corresponding L and U in the form

$$\begin{bmatrix} u_{11} & u_{12} & u_{13} \\ l_{21} & u_{22} & u_{23} \\ l_{31} & l_{32} & u_{33} \end{bmatrix}.$$

Moreover, it is quite natural to create this form as we proceed through the row reduction of A.

For example, for the 3 by 3 matrix A, let $l_{21} = (-a_{21}/a_{11}), l_{31} = (-a_{31}/a_{11})$. Then,

$$\begin{bmatrix} a_{11} & a_{12} & a_{13} \\ a_{21} & a_{22} & a_{23} \\ a_{31} & a_{32} & a_{33} \end{bmatrix} \begin{matrix} \\ R_2 + l_{21}R_1 \\ R_3 + l_{31}R_1 \end{matrix} \rightarrow \begin{bmatrix} u_{11} & u_{12} & u_{13} \\ 0 & u_{22} & u_{23} \\ 0 & c_{32} & c_{33} \end{bmatrix},$$

where

$$u_{11} = a_{11},\ u_{12} = a_{12},\ u_{13} = a_{13},\ u_{22} = a_{22} - l_{21}a_{12}, u_{23} = a_{23} - l_{21}a_{13},$$
$$c_{32} = a_{32} - l_{31}a_{12},\ c_{33} = a_{33} - l_{31}a_{13}.$$

In the computer, instead of overwriting A by the matrix $\begin{bmatrix} u_{11} & u_{12} & u_{13} \\ 0 & u_{22} & u_{23} \\ 0 & c_{32} & c_{33} \end{bmatrix}$, replace A by $\begin{bmatrix} u_{11} & u_{12} & u_{13} \\ l_{21} & u_{22} & u_{23} \\ l_{31} & c_{32} & c_{33} \end{bmatrix}$. Now let $l_{32} = (-c_{32}/u_{22})$, and the elementary row operation described by $R_3 + l_{32}R_2$ brings the matrix into the desired LU form.

To find out more about the LU-decomposition and related issues of efficiency and accuracy of procedures for solving systems of linear equations, consult **Numerical Analysis: A Practical Approach** (3rd Edition), by M.J. Maron and R.J. Lopez, published by Wadsworth, or other books on the numerical analysis of systems of linear equations.

EXERCISES 3–6

A1 Write 3 by 3 elementary matrices that correspond to each of the following elementary row operations, AND multiply each of these elementary matrices by $A = \begin{bmatrix} 1 & 2 & 3 \\ -1 & 3 & 4 \\ 4 & 2 & 0 \end{bmatrix}$ and verify that the product EA is indeed the matrix obtained from A by the elementary row operation.

(a) add (-5) times the second row to the first row
(b) interchange the second and third rows
(c) multiply the last row by (-1)
(d) multiply the second row by 6
(e) add 4 times the first row to the last row

A2 Write 4 by 4 elementary matrices that correspond to each of the following elementary row operations.

(a) add (-3) times the third row to the fourth row
(b) interchange the second and fourth rows
(c) multiply the third row by (-3)

A3 For each of the following matrices, EITHER state that it is elementary and state the corresponding elementary row operation OR explain why it is not elementary.

(a) $A = \begin{bmatrix} 1 & 0 & 0 \\ 0 & 1 & 0 \\ 0 & -4 & 1 \end{bmatrix}$ (b) $B = \begin{bmatrix} -1 & 0 & 0 \\ 0 & 1 & 0 \\ 0 & 0 & -1 \end{bmatrix}$ (c) $C = \begin{bmatrix} 3 & 0 & 1 \\ 0 & 1 & 0 \\ 0 & 0 & 1 \end{bmatrix}$

(d) $D = \begin{bmatrix} 0 & 0 & 1 \\ 0 & 1 & 0 \\ 1 & 0 & 0 \end{bmatrix}$ (e) $E = \begin{bmatrix} 0 & 1 & 0 \\ 0 & 0 & 1 \\ 1 & 0 & 0 \end{bmatrix}$

A4 Let $A = \begin{bmatrix} 1 & 3 & 4 \\ 0 & 0 & 2 \\ 0 & 1 & 0 \end{bmatrix}$.

(a) Determine a sequence of elementary row operations that reduces A to I.
(b) Write the corresponding elementary matrices (such that $E_k \dots E_2 E_1 A = I$).
(c) Determine A^{-1} by computing the product $E_k \dots E_2 E_1$.
(d) Determine the inverse of each of the elementary matrices and hence represent A as a product of elementary matrices.

A5 Let $A = \begin{bmatrix} 1 & 2 & 2 \\ 0 & 1 & 3 \\ 2 & 4 & 5 \end{bmatrix}$, and carry out the same instructions as in A4.

B1 Write 3 by 3 elementary matrices that correspond to each of the following elementary row operations, AND multiply each of these elementary matrices by $A = \begin{bmatrix} 2 & 1 & -1 \\ 2 & 0 & 5 \\ 1 & -3 & -2 \end{bmatrix}$ and verify that the product EA is indeed the matrix obtained from A by the elementary row operation.

(a) add 4 times the third row to the second row
(b) interchange the first and third rows
(c) multiply the second row by (-3)
(d) add (-2) times the first row to the third row

B2 Write 4 by 4 elementary matrices that correspond to each of the following elementary row operations.

(a) add 6 times the fourth row to the second row
(b) multiply the second row by 5
(c) interchange the first and fourth rows

B3 For each of the following matrices, EITHER state that it is elementary and state the corresponding elementary row operation OR explain why it is not elementary.

(a) $A = \begin{bmatrix} 0 & 0 & 1 \\ 1 & 0 & 0 \\ 0 & 1 & 0 \end{bmatrix}$ (b) $B = \begin{bmatrix} 1 & 0 & 0 \\ 0 & 1 & 0 \\ 0 & 3 & 1 \end{bmatrix}$

(c) $C = \begin{bmatrix} -1 & 0 & 0 \\ 0 & 0 & 1 \\ 0 & 1 & 0 \end{bmatrix}$ (d) $D = \begin{bmatrix} 1 & 0 & 1 \\ 0 & 1 & 0 \\ 1 & 0 & 1 \end{bmatrix}$

B4 Let $A = \begin{bmatrix} 1 & 2 & -1 \\ 0 & 1 & 2 \\ 2 & 4 & 0 \end{bmatrix}$.

(a) Determine a sequence of elementary row operations that reduces A to I.

(b) Write the corresponding elementary matrices (such that $E_k \dots E_2E_1A = I$).

(c) Determine A^{-1} by computing the product $E_k \dots E_2E_1$.

(d) Determine the inverse of each of the elementary matrices and hence represent A as a product of elementary matrices.

B5 Let $A = \begin{bmatrix} 1 & 0 & 0 \\ 2 & 3 & 0 \\ 1 & 4 & 1 \end{bmatrix}$, and carry out the same instructions as in B4.

CONCEPTUAL EXERCISES

D1 Interpret all three kinds of 2 by 2 elementary matrices as geometrical linear mappings $\mathbb{R}^2 \to \mathbb{R}^2$.

D2 Interpret all three kinds of 3 by 3 elementary matrices as geometrical linear mappings $\mathbb{R}^3 \to \mathbb{R}^3$.

Chapter Review

Suggestions for Student Review

Try to answer all of these questions before checking answers at the suggested locations. In particular, try to invent your own examples. These review suggestions are intended to help you carry out your review—they may not cover every idea you need to master. Remember that working in small groups may improve the efficiency of your learning.

1 (Section 3–1) State the rule for determining the product of two matrices A and B. What condition(s) must be satisfied by the sizes of A and B for the product to be defined? For the case where A and B are 2 by 2 matrices, show what the product looks like if we treat the columns of B as blocks.

2 (Section 3–2) Explain clearly the relation between a matrix and the corresponding matrix mapping (Theorem 1). Explain how you determine the matrix of a given linear mapping (Theorem 2). Pick some vector $\mathbf{n}$ in $\mathbb{R}^2$ and determine the standard matrices of the linear mappings $\mathbf{proj_n}$, $\mathbf{perp_n}$, and $\mathbf{refl_n}$. (Section 3–2 Example 13, Exercises A5–7, Section 3–3 Example 17) Check that your answers are correct by using each matrix to determine the image under the mapping of $\mathbf{n}$ and of a vector orthogonal to $\mathbf{n}$.

3 (Section 3–3) Determine the image of the vector $(1, 1)$ under the rotation of the plane through angle $\frac{\pi}{6}$. Check that the image has the same length as the original vector $(1, 1)$.

4 (Section 3–4) (a) State the definition of "subspace of $\mathbb{R}^n$". Indicate how you can check whether a given set V is a subspace.
(b) Because of theorems, certain subsets that arise naturally turn out to be automatically subspaces. Make a list of these kinds of subsets. (Section 3–4 Theorem 5 and Corollary; Theorems 6, 1, 3, 7, 8)
(c) Make a list of some subsets of $\mathbb{R}^n$ that are certainly *not* subspaces. (Section 3–4 Theorem 5 (2), Examples 31, 32)
(d) Give an example of a subspace containing exactly one vector. Is there an example consisting of exactly three vectors? Why or why not?

5 Outline relations between the solution set for the homogeneous system $A\mathbf{x} = \mathbf{0}$ and solutions for the corresponding nonhomogeneous system $A\mathbf{x} = \mathbf{b}$. Illustrate by giving some specific examples where A is 2 by 3 and in reduced row echelon form. Also discuss connections between these solution sets and special subsets or subspaces for linear mappings. (Section 3–4, Theorems 2, 1, 3, 4)

6 (Section 3–4) (a) Give an example of a spanning set for some simple subspace (for example, some 2-dimensional subspace of $\mathbb{R}^3$). Now give a different spanning set for the same subspace. Pick any vector in your subspace and express it as a linear combination of the vectors of your first spanning set. (See Examples 35 and 36.)
(b) Explain what is meant by a spanning set.

7 (Section 3–5) (a) Outline the procedure for determining the inverse of a matrix, and indicate why it might not produce an inverse for some matrix A. Use the matrices of some geometric linear mappings to give two or three examples of matrices that do have inverses and two examples of square matrices that do not have inverses.
(b) Pick a fairly simple 3 by 3 matrix (not too many zeros) and try to find its inverse. If it is not invertible, try another. When you have an inverse, check its correctness by multiplication.

8 (Section 3–6) For 3 by 3 matrices, choose one elementary row operation of each of the three types; call these E_1, E_2, E_3. Choose an arbitrary 3 by 3 matrix A and check that E_iA is the matrix obtained from A by the appropriate elementary row operation.

Chapter Quiz

1 Let $A = \begin{bmatrix} 2 & -5 & -3 \\ -3 & 4 & -7 \end{bmatrix}$ and $B = \begin{bmatrix} 2 & -1 & 4 \\ 3 & 0 & 2 \\ 1 & -1 & 5 \end{bmatrix}$. Either determine the following products or explain why they are not defined. (a) AB (b) BA (c) BA^T

2 (a) Let $A = \begin{bmatrix} -3 & 0 & 4 \\ 2 & -4 & -1 \end{bmatrix}$, let f_A be the matrix mapping with matrix A, and let $\mathbf{c} = (1, 1, -2)$ and $\mathbf{d} = (4, -2, -1)$. Determine $f_A(\mathbf{c})$ and $f_A(\mathbf{d})$.

(b) Use the result of part (a) to calculate $A \begin{bmatrix} 4 & 1 \\ -2 & 1 \\ -1 & -2 \end{bmatrix}$.

3 Let L be rotation through angle $\frac{\pi}{3}$ about the z-axis in $\mathbb{R}^3$, and let M be reflection (in $\mathbb{R}^3$) in the plane with equation $-x - y + 2z = 0$. Determine (a) the matrix of L; (b) the matrix of M; (c) the matrix $[L \circ M]$.

4 Let $A = \begin{bmatrix} 1 & 0 & 2 & 1 & 0 \\ 2 & 1 & 3 & 2 & 0 \\ 1 & 1 & 1 & 1 & 1 \end{bmatrix}$, and let $[\mathbf{b}] = \begin{bmatrix} 5 \\ 16 \\ 18 \end{bmatrix}$. Determine the solution set of $A\mathbf{x} = \mathbf{b}$ and the solution space of $A\mathbf{x} = \mathbf{0}$, and discuss the relation between these two sets.

5 Let $B = \begin{bmatrix} 1 & 2 & 0 \\ -1 & -1 & -1 \\ 1 & 3 & 0 \\ 0 & 2 & -1 \end{bmatrix}$, $\mathbf{c} = (4, -3, 5, 3)$, $\mathbf{d} = (-5, 6, -7, -1)$.

(a) With one row reduction (that is, one sequence of row operations), determine whether $\mathbf{c}$ is in the columnspace of B and whether $\mathbf{d}$ is in the range of the linear mapping f_B with matrix B.

(b) Determine from your calculation in part (a) a vector $\mathbf{x}$ such that $f_B(\mathbf{x}) = \mathbf{d}$.

(c) Determine a vector $\mathbf{y}$ such that $f_B(\mathbf{y}) = (2, -1, 3, 2)$ (the second column of B).

6 State whether the following sets are subspaces of some $\mathbb{R}^n$, indicating n in each case, and give reasons for your conclusion.

(a) the set of vectors that are orthogonal to both $\mathbf{a} = (3, -4, 5, -6)$ and $\mathbf{b} = (5, 4, -3, -2)$;

(b) the set of all vectors of the form $\mathbf{x} = (1, 1, 0, 0) + s(0, 0, 1, 2) + t(0, 0, 0, -3)$, $s \in \mathbb{R}, t \in \mathbb{R}$;

(c) $\mathrm{Sp}\big(\{(1, 0, 2, 0, -1), (2, 3, 4, 0, 0), (0, 0, 0, 1, 2)\}\big)$.

7 Determine the inverse of the matrix $A = \begin{bmatrix} 1 & 0 & 0 & -1 \\ 0 & 0 & 1 & 0 \\ 0 & 2 & 0 & 1 \\ 1 & 0 & 0 & 2 \end{bmatrix}$. Check your answer.

8 Determine all values of the parameter p such that the matrix $\begin{bmatrix} 1 & 0 & p \\ 1 & 1 & 0 \\ 2 & 1 & 1 \end{bmatrix}$ is invertible, and determine its inverse.

9 Prove that the range of a linear mapping $L : \mathbb{R}^n \to \mathbb{R}^m$ is a subspace of the codomain.

10 For each of the following, either give an example or explain (in terms of theorems or definitions) why no such example can exist.

(a) A matrix K such that $KM = MK$ for all 3 by 3 matrices M.

(b) A matrix K such that $KM = MK$ for all 3 by 4 matrices M.

(c) The matrix of a linear map $L : \mathbb{R}^2 \to \mathbb{R}^3$ whose range is $\text{Sp}(\{(1,1)\})$ and whose nullspace is $\text{Sp}(\{(2,3)\})$.

(d) The matrix of a linear map $L : \mathbb{R}^2 \to \mathbb{R}^3$ whose range is $\text{Sp}(\{(1,1,2)\})$ and whose nullspace is $\text{Sp}(\{(2,3)\})$.

(e) A linear mapping $L : \mathbb{R}^3 \to \mathbb{R}^3$ such that the range of L is all of $\mathbb{R}^3$ and the nullspace of L is $\text{Sp}(\{(1,-1,1)\})$.

(f) An invertible 4 by 4 matrix of rank 3.

11 Let $A = \begin{bmatrix} 1 & 0 & -2 \\ 0 & 2 & -3 \\ 0 & 0 & 4 \end{bmatrix}$.

(a) Determine a sequence of elementary matrices, $E_1, E_2, \ldots, E_k$, such that $E_k E_{k-1} \ldots E_2 E_1 A = I$.

(b) By inverting the elementary matrices of part (a), write A as a product of elementary matrices.

Further Exercises

F1 We say that matrix C commutes with matrix D if $CD = DC$. Show that the set of matrices that commute with $A = \begin{bmatrix} 3 & 2 \\ 0 & 1 \end{bmatrix}$ is the set of matrices of the form $pI + qA$, where p and q are arbitrary scalars.

F2 Let A be some fixed n by n matrix. Show that the set $C(A)$ of matrices that commutes with A is closed under addition, multiplication by scalars, and under multiplication. (Here "closed under multiplication" means that if C and D are in the set, so are CD and DC. Exercise F1 is a simple illustration of the general case.)

F3 A square matrix A is said to be **nilpotent** if some power of A is equal to the zero matrix. Show that the matrix $\begin{bmatrix} 0 & a_{12} & a_{13} \\ 0 & 0 & a_{23} \\ 0 & 0 & 0 \end{bmatrix}$ is nilpotent. Generalize.

F4 (a) Suppose that ℓ is a line in $\mathbb{R}^2$ passing through the origin and making an angle θ with the positive x_1-axis. Let $\mathbf{refl}_\theta$ denote reflection in this line. Determine the matrix $[\mathbf{refl}_\theta]$ in terms of functions of θ.

(b) Let $\mathbf{refl}_\alpha$ denote reflection in a second line, and by considering the matrix $[\mathbf{refl}_\alpha \circ \mathbf{refl}_\theta]$, show that the composition of two reflections in the plane is a rotation. Express the angle of the rotation in terms of α and θ. (Exercise D4 of Section 3–3 is a particular case of the general result.)

F5 (ISOMETRIES OF $\mathbb{R}^2$) A linear transformation $L : \mathbb{R}^2 \to \mathbb{R}^2$ is an **isometry of** $\mathbb{R}^2$ if L preserves lengths (that is, if $\|L(\mathbf{x})\| = \|\mathbf{x}\|$ for every $\mathbf{x}$ in $\mathbb{R}^2$).
(a) Show that an isometry preserves inner products (that is, $(L(\mathbf{x})) \cdot L(\mathbf{y}) = \mathbf{x} \cdot \mathbf{y}$ for every $\mathbf{x}$ and $\mathbf{y}$ in $\mathbb{R}^2$). [Hint: Consider $L(\mathbf{x}+\mathbf{y})$.]
(b) Show that the columns of the matrix $[L]$ must be orthogonal to each other, and of length 1, and hence, deduce that any isometry of $\mathbb{R}^2$ must be the composition of a reflection and a rotation. [Hint: You may find it helpful to use the result of Exercise F4(a).]

F6 In Section 3–5, two proofs were given for Theorem 3. Give still another proof for this theorem by filling in the details in the following steps.
(a) If $AB = I$, then B and A have rank n.
(b) Since A has rank n, the matrix linear mapping $f_A : \mathbb{R}^n \to \mathbb{R}^n$ defined by $[f_A(\mathbf{x})] = A[\mathbf{x}]$ has the property that for every $\mathbf{y}$ in $\mathbb{R}^n$ there is exactly one $\mathbf{x}$ in $\mathbb{R}^n$ such that $f_A(\mathbf{x}) = \mathbf{y}$. (We often say that such a mapping is "one–to–one".)
(c) Define an inverse *mapping* $f_A^{-1} : \mathbb{R}^n \to \mathbb{R}^n$ by $f_A^{-1}(\mathbf{y}) = \mathbf{x}$ if and only if $f_A(\mathbf{x}) = \mathbf{y}$. Is this mapping well defined? Is it linear? (See Exercise 3–5–D3.)
(d) Since f_A^{-1} is linear, it has a standard matrix; call this matrix B and explain why it must be true that $AB = BA = I$.

F7 (a) Suppose that A and B are n by n matrices such that $A+B$ and $A-B$ are invertible, and that C and D are arbitrary n by n matrices. Show that there are n by n matrices X and Y satisfying the system

$$\begin{aligned} AX + BY &= C \\ BX + AY &= D. \end{aligned}$$

(b) With the same assumptions as in part (a), give a careful explanation of why the matrix $\begin{bmatrix} A & B \\ B & A \end{bmatrix}$ must be invertible, and obtain an expression for its inverse in terms of $(A+B)^{-1}$ and $(A-B)^{-1}$.

F8 **Postscript®** is a computer graphics and typesetting system. It uses a "Current Transformation Matrix" $[\, a \quad b \quad c \quad d \quad T_x \quad T_y \,]$ to determine how images on a "user's" screen or page are positioned on a "device" (or output) page. The coordinates on a user's page are transformed coordinates on the device page by the equations

$$\begin{aligned} x_{\text{device}} &= a x_{\text{user}} + c y_{\text{user}} + T_x \\ y_{\text{device}} &= b x_{\text{user}} + d y_{\text{user}} + T_y, \end{aligned}$$

where $\begin{bmatrix} a & c \\ b & d \end{bmatrix}$ is an invertible matrix describing "scaling" (stretches), shear, rotation, and reflection in the plane, and (T_x, T_y) is a vector determining how the user's origin is translated to a suitable position on the device page.

In order to explore these transformations, we represent them as matrix mappings $T_{A,\mathbf{b}} : \mathbb{R}^3 \to \mathbb{R}^3$. With a slight abuse of notation, we use $T_{A,\mathbf{b}}$ to denote both the transformation and the corresponding matrix. Then we write the matrix, and rewrite it using block form:

$$T_{A,\mathbf{b}} = \begin{bmatrix} a_{11} & a_{12} & b_1 \\ a_{21} & a_{22} & b_2 \\ 0 & 0 & 1 \end{bmatrix} = \begin{bmatrix} A & \mathbf{b} \\ 0_2 & 1 \end{bmatrix}.$$

Here, $A = \begin{bmatrix} a_{11} & a_{12} \\ a_{21} & a_{22} \end{bmatrix} = \begin{bmatrix} a & c \\ b & d \end{bmatrix}$, $\mathbf{b} = \begin{bmatrix} b_1 \\ b_2 \end{bmatrix} = \begin{bmatrix} T_x \\ T_y \end{bmatrix}$, and $0_2 = [\,0 \quad 0\,]$.

(a) Check that the transformation $T_{A,\mathbf{b}}$ leaves the plane $x_3 = 1$ invariant. (That is, if $(\mathbf{x})_3 = 1$, then $(T_{A,\mathbf{b}}(\mathbf{x}))_3 = 1$.)

(b) We may think of the plane $x_3 = 1$ as a "copy" of $\mathbb{R}^2$. Then the "action" of $T_{A,\mathbf{b}}$ on $\mathbb{R}^3$ defined by

$$(x_1, x_2, x_3) \mapsto T_{A,\mathbf{b}}(x_1, x_2, x_3)$$

determines an "action" of a corresponding map $\tilde{T}_{A,\mathbf{b}}$ on $\mathbb{R}^2$ defined by

$$(x_1, x_2) \mapsto \tilde{T}_{A,\mathbf{b}}(x_1, x_2) = A \begin{bmatrix} x_1 \\ x_2 \end{bmatrix} + \begin{bmatrix} b_1 \\ b_2 \end{bmatrix}.$$

$\tilde{T}_{A,\mathbf{b}}$ is the Postscript® transformation described at the beginning of this exercise.

More formally, introduce $IN : \mathbb{R}^2 \to \mathbb{R}^3$ by $IN(x_1, x_2) = (x_1, x_2, 1)$ and $P : \mathbb{R}^3 \to \mathbb{R}^2$ by $P(x_1, x_2, x_3) = (x_1, x_2)$. Check that

$$\tilde{T}_{A,\mathbf{b}} = P \circ T_{A,\mathbf{b}} \circ IN.$$

Explain why $T_{A,\mathbf{b}}$ is a linear mapping but $\tilde{T}_{A,\mathbf{b}}$ is not linear. Also explain why IN is not linear.

> *REMARK*
>
> For use in Postscript®, the important issue is how to choose a, b, c, d, T_x, T_y to achieve the desired output pages, and how the choices are described by Postscript® commands. Readers who wish to pursue this may consult *Postscript® by Example* by Henry McGitton and Mary Campione, Addison-Wesley Publishing Company, 1992, particularly Chapter 4.

We proceed to explore the set of transformations of $\mathbb{R}^3$ of the form $T_{A,\mathbf{b}}$. Denote the set of such transformations by $\mathcal{A}$, and let $\tilde{\mathcal{A}}$ be the set of corresponding transformations of $\mathbb{R}^2$.

(c) Show that if $T_{A,\mathbf{b}}$ and $T_{C,\mathbf{d}}$ are in $\mathcal{A}$, then so is $T_{A,\mathbf{b}} \circ T_{C,\mathbf{d}}$. Give an expression for the matrix product $T_{A,\mathbf{b}}\, T_{C,\mathbf{d}}$.

(d) Observe that the product of matrices of $\mathcal{A}$ is associative, that is

$$T_{A,\mathbf{b}}\ \left(T_{C,\mathbf{d}}\, T_{E,\mathbf{f}}\right) = \left(T_{A,\mathbf{b}}\, T_{C,\mathbf{d}}\right)\ T_{E,\mathbf{f}}.$$

(e) Observe that $I_3 = \begin{bmatrix} I_2 & 0 \\ 0_2 & 1 \end{bmatrix}$ is an element of $\mathcal{A}$ such that $I_3\, T_{A,\mathbf{b}} = T_{A,\mathbf{b}}\, I_3 = T_{A,\mathbf{b}}$ for all A, $\mathbf{b}$.

(f) Show that if $T_{A,\mathbf{b}}$ is in $\mathcal{A}$, then so is $\left(T_{A,\mathbf{b}}\right)^{-1}$. Determine the matrix of this inverse in terms of A, $\mathbf{b}$.

REMARK

A set of transformations that satisfies (c), (d), (e), (f) is called a **group** of transformations. The fact that $\mathcal{A}$ is a group implies that $\tilde{\mathcal{A}}$ is also a group; $\tilde{\mathcal{A}}$ is called the **group of affine transformations of the plane.** The group $\mathcal{A}$ of linear transformations of $\mathbb{R}^3$ is called a "representation of $\tilde{\mathcal{A}}$".

An Essay on Linearity and Superposition in Physics

The material of this essay is not really linear algebra, and it is not essential to what follows; there are no exercises. The purpose is to convince you that there are good reasons for setting up a proper general framework for discussing linearity. Some of the questions that will be considered later in this book are introduced. The material does not assume much previous knowledge of physics, but some knowledge of calculus is used. In reading this material, you should not worry about understanding every detail. Instead, try to see that the idea of linearity shows up in many important settings, so that it is worth investing the effort to learn how to take advantage of linearity.

The Simple Spring

Hooke's Law

Consider a spring attached at one end to a wall and resting on a smooth horizontal surface. Since the surface is smooth, it is assumed that friction is zero and there is no damping. Suppose that a ball of mass m is attached to the free end of the spring; it is assumed that the mass can move in only one direction, the direction of the axis of the spring. To describe the motion of this mass, a coordinate is introduced: when the mass is moved, the displacement of the mass from its rest position is measured by the coordinate x (Figure E–1). x is positive when the spring is stretched, negative when the spring is compressed. When the mass is moved from its rest position, the spring exerts a force on the mass; the force of the spring when the displacement is x will be denoted $F(x)$. Observations led Hooke to say that if the displacement is small, the force (in most springs) is directly proportional to the displacement, but opposite in direction to the displacement: $F(x) = -Kx$, where K is a positive constant that depends on the spring (the "spring constant"). Thus, $F : \mathbb{R} \to \mathbb{R}$ is a **linear** function.

Now suppose that an **external** force G is applied to the mass. Then according to the laws of physics, the mass is at rest only when the sum of the spring force and the external force is zero. This means that $G = -F$, and hence, the mass will be at rest with displacement $x_G = (1/K)G$. Suppose that the first force is removed and a second force H is applied: then the mass will be at rest with displacement $x_H = (1/K)H$. Now suppose that H is removed and a force $(G + H)$ is applied to the mass. Then the resulting displacement is $x_{(G+H)} = (1/K)(G + H) = (1/K)G + (1/K)H = x_G + x_H$. In situations such as this, physicists sometimes speak of **superposition**: the displacement for the compound problem (with force $G + H$) is obtained by **superposing** (adding) the displacements for the two simpler problems (force G, then force H) that make up the compound problem.

It should be apparent that "the principle of superposition" in this case is really just an expression of the fact that Hooke's Law is **linear**. We shall consider the spring further below, but first we sketch the situation for deformation of solids.

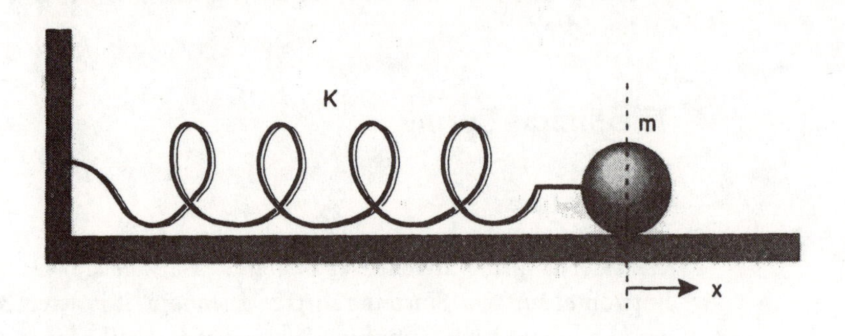

FIGURE E–1. *A simple spring-mass system.*

Elastic Deformation in Three Dimensions

The deformations of a solid can be more complicated than the one-dimensional stretching of a spring. In addition to being stretched, a line segment in the x_1-direction can experience shearing in the x_2- and x_3-directions; line segments in other directions experience similar changes. It turns out that for small deformations, the best way to describe this is to introduce the **infinitesimal strain**, which is a 3 by 3 symmetric matrix defined at each point in the solid. (A matrix is **symmetric** if it is equal to its own transpose; in terms of entries, this means that $a_{jk} = a_{kj}$.) When a body is deformed, internal forces analogous to the restoring force in a spring arise, and it turns out that these can be described by another 3 by 3 symmetric matrix called the **stress** matrix. Since the stress depends on the strain, for each material under small deformations, there is a function that we denote by

$$C : \{\text{symmetric matrices}\} \rightarrow \{\text{symmetric matrices}\},$$

such that $C(\text{strain}) = \text{stress}$, for that material. For many materials under small deformations, it is observed that superposition holds:

if $C(\text{strain}\#1) = \text{stress}\#1$ and $C(\text{strain}\#2) = \text{stress}\#2$,

then $C(\text{strain}\#1 + \text{strain}\#2) = \text{stress}\#1 + \text{stress}\#2$.

Moreover, twice the strain gives twice the stress; in general,

for a real number k, $C(k\ \text{strain}\#1) = k\ \text{stress}\#1$.

Thus in this case, called **linear elasticity**, the function C satisfies the **linearity properties**. We had earlier defined the linearity properties only for mappings from $\mathbb{R}^n$ to $\mathbb{R}^m$ but they seem to hold for this mapping from symmetric matrices to symmetric matrices.

Question #1: *What is the general setting in which it makes sense to talk about linear mappings?* (The answer is, roughly speaking, that you must be working with spaces in which addition and multiplication by scalars is defined. See Section 4–1.)

In addition to strain and stress there are many other physical quantities that are described by matrices. The inertia tensor used in describing rotating motion is another example (described briefly in Section 8–5). Matrices also arise naturally when we try to determine maximum and minimum values of functions of two or more variables (Section 8–4). Trying to understand in an intuitive way the information contained in a matrix (such as the infinitesimal strain matrix) appears to be a difficult task. It is helpful that in many important applications the matrices are square, and often symmetric.

Question #2: *Is there some way of getting a good geometric feeling for the information contained in a square matrix? Is this question easier if the matrix is symmetric?*

The answer for symmetric matrices is sometimes called the Principal Axis Theorem: if you pick the right coordinate system, the matrix simply describes "stretches" and reflections in certain preferred directions. For general square matrices, this "diagonalization" is a little more complicated. The ideas and techniques required to understand diagonalization take up much of the rest of this book.

The Differential Equation of a Spring and Its Solution Space

The most familiar of Newton's laws is that the force on an object is equal to its mass times its acceleration: $\mathbf{F} = m\mathbf{a}$. For the one-dimensional spring problem with no damping, this results in the equation:

$$m\frac{d^2x}{dt^2} = -Kx \quad \text{or} \quad m\frac{d^2x}{dt^2} + Kx = 0.$$

It is convenient to introduce the notation $D = \dfrac{d}{dt}$. Some of the most basic properties of differentiation are that for differentiable functions f and g, and constant c, $D(f + g) = Df + Dg$, and $D(cf) = cDf$. The differentiation operator D satisfies the linearity properties, and we could say (if Question #1 has been answered) that

$$D : \{\text{differentiable functions}\} \to \{\text{functions}\}$$

is a linear mapping. It is common to use the phrase **linear operator** when the objects being mapped are functions.

In this notation, the spring equation can be rewritten:

$$\left(D^2 + (K/m)\right)x = 0.$$

This is a differential equation for the function $x(t)$. The operator $\left(D^2 + (K/m)\right)$ is a linear operator because

$$\left(D^2 + (K/m)\right)(ax(t) + by(t)) = a\left(D^2 + (K/m)\right)x(t) + b\left(D^2 + (K/m)\right)y(t)$$

for any real numbers a and b, and for any twice differentiable functions x and y. Therefore, the spring equation is called a **linear differential equation**. Moreover, it is a **homogeneous** equation, because the right-hand side is zero. (You should think of the right-hand side as the zero function, that is, the right-hand side is zero for every t.)

The theory of linear differential equations has strong similarities to the theory of algebraic linear systems as developed in Chapters 2 and 3. For example, for homogeneous linear algebraic systems it was true that "a linear combination of solutions is a solution"; for a homogeneous linear differential equation such as the spring equation it is easy to see that if $x(t)$ and $y(t)$ are solutions, then so is any linear combination $ax(t) + by(t)$. For homogenous linear systems, we saw that the solution space could always be spanned by an appropriate set of vectors. There are similar results for the linear differential equations; for the case of the homogeneous spring equation it turns out that the solution space is the space of functions spanned by $\cos \omega t$ and $\sin \omega t$, where $\omega^2 = K/m$.

Now consider the "forced spring": suppose that a time-varying force $G(t)$ is applied to the mass. Then the equation becomes

$$\left(D^2 + (K/m)\right)x(t) = G(t).$$

This is now a nonhomogeneous linear differential equation for the position $x(t)$ of the mass at time t.

There are also parallels between the theory of nonhomogeneous linear algebraic systems (discussed in Section 3–4) and nonhomogeneous linear differential equations. Recall that for the linear system $A\mathbf{x} = \mathbf{b}$, the difference between two solutions $\mathbf{x}^{(1)}$ and $\mathbf{x}^{(2)}$ must be a solution of the corresponding homogeneous system because

$$A\left(\mathbf{x}^{(1)} - \mathbf{x}^{(2)}\right) = A\mathbf{x}^{(1)} - A\mathbf{x}^{(2)} = \mathbf{b} - \mathbf{b} = \mathbf{0}.$$

In exactly similar fashion one can show that if $x^{(1)}(t)$ and $x^{(2)}(t)$ are solutions of the nonhomogeneous spring equation, their difference is a solution of the corresponding homogeneous spring equation:

$$\left(D^2 + (K/m)\right)\left(x^{(1)}(t) - x^{(2)}(t)\right) = G(t) - G(t) = 0.$$

In the case of algebraic systems, it followed from arguments like these that the general solution of the nonhomogeneous system could be written in the form $\mathbf{x} = \mathbf{x}_p + \mathbf{x}_h$, where $\mathbf{x}_p$ is a "particular" solution of the nonhomogeneous system, and $\mathbf{x}_h$ is the general solution for the corresponding homogeneous system. For the nonhomogeneous linear spring equation, the general solution can be written $x(t) = x_p(t) + x_h(t)$, where again $x_p(t)$ is a "particular solution" of the nonhomogeneous equation, and $x_h(t)$ is the general solution of the corresponding homogeneous equation. The word "superposition" is sometimes used to describe this addition of a particular solution and a solution of the homogeneous problem.

The analogy between linear algebraic systems and linear differential equations has been exhibited here only in terms of the spring equation, but the parallels are true for general linear differential equations. In any situation where one is dealing with linear equations, the linearity will dictate important features of the solution sets, and general solutions will be produced by superposition. This raises an important question. In many homogeneous linear problems, the solution is given as a linear combination of the vectors (functions) in some spanning set. It would be nice to be sure that everybody producing a solution in this way will get the same answer.

Question #3: *Can an element in a space be represented in a unique way as a linear combination of a spanning set? (The answer requires the introduction of the concepts of linear independence and basis in Chapter 4.)*

Coupled Linear Oscillators

In many systems, it is important to understand how the changes in one variable interact with changes in another variable. This is particularly important in structures that undergo motion; bridges have collapsed because of unanticipated interactions of this kind. This is also a central issue in the design of devices such as amplifiers.

One of the simplest interesting models that exhibits some of this behaviour is illustrated in Figure E-2. There are three springs and two balls. One spring, with spring constant K_1, is attached to the wall at the left and to the first ball of mass m; the second spring, with constant K_2, joins the first ball to the second, also of mass m; the third spring, with constant K_3, joins the second ball to the wall on the right. It is assumed that when the system is in its rest position there is no force in any spring. Again it will be assumed that there is no damping force due to friction. Displacements of the first mass from its rest position are measured by x, displacements of the second mass from its rest position are measured by y; note that with such displacements, the centre spring is extended by amount $y - x$.

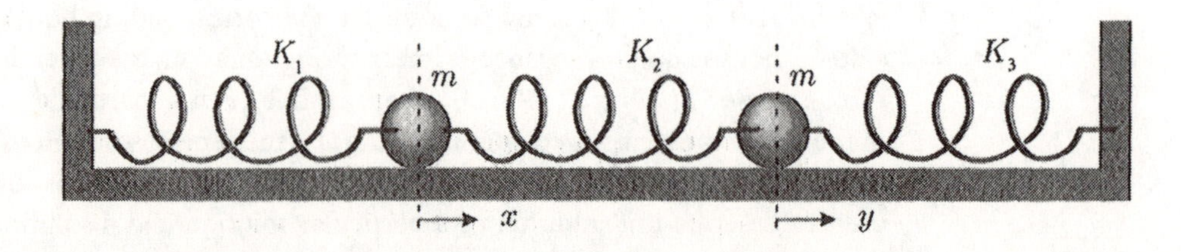

FIGURE E–2. *A system with three springs and two masses.*

The problem is to determine the positions $x(t)$ and $y(t)$ of the masses at time t. Let us first review the case of a simple spring. In that case, if the single mass is displaced from rest and then released, it is observed that the mass moves back and forth in a regular periodic way: in fact, such a motion can be described by the equation $x(t) = A\cos(\omega t + \phi)$, where ϕ is the initial phase and A is the amplitude of the oscillation.

Now suppose that in our three-spring system, each mass is displaced by a small random amount, and both masses are released. What you will probably see is that each mass moves back and forth, apparently independently; the motion looks very irregular and is probably not periodic. The compound system seems much more complicated than the simple spring. However, if you were a **very** patient and lucky experimenter, you might find that if you moved each mass to the right by exactly the correct amount (and the amounts are usually different for the two masses — the magic initial displacements depend on the spring constants), you would see the two masses start moving together to the left, then together to the right in a regular periodic way. This is one of the two **normal modes of oscillation** of the system. In the second normal mode, the motion is again regular and periodic, but at any given time the two masses are moving in opposite directions. (Normal modes are particularly important in understanding the behaviour of compound systems when external forces are applied, but this will not be considered here.)

Remarkable Fact: *Any motion of the two masses can be described as a superposition (that is, a linear combination) of the two normal modes.*

This is not the place to tell the full story about this system, but if we go into a little more detail we can get some idea of why superposition applies, and as a bonus, see that Question #2 arises in an essential way in this problem.

Newton's law for the two masses gives the following differential equations:

$$mD^2x = -K_1x + K_2(y - x) = (-K_1 - K_2)x + K_2y,$$
$$mD^2y = -K_2(y - x) - K_3y = K_2x + (-K_2 - K_3)y.$$

It is convenient to write this in terms of the state vector function $(x(t), y(t))$:

$$D^2 \begin{bmatrix} x(t) \\ y(t) \end{bmatrix} - (1/m) \begin{bmatrix} -K_1 - K_2 & K_2 \\ K_2 & -K_2 - K_3 \end{bmatrix} \begin{bmatrix} x(t) \\ y(t) \end{bmatrix} = \begin{bmatrix} 0 \\ 0 \end{bmatrix}.$$

This is easily seen to be a homogeneous linear differential equation for the state vector function, so it is expected that the solution space for the problem consists of linear combinations of a suitable spanning set (and the spanning set turns out to be the two solutions describing the normal modes). Thus, the principle of superposition applies to this situation.

It is interesting to see if the normal modes can be found. From the description above, the state vector function for a normal mode should be of the form $\begin{bmatrix} x(t) \\ y(t) \end{bmatrix} = \cos \omega t \begin{bmatrix} a \\ b \end{bmatrix}$, for some real numbers ω, a, b. (ω is 2π times the frequency of the oscillation, and a and b determine the initial displacements of the two balls.) Substitute this into the system of differential equations; since $D^2 \cos \omega t = -\omega^2 \cos \omega t$, this substitution gives

$$\begin{bmatrix} -K_1 - K_2 & K_2 \\ K_2 & -K_2 - K_3 \end{bmatrix} \begin{bmatrix} a \\ b \end{bmatrix} = -m\omega^2 \begin{bmatrix} a \\ b \end{bmatrix}.$$

Therefore, this procedure will produce a normal mode only if there is a number ω and a vector (a, b) with the very special property that the image of (a, b) under the matrix mapping with matrix $\begin{bmatrix} -K_1 - K_2 & K_2 \\ K_2 & -K_2 - K_3 \end{bmatrix}$ is simply a scalar multiple of itself. Such vectors are called **eigenvectors** of the matrix, and the scalar factor ($-m\omega^2$ in this case) is called the corresponding eigenvalue. Finding the eigenvectors of a matrix will be the key (in Chapter 6) to answering Question #2 of this essay.

To conclude the discussion of the coupled spring system, consider the particular example where $m = 1$ and $K_1 = K_2 = K_3 = 1$. Then the matrix is $\begin{bmatrix} -2 & 1 \\ 1 & -2 \end{bmatrix}$. Can ω^2 and (a, b) be found to satisfy the eigenvector condition? Later we shall learn a method for finding eigenvectors, but for now we simply verify the answer.

(1) It is easy to verify that $\begin{bmatrix} -2 & 1 \\ 1 & -2 \end{bmatrix} \begin{bmatrix} 1 \\ -1 \end{bmatrix} = -3 \begin{bmatrix} 1 \\ -1 \end{bmatrix}$, so $\omega^2 = 3$ and one normal mode is $\begin{bmatrix} x(t) \\ y(t) \end{bmatrix} = A \cos \sqrt{3}t \begin{bmatrix} 1 \\ -1 \end{bmatrix}$. This is the normal mode where the displacements x and y are in opposite directions at any given time, since the components of the eigenvector have opposite signs.

(2) It is also easy to verify that $\begin{bmatrix} -2 & 1 \\ 1 & -2 \end{bmatrix} \begin{bmatrix} 1 \\ 1 \end{bmatrix} = -1 \begin{bmatrix} 1 \\ 1 \end{bmatrix}$, so $\omega^2 = 1$ and the other normal mode in this case is $\begin{bmatrix} x(t) \\ y(t) \end{bmatrix} = B \cos t \begin{bmatrix} 1 \\ 1 \end{bmatrix}$. This is the normal mode in which both displacements have the same sign for all t.

As a final remark on the coupled spring system, if you consider linear combinations of the two normal modes for the particular example, you can see that in general the motion is not periodic since $(\cos t + \cos \sqrt{3}t)$ is not periodic.

Concluding Remarks

There are many other phenomena in physics that are described by linear laws. To such phenomena, the principle of superposition will apply because of the linearity. Some important examples are the wave equation, which describes a vibrating string or the propagation of electromagnetic waves in space; the heat equation, which describes the distribution of heat in a solid; and Laplace's equation, which is satisfied by gravitational and electrostatic potentials. The details of linear algebra do not apply to exact solution methods for these problems, but the concepts of linearity and linear combinations are very important in solving them. Numerical methods for solving these problems make extensive use of linear algebra.

One of our earlier questions does arise in the solution methods for problems such as the heat equation. In the early 1800s, Fourier worked on the heat equation and gave a solution based on the very non-obvious idea that every periodic function can be represented as a linear combination of a spanning set consisting of suitable cosine and sine functions. Question #3 is relevant here. Fourier's method is beyond the scope of this book, but there is an introduction to Fourier series in Chapter 7.

Truth in Advertising

Linearity and linear algebra are important. However, it is usually an idealization to assume that a linear law applies, and often it is important to take account of non-linearities as well. In fact, in some phenomena, the non-linear effects are the most important and interesting. Generally speaking, non-linear problems are difficult to solve exactly, so for centuries, mathematicians, physicists, and engineers focussed on the linear problems. Now, with computers, we can tackle non-linear problems directly, but the understanding and tools from the analysis of linear problems continue to be essential.

CHAPTER 4

Vector Spaces, Bases, and Change of Basis

This chapter explores some of the most important ideas in linear algebra. Some of these ideas have appeared in special cases before, but here we give definitions and examine them in more general settings.

SECTION 4–1

Vector Spaces and Linear Mappings

Linearity of mappings makes sense only in spaces in which objects can be added and multiplied by scalars. (See Chapter 3 and the Essay on Linearity and Superposition in Physics for examples.) The following definition of vector spaces captures these properties. The most important properties are that the space is **closed under addition** ((1) below) **and under multiplication by scalars** ((6)), and that it possesses a **zero element** (or zero vector, (3)). The other properties simply say that addition and multiplication by scalars behave in the way we expect them to behave; it is necessary to spell them out because the idea of vector space is used in settings where these properties are not so obvious as they are in $\mathbb{R}^n$.

DEFINITION

A **vector space over the real numbers** $\mathbb{R}$ is a set $\mathbb{V}$ together with an operation of **addition**, denoted $\mathbf{x}+\mathbf{y}$, and an operation of **multiplication by scalars**, denoted by $a\mathbf{x}$, such that for any $\mathbf{x}, \mathbf{y}, \mathbf{z}$ in $\mathbb{V}$, and any a and b in $\mathbb{R}$:

(1) $\mathbf{x}+\mathbf{y}$ is defined and in $\mathbb{V}$;

(2) $(\mathbf{x}+\mathbf{y})+\mathbf{z} = \mathbf{x}+(\mathbf{y}+\mathbf{z})$; (addition is *associative*)

(3) there is a zero element in $\mathbb{V}$, denoted $\mathbf{0}$, such that

$$\mathbf{x}+\mathbf{0} = \mathbf{0}+\mathbf{x} = \mathbf{x};$$

(4) for each $\mathbf{x}$, there is an "additive inverse", denoted $-\mathbf{x}$, such that

$$\mathbf{x} + (-\mathbf{x}) = \mathbf{0};$$

(5) $\mathbf{x} + \mathbf{y} = \mathbf{y} + \mathbf{x}$; (addition is *commutative*)

(6) $a\mathbf{x}$ is defined and in V;

(7) $a(b\mathbf{x}) = (ab)\mathbf{x}$;

(8) $(a + b)\mathbf{x} = a\mathbf{x} + b\mathbf{x}$; (a *distributive* law)

(9) $a(\mathbf{x} + \mathbf{y}) = a\mathbf{x} + a\mathbf{y}$; (another *distributive* law)

(10) $1\mathbf{x} = \mathbf{x}$.

Some additional properties (consequences of the axioms).

(11) $0\mathbf{x} = \mathbf{0}$ (the zero vector), for any $\mathbf{x}$ in V.

(12) $(-1)\mathbf{x} = -\mathbf{x}$, for any $\mathbf{x}$ in V.

Proof of (11).

By (8) and (10), $\mathbf{x} + 0\mathbf{x} = (1 + 0)\mathbf{x} = 1\mathbf{x} = \mathbf{x}$. Add the negative of $\mathbf{x}$ to both sides, and use (2) and (4) and (3); the left-hand side gives

$$(-\mathbf{x}) + (\mathbf{x} + 0\mathbf{x}) = ((-\mathbf{x}) + \mathbf{x}) + 0\mathbf{x} = \mathbf{0} + 0\mathbf{x} = 0\mathbf{x},$$

while the right-hand side gives $(-\mathbf{x}) + \mathbf{x} = \mathbf{0}$, so $0\mathbf{x} = \mathbf{0}$, as claimed. ■

You are asked to prove Property (12) in Exercise D1.

REMARKS

(1) Since every vector space contains a zero element, the empty set cannot be a vector space. (You are not likely ever to try to use it as a vector space, but it is helpful to know that there is one special case we do not have to worry about when we make general statements about vector spaces.)

(2) Vector spaces can be defined with other number systems as the scalars. For example, note that the definition makes perfect sense if complex

numbers are used instead of the real numbers, provided that the rules of complex arithmetic are known; vector spaces over the complex numbers (that is, with complex scalars) are discussed in Section 4–8. Vector spaces over "finite number fields" have applications in coding theory and communications engineering; they are also of considerable interest to mathematicians. In this book, "vector space" means "vector space over the reals", except where it is stated that the scalars are complex numbers.

Once general vector spaces are defined, it is easy to define general linear mappings.

If U and V are vector spaces over the real numbers, a function $f : \mathrm{U} \to \mathrm{V}$ is a **linear mapping** if it satisfies the linearity properties: for any $\mathbf{x}$ and $\mathbf{y}$ in U, and for any real number a,

(L1) $f(\mathbf{x}+\mathbf{y}) = f(\mathbf{x}) + f(\mathbf{y})$;

(L2) $f(a\mathbf{x}) = af(\mathbf{x})$.

As before, the two properties can be combined into one statement:

$$f(a\mathbf{x} + b\mathbf{y}) = af(\mathbf{x}) + bf(\mathbf{y}), \text{ for all } \mathbf{x}, \mathbf{y} \text{ in U, and all } a, b \text{ in } \mathbb{R}.$$

Other definitions from Chapter 3, such as the definitions of nullspace and range in Section 3–4, make sense in general vector spaces with minor changes.

EXAMPLE 1

$\mathbb{R}^n$ is clearly a vector space, for any n. Chapter 3 contains many examples of linear mappings from $\mathbb{R}^n$ to $\mathbb{R}^m$.

EXAMPLE 2

Let $\mathbb{Z}$ denote the set of integers $\{\dots, -2, -1, 0, 1, 2, \dots\}$. Is this a vector space over the reals?

SOLUTION: If x and y are integers, $x + y$ is an integer, so $\mathbb{Z}$ is closed under addition. However, if r is a real number and x is an integer, usually rx is *not* an integer. For example, $(\sqrt{2})2$ is not an integer, so $\mathbb{Z}$ is not closed under multiplication by real scalars and is not a vector space over the real numbers.

Subspaces

Recall from Section 3–4 that $\mathbb{U}$ is a subspace of $\mathbb{R}^n$ if it is a non-empty subset of $\mathbb{R}^n$ and is closed under addition and multiplication by scalars. Because $\mathbb{R}^n$ satisfies property (11), and $\mathbb{U}$ is closed under multiplication by scalars, $\mathbb{U}$ automatically contains the zero vector $\mathbf{0}$. Because addition and multiplication by scalars in $\mathbb{U}$ are inherited from $\mathbb{R}^n$, these operations when performed on objects in $\mathbb{U}$ automatically satisfy all of the other properties for a vector space. Hence, any subspace $\mathbb{U}$ of $\mathbb{R}^n$ is itself a vector space. These ideas can easily be generalized to general vector spaces.

DEFINITION

Suppose that $\mathbb{V}$ is a vector space. A non-empty subset $\mathbb{U}$ of $\mathbb{V}$ is a **subspace** of $\mathbb{V}$ if $\mathbb{U}$ is closed under addition and multiplication by scalars (where these operations in $\mathbb{U}$ are inherited from $\mathbb{V}$).

Equivalent Definition. $\mathbb{U}$ is a subspace of the vector space $\mathbb{V}$ if $\mathbb{U}$ is a non-empty subset of $\mathbb{V}$ and $\mathbb{U}$ is itself a vector space with the operations inherited from $\mathbb{V}$.

It is easy to prove that these definitions are equivalent by a discussion just like the discussion above for subspaces of $\mathbb{R}^n$.

For the questions to be considered in the rest of this book, the important examples of vector spaces will continue to be $\mathbb{R}^n$ and subspaces of $\mathbb{R}^n$. We consider a few other examples, partly because they are important in the study of differential equations and many other topics in applied and pure mathematics, and partly because it helps to illustrate the idea of a vector space.

EXAMPLE 3

The vector space $\mathcal{M}(2,3)$ of 2 by 3 matrices

Let $\mathcal{M}(2,3)$ be the set of 2 by 3 matrices. With the usual rules for addition of matrices and multiplication of matrices by real numbers (see Section 3–1), it is easy to see that $\mathcal{M}(2,3)$ is a vector space; you should check this. Note that the zero element of $\mathcal{M}(2,3)$ is the 2 by 3 zero matrix $0_{2,3}$.

Some linear mappings of $\mathcal{M}(2,3)$

Let A be any 2 by 2 matrix; define the mapping $M_A : \mathcal{M}(2,3) \to \mathcal{M}(2,3)$ by $M_A(B) = AB$.

(i) Since $A(B+C) = AB + AC$, $M_A(B+C) = M_A(B) + M_A(C)$.

(ii) Since $A(cB) = cAB$, $M_A(cB) = cM_AB$.

It follows that M_A is a linear mapping. We consider two particular examples.

(1) Let E be the 2 by 2 matrix, $\begin{bmatrix} 0 & 1 \\ 1 & 0 \end{bmatrix}$, and define the mapping $M_E : \mathcal{M}(2,3) \to \mathcal{M}(2,3)$ by $M_E(B) = EB$. Then, for a general matrix B,

$$EB = \begin{bmatrix} 0 & 1 \\ 1 & 0 \end{bmatrix} \begin{bmatrix} a & b & c \\ d & e & f \end{bmatrix} = \begin{bmatrix} d & e & f \\ a & b & c \end{bmatrix},$$

so M_E exchanges the two rows of B. Note that the nullspace of M_E is $\{0_{2,3}\}$ and the range of M_E is $\mathcal{M}(2,3)$.

(2) Consider the case $F = \begin{bmatrix} 0 & 1 \\ 0 & 0 \end{bmatrix}$. Define M_F by $M_F(B) = FB$; then,

$$FB = \begin{bmatrix} 0 & 1 \\ 0 & 0 \end{bmatrix} \begin{bmatrix} a & b & c \\ d & e & f \end{bmatrix} = \begin{bmatrix} d & e & f \\ 0 & 0 & 0 \end{bmatrix}.$$

The range of M_F consists of 2 by 3 matrices with zero second row. The nullspace of M_F is also the set of 2 by 3 matrices with zero second row because $M_F(B) = \begin{bmatrix} 0 & 0 & 0 \\ 0 & 0 & 0 \end{bmatrix}$ only if $d = e = f = 0$. The range and the nullspace of this linear mapping coincide!

This example can be generalized in many ways, some of which are important in advanced algebra.

Spaces of Polynomials

Recall that a **polynomial** (over the reals, or with real coefficients) is an expression of the form

$$p(x) = a_0 + a_1x + a_2x^2 + \cdots + a_nx^n,$$

where $a_0, a_1, a_2, \ldots, a_n$ are real numbers and x is an "indeterminate" (that means that x stands for nothing in particular; when dealing with polynomials in algebra, you should not assume that x necessarily represents one of the numbers in the number field—sometimes you substitute all sorts of things for x, for example, a matrix).

It is a convention that when a polynomial is written in this form, the coefficient a_n of the highest power of x is **non-zero**; the number n is called the **degree** of the polynomial. Note that the polynomials of degree 0 are the non-zero constants. The zero polynomial (denoted 0) is a special case: although it has no non-zero coefficient, it is considered to be of degree zero. (Many statements about the degree of polynomials must be made only for **non-zero** polynomials: for example, for **non-zero** $p(x)$ and $q(x)$, the degree of the product is the sum of the degrees.)

The sum of two polynomials is defined in an obvious way; if $m < n$,

$$\begin{aligned}&(a_0 + a_1x + a_2x^2 + \cdots + a_nx^n) + (b_0 + b_1x + b_2x^2 + \cdots + b_mx^m)\\ &= (a_0 + b_0) + (a_1 + b_1)x + \cdots + (a_m + b_m)x^m + a_{m+1}x^{m+1} + \cdots + a_nx^n.\end{aligned}$$

Similarly, multiplication by a scalar c is defined by

$$c(a_0 + a_1x + a_2x^2 + \cdots + a_nx^n) = ca_0 + ca_1x + ca_2x^2 + \cdots + ca_nx^n.$$

EXAMPLE 4

Consider the set of polynomials of degree n (for some fixed $n > 0$). Is this a vector space? Since it does not contain the zero polynomial, the answer is no. Note also that the sum of two polynomials of degree n may be of degree lower than n: for example, $(1 + x^n) + (1 - x^n) = 2$, which is of degree 0. Since the set is not closed under addition, it is not a vector space.

EXAMPLE 5

Let P_n be the set of all polynomials of degree less than or equal to n, for some positive integer n. Is P_n a vector space? This set does contain 0, and it is easy to see that it is closed under addition and multiplication by scalars. Note that there is an obvious spanning set for this vector space: every polynomial of degree less than or equal to n can be written as a linear combination of the polynomials in the set $\{1, x, x^2, \ldots, x^n\}$. These spaces of polynomials are important in solving some differential equations.

Let D denote differentiation: D satisfies $D(x^n) = nx^{n-1}$ if n is an integer greater than or equal to 1. If $p(x)$ is a polynomial of degree zero (that is, a constant) $D(p(x)) = 0$. By standard calculus theorems, D also satisfies the linearity properties:

$$D(p(x) + q(x)) = D(p(x)) + D(q(x)) \text{ and } D(cp(x)) = cD(p(x)).$$

Thus, for any $n \geq 1$, $D : P_n \to P_{n-1}$ is a linear mapping. The nullspace of D is the set of polynomials of degree zero (that is, the space of constant polynomials). We can also define $D : P_0 \to P_0$ to be the zero linear mapping (that is, the mapping that maps any constant polynomial to the zero polynomial).

EXAMPLE 6

Let $\mathcal{P}$ denote the set of **all** polynomials with real coefficients. Verify that $\mathcal{P}$ is closed under addition and multiplication by scalars, and contains 0, so $\mathcal{P}$ is a vector space. The most obvious spanning set for $\mathcal{P}$ is $\{1, x, x^2, \ldots, x^n, \ldots\}$: every polynomial can be written as a linear combination of these terms. Note that while the spanning set has infinitely many members, any polynomial in fact has only finitely many non-zero coefficients. (Infinite power series, that is, expressions with infinitely many non-zero coefficients, are beyond the scope of this book.) It seems very unlikely that $\mathcal{P}$ has any spanning set with only finitely many members; this fact can be proved, so $\mathcal{P}$ is called an "infinite-dimensional" vector space.

Spaces of Functions

We shall consider here only real-valued functions of a real variable, that is, functions $f\colon \mathbb{R} \to \mathbb{R}$. Examples of such functions are polynomial functions (that is, polynomials with real coefficients where x is now to be considered as a real variable rather than an indeterminate) and such familiar functions from calculus as $\cos x$ or e^x. Some functions such as $\log x$ are not defined for every x, so sometimes we may be interested in functions defined on an interval $(a, b) = \{x \mid a < x < b\}$. Let $\mathcal{F}(a, b)$ denote the set of functions $f\colon (a, b) \to \mathbb{R}$. If f and g are elements of $\mathcal{F}(a, b)$, then their sum is defined by $(f+g)(x) = f(x)+g(x)$, and multiplication by a real number c is defined by $(cf)(x) = c(f(x))$.

With these definitions, $\mathcal{F}(a, b)$ is a vector space. There are some subspaces that turn out to be particularly important in more advanced applications of calculus.

EXAMPLE 7

Let $C(a, b)$ be the set of all functions that are continuous on the interval (a, b). Since the sum of continuous functions is continuous and a scalar (real) multiple of a continuous function is continuous, $C(a, b)$ is a vector space. See Figure 4–1–1.

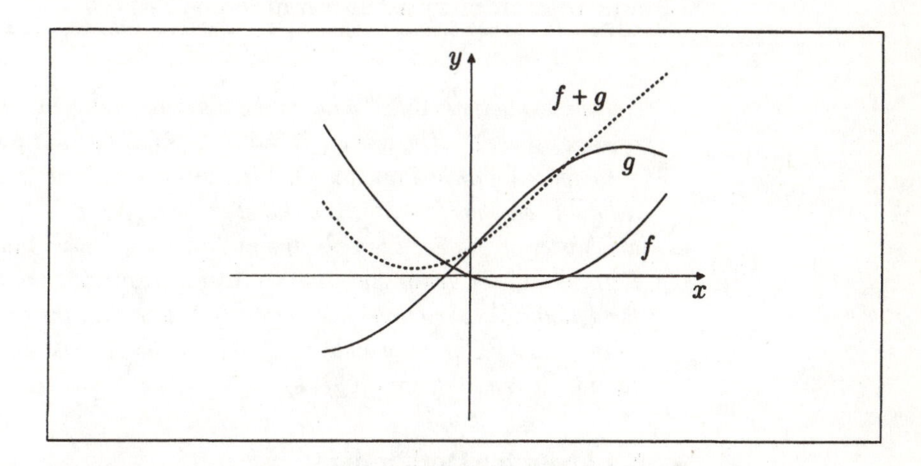

FIGURE 4-1-1. *$C(a, b)$ is closed under addition: on the interval (a, b), the sum $(f + g)$ of continuous functions f and g is a continuous function.*

Let $C^1(a, b)$ be the set of functions that have a continuous first derivative on the interval (a, b). It is easy to see that this is also a vector space, and that if D denotes differentiation, $D : C^1(a, b) \to C(a, b)$ is a linear mapping with nullspace equal to the set of constant functions. (Part of the Fundamental Theorem of Calculus says that the range of this mapping is all of $C(a, b)$.) In some calculus problems, we work over a closed interval $[a, b] = \{x \mid a \leq x \leq b\}$. In such cases we may work with the spaces $C[a, b]$ or $C^1[a, b]$, which are defined similarly to $C(a, b)$ and $C^1(a, b)$.

EXAMPLE 8

Let $\mathcal{PF}(2\pi)$ be the set of continuous functions that are periodic with period 2π, that is, functions $f : \mathbb{R} \to \mathbb{R}$ that satisfy $f(x + 2\pi) = f(x)$ for every x. It is easy to check that a scalar multiple of a periodic function with period 2π

also has period 2π, and that the sum of periodic functions with period 2π also has period 2π, and that the zero function is periodic with period 2π (in fact, all constant functions are periodic). Hence, $\mathcal{PF}(2\pi)$ is a vector space. This is exactly the space of interest in Fourier series, which play a very important role in the analysis of heat flow and of wave propagation. Fourier's method depends on understanding the remarkable fact that although the set

$$\{1, \cos x, \sin x, \cos 2x, \sin 2x, \ldots, \cos nx, \sin nx, \ldots\}$$

is not quite a spanning set for $\mathcal{PF}(2\pi)$, nevertheless, any function in $\mathcal{PF}(2\pi)$ can be represented as a *limit* of linear combinations of these cosines and sines. Fourier series are discussed further in Section 7–4.

Students sometimes ask, "What is the difference between 'subspace' and a 'vector space'?" The answer is that it depends on your point of view. A subspace (of any vector space $\mathbb{V}$*) is a vector space, but it is sometimes helpful to emphasize that it is a subset of the bigger space* $\mathbb{V}$ *by calling it a subspace of* $\mathbb{V}$*. For example, the nullspace of a linear transformation* $L : \mathbb{R}^3 \to \mathbb{R}^4$ *is a vector space, but it is useful to consider it as a subspace of the domain* $\mathbb{R}^3$*. Sometimes it is most useful to consider the set of functions* $C^1(a, b)$ *as a vector space all by itself—but at other times we may wish to consider it as a subspace of* $C(a, b)$*.*

What About the Dot Product?

Notice that the definition of a vector space makes no mention of the dot product. It is quite possible to develop most of the theory of Chapters 2–6 without any mention of dot products, lengths, or angles. If, in addition to the vector space axioms, we assume that a dot product (or in general cases, an "inner product") is defined, then we have an "inner product space". This is discussed further in Section 7–4. Some other books use notations to make clear the distinction: for example, $\mathbb{R}^3$ may denote the 3-dimensional vector space *without* any inner product, while $\mathbb{E}^3$ denotes the 3-dimensional vector space *with* the usual Euclidean inner product.

In this book, I have chosen to assume throughout that we have the usual inner product in $\mathbb{R}^3$ and $\mathbb{R}^n$, because this allows us to develop ideas about linear mappings in a familiar geometric setting, and allows us to introduce interesting and important geometric mappings such as rotations, projections, and reflections as examples. If and when it becomes necessary, it is easy to sort out those things that only make sense when you have an inner product. We shall continue to use examples that make use of the dot product, but you should check from time to time that the definitions about bases, coordinates, and (in Chapter 6) eigenvectors do not really require the use of the dot product.

A careful reader might be concerned that we used the dot product in defining the matrix product. However, we used the algebraic formula for dot product in the matrix product only as a convenient way to remember the rule for the matrix product; ideas such as length of a vector and orthogonality, which are essential features of an inner product, are not part of the definition of matrix product.

EXERCISES 4–1

A1 Let $\mathcal{M}(n, n)$ denote the space of n by n matrices, with addition of matrices and multiplication of matrices by real numbers defined as in Section 3–1. Check that $\mathcal{M}(n, n)$ is a vector space over the reals. (See if you can check the properties in your head.) Determine whether the following subsets of $\mathcal{M}(n, n)$ are subspaces. If the subset is not a subspace, indicate clearly why not (an example is usually the best way).

(a) The subset of **diagonal** matrices. (A matrix is **diagonal** if all entries that are not on the main diagonal are zero; that is, $a_{jk} = 0$ if $j \neq k$. Note that $a_{jj} = 0$ is allowed.)

(b) The subset of matrices that are in row echelon form.

(c) The subset of **symmetric** matrices. (*A* matrix is **symmetric** if $A^T = A$, or equivalently, if $a_{jk} = a_{kj}$ for all j and k.)

(d) The subset of contraction matrices. (See Section 3–3.)

(e) The set of **upper triangular** matrices. (A matrix is **upper triangular** if all the entries below the main diagonal are zero; that is, $a_{jk} = 0$ if $j > k$.)

A2 Each of the following sets of polynomials is a subset of P_5. In each case, either show that the subset is a subspace of P_5 or show (by example or explanation) that it is not a subspace of P_5.

(a) $E = \{p(x) \mid p(x) \in P_5, p(-x) = p(x)\}$. (The subset of **even polynomials**; a polynomial p is even if $p(-x) = p(x)$, so for example, $p(x) = x^2 + x^4$ is even.)

(b) $F = \{(1 + x^2)p(x) \mid p(x) \in P_3\}$. (Polynomials of the form $(1 + x^2)p(x)$, where p is of degree less than or equal to 3.)

(c) $G = \{p(x) = a_0 + a_1x + \ldots + a_4x^4 \mid a_0 = a_4, a_1 = a_3\}$

(d) $H = \{x^3p(x) \mid p(x) \in P_2\}$

A3 Let $\mathcal{F}$ be the set of all real-valued functions of a real variable. For each of the following subsets of $\mathcal{F}$, either prove that the set is a subspace of $\mathcal{F}$ or give an example to show that it is not.

(a) $\{f \mid f(3) = 0\}$

(b) $\{f \mid f(3) = 1\}$

(c) the set of **even** functions (a function is even if $f(-x) = f(x)$ for all real numbers x)
(d) the set of non-negative functions $\{f \mid f(x) \geq 0, x \in \mathbb{R}\}$

B1 Let $\mathcal{M}(3,3)$ be the vector space of 3 by 3 matrices (see Exercise A1). Determine whether the following subsets of $\mathcal{M}(3,3)$ are subspaces. If it is not, indicate clearly why it is not.
(a) $\{A \mid A \in \mathcal{M}(3,3), a_{11} + a_{22} + a_{33} = 0\}$ (the sum $a_{11} + a_{22} + a_{33}$ is called the **trace** of A, so the subset here is called the set of trace-free matrices).
(b) The subset of invertible 3 by 3 matrices.
(c) The subset of matrices A such that $A \begin{bmatrix} 1 \\ 2 \\ 3 \end{bmatrix} = \begin{bmatrix} 0 \\ 0 \\ 0 \end{bmatrix}$.
(d) The subset of matrices A such that $A \begin{bmatrix} 4 \\ 5 \\ 6 \end{bmatrix} = \begin{bmatrix} 1 \\ 2 \\ 3 \end{bmatrix}$.
(e) The subset of **skew-symmetric** (or anti-symmetric) matrices. (A matrix is skew-symmetric if $A^T = -A$, or if $a_{jk} = -a_{jk}$ for all j and k.)

B2 Each of the following sets of polynomials is a subset of P_5. In each case, either show that the subset is a subspace of P_5 or show (by example or explanation) that it is not a subspace of P_5.
(a) $E = \{p(x) \mid p(x) \in P_5, p(-x) = -p(x)\}$ (the subset of **odd** polynomials)
(b) $F = \{(p(x))^2 \mid p(x) \in P_2\}$
(c) $G = \{p(x) = a_0 + a_1 x + \cdots + a_4 x^4 \mid a_1 a_4 = 1\}$
(d) $H = \{(x + x^3)p(x) \mid p(x) \in P_2\}$

B3 Let $\mathcal{F}$ be the set of all real-valued functions of a real variable. For each of the following subsets of $\mathcal{F}$, either prove that the set is a subspace of $\mathcal{F}$ or give an example to show that it is not.
(a) $\{f \mid f(3) + f(5) = 0\}$
(b) $\{f \mid f(1) + f(2) = 1\}$
(c) $\{f \mid |f(x)| \leq 1\}$ (the set of functions whose absolute value is bounded by 1)
(d) $\{f \mid f \text{ is increasing on } \mathbb{R}\}$ (f is increasing on $\mathbb{R}$ if $f(x_1) < f(x_2)$ whenever $x_1 < x_2$)

CONCEPTUAL EXERCISES

***D1** Use properties (1) – (11) of vector spaces to prove that $(-1)\mathbf{x} = -\mathbf{x}$. (Property (12))

D2 For linear mappings from $\mathbb{R}^n$ to $\mathbb{R}^m$, addition and multiplication by scalars were defined in Section 3–3. Show that the set of linear mappings from $\mathbb{R}^n$ to $\mathbb{R}^m$ is a vector space over the reals. (Part of what you need to prove is in Exercise 3–3 D4.)

D3 (a) Let $\mathcal{A}f$ be the antiderivative (or indefinite integral) of the continuous function f such that $(\mathcal{A}f)(0) = 0$. Show that $\mathcal{A}$ is a linear mapping: $C(\mathbb{R}) \to C(\mathbb{R})$.
(b) For fixed real numbers a and b, let $I_a^b(f)$ be the definite integral of the continuous function f from a to b. Show that I_a^b is a linear mapping. What is its domain and codomain?

D4 Suppose that U and V are vector spaces over the real numbers. The Cartesian product of U and V is defined to be

$$\mathrm{U} \times \mathrm{V} = \{(\mathbf{u}, \mathbf{v}) \mid \mathbf{u} \in \mathrm{U}, \mathbf{v} \in \mathrm{V}\}.$$

(a) In $\mathrm{U} \times \mathrm{V}$, define addition by

$$(\mathbf{u}_1, \mathbf{v}_1) + (\mathbf{u}_2, \mathbf{v}_2) = (\mathbf{u}_1 + \mathbf{u}_2, \mathbf{v}_1 + \mathbf{v}_2)$$

and multiplication by real scalars by

$$c(\mathbf{u}, \mathbf{v}) = (c\mathbf{u}, c\mathbf{v}).$$

Verify that with these operations, $\mathrm{U} \times \mathrm{V}$ is closed under addition and multiplication by scalars; in fact, it is a vector space—verify the other vector space properties in $\mathrm{U} \times \mathrm{V}$.
(b) Verify that $\mathrm{U} \times \{\mathbf{0}\}$ is a subspace of $\mathrm{U} \times \mathrm{V}$.
(c) Suppose instead that multiplication by scalars is defined by $c(\mathbf{u}, \mathbf{v}) = (c\mathbf{u}, \mathbf{v})$, while addition is defined as in part (a). Is $\mathrm{U} \times \mathrm{V}$ a vector space with these operations?

SECTION

4–2 Linear Independence and Bases

The material of this section is crucial to everything that follows. We first emphasize why linear independence is important, and then we define linear independence by a standard computational test. A common geometrical description is given at the end of the section.

In Chapters 1 and 3, much of the discussion depended on the use of the standard basis in $\mathbb{R}^n$. To give just two examples, the dot product of two vectors $\mathbf{a}$ and $\mathbf{b}$ was defined in terms of the standard components of the vectors, and the matrix $[L]$ of a linear mapping L was determined by calculating the images of the standard basis vectors. That means that the standard basis vectors (the directions of the coordinate axes) enjoy a special preferred status in the discussion.

There are important problems where it would be much more convenient if we could base our discussion on different preferred vectors. For example, in $\mathbb{R}^2$, a stretch by factor 3 in the direction of the vector $(1, 2)$ is geometrically easy to understand; it must leave the orthogonal vector $(-2, 1)$ fixed (or unchanged). With the tools we have so far, it would be awkward to determine the standard matrix of this stretch, and then determine its effect on any other vector. It would be better to have a description that takes advantage of the preferred directions $(1, 2)$ and $(-2, 1)$ in this example.

Or, consider the reflection in the plane $x_1 + 2x_2 - 3x_3 = 0$ in $\mathbb{R}^3$. It is easy to describe this by saying that it reverses the normal vector $(1, 2, -3)$ to $(-1, -2, 3)$ and leaves unchanged any vectors lying in the plane (such as $(2, -1, 0)$ and $(3, 0, 1)$—check that these satisfy the equation of the plane). See Figure 4-2-1. Describing this reflection in terms of these vectors gives more geometric information than describing it in terms of the standard basis vectors.

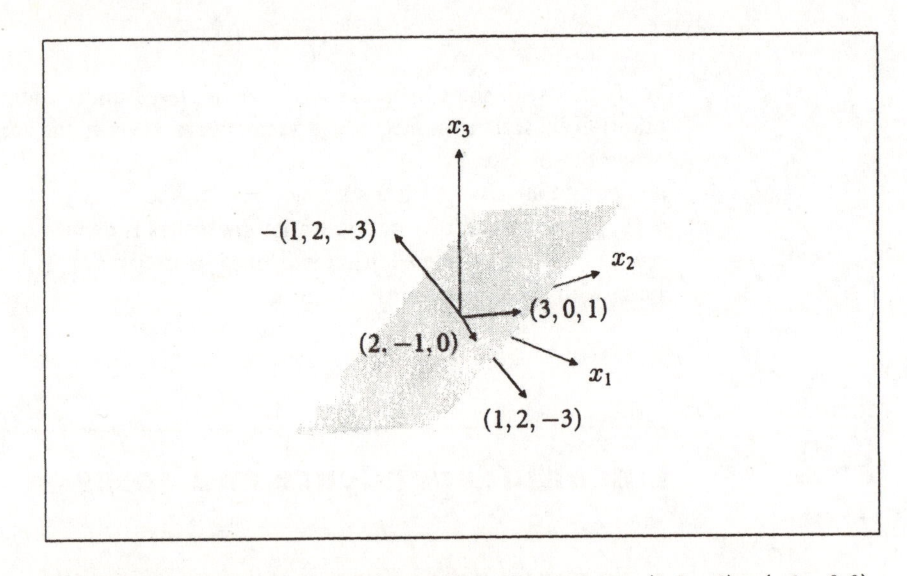

FIGURE 4-2-1. *Reflection in the plane* $x_1 + 2x_2 - 3x_3 = 0$ *maps* $(1, 2, -3)$ *to* $(-1, -2, 3)$*, and leaves* $(2, -1, 0)$ *and* $(3, 0, 1)$ *unchanged.*

Notice that in these examples the geometry itself provides us with a preferred set of vectors that make up a **spanning** set for the appropriate space: in the case of the stretch, $\{(1, 2), (-2, 1)\}$ is a spanning set for $\mathbb{R}^2$, and in the case of the reflection, $\{(1, 2, -3), (2, -1, 0), (3, 0, 1)\}$ is a spanning set for $\mathbb{R}^3$. (It is easy to check that these are spanning sets; the set in $\mathbb{R}^3$ will be checked below.)

We shall usually require that our spanning sets have an additional property, called linear independence. The following example illustrates why we need this extra property.

EXAMPLE 9

Let $\mathbb{V}$ be the rowspace of the matrix $A = \begin{bmatrix} 1 & 1 & 1 \\ 0 & 1 & 1 \\ 2 & 3 & 3 \end{bmatrix}$. $\mathbb{V}$ is the subspace of $\mathbb{R}^3$ spanned by $\mathbf{v}_1 = (1,1,1)$, $\mathbf{v}_2 = (0,1,1)$, and $\mathbf{v}_3 = (2,3,3)$. By Theorem 8 of Section 3–4, $\mathbb{V}$ is also the rowspace of the reduced row echelon form $R = \begin{bmatrix} 1 & 0 & 0 \\ 0 & 1 & 1 \\ 0 & 0 & 0 \end{bmatrix}$ of A. Thus, we have a second spanning set for $\mathbb{V}$ consisting of $\mathbf{w}_1 = (1,0,0)$ and $\mathbf{w}_2 = (0,1,1)$.

Now try to write a vector in $\mathbb{V}$ in terms of these spanning sets. For example, we write $(3,5,5)$ in terms of the first spanning set: $(3,5,5) = 3\mathbf{v}_1 + 2\mathbf{v}_2 + 0\mathbf{v}_3$, OR $(3,5,5) = 1\mathbf{v}_1 + 1\mathbf{v}_2 + 1\mathbf{v}_3$, OR $(3,5,5) = (-1)\mathbf{v}_1 + 0\mathbf{v}_2 + 2\mathbf{v}_3$; in fact, there are infinitely many ways to write $(3,5,5)$ in terms of $\{\mathbf{v}_1, \mathbf{v}_2, \mathbf{v}_3\}$. On the other hand, there is only one way to write $(3,5,5)$ in terms of the second spanning set: if

$$(3,5,5) = c_1\mathbf{w}_1 + c_2\mathbf{w}_2 = c_1(1,0,0) + c_2(0,1,1) = (c_1, c_2, c_2),$$

then $c_1 = 3$ and $c_2 = 5$, and there are no other possibilities. A similar situation occurs for any vector $\mathbf{v}$ in $\mathbb{V}$ (*try it for* $(4,-7,-7)$). For many purposes this makes $\{\mathbf{w}_1, \mathbf{w}_2\}$ a "better" spanning set than $\{\mathbf{v}_1, \mathbf{v}_2, \mathbf{v}_3\}$. We do not explore this example further, but instead turn to the general case.

We shall say that a spanning set for a vector space $\mathbb{V}$ has the **unique representation property** if every vector in $\mathbb{V}$ can be written in a **unique** way as a linear combination of the spanning vectors. In Example 9, $\{\mathbf{w}_1, \mathbf{w}_2\}$ has this property, but $\{\mathbf{v}_1, \mathbf{v}_2, \mathbf{v}_3\}$ does not. This property is so obvious when our spanning set consists of standard basis vectors that we take it for granted, but it is really quite essential in many calculations. If spanning sets other than the standard basis are to be used, we prefer spanning sets that possess this property.

How can we determine whether a spanning set has this property? Suppose that $S = \{\mathbf{v}_1, \mathbf{v}_2, \ldots, \mathbf{v}_k\}$ is a spanning set for $\mathbb{V}$ ($\mathbb{V}$ could be a subspace of $\mathbb{R}^n$ or a space of functions, so this discussion is quite general). To explore whether this set S has the unique representation property, consider the possibility that there are two representations for some $\mathbf{x}$ in V:

$$\mathbf{x} = a_1\mathbf{v}_1 + a_2\mathbf{v}_2 + \cdots + a_k\mathbf{v}_k = b_1\mathbf{v}_1 + b_2\mathbf{v}_2 + \cdots + b_k\mathbf{v}_k.$$

These representations coincide and the vector $\mathbf{x}$ is **uniquely** represented as a linear combination of the $\mathbf{v}$'s if and only if $a_j = b_j$ for $j = 1, 2, \ldots, k$. We want this to be true for **every** $\mathbf{x}$ in V, so S has the unique representation property if and only if

> **whenever** $(a_1 - b_1)\mathbf{v}_1 + (a_2 - b_2)\mathbf{v}_2 + \cdots + (a_k - b_k)\mathbf{v}_k = \mathbf{0}$,
> it follows that $(a_j - b_j) = 0$ for $j = 1, 2, \ldots, k$.

This leads to the following definition, which is one of the most important in linear algebra.

DEFINITION

A set of vectors $S = \{\mathbf{v}_1, \mathbf{v}_2, \ldots, \mathbf{v}_k\}$ is **linearly independent** if

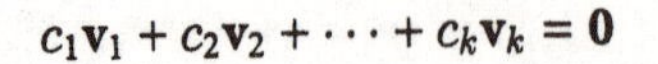

$$c_1\mathbf{v}_1 + c_2\mathbf{v}_2 + \cdots + c_k\mathbf{v}_k = \mathbf{0}$$

implies that $c_j = 0$ for $j = 1, 2, \ldots, k$. This is often stated in the equivalent form: S is linearly independent if the only linear combination of vectors of S that is equal to the zero vector is the **trivial** linear combination. (The trivial linear combination is the linear combination with all coefficients equal to zero.)

Although the definition applies to a **set** of vectors, we shall often be a little sloppy and say "the vectors $\mathbf{v}_1, \mathbf{v}_2, \ldots, \mathbf{v}_k$ are linearly independent."

EXAMPLE 10

The standard basis vectors $\mathbf{e}_1, \mathbf{e}_2, \ldots, \mathbf{e}_n$ in $\mathbb{R}^n$ form a linearly independent set. To test this, consider the equation

$$c_1\mathbf{e}_1 + c_2\mathbf{e}_2 + \cdots + c_n\mathbf{e}_n = \mathbf{0};$$

rewrite this in column form:

$$c_1[\mathbf{e}_1] + c_2[\mathbf{e}_2] + \cdots + c_n[\mathbf{e}_n] = [\mathbf{0}]$$

or

$$\begin{bmatrix} [\mathbf{e}_1] & [\mathbf{e}_2] & \ldots & [\mathbf{e}_n] \end{bmatrix} \begin{bmatrix} c_1 \\ c_2 \\ \vdots \\ c_n \end{bmatrix} = [\mathbf{0}].$$

Because of the special form of the standard basis vectors, the coefficient matrix in this system is the identity matrix. Therefore, $\mathbf{c} = \mathbf{0}$ (the trivial solution) is the only solution and the vectors are linearly independent.

EXAMPLE 11

The set $\{\mathbf{v}_1 = (1,1,1), \mathbf{v}_2 = (0,1,1), \mathbf{v}_3 = (2,3,3)\}$ of Example 9 was an example of a spanning set that did not have the unique representation property. Let us check that it is not linearly independent according to the definition. Consider the equation

$$c_1(1,1,1) + c_2(0,1,1) + c_3(2,3,3) = (0,0,0),$$

or in column form

$$c_1 \begin{bmatrix} 1 \\ 1 \\ 1 \end{bmatrix} + c_2 \begin{bmatrix} 0 \\ 1 \\ 1 \end{bmatrix} + c_3 \begin{bmatrix} 2 \\ 3 \\ 3 \end{bmatrix} = \begin{bmatrix} 0 \\ 0 \\ 0 \end{bmatrix}$$

or

$$\begin{bmatrix} 1 & 0 & 2 \\ 1 & 1 & 3 \\ 1 & 1 & 3 \end{bmatrix} [\mathbf{c}] = [\mathbf{0}].$$

The reduced row echelon form of the coefficient matrix is $\begin{bmatrix} 1 & 0 & 2 \\ 0 & 1 & 1 \\ 0 & 0 & 0 \end{bmatrix}$. The homogeneous system has 3 variables and rank 2, so there are non-trivial solutions for c_1, c_2, c_3, and $\{\mathbf{v}_1, \mathbf{v}_2, \mathbf{v}_3\}$ is not a linearly independent set.

If a set is **not** linearly independent we shall call it a linearly **dependent** set. It is worth restating the definition for this case.

DEFINITION

A set S of vectors is **linearly dependent** if there is a non-trivial linear combination of the vectors of S that is equal to the zero vector.

EXAMPLE 12

Are the vectors $(3, 2, 2, 2)$, $(1, 0, 1, 1)$, $(1, 1, 1, 0)$ linearly independent? Consider

$$c_1 \begin{bmatrix} 3 \\ 2 \\ 2 \\ 2 \end{bmatrix} + c_2 \begin{bmatrix} 1 \\ 0 \\ 1 \\ 1 \end{bmatrix} + c_3 \begin{bmatrix} 1 \\ 1 \\ 1 \\ 0 \end{bmatrix} = \begin{bmatrix} 3 & 1 & 1 \\ 2 & 0 & 1 \\ 2 & 1 & 1 \\ 2 & 1 & 0 \end{bmatrix} \begin{bmatrix} c_1 \\ c_2 \\ c_3 \end{bmatrix} = \begin{bmatrix} 0 \\ 0 \\ 0 \\ 0 \end{bmatrix}.$$

Then,

$$\begin{bmatrix} 3 & 1 & 1 \\ 2 & 0 & 1 \\ 2 & 1 & 1 \\ 2 & 1 & 0 \end{bmatrix} \to \begin{bmatrix} 1 & 1 & 0 \\ 2 & 0 & 1 \\ 2 & 1 & 1 \\ 2 & 1 & 0 \end{bmatrix} \to \begin{bmatrix} 1 & 1 & 0 \\ 0 & -2 & 1 \\ 0 & -1 & 1 \\ 0 & -1 & 0 \end{bmatrix} \to \cdots \to \begin{bmatrix} 1 & 0 & 0 \\ 0 & 1 & 0 \\ 0 & 0 & 1 \\ 0 & 0 & 0 \end{bmatrix},$$

so the rank of the coefficient matrix is 3, equal to the number of variables, and the system has a unique solution; the trivial solution is the only solution to this homogeneous system, and the given vectors are linearly independent.

General Procedure for Determining Linear Independence of a Set of Vectors $\{\mathbf{v}_1, \mathbf{v}_2, \ldots, \mathbf{v}_k\}$ in $\mathbb{R}^n$

Write the equation $c_1[\mathbf{v}_1] + c_2[\mathbf{v}_2] + \cdots + c_k[\mathbf{v}_k] = [\mathbf{0}]$. This gives a homogeneous linear system for $\mathbf{c}$ with coefficient matrix $\Big[[\mathbf{v}_1], [\mathbf{v}_2], [\mathbf{v}_3], \ldots, [\mathbf{v}_k]\Big]$. Row reduce this matrix to determine the nature of the solution set. If the only solution is the trivial solution, the vectors are linearly independent; otherwise, they are linearly dependent.

Note that this procedure would allow us to determine the coefficients $c_1, c_2, \ldots, c_k$ if there were non-trivial solutions. In this case, we could thus write an explicit linear dependence relation among the vectors. This is rarely necessary.

Suppose that a set S in $\mathbb{R}^n$ contains k vectors: $S = \{\mathbf{v}_1, \mathbf{v}_2, \ldots, \mathbf{v}_k\}$; suppose also that $k > n$. Then the coefficient matrix $\big[[\mathbf{v}_1], [\mathbf{v}_2], [\mathbf{v}_3], \ldots, [\mathbf{v}_k]\big]$ is n by k; since $k > n$, it follows from Theorem 4 in Section 2–2 that the homogeneous system always has non-trivial solutions. Hence, we have proved the following theorem.

THEOREM 1. If the number of vectors in a set S in $\mathbb{R}^n$ is greater than the dimension n of $\mathbb{R}^n$, then the set S is necessarily linearly dependent.

Let us now consider how the concept of linear independence applies to elements in spaces other than $\mathbb{R}^n$.

EXAMPLE 13

In Section 4–1, the vector space P_n of polynomials of degree less than or equal to n was introduced. Is the set $\{1, x, x^2, \ldots, x^n\}$ a linearly independent spanning set for P_n? The question we need to ask is: for what values of $c_0, c_1, c_2, \ldots, c_n$ is the polynomial $c_0 + c_1 x + c_2 x^2 + \cdots + c_n x^n$ equal to the *zero polynomial*? The answer is that to have the zero polynomial we require $c_0 = c_1 = c_2 = \cdots = c_n = 0$, so the set is linearly independent.

EXAMPLE 14

In the space of continuous functions, consider the set of functions $\{\sin x, \cos x, \sin(x + \pi/4)\}$. From the identity

$$\sin(x + \pi/4) = (\sqrt{2}/2)\sin x + (\sqrt{2}/2)\cos x,$$

it follows that

$$(\sqrt{2}/2)\sin x + (\sqrt{2}/2)\cos x - \sin(x + \pi/4) = 0 \quad \text{for all } x \text{ in } \mathbb{R},$$

so this is a linearly dependent set of functions. (This turns out to be quite important in studying the equation of an oscillating spring.) On the other hand, there are no non-zero constants c_1, c_2, c_3 such that

$$c_1 \sin x + c_2 x^2 + c_3 e^x = 0 \quad \text{for all } x \text{ in } \mathbb{R},$$

so the set of functions $\{\sin x, x^2, e^x\}$ is linearly independent.

The idea of linear independence was introduced in order to check whether a spanning set had the unique representation property. It is worth stating the conclusion as a theorem; the proof of this theorem is essentially the discussion leading to the definition of linear independence.

THEOREM 2. Let S be a spanning set for a vector space V. Then every vector in V can be expressed in a **unique** way as a linear combination of the vectors of S if and only if the set S is linearly independent.

Bases

Since linearly independent spanning sets are particularly important, they are given a special name.

DEFINITION

A set $\mathcal{B}$ of vectors in a vector space V is a **basis** for V if

(1) $\mathcal{B}$ is a spanning set for V, and

(2) $\mathcal{B}$ is linearly independent.

Note that the plural of "basis" is "bases".

For the zero space $\{\mathbf{0}\}$, there is no linearly independent spanning set, so there is no basis for $\{\mathbf{0}\}$.

EXAMPLE 15

The standard basis vectors $\mathbf{e}_1, \mathbf{e}_2, \ldots, \mathbf{e}_n$ of Example 10 form a basis of $\mathbb{R}^n$.

EXAMPLE 16

In Example 12, the set $\{(3,2,2,2),(1,0,1,1),(1,1,1,0)\}$ is linearly independent; it is a basis for the subspace it spans. (This set does **not** span all of $\mathbb{R}^4$, so it is not a basis for $\mathbb{R}^4$.)

EXAMPLE 17

The set $\{1, x, x^2, \ldots, x^n\}$ is a basis for P_n. (See Example 13.)

EXAMPLE 18

Is the set $S = \{(1,2,-1,1),(0,1,3,1),(2,5,1,3)\}$ a basis for the subspace $\mathrm{Sp}(S)$? S is by definition a spanning set for $\mathrm{Sp}(S)$, so we need only check whether S is linearly independent. Consider the vector equation

$$c_1\begin{bmatrix}1\\2\\-1\\1\end{bmatrix}+c_2\begin{bmatrix}0\\1\\3\\1\end{bmatrix}+c_3\begin{bmatrix}2\\5\\1\\3\end{bmatrix}=\begin{bmatrix}1&0&2\\2&1&5\\-1&3&1\\1&1&3\end{bmatrix}[\mathbf{c}]=[\mathbf{0}].$$

Then,

$$\begin{bmatrix}1&0&2\\2&1&5\\-1&3&1\\1&1&3\end{bmatrix}\to\begin{bmatrix}1&0&2\\0&1&1\\0&3&3\\0&1&1\end{bmatrix}\to\begin{bmatrix}1&0&2\\0&1&1\\0&0&0\\0&0&0\end{bmatrix}$$

and it is clear that there are non-trivial solutions $\mathbf{c}$, so the given vectors are linearly dependent and do not form a basis for $\text{Sp}(S)$.

EXAMPLE 19

Reflection in the plane $x_1 + 2x_2 - 3x_3 = 0$ was used as an example in the introduction to this section. Consider the set $\mathcal{B} = \{(1,2,-3),(2,-1,0),(3,0,1)\}$ consisting of the normal vector to the plane and two vectors lying in the plane. Is $\mathcal{B}$ a basis for $\mathbb{R}^3$? We must check both whether it is a spanning set and whether it is linearly independent.

Does $\mathcal{B}$ span $\mathbb{R}^3$? Can the system $c_1\begin{bmatrix}1\\2\\-3\end{bmatrix}+c_2\begin{bmatrix}2\\-1\\0\end{bmatrix}+c_3\begin{bmatrix}3\\0\\1\end{bmatrix}=\mathbf{b}$ be solved for arbitrary $\mathbf{b}$ in $\mathbb{R}^3$? The answer depends on the row echelon form of the coefficient matrix:

$$\begin{bmatrix}1&2&3\\2&-1&0\\-3&0&1\end{bmatrix}\to\begin{bmatrix}1&2&3\\0&-5&-6\\0&6&10\end{bmatrix}\to\begin{bmatrix}1&2&3\\0&1&6/5\\0&0&14/5\end{bmatrix}\to\begin{bmatrix}1&2&3\\0&1&6/5\\0&0&1\end{bmatrix}.$$

It follows that the nonhomogeneous system for $\mathbf{c}$ has a unique solution for any $\mathbf{b}$, so the given set of vectors is a spanning set for $\mathbb{R}^3$.

Is $\mathcal{B}$ linearly independent? Are there non-trivial solutions for the system of equations

$$c_1\begin{bmatrix}1\\2\\-3\end{bmatrix}+c_2\begin{bmatrix}2\\-1\\0\end{bmatrix}+c_3\begin{bmatrix}3\\0\\1\end{bmatrix}=\mathbf{0}?$$

Notice that the coefficient matrix for this system is exactly the same as the coefficient matrix for the system above. Since the rank of the coefficient matrix is 3 (as determined from the row echelon form above), this system has a unique solution: the trivial solution $\mathbf{c} = \mathbf{0}$ is the only solution, so $\mathcal{B}$ is a linearly independent set.

Since $\mathcal{B}$ is both a spanning set for $\mathbb{R}^3$ and a linearly independent set, $\mathcal{B}$ is a basis for $\mathbb{R}^3$.

Example 19 exhibits an important and convenient fact: **for a set of vectors in $\mathbb{R}^n$ (or in a subspace of $\mathbb{R}^n$) the same coefficient matrix appears in the system of equations that determines whether the set is a spanning set and in the system that determines whether the set is linearly independent.** If you set up your solution procedure carefully, you will have to do only one row reduction.

An Intuitive Description of Linear Independence

Suppose that two vectors $\mathbf{u}$ and $\mathbf{v}$ in a vector space $\mathbb{V}$ are linearly **dependent.** Then, for some c and d, not both zero, $c\mathbf{u} + d\mathbf{v} = \mathbf{0}$. If $c \neq 0$, then we may write $\mathbf{u} = (-d/c)\mathbf{v}$, so $\mathbf{u}$ is a scalar multiple of $\mathbf{v}$, and it is natural to say that "$\mathbf{u}$ depends on $\mathbf{v}$". On the other hand, if $d \neq 0$, then $\mathbf{v} = (-c/d)\mathbf{u}$, and $\mathbf{v}$ depends on $\mathbf{u}$. Since at least one of c and d must be non-zero, the fact that $\mathbf{u}$ and $\mathbf{v}$ are linearly dependent means that one vector is dependent on the other. Geometrically, the vectors must point in the same direction: they are **collinear** (along the same line). Turning this around, we may say that two vectors are linearly **independent** if they are **not** collinear.

Note that it is possible that one of c or d is zero in this case. For example, in $\mathbb{R}^2$, $0(1,1) + 1(0,0) = (0,0)$, so that the set $\{(1,1),(0,0)\}$ is linearly dependent. In fact, in any space, any set of vectors that includes the zero vector must be linearly dependent. Note that in this case, the statement that one vector depends on the other is $(0,0) = 0(1,1)$.

Next consider a linearly dependent set of three vectors $\{\mathbf{u},\mathbf{v},\mathbf{w}\}$: for some c, d, e, not all zero, $c\mathbf{u} + d\mathbf{v} + e\mathbf{w} = \mathbf{0}$. By arguments just like those above, we see that it must be possible to express at least one of $\mathbf{u}, \mathbf{v}, \mathbf{w}$ as a linear combination of the other two. However, without knowing which of the coefficients is non-zero, we cannot be sure before we start which of the three can be expressed in terms of the other two. Geometrically, in this case, linear dependence means that one of the vectors lies in the subspace spanned by the other two, so the three vectors are **coplanar** (in the same plane). Notice that "coplanar" includes degenerate cases where one of the vectors is the zero vector, or the three vectors are in fact collinear. Three vectors are linearly **independent** if and only if they are **not** coplanar.

Generally, if $\{\mathbf{v}_1, \mathbf{v}_2, \ldots, \mathbf{v}_k\}$ is a linearly dependent set, by the same arguments as above, at least one of the vectors can be expressed as a linear combination of the others. Unfortunately, we do not know when we start which of the vectors can be expressed in terms of the others. If we try to build a test of linear independence around this idea, we would first have to check whether $\mathbf{v}_1$ can be written as a linear combination of the others; if not, then whether $\mathbf{v}_2$ can be written as a linear combination of the others, and so on. This would be a very inefficient test. In fact, if you started this way, you would eventually be led to the definition of linear independence given earlier. The earlier approach also has the advantage of emphasizing the crucial property of unique representation.

Although this geometrical approach does not immediately lead to the best definition, it can be very helpful in thinking about linear independence, particularly in low-dimensional cases.

EXAMPLE 20

Let us apply this intuitive approach to the set $\mathcal{B} = \{(1, 2, -3), (2, -1, 0), (3, 0, 1)\}$ introduced in Example 19. Clearly $(2, -1, 0)$ and $(3, 0, 1)$ are linearly independent, since neither is a multiple of the other (this is easy to check since each vector has one component equal to zero where the other vector does not). The normal vector $(1, 2, -3)$ obviously does not lie in the plane spanned by the other two vectors. Intuitively, it seems clear that the three vectors must be linearly independent. An argument such as this does not constitute a proof of linear independence (unless you prove some theorems that justify such an approach), but it can be a very valuable help in choosing a set of vectors to use as a basis.

There is an issue concerning the definition of linear independence that may worry some readers. The correct use of the language and notation of sets requires that in the set $\{\mathbf{v}_1, \mathbf{v}_2, \ldots, \mathbf{v}_k\}$ there are no repetitions *— the elements in the set must be distinct. It follows that, strictly speaking, our definition of linear independence ought to be applied only when we have k distinct vectors. However, we shall need to ask questions such as: "Are the columns of a given matrix linearly independent?" — and the columns may not be distinct. For the questions we consider in this book, no practical difficulties arise if we are a little sloppy with language at this point and apply the definition of linear independence to "sets with repetitions allowed", instead of introducing additional formal vocabulary. Thus, for example, we shall say that the columns of* $\begin{bmatrix} 1 & 1 \\ 2 & 2 \end{bmatrix}$ *are linearly dependent, because*

$c_1(1,2) + c_2(1,2) = (0,0)$ *has non-trivial solutions* $c_1 = -c_2$, *although the columns are not distinct vectors in* $\mathbb{R}^2$.

EXERCISES 4–2

A1 Consider the vectors $\mathbf{v}_1 = (1,1,1)$, $\mathbf{v}_2 = (0,1,2)$, $\mathbf{v}_3 = (1,2,4)$. Suppose that $\mathbf{b}$ is a vector in the subspace spanned by $\mathbf{v}_1, \mathbf{v}_2, \mathbf{v}_3$; equivalently, there are numbers c_1, c_2, c_3 such that

$$c_1[\mathbf{v}_1] + c_2[\mathbf{v}_2] + c_3[\mathbf{v}_3] = [\mathbf{b}].$$

(a) The given information says that the system of equations for c_1, c_2, c_3 is consistent. By row reduction, determine whether the system has a unique solution.
(b) Does the spanning set $\{\mathbf{v}_1, \mathbf{v}_2, \mathbf{v}_3\}$ have the unique representation property?
(c) Consider the case $\mathbf{b} = \mathbf{0}$; the system certainly has $c_1 = c_2 = c_3 = 0$ as one solution. Use your answers to (a) and (b) to decide whether $\{\mathbf{v}_1, \mathbf{v}_2, \mathbf{v}_3\}$ is linearly independent.
(d) Next consider vectors $\mathbf{w}_1 = (1,1,-1)$, $\mathbf{w}_2 = (0,1,-2)$, $\mathbf{w}_3 = (1,1,-3)$. Suppose now that the vector $\mathbf{b}$ is in $\mathrm{Sp}(\{\mathbf{w}_1, \mathbf{w}_2, \mathbf{w}_3\})$, so that the system

$$d_1[\mathbf{w}_1] + d_2[\mathbf{w}_2] + d_3[\mathbf{w}_3] = \mathbf{b}$$

is consistent. By a procedure similar to (a), (b), (c), decide whether $\{\mathbf{w}_1, \mathbf{w}_2, \mathbf{w}_3\}$ has the unique representation property and whether it is linearly independent.

A2 Repeat Exercise A1 with $\mathbf{v}_1 = (1,0,-1,1)$, $\mathbf{v}_2 = (1,1,1,2)$, $\mathbf{v}_3 = (2,-1,-4,1)$, and $\mathbf{w}_1 = (1,1,-1,0)$, $\mathbf{w}_2 = (1,2,1,1)$, $\mathbf{w}_3 = (2,1,4,1)$.

A3 Determine whether the following sets are linearly independent. If the set is linearly dependent, find all linear combinations of the vectors that are $\mathbf{0}$.
(a) $\{(1,2,1,-1), (1,2,3,1), (1,-3,2,1)\}$
(b) $\{(1,0,1,0), (0,1,1,1), (0,0,1,1), (3,2,6,3)\}$
(c) $\{(1,1,0,1,1), (2,3,1,3,3), (0,1,1,1,1)\}$
(d) in the space of functions, $\{\cos^2 x, \sin^2 x, \cos 2x\}$

A4 Determine all values of k such that the given set is linearly independent. Explain.
(a) $\{(1,0,1,0), (0,1,1,1), (2,-3,-1,k)\}$
(b) $\{(1,1,1,2), (1,-1,2,0), (-1,2,k,1)\}$

A5 Determine whether the given set is a basis for $\mathbb{R}^3$. Explain briefly.
(a) $\{(1,1,2),(1,-1,-1),(2,1,1)\}$
(b) $\{(1,0,1),(-1,2,1),(1,3,5),(2,-1,-4)\}$
(c) $\{(1,-1,1)(1,2,-1),(3,0,1)\}$

A6 Let $S = \{(1,1,1,2),(2,3,1,3),(3,3,4,11),(2,3,1,4)\}$. Showing your reasoning clearly, prove that S is a basis for $\mathbb{R}^4$.

A7 By arguments similar to Example 20, choose a basis for $\mathbb{R}^3$ that is well adapted to describing the following linear mappings.
(a) $\mathbf{perp}_{(-2,1,1)}$
(b) reflection in the plane $x_1 - 3x_2 - 5x_3 = 0$

A8 Determine whether the following sets are linearly independent in P_5.
(a) $\{1+x+x^3, x+x^3+x^5, 1-x^5\}$
(b) $\{1-2x+x^4, x-2x^2+x^5, 1-3x+x^3\}$

B1 Consider the vectors $\mathbf{v}_1 = (1,1,1)$, $\mathbf{v}_2 = (1,2,1)$, $\mathbf{v}_3 = (1,-2,1)$. Suppose that $\mathbf{b}$ is a vector in the subspace spanned by $\mathbf{v}_1$, $\mathbf{v}_2$, $\mathbf{v}_3$; equivalently, there are numbers c_1, c_2, c_3 such that

$$c_1[\mathbf{v}_1] + c_2[\mathbf{v}_2] + c_3[\mathbf{v}_3] = [\mathbf{b}].$$

(a) The given information says that the system of equations for c_1, c_2, c_3 is consistent. By row reduction, determine whether the system has a unique solution.
(b) Does the spanning set $\{\mathbf{v}_1, \mathbf{v}_2, \mathbf{v}_3\}$ have the unique representation property?
(c) Consider the case $\mathbf{b} = \mathbf{0}$; the system certainly has $c_1 = c_2 = c_3 = 0$ as one solution. Use your answers to (a) and (b) to decide whether $\{\mathbf{v}_1, \mathbf{v}_2, \mathbf{v}_3\}$ is linearly independent.
(d) Next consider vectors $\mathbf{w}_1 = (1,-1,1)$, $\mathbf{w}_2 = (1,2,-1)$, $\mathbf{w}_3 = (-1,-8,0)$. Suppose now that the vector $\mathbf{b}$ is in $\mathrm{Sp}(\{\mathbf{w}_1, \mathbf{w}_2, \mathbf{w}_3\})$, so that the system

$$d_1[\mathbf{w}_1] + d_2[\mathbf{w}_2] + d_3[\mathbf{w}_3] = \mathbf{b}$$

is consistent. By a procedure similar to (a), (b), (c), decide whether $\{\mathbf{w}_1, \mathbf{w}_2, \mathbf{w}_3\}$ has the unique representation property, and whether it is linearly independent.

B2 Repeat Exercise B1 with $\mathbf{v}_1 = (1,0,1,0)$, $\mathbf{v}_2 = (0,1,1,1)$, $\mathbf{v}_3 = (1,2,2,2)$, and $\mathbf{w}_1 = (1,1,1,1)$, $\mathbf{w}_2 = (1,1,2,-2)$, $\mathbf{w}_3 = (1,1,-1,7)$.

B3 Determine whether the following sets are linearly independent. If the set is linearly dependent, find all linear combinations of the vectors that are $\mathbf{0}$.
(a) $\{(1,0,1,0),(1,2,0,1),(0,1,2,3)\}$
(b) $\{(1,0,1,1,0),(1,2,0,0,1),(0,1,1,2,-2),(1,-3,1,0,1)\}$

B4 Determine all values of k such that the given set is linearly independent. Explain.

(a) $\{(1,1,2,1),(-1,1,-1,2),(3,1,5,k)\}$

(b) $\{(1,-1,3,1),(-1,1,2,1),(-1,1,k,5)\}$

B5 Determine whether the given set is a basis for $\mathbb{R}^3$. Explain briefly.

(a) $\{1,-1,3),(2,1,-1),(1,2,-4)\}$

(b) $\{(1,3,1),(-1,0,1),(5,1,0),(0,0,3)\}$

(c) $\{(1,2,1),(-1,-1,0),(1,1,7)\}$

B6 Determine whether the given set is a basis for $\mathbb{R}^3$. Explain briefly.

(a) $\{3,2,1),(2,1,5),(1,0,1)\}$

(b) $\{(2,3,1),(1,5,0),(6,8,-1),(30,0,0)\}$

(c) $\{(1,2,3),(3,2,1),(5,6,7)\}$

B7 Let $S = \{(1,-2,1,1),(2,-3,3,4),(3,-6,4,5),(3,-6,4,6)\}$. Showing your reasoning clearly, prove that S is a basis for $\mathbb{R}^4$.

B8 By arguments similar to Example 20, choose a basis for $\mathbb{R}^3$ that is well adapted to describing the following linear mappings.

(a) reflection in the plane $2x_1 - x_2 - x_3 = 0$

(b) $\mathbf{perp}_{(1,1,3)}$

B9 Determine whether the following sets are linearly independent in P_4.

(a) $\{1+\frac{x^2}{2}, 1-\frac{x^2}{2}, x+\frac{x^3}{6}, x-\frac{x^3}{6}\}$

(b) $\{x^2, x^3, x^2+x^3+x^4\}$

B10 In calculus, the hyperbolic functions are defined by

$$\cosh x = \frac{e^x + e^{-x}}{2}, \quad \sinh x = \frac{e^x - e^{-x}}{2}.$$

Prove that $\{\cosh x, \sinh x\}$ is a basis for $\text{Sp}(\{e^x, e^{-x}\})$.

CONCEPTUAL EXERCISES

D1 In any vector space, show that a set of vectors that includes the zero vector must be linearly dependent.

***D2** Suppose that $V = \{\mathbf{v}_1, \mathbf{v}_2, \ldots, \mathbf{v}_k\}$ is a linearly independent set in some vector space. Show that any non-empty subset of V is linearly independent.

D3 Show that if $\{\mathbf{v}_1, \mathbf{v}_2\}$ is a basis for a vector space V, then for any real number c, $\{\mathbf{v}_1, \mathbf{v}_2 + c\mathbf{v}_1\}$ is also a basis for V. Show that if $\{\mathbf{v}_1, \mathbf{v}_2, \mathbf{v}_3\}$ is a basis for a vector space W, then for any real numbers c and d, $\{\mathbf{v}_1, \mathbf{v}_2, \mathbf{v}_3 + c\mathbf{v}_1 + d\mathbf{v}_2\}$ is also a basis for W. (Compare the proof of Theorem 8 in Section 3–4. These results are used in the Gram-Schmidt procedure in Chapter 7.)

D4 Suppose that $L : \mathbb{R}^4 \to \mathbb{R}^5$ is a linear mapping such that the nullspace of L is $\{\mathbf{0}\}$. Suppose that $\{\mathbf{v}_1, \mathbf{v}_2, \mathbf{v}_3\}$ is a linearly independent set in $\mathbb{R}^4$. Prove that $\{L(\mathbf{v}_1), L(\mathbf{v}_2), L(\mathbf{v}_3)\}$ is linearly independent.

D5 Suppose that the vectors $\mathbf{v}_1$, $\mathbf{v}_2$, and $\mathbf{v}_3$ in $\mathbb{R}^3$ are non-zero and mutually orthogonal (that is, each is orthogonal to both of the others).

(a) Argue intuitively that they must form a basis for $\mathbb{R}^3$.

(b) Prove that they must be linearly independent. (Consider $c_1\mathbf{v}_1 + c_2\mathbf{v}_2 + c_3\mathbf{v}_3 = \mathbf{0}$, and take suitable dot products.)

***D6** Invent a "standard" basis for $\mathcal{M}(2,2)$, the vector space of 2 by 2 matrices with real entries. Verify that it really is a basis.

D7 Invent a basis for each of the following subspaces of $\mathcal{M}(3,3)$. (See Exercises 4–1–A1 and B1 for some of the definitions.)

(a) the diagonal 3 by 3 matrices

(b) upper triangular 3 by 3 matrices

(c) 3 by 3 diagonal matrices that are trace-free

SECTION

4–3 Coordinates with Respect to a Basis

It was suggested in the introduction of Section 4–2 that for certain problems such as reflections, it might be helpful to base our description on a non-standard basis. To develop this approach, we need to know how to represent an arbitrary vector in a vector space $\mathbb{V}$ in terms of a basis for $\mathbb{V}$.

DEFINITION

Suppose that $\mathcal{B} = \{\mathbf{v}_1, \mathbf{v}_2, \ldots, \mathbf{v}_k\}$ is a basis for the vector space $\mathbb{V}$. For any $\mathbf{x}$ in $\mathbb{V}$, the unique k-tuple $(\tilde{x}_1, \tilde{x}_2, \ldots, \tilde{x}_k)$ of real numbers such that $\mathbf{x} = \tilde{x}_1\mathbf{v}_1 + \tilde{x}_2\mathbf{v}_2 + \cdots + \tilde{x}_k\mathbf{v}_k$ is called the **coordinate vector of x with respect to the basis** $\mathcal{B}$ (or **relative to the basis** $\mathcal{B}$). The notation used to describe this is $(\mathbf{x})_{\mathcal{B}} = (\tilde{x}_1, \tilde{x}_2, \ldots, \tilde{x}_k)$.

REMARKS

(1) Note that this definition makes sense because of the unique representation property.

(2) The statement $(\mathbf{x})_{\mathcal{B}} = (\tilde{x}_1, \tilde{x}_2, \ldots, \tilde{x}_k)$ cannot be interpreted properly unless there is an agreement on the **order** in which the basis vectors appear. In this book, "basis" will always mean **ordered basis**; that is, it is always assumed that it is specified which basis vector is first, which is second, and so on.

(3) We often speak of "the coordinates of $\mathbf{x}$ relative to $\mathcal{B}$" (instead of the more correct "coordinate vector"), or even of "the $\mathcal{B}$-coordinates of $\mathbf{x}$". Notice that this is a switch in language, because in Chapter 1 we called such numbers "components" of the vector $\mathbf{x}$.

(4) The wavy ˜ is called "tilde". Many people prefer to use a prime (′) to distinguish the $\mathcal{B}$-coordinates, but hand-written primes often get lost, so we prefer the tilde. Others prefer to use a completely different letter (for example, $(c_1, c_2, \ldots, c_k)$) to denote the $\mathcal{B}$-coordinates, but in many situations it is helpful to use a notation that suggests the connection to the vector $\mathbf{x}$. In our notation, the $\mathcal{B}$-coordinate vector of $\mathbf{y}$ is $(\tilde{y}_1, \tilde{y}_2, \ldots, \tilde{y}_k)$.

(5) When it is necessary to indicate that the coordinate vector is a column vector, we shall write $[\mathbf{x}]_{\mathcal{B}}$ to denote $[\tilde{\mathbf{x}}] = \begin{bmatrix} \tilde{x}_1 \\ \tilde{x}_2 \\ \vdots \\ \tilde{x}_k \end{bmatrix}$.

(6) $\mathcal{S}$ will always denote the standard basis in $\mathbb{R}^n$. Since $(\mathbf{x})_{\mathcal{S}} = \mathbf{x}$ (make sure you understand why), it is usual to omit the subscript $\mathcal{S}$ in this case unless it is needed for emphasis.

EXAMPLE 21

The set $\mathcal{B} = \{(1,0), (1,1)\}$ is a basis for $\mathbb{R}^2$. (Check this for yourself.) Find the coordinates of $\mathbf{a} = (3,2)$ and $\mathbf{b} = (1,-2)$ with respect to this basis.

SOLUTION: For $\mathbf{a}$, we must find $(\tilde{a}_1, \tilde{a}_2)$ such that $\tilde{a}_1 \begin{bmatrix} 1 \\ 0 \end{bmatrix} + \tilde{a}_2 \begin{bmatrix} 1 \\ 1 \end{bmatrix} = \begin{bmatrix} 3 \\ 2 \end{bmatrix}$.

The augmented matrix for this system is $\begin{bmatrix} 1 & 1 & 3 \\ 0 & 1 & 2 \end{bmatrix}$, and it is easy to see that the solution is $(\tilde{a}_1, \tilde{a}_2) = (1, 2)$; thus,

$$(\mathbf{a})_{\mathcal{B}} = (3,2)_{\mathcal{B}} = (1,2);$$

in column form, $[\mathbf{a}]_{\mathcal{B}} = [(3,2)]_{\mathcal{B}} = \begin{bmatrix} 1 \\ 2 \end{bmatrix}$.

By a similar calculation, the $\mathcal{B}$-coordinates of **b** are found as the solution of the system with augmented matrix $\begin{bmatrix} 1 & 1 & 1 \\ 0 & 1 & -2 \end{bmatrix}$. It follows that $(\tilde{b}_1, \tilde{b}_2) = (\mathbf{b})_{\mathcal{B}} = (1,-2)_{\mathcal{B}} = (3,-2)$, and $[\mathbf{b}]_{\mathcal{B}} = [(1,-2)]_{\mathcal{B}} = \begin{bmatrix} 3 \\ -2 \end{bmatrix}$.

Figure 4–3–1 shows $\mathbb{R}^2$ with this basis and **a**. Notice that the use of the basis $\mathcal{B}$ means that the space is covered with two families of parallel coordinate lines, one with direction vector $(1, 0)$ and the other with direction vector $(1, 1)$. Coordinates are established relative to these two families. The axes of this new coordinate system are obviously not orthogonal to each other. Such non-orthogonal coordinate systems arise naturally in the study of some crystalline structures in materials science.

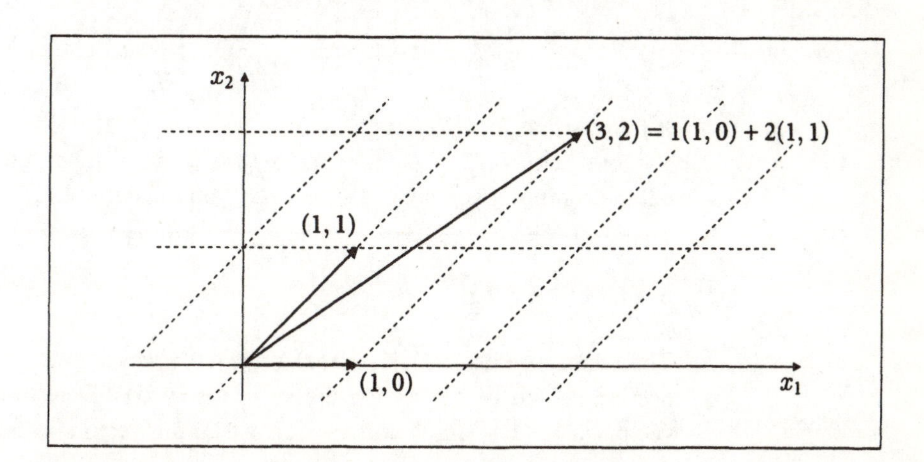

FIGURE 4-3-1. *The basis* $\mathcal{B} = \{(1,0),(1,1)\}$ *in* $\mathbb{R}^2$*; the* $\mathcal{B}$*-coordinate vector of* $\mathbf{a} = (3,2)$ *is* $(3,2)_{\mathcal{B}} = (1,2)$.

Since polynomial spaces are less familiar to us, they help to emphasize the meaning of coordinates, as in the following example.

EXAMPLE 22

Suppose that you have written a computer program to perform certain operations with polynomials in P_2 (polynomials of degree less than or equal to 2). You need to include some way of inputting which polynomial you are considering. If you use the standard basis $\{1, x, x^2\}$ for P_2, to input the polynomial $3 - 5x + 2x^2$, you would surely write your program in such a way that you would type $(3, -5, 2)$ as the input. $(3, -5, 2)$ is the standard coordinate vector for this element of P_2.

On the other hand, for some problems in differential equations you might prefer the basis $\mathcal{B} = \{1 - x^2, x, 1 + x^2\}$. (We could verify that this is a basis by checking that there are unique $\mathcal{B}$-coordinates for any vector in P_2.) To find the $\mathcal{B}$-coordinates of $3 - 5x + 2x^2$, we must find c_1, c_2, c_3 such that

$$c_1(1 - x^2) + c_2 x + c_3(1 + x^2) = 3 - 5x + 2x^2.$$

By considering the different powers $1, x, x^2$, we get equations for the coefficients c_1, c_2, c_3; since the arithmetic is simple, we write the augmented matrix of the system for c_1, c_2, c_3 and row reduce.

$$\begin{bmatrix} 1 & 0 & 1 & 3 \\ 0 & 1 & 0 & -5 \\ -1 & 0 & 1 & 2 \end{bmatrix} \to \begin{bmatrix} 1 & 0 & 1 & 3 \\ 0 & 1 & 0 & -5 \\ 0 & 0 & 2 & 5 \end{bmatrix} \to \begin{bmatrix} 1 & 0 & 0 & 1/2 \\ 0 & 1 & 0 & -5 \\ 0 & 0 & 1 & 5/2 \end{bmatrix}$$

It follows that $(c_1, c_2, c_3) = (3 - 5x + 2x^2)_{\mathcal{B}} = (1/2, -5, 5/2)$, which means $3 - 5x + 2x^2 = (1/2)(1 - x^2) - 5(x) + (5/2)(1 + x^2)$. If you use the basis $\mathcal{B}$, to tell the computer you mean $3 - 5x + 2x^2$ you must input $(1/2, -5, 5/2)$.

EXAMPLE 23

The set $\mathcal{B} = \{(3, 2, 2, 2), (1, 0, 1, 1), (1, 1, 1, 0)\}$ was shown in Example 12 of Section 4–2 to be a basis for the subspace $\text{Sp}(\mathcal{B})$. Determine whether the vectors $\mathbf{u} = (1, -1, 0, 3)$ and $\mathbf{v} = (1, 4, 0, 3)$ are vectors in $\text{Sp}(\mathcal{B})$, and if so, determine their $\mathcal{B}$-coordinate vectors.

SOLUTION: We are required to determine whether there are numbers $\tilde{u}_1, \tilde{u}_2, \tilde{u}_3$ such that

$$\tilde{u}_1 \begin{bmatrix} 3 \\ 2 \\ 2 \\ 2 \end{bmatrix} + \tilde{u}_2 \begin{bmatrix} 1 \\ 0 \\ 1 \\ 1 \end{bmatrix} + \tilde{u}_3 \begin{bmatrix} 1 \\ 1 \\ 1 \\ 0 \end{bmatrix} = \begin{bmatrix} 1 \\ -1 \\ 0 \\ 3 \end{bmatrix},$$

and numbers $\tilde{v}_1, \tilde{v}_2, \tilde{v}_3$ satisfying the system with the same coefficient matrix, but with right-hand side $[\mathbf{v}] = [(1, 4, 0, 3)]$. Since the two systems have the same coefficient matrix, augment the coefficient matrix twice by adjoining **both** right-hand sides, and row reduce:

$$\left[\begin{array}{ccc|c|c} 3 & 1 & 1 & 1 & 1 \\ 2 & 0 & 1 & -1 & 4 \\ 2 & 1 & 1 & 0 & 0 \\ 2 & 1 & 0 & 3 & 3 \end{array}\right] \overset{R_1 - R_2}{\longrightarrow} \left[\begin{array}{ccc|c|c} 1 & 1 & 0 & 2 & -3 \\ 2 & 0 & 1 & -1 & 4 \\ 2 & 1 & 1 & 0 & 0 \\ 2 & 1 & 0 & 3 & 3 \end{array}\right]$$

$$\longrightarrow \left[\begin{array}{ccc|c|c} 1 & 1 & 0 & 2 & -3 \\ 0 & -2 & 1 & -5 & 10 \\ 0 & -1 & 1 & -4 & 6 \\ 0 & -1 & 0 & -1 & 9 \end{array}\right] \longrightarrow \left[\begin{array}{ccc|c|c} 1 & 0 & 0 & 1 & 6 \\ 0 & 0 & 1 & -3 & -8 \\ 0 & 0 & 1 & -3 & -3 \\ 0 & 1 & 0 & 1 & -9 \end{array}\right]$$

$$\longrightarrow \left[\begin{array}{ccc|c|c} 1 & 0 & 0 & 1 & 6 \\ 0 & 0 & 1 & -3 & -8 \\ 0 & 0 & 0 & 0 & 5 \\ 0 & 1 & 0 & 1 & -9 \end{array}\right] \longrightarrow \left[\begin{array}{ccc|c|c} 1 & 0 & 0 & 1 & 6 \\ 0 & 1 & 0 & 1 & -9 \\ 0 & 0 & 1 & -3 & -8 \\ 0 & 0 & 0 & 0 & 5 \end{array}\right]$$

It is now clear that the system with the first vector $(1, -1, 0, 3)$ on the right-hand side does have a solution $\tilde{u}_1 = 1$, $\tilde{u}_2 = 1$, $\tilde{u}_3 = -3$, so that $(1, -1, 0, 3)_{\mathcal{B}} = (1, 1, -3)$. Note that there are only three $\mathcal{B}$-coordinates because the basis $\mathcal{B}$ has only three vectors. Note also that there is an immediate check available since it must be true that

$$(1, -1, 0, 3) = 1(3, 2, 2, 2) + 1(1, 0, 1, 1) - 3(1, 1, 1, 0).$$

On the other hand, by looking at the reduced row echelon form of the coefficient matrix and the second augmenting column, we see that this system is not consistent, so that $(1, 4, 0, 3)$ is not a vector in the subspace spanned by $\mathcal{B}$.

These ideas and the procedure for determining $\mathcal{B}$-coordinates will play an important part in the discussion of change of basis in Section 4–6, so we summarize the procedure here.

A Procedure for Finding $\mathcal{B}$-coordinates of a Vector in a Subspace of $\mathbb{R}^n$

Suppose that $\mathcal{B} = \{\mathbf{v}_1, \mathbf{v}_2, \ldots, \mathbf{v}_k\}$ is the basis of a subspace V in $\mathbb{R}^n$ and that $\mathbf{x}$ is a vector in V. Since $[\mathbf{x}]_{\mathcal{B}} = [\tilde{\mathbf{x}}]$,

$$\mathbf{x} = \tilde{x}_1\mathbf{v}_1 + \tilde{x}_2\mathbf{v}_2 + \cdots + \tilde{x}_k\mathbf{v}_k$$

or

$$\begin{bmatrix}[\mathbf{v}_1] & [\mathbf{v}_2] & \dots & [\mathbf{v}_k]\end{bmatrix}[\tilde{\mathbf{x}}] = [\mathbf{x}].$$

Thus, for any $\mathbf{x}$, $[\mathbf{x}]_{\mathcal{B}} = [\tilde{\mathbf{x}}]$ is determined as the solution of the matrix-vector equation $\begin{bmatrix}[\mathbf{v}_1] & [\mathbf{v}_2] & \dots & [\mathbf{v}_k]\end{bmatrix}[\mathbf{x}]_{\mathcal{B}} = [\mathbf{x}]$.

REMARK

Temporarily denote the matrix $\begin{bmatrix}[\mathbf{v}_1] & [\mathbf{v}_2] & \dots & [\mathbf{v}_k]\end{bmatrix}$ by Q. If $Q[\mathbf{x}]_{\mathcal{B}} = [\mathbf{x}]$ and $Q[\mathbf{y}]_{\mathcal{B}} = [\mathbf{y}]$, then $Q([\mathbf{x}]_{\mathcal{B}} + [\mathbf{y}]_{\mathcal{B}}) = [\mathbf{x}] + [\mathbf{y}] = [\mathbf{x}+\mathbf{y}]$. It follows that $[\mathbf{x}+\mathbf{y}]_{\mathcal{B}} = [\mathbf{x}]_{\mathcal{B}} + [\mathbf{y}]_{\mathcal{B}}$. Similarly, $[k\mathbf{x}]_{\mathcal{B}} = k[\mathbf{x}]_{\mathcal{B}}$ for any real number k. The operation of taking $\mathcal{B}$-coordinates is linear.

The case where $k = n$ (so that the subspace is all of $\mathbb{R}^n$) will be particularly important. In this case the matrix $\begin{bmatrix}[\mathbf{v}_1] & [\mathbf{v}_2] & \dots & [\mathbf{v}_n]\end{bmatrix}$ will be square; it will be denoted P:

$$P = \begin{bmatrix}[\mathbf{v}_1] & [\mathbf{v}_2] & \dots & [\mathbf{v}_n]\end{bmatrix}.$$

Since the columns of P form a basis for $\mathbb{R}^n$, the columnspace of P is $\mathbb{R}^n$, so P is invertible by Theorem 5 of Section 3–5.

EXAMPLE 24

Find the coordinates of $(-2,-4,6)$, $(4,5,6)$, and $(2,3,4)$ with respect to the basis $\mathcal{B} = \{(1,2,-3),(2,-1,0),(3,0,1)\}$ of $\mathbb{R}^3$ (the basis of Example 19 in Section 4–2).

SOLUTION: By inspection, since $(-2,-4,6) = -2(1,2,-3)$, we have

$$(-2,-4,6)_{\mathcal{B}} = (-2,0,0) \text{ or } [(-2,-4,6)]_{\mathcal{B}} = \begin{bmatrix}-2\\0\\0\end{bmatrix}.$$

By the discussion preceding the example,

$$\begin{bmatrix}[\mathbf{v}_1] & [\mathbf{v}_2] & [\mathbf{v}_3]\end{bmatrix}[(4,5,6)]_{\mathcal{B}} = \begin{bmatrix}1 & 2 & 3\\2 & -1 & 0\\-3 & 0 & 1\end{bmatrix}[(4,5,6)]_{\mathcal{B}} = \begin{bmatrix}4\\5\\6\end{bmatrix}$$

and

$$\begin{bmatrix} 1 & 2 & 3 \\ 2 & -1 & 0 \\ -3 & 0 & 1 \end{bmatrix} [(2,3,4)]_{\mathcal{B}} = \begin{bmatrix} 2 \\ 3 \\ 4 \end{bmatrix}.$$

Denote the matrix $\begin{bmatrix} 1 & 2 & 3 \\ 2 & -1 & 0 \\ -3 & 0 & 1 \end{bmatrix}$ by P, and find P^{-1}:

$$\left[\begin{array}{ccc|ccc} 1 & 2 & 3 & 1 & 0 & 0 \\ 2 & -1 & 0 & 0 & 1 & 0 \\ -3 & 0 & 1 & 0 & 0 & 1 \end{array}\right] \to \dots \to \left[\begin{array}{ccc|ccc} 1 & 0 & 0 & 1/14 & 2/14 & -3/14 \\ 0 & 1 & 0 & 2/14 & -10/14 & -6/14 \\ 0 & 0 & 1 & 3/14 & 6/14 & 5/14 \end{array}\right].$$

Thus,

$$[(4,5,6)]_{\mathcal{B}} = P^{-1} \begin{bmatrix} 4 \\ 5 \\ 6 \end{bmatrix} = \begin{bmatrix} 1/14 & 2/14 & -3/14 \\ 2/14 & -10/14 & -6/14 \\ 3/14 & 6/14 & 5/14 \end{bmatrix} \begin{bmatrix} 4 \\ 5 \\ 6 \end{bmatrix} = \begin{bmatrix} -4/14 \\ -78/14 \\ 72/14 \end{bmatrix}$$

and

$$[(2,3,4)]_{\mathcal{B}} = P^{-1} \begin{bmatrix} 2 \\ 3 \\ 4 \end{bmatrix} = \begin{bmatrix} 1/14 & 2/14 & -3/14 \\ 2/14 & -10/14 & -6/14 \\ 3/14 & 6/14 & 5/14 \end{bmatrix} \begin{bmatrix} 2 \\ 3 \\ 4 \end{bmatrix} = \begin{bmatrix} -4/14 \\ -50/14 \\ 44/14 \end{bmatrix}.$$

EXERCISES 4-3

A1 In each case, check that the given vectors in $\mathcal{B}$ are linearly independent (and therefore form a basis for the subspace that they span); then, determine the coordinates of $\mathbf{x}$ and $\mathbf{y}$ with respect to $\mathcal{B}$.

(a) $\mathcal{B} = \{(1,0,1,1), (-1,1,-1,0)\}$; $\mathbf{x} = (5,-2,5,3)$, $\mathbf{y} = (-1,3,-1,2)$

(b) $\mathcal{B} = \{(1,1,1,0), (0,1,1,1), (2,0,0,-1)\}$; $\mathbf{x} = (0,1,1,2)$, $\mathbf{y} = (-4,1,1,4)$

(c) $\mathcal{B} = \{(1,0,1,0,1), (1,1,2,1,1), (0,1,-1,1,-2)\}$; $\mathbf{x} = (2,1,-5,1,-6)$, $\mathbf{y} = (1,1,5,1,3)$

A2 (a) Verify that $\mathcal{B} = \{(1,2,0), (0,-2,1)\}$ is a basis for the plane $2x_1 - x_2 - 2x_3 = 0$. (Hint: To check whether the set spans the plane, solve the system that arises from asking what condition(s) an arbitrary $\mathbf{b}$ must satisfy to be a linear combination of the given vectors.)

(b) For each of the following vectors, determine whether it lies in the plane of part (a), and if so, find its $\mathcal{B}$-coordinates.

(i) $(3,2,1)$ (ii) $(3,2,2)$ (iii) $(5,2,3)$

A3 (a) Verify that $\mathcal{B} = \{(1,0,1), (1,-1,2), (-1,-1,1)\}$ is a basis for $\mathbb{R}^3$.

(b) Determine the coordinates relative to $\mathcal{B}$ of the vectors
(i) $(1,0,0)$; (ii) $(4,-2,7)$; (iii) $(-2,-2,3)$. (You may want to use a computer or calculator to invert the coefficient matrix in this calculation.)
(c) Determine $(2,-4,10)_{\mathcal{B}}$ and use your answers to part (b) to check that $(4,-2,7)_{\mathcal{B}}+(-2,-2,3)_{\mathcal{B}}=(4-2,-2-2,7+3)_{\mathcal{B}}$.

A4 In each case, (i) determine whether the given vector $\mathbf{w}$ belongs to Sp($\mathcal{V}$); (ii) use your row reduction from (i) to explain whether or not $\mathcal{V}$ is a basis for Sp($\mathcal{V}$); (iii) if $\mathcal{V}$ is a basis for Sp($\mathcal{V}$) and $\mathbf{w}$ is in Sp($\mathcal{V}$), determine $(\mathbf{w})_{\mathcal{V}}$.
(a) $\mathcal{V}=\{(1,2,1,3),(2,1,-1,2),(-2,2,4,10)\}$; $\mathbf{w}=(-1,1,2,7)$
(b) $\mathcal{V}=\{(1,3,2,3),(1,-2,1,2),(0,1,-1,1)\}$; $\mathbf{w}=(4,-1,9,3)$

B1 In each case, check that the given vectors in $\mathcal{B}$ are linearly independent (and therefore form a basis for the subspace that they span); then, determine the coordinates of $\mathbf{x}$ and $\mathbf{y}$ with respect to $\mathcal{B}$.
(a) $\mathcal{B}=\{(1,1,2,1),(2,1,1,1)\}$; $\mathbf{x}=(5,1,-2,1), \mathbf{y}=(0,1,3,1)$
(b) $\mathcal{B}=\{(1,1,0,-1),(1,0,1,1),(0,1,1,2)\}$; $\mathbf{x}=(1,-3,2,3), \mathbf{y}=(-1,0,3,7)$
(c) $\mathcal{B}=\{(1,1,0,1,0),(1,0,2,1,1),(0,0,1,1,3)\}$; $\mathbf{x}=(2,-2,5,-1,-5)$, $\mathbf{y}=(-1,-3,3,-2,-1)$

B2 (a) Verify that $\mathcal{B}=\{(3,2,0),(0,2,3)\}$ is a basis for the plane $2x_1-3x_2+2x_3=0$. (Hint: To check whether the set spans the plane, solve the system that arises from asking what condition(s) an arbitrary $\mathbf{b}$ must satisfy to be a linear combination of the given vectors.)
(b) For each of the following vectors, determine whether it lies in the plane of part (a), and if so, find its $\mathcal{B}$-coordinates.
(i) $(1,2,2)$ (ii) $(4,2,2)$ (iii) $(5,4,1)$

B3 (a) Verify that $\mathcal{B}=\{(1,0,1),(1,1,0),(0,1,1)\}$ is a basis for $\mathbb{R}^3$.
(b) Determine the coordinates relative to $\mathcal{B}$ of the vectors
(i) $(3,4,5)$; (ii) $(4,5,-7)$; (iii) $(1,1,1)$. (You may want to use a computer or calculator to invert the coefficient matrix in this calculation.)
(c) Determine $(4,5,6)_{\mathcal{B}}$ and use your answers to part (b) to check that $(3,4,5)_{\mathcal{B}}+(1,1,1)_{\mathcal{B}}=(3+1,4+1,5+1)_{\mathcal{B}}$.

B4 (a) Verify that $\mathcal{B}=\{(1,1,-1),(1,0,2),(2,2,1)\}$ is a basis for $\mathbb{R}^3$.
(b) Determine the coordinates relative to $\mathcal{B}$ of the vectors
(i) $(6,5,3)$; (ii) $(4,5,-4)$; (iii) $(5,4,1)$. (You may want to use a computer or calculator to invert the coefficient matrix in this calculation.)
(c) Determine $(9,9,-3)_{\mathcal{B}}$ and use your answers to part (b) to check that $(4,5,-4)_{\mathcal{B}}+(5,4,1)_{\mathcal{B}}=(9,9,-3)_{\mathcal{B}}$.

B5 In each case, (i) determine whether the given vector $\mathbf{w}$ belongs to Sp($\mathcal{V}$); (ii) use your row reduction from (i) to explain whether or not $\mathcal{V}$ is a basis for Sp($\mathcal{V}$); (iii) if $\mathcal{V}$ is a basis for Sp($\mathcal{V}$), and $\mathbf{w}$ is in Sp($\mathcal{V}$), determine, $(\mathbf{w})_{\mathcal{V}}$.
(a) $\mathcal{V}=\{(1,1,3,2),(-1,2,1,-4),(0,3,4,-2)\}$; $\mathbf{w}=(-1,5,13,-9)$
(b) $\mathcal{V}=\{(1,1,1,0),(0,1,1,1),(1,0,1,1)\}$; $\mathbf{w}=(1,1,-1,-4)$

CONCEPTUAL EXERCISES

***D1** Extend Figure 4–3–1, and use the grid based on the basis $\mathcal{B} = \{(1,0),(1,1)\}$ to plot the vectors whose $\mathcal{B}$-coordinates are $(\mathbf{c})_{\mathcal{B}} = (3,-2)$, $(\mathbf{d})_{\mathcal{B}} = (-2,-3)$, $(\mathbf{f})_{\mathcal{B}} = (4,2)$. Then use the figure to determine the standard coordinates of these vectors.

D2 Suppose that $\mathcal{B} = \{\mathbf{v}_1, \mathbf{v}_2, \ldots, \mathbf{v}_k\}$ is an ordered basis for vector space V, and that $\mathcal{C} = \{\mathbf{w}_1, \mathbf{w}_2, \ldots, \mathbf{w}_k\}$ is another ordered basis, and that for every $\mathbf{x}$ in V, $(\mathbf{x})_{\mathcal{B}} = (\mathbf{x})_{\mathcal{C}}$. Must it be true that $\mathbf{v}_i = \mathbf{w}_i$ for each $i = 1, 2, \ldots, k$? Explain or prove your conclusion.

SECTION

4–4 Dimension

The idea of **dimension** has been used informally earlier. Certainly, we expect that the dimension of $\mathbb{R}^n$ is n, and we expect (for example) that the dimension of a plane in $\mathbb{R}^3$ is two, or that the dimension of the solution space of a homogeneous system is determined by the number of non-leading variables in its reduced row echelon form.

The formal definition of the dimension of a general vector space depends on the following theorem about bases.

THEOREM 1. If $\{\mathbf{v}_1, \mathbf{v}_2, \ldots, \mathbf{v}_k\}$ and $\{\mathbf{u}_1, \mathbf{u}_2, \ldots, \mathbf{u}_n\}$ are both bases of a vector space V, then $k = n$. That is, if V has one basis containing n elements, then every basis of V contains exactly n elements.

The proof depends on the following Lemma.

LEMMA 1. Suppose that $\{\mathbf{v}_1, \mathbf{v}_2, \ldots, \mathbf{v}_k\}$ is a basis for a vector space V and $\{\mathbf{u}_1, \mathbf{u}_2, \ldots, \mathbf{u}_n\}$ is a linearly independent set in V. Then $n \leq k$.

The proof is really a generalization of the proof of Theorem 1 of Section 4–2, but since the details are a little technical, we defer it until the end of this section.

EXAMPLE 25

(a) In $\mathbb{R}^3$, the standard basis consists of three vectors. By Lemma 1, any linearly independent set in $\mathbb{R}^3$ must consist of not more than three vectors.

A set such as $\{(1,1,2),(2,1,-3),(1,-5,-2),(2,1,1)\}$ must be linearly dependent because it consists of more than three vectors.

(b) In P_3 (polynomials of degree less than or equal to 3), the usual basis $\{1, x, x^2, x^3\}$ has four elements. A set in P_3 with more than four elements, such as $\{1+x^2+x^4, 1-x^2+x^3, x-x^4, x+2x^3, 2x-x^2\}$, must therefore be linearly dependent by Lemma 1.

We can now prove Theorem 1.

Proof of Theorem 1. Since $\{\mathbf{v}_1, \mathbf{v}_2, \ldots, \mathbf{v}_k\}$ is a basis and $\{\mathbf{u}_1, \mathbf{u}_2, \ldots, \mathbf{u}_n\}$ (as a basis) is linearly independent, $n \leq k$. Since $\{\mathbf{u}_1, \mathbf{u}_2, \ldots, \mathbf{u}_n\}$ is a basis and $\{\mathbf{v}_1, \mathbf{v}_2, \ldots, \mathbf{v}_k\}$ (as a basis) is linearly independent, $k \leq n$. Therefore, $k = n$. ■

Theorem 1 justifies the following definition of dimension.

DEFINITION

If a vector space V has a basis with n elements, then the **dimension** of V is n. If a vector space V has no basis with finitely many elements, V is **infinite-dimensional.** The dimension of the trivial vector space (consisting of only the zero vector) is defined to be zero.

EXAMPLE 26

(a) $\mathbb{R}^n$ is n-dimensional because the standard basis contains n vectors.

(b) A plane through the origin in $\mathbb{R}^3$ is a two-dimensional subspace because a basis consisting of two vectors in the plane can be found for it (take any two non-collinear vectors that are orthogonal to the normal to the plane). Similarly, it can be shown that a hyperplane through the origin in $\mathbb{R}^n$ is $(n-1)$-dimensional. (See Exercise 4–5–D4.)

(c) The space P_n of polynomials of degree less than or equal to n is $(n+1)$-dimensional (the usual basis is $\{1, x, x^2, \ldots, x^n\}$).

(d) The space of all polynomials is infinite-dimensional, and so are most function spaces. (However, it turns out that in the theory of ordinary differential equations, solution spaces are often finite-dimensional.) Infinite-dimensional spaces are beyond the scope of this book.

Procedure for Obtaining a Basis as a Subset of an Arbitrary Finite Spanning Set

Suppose that $T = \{\mathbf{v}_1, \mathbf{v}_2, \ldots, \mathbf{v}_k\}$ is a spanning set for a finite-dimensional vector space V; to avoid uninteresting special cases, suppose that $\mathbf{0}$ is not an element of T. Let us choose a subset of T to use as a basis for V. If T is a linearly independent set, it is already a basis for V, so suppose that T is linearly dependent. Then, for some non-trivial **c**, $c_1\mathbf{v}_1 + c_2\mathbf{v}_2 + \cdots + c_k\mathbf{v}_k = \mathbf{0}$. At least one of the coefficients is non-zero, say $c_j \neq 0$. Then we may "solve" the equation to get

$$\mathbf{v}_j = (-1/c_j)(c_1\mathbf{v}_1 + c_2\mathbf{v}_2 + \cdots + c_{j-1}\mathbf{v}_{j-1} + c_{j+1}\mathbf{v}_{j+1} + \cdots + c_k\mathbf{v}_k);$$

note particularly that $\mathbf{v}_j$ does not appear on the right-hand side. Now if $\mathbf{x}$ is any vector in V, it can be written as a linear combination of the elements of T: $\mathbf{x} = a_1\mathbf{v}_1 + a_2\mathbf{v}_2 + \cdots + a_k\mathbf{v}_k$. In this expression, replace $\mathbf{v}_j$ by its representation in terms of the other vectors of T:

$$\mathbf{x} = a_1\mathbf{v}_1 + a_2\mathbf{v}_2 + \cdots + a_{j-1}\mathbf{v}_{j-1} + a_j(-1/c_j)\big(c_1\mathbf{v}_1 + c_2\mathbf{v}_2 + \cdots + c_{j-1}\mathbf{v}_{j-1}$$
$$+ c_{j+1}\mathbf{v}_{j+1} + \cdots + c_k\mathbf{v}_k\big) + a_{j+1}\mathbf{v}_{j+1} + \cdots + a_k\mathbf{v}_k.$$

Thus, $\mathbf{x}$ is expressed as a linear combination of the vectors of the set obtained from T by omitting $\mathbf{v}_j$; this set is denoted $T\backslash\{\mathbf{v}_j\}$. Since $\mathbf{x}$ was an arbitrary vector in V, this argument shows that $T\backslash\{\mathbf{v}_j\}$ is a spanning set for V. If $T\backslash\{\mathbf{v}_j\}$ is linearly independent, it is the desired basis of V, and the procedure is finished. If $T\backslash\{\mathbf{v}_j\}$ is not linearly independent, repeat the procedure to omit a second vector, say $\mathbf{v}_k$, and get $T\backslash\{\mathbf{v}_j, \mathbf{v}_k\}$, which still spans V. In this fashion, we must eventually get to a linearly independent set (certainly, if there is only one vector left, it forms a linearly independent set), and thus, we obtain a subset of T that is a basis for V.

EXAMPLE 27

If $T = \{\mathbf{v}_1 = (1, 1, -2), \mathbf{v}_2 = (2, -1, 1), \mathbf{v}_3 = (1, -2, 3), \mathbf{v}_4 = (1, 5, 3)\}$, determine a subset of T that is a basis for Sp(T).

SOLUTION: Consider

$$c_1\begin{bmatrix} 1 \\ 1 \\ -2 \end{bmatrix} + c_2\begin{bmatrix} 2 \\ -1 \\ 1 \end{bmatrix} + c_3\begin{bmatrix} 1 \\ -2 \\ 3 \end{bmatrix} + c_4\begin{bmatrix} 1 \\ 5 \\ 3 \end{bmatrix} = \begin{bmatrix} 0 \\ 0 \\ 0 \end{bmatrix}.$$

We row reduce the corresponding coefficient matrix

$$\begin{bmatrix} 1 & 2 & 1 & 1 \\ 1 & -1 & -2 & 5 \\ -2 & 1 & 3 & 3 \end{bmatrix} \to \begin{bmatrix} 1 & 0 & -1 & 0 \\ 0 & 1 & 1 & 0 \\ 0 & 0 & 0 & 1 \end{bmatrix}.$$

The general solution for **c** is $(c_3, -c_3, c_3, 0)$, where c_3 is arbitrary, so

$$\mathbf{v}_1 - \mathbf{v}_2 + \mathbf{v}_3 = 0, \quad \text{or} \quad \mathbf{v}_3 = -\mathbf{v}_1 + \mathbf{v}_2,$$

and $\mathbf{v}_3$ depends on $\mathbf{v}_1$ and $\mathbf{v}_2$. We omit $\mathbf{v}_3$ and consider $T \backslash \{\mathbf{v}_3\} = \{\mathbf{v}_1, \mathbf{v}_2, \mathbf{v}_4\}$.

Now consider

$$c_1 \begin{bmatrix} 1 \\ 1 \\ -2 \end{bmatrix} + c_2 \begin{bmatrix} 2 \\ -1 \\ 1 \end{bmatrix} + c_4 \begin{bmatrix} 1 \\ 5 \\ 3 \end{bmatrix} = \begin{bmatrix} 0 \\ 0 \\ 0 \end{bmatrix}.$$

The matrix is as above except that the third column is omitted, so the same row operations give

$$\begin{bmatrix} 1 & 2 & 1 \\ 1 & -1 & 5 \\ -2 & 1 & 3 \end{bmatrix} \to \begin{bmatrix} 1 & 0 & 0 \\ 0 & 1 & 0 \\ 0 & 0 & 1 \end{bmatrix},$$

and we conclude that $\{\mathbf{v}_1, \mathbf{v}_2, \mathbf{v}_4\}$ is linearly independent and thus a basis for $\mathrm{Sp}(T)$. It will follow from Corollary 2 to Theorem 2 that $\mathrm{Sp}(T)$ is in fact equal to $\mathbb{R}^3$ in this example.

In practice, you can often short-circuit this step-by-step procedure for selecting a basis by appealing to Theorem 2 below and simply selecting a linearly independent subset of T that contains a number of vectors equal to the dimension of V.

Lemma 1 tells us that the number of elements in a *linearly independent* set in a k-dimensional vector space must be *less* than or equal to k. It follows from Theorem 1 and the procedure just discussed that there is an analogous fact about *spanning* sets.

COROLLARY TO THEOREM 1 If T is a spanning set in a k-dimensional vector space V, the number of elements in T must be *greater* than or equal to k.

Proof. By the procedure, we can select a subset of T that is a basis for V, and by Theorem 1, the number of elements in T must thus be at least as great as k. ■

Procedure for Extending a Linearly Independent Set to Get a Basis

Sometimes a linearly independent set $T = \{\mathbf{v}_1, \mathbf{v}_2, \ldots, \mathbf{v}_k\}$ is given in an n-dimensional space V, where $k \leq n$, and it is necessary to include these vectors in a basis for V. If Sp(T) is not equal to all of V, then there exists some vector $\mathbf{w}_{k+1}$ that is in V, but not in Sp(T). Then $T \cup \{\mathbf{w}_{k+1}\} = \{\mathbf{v}_1, \mathbf{v}_2, \ldots, \mathbf{v}_k, \mathbf{w}_{k+1}\}$; it is easy to check that $T \cup \{\mathbf{w}_{k+1}\}$ is linearly independent. If $\text{Sp}(T \cup \{\mathbf{w}_{k+1}\}) = \text{V}$, $T \cup \{\mathbf{w}_{k+1}\}$ is thus a basis for V; if not, repeat the step to get $T \cup \{\mathbf{w}_{k+1}, \mathbf{w}_{k+2}\} = \{\mathbf{v}_1, \mathbf{v}_2, \ldots, \mathbf{v}_k, \mathbf{w}_{k+1}, \mathbf{w}_{k+2}\}$, which is linearly independent. In this fashion, we must eventually get a basis, since there cannot be more than n linearly independent vectors in an n-dimensional space.

Again, in practice, you can often omit the step-by-step procedure and simply add the appropriate number of extra vectors all at once by inspection.

EXAMPLE 28

(a) Produce a basis for the plane P in $\mathbb{R}^3$ with equation $x_1 + 2x_2 - x_3 = 0$, and (b) extend this basis to obtain a basis for $\mathbb{R}^3$.

SOLUTION: (a) First, choose any vector in P. One easy way to do this is to set one coordinate equal to zero — say, $x_3 = 0$ — and choose a second coordinate in any convenient way — say, $x_2 = 1$. Then solve for the remaining coordinate, and we have $\mathbf{v}_1 = (-2, 1, 0)$ in P.

Next, chose a second vector by taking a different coordinate equal to zero — say $x_2 = 0$. Thus we get $\mathbf{v}_2 = (1, 0, 1)$. Then,

$$c_1\mathbf{v}_1 + c_2\mathbf{v}_2 = (-2c_1 + c_2, c_1, c_2) = (0, 0, 0)$$

if and only if $c_1 = c_2 = 0$, so $\mathbf{v}_1$ and $\mathbf{v}_2$ are linearly independent.

Is $\{\mathbf{v}_1, \mathbf{v}_2\}$ a basis for P? We could show by methods of Section 3–4 that $\{\mathbf{v}_1, \mathbf{v}_2\}$ spans P and is thus a basis. However, we shall see in Section 4–5 that a plane is two-dimensional, so by Corollary 1 to Theorem 2 below, $\{\mathbf{v}_1, \mathbf{v}_2\}$ is a basis for P.

(b) We need to add a vector $\mathbf{w}_3$ so that $\{\mathbf{v}_1, \mathbf{v}_2, \mathbf{w}_3\}$ is a basis for $\mathbb{R}^3$. One simple way would be to take $\mathbf{w}_3 = \mathbf{v}_1 \times \mathbf{v}_2$. However, the cross-product is not available in most vector spaces (in fact, it is available only in $\mathbb{R}^3$), so let us use another approach here. Let us try the standard basis vectors: let $\mathbf{w}_3 = \mathbf{e}_1 = (1,0,0)$. $(1,0,0)$ is not in $\mathbb{P}$, so $\{\mathbf{v}_1, \mathbf{v}_2, \mathbf{w}_3\}$ is linearly independent and (by Theorem 2 below) we have a basis for $\mathbb{R}^3$. If $\mathbf{e}_1$ had not worked, we would have tried $\mathbf{e}_2$, then $\mathbf{e}_3$. *Why can we be sure that at least one of* $\mathbf{e}_1, \mathbf{e}_2, \mathbf{e}_3$ *will provide the required third basis vector?*

REMARK

It is easy to prove that if V is a subspace of W, then the dimension of V is less than or equal to the dimension of W. Example 28 provides one illustration of this fact.

It is tiresome to have to check both, whether a set spans V and whether the set is linearly independent. It is very convenient that *if the dimension of* V *is known*, it is sufficient to check only one of the properties.

THEOREM 2. Suppose that V is a vector space of dimension n. Then a set $T = \{\mathbf{v}_1, \mathbf{v}_2, \ldots, \mathbf{v}_n\}$ with n elements is a spanning set for V if and only if it is linearly independent.

Proof for the Special Case $V = \mathbb{R}^n$**.** By definition, T spans $\mathbb{R}^n$ if and only if the system $\big[[\mathbf{v}_1] \ \ [\mathbf{v}_2] \ \ \ldots \ \ [\mathbf{v}_n]\big][\mathbf{x}] = [\mathbf{b}]$ is consistent for every $\mathbf{b}$ in $\mathbb{R}^n$. Since the coefficient matrix is n by n, this is true if and only if the rank of the coefficient matrix is n. However, the rank of the coefficient matrix is n if and only if the system $\big[[\mathbf{v}_1] \ \ [\mathbf{v}_2] \ \ \ldots \ \ [\mathbf{v}_n]\big][\mathbf{x}] = [\mathbf{0}]$ has only the trivial solution, and by definition, this is true if and only if T is linearly independent. Thus, T spans $\mathbb{R}^n$ if and only if T is linearly independent.

General Proof. If T spans V, then by the procedure above, there is a linearly independent subset of T that spans V. Since the dimension of V is n, any linearly independent spanning set must contain n elements, so the linearly independent subset of T that spans V must be T itself. Conversely, if T is linearly independent, it can be extended to get a basis of V. However, any basis of V must contain exactly n elements, so T must already be a basis, and therefore T spans V, as claimed.

■

COROLLARY 1 TO THEOREM 2

(a) If $T = \{\mathbf{v}_1, \mathbf{v}_2, \ldots, \mathbf{v}_n\}$ is a set with n elements that spans $\mathbb{V}$, it is necessarily a basis for $\mathbb{V}$.

(b) If $T = \{\mathbf{v}_1, \mathbf{v}_2, \ldots, \mathbf{v}_n\}$ is a linearly independent set in $\mathbb{V}$ with n elements, it is necessarily a basis for $\mathbb{V}$.

Proof. These statements follow immediately from Theorem 2.

■

COROLLARY 2 TO THEOREM 2 If $\mathbb{V}$ is an n-dimensional subspace of $\mathbb{R}^n$, then $\mathbb{V} = \mathbb{R}^n$.

The proof is left as Exercise D1.

The ideas of basis and dimension are essential in the development of the theory of general finite-dimensional vector spaces. Another key concept is "isomorphism". Some of this theory is introduced briefly in Section 4–7.

To conclude this section, we give the proof of Lemma 1.

Proof of Lemma 1.

Step 1. Since $\mathbf{u}_j$ is a vector in $\mathbb{V} = \text{Sp}(\{\mathbf{v}_1, \mathbf{v}_2, \ldots, \mathbf{v}_k\})$, it can be written as a linear combination of the $\mathbf{v}$'s. We label the coefficients in what may seem to be an unusual way:

$$\begin{aligned}
\mathbf{u}_1 &= a_{11}\mathbf{v}_1 + a_{21}\mathbf{v}_2 + \cdots + a_{k1}\mathbf{v}_k \\
\mathbf{u}_2 &= a_{12}\mathbf{v}_1 + a_{22}\mathbf{v}_2 + \cdots + a_{k2}\mathbf{v}_k \\
&\cdots \\
\mathbf{u}_n &= a_{1n}\mathbf{v}_1 + a_{2n}\mathbf{v}_2 + \cdots + a_{kn}\mathbf{v}_k.
\end{aligned}$$

Step 2. Since the $\mathbf{u}$'s are linearly independent,

$$c_1\mathbf{u}_1 + c_2\mathbf{u}_2 + \cdots + c_n\mathbf{u}_n = \mathbf{0} \text{ if and only if } c_j = 0 \text{ for } j = 1, 2, \ldots, n.$$

Step 3. Substitute the equations obtained in Step 1 into the statement in Step 2.

$$\begin{aligned}
c_1(a_{11}\mathbf{v}_1 + a_{21}\mathbf{v}_2 + \cdots + a_{k1}\mathbf{v}_k) + c_2(a_{12}\mathbf{v}_1 + a_{22}\mathbf{v}_2 + \cdots + a_{k2}\mathbf{v}_k) \\
+ \cdots + c_n(a_{1n}\mathbf{v}_1 + a_{2n}\mathbf{v}_2 + \cdots + a_{kn}\mathbf{v}_k) = \mathbf{0}
\end{aligned}$$

if and only if each $c_j = 0$.

Expand, and collect terms in $\mathbf{v}_1, \mathbf{v}_2, \ldots, \mathbf{v}_k$:

$$(a_{11}c_1 + a_{12}c_2 + \cdots + a_{1n}c_n)\mathbf{v}_1 + (a_{21}c_1 + a_{22}c_2 + \cdots + a_{2n}c_n)\mathbf{v}_2 + \cdots + (a_{k1}c_1 + a_{k2}c_2 + \cdots + a_{kn}c_n)\mathbf{v}_k = \mathbf{0}$$

if and only if each $c_j = 0$.

Step 4. Since the v's form a basis, they are linearly independent, and therefore, the coefficients of the v's in this last equation must be zero:

$$\begin{aligned} a_{11}c_1 + a_{12}c_2 + \cdots + a_{1n}c_n &= 0 \\ a_{21}c_1 + a_{22}c_2 + \cdots + a_{2n}c_n &= 0 \\ &\cdots \\ a_{k1}c_1 + a_{k2}c_2 + \cdots + a_{kn}c_n &= 0 \end{aligned}$$

if and only if each $c_j = 0$. Thus, in an obvious notation, $A\mathbf{c} = \mathbf{0}$ if and only $\mathbf{c} = \mathbf{0}$. But if $k < n$, this homogeneous system must have non-trivial solutions. Since we are told that it has only the trivial solution (each $c_j = 0$), it follows that $k \geq n$, or $n \leq k$, as claimed.

■

EXERCISES 4–4

A1 Use the procedure illustrated in Example 27 to select a basis for Sp(S) from each of the following sets S, and hence, determine the dimension of Sp(S).
(a) $S = \{(1,-2,1), (0,1,2), (2,0,10), (1,1,7)\}$
(b) $S = \{(1,3,2), (-2,-6,-4), (-1,-1,2), (0,4,8), (0,1,1)\}$

A2 Same instructions as in A1.
(a) $S = \{(1,1,-1,1), (0,1,3,-1), (1,-1,2,-3), (2,1,4,-3)\}$
(b) $S = \{(1,-1,-1,-1), (1,1,1,-1), (0,1,2,1), (3,0,1,-2), (0,1,0,1)\}$
(c) $S = \{(1,0,0,1), (0,1,1,0), (0,1,0,-1), (1,1,1,0)\}$

A3 Determine the dimension of the vector space of polynomials spanned by S.
(a) $S = \{1+x, 1+x+x^2, 1+x^3\}$
(b) $S = \{1+x, 1-x, 1+x^3, 1-x^3\}$

A4 (a) By the method of Example 28, determine a basis for the plane $2x_1 - x_2 - x_3 = 0$ in $\mathbb{R}^3$.
(b) Extend the basis of part (a) to obtain a basis for $\mathbb{R}^3$.

A5 (a) By the method of Example 28, determine a basis for the hyperplane $x_1 - x_2 + x_3 - x_4 = 0$ in $\mathbb{R}^4$.
(b) Extend the basis of part (a) to obtain a basis for $\mathbb{R}^4$.

B1 Same instructions as in A1.
(a) $S = \{(1,2,1),(2,3,-2),(-1,0,7)\}$
(b) $S = \{(1,1,-1),(1,-1,2),(2,0,1),(0,2,-3),(3,-3,-2)\}$

B2 Same instructions as in A1.
(a) $S = \{(1,1,2,1),(-2,-2,-4,-2),(1,-1,1,0),(3,1,5,2)\}$
(b) $S = \{(1,2,-1,1),(1,2,2,1),(3,6,0,3),(0,1,0,-2),(1,3,2,-1)\}$

B3 Determine the dimension of the vector space of polynomials spanned by S.
(a) $S = \{1+x+x^2, x+x^2+x^3, 1-x^3\}$
(b) $S = \{1+x+x^2, x+x^2+x^3, 1+x^2+x^3\}$

B4 (a) By the method of Example 28, determine a basis for the plane $x_1 + 3x_2 + 4x_3 = 0$ in $\mathbb{R}^3$.
(b) Extend the basis of part (a) to obtain a basis for $\mathbb{R}^3$.

B5 (a) By the method of Example 28, determine a basis for the hyperplane $x_1 + x_2 + 2x_3 + x_4 = 0$ in $\mathbb{R}^4$.
(b) Extend the basis of part (a) to obtain a basis for $\mathbb{R}^4$.

CONCEPTUAL EXERCISES

D1 Prove Corollary 2 to Theorem 2.

***D2** (a) It is sometimes said that "a basis for V is a maximal (largest possible) linearly independent set in V." Explain why this makes sense in terms of statements in this section. (Assume V is finite-dimensional.)
(b) Similarly, explain the statement "a basis for V is a minimal spanning set for V." (Again, assume V is finite-dimensional.)

D3 Determine the dimension of $\mathcal{M}(m,n)$, the vector space of m by n matrices. (In Exercise D6 of Section 4–2 you were asked to provide a basis for $\mathcal{M}(2,2)$.)

***D4** Determine the dimensions of the following subspaces of $\mathcal{M}(3,3)$. (See Section 4–1, Exercises A1 and B1 and Section 4–2, Exercise D7.)
(a) The subspace of diagonal matrices.
(b) The subspace of symmetric matrices.
(c) The subspace of upper triangular matrices.

D5 Generalize Exercise D4 for the case of n by n matrices.

D6 Determine the dimensions of the following subspaces of P_5.
(a) $V = \{(1+x^2)p(x) \mid p(x) \in P_3\}$
(b) The even polynomials (polynomials such that $p(-x) = p(x)$).

SECTION

4–5 The Rank of a Linear Mapping

The Dimension of the Rowspace of a Matrix Equals Its Rank

The rank of a matrix has played an important role in Section 3–5 and in this Chapter. Recall that in Section 2–2, the rank of a matrix A was defined to be the number of leading 1's in the reduced row echelon form of A. This definition has two attractive features: it provides an easy way of computing the rank of a matrix, and it is closely related to statements about solutions of systems of linear equations. However, some people prefer a definition that is more related to the geometry of linear mappings. Such a definition will have to be equivalent to our original definition. The rest of this section is aimed at obtaining such a definition. On the way, we shall learn how to determine bases for the rowspace and columnspace of a matrix and for the nullspace and range of a linear mapping.

First, recall from Section 3–4, Theorem 8, that the rowspace of a matrix is left unchanged by elementary row operations, so that if B is the reduced row echelon form of a matrix A, then the rowspace of A is equal to the rowspace of B. Moreover, the non-zero rows of B form a spanning set for this common rowspace. Are these non-zero rows linearly independent?

THEOREM 1.

(1) If the matrix B is in reduced row echelon form, the non-zero rows of B are linearly independent, so they actually form a basis for the rowspace of B; if B is the reduced row echelon form of A, the non-zero rows of B also form a basis for the rowspace of A.

(2) The dimension of the rowspace of A is equal to the rank of A.

The idea of the proof will first be illustrated by an example.

EXAMPLE 29

The matrix $B = \begin{bmatrix} 1 & 2 & 0 & 3 \\ 0 & 0 & 1 & 4 \\ 0 & 0 & 0 & 0 \end{bmatrix}$ is in reduced row echelon form. To test the non-zero rows for linear independence, consider

$$c_1(1,2,0,3) + c_2(0,0,1,4) = (0,0,0,0) \text{ or } (c_1, *, c_2, **) = (0,0,0,0)$$

(where $*$ and $**$ can easily be calculated but do not matter). Clearly the linear combination is the zero vector only if $c_1 = c_2 = 0$, so the non-zero rows are linearly independent.

Proof of Theorem 1. (1) Suppose that B has r non-zero rows (this means that the rank of B is r). Which linear combinations of these rows of B give the zero vector? That is, for what $\mathbf{c}$ is $c_1 b_{1\rightarrow} + c_2 b_{2\rightarrow} + \cdots + c_r b_{r\rightarrow} = \mathbf{0}$? Consider the coordinate of the vector $c_1 b_{1\rightarrow} + c_2 b_{2\rightarrow} + \cdots + c_r b_{r\rightarrow}$ that corresponds to the first leading 1 in B; since all rows except $b_{1\rightarrow}$ have zeros in this column, this coordinate must be c_1, so this coordinate is 0 only if $c_1 = 0$. Similarly, by considering the coordinate corresponding to the leading 1 in the jth row, we see that it is zero if and only if $c_j = 0$. It follows that the rows are linearly independent. Since they were shown to be a spanning set in Section 3–4, they form a basis for the rowspace of B and therefore (by Theorem 8 of Section 3–4) a basis for the rowspace of A.

(2) The equality of the dimension of the rowspace of A and the rank of A follows from the discussion above and the definition of dimension.

■

> *REMARK*
>
> This theorem provides an alternative to the procedure described in Section 4–3 for finding a basis of V when a spanning set is given, at least when V is a subspace of $\mathbb{R}^n$. The new procedure is to write the given spanning vectors as the rows of a matrix, and then find the reduced row echelon form of this matrix. The resulting non-zero rows are a basis for V.

EXAMPLE 30

Let $T = \{(1,1,1), (1,-1,3), (3,1,5)\}$, and let $\mathrm{V} = \mathrm{Sp}(T)$. Find a basis for V.

SOLUTION 1: (The new alternative procedure.) Write the vectors of T as rows of a matrix, and reduce.

$$\begin{bmatrix} 1 & 1 & 1 \\ 1 & -1 & 3 \\ 3 & 1 & 5 \end{bmatrix} \rightarrow \cdots \rightarrow \begin{bmatrix} 1 & 0 & 2 \\ 0 & 1 & -1 \\ 0 & 0 & 0 \end{bmatrix}$$

Therefore, a basis for V is $\{(1,0,2), (0,1,-1)\}$. Notice that neither of these basis vectors is a member of the original spanning set. In fact, it is generally true that this solution method produces basis vectors that are not members of the original spanning set, and for some purposes, this is not satisfactory.

SOLUTION 2: (Based on ideas in Section 4–4.) Check T for linear independence, and find that it is a linearly dependent set. However, by inspection (using the geometrical ideas about independence), the first two vectors form a linearly independent set. Hence $\mathbb{V}$ must be 2-dimensional and a basis for $\mathbb{V}$ is $\{(1,1,1),(1,-1,3)\}$.

The Dimension of the Columnspace of a Matrix Equals Its Rank

What about a basis for the columnspace of a matrix A? It is remarkable that the same row reduction that gives the basis for the rowspace of A also indicates how to find a basis for the columnspace of A. However, the method is a little more subtle and requires a little more attention.

Again let B be the reduced row echelon form of the m by n matrix A.

The whole point of the method of elimination is that a vector $\mathbf{x}$ satisfies $A\mathbf{x} = \mathbf{0}$ if and only if it satisfies $B\mathbf{x} = \mathbf{0}$. Let $\mathbf{c}$ be a vector that does satisfy both systems, and write the products using the columns of A or B as blocks (as in Section 3–1 and in the Proof of Theorem 3, Section 3–4). It follows that

$$c_1[a_{\downarrow 1}] + c_2[a_{\downarrow 2}] + \cdots + c_n[a_{\downarrow n}] = [\mathbf{0}]$$

if and only if

$$c_1[b_{\downarrow 1}] + c_2[b_{\downarrow 2}] + \cdots + c_n[b_{\downarrow n}] = [\mathbf{0}].$$

If $\mathbf{c} \neq 0$, these statements indicate that the columns of A are linearly dependent, and that the columns of B are linearly dependent. Or suppose that $c_1 = 0$, but that the other c's are not zero; then the set of columns $\{[a_{\downarrow 2}],[a_{\downarrow 3}],\ldots,[a_{\downarrow n}]\}$ is linearly dependent, and the corresponding set $\{[b_{\downarrow 2}],[b_{\downarrow 3}],\ldots,[b_{\downarrow n}]\}$ is linearly dependent. In fact, **any statement about linear dependence of columns of A is true if and only if the same statement is true for the corresponding columns of B.** Since a basis for a space is a largest possible linearly independent set in the space (see Exercise D2 in Section 4–4), the following theorem is an immediate consequence of this fact.

THEOREM 2. Suppose that B is the reduced row echelon form of A. Then certain columns of A form a basis of the columnspace of A if and only if the *corresponding* columns of B form a basis of the columnspace of B.

Proof. Preceding discussion.

■

REMARK

Be careful! In general, the two columnspaces are not equal, so columns of B do not form a basis for the columnspace of A. (See Example 31 below.)

EXAMPLE 31

Let $A = \begin{bmatrix} 1 & 2 & 1 & 1 \\ 1 & 2 & 2 & 1 \\ 2 & 4 & 2 & 3 \\ 3 & 6 & 4 & 3 \end{bmatrix}$. Find a basis for the columnspace of A.

SOLUTION: The reduced row echelon form of A is easily found to be $B = \begin{bmatrix} 1 & 2 & 0 & 0 \\ 0 & 0 & 1 & 0 \\ 0 & 0 & 0 & 1 \\ 0 & 0 & 0 & 0 \end{bmatrix}$. The first, third, and fourth columns of B form a linearly independent set, and hence a basis for the columnspace of B. Therefore, by Theorem 2, the first, third, and fourth columns of the matrix A form a basis for the columnspace of A. Thus, $\{(1,1,2,3),(1,2,2,4),(1,1,3,3)\}$ is a basis for the columnspace of A.

Notice in this example that every vector in the columnspace of B has last coordinate 0, which is not true in the columnspace of A, so the two columnspaces are not equal. Thus, the first, third, and fourth columns of B do *not* form a basis for the columnspace of A.

COROLLARY TO THEOREM 2

(1) If B is the reduced row echelon form of A, then the columns of A that *correspond to* the columns of B with leading 1's form a basis of the columnspace of A.

(2) The dimension of the columnspace of A is equal to the rank of A.

Proof. (1) For an m by n matrix B in reduced row echelon form, the set of columns containing leading 1's is linearly independent (they are standard basis vectors from $\mathbb{R}^m$), and no larger set of columns of B can be linearly independent. Hence, these columns form a basis of the columnspace of B.

Hence, the corresponding columns of A form a basis for the columnspace of B.

(2) Since the number of leading 1's is the maximum number of linearly independent columns, the rank of A is the dimension of the columnspace of A. ■

There is an alternative procedure for finding a basis for the columnspace of a matrix A, which uses the transpose of a matrix. Since the columnspace of A is equal to the rowspace of A^T, simply use the procedure of Example 30 to produce a basis for the rowspace of the transposed matrix A^T and it will be a basis for the columnspace of A. This basis is sometimes not as useful as the basis produced by the method of Theorem 2, since in general it does not consist of columns of A.

EXAMPLE 32

Apply the procedure just described to find a basis for the columnspace of the matrix A of Example 31.

SOLUTION: $A = \begin{bmatrix} 1 & 2 & 1 & 1 \\ 1 & 2 & 2 & 1 \\ 2 & 4 & 2 & 3 \\ 3 & 6 & 4 & 3 \end{bmatrix}$, so $A^T = \begin{bmatrix} 1 & 1 & 2 & 3 \\ 2 & 2 & 4 & 6 \\ 1 & 2 & 2 & 4 \\ 1 & 1 & 3 & 3 \end{bmatrix}$. It is easy to see that $A^T \to \begin{bmatrix} 1 & 0 & 0 & 2 \\ 0 & 1 & 0 & 1 \\ 0 & 0 & 1 & 0 \\ 0 & 0 & 0 & 0 \end{bmatrix}$, so that $\{(1, 0, 0, 2), (0, 1, 0, 1), (0, 0, 1, 0)\}$ is a basis for the rowspace of A^T, and therefore a basis for the columnspace of A. However, these basis vectors do not appear as original columns of A, so for some purposes this basis is less satisfactory than the basis found in Example 31.

Rank Is the Dimension of Range

THEOREM 3. Let $L : \mathbb{R}^n \to \mathbb{R}^m$ be a linear mapping. Then the dimension of the range of L is equal to the rank of the matrix $[L]$ of L. (Here, as in Chapter 3, $[L]$ denotes the matrix of L with respect to the standard basis in $\mathbb{R}^n$ and the standard basis in $\mathbb{R}^m$.)

Proof. By Theorem 3 of Section 3–4, the range of L is the columnspace of its matrix $[L]$, so the present theorem is an immediate consequence of the Corollary to Theorem 2 above. ■

The range of a linear mapping L is a subspace that is independent of the choice of basis, and its dimension is also independent of the choice of basis. Therefore, the dimension of the range of L is a more geometrical idea than the number of leading 1's in the row echelon form of $[L]$. For this reason, many mathematicians feel that it is better to use the following definition as the starting point for the discussion of rank.

DEFINITION

The rank of a linear mapping $L : \mathbb{R}^n \to \mathbb{R}^m$ is the dimension of its range.

Because of Theorem 3, this definition of rank of a linear mapping L fits with the previous computational definition of the rank of its matrix $[L]$, and we shall use whichever version of the definition is most useful for the problems we are considering.

Dimension of the Solution Space and the Nullspace

The ideas of linear independence and dimension make it possible to clarify a statement in Section 3–4 about the solution space of a homogeneous system of m equations in n variables, $A\mathbf{x} = \mathbf{0}$. We saw that if the rank of A was r, the general solution of the system was automatically expressed as a linear combination of $n - r$ spanning vectors. We can now show that these spanning vectors are linearly independent, so that the dimension of the solution space is $n - r$. We first consider a simple example.

EXAMPLE 33

Consider the homogeneous system $A\mathbf{x} = \mathbf{0}$, where the coefficient matrix

$$A = \begin{bmatrix} 1 & 2 & 0 & 3 & 4 \\ 0 & 0 & 1 & 5 & 6 \end{bmatrix}$$

is already in reduced row echelon form. Determine a basis for the solution space; hence, determine the dimension of the solution space and relate it to the rank of A.

SOLUTION: By back-substitution, as in Chapter 2, the general solution is

$$\mathbf{x} = c_1(-4, 0, -6, 0, 1) + c_2(-3, 0, -5, 1, 0) + c_3(-2, 1, 0, 0, 0).$$

As we saw in Section 3–4, this means that $\{(-4,0,-6,0,1),(-3,0,-5,1,0),(-2,1,0,0,0)\}$ is a spanning set for the solution space. Is this a basis for the solution space? We must check for linear independence.

Let us look closely at the coordinates of the general solution $\mathbf{x}$ corresponding to the **non-leading** variables (x_2, x_4, and x_5 in this example).

$$\begin{aligned}\mathbf{x} &= c_1(*,0,*,0,1) + c_2(*,0,*,1,0) + c_3(*,1,*,0,0)\\ &= (*,c_3,*,c_2,c_1)\end{aligned}$$

(We could easily determine the entries denoted by $*$'s, but we want to focus our attention on the others.) Clearly this linear combination is the zero vector only if $c_1 = c_2 = c_3 = 0$, so the spanning vectors are linearly independent and form a basis for the solution space.

It follows that the dimension of the solution space is 3. The number of variables in the system is 5, and the rank of A is 2 (the number of non-zero rows). Thus, we see that the dimension 3 is equal to $n - r = 5 - 2$, as we expected from the general discussion.

We could prove a general result about rank and the dimension of the solution space by developing ideas in Example 33. The essential point is that the solution method automatically provides us with spanning vectors for the solution space with the following property: in each slot corresponding to a non-leading variable (in the example, the second, fourth, and fifth slots), the corresponding spanning vector has a 1 while all the other spanning vectors have a 0. However, it is awkward to write such a proof because it is awkward to describe the positions of leading and non-leading variables in a general case. Instead, we prove the general result by proving the corresponding result for the rank and the dimension of the nullspace of a linear mapping.

Recall that if $L : \mathbb{R}^n \to \mathbb{R}^m$ is linear, with matrix $[L] = A$, then the nullspace of L is identical to the solution space of $A\mathbf{x} = \mathbf{0}$ (Section 3–4, Theorem 1). Thus, our claim that the dimension of the solution space is $n -$ (rank of A) will be proved if we prove that the dimension of the nullspace of L is $n -$ (rank of L). It is easiest to prove this in a general setting. To shorten statements, we call the dimension of the nullspace of L the **nullity** of L.

THEOREM 4. **Rank Theorem (or Dimension Theorem)** Suppose that $\mathbb{V}$ is an n-dimensional vector space and that $L : \mathbb{V} \to \mathbb{W}$ is a linear mapping into a vector space $\mathbb{W}$. Then, (rank of L) + (nullity of L) = (dimension of $\mathbb{V}$). In the special case $L : \mathbb{R}^n \to \mathbb{R}^m$, (rank of L) + (nullity of L) = n.

Proof. The idea of the proof is to find bases for the nullspace and the range so that we can determine their dimensions. To simplify the argument, suppose that the nullspace of L is not trivial; the case where it is trivial can easily be dealt with by adapting the following argument.

Let the nullity of L be k. Then by Section 4–4 there is a basis $\{\mathbf{v}_1, \mathbf{v}_2, \ldots, \mathbf{v}_k\}$ for the nullspace. Since $\mathbb{V}$ is n-dimensional, we can extend the basis for the nullspace to a basis for $\mathbb{V}$ by adding vectors $\mathbf{u}_{k+1}, \ldots, \mathbf{u}_n$ (again by the methods of Section 4–4).

Now consider any element $\mathbf{w}$ of the range of L. Then $\mathbf{w} = L(\mathbf{x})$ for some $\mathbf{x}$ in $\mathbb{V}$. But any $\mathbf{x}$ in $\mathbb{V}$ can be written as a linear combination of vectors in the basis $\{\mathbf{v}_1, \mathbf{v}_2, \ldots, \mathbf{v}_k, \mathbf{u}_{k+1}, \ldots, \mathbf{u}_n\}$, so

$$\begin{aligned}\mathbf{w} &= L(c_1\mathbf{v}_1 + c_2\mathbf{v}_2 + \cdots + c_k\mathbf{v}_k + c_{k+1}\mathbf{u}_{k+1} + \cdots + c_n\mathbf{u}_n)\\ &= c_1L(\mathbf{v}_1) + c_2L(\mathbf{v}_2) + \cdots + c_{k+1}L(\mathbf{u}_{k+1}) + \cdots + c_nL(\mathbf{u}_n).\end{aligned}$$

But each $\mathbf{v}_i$ is in the nullspace of L, so $L(\mathbf{v}_i) = \mathbf{0}$, and $\mathbf{w}$ is expressed as a linear combination of vectors in the set $\{L(\mathbf{u}_{k+1}), L(\mathbf{u}_{k+2}), \ldots, L(\mathbf{u}_n)\}$. This set is therefore a spanning set for the range of L. Is it a basis for the range?

To test for linear independence, consider the equation

$$c_{k+1}L(\mathbf{u}_{k+1}) + c_{k+2}L(\mathbf{u}_{k+2}) + \cdots + c_nL(\mathbf{u}_n) = \mathbf{0}.$$

By the linearity of L, this is equivalent to

$$L(c_{k+1}\mathbf{u}_{k+1} + \cdots + c_n\mathbf{u}_n) = \mathbf{0},$$

so if this is true, $(c_{k+1}\mathbf{u}_{k+1} + c_{k+2}\mathbf{u}_{k+2} + \cdots + c_n\mathbf{u}_n)$ is a vector in the nullspace of L. Hence, for some $d_1, d_2, \ldots, d_k$,

$$c_{k+1}\mathbf{u}_{k+1} + c_{k+2}\mathbf{u}_{k+2} + \cdots + c_n\mathbf{u}_n = d_1\mathbf{v}_1 + \cdots + d_k\mathbf{v}_k.$$

But this is impossible unless all the c's and d's are zero, because $\{\mathbf{v}_1, \ldots, \mathbf{v}_k, \mathbf{u}_{k+1}, \ldots, \mathbf{u}_n\}$ is a basis for $\mathbb{V}$. It follows that $\{L(\mathbf{u}_{k+1}), L(\mathbf{u}_{k+2}), \ldots, L(\mathbf{u}_n)\}$ is a linearly independent set, and thus a basis for the range of L.

Since we have a basis for the range of L consisting of $n - k$ vectors, the dimension of the range is $n - k$. But this dimension is the rank of L, so (rank of L) $= n - k = n -$(nullity of L), and the proof is complete.

■

A Summary of Facts About Rank

For an m by n matrix A: the rank of A

= the number of leading 1's in the reduced row echelon form of A

= the number of non-zero rows in any row echelon form of A

= the dimension of the rowspace of A

= the dimension of the columnspace of A

$= n-$ (the dimension of the solution space of $A\mathbf{x} = \mathbf{0}$).

For a linear mapping $L : \mathbb{R}^n \to \mathbb{R}^m$: the rank of L

= the dimension of the range of L

$= n-$ (dimension of the nullspace of L)

= the rank of the matrix $[L]$.

An Application of Rank in Control Engineering

In various kinds of engineering, it is important to be able to control the **state** of some system. The state at time t is represented by the state vector function

$$\mathbf{x}(t) = \Big(x_1(t), x_2(t), \ldots, x_n(t)\Big).$$

In electrical engineering applications, the components of $\mathbf{x}$ might be the voltages and currents in some device; in chemical engineering, they could be the temperature and chemical concentrations; or they could be the angular coordinates and angular velocities of a satellite. A standard simple model for describing the control of such systems is the vector differential equation

$$\frac{d}{dt}\mathbf{x}(t) = A\mathbf{x}(t) + u(t)\mathbf{b},$$

where A is an n by n matrix, $u(t)$ is a real-valued function called the **control,** and $\mathbf{b}$ is a vector in $\mathbb{R}^n$.

One of the first important questions is whether the system is "controllable": can the system be driven from any initial state $\mathbf{x}_0$ to any desired final state $\mathbf{x}_F$ by appropriate choice of $u(t)$? One of the basic theorems of control theory is that the system is controllable if and only if the rank of the n by n matrix $\Big[[\mathbf{b}] \quad A[\mathbf{b}] \quad A^2[\mathbf{b}] \quad \ldots \quad A^{n-1}[\mathbf{b}]\Big]$ is n, the dimension of the state space.

A somewhat more general model uses the equation

$$\frac{d}{dt}\mathbf{x}(t) = A\mathbf{x}(t) + B\mathbf{u}(t),$$

where A is an n by n matrix, B is an n by k matrix, and $\mathbf{u}$ is a vector in $\mathbb{R}^k$. The theorem on controllability is essentially the same except that one must consider the rank of the n by nk matrix $[\, B \quad AB \quad A^2B \quad \dots \quad A^{n-1}B \,]$.

EXERCISES 4-5

A1 Each of the following matrices is in row echelon form. Suppose that the matrix is the coefficient matrix of a homogeneous system of equations. State:
(i) the number of variables in the system;
(ii) the rank of the matrix; and
(iii) the dimension of the solution space.

(a) $A = \begin{bmatrix} 1 & 2 & 1 & 3 \\ 0 & 0 & 1 & -2 \end{bmatrix}$ (b) $B = \begin{bmatrix} 1 & -2 & 0 & 0 & 5 \\ 0 & 1 & 3 & 4 & -1 \\ 0 & 0 & 0 & 1 & 2 \end{bmatrix}$

(c) $C = \begin{bmatrix} 1 & 0 & 2 & 1 & 1 \\ 0 & 0 & 1 & 5 & -2 \\ 0 & 0 & 0 & 0 & 0 \end{bmatrix}$ (d) $D = \begin{bmatrix} 1 & 0 & 3 & -5 & 1 & -1 \\ 0 & 1 & 2 & -4 & 2 & 1 \\ 0 & 0 & 0 & 1 & 1 & 4 \\ 0 & 0 & 0 & 0 & 0 & 0 \end{bmatrix}$

A2 For each of the following matrices, determine a basis for the rowspace, and determine a subset of the columns that forms a basis for the columnspace. Also, determine a basis for the nullspace of the linear mapping that has the given matrix. Hence, determine the dimension of the columnspace and the rank of the matrix.

(a) $\begin{bmatrix} 1 & 2 & 8 \\ 1 & 1 & 5 \\ 1 & 0 & -2 \end{bmatrix}$ (b) $\begin{bmatrix} 1 & 1 & -3 & 1 \\ 2 & 3 & -8 & 4 \\ 0 & 1 & -2 & 3 \end{bmatrix}$

(c) $\begin{bmatrix} 1 & 2 & 0 & 3 & 0 \\ 1 & 2 & 1 & 7 & 1 \\ 2 & 4 & 0 & 6 & 1 \\ 3 & 6 & 1 & 13 & 2 \end{bmatrix}$

A3 By geometrical arguments, give a basis for the nullspace and a basis for the range of the following linear transformations, and hence, determine the rank of the transformation.
(a) $\mathbf{proj}_{(1,-2,2)} : \mathbb{R}^3 \to \mathbb{R}^3$
(b) $\mathbf{perp}_{(3,1,2)} : \mathbb{R}^3 \to \mathbb{R}^3$
(c) $\mathbf{refl}_{(0,1,0)} : \mathbb{R}^3 \to \mathbb{R}^3$

A4 The matrix $A = \begin{bmatrix} 1 & 1 & 1 & 1 & 5 \\ 2 & 3 & 1 & 2 & 11 \\ 1 & 1 & 1 & 3 & 7 \\ 1 & 2 & 0 & -1 & 4 \end{bmatrix}$ has reduced row echelon form
$$R = \begin{bmatrix} 1 & 0 & 2 & 0 & 3 \\ 0 & 1 & -1 & 0 & 1 \\ 0 & 0 & 0 & 1 & 1 \\ 0 & 0 & 0 & 0 & 0 \end{bmatrix}.$$
(a) The rowspace of A is a subspace of some $\mathbb{R}^n$, what is n?
(b) Without calculation, give a basis for the rowspace of A; outline the theory that explains why this is a basis.
(c) The columnspace of A is a subspace of some $\mathbb{R}^m$, what is m?
(d) Without calculation, give a basis for the columnspace of A. Indicate clearly why this is the required basis. Call this basis $\mathcal{B}$, and determine $(5, 11, 7, 4)_{\mathcal{B}}$ (the $\mathcal{B}$-coordinates of the last column of A); you can answer this without calculation by considering the matrix R.
(e) Determine the general solution of the system $A\mathbf{x} = \mathbf{0}$, and hence, obtain a spanning set for the solution space.
(f) Explain why this spanning set is in fact a basis for the solution space.
(g) Verify that the rank of A plus the dimension of the solution space of the system $A\mathbf{x} = \mathbf{0}$ is equal to the number of variables in the system.
(h) Use block multiplication to write the system $A\mathbf{x} = \mathbf{0}$ in a way that makes it clear that solutions of the system are orthogonal to rows of A. Use this fact to explain why the set consisting of non-zero rows of R together with a basis for the solution space forms a basis for $\mathbb{R}^n$.

B1 Each of the following matrices is in row echelon form. Suppose that the matrix is the coefficient matrix of a homogeneous system of equations. State:
(i) the number of variables in the system;
(ii) the rank of the matrix; and
(iii) the dimension of the solution space.

(a) $A = \begin{bmatrix} 1 & 2 & 0 & 1 & -2 \\ 0 & 1 & 1 & 2 & 0 \\ 0 & 0 & 0 & 1 & 0 \end{bmatrix}$ (b) $B = \begin{bmatrix} 1 & 0 & 2 & 0 \\ 0 & 1 & -1 & 3 \\ 0 & 0 & 0 & 1 \\ 0 & 0 & 0 & 0 \end{bmatrix}$

(c) $C = \begin{bmatrix} 1 & 5 & 0 & 2 & -3 \\ 0 & 1 & 3 & -2 & 1 \\ 0 & 0 & 1 & 4 & 2 \\ 0 & 0 & 0 & 0 & 1 \end{bmatrix}$ (d) $D = \begin{bmatrix} 1 & 6 & 0 & 2 & -1 & 0 \\ 0 & 0 & 1 & -2 & 1 & 2 \\ 0 & 0 & 0 & 0 & 1 & 3 \\ 0 & 0 & 0 & 0 & 0 & 0 \end{bmatrix}$

B2 For each of the following matrices, determine a basis for the rowspace, and determine a subset of the columns that forms a basis for the columnspace. Also, determine a basis for the nullspace of the linear mapping that has the given matrix. Hence, determine the dimension of the columnspace and the rank of the matrix.

(a) $\begin{bmatrix} 3 & 6 & 1 \\ 2 & 4 & 1 \\ 1 & 2 & 0 \end{bmatrix}$ (b) $\begin{bmatrix} 0 & 1 & 0 & -2 \\ 1 & 2 & 1 & -1 \\ 2 & 4 & 3 & -1 \end{bmatrix}$

(c) $\begin{bmatrix} 1 & 1 & 1 & 1 & 1 \\ 0 & 1 & 2 & 3 & 4 \\ 1 & 0 & 1 & 3 & 3 \\ 1 & 1 & 3 & 6 & 8 \end{bmatrix}$

B3 By geometrical arguments, give a basis for the nullspace and a basis for the range of the following linear transformations; determine the rank of each transformation.

(a) $\mathbf{proj}_{(1,0,-1)} : \mathbb{R}^3 \to \mathbb{R}^3$
(b) $\mathbf{perp}_{(-1,1,3)} : \mathbb{R}^3 \to \mathbb{R}^3$
(c) $\mathbf{refl}_{(1,1,1)} : \mathbb{R}^3 \to \mathbb{R}^3$

B4 The matrix $A = \begin{bmatrix} 1 & 2 & 0 & 0 & 3 & 0 & -1 \\ 1 & 2 & 1 & 0 & 2 & 0 & -3 \\ 1 & 2 & 0 & 1 & 1 & 1 & 4 \\ 1 & 2 & 1 & 0 & 2 & 1 & 1 \\ 3 & 6 & 2 & 1 & 5 & 2 & 2 \end{bmatrix}$ has reduced row echelon form

$$R = \begin{bmatrix} 1 & 2 & 0 & 0 & 3 & 0 & -1 \\ 0 & 0 & 1 & 0 & -1 & 0 & -2 \\ 0 & 0 & 0 & 1 & -2 & 0 & 1 \\ 0 & 0 & 0 & 0 & 0 & 1 & 4 \\ 0 & 0 & 0 & 0 & 0 & 0 & 0 \end{bmatrix}.$$

(a) The rowspace is a subspace of some $\mathbb{R}^n$, what is n?
(b) Without calculation, give a basis for the rowspace of A. Explain briefly.
(c) The columnspace of A is a subspace of some $\mathbb{R}^m$, what is m?
(d) Without calculation, give a basis for the columnspace of A. Call this basis $\mathcal{B}$, and determine $(-1, -3, 4, 1, 2)_{\mathcal{B}}$; you can answer this without calculation by considering the matrix R.
(e) Determine the general solution of $A\mathbf{x} = \mathbf{0}$, and hence, determine a basis for the solution space.
(f) Verify that the rank of A plus the dimension of the solution space is n (where n is as in part (a)).
(g) Verify that the basis for the rowspace and the basis for the solution space together make up a basis for $\mathbb{R}^n$.

CONCEPTUAL EXERCISES

D1 Suppose that $L : \mathbb{R}^n \to \mathbb{R}^m$ is a linear mapping where some information is given about the dimension of the range, or the nullspace of L, or about the rank of L. Answer the following questions, giving a brief explanation in each case.

(a) If $m = 5$, $n = 7$, and rank $(L) = 4$, what is the dimension of the nullspace of L and what is the dimension of the range of L ?

(b) If $m = 5$, $n = 4$, what is the largest possible dimension of the nullspace of L? What is the largest possible rank of L?
(c) If $m = n = 5$, what are the largest possible dimensions of the nullspace and the range of L?
(d) If $m = 4$, $n = 5$, and rank $(L) = 3$, what is the dimension of the range of L? of the nullspace of L?

***D2** (a) Show that if $L : \mathbb{R}^n \to \mathbb{R}^m$ and $M : \mathbb{R}^m \to \mathbb{R}^p$ are linear transformations, the rank of $(M \circ L)$ is less than or equal to the rank of M. (Hint: Consider the dimension of the ranges.)
(b) For L and M as in part (a), show that the rank of $M \circ L$ is less than or equal to the rank of L. (Hint: Use nullspaces and Theorem 4.)
(c) Construct an example such that the rank of the composition is strictly less than the maximum of the ranks. (Hint: Can you find two 2 by 2 matrices each of rank 1 whose product is of rank 0?)

D3 Suppose that $L : \mathbb{R}^n \to \mathbb{R}^m$ and $M : \mathbb{R}^m \to \mathbb{R}^m$ are linear transformations. If M is invertible, show that the rank of $(M \circ L)$ is equal to the rank of L.

D4 (a) By considering the hyperplane $\mathbf{n} \cdot \mathbf{x} = 0$ in $\mathbb{R}^n$ as the nullspace of a linear mapping, prove that it is $(n-1)$-dimensional.
(b) Show that the intersection of two-hyperplanes $\mathbf{n} \cdot \mathbf{x} = 0$ and $\mathbf{m} \cdot \mathbf{x} = 0$ in $\mathbb{R}^n$ is $(n-2)$-dimensional if and only if $\{\mathbf{m}, \mathbf{n}\}$ is linearly independent.
(c) Generalize.

D5 By considering the dimension of the range or nullspace, determine the rank and the nullity of the following linear maps.
(a) $D : P_n \to P_{n-1}$, where P_n is the space of polynomials of degree less than n, and $D(x^k) = kx^{k-1}$.
(b) $D : P_n \to P_n$, where P_n and D are as in (a).
(c) $L : \mathcal{M}(2,3) \to \mathcal{M}(2,3)$, where L is defined by

$$L\left(\begin{bmatrix} a & b & c \\ d & e & f \end{bmatrix}\right) = \begin{bmatrix} d & e & f \\ 0 & 0 & 0 \end{bmatrix}.$$

(d) $Tr : \mathcal{M}(3,3) \to \mathbb{R}$, where $Tr(A) = a_{11} + a_{22} + a_{33}$ (the trace of A).
(e) $S : \mathcal{M}(3,3) \to \mathcal{M}(3,3)$, where S is defined by $S(A) = \frac{1}{2}(A + A^T)$ and A^T is the transpose of the 3 by 3 matrix A. ($S(A)$ is called "the symmetric part of A".)

SECTION

4–6 Matrices with Respect to an Arbitrary Basis; Change of Basis in $\mathbb{R}^n$

The ideas and techniques of this section are essential to achieving one of the main goals of the rest of this book, namely, the "diagonalization" of matrices. The ideas of Section 4–3 play an essential role.

The Matrix of L with Respect to Basis $\mathcal{B}$

In Chapter 3, the matrix of a linear transformation $L : \mathbb{R}^n \rightarrow \mathbb{R}^n$ was defined to be the square matrix whose columns were the images of the **standard** basis vectors of $\mathbb{R}^n$ under the transformation L. In this section, coordinates of vectors with respect to the standard basis will be indicated with the subscript $\mathcal{S}$. Moreover, since we shall soon be discussing the matrix of L with respect to other bases, we shall also use the subscript $\mathcal{S}$ to distinguish the matrix of L with respect to the standard basis. Thus, the **standard matrix** of L is

$$[L]_{\mathcal{S}} = \begin{bmatrix} [L(\mathbf{e}_1)]_{\mathcal{S}} & [L(\mathbf{e}_2)]_{\mathcal{S}} & \ldots & [L(\mathbf{e}_n)]_{\mathcal{S}} \end{bmatrix},$$

and the standard coordinates of the image under L of a vector $\mathbf{x}$ are given by the equation

$$[L(\mathbf{x})]_{\mathcal{S}} = [L]_{\mathcal{S}}[\mathbf{x}]_{\mathcal{S}}.$$

These equations are exactly the same as the equations in Chapter 3 except that the notation is fancier, so that it will be possible to compare this standard description with a description with respect to some other basis. We follow the same pattern in defining the matrix of L with respect to another basis $\mathcal{B}$.

DEFINITION

Suppose that $\mathcal{B} = \{\mathbf{v}_1, \mathbf{v}_2, \ldots, \mathbf{v}_n\}$ is any basis for $\mathbb{R}^n$. Define **the matrix of** L **with respect to** $\mathcal{B}$ (or the $\mathcal{B}$-matrix of L) to be the matrix

$$[L]_{\mathcal{B}} = \begin{bmatrix} [L(\mathbf{v}_1)]_{\mathcal{B}} & [L(\mathbf{v}_2)]_{\mathcal{B}} & \ldots & [L(\mathbf{v}_n)]_{\mathcal{B}} \end{bmatrix}.$$

Note that the columns of $[L]_{\mathcal{B}}$ are the $\mathcal{B}$-coordinate vectors of the images of the $\mathcal{B}$-basis vectors under the linear transformation L. Note that the pattern is exactly the same as above, but that here everything is done in terms of the vectors in $\mathcal{B}$. It is important to emphasize that by basis we always mean ordered basis; the order of the basis elements determines the order of the columns of the matrix $[L]_\mathcal{B}$.

Recall from Section 4–3 that the $\mathcal{B}$-coordinate vector of $\mathbf{x}$ is $(\mathbf{x})_\mathcal{B} = (\tilde{x}_1, \tilde{x}_2, \ldots, \tilde{x}_n)$, so that

$$\mathbf{x} = \tilde{x}_1\mathbf{v}_1 + \tilde{x}_2\mathbf{v}_2 + \cdots + \tilde{x}_n\mathbf{v}_n;$$

also, $[\mathbf{x}]_\mathcal{B} = [\,\tilde{\mathbf{x}}\,]$ is the corresponding $\mathcal{B}$-coordinate column.

The $\mathcal{B}$-coordinates of $L(\mathbf{x})$ are determined with the help of the linearity properties, just as the standard coordinates were determined in Chapter 3. First, note that by linearity,

$$L(\mathbf{x}) = L(\tilde{x}_1\mathbf{v}_1 + \tilde{x}_2\mathbf{v}_2 + \cdots + \tilde{x}_n\mathbf{v}_n) = \tilde{x}_1 L(\mathbf{v}_1) + \tilde{x}_2 L(\mathbf{v}_2) + \cdots + \tilde{x}_n L(\mathbf{v}_n).$$

By a Remark in Section 4–3, taking $\mathcal{B}$-coordinates is a linear operation, so this equation is still true if we replace each vector by its $\mathcal{B}$-coordinate column, and use block multiplication:

$$\begin{aligned}[L(\mathbf{x})]_\mathcal{B} &= \tilde{x}_1[L(\mathbf{v}_1)]_\mathcal{B} + \tilde{x}_2[L(\mathbf{v}_2)]_\mathcal{B} + \cdots + \tilde{x}_n[L(\mathbf{v}_n)]_\mathcal{B} \\ &= \begin{bmatrix} [L(\mathbf{v}_1)]_\mathcal{B} & [L(\mathbf{v}_2)]_\mathcal{B} & \cdots & [L(\mathbf{v}_n)]_\mathcal{B} \end{bmatrix} \begin{bmatrix} \tilde{x}_1 \\ \vdots \\ \tilde{x}_n \end{bmatrix}.\end{aligned}$$

It follows that

$$[L(\mathbf{x})]_\mathcal{B} = [L]_\mathcal{B}[\mathbf{x}]_\mathcal{B}.$$

The definition of the $\mathcal{B}$-matrix of L was exactly what was needed in order for the pattern for $\mathcal{B}$-coordinates and the $\mathcal{B}$-matrix of L to be exactly parallel to the pattern for standard coordinates and the standard matrix.

EXAMPLE 34

In Section 3–2, Example 13, the standard matrix of the linear transformation $\mathbf{proj}_{(3,4)} : \mathbb{R}^2 \to \mathbb{R}^2$ was found to be

$$[\mathbf{proj}_{(3,4)}]_\mathcal{S} = \begin{bmatrix} 9/25 & 12/25 \\ 12/25 & 16/25 \end{bmatrix}.$$

Find the matrix of this linear transformation with respect to a basis that shows the geometry of the transformation more clearly.

SOLUTION: For this linear transformation, it is natural to use a basis for $\mathbb{R}^2$ consisting of the vector $\mathbf{v}_1 = (3,4)$, which is the direction vector for the projection, and a second vector orthogonal to $(3,4)$, say $\mathbf{v}_2 = (-4,3)$. Then, with $\mathcal{B} = \{\mathbf{v}_1, \mathbf{v}_2\}$, by geometry,

$$\mathbf{proj}_{(3,4)}(\mathbf{v}_1) = \mathbf{proj}_{(3,4)}(3,4) = (3,4) = 1\mathbf{v}_1 + 0\mathbf{v}_2,$$

so the $\mathcal{B}$-coordinate vector is $(\mathbf{proj}_{(3,4)}(\mathbf{v}_1))_\mathcal{B} = (1,0)$, and

$$\mathbf{proj}_{(3,4)}(\mathbf{v}_2) = \mathbf{proj}_{(3,4)}(-4,3) = (0,0) = 0\mathbf{v}_1 + 0\mathbf{v}_2,$$

so the $\mathcal{B}$-coordinate vector is $(\mathbf{proj}_{(3,4)}(\mathbf{v}_2))_\mathcal{B} = (0,0)$.

To obtain the $\mathcal{B}$-matrix of $\mathbf{proj}_{(3,4)}$, take these two vectors as the columns of the matrix, so that

$$[\mathbf{proj}_{(3,4)}]_\mathcal{B} = \begin{bmatrix} 1 & 0 \\ 0 & 0 \end{bmatrix}.$$

Now consider $\mathbf{proj}_{(3,4)}(\mathbf{a})$ for an arbitrary $\mathbf{a}$ in $\mathbb{R}^2$. By the general rule $[L(\mathbf{a})]_\mathcal{B} = [L]_\mathcal{B}[\mathbf{a}]_\mathcal{B}$, we have

$$\begin{aligned}[\mathbf{proj}_{(3,4)}(\mathbf{a})]_\mathcal{B} &= [\mathbf{proj}_{(3,4)}][\mathbf{a}]_\mathcal{B} \\ &= \begin{bmatrix} 1 & 0 \\ 0 & 0 \end{bmatrix}\begin{bmatrix} \tilde{a}_1 \\ \tilde{a}_2 \end{bmatrix} = \begin{bmatrix} \tilde{a}_1 \\ 0 \end{bmatrix}.\end{aligned}$$

In terms of $\mathcal{B}$-coordinates, $\mathbf{proj}_{(3,4)}$ is described as the linear mapping that sends $(\tilde{a}_1, \tilde{a}_2)$ to $(\tilde{a}_1, 0)$.

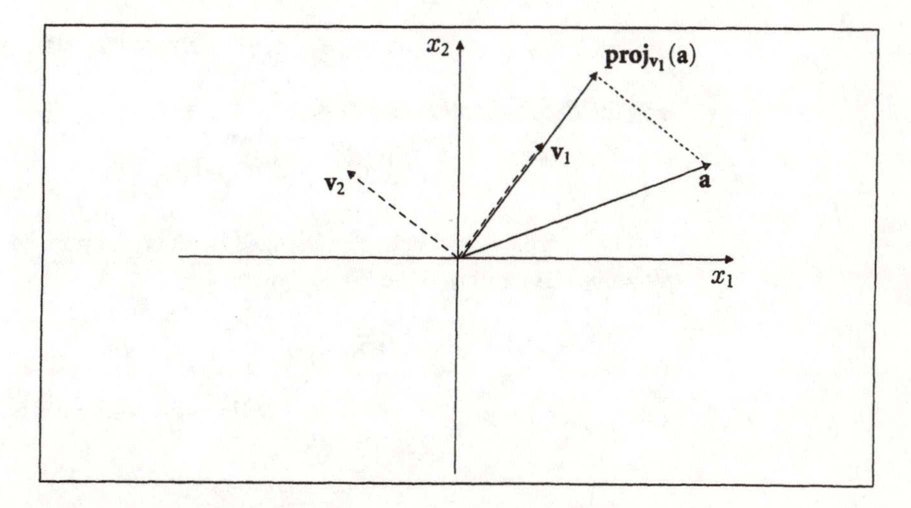

FIGURE 4-6-1. $\mathbf{proj}_{\mathbf{v}_1} = \tilde{a}_1\mathbf{v}_1 + 0\mathbf{v}_2$, *so* $(\mathbf{proj}_{\mathbf{v}_1}(\mathbf{a}))_\mathcal{B} = (\tilde{a}_1, 0)$.

This simple geometrical description is obtained when we use a basis $\mathcal{B}$ that is adapted to the geometry of the transformation. See Figure 4–6–1. This example will be discussed further below.

EXAMPLE 35

In Section 3–3, Example 17, it was found that

$$[\mathbf{refl}_{(1,-1,2)}]_{\mathcal{S}} = \begin{bmatrix} 2/3 & 1/3 & -2/3 \\ 1/3 & 2/3 & 2/3 \\ -2/3 & 2/3 & -1/3 \end{bmatrix}.$$

Find the matrix of this linear transformation with respect to a basis that shows the geometry of the transformation more clearly.

SOLUTION: For this linear transformation of $\mathbb{R}^3$, it seems natural to use as a basis the vector $\mathbf{v}_1 = (1,-1,2)$, which is the normal vector to the plane of the reflection, and two vectors that lie in this plane (and are therefore orthogonal to $\mathbf{v}_1$). We take $\mathbf{v}_2 = (1,1,0)$ and $\mathbf{v}_3 = (0,2,1)$, because they can easily be found by inspection to be orthogonal to $\mathbf{v}_1$. Now let $\mathcal{B} = \{\mathbf{v}_1, \mathbf{v}_2, \mathbf{v}_3\}$; by arguments given at the end of Section 4–2, this is a basis for $\mathbb{R}^3$.

To determine the matrix of $\mathbf{refl}_{(1,-1,2)}$ with respect to this basis, it is necessary to calculate the $\mathcal{B}$-coordinates of the images of the basis vectors. From the geometry of the transformation, it is easy to calculate the images, and because the basis is adapted to the geometry, it is easy to calculate $\mathcal{B}$-coordinates. The vector $\mathbf{v}_1$ is reversed by the reflection, so

$$\mathbf{refl}_{(1,-1,2)}(\mathbf{v}_1) = -\mathbf{v}_1 = (-1)\mathbf{v}_1 + 0\mathbf{v}_2 + 0\mathbf{v}_3,$$

and the $\mathcal{B}$-coordinates are

$$(\mathbf{refl}_{(1,-1,2)}(\mathbf{v}_1))_{\mathcal{B}} = (-1,0,0).$$

The vectors $\mathbf{v}_2$ and $\mathbf{v}_3$ are unchanged by the reflection, so $\mathbf{refl}_{(1,-1,2)}(\mathbf{v}_2) = \mathbf{v}_2 = 0\mathbf{v}_1 + 1\mathbf{v}_2 + 0\mathbf{v}_3$, so the $\mathcal{B}$-coordinates are

$$(\mathbf{refl}_{(1,-1,2)}(\mathbf{v}_2))_{\mathcal{B}} = (0,1,0);$$

$\mathbf{refl}_{(1,-1,2)}(\mathbf{v}_3) = \mathbf{v}_3 = 0\mathbf{v}_1 + 0\mathbf{v}_2 + 1\mathbf{v}_3$, and the $\mathcal{B}$-coordinates are

$$(\mathbf{refl}_{(1,-1,2)}(\mathbf{v}_3))_{\mathcal{B}} = (0,0,1).$$

Taking these three $\mathcal{B}$-coordinate vectors as columns, we find that the matrix of $\mathbf{refl}_{(1,-1,2)}$ with respect to $\mathcal{B}$ is

$$[\mathbf{refl}_{(1,-1,2)}]_{\mathcal{B}} = \begin{bmatrix} -1 & 0 & 0 \\ 0 & 1 & 0 \\ 0 & 0 & 1 \end{bmatrix}.$$

Again, it is much easier to understand the geometry of this transformation from the matrix with respect to the adapted basis $\mathcal{B}$ than from the matrix with respect to the standard basis: the transformation takes the vector with $\mathcal{B}$-coordinates $(\tilde{a}_1, \tilde{a}_2, \tilde{a}_3)$ to the vector with $\mathcal{B}$-coordinates $(-\tilde{a}_1, \tilde{a}_2, \tilde{a}_3)$.

Change of Basis

In Examples 34 and 35, we used special geometrical properties of the linear transformation and of the chosen basis $\mathcal{B}$ to determine the $\mathcal{B}$-coordinate columns that make up the $\mathcal{B}$-matrix $[L]_{\mathcal{B}}$ of the linear transformation L. In some applications of these ideas, the geometry does not provide us with such a simple way of determining $[L]_{\mathcal{B}}$. We need a general method for determining $[L]_{\mathcal{B}}$, given the standard matrix of L and a new basis $\mathcal{B}$.

The starting point for such a method is the general procedure given in Section 4–3 for finding the $\mathcal{B}$-coordinates of a vector in $\mathbb{R}^n$, as illustrated in Example 24 of that section. Recall that if the basis $\mathcal{B} = \{\mathbf{v}_1, \mathbf{v}_2, \ldots, \mathbf{v}_n\}$ is given, then

$$\Big[[\mathbf{v}_1] \quad [\mathbf{v}_2] \quad \ldots \quad [\mathbf{v}_n] \Big] [\mathbf{x}]_{\mathcal{B}} = [\mathbf{x}].$$

The matrix in this equation is particularly important: it is called **the change of basis matrix** (from the standard basis $\mathcal{S}$ to the new basis $\mathcal{B}$). Denote this matrix by P; recall that the columns of this matrix were the standard coordinates of the new basis vectors, so that

$$P = \Big[[\mathbf{v}_1]_{\mathcal{S}} \quad [\mathbf{v}_2]_{\mathcal{S}} \quad \ldots \quad [\mathbf{v}_n]_{\mathcal{S}} \Big].$$

Notice that the order of the vectors in $\mathcal{B}$ determines the order of the columns of P.

LEMMA 1. The change of basis matrix P is invertible.

Proof. The columns of P are linearly independent, since they are the vectors of a basis, so the rank of P is n by the discussion in Section 4–5. Therefore, P is invertible by Theorem 4 of Section 3–5. ∎

To emphasize the role of the bases, we use the fact that $[\mathbf{x}] = [\mathbf{x}]_S$ (see Remark (6) following the definition of coordinates in Section 4–3). We also take advantage of the fact that P is invertible. The **change of coordinates equation** relating the $\mathcal{B}$-coordinates of **x** to the $\mathcal{S}$-coordinates of **x** may thus be rewritten

$$P[\mathbf{x}]_{\mathcal{B}} = [\mathbf{x}]_{\mathcal{S}} \qquad \text{or} \qquad [\mathbf{x}]_{\mathcal{B}} = P^{-1}[\mathbf{x}]_{\mathcal{S}}.$$

Next, apply the change of coordinates equation to the vector $L(\mathbf{x})$, and obtain

$$[L(\mathbf{x})]_{\mathcal{B}} = P^{-1}[L(\mathbf{x})]_{\mathcal{S}}.$$

From the discussion earlier in this section, substitute for $[L(\mathbf{x})]_{\mathcal{B}}$ and $[L(\mathbf{x})]_{\mathcal{S}}$ to get

$$[L]_{\mathcal{B}}[\mathbf{x}]_{\mathcal{B}} = [L(\mathbf{x})]_{\mathcal{B}} = P^{-1}[L(\mathbf{x})]_{\mathcal{S}} = P^{-1}[L]_{\mathcal{S}}[\mathbf{x}]_{\mathcal{S}}.$$

Finally, use the change of coordinates equation to substitute for the $[\mathbf{x}]_{\mathcal{S}}$ in the last expression:

$$[L]_{\mathcal{B}}[\mathbf{x}]_{\mathcal{B}} = P^{-1}[L]_{\mathcal{S}}P[\mathbf{x}]_{\mathcal{B}}.$$

Since this is true for every $[\mathbf{x}]_{\mathcal{B}}$, it must be true (by Theorem 2 of Section 3–1) that the matrices multiplying $[\mathbf{x}]_{\mathcal{B}}$ are equal:

$$\boxed{[L]_{\mathcal{B}} = P^{-1}[L]_{\mathcal{S}}P.}$$

This is the solution to the problem of determining the $\mathcal{B}$-matrix of L, given the standard matrix of L and a new basis $\mathcal{B}$.

We shall first apply this change of basis method to Examples 34 and 35 to make sure that things work out as we expect.

EXAMPLE 36

In Example 34, we determined the matrix of $\mathbf{proj}_{(3,4)}$ with respect to a geometrically adapted basis $\mathcal{B}$. Let us verify that the change of basis method just described really does transform the standard matrix $[\mathbf{proj}_{(3,4)}]_S$ to the $\mathcal{B}$-matrix $[\mathbf{proj}_{(3,4)}]_\mathcal{B}$.

The matrix $[\mathbf{proj}_{(3,4)}]_S = \begin{bmatrix} 9/25 & 12/25 \\ 12/25 & 16/25 \end{bmatrix}$. The basis $\mathcal{B} = \{(3,4),(-4,3)\}$, so that the change of basis matrix is $P = \begin{bmatrix} 3 & -4 \\ 4 & 3 \end{bmatrix}$. The inverse is found to be $P^{-1} = \begin{bmatrix} 3/25 & 4/25 \\ -4/25 & 3/25 \end{bmatrix}$. Hence, the $\mathcal{B}$-matrix of $\mathbf{proj}_{(3,4)}$ is given by

$$\begin{aligned} P^{-1}[\mathbf{proj}_{(3,4)}]_S P &= \begin{bmatrix} 3/25 & 4/25 \\ -4/25 & 3/25 \end{bmatrix} \begin{bmatrix} 9/25 & 12/25 \\ 12/25 & 16/25 \end{bmatrix} \begin{bmatrix} 3 & -4 \\ 4 & 3 \end{bmatrix} \\ &= \begin{bmatrix} 3/25 & 4/25 \\ -4/25 & 3/25 \end{bmatrix} \begin{bmatrix} 75/25 & 0/25 \\ 100/25 & 0/25 \end{bmatrix} \\ &= \frac{1}{25} \begin{bmatrix} 3 & 4 \\ -4 & 3 \end{bmatrix} \begin{bmatrix} 3 & 0 \\ 4 & 0 \end{bmatrix} = \begin{bmatrix} 1 & 0 \\ 0 & 0 \end{bmatrix}. \end{aligned}$$

Thus, we obtain exactly the same $\mathcal{B}$-matrix $[\mathbf{proj}_{(3,4)}]_\mathcal{B}$ as we obtained by the earlier geometric argument.

To make sure we understand what this means, let us calculate the $\mathcal{B}$-coordinates of the image of the vector (5, 2) under $\mathbf{proj}_{(3,4)}$. We can do this in two ways.

Method 1. Use the fact that $[\mathbf{proj}_{(3,4)}(5,2)]_\mathcal{B} = [\mathbf{proj}_{(3,4)}]_\mathcal{B}[(5,2)]_\mathcal{B}$. We need the $\mathcal{B}$-coordinates of $(5,2)$:

$$[(5,2)]_\mathcal{B} = P^{-1}[(5,2)]_S = \begin{bmatrix} 3/25 & 4/25 \\ -4/25 & 3/25 \end{bmatrix} \begin{bmatrix} 5 \\ 2 \end{bmatrix} = \begin{bmatrix} 23/25 \\ -14/25 \end{bmatrix}.$$

Hence,

$$\begin{aligned} [\mathbf{proj}_{(3,4)}(5,2)]_\mathcal{B} &= [\mathbf{proj}_{(3,4)}]_\mathcal{B}[(5,2)]_\mathcal{B} \\ &= \begin{bmatrix} 1 & 0 \\ 0 & 0 \end{bmatrix} \begin{bmatrix} 23/25 \\ -14/25 \end{bmatrix} = \begin{bmatrix} 23/25 \\ 0 \end{bmatrix}. \end{aligned}$$

Method 2. Use the fact that $[\mathbf{proj}_{(3,4)}(5,2)]_\mathcal{B} = P^{-1}[\mathbf{proj}_{(3,4)}(5,2)]_S$.

$$\begin{aligned} [\mathbf{proj}_{(3,4)}(5,2)]_S = [\mathbf{proj}_{(3,4)}]_S \begin{bmatrix} 5 \\ 2 \end{bmatrix} &= \begin{bmatrix} 9/25 & 12/25 \\ 12/25 & 16/25 \end{bmatrix} \begin{bmatrix} 5 \\ 2 \end{bmatrix} \\ &= \begin{bmatrix} 69/25 \\ 92/25 \end{bmatrix}. \end{aligned}$$

Therefore,

$$[\mathbf{proj}_{(3,4)}(5,2)]_{\mathcal{B}} = \begin{bmatrix} 3/25 & 4/25 \\ -4/25 & 3/25 \end{bmatrix} \begin{bmatrix} 69/25 \\ 92/25 \end{bmatrix}$$

$$= \begin{bmatrix} 575/625 \\ 0 \end{bmatrix} = \begin{bmatrix} 23/25 \\ 0 \end{bmatrix}.$$

The calculation is probably slightly easier if we use the first method and the matrix $[\mathbf{proj}_{(3,4)}]_{\mathcal{B}}$, but that really is not the point of change of basis. What is really important is that it is easy to get a geometrical understanding of what happens to vectors if you multiply by $\begin{bmatrix} 1 & 0 \\ 0 & 0 \end{bmatrix}$ (the $\mathcal{B}$-matrix), and much more difficult to understand what happens if you multiply by $\begin{bmatrix} 9/25 & 12/25 \\ 12/25 & 16/25 \end{bmatrix}$ (the standard matrix). Using a non-standard basis may make it much easier to understand the geometry of a linear transformation.

EXAMPLE 37

The standard matrix of the reflection in Example 35 is

$$[\mathbf{refl}_{(1,-1,2)}]_{\mathcal{S}} = \begin{bmatrix} 2/3 & 1/3 & -2/3 \\ 1/3 & 2/3 & 2/3 \\ -2/3 & 2/3 & -1/3 \end{bmatrix}.$$

We want to check that a change of basis to the basis $\mathcal{B} = \{(1,-1,2), (1,1,0), (0,2,1)\}$ leads to the $\mathcal{B}$-matrix obtained in Example 35.

The change of basis matrix is $P = \begin{bmatrix} 1 & 1 & 0 \\ -1 & 1 & 2 \\ 2 & 0 & 1 \end{bmatrix}$. Calculate that

$P^{-1} = 1/6 \begin{bmatrix} 1 & -1 & 2 \\ 5 & 1 & -2 \\ -2 & 2 & 2 \end{bmatrix}$. It follows that

$$[\mathbf{refl}_{(1,-1,2)}]_{\mathcal{B}} = P^{-1}[\mathbf{refl}_{(1,-1,2)}]_{\mathcal{B}}P$$

$$= (1/6) \begin{bmatrix} 1 & -1 & 2 \\ 5 & 1 & -2 \\ -2 & 2 & 2 \end{bmatrix} \begin{bmatrix} 2/3 & 1/3 & -2/3 \\ 1/3 & 2/3 & 2/3 \\ -2/3 & 2/3 & -1/3 \end{bmatrix} \begin{bmatrix} 1 & 1 & 0 \\ -1 & 1 & 2 \\ 2 & 0 & 1 \end{bmatrix}$$

$$= (1/6)\begin{bmatrix} 1 & -1 & 2 \\ 5 & 1 & -2 \\ -2 & 2 & 2 \end{bmatrix}\begin{bmatrix} -3/3 & 3/3 & 0/3 \\ 3/3 & 3/3 & 6/3 \\ -6/3 & 0/3 & 3/3 \end{bmatrix}$$

$$= (1/6)\begin{bmatrix} 1 & -1 & 2 \\ 5 & 1 & -2 \\ -2 & 2 & 2 \end{bmatrix}\begin{bmatrix} -1 & 1 & 0 \\ 1 & 1 & 2 \\ -2 & 0 & 1 \end{bmatrix} = \begin{bmatrix} -1 & 0 & 0 \\ 0 & 1 & 0 \\ 0 & 0 & 1 \end{bmatrix}.$$

Again, this $\mathcal{B}$-matrix agrees with the $\mathcal{B}$-matrix of this linear transformation determined earlier by geometric arguments.

Now that we have checked that the method produces the correct answers in cases where the answer is known in advance, let us consider some other examples.

EXAMPLE 38

Let L be the linear transformation with matrix $A = \begin{bmatrix} 2 & 3 \\ 4 & 5 \end{bmatrix}$. Let $\mathcal{B} = \{(3, 1), (-1, 1)\}$; it is easy to check that $\mathcal{B}$ is a basis for $\mathbb{R}^2$. Find the matrix of L with respect to the basis $\mathcal{B}$.

SOLUTION: The change of basis matrix is $P = \begin{bmatrix} 3 & -1 \\ 1 & 1 \end{bmatrix}$. By a standard calculation, $P^{-1} = (1/4)\begin{bmatrix} 1 & 1 \\ -1 & 3 \end{bmatrix}$. It follows that the $\mathcal{B}$-matrix of the linear transformation is

$$[L]_{\mathcal{B}} = P^{-1}AP = (1/4)\begin{bmatrix} 1 & 1 \\ -1 & 3 \end{bmatrix}\begin{bmatrix} 2 & 3 \\ 4 & 5 \end{bmatrix}\begin{bmatrix} 3 & -1 \\ 1 & 1 \end{bmatrix}$$

$$= (1/4)\begin{bmatrix} 6 & 8 \\ 10 & 12 \end{bmatrix}\begin{bmatrix} 3 & -1 \\ 1 & 1 \end{bmatrix} = \begin{bmatrix} 13/2 & 1/2 \\ 21/2 & 1/2 \end{bmatrix}.$$

This calculation correctly illustrates the procedure of the change of basis, but the result is not very interesting. The result of the next example raises more interesting questions.

EXAMPLE 39

Let L be a linear transformation with standard matrix

$$A = \begin{bmatrix} -3 & 5 & -5 \\ -7 & 9 & -5 \\ -7 & 7 & -3 \end{bmatrix}.$$

Let $\mathcal{B}$ be the basis $\{(1,1,0),(1,1,1),(0,1,1)\}$. (The fact that this is a basis will in fact be checked when we calculate P^{-1}, since P has an inverse if and only if its columns are linearly independent.) Determine the matrix of the linear transformation with respect to the basis $\mathcal{B}$.

SOLUTION:

$P = \begin{bmatrix} 1 & 1 & 0 \\ 1 & 1 & 1 \\ 0 & 1 & 1 \end{bmatrix}$. Calculate $P^{-1} = \begin{bmatrix} 0 & 1 & -1 \\ 1 & -1 & 1 \\ -1 & 1 & 0 \end{bmatrix}$.

Thus, the $\mathcal{B}$-matrix corresponding to the standard matrix A is

$$\begin{aligned} [L]_{\mathcal{B}} &= P^{-1}AP \\ &= \begin{bmatrix} 0 & 1 & -1 \\ 1 & -1 & 1 \\ -1 & 1 & 0 \end{bmatrix} \begin{bmatrix} -3 & 5 & -5 \\ -7 & 9 & -5 \\ -7 & 7 & -3 \end{bmatrix} \begin{bmatrix} 1 & 1 & 0 \\ 1 & 1 & 1 \\ 0 & 1 & 1 \end{bmatrix} \\ &= \begin{bmatrix} 0 & 2 & -2 \\ -3 & 3 & -3 \\ -4 & 4 & 0 \end{bmatrix} \begin{bmatrix} 1 & 1 & 0 \\ 1 & 1 & 1 \\ 0 & 1 & 1 \end{bmatrix} \\ &= \begin{bmatrix} 2 & 0 & 0 \\ 0 & -3 & 0 \\ 0 & 0 & 4 \end{bmatrix}. \end{aligned}$$

The resulting matrix is said to be **in diagonal form** or to be a **diagonal** matrix, since all entries not on the main diagonal are zero. What does this mean in terms of the geometry of the linear transformation? From the definition of $[L]_{\mathcal{B}}$ and the definition of $\mathcal{B}$-coordinates of a vector, we see that the linear transformation stretches the first basis vector $(1,1,0)$ by a factor of 2, it reflects (because of the minus sign) the second basis vector $(1,1,1)$ in the origin and stretches it by a factor of 3, and it stretches the third basis vector $(0,1,1)$ by a factor of 4. This gives a very clear geometrical picture of how the linear transformation maps vectors — a picture that is certainly not obvious from looking at the standard matrix A.

Big Questions. An arbitrary square matrix A can always be thought of as the matrix of a linear transformation; is there a basis $\mathcal{B}$ such that the $\mathcal{B}$-matrix of this linear transformation is in diagonal form? How can we find such a basis if it exists?

The answer to these questions is to be found in Chapter 6 (although one of the exercises invites you to try to discover the answers for yourself, at least for 2 by 2 matrices). In order to deal with these questions in the general case, one more computational tool is needed: the determinant, which is discussed in Chapter 5.

We have been using the machinery of change of basis to change from the standard basis $\mathcal{S}$ to another basis $\mathcal{B}$. The machinery can be reversed to compute an $\mathcal{S}$-matrix, given a $\mathcal{B}$-matrix and the basis $\mathcal{B}$, as the next example shows.

EXAMPLE 40

Consider the linear transformation $L = \mathbf{perp}_{(1,2,2)}$. Introduce a basis $\mathcal{B}$ consisting of the vector $\mathbf{v}_1 = (1, 2, 2)$ and two vectors orthogonal to it, say $\mathbf{v}_2 = (2, -1, 0)$ and $\mathbf{v}_3 = (0, 1, -1)$. By geometrical arguments we see that this is a basis, and by methods similar to those of Example 35, we see that

$$[\mathbf{perp}_{(1,2,2)}]_{\mathcal{B}} = \begin{bmatrix} 0 & 0 & 0 \\ 0 & 1 & 0 \\ 0 & 0 & 1 \end{bmatrix}.$$

Determine the standard matrix of this linear transformation by a change of basis.

SOLUTION: The change of basis matrix (from $\mathcal{S}$ to $\mathcal{B}$) is $P = \begin{bmatrix} 1 & 2 & 0 \\ 2 & -1 & 1 \\ 2 & 0 & -1 \end{bmatrix}$.

By straightforward calculation, $P^{-1} = (1/9)\begin{bmatrix} 1 & 2 & 2 \\ 4 & -1 & -1 \\ 2 & 4 & -5 \end{bmatrix}$. Since $[L]_{\mathcal{B}} = P^{-1}[L]_{\mathcal{S}}P$, it follows that $P[L]_{\mathcal{B}}P^{-1} = [L]_{\mathcal{S}}$. Hence,

$$\begin{aligned}
[\mathbf{perp}_{(1,2,2)}]_{\mathcal{S}} &= P[\mathbf{perp}_{(1,2,2)}]_{\mathcal{B}}P^{-1} \\
&= \begin{bmatrix} 1 & 2 & 0 \\ 2 & -1 & 1 \\ 2 & 0 & -1 \end{bmatrix} \begin{bmatrix} 0 & 0 & 0 \\ 0 & 1 & 0 \\ 0 & 0 & 1 \end{bmatrix} (1/9) \begin{bmatrix} 1 & 2 & 2 \\ 4 & -1 & -1 \\ 2 & 4 & -5 \end{bmatrix}
\end{aligned}$$

$$= (1/9)\begin{bmatrix} 0 & 2 & 0 \\ 0 & -1 & 1 \\ 0 & 0 & -1 \end{bmatrix}\begin{bmatrix} 1 & 2 & 2 \\ 4 & -1 & -1 \\ 2 & 4 & -5 \end{bmatrix}$$

$$= (1/9)\begin{bmatrix} 8 & -2 & -2 \\ -2 & 5 & -4 \\ -2 & -4 & 5 \end{bmatrix}.$$

It is easy to calculate the standard matrix of $\mathbf{perp}_{(1,2,2)}$ by the methods of Section 3–2 to verify that this matrix is correct. In fact, for this linear transformation, the methods of Section 3–2 are easier than the method we have used.

EXERCISES 4–6

A1 Determine the matrix of the linear mapping L with respect to the basis $\mathcal{B}$ in the following cases, and determine $(L(\mathbf{x}))_{\mathcal{B}}$ for the given $(\mathbf{x})_{\mathcal{B}}$.
(a) In $\mathbb{R}^2$, $\mathcal{B} = \{\mathbf{v}_1, \mathbf{v}_2\}$ and $L(\mathbf{v}_1) = \mathbf{v}_2$, $L(\mathbf{v}_2) = 2\mathbf{v}_1 - \mathbf{v}_2$; $(\mathbf{x})_{\mathcal{B}} = (4, 3)$
(b) In $\mathbb{R}^3$, $\mathcal{B} = \{\mathbf{v}_1, \mathbf{v}_2, \mathbf{v}_3\}$ and $L(\mathbf{v}_1) = 2\mathbf{v}_1 - \mathbf{v}_3$, $L(\mathbf{v}_2) = 2\mathbf{v}_1 - \mathbf{v}_3$, $L(\mathbf{v}_3) = 4\mathbf{v}_2 + 5\mathbf{v}_3$; $(\mathbf{x})_{\mathcal{B}} = (3, 3, -1)$

A2 Consider the basis $\mathcal{B} = \{\mathbf{v}_1 = (1,1), \mathbf{v}_2 = (-1,2)\}$ of $\mathbb{R}^2$. In each of the following cases, L is a linear mapping. Determine $[L(\mathbf{v}_1)]_{\mathcal{B}}$ and $[L(\mathbf{v}_2)]_{\mathcal{B}}$ **by inspection**, and hence, determine $[L]_{\mathcal{B}}$.
(a) $L(1,1) = (-3,-3)$ and $L(-1,2) = (-4,8)$
(b) $L(1,1) = (-1,2)$ and $L(-1,2) = (2,2)$

A3 Consider the basis $\mathcal{B} = \{\mathbf{v}_1 = (1,1,1), \mathbf{v}_2 = (-1,2,0), \mathbf{v}_3 = (0,-1,4)\}$ of $\mathbb{R}^3$. In each of the following cases, L is a linear mapping. Determine $[L(\mathbf{v}_1)]_{\mathcal{B}}$, $[L(\mathbf{v}_2)]_{\mathcal{B}}$, and $[L(\mathbf{v}_3)]_{\mathcal{B}}$ **by inspection**, and hence, determine $[L]_{\mathcal{B}}$.
(a) $L(1,1,1) = (-1,2,0)$, $L(-1,2,0) = (1,1,1)$, $L(0,-1,4) = (5,5,5)$
(b) $L(1,1,1) = (0,-1,4)$, $L(-1,2,0) = (1,1,1)$, $L(0,-1,4) = (-1,2,0)$

A4 For each of the following linear transformations, determine a geometrically natural basis $\mathcal{B}$ (as in Examples 34 and 35), and determine (by inspection) the $\mathcal{B}$-matrix of the transformation.
(a) $\mathbf{refl}_{(1,-2)}$ (b) $\mathbf{proj}_{(2,1,-1)}$ (c) $\mathbf{refl}_{(-1,-1,1)}$

A5 (a) Find the coordinates of $(1,2,4)$ with respect to the basis $\mathcal{B} = \{(1,0,1), (1,-1,0), (0,1,2)\}$ in $\mathbb{R}^3$.
(b) Suppose that $L : \mathbb{R}^3 \to \mathbb{R}^3$ is a linear transformation such that $L(1,0,1) = (1,2,4)$, $L(1,-1,0) = (0,1,2)$, $L(0,1,2) = (2,-2,0)$. Without doing further row reduction, determine the $\mathcal{B}$-matrix of L.
(c) Determine $[L(1,2,4)]_{\mathcal{B}}$, and use it to determine $[L(1,2,4)]_{\mathcal{S}}$.

A6 (a) Find the coordinates of $(5, 3, -5)$ with respect to the basis $\mathcal{B} = \{(1, 0, -1), (1, 2, 0), (0, 1, 1)\}$ in $\mathbb{R}^3$.

(b) Suppose that $L : \mathbb{R}^3 \to \mathbb{R}^3$ is a linear transformation such that $L(1, 0, -1) = (0, 1, 1)$, $L(1, 2, 0) = (-2, 0, 2)$, $L(0, 1, 1) = (5, 3, -5)$. Without doing further row reduction, determine the $\mathcal{B}$-matrix of L.

(c) Determine $[L(5, 3, -5)]_{\mathcal{B}}$, and use it to determine $[L(5, 3, -5)]_{\mathcal{S}}$.

A7 Assume that each of the following matrices is the matrix of some linear transformation with respect to the standard basis. Determine the matrix of the linear transformation with respect to the given basis $\mathcal{B}$. You may find it helpful to use a computer to find inverses and to multiply matrices.

(a) $\begin{bmatrix} 1 & 3 \\ -8 & 7 \end{bmatrix}$, $\mathcal{B} = \{(1, 2), (1, 4)\}$

(b) $\begin{bmatrix} 1 & -6 \\ -4 & -1 \end{bmatrix}$, $\mathcal{B} = \{(3, -2), (1, 1)\}$

(c) $\begin{bmatrix} 4 & -6 \\ 2 & 8 \end{bmatrix}$, $\mathcal{B} = \{(3, 1), (7, 3)\}$

(d) $\begin{bmatrix} 16 & -20 \\ 6 & -6 \end{bmatrix}$, $\mathcal{B} = \{(5, 3), (4, 2)\}$

(e) $\begin{bmatrix} 3 & 1 & 1 \\ 0 & 4 & 2 \\ 1 & -1 & 5 \end{bmatrix}$, $\mathcal{B} = \{(1, 1, 0), (0, 1, 1), (1, 0, 1)\}$

(f) $\begin{bmatrix} 4 & 1 & -3 \\ 16 & 4 & -18 \\ 6 & 1 & -5 \end{bmatrix}$, $\mathcal{B} = \{(1, 1, 1), (0, 3, 1), (1, 2, 1)\}$

B1 Determine the matrix of the linear mapping L with respect to the basis $\mathcal{B}$ in the following cases, and determine $(L(\mathbf{x}))_{\mathcal{B}}$ for the given $(\mathbf{x})_{\mathcal{B}}$.

(a) In $\mathbb{R}^2$, $\mathcal{B} = \{\mathbf{v}_1, \mathbf{v}_2\}$ and $L(\mathbf{v}_1) = \mathbf{v}_1 + 3\mathbf{v}_2$, $L(\mathbf{v}_2) = 5\mathbf{v}_1 - 7\mathbf{v}_2$; $(\mathbf{x})_{\mathcal{B}} = (4, -2)$

(b) In $\mathbb{R}^3$, $\mathcal{B} = \{\mathbf{v}_1, \mathbf{v}_2, \mathbf{v}_3\}$ and $L(\mathbf{v}_1) = 2\mathbf{v}_1 - 3\mathbf{v}_2$, $L(\mathbf{v}_2) = 3\mathbf{v}_1 + 4\mathbf{v}_2 - \mathbf{v}_3$, $L(\mathbf{v}_3) = -\mathbf{v}_1 + 2\mathbf{v}_2 + 6\mathbf{v}_3$; $(\mathbf{x})_{\mathcal{B}} = (5, -3, 1)$

B2 Consider the basis $\mathcal{B} = \{\mathbf{v}_1 = (1, 2), \mathbf{v}_2 = (1, -2)\}$ of $\mathbb{R}^2$. In each of the following cases, determine $[L(\mathbf{v}_1)]_{\mathcal{B}}$ and $[L(\mathbf{v}_2)]_{\mathcal{B}}$ **by inspection**, and hence, determine $[L]_{\mathcal{B}}$.

(a) $L(1, 2) = (1, -2)$ and $L(1, -2) = (4, 8)$

(b) $L(1, 2) = (5, 10)$ and $L(1, -2) = (-3, 6)$

B3 Consider the basis $\mathcal{B} = \{\mathbf{v}_1 = (1, 1, 0), \mathbf{v}_2 = (1, -1, 1), \mathbf{v}_3 = (3, 0, 1)\}$ of $\mathbb{R}^3$. In each of the following cases, determine $[L(\mathbf{v}_1)]_{\mathcal{B}}$, $[L(\mathbf{v}_2)]_{\mathcal{B}}$, and $[L(\mathbf{v}_3)]_{\mathcal{B}}$ **by inspection**, and hence, determine $[L]_{\mathcal{B}}$.

(a) $L(1, 1, 0) = (3, 0, 1)$, $L(1, -1, 1) = (1, 1, 0)$, $L(3, 0, 1) = (2, -2, 2)$

(b) $L(1, 1, 0) = (4, 4, 0)$, $L(1, -1, 1) = (6, 0, 2)$, $L(3, 0, 1) = (-1, 1, -1)$

B4 For each of the following linear transformations, determine a geometrically natural basis $\mathcal{B}$ (as in Examples 34 and 35), and determine (by inspection) the $\mathcal{B}$-matrix of the transformation.

(a) $\mathbf{perp}_{(3,2)}$ (b) $\mathbf{perp}_{(2,1,-2)}$ (c) $\mathbf{refl}_{(1,2,2)}$

B5 (a) Find the coordinates of $(5, 2, 1)$ with respect to the basis $\mathcal{B} = \{(1, 1, 0), (0, 1, 1), (1, 0, 1)\}$ in $\mathbb{R}^3$.
(b) Suppose that $L : \mathbb{R}^3 \to \mathbb{R}^3$ is a linear transformation such that $L(1, 1, 0) = (0, 5, 5)$, $L(0, 1, 1) = (2, 0, 2)$, $L(1, 0, 1) = (5, 2, 1)$. Without doing further row reduction, determine the $\mathcal{B}$-matrix of L.
(c) Determine $[L(5, 2, 1)]_{\mathcal{B}}$, and use it to determine $[L(5, 2, 1)]_{\mathcal{S}}$.

B6 (a) Find the coordinates of $(1, 4, 4)$ with respect to the basis $\mathcal{B} = \{(2, 1, 0), (-1, 0, 1), (1, 1, 0)\}$ in $\mathbb{R}^3$.
(b) Suppose that $L : \mathbb{R}^3 \to \mathbb{R}^3$ is a linear transformation such that $L(2, 1, 0) = (3, 3, 0)$, $L(-1, 0, 1) = (1, 4, 4)$, $L(1, 1, 0) = (-2, 0, 2)$. Without doing further row reduction, determine the $\mathcal{B}$-matrix of L.
(c) Determine $[L(1, 4, 4)]_{\mathcal{B}}$, and use it to determine $[L(1, 4, 4)]_{\mathcal{S}}$.

B7 Assume that each of the following matrices is the matrix of some linear transformation with respect to the standard basis. Determine the matrix of the linear transformation with respect to the given basis $\mathcal{B}$. You may find it helpful to use a computer to find inverses and to multiply matrices.

(a) $\begin{bmatrix} -3 & 1 \\ -16 & 5 \end{bmatrix}$, $\mathcal{B} = \{(1, 4), (1, 5)\}$

(b) $\begin{bmatrix} 6 & -10 \\ 2 & -6 \end{bmatrix}$, $\mathcal{B} = \{(5, 1), (1, 1)\}$

(c) $\begin{bmatrix} -6 & -2 & 9 \\ -5 & -1 & 7 \\ -7 & -2 & 10 \end{bmatrix}$, $\mathcal{B} = \{(1, 1, 1), (1, 0, 1), (1, 3, 2)\}$

(d) $\begin{bmatrix} -7 & -3 & 3 \\ 2 & 2 & -1 \\ -16 & -6 & 7 \end{bmatrix}$, $\mathcal{B} = \{(1, 0, 2), (0, 1, 1), (1, -1, 2)\}$

B8 Assume that each of the following matrices is the matrix of some linear transformation with respect to the standard basis. Determine the matrix of the linear transformation with respect to the given basis $\mathcal{B}$. You may find it helpful to use a computer to find inverses and to multiply matrices.

(a) $\begin{bmatrix} 12 & -15 \\ -16 & -7 \end{bmatrix}$, $\mathcal{B} = \{(5, 3), (3, 2)\}$

(b) $\begin{bmatrix} 6 & -2 \\ 36 & -7 \end{bmatrix}$, $\mathcal{B} = \{(1, 5), (1, 2)\}$

(c) $\begin{bmatrix} 10 & -20 & -24 \\ 5 & -10 & -15 \\ -2 & 4 & 8 \end{bmatrix}$, $\mathcal{B} = \{(2, 1, 0), (1, -1, 1), (1, 1, -1)\}$

(d) $\begin{bmatrix} 3 & 6 & 1 \\ 5 & -4 & 5 \\ 3 & -6 & 5 \end{bmatrix}$, $\mathcal{B} = \{(1, 0, -1), (-1, 1, 1), (2, 1, 0)\}$

CONCEPTUAL EXERCISES

D1 When can a 2 by 2 matrix be diagonalized? Suppose A is a 2 by 2 matrix; interpret it as the matrix of a linear transformation. What conditions will have to be satisfied by the vectors $\mathbf{v}_1$ and $\mathbf{v}_2$ of a basis $\mathcal{B}$ in order for $P^{-1}AP = \begin{bmatrix} d_1 & 0 \\ 0 & d_2 \end{bmatrix} = D$, for some real numbers d_1 and d_2? (Hint: Consider the equation $AP = PD$, or $A\big[[\mathbf{v}_1]\ [\mathbf{v}_2]\big] = \big[[\mathbf{v}_1]\ [\mathbf{v}_2]\big]D$.)

***D2** Suppose that P is the change of basis matrix from the standard basis $\mathcal{S}$ to basis $\mathcal{B}$, and that Q is the change of basis matrix from standard basis $\mathcal{S}$ to basis $\mathcal{C}$. Express the matrix $[L]_{\mathcal{C}}$ in terms of $[L]_{\mathcal{B}}$, P, and Q.

D3 Let $D : P_3 \to P_3$ be differentiation of polynomials as in Section 4–1, Example 3.

(a) Determine the matrix of D with respect to the standard basis $\{1, x, x^2, x^3\}$ of P_3.

(b) It might be more natural to think of D as a map from P_3 to P_2. Since the codomain is now a different space from the domain, we will have to introduce one basis in the domain and one in the codomain. Using the standard basis in each case, determine the matrix of D (it is 3 by 4).

D4 (This is a sequel to D3.) Give a general prescription for finding the matrix of a linear transformation $L : \mathbb{V} \to \mathbb{W}$.

SECTION

4–7 Isomorphism of Vector Spaces

Some of the ideas of Chapters 3 and 4 lead to generalizations that are important in the further development of linear algebra (and also in abstract algebra). In this section, some of these generalizations are outlined. Most of the proofs are easy or simple variations on proofs given earlier, so they will be left as exercises. Throughout this section, it will be assumed that $\mathbb{U}$, $\mathbb{V}$, and $\mathbb{W}$ are vector spaces over the real numbers and that $L : \mathbb{U} \to \mathbb{V}$ and $M : \mathbb{V} \to \mathbb{W}$ are linear mappings.

DEFINITION

L is a **one-to-one** mapping if for each $\mathbf{v}$ in the range of L, there is exactly one $\mathbf{u}$ in $\mathbb{U}$ such that $L(\mathbf{u}) = \mathbf{v}$.

LEMMA 1. *L* is one-to-one if and only if the nullspace of *L* is trivial, that is, the nullspace consists only of the zero vector. (Compare Section 3–5, Theorem 5 statement (8).)

EXERCISE D1 Prove Lemma 1. [Hint: Suppose that L is one-to-one. What is the unique $\mathbf{u}$ such that $L(\mathbf{u}) = \mathbf{0}$? Conversely, suppose that the nullspace is trivial. If $L(\mathbf{u}_1) = L(\mathbf{u}_2)$, then $L(\mathbf{u}_1 - \mathbf{u}_2) = ?$]

EXAMPLES

Every invertible linear transformation from $\mathbb{R}^n$ to $\mathbb{R}^n$ is one-to-one; it is the fact that such a transformation is one-to-one that allows the definition of the inverse. The mapping *inj* of Example 43 of Section 3–5 is a one-to-one mapping that is not invertible. For any **n**, $\mathbf{proj_n} : \mathbb{R}^3 \to \mathbb{R}^3$ is **not** one-to-one, since many elements in the domain are mapped to the same vector in the range; the mapping P of Example 43 of Section 3–5 also fails to be one-to-one. (Such mappings are sometimes said to be "many-to-one".)

EXERCISE D2 Suppose that $\{\mathbf{u}_1, \mathbf{u}_2, \ldots, \mathbf{u}_n\}$ is a linearly independent set in U and that L is one-to-one. Prove that $\{L(\mathbf{u}_1), L(\mathbf{u}_2), \ldots, L(\mathbf{u}_n)\}$ is linearly independent.

EXERCISE D3 (a) Prove that if L and M are one-to-one, then $M \circ L$ is one-to-one.

(b) Give an example where M is not one-to-one, but $M \circ L$ is one-to-one.

(c) Is it possible to give an example where L is not one-to-one but $M \circ L$ is one-to-one?

$L : \mathrm{U} \to \mathrm{V}$ is **onto** if every $\mathbf{v}$ in V is the image of some $\mathbf{u}$ in U under the linear mapping L; that is, for each $\mathbf{v}$ in V, there is some $\mathbf{u}$ in U such that $L(\mathbf{u}) = \mathbf{v}$. (Note that "onto" is used as an adjective here.)

LEMMA 2. *L* is onto if and only if the range of *L* is V, the codomain of the mapping *L*. (Compare statement (7) of Theorem 5, Section 3–5.)

Proof. Restatement of the definition. ■

EXAMPLES

Invertible linear transformations of $\mathbb{R}^n$ are all onto mappings. The mapping P of Example 43, Section 3–5 is an onto mapping. Every linear mapping maps **onto its range.**

EXERCISE D4 Suppose that $\{\mathbf{u}_1, \mathbf{u}_2, \ldots, \mathbf{u}_n\}$ is a spanning set for U and that $L : \mathrm{U} \to \mathrm{V}$ is an onto linear mapping. Show that $\{L(\mathbf{u}_1), L(\mathbf{u}_2), \ldots, L(\mathbf{u}_n)\}$ is a spanning set for V.

EXERCISE D5 Prove that if L and M are onto, then $M \circ L$ is onto.

THEOREM 1. The linear mapping $L : \mathrm{U} \to \mathrm{V}$ has an inverse linear mapping $L^{-1} : \mathrm{V} \to \mathrm{U}$ if and only if L is one-to-one and onto.

EXERCISE D6 Prove Theorem 1.

DEFINITION

If U and V are vector spaces over the real numbers, and $L : \mathrm{U} \to \mathrm{V}$ is a one-to-one linear mapping from U onto V, L is called an **isomorphism** (or a vector space isomorphism), and U is said to be **isomorphic** to V.

By Theorem 1, the isomorphism L has an inverse and V is isomorphic to U, so we often say that "U and V are isomorphic".

The word "isomorphism" comes from Greek words meaning "same form". The concept of isomorphism is a very powerful and important one. It implies that the essential structure of the vector spaces is the same, so that a vector space statement that is true in one space is immediately true in any isomorphic space. (Of course, some vector spaces such as the space of m by n matrices or spaces of polynomials have some features that are not purely vector space properties, and these particular features cannot automatically be transferred from these spaces to spaces that are isomorphic as vector spaces.)

EXERCISE D7 Use Exercises D2 and D4 to show that if $L : \mathrm{U} \to \mathrm{V}$ is an isomorphism and $\{\mathbf{u}_1, \mathbf{u}_2, \ldots, \mathbf{u}_n\}$ is a basis for U, then $\{L(\mathbf{u}_1), L(\mathbf{u}_2), \ldots, L(\mathbf{u}_n)\}$ is a basis for V.

THEOREM 2. Suppose that U and V are finite-dimensional vector spaces over the real numbers. U and V are isomorphic if and only if they are of the same dimension.

EXAMPLES

(a) Every plane in $\mathbb{R}^3$ is isomorphic to $\mathbb{R}^2$.

(b) The space $\mathcal{M}(m, n)$ of m by n matrices with real entries is isomorphic to $\mathbb{R}^{mn}$.

(c) The space P_n of polynomials of degree less than or equal to n is isomorphic to $\mathbb{R}^{n+1}$.

EXERCISE D8 Prove Theorem 2. To prove "isomorphic $\Rightarrow$ same dimension", use Exercise D7; to prove "same dimension $\Rightarrow$ isomorphic", take a basis in each space and define an isomorphism by taking $L(\mathbf{u}_j) = \mathbf{v}_j$ for each j and requiring that L be linear (you must make sure that this is an isomorphism).

REMARK

Theorem 2 is sometimes restated as follows: over the real numbers, every n-dimensional vector space is isomorphic to $\mathbb{R}^n$, or *up to isomorphism, there is only one vector space of dimension n over the real numbers.* This fact explains why, for many users, it is sufficient to develop the theory of linear algebra in $\mathbb{R}^n$.

THEOREM 3. If U and V are real vector spaces of dimension n, then a linear mapping $L : \mathrm{U} \to \mathrm{V}$ is one-to-one if and only if it is onto. (Compare Theorem 5 in Section 3–5, also Theorem 2 in Section 4–4.)

EXERCISE D9 Prove Theorem 3.

EXERCISES 4–7

A1 For each of the following pairs of vector spaces, define an explicit isomorphism to establish that the spaces are isomorphic. Indicate clearly why the map is an isomorphism — is it linear? one-to-one? onto?

(a) P_3 and $\mathbb{R}^4$

(b) $\mathcal{M}(2, 2)$ and $\mathbb{R}^4$

(c) P_3 and $\mathcal{M}(2, 2)$

B1 For each of the following pairs of vector spaces, define an explicit isomorphism to establish that the spaces are isomorphic. Indicate clearly why the map is an isomorphism —is it linear? one-to-one? onto?
(a) P_5 and $\mathbb{R}^6$
(b) $\mathcal{M}(2,3)$ and $\mathbb{R}^6$
(c) P_5 and $\mathcal{M}(2,3)$

CONCEPTUAL EXERCISES

Exercises D1–D9 appear throughout this section.

***D10** Prove that any plane through the origin in $\mathbb{R}^3$ is isomorphic to $\mathbb{R}^2$.

D11 Recall the definition of the Cartesian product from Exercise 4–1–D4. Prove that $\mathrm{U} \times \{\mathbf{0}_\mathrm{V}\}$ is a subspace of $\mathrm{U} \times \mathrm{V}$ that is isomorphic to U.

***D12** (a) Prove that $\mathbb{R}^2 \times \mathbb{R}$ is isomorphic to $\mathbb{R}^3$.
(b) Prove that $\mathbb{R}^n \times \mathbb{R}^m$ is isomorphic to $\mathbb{R}^{n+m}$.

D13 Suppose that $L : \mathrm{U} \to \mathrm{V}$ is a vector space isomorphism and that $M : \mathrm{V} \to \mathrm{V}$ is a linear mapping. Prove that $L^{-1} \circ M \circ L$ is a linear mapping from U to U, and describe the nullspace and range of $L^{-1} \circ M \circ L$ in terms of the nullspace and range of M.

SECTION 4–8 Vector Spaces Over the Complex Numbers

The definition of a vector space in Section 4–1 is given in the case where the scalars are the real numbers $\mathbb{R}$. In fact, the definition makes sense when the scalar multipliers are taken from any one system of numbers such that addition, subtraction, multiplication, and division are defined for any pair of numbers (excluding division by 0), and satisfy the usual commutative, associative, and distributive rules for doing arithmetic. Thus, the vector space axioms make sense if we allow the scalar multipliers to be the set of complex numbers C. (The complex numbers are reviewed in the Appendix.) In such cases, we say that we have a *vector space over the complex numbers* or a *complex vector space.*

EXAMPLE 41

Let C^2 denote the set of pairs of complex numbers:

$$\mathrm{C}^2 = \{ (z_1, z_2) | z_1, z_2 \in \mathrm{C} \},$$

with addition of elements of $\mathbf{C}^2$ defined (as we might expect) by

$$(z_1, z_2) + (w_1, w_2) = (z_1 + w_1, z_2 + w_2),$$

and multiplication by a *complex* scalar α by

$$\alpha(z_1, z_2) = (\alpha z_1, \alpha z_2).$$

It is easy to check that $\mathbf{C}^2$ satisfies all the vector space axioms, provided that we revise these axioms to allow complex scalars. Thus, $\mathbf{C}^2$ is a *vector space over the complex numbers.* We have used a Greek letter to denote the complex scalar because in some cases we shall want to make a distinction between a complex scalar and a real scalar.

It is instructive to look carefully at the ideas of basis and dimension for complex vector spaces. We begin by considering the set of complex numbers $\mathbf{C}$ itself as a vector space.

EXAMPLE 42

The vector space C

As a vector space over the complex numbers, $\mathbf{C}$ has a basis consisting of a single element: $\{1\}$. Every complex number can be written in the form

$$z = z\,1,$$

where z is a complex number. Thus, with respect to this basis, the coordinate of the complex number z is z itself. Alternatively, we could choose to use the basis $\{i\}$ consisting of the single complex number i. Then an arbitrary element z of the vector space $\mathbf{C}$ must be written in the form

$$z = (-i\,z)\,i.$$

In either case, we see that $\mathbf{C}$ has a basis consisting of one element, so $\mathbf{C}$ is a one-dimensional complex vector space.

Another way of looking at this is to observe that *when we use complex scalars* any two non-zero elements of the space $\mathbf{C}$ are linearly dependent. That is, given z_1 and z_2 in $\mathbf{C}$, there exist complex scalars α_1 and α_2, not both zero, such that $\alpha_1 z_1 + \alpha_2 z_2 = 0$. For example, we may take $\alpha_1 = 1$, $\alpha_2 = -z_1/z_2$, since we have assumed that $z_2 \neq 0$. It follows that with respect to complex scalars, a basis for $\mathbf{C}$ must have dimension less than two.

However, we could also view C as a vector space over the *real* scalars. Addition of complex numbers is defined as usual, and multiplication of $z = x + iy = x1 + yi$ by a *real* scalar k gives

$$kz = kx\,1 + ky\,i.$$

If we use real scalars, then the elements 1 and i in C are linearly *independent* because the only way to get $k_1 1 + k_2\, i = 0$ with k_1 and k_2 real is to take $k_1 = 0 = k_2$. We know that any complex number can be written as a *real* linear combination of 1 and i :

$$z = x\,1 + y\,i, \quad \text{where} \quad x, y \in \mathbb{R}.$$

Thus, viewed as a vector space over the real scalars, the set of complex numbers is two-dimensional with "standard" basis $\{1, i\}$.

We sometimes write complex numbers in a way that exhibits the property that C is a two-dimensional real vector space: we write the complex number z as an "ordered pair" in the form

$$z = x + i\,y = (x, y) = x(1, 0) + y(0, 1).$$

Note that $(1, 0)$ denotes the *complex number* 1 and that $(0, 1)$ denotes the *complex number* i. With this notation, *as a vector space over the reals*, the set of complex numbers looks just like the vector space $\mathbb{R}^2$, and we speak of the "complex plane". Notice that this representation of the complex numbers as a real vector space does not yet include multiplication by *complex* scalars. We shall describe multiplication by a complex scalar as a linear mapping of the two-dimensional real vector space below.

By arguments similar to those in Example 42, we see that C^2 is a two-dimensional vector space over the complex numbers. However, we may view the set of vectors in C^2 as a *real* vector space of dimension *four.*

The vector space C^n is defined to be the set

$$\{(z_1, z_2, \ldots, z_n) \mid z_k \in \mathrm{C},\ k = 1, 2, \ldots, n\},$$

with addition of vectors and multiplication by complex scalars defined similarly to the definitions in Example 41. These vector spaces play an important role in much modern mathematics. Vector spaces of functions over the complex numbers can also be defined; they are important in transform methods for solving differential equations.

Linear Mappings and Subspaces

The concept of a linear mapping $L : \mathrm{V} \to \mathrm{W}$ makes sense if the vector spaces V and W are both vector spaces over the complex numbers. Then,

$$L(\alpha \mathbf{v}_1 + \beta \mathbf{v}_2) = \alpha L(\mathbf{v}_1) + \beta L(\mathbf{v}_2)$$

makes sense, because on both sides of the equation we are using complex scalars. We say in this case that "L is linear over the complexes".

We can also define subspaces just as we did in Section 4–1, and the range and nullspace of a linear mapping will be subspaces of the appropriate vector spaces, as before.

EXAMPLE 43

Let $L : \mathbb{C}^3 \to \mathbb{C}^2$ be the linear mapping such that

$$L(1,0,0) = (1+i, 2), \quad L(0,1,0) = (-2i, 1-i), \quad L(0,0,1) = (1+2i, 3+i).$$

Then, by arguments exactly like those in Sections 3–2 and 4–6, the standard matrix of L is

$$[L] = A = \begin{bmatrix} 1+i & -2i & 1+2i \\ 2 & 1-i & 3+i \end{bmatrix}.$$

To determine the image of some vector, say for example $L(1, 2i, 1-i)$, we calculate

$$\begin{bmatrix} 1+i & -2i & 1+2i \\ 2 & 1-i & 3+i \end{bmatrix} \begin{bmatrix} 1 \\ 2i \\ 1-i \end{bmatrix} = \begin{bmatrix} 8+2i \\ 8 \end{bmatrix}.$$

The range of L is the subspace of the codomain $\mathbb{C}^2$ spanned by the columns of A, and the nullspace of L is the solution space of the system of linear equations $A\mathbf{z} = \mathbf{0}$, just as in Section 3–4.

Complex Multiplication as a Matrix Mapping

In Example 42, we saw that the set of complex numbers C could be regarded as a *real* two-dimensional vector space. In this picture, how can we represent multiplication by a complex number? We first consider a special case.

EXAMPLE 44

Muliplication by i

As in Example 42, we write

$$z = x + iy = (x, y)$$

so that we may regard $\mathbb{C}$ as a real vector space. Let us consider multiplication by i :

$$iz = i(x + iy) = ix - y = (-y, x).$$

It is easy to see that this corresponds to a linear mapping:

$$M_i : \mathbb{R}^2 \to \mathbb{R}^2 \quad \text{defined by} \quad M_i(x, y) = (-y, x).$$

The standard matrix of M_i is $\begin{bmatrix} 0 & -1 \\ 1 & 0 \end{bmatrix}$. Thus, multiplication of z by i corresponds to multiplication of the column vector $\begin{bmatrix} x \\ y \end{bmatrix}$ by the matrix as follows:

$$iz \longleftrightarrow \begin{bmatrix} 0 & -1 \\ 1 & 0 \end{bmatrix} \begin{bmatrix} x \\ y \end{bmatrix}.$$

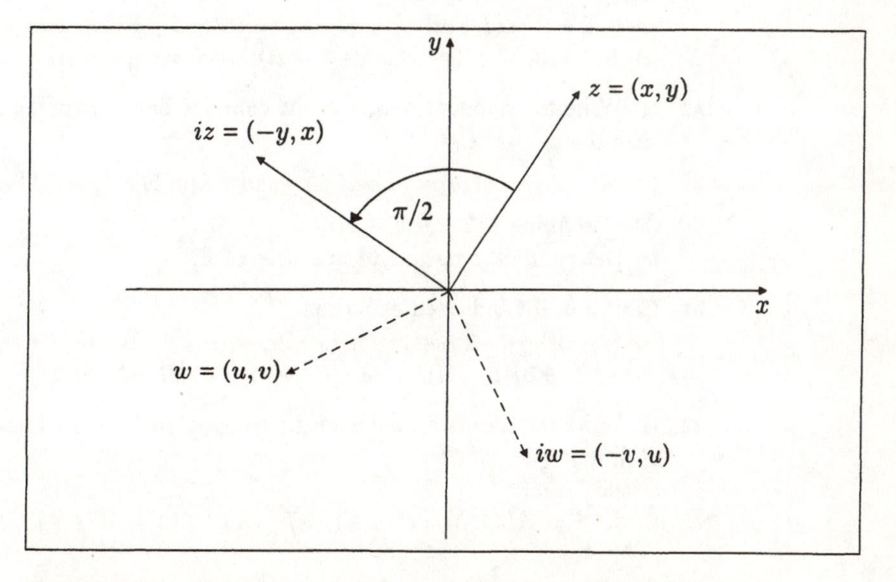

FIGURE 4-8-1. *Multiplication of $\mathbb{C}$ by i corresponds to rotation by angle $\frac{\pi}{2}$.*

By considering rotations of the plane (Section 3–3), we see that multiplication by i corresponds to rotation of the complex plane by an angle of $\frac{\pi}{2}$ (Figure 4–8–1). In the language of complex numbers, we can also describe this mapping as "increasing the argument of each complex number by $\frac{\pi}{2}$". This is very useful in applications to electrical circuits, where the argument of a complex number may be interpreted as the "phase" of a current or voltage, so multiplication by i corresponds to "advancing the phase by $\frac{\pi}{2}$". (See Section 2–4.)

More generally, we may consider multiplication of complex numbers by an arbitrary complex number $\alpha = a + ib$.

CLAIM Multiplication of numbers in $\mathbb{C}$ by the complex number $\alpha = a + ib$ may be represented as a linear mapping of $\mathbb{R}^2$ as follows:

$$\alpha z \longleftrightarrow \begin{bmatrix} a & -b \\ b & a \end{bmatrix} \begin{bmatrix} x \\ y \end{bmatrix}.$$

The verification of this claim is left as Exercise D1.

EXERCISES 4–8

A1 Carry out the indicated operation.
(a) $(-2 + i) - (3 + 4i)$ (b) $(2 - i, 3 + i, 2 - 5i) + (3 - 2i, 4 + 7i, -3 - 4i)$
(c) $2i(2 + 5i, 3 - 2i)$ (d) $(-1 - 2i)(2 - i, 3 + i, 2 - 5i)$

A2 (a) Write the standard matrix of the complex linear mapping $L : \mathbb{C}^2 \to \mathbb{C}^2$ such that

$$L(1,0) = (1 + 2i, 1) \quad \text{and} \quad L(0,1) = (3 + i, 1 - i).$$

(b) Determine $L(2 + 3i, 1 - 4i)$.
(c) Determine the range and nullspace of L.

B1 Carry out the indicated operation.
(a) $(4 - 3i) - (-2 - 4i)$ (b) $(3 + 2i, -2 - i, 1 + 3i) - (2 + i, 3 + i, -4i)$
(c) $-3i(1 + 3i, 5 - 3i)$ (d) $(-1 - i)(i, 2 + 3i, -2 - 7i)$

B2 (a) Write the standard matrix of the complex linear mapping $L : \mathbb{C}^2 \to \mathbb{C}^2$ such that

$$L(1,0) = (-i, -1 + i) \quad \text{and} \quad L(0,1) = (1 + i, -2i).$$

(b) Determine $L(2 - i, -4 + i)$.
(c) Determine the range and nullspace of L.

CONCEPTUAL EXERCISES

*D1 (a) Check the correctness of the Claim at the end of this section.
(b) Interpret multiplication by an arbitrary complex number as a composition of a contraction or dilatation, and a rotation in the plane $\mathbb{R}^2$.
(c) Check the correctness of the Claim when $\alpha = 3 - 4i$, and interpret as in part (b).

D2 Verify that for a general 2 by 2 matrix A with complex entries,
$A = \begin{bmatrix} a_{11} & a_{12} \\ a_{21} & a_{22} \end{bmatrix}$, and vector $\mathbf{z} = (z_1, z_2)$,

$$\overline{A\mathbf{z}} = \overline{A}\,\overline{\mathbf{z}}.$$

D3 Show that the space of 2 by 2 matrices with complex entries, with the usual rules for addition and multiplication by scalars is isomorphic to $\mathbb{C}^4$.

D4 Define isomorphism of complex vector spaces, and check that the arguments and results of Section 4–7 are correct, provided that the scalars are always taken to be complex numbers.

Chapter Review

Suggestions for Student Review

Remember that if you have understood the ideas of Chapter 4 you should be able to give answers to these questions without looking them up. Try hard to answer them from your own understanding.

1 State the essential properties of a vector space with real scalars. Why is the empty set not a vector space? Describe two or three examples of vector spaces that are not subspaces of $\mathbb{R}^n$ and linear mappings defined with these vector spaces as their domains. (Section 4–1)

2 State the formal definition of linear independence. Explain the connection between the formal definition of linear dependence and an intuitive geometric understanding of linear dependence. What is a basis? Why is linear independence of a spanning set important when we define coordinates with respect to the spanning set? (Sections 4–2 and 4–3)

3 Invent and analyze an example as follows.
(a) Give three vectors in $\mathbb{R}^5$ that are linearly independent. (Don't make it too easy by choosing the three to be standard basis vectors, but don't make it too hard by choosing completely random components.) (Sections 4–2 and 4–3)
(b) Determine the standard coordinates in $\mathbb{R}^5$ of the vector that has coordinates $(2, -3, 4)$ with respect to the basis consisting of these three vectors.
(c) Take the vector you found in part (b) and carry out the standard proce-

dure to determine its coordinates with respect to the basis of three vectors. Did you get the right answer $(2, -3, 4)$?

(d) Pick any two vectors in $\mathbb{R}^5$ and determine whether they lie in the subspace spanned by your basis of three vectors.

4 Write a short explanation of how you use information about consistency of systems and uniqueness of solutions in testing for linear independence and in determining whether a vector belongs to a given subspace. (Sections 4–2 and 4–3)

5 (a) Imagine that you are returning to your high school and are supposed to give a seminar to the senior mathematics class. Explain the concept of dimension, with liberal examples using lines and planes in $\mathbb{R}^3$. (Section 4–4)

(b) Explain how knowing the dimension of a space (or subspace) is helpful when you have to find a basis for the space. (4–4, Theorem 2 and Corollary)

6 (a) How many ways can you give for recognizing the rank of a matrix or linear mapping? State them all. (Section 4–5)

(b) State the connection between the rank of a matrix A and the dimension of the solution space of $A\mathbf{x} = \mathbf{0}$.

(c) Illustrate your answers to (a) and (b) by constructing examples of 4 by 5 matrices in row echelon form of (i) rank 4; (ii) rank 3; (iii) rank 2. In each case, actually determine the general solution of the system $A\mathbf{x} = \mathbf{0}$ and check that the solution space has the correct dimension.

7 State how to determine the standard matrix and the $\mathcal{B}$-matrix of a linear mapping $L : \mathbb{R}^3 \to \mathbb{R}^3$, and explain how $[L(\mathbf{x})]_\mathcal{B}$ is determined in terms of $[L]_\mathcal{B}$. Explain how the $\mathcal{B}$-coordinates of a vector are determined, and show how this leads to the change of basis matrix P. Use the answers to these questions to express $[L]_\mathcal{B}$ in terms of $[L]_\mathcal{S}$ and P. (Section 4–6)

8 State the definition of isomorphism of vector spaces, and give some examples. Explain why a finite-dimensional vector space cannot be isomorphic to a *proper* subspace of itself. (A proper subspace of $\mathbb{V}$ is a subspace that is not equal to $\mathbb{V}$ itself.) (Section 4–7)

Chapter Quiz

1 Determine whether the following sets are vector spaces; explain briefly.

(a) The set of 4 by 3 matrices such that the sum of the entries in the first row is zero. $(a_{11} + a_{12} + a_{13} = 0.)$

(b) The set of differentiable functions $\mathbb{R} \to \mathbb{R}$ such that $Df(1) = 0$, $Df(2) = 0$.

(c) The set of 2 by 2 matrices such that all entries are integers.

2 In each of the following cases, determine whether the given set of vectors is linearly independent.

(a) $\{(1, 1, 2, 1), (0, 1, 1, -1), (0, 1, 1, 3), (2, 2, 4, -2), (0, 2, 3, 0)\}$

(b) $\{(1, 1, 2, 1), (0, 1, 1, -1), (0, 1, 1, 3), (2, 2, 4, -2)\}$

(c) $\{(1,1,2,1),(0,1,1,-1),(0,1,1,3)\}$

3 (a) Determine the dimension of the subspace $\mathbb{S}$ of $\mathbb{R}^5$ spanned by the vectors $\mathbf{v}_1 = (1,0,1,1,3), \mathbf{v}_2 = (1,1,0,1,1), \mathbf{v}_3 = (3,3,1,0,2), \mathbf{v}_4 = (1,1,1,-2,0)$.
(b) Pick a subset of the given spanning set for $\mathbb{S}$ which forms a basis $\mathcal{B}$ for $\mathbb{S}$, and determine the coordinates of $(0,2,-1,-3,-5)$ with respect to this basis.

4 You are given the matrix A below and a row echelon form corresponding to A. Determine a basis for the range of the linear mapping corresponding to A and a basis for the nullspace of the linear mapping. Explain briefly.

$$A = \begin{bmatrix} 1 & 0 & 1 & 1 & 1 \\ 2 & 1 & 1 & 2 & 5 \\ 0 & 2 & -2 & 1 & 8 \\ 3 & 3 & 0 & 4 & 14 \end{bmatrix} \rightarrow \begin{bmatrix} 1 & 0 & 1 & 1 & 1 \\ 0 & 1 & -1 & 0 & 3 \\ 0 & 0 & 0 & 1 & 2 \\ 0 & 0 & 0 & 0 & 0 \end{bmatrix}$$

5 Let $L : \mathbb{R}^3 \rightarrow \mathbb{R}^3$ be a linear mapping with standard matrix $\begin{bmatrix} 1 & -1 & 2 \\ -1 & 0 & 1 \\ -2 & 1 & 0 \end{bmatrix}$ and let $\mathcal{B}$ be the basis $\{(1,1,0),(0,1,1),(1,-1,1)\}$. Determine the matrix $[L]_{\mathcal{B}}$.

6 Let $L : \mathbb{R}^3 \rightarrow \mathbb{R}^3$ be reflection in the plane with equation $x_1 - x_3 = 0$. Choose a basis $\mathcal{B}$ consisting of a normal to the plane and two vectors lying in the plane. Use geometric arguments to determine $[L]_{\mathcal{B}}$, and then by a change of basis, determine the standard matrix of L.

7 Suppose that $L : \mathbb{R}^n \rightarrow \mathbb{R}^m$ is a linear mapping with nullspace equal to $\{\mathbf{0}\}$, and suppose that $\{\mathbf{v}_1, \mathbf{v}_2, \dots, \mathbf{v}_k\}$ is a linearly independent set in $\mathbb{R}^n$. Prove that $\{L(\mathbf{v}_1), L(\mathbf{v}_2), \dots, L(\mathbf{v}_k)\}$ is linearly independent.

8 For each of the following statements decide whether it is true or false. If it is true, explain *briefly*; if it is false, give an example to show that it is false.
(a) A subspace of $\mathbb{R}^n$ must have dimension less than n.
(b) A set of 4 vectors in $\mathbb{R}^3$ must be linearly dependent.
(c) If $\{\mathbf{v}_1, \mathbf{v}_2, \dots, \mathbf{v}_k\}$ is a linearly independent set of vectors in $\mathbb{R}^n$, then the set $\{\mathbf{v}_1, \mathbf{v}_2, \mathbf{v}_3\}$ is linearly independent.
(d) If $\mathcal{B}$ is a basis for a subspace of $\mathbb{R}^5$, the $\mathcal{B}$-coordinate vector of some vector $\mathbf{x}$ in $\mathbb{R}^5$ has five components.
(e) For any linear mapping $L : \mathbb{R}^n \rightarrow \mathbb{R}^n$ and any basis $\mathcal{B}$ of $\mathbb{R}^n$, the rank of the matrix $[L]_{\mathcal{B}}$ is the same as the rank of the matrix $[L]_{\mathcal{S}}$.

Further Exercises

F1 MAGIC SQUARES—AN EXPLORATION OF THEIR VECTOR SPACE PROPERTIES

Let $\mathcal{M}(3,3)$ denote the set of all 3 by 3 matrices. If A and B are elements of $\mathcal{M}(3,3)$, we define their sum by $(A+B)_{jk} = a_{jk} + b_{jk}$; and for c in $\mathbb{R}$, we define scalar multiplication by $(cA)_{jk} = ca_{jk}$. With these definitions, $\mathcal{M}(3,3)$ is a vector space over the reals (see Section 4–1). We consider the three **row sums** of a 3 by 3 matrix A (each row sum is the sum of the entries in one row of A), the three **column sums** of A, and the **two diagonal sums** of A ($a_{11} + a_{22} + a_{33}$ and $a_{13} + a_{22} + a_{31}$).

We define an element A in $\mathcal{M}(3,3)$ to be a 3 by 3 **magic square** if all row sums, all column sums, and both diagonal sums have the same value k; the common sum k is called the **weight** of the magic square A and denoted by $wt(A) = k$.

For example, if $A = \begin{bmatrix} 2 & 2 & -1 \\ -2 & 1 & 4 \\ 3 & 0 & 0 \end{bmatrix}$, A is a magic square, with $wt(A) = 3$.

The aim of this exploration is to find all 3 by 3 magic squares. The subset of $\mathcal{M}(3,3)$ consisting of magic squares is denoted by MS_3.

(a) Show that MS_3 is a subspace of $\mathcal{M}(3,3)$.

(b) Observe that **weight** determines a map $wt\colon MS_3 \to \mathbb{R}$. Show that wt is a linear mapping.

(c) Compute the nullspace of wt: suppose that

$$\underline{X}_1 = \begin{bmatrix} 1 & 0 & a \\ b & c & d \\ e & f & g \end{bmatrix}, \quad \underline{X}_2 = \begin{bmatrix} 0 & 1 & h \\ i & j & k \\ l & m & n \end{bmatrix},$$

and

$$\underline{0} = \begin{bmatrix} 0 & o & p \\ q & r & s \\ t & u & v \end{bmatrix}$$

are all in the nullspace, where $a, b, c, \ldots$ denote unknown entries. Determine these unknown entries, and prove that $\underline{X}_1$ and $\underline{X}_2$ form a basis for the nullspace of wt. [Hint: If A is in the nullspace, consider $A - a_{11}\underline{X}_1 - a_{12}\underline{X}_2$.]

(d) If $\underline{J} = \begin{bmatrix} 1 & 1 & 1 \\ 1 & 1 & 1 \\ 1 & 1 & 1 \end{bmatrix}$, show that $\underline{J}$ is a magic square, and compute $wt(\underline{J})$.

Show that all A in MS_3 that have weight k are of the form

$$(k/3)\underline{J} + p\underline{X}_1 + q\underline{X}_2 \text{ for some } p, q \text{ in } \mathbb{R}.$$

(e) Show that $\underline{J}$, $\underline{X}_1$, and $\underline{X}_2$ form a basis for MS_3.

(f) As an example, find all 3 by 3 magic squares of weight 1.

(g) Find the coordinates of $A = \begin{bmatrix} 3 & 1 & 2 \\ 1 & 2 & 3 \\ 2 & 3 & 1 \end{bmatrix}$ with respect to the basis $\mathcal{B} = \{\underline{J}, \underline{X}_1, \underline{X}_2\}$ of MS_3. [Answer: $(2, 1, -1)$]

Conclusion: MS_3 is a three-dimensional subspace of the nine-dimensional vector space $\mathcal{M}(3,3)$ of real 3 by 3 matrices.

Exercises F2–F5 require the following definitions.
If S and T are subspaces of the vector space V, we define

$$\mathrm{S} + \mathrm{T} = \{p\mathbf{s} + q\mathbf{t} \mid p, q \in \mathbb{R}, \mathbf{s} \in \mathrm{S}, \mathbf{t} \in \mathrm{T}\}$$

(that is, the subspace of V spanned by the vectors of S and T).
If S and T are subspaces of the vector space V such that $\mathrm{S} + \mathrm{T} = \mathrm{V}$ and $\mathrm{S} \cap \mathrm{T} = \{\mathbf{0}\}$, we say that S is a *complement* of T (and T is a *complement* of S). In general, given a subspace S of V, one can choose a complement in many ways: the complement of S is not unique. For example, in $\mathbb{R}^2$, we may take a complement of Sp$(\{(1,0)\})$, to be Sp$(\{(0,1)\})$, or Sp$(\{(1,1)\})$. (In Exercise D2 of Section 3–4, we met the special case of the *orthogonal* complement, which is unique. In our present exercises, we do not assume that a dot product is defined.)

F2 In the vector space of continuous real-valued functions of a real variable, show that the even functions and the odd functions form subspaces such that each is the complement of the other. (A function is **even** if $f(-x) = f(x)$ for all x and **odd** if $f(-x) = -f(x)$ for all x.)

F3 (a) If S is a k-dimensional subspace of $\mathbb{R}^n$, show that any complement of S must be of dimension $n - k$.
(b) Suppose that S is a subspace of $\mathbb{R}^n$, that has a unique complement. Must it be true that S is either $\{\mathbf{0}\}$ or $\mathbb{R}^n$?

F4 Suppose that **a** and **b** are vectors in a vector space V and that S is a subspace of V. Let T be the subspace of V spanned by **a** and S and let U be the subspace spanned by **b** and S. Prove that if **b** is in T but not in S, then **a** is in U.

F5 Show that if S and T are finite-dimensional subspaces of V, then

$$\dim \mathrm{S} + \dim \mathrm{T} = \dim(\mathrm{S} + \mathrm{T}) + \dim(\mathrm{S} \cap \mathrm{T}),$$

where dim S denotes the dimension of S.

F6 Use the ideas of Section 4–4 to prove the uniqueness of the reduced row echelon form for a given matrix A (that is, given A, there is only one matrix in reduced row echelon form that is row equivalent to A). Hint: Begin by assuming that there are two such matrices R and S. What can you say about the columns with leading 1's in the two matrices?

F7 (a) Let $L : \mathbb{R}^n \to \mathbb{R}^m$ be a linear mapping, and let $\mathcal{S} = \{\mathbf{e}_1, \mathbf{e}_2, \ldots, \mathbf{e}_n\}$ denote the standard basis in $\mathbb{R}^n$ and $\mathcal{T} = \{\mathbf{f}_1, \mathbf{f}_2, \ldots, \mathbf{f}_m\}$ the standard basis in $\mathbb{R}^m$. Satisfy yourself that the correct way to define the standard matrix of L is

$$_{\mathcal{T}}[L]_{\mathcal{S}} = \Big[[L(\mathbf{e}_1)]_{\mathcal{T}}[L(\mathbf{e}_2)]_{\mathcal{T}} \cdots [L(\mathbf{e}_n)]_{\mathcal{T}}\Big].$$

(b) Now suppose that P is the change of basis matrix from basis $\mathcal{S}$ to basis $\mathcal{B}$ in $\mathbb{R}^n$ and Q is the change of basis matrix from basis $\mathcal{T}$ to $\mathcal{C}$ in $\mathbb{R}^m$. Determine the matrix that represents L with respect to the bases $\mathcal{B}$ and $\mathcal{C}$ in terms of P, Q, and $_{\mathcal{S}}[L]_{\mathcal{T}}$.

(c) Adapt the theory of this question to the case where L is a linear mapping from an n-dimensional vector space $\mathbb{V}$ to an m-dimensional vector space $\mathbb{W}$.

F8 Suppose that $\mathbb{V}$ is an n-dimensional vector space over the real numbers. Let $\mathbb{V}^*$ be the set of all linear mappings from $\mathbb{V}$ to $\mathbb{R}$.

(a) Prove that $\mathbb{V}^*$ is a vector space over the reals. It is called the **dual space of** $\mathbb{V}$.

(b) Let $\mathcal{B} = \{\mathbf{v}_1, \mathbf{v}_2, \ldots, \mathbf{v}_n\}$ be any basis for $\mathbb{V}$. For each $j = 1, 2, \ldots, n$, define $\mathbf{d}_j : \mathbb{V} \to \mathbb{R}$ to be the mapping such that

$$\mathbf{d}_j(\mathbf{v}_k) = \begin{cases} 1 & \text{if } j = k \\ 0 & \text{if } j \neq k, \end{cases}$$

and for any vectors $\mathbf{x}$ and $\mathbf{y}$ in $\mathbb{V}$, and real numbers a and b,

$$\mathbf{d}_j(a\mathbf{x} + b\mathbf{y}) = a\mathbf{d}_j(\mathbf{x}) + b\mathbf{d}_j(\mathbf{y}),$$

so that $\mathbf{d}_j$ is linear. Prove that $\mathcal{B}^* = \{\mathbf{d}_1, \mathbf{d}_2, \ldots, \mathbf{d}_n\}$ is a basis for $\mathbb{V}^*$ (called the **dual basis**).

(c) Conclude that $\mathbb{V}^*$ is isomorphic to $\mathbb{V}$.

(d) Prove that $(\mathbb{V}^*)^*$ is isomorphic to $\mathbb{V}$. [Hint: If $\mathbf{x} \in \mathbb{V}$, define $\mathbf{x}^{**} \in (\mathbb{V}^*)^*$ by $\mathbf{x}^{**}(\mathbf{d}) = \mathbf{d}(\mathbf{x})$ for every $\mathbf{d}$ in $\mathbb{V}^*$. Is this a well-defined mapping from $\mathbb{V}$ to $\mathbb{V}^{**}$? Can you use dimension to help prove it must be an isomorphism?]

REMARKS

(1) Part (d) is true only for finite-dimensional vector spaces.

(2) The isomorphism in (d) is said to be "natural" or "canonical", because it is defined in a direct natural way; in contrast, the isomorphism in (c) is only established by an abstract theorem, or by some basis-dependent definition.

F9 Consider $\mathbb{R}^n$ with the usual dot product. Let $(\mathbb{R}^n)^*$ be the dual space of $\mathbb{R}^n$ as defined in Exercise F8. Define a mapping $Dot : \mathbb{R}^n \to (\mathbb{R}^n)^*$ by letting $Dot(\mathbf{a})$ be the mapping $\mathbb{R}^n \to \mathbb{R}$:

$$Dot(\mathbf{a})(\mathbf{x}) = \mathbf{a} \cdot \mathbf{x} \text{ for any } \mathbf{x} \text{ in } \mathbb{R}^n.$$

Prove that Dot is an isomorphism.

REMARK

In the language of F8, we say that Dot provides a "natural" or "canonical" isomorphism from $\mathbb{R}^n$ to $(\mathbb{R}^n)^*$. In practice, this means that if we have a dot product, we may "identify" $(\mathbb{R}^n)^*$ with $\mathbb{R}^n$, and it is really not necessary to consider $(\mathbb{R}^n)^*$ as a distinct vector space.

CHAPTER

The Determinant

For each square matrix A, we define a number called the determinant of A. The determinant of A provides a second method for finding the inverse of A; it plays an important role in the discussion of volume; and it is an important tool for finding eigenvalues in Chapter 6.

SECTION

5–1 Determinant Defined in Terms of Cofactors

Consider the system of two linear equations in two variables:

$$\begin{aligned} a_{11}x_1 + a_{12}x_2 &= b_1 \\ a_{21}x_1 + a_{22}x_2 &= b_2. \end{aligned}$$

By the standard procedure of elimination, the solution is found to be

$$x_1 = \frac{b_1a_{22} - b_2a_{12}}{a_{11}a_{22} - a_{12}a_{21}}, \qquad x_2 = \frac{a_{11}b_2 - a_{21}b_1}{a_{11}a_{22} - a_{12}a_{21}},$$

provided that the denominator is not zero. Ignore the numerators for now (we shall examine them when we discuss Cramer's Rule in Section 5–3). The denominator is the same in the two expressions, and this fact prompts the following definition.

For a 2 by 2 matrix A, the **determinant** of A (denoted $\det(A)$ or $\det A$) is defined by

$$\det \begin{bmatrix} a_{11} & a_{12} \\ a_{21} & a_{22} \end{bmatrix} = a_{11}a_{22} - a_{12}a_{21}.$$

EXAMPLE 1

$$\det\begin{bmatrix} 2 & 7 \\ 8 & -5 \end{bmatrix} = 2(-5) - 7(8) = -10 - 56 = -66$$

An Alternative Notation

The determinant is often denoted by vertical straight lines:

$$\begin{vmatrix} a_{11} & a_{12} \\ a_{21} & a_{22} \end{vmatrix} = \det\begin{bmatrix} a_{11} & a_{12} \\ a_{21} & a_{22} \end{bmatrix} = a_{11}a_{22} - a_{12}a_{21}.$$

For example, $\begin{vmatrix} 2 & 7 \\ 8 & -5 \end{vmatrix} = -10 - 56 = -66$. One risk with this notation is that one may fail to distinguish between a matrix and the determinant of the matrix (a rather gross error).

The 3 by 3 Case

If the general system $A\mathbf{x} = \mathbf{b}$ of three linear equations in three variables is solved by elimination, and it has a unique solution, the solution appears in the form $x_j = \dfrac{j^{\text{th}}\text{ numerator}}{\text{denominator}}$, $j = 1, 2, 3$, where the numerators will be discussed later, and the common denominator D in all three expressions is

$$D = a_{11}a_{22}a_{33} - a_{11}a_{23}a_{32} - a_{12}a_{21}a_{33} + a_{12}a_{23}a_{31} + a_{13}a_{21}a_{32} - a_{13}a_{22}a_{31}.$$

We would like to simplify or reorganize this expression, so that we can remember it more easily, and so that we can guess how to generalize to the 4 by 4 case, and to the n by n case.

Notice that in the first pair of terms in D, a_{11} is a common factor, similarly in the second pair a_{12}, and in the third pair a_{13}. Thus D can be rewritten

$$D = a_{11}(a_{22}a_{33} - a_{23}a_{32}) - a_{12}(a_{21}a_{33} - a_{23}a_{31}) + a_{13}(a_{21}a_{32} - a_{22}a_{31}).$$

Notice that each expression in parentheses can be regarded as the determinant of a 2 by 2 **submatrix** of A, selected according to some rule. Notice that the first and third expressions have plus signs, and the middle expression has a minus sign. The following set of rules turns out to be one of the best ways to organize the information in D.

DEFINITION

Let A be a 3 by 3 matrix. Let $A(\hat{\jmath}, \hat{k})$ be the 2 by 2 matrix obtained from A by deleting the jth row and the kth column.

Define the **cofactor** of a_{jk} to be $C_{jk} = (-1)^{(j+k)} \det A(\hat{\jmath}, \hat{k})$.

Define the **determinant** of A to be

$$\det A = a_{11}C_{11} + a_{12}C_{12} + a_{13}C_{13}.$$

First, let us check that this definition really does produce the value D from the 3 by 3 matrix A. First calculate C_{11} : the 2 by 2 matrix $A(\hat{1}, \hat{1})$ obtained by deleting the first row and the first column is $\begin{bmatrix} a_{22} & a_{23} \\ a_{32} & a_{33} \end{bmatrix}$, and its determinant is $(a_{22}a_{33} - a_{23}a_{32})$. Thus,

$$C_{11} = (-1)^{(1+1)}(a_{22}a_{33} - a_{23}a_{32}) = (a_{22}a_{33} - a_{23}a_{32}).$$

Similarly,

$$C_{12} = (-1)^{(1+2)} \det \begin{bmatrix} a_{21} & a_{23} \\ a_{31} & a_{33} \end{bmatrix} = (-1)(a_{21}a_{33} - a_{23}a_{31}),$$

and

$$C_{13} = (-1)^{(1+3)} \det \begin{bmatrix} a_{21} & a_{22} \\ a_{31} & a_{32} \end{bmatrix} = (a_{21}a_{32} - a_{22}a_{31}).$$

It follows that $a_{11}C_{11} + a_{11}C_{12} + a_{13}C_{13}$ does give the value of D as required.

The usual brief way of showing this calculation is as follows:

$$\det \begin{bmatrix} a_{11} & a_{12} & a_{13} \\ a_{21} & a_{22} & a_{23} \\ a_{31} & a_{32} & a_{33} \end{bmatrix} = a_{11} \det \begin{bmatrix} a_{22} & a_{23} \\ a_{32} & a_{33} \end{bmatrix} + a_{12}(-1) \det \begin{bmatrix} a_{21} & a_{23} \\ a_{31} & a_{33} \end{bmatrix}$$
$$+ a_{13} \det \begin{bmatrix} a_{21} & a_{22} \\ a_{31} & a_{32} \end{bmatrix}.$$

Alternatively, in the vertical line notation,

$$\begin{vmatrix} a_{11} & a_{12} & a_{13} \\ a_{21} & a_{22} & a_{23} \\ a_{31} & a_{32} & a_{33} \end{vmatrix} = a_{11} \begin{vmatrix} a_{22} & a_{23} \\ a_{32} & a_{33} \end{vmatrix} + a_{12}(-1) \begin{vmatrix} a_{21} & a_{23} \\ a_{31} & a_{33} \end{vmatrix}$$
$$+ a_{13} \begin{vmatrix} a_{21} & a_{22} \\ a_{31} & a_{32} \end{vmatrix}.$$

REMARKS

(1) This definition of a 3 by 3 determinant is called **expansion of the determinant along the first row.** As we shall see below, a determinant can be expanded along any row or any column.

(2) The signs attached to the cofactors can cause trouble if you are not careful. One helpful way to remember which sign to attach to which position is to take a blank matrix and put a + in the top left corner and then alternate – and + both across and down. This is shown for a 3 by 3 matrix, but it works for a square matrix of any size: $\begin{bmatrix} + & - & + \\ - & + & - \\ + & - & + \end{bmatrix}$.

(3) C_{jk} is called the cofactor of a_{jk} because it is the "factor with a_{jk}" in the expansion of the determinant. Note that each C_{jk} is a number, not a matrix. However, there is a **matrix** of cofactors; it will be denoted Cof(A), with typical entry C_{jk}.

(4) Notice that the determinant of A is the dot product of the first **row** of A with the corresponding first **row** of Cof(A):

$$\det A = a_{11}C_{11} + a_{12}C_{12} + a_{13}C_{13} = a_{1\rightarrow} \cdot C_{1\rightarrow}.$$

EXAMPLE 2

The following determinant is calculated by expansion along the first row.

$$\det \begin{bmatrix} 4 & -1 & 1 \\ 2 & 3 & 5 \\ 1 & 0 & 6 \end{bmatrix} = 4 \det \begin{bmatrix} 3 & 5 \\ 0 & 6 \end{bmatrix} - 1(-1) \det \begin{bmatrix} 2 & 5 \\ 1 & 6 \end{bmatrix} + 1 \det \begin{bmatrix} 2 & 3 \\ 1 & 0 \end{bmatrix}$$

$$= 4(18 - 0) + (12 - 5) + (0 - 3) = 72 + 7 - 3 = 76.$$

Calculating determinants can involve a lot of arithmetic, so it is important to know rules that simplify the calculations.

THEOREM 1. **(3 by 3 case)** The determinant of a 3 by 3 matrix A can be expanded along any row or any column, with the same result. That is,

$$D = a_{1\to} \cdot C_{1\to} = a_{2\to} \cdot C_{2\to} = a_{3\to} \cdot C_{3\to}$$
$$= a_{\downarrow 1} \cdot C_{\downarrow 1} = a_{\downarrow 2} \cdot C_{\downarrow 2} = a_{\downarrow 3} \cdot C_{\downarrow 3}.$$

To prove Theorem 1, simply calculate D in each of the six ways indicated and verify that the result is the same in each case. (Inelegant proofs like this explain why some people prefer alternative definitions for the determinant.)

EXAMPLE 3

Calculate the determinant of the matrix in Example 2 by expanding along the second column to take advantage of the fact that one entry in this column is 0; make sure that you give each cofactor the correct sign:

$$\det \begin{bmatrix} 4 & -1 & 1 \\ 2 & 3 & 5 \\ 1 & 0 & 6 \end{bmatrix} = -1(-1) \det \begin{bmatrix} 2 & 5 \\ 1 & 6 \end{bmatrix} + 3 \det \begin{bmatrix} 4 & 1 \\ 1 & 6 \end{bmatrix} + 0(-1) \det \begin{bmatrix} 4 & 1 \\ 2 & 5 \end{bmatrix}$$
$$= 1(12 - 5) + 3(24 - 1) + 0 = 7 + 69 = 76,$$

as in Example 2. Check by expanding along the third row.

THEOREM 2. **(3 by 3 case)** If one row (or column) of a 3 by 3 matrix A is a zero row (or column), then $\det A = 0$.

Proof. If the ith row consists of zeros, then expansion along the ith row gives

$$\det A = a_{i\to} \cdot C_{i\to} = (0, 0, 0) \cdot C_{i\to} = 0.$$

■

THEOREM 3. **(3 by 3 case)** If two rows of the 3 by 3 matrix A are equal to each other, then $\det A = 0$. Similarly, if two columns are equal, $\det A = 0$.

Before proving Theorem 3, consider an example that illustrates why it must be true.

EXAMPLE 4

Evaluate $\det \begin{bmatrix} 1 & 2 & 3 \\ 4 & 5 & 6 \\ 1 & 2 & 3 \end{bmatrix}$.

SOLUTION: Since the first and third rows of the matrix are equal, expand the determinant along the *second* row, and observe that the corresponding cofactors are all zero:

$$\det\begin{bmatrix}1 & 2 & 3\\4 & 5 & 6\\1 & 2 & 3\end{bmatrix} = -4\det\begin{bmatrix}2 & 3\\2 & 3\end{bmatrix} + 5\det\begin{bmatrix}1 & 3\\1 & 3\end{bmatrix} - 6\det\begin{bmatrix}1 & 2\\1 & 2\end{bmatrix}$$

$$= -4(0) + 5(0) - 6(0) = 0.$$

Proof of Theorem 3. Suppose (to be definite) that the first and third rows of A are equal; expand along the second row. Then each cofactor in the expansion is (plus or minus) the determinant of a 2 by 2 matrix with equal rows, and clearly such cofactors are all zero. A similar argument works if two columns are equal.

■

Further general properties of the determinant will be considered after the definition is given for the n by n case. Note that this definition follows exactly the pattern of the definition for the 3 by 3 case.

DEFINITION

Let A be an n by n matrix. Let $A(\hat{j}, \hat{k})$ be the $(n-1)$ by $(n-1)$ matrix obtained from A by deleting the jth row and the kth column.

Define the **cofactor** of a_{jk} to be $C_{jk} = (-1)^{j+k} \det A(\hat{j}, \hat{k})$.

Define the **determinant** of A to be

$$a_{11}C_{11} + a_{12}C_{12} + \cdots + a_{1n}C_{1n} = a_{1\rightarrow} \cdot C_{1\rightarrow}.$$

REMARK

This is a **recursive** definition: the result for the n by n case is defined in terms of the $(n-1)$ by $(n-1)$ case, which in turn must be calculated in terms of the $(n-2)$ by $(n-2)$ case, and so on, until we get back to the 2 by 2 case for which the result is given explicitly. With such a definition, we may expect proof by induction to play a role.

EXAMPLE 5

We calculate the following determinant by using the definition. Expand along the first row to take advantage of the zeros. Note that $*$ and $**$ represent cofactors whose values are irrelevant.

$$\det\begin{bmatrix} 0 & 2 & 3 & 0 \\ 1 & 5 & 6 & 7 \\ -2 & 3 & 0 & 4 \\ -5 & 1 & 2 & 3 \end{bmatrix} = 0(*) + 2(-1)\det\begin{bmatrix} 1 & 6 & 7 \\ -2 & 0 & 4 \\ -5 & 2 & 3 \end{bmatrix}$$

$$+ 3(1)\det\begin{bmatrix} 1 & 5 & 7 \\ -2 & 3 & 4 \\ -5 & 1 & 3 \end{bmatrix} + 0(**)$$

$$= -2\Big(1(0-8) - 6(-6+20) + 7(-4-0)\Big)$$

$$+ 3\Big(1(9-4) - 5(-6+20) + 7(-2+15)\Big)$$

$$= -2(-8 - 84 - 28) + 3(5 - 70 + 91)$$

$$= -2(-120) + 3(26) = 318.$$

It is apparent that evaluating the determinant of a 4 by 4 matrix is a fairly lengthy calculation, and things will get worse for larger matrices, so it is important to have results that simplify the evaluation of the determinant. The theorems stated earlier for 3 by 3 matrices are all true in general.

THEOREM 1. *(n by n case)* Suppose that A is an n by n matrix. Then $\det A$ may be obtained by a cofactor expansion along any row or any column of A.

We omit a proof here since there is no conceptually helpful proof, and it would be a bit grim to verify the result in the general case.

Theorem 1 is a very practical result. It allows us to **choose** the row or column from A we are going to expand along. If one row or column has many zeros, it is sensible to expand along it, since we shall then not have to evaluate the cofactors of these zero entries. In Example 5, it was necessary to evaluate only two of the cofactors, since the expansion was along the first row, and two entries in the first row were zero.

THEOREM 2. *(n by n case)* If one row (or column) of a square matrix A is the zero row (or column), then $\det A = 0$.

Proof. The same as for the 3 by 3 case—expand along the zero row (or column).

■

THEOREM 3. *(n by n case)* If A is an n by n matrix, and if two rows (or columns) of A are equal to each other, then $\det A = 0$.

Proof. (By induction—if you have forgotten how this works, see the review below.) We consider only the case of equal rows, since the case of columns is similar. The statement is certainly true if $n = 2$. As the Induction Hypothesis, suppose that the statement is true for all $(n - 1)$ by $(n - 1)$ matrices; we must show that it then follows for n by n matrices. Expand the n by n matrix A along any row that is **not** one of the two equal rows (possible if $n > 2$). Each cofactor in this expansion is (plus or minus) the determinant of an $(n - 1)$ by $(n - 1)$ matrix with two equal rows, and therefore each cofactor is zero by the Induction Hypothesis. Hence $\det A = 0$ for the n by n case and the proof is complete.

■

A BRIEF REVIEW OF PROOF BY INDUCTION

Given a statement that depends on the natural number n,
if

(i) the statement is true for some initial value (usually $n = 1$, but for determinant proofs, usually $n = 2$), and

(ii) whenever the statement is true for $n - 1$ (the Induction Hypothesis), it follows that it is true for n,

then *the statement is true for all n greater than or equal to the initial value.*

The determinant of the transpose of a matrix will be used in Chapters 7 and 8. The next theorem tells us about the value of $\det A^T$.

THEOREM 4. If A is an n by n matrix, then $\det A = \det A^T$.

Proof. Expand $\det A$ along its first row, then each $(n - 1)$ by $(n - 1)$ cofactor along its first row, and so on. This will give exactly the same expression as expanding $\det A^T$ along its first **column**, then each of its $(n-1)$ by $(n-1)$ cofactors along its first column, and so on. More formally, we prove this theorem by induction.

The statement is easy to verify for 2 by 2 matrices. Suppose that it is true for $(n-1)$ by $(n-1)$ matrices. In writing this proof, it is convenient to use a notation introduced in Section 3–1: $(A^T)_{ij}$ means the entry in the ith row and j^{th} column of A^T.

Expand $\det(A^T)$ along its first column:

$$\det(A^T) = (A^T)_{11}\det(A^T(\hat{1},\hat{1})) + (A^T)_{21}(-1)\det(A^T(\hat{2},\hat{1})) + \cdots \\ + (A^T)_{n1}(-1)^{n+1}\det(A^T(\hat{n},\hat{1})).$$

We wish to rewrite the right-hand side in terms of entries and cofactors of A. Note that $(A^T)_{ij} = (A)_{ji} = a_{ji}$. Also, observe that $A^T(\hat{j},\hat{k}) = (A(\hat{k},\hat{j}))^T$; since these are $(n-1)$ by $(n-1)$ matrices, by this equation and the induction hypothesis, $\det(A^T(\hat{j},\hat{k})) = \det((A(\hat{k},\hat{j}))^T) = \det(A(\hat{k},\hat{j}))$. Thus,

$$\begin{aligned}\det(A^T) &= a_{11}\det(A(\hat{1},\hat{1})) + a_{12}(-1)\det(A(\hat{1},\hat{2})) + \cdots \\ &\quad + a_{1n}(-1)^{1+n}\det(A(\hat{1},\hat{n})) \\ &= \det(A).\end{aligned}$$

Since this is true for any n, the theorem is proved.

■

With the tools we have so far, evaluation of determinants is still a very tedious business. Properties of the determinant with respect to elementary row operations make the evaluation much easier. These properties are discussed in the next section. This section concludes with a sketch of an alternative definition of the determinant, and with a special method that only works for 3 by 3 matrices.

An Alternative Approach to the Determinant

Consider again the explicit expression for the determinant of a 3 by 3 matrix A:

$$a_{11}a_{22}a_{33} - a_{11}a_{23}a_{32} - a_{12}a_{21}a_{33} + a_{12}a_{23}a_{31} + a_{13}a_{21}a_{32} - a_{13}a_{22}a_{31}.$$

It is the sum of six products; each product has three factors. Check that each product has one factor from each row of A, and one factor from each column of A. In fact, a typical product can be written $a_{1j_1}a_{2j_2}a_{3j_3}$, where the column indices (j_1, j_2, j_3) are a rearrangement or **permutation** of $(1, 2, 3)$. Each product also has a sign attached to it; the rule for the sign is a little complicated. (If you really want to know, the sign is $+1$ if (j_1, j_2, j_3) can be brought into the natural order $(1, 2, 3)$ by an even number of interchanges of adjacent indices, and -1 if it takes an odd number of adjacent interchanges. It takes a bit of work to show that this definition of the sign really makes sense in the general case.) Once the sign associated to each permutation is known, the determinant of the 3 by 3 matrix A can be defined to be

$$\det A = \sum \operatorname{sign}(j_1, j_2, j_3) a_{1j_1} a_{2j_2} a_{3j_3},$$

where the sum is taken over all permutations (j_1, j_2, j_3) of $(1, 2, 3)$.

This definition can easily be generalized for n by n matrices. This approach is helpful because it emphasizes that each product in the sum has one factor from each row, and one from each column of A. Some of the properties of the determinant then become very obvious — for example, if one row is zero, the determinant is obviously zero. This approach also leads particularly well into some more advanced mathematics such as "exterior algebra". The difficulty with this approach is that it does not give an immediately effective method of computation, and that it requires a modest digression into the signs of permutations.

A Special Method for the Determinants of 3 by 3 Matrices

Perhaps the most important thing for you to know about this method is that it works only for the 3 by 3 case. There are no similar rules for $n = 4$ or greater. This method is based on "diagonals": the three **forward** diagonals (the **main** diagonal and two diagonals "parallel" to it) and the three **back** diagonals. Take the products along the forward diagonals with a plus sign, and the products along the back diagonals with a minus sign, and add to get the determinant.

Forward diagonals

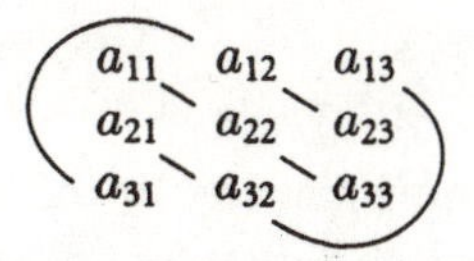

give $a_{11}a_{22}a_{33} + a_{12}a_{23}a_{31} + a_{13}a_{21}a_{32}$.

Back diagonals

$$\begin{matrix} a_{11} & a_{12} & a_{13} \\ a_{21} & a_{22} & a_{23} \\ a_{31} & a_{32} & a_{33} \end{matrix}$$

give $-a_{13}a_{22}a_{31} - a_{12}a_{21}a_{33} - a_{11}a_{23}a_{32}$.

EXAMPLE 6

The following determinant is evaluated by the special 3 by 3 method.

$$\begin{vmatrix} 1 & 2 & 0 \\ -1 & 3 & 5 \\ 4 & 1 & 7 \end{vmatrix} = 1 \cdot 3 \cdot 7 + 2 \cdot 5 \cdot 4 + 0(-1)1 - 0 \cdot 3 \cdot 4 - 2(-1)7 - 1 \cdot 5 \cdot 1$$

$$= 21 + 40 + 0 - 0 - (-14) - 5 = 70.$$

A Comment on the Cross-product as Determinant

Recall that in Chapter 1, the cross-product of two vectors **a** and **b** was defined by $\mathbf{a} \times \mathbf{b} = \begin{vmatrix} \mathbf{i} & \mathbf{j} & \mathbf{k} \\ a_1 & a_2 & a_3 \\ b_1 & b_2 & b_3 \end{vmatrix}$. This formula helps us to remember how to calculate the cross-product, but it is not an "honest" determinant, because some of the entries are vectors and not real numbers. It should be regarded as a "formal" determinant; expansion along the first row gives results that we can interpret sensibly, but some of the statements made about determinants in the next section would make no sense for this formal determinant.

EXERCISES 5–1

A1 Explain in terms of the theorems of this section why each of the following matrices has determinant equal to zero.

(a) $A = \begin{bmatrix} 3 & 2 & 1 \\ -1 & 4 & 5 \\ 3 & 2 & 1 \end{bmatrix}$ (b) $B = \begin{bmatrix} 3 & 5 & 0 \\ -2 & 6 & 0 \\ 4 & 1 & 0 \end{bmatrix}$

(c) $C = \begin{bmatrix} 0 & 1 & 5 & 1 \\ 2 & -1 & 1 & -1 \\ 0 & 1 & 0 & 1 \\ 3 & -2 & 4 & -2 \end{bmatrix}$

A2 Evaluate the determinants of the following matrices by cofactor expansion or by one of the theorems. Check the 3 by 3 cases by the special diagonal method.

(a) $A = \begin{bmatrix} 2 & 1 & 5 \\ 4 & 3 & -1 \\ 0 & 1 & -2 \end{bmatrix}$ (b) $B = \begin{bmatrix} 7 & 4 & 0 \\ 2 & -3 & 2 \\ 3 & 5 & 0 \end{bmatrix}$

(c) $C = \begin{bmatrix} 1 & 2 & 3 & 4 \\ -2 & 0 & 2 & 5 \\ 3 & 0 & 1 & 4 \\ 4 & 5 & 1 & -2 \end{bmatrix}$ (d) $D = \begin{bmatrix} 2 & 1 & -1 & 3 \\ 3 & 5 & 7 & 9 \\ -4 & 2 & 2 & 6 \\ 2 & 1 & -1 & 3 \end{bmatrix}$

(e) $E = \begin{bmatrix} 1 & 3 & 4 & -5 & 7 \\ 0 & 3 & 1 & 2 & 3 \\ 0 & 0 & 4 & 1 & 0 \\ 0 & 0 & 0 & -1 & 8 \\ 0 & 0 & 0 & 4 & 3 \end{bmatrix}$

A3 Write the matrix of cofactors, Cof (A), for each of the following matrices. Make sure that each entry has the correct sign. Then transpose Cof (A) and calculate the product $A\,(\mathrm{Cof}(A))^T$. You should get $(\det A)\,I$; we shall see why this is true in Section 5–3.

(a) $A = \begin{bmatrix} 2 & -4 \\ 7 & 5 \end{bmatrix}$ (b) $A = \begin{bmatrix} 2 & 3 & 1 \\ 1 & -1 & 4 \\ 3 & 2 & 5 \end{bmatrix}$

(c) $A = \begin{bmatrix} 1 & -5 & -3 \\ 1 & -6 & 2 \\ 4 & 1 & -6 \end{bmatrix}$

B1 Explain in terms of the theorems of this section why each of the following matrices has determinant equal to zero.

(a) $A = \begin{bmatrix} -2 & 0 & 2 \\ 4 & 1 & 5 \\ 0 & 0 & 0 \end{bmatrix}$ (b) $B = \begin{bmatrix} 1 & 5 & 1 \\ 1 & 6 & 1 \\ 7 & 1 & 7 \end{bmatrix}$

(c) $C = \begin{bmatrix} 1 & 2 & 1 & -5 \\ 2 & 4 & 0 & 1 \\ 3 & 0 & 1 & 6 \\ 1 & 2 & 1 & -5 \end{bmatrix}$

B2 Evaluate the determinants of the following matrices by cofactor expansion or by one of the theorems. Check the 3 by 3 cases by the special diagonal method.

(a) $A = \begin{bmatrix} 3 & -2 & 4 \\ 3 & 0 & 0 \\ 0 & 8 & -3 \end{bmatrix}$ (b) $B = \begin{bmatrix} 2 & 5 & 4 \\ -3 & 2 & 1 \\ 3 & -6 & 0 \end{bmatrix}$

(c) $C = \begin{bmatrix} 2 & 3 & -3 & 1 \\ -3 & 1 & 0 & 5 \\ 1 & 2 & 1 & -2 \\ 3 & 0 & 1 & 0 \end{bmatrix}$ (d) $D = \begin{bmatrix} 3 & 4 & 2 & 4 \\ 2 & 1 & -7 & 1 \\ -3 & 2 & 1 & 2 \\ 2 & 3 & -1 & 3 \end{bmatrix}$

(e) $E = \begin{bmatrix} 1 & 3 & 4 & -5 & 7 \\ 3 & 3 & 1 & 2 & 0 \\ 2 & -1 & 4 & 0 & 0 \\ 5 & 3 & 0 & 0 & 0 \\ -2 & 0 & 0 & 0 & 0 \end{bmatrix}$

B3 Write the matrix of cofactors, Cof (A), for each of the following matrices. Make sure that each entry has the correct sign. Then transpose Cof (A) and calculate the product A (Cof$(A))^T$. You should get (det A) I; we shall see why this is true in Section 5–3.

(a) $A = \begin{bmatrix} 3 & -2 \\ 2 & 5 \end{bmatrix}$ (b) $A = \begin{bmatrix} 4 & 6 \\ -1 & -1 \end{bmatrix}$

(c) $A = \begin{bmatrix} 5 & -1 & 1 \\ 1 & 4 & 6 \\ -3 & 2 & 1 \end{bmatrix}$ (d) $A = \begin{bmatrix} 2 & -1 & -2 \\ 1 & 0 & 3 \\ 3 & -1 & 0 \end{bmatrix}$

COMPUTER RELATED EXERCISES

C1 Use a computer or calculator to evaluate the determinants of the following matrices.

(a) $\begin{bmatrix} 1.09 & 4.83 & 2.95 \\ 2.13 & -3.25 & 1.57 \\ 1.72 & 2.15 & -0.89 \end{bmatrix}$ (b) $\begin{bmatrix} 1.23 & 2.35 & 4.19 & -1.28 \\ -2.09 & 0.17 & 3.89 & 22.1 \\ 0.78 & 2.15 & -3.55 & 4.15 \\ 1.58 & -2.59 & 1.01 & 0.00 \end{bmatrix}$

CONCEPTUAL EXERCISES

D1 (a) Consider the points (a_1, a_2) and (b_1, b_2) in $\mathbb{R}^2$. Show that

$\det \begin{bmatrix} x_1 & x_2 & 1 \\ a_1 & a_2 & 1 \\ b_1 & b_2 & 1 \end{bmatrix} = 0$ is the equation of the line containing the two points.

[Hint: Is it a linear equation? Do the points satisfy it? What if the points are identical?]

(b) Write a determinantal equation analagous to the equation found in part (a) for the plane in $\mathbb{R}^3$ that contains the points (a_1, a_2, a_3), (b_1, b_2, b_3), (c_1, c_2, c_3). What happens if the points are collinear?

(c) Describe geometrically the set of points in the xy-plane that satisfy the equation

$$\det \begin{bmatrix} y & x^2 & x & 1 \\ q & p^2 & p & 1 \\ s & r^2 & r & 1 \\ u & t^2 & t & 1 \end{bmatrix} = 0,$$

where (p, q), (r, s), and (t, u) are given points in $\mathbb{R}^2$. What if $p = r$? Are there other degenerate cases?

Remark Although determinantal equations such as these are not always computationally easy to work with, it is sometimes helpful to see that, in principle, one can easily write the equations of some line, or plane, or other set in terms of the determinant.

SECTION

5-2 Elementary Row Operations and the Determinant

Calculation of the determinant of a matrix can be lengthy, so it is helpful to know that the calculation can often be simplified by applying elementary row operations to the matrix, provided that suitable adjustments are made to the determinant.

To see what happens to the determinant of A when we multiply a row of A by a constant, we first consider a 3 by 3 example. Following the example, we state and prove a general result.

EXAMPLE 7

Let $A = \begin{bmatrix} a & b & c \\ d & e & f \\ g & h & k \end{bmatrix}$, and let $B = \begin{bmatrix} a & b & c \\ d & e & f \\ rg & rh & rk \end{bmatrix}$, so that B is obtained from A by multiplying the third row by r. Calculate and compare $\det A$ and $\det B$.

SOLUTION: By expansion along the third row,

$$\det A = g\begin{vmatrix} b & c \\ e & f \end{vmatrix} - h\begin{vmatrix} a & c \\ d & f \end{vmatrix} + k\begin{vmatrix} a & b \\ d & e \end{vmatrix}.$$

Similarly,

$$\det B = rg\begin{vmatrix} b & c \\ e & f \end{vmatrix} - rh\begin{vmatrix} a & c \\ d & f \end{vmatrix} + rk\begin{vmatrix} a & b \\ d & e \end{vmatrix}.$$

Thus, $\det \begin{bmatrix} a & b & c \\ d & e & f \\ rg & rh & rk \end{bmatrix} = r \det \begin{bmatrix} a & b & c \\ d & e & f \\ g & h & k \end{bmatrix}$.

THEOREM 1. Let A be an n by n matrix, and let B be the matrix obtained from A by multiplying the jth row of A by the real number r. Then $\det B = r \det A$.

Proof. Expand the determinant of B along the jth row. Notice that the cofactors of the elements in this row are exactly the cofactors of the jth row of A, since B agrees with A except in the jth row. Therefore,

$$\det B = b_{j\rightarrow} \cdot C_{j\rightarrow} = (ra_{j\rightarrow}) \cdot C_{j\rightarrow} = r(a_{j\rightarrow} \cdot C_{j\rightarrow}) = r \det A.$$

■

EXAMPLE 8

$$\det \begin{bmatrix} 1 & 3 & 4 \\ 5 & 10 & 15 \\ 2 & -1 & 0 \end{bmatrix} = 5 \det \begin{bmatrix} 1 & 3 & 4 \\ 1 & 2 & 3 \\ 2 & -1 & 0 \end{bmatrix} = 5\Big(2(9-8) - (-1)(3-4)\Big)$$
$$= 5(2-1) = 5.$$

For some reason, some students get confused with this theorem and introduce the reciprocal $(1/r)$. It may help to think about this theorem in terms of the permutation approach to the determinant described in the preceding section: since each product in the sum making up $\det B$ has one factor from the jth row, a common factor r in the jth row becomes a common factor in $\det B$.

Next we consider the effect of interchanging two rows.

EXAMPLE 9

The following calculations illustrate the fact that an interchange of rows causes a change of sign of the determinant.

(a) $\det \begin{bmatrix} c & d \\ a & b \end{bmatrix} = cb - da = -(ad - bc) = -\det \begin{bmatrix} a & b \\ c & d \end{bmatrix}$.

(b) We evaluate the determinant of a matrix whose first and third rows have been interchanged by expanding along the *second* row, and using part (a):

$$\det\begin{bmatrix} c_1 & c_2 & c_3 \\ b_1 & b_2 & b_3 \\ a_1 & a_2 & a_3 \end{bmatrix}$$

$$= -b_1\begin{vmatrix} c_2 & c_3 \\ a_2 & a_3 \end{vmatrix} + b_2\begin{vmatrix} c_1 & c_3 \\ a_1 & a_3 \end{vmatrix} - b_3\begin{vmatrix} c_1 & c_2 \\ a_1 & a_2 \end{vmatrix}$$

$$= -b_1\left(-\begin{vmatrix} a_2 & a_3 \\ c_2 & c_3 \end{vmatrix}\right) + b_2\left(-\begin{vmatrix} a_1 & a_3 \\ c_1 & c_3 \end{vmatrix}\right) - b_3\left(-\begin{vmatrix} a_1 & a_2 \\ c_1 & c_2 \end{vmatrix}\right)$$

$$= -\det\begin{bmatrix} a_1 & a_2 & a_3 \\ b_1 & b_2 & b_3 \\ c_1 & c_2 & c_3 \end{bmatrix}.$$

THEOREM 2. Suppose that A is an n by n matrix, and that B is the matrix obtained from A by interchanging two rows. Then $\det B = -\det A$.

Proof (by induction). The statement can easily be verified directly if $n = 2$, so let $n > 2$. Suppose that the statement of the theorem is true for all $(n-1)$ by $(n-1)$ matrices; we must show that it follows for n by n matrices. Expand the determinant of B along some row **other than** the two rows that are interchanged, and compare with the expansion of A along the same row. This row of A has exactly the same entries as the corresponding row of B. The cofactors in this expansion of B are (plus or minus) the determinants of $(n-1)$ by $(n-1)$ matrices, and these $(n-1)$ by $(n-1)$ matrices are the same as the matrices giving the cofactors of A **except** that those in the expansion of B have two rows interchanged. By the induction hypothesis, this means that the cofactors in the expansion of $\det B$ are (-1) times the cofactors in the corresponding expansion of $\det A$. Hence $\det B = -\det A$ for the n by n case, and the theorem is true by induction.

■

COROLLARY If two rows of A are equal, $\det A = 0$. (Section 5–1, Theorem 3)

Proof. Let B be the matrix obtained from A by interchanging the two equal rows. Obviously $B = A$, so $\det B = \det A$. But by Theorem 2, $\det B = -\det A$, so $\det A = 0$.

■

Further examples of Theorem 2 will be given below.

In order to see how the third type of elementary row operation affects the determinant of a matrix, we require a lemma. This lemma will play an important role in other calculations as well as in the next theorem. It is useful to look at a 3 by 3 example before considering the general case.

EXAMPLE 10

By expansion along the second row,

$$\begin{vmatrix} a_{11} & a_{12} & a_{13} \\ b_1 & b_2 & b_3 \\ a_{31} & a_{32} & a_{33} \end{vmatrix} = b_1(-1)\begin{vmatrix} a_{12} & a_{13} \\ a_{32} & a_{33} \end{vmatrix} + b_2\begin{vmatrix} a_{11} & a_{13} \\ a_{31} & a_{33} \end{vmatrix}$$

$$+ b_3(-1)\begin{vmatrix} a_{11} & a_{12} \\ a_{31} & a_{32} \end{vmatrix} = b_1C_{21} + b_2C_{22} + b_3C_{23}$$

$$= \mathbf{b} \cdot C_{2\rightarrow}.$$

The content of this calculation is described in the next lemma.

LEMMA. **(Substitution Lemma)** Suppose that A is an n by n matrix with jk^{th} cofactor C_{jk}. Then if $\mathbf{b}$ is any n-vector, the dot product $\mathbf{b} \cdot C_{j\rightarrow}$ is equal to the determinant of the matrix obtained from A by substituting $\mathbf{b}$ in place of the jth row of A.

Proof. Let B be the matrix obtained from A by substituting $\mathbf{b}$ in place of $a_{j\rightarrow}$. Expand $\det B$ along its jth row: then b_1 is multiplied by the cofactor C_{j1} of a_{j1}, b_2 by the cofactor C_{j2} of a_{j2}, and so on, so $\det B = b_1C_{j1} + b_2C_{j2} + \cdots + b_nC_{jn}$, as claimed. ■

One immediate application of this lemma is the following theorem.

THEOREM. **(False Expansion Theorem)** If A is an n by n matrix and $j \neq k$, then $a_{j\rightarrow} \cdot C_{k\rightarrow} = 0$. That is, the dot product of the jth row of A with the row of cofactors belonging to a **different** row is zero.

Proof. By the Substitution Lemma, $a_{j\rightarrow} \cdot C_{k\rightarrow}$ is equal to the determinant of the matrix obtained from A by replacing the kth row by $a_{j\rightarrow}$, that is, by the jth row. But this means that in the new matrix the jth and k^{th} rows are equal, so the determinant is zero. ■

This theorem is one key to the cofactor method of finding inverses described in the next section. A more immediate application is the next theorem.

THEOREM 3. Suppose that A is an n by n matrix, and that B is obtained from A by adding r times the kth row of A to the jth row. Then $\det B = \det A$.

Proof. This elementary row operation can also be described as substituting $(a_{j\rightarrow} + ra_{k\rightarrow})$ in place of the original j^{th} row. Therefore, by the Substitution Lemma,

$$\det B = (a_{j\rightarrow} + ra_{k\rightarrow}) \cdot C_{j\rightarrow} = a_{j\rightarrow} \cdot C_{j\rightarrow} + ra_{k\rightarrow} \cdot C_{j\rightarrow} = \det A + 0,$$

where the last term is zero by the False Expansion Theorem.

■

To take full advantage of the ways that elementary row operations affect the determinant of a matrix, we need one more definition and one simple theorem.

DEFINITION

A matrix A is **upper triangular** if all entries below the main diagonal are zero, that is, if $a_{jk} = 0$ for $j > k$.

EXAMPLE 11

$$\begin{bmatrix} 2 & 6 & -1 \\ 0 & -1 & 0 \\ 0 & 0 & 3 \end{bmatrix} \text{ and } \begin{bmatrix} 1 & 2 & 3 & -1 \\ 0 & 5 & 1 & 0 \\ 0 & 0 & 0 & 2 \\ 0 & 0 & 0 & 1 \end{bmatrix} \text{ are upper triangular.}$$

THEOREM 4. If A is a square upper triangular matrix, $\det A$ is the product of the diagonal entries of A, that is,

$$\det A = a_{11}a_{22}\ldots a_{nn}.$$

Proof. Expand repeatedly along the first column.

$$\begin{vmatrix} a_{11} & a_{12} & a_{13} & \cdots & \\ 0 & a_{22} & a_{23} & \cdots & \\ 0 & 0 & a_{33} & \cdots & \\ \vdots & \vdots & \vdots & \ddots & \\ 0 & 0 & 0 & \cdots & a_{nn} \end{vmatrix} = a_{11} \begin{vmatrix} a_{22} & a_{23} & \cdots & \\ 0 & a_{33} & \cdots & \\ \vdots & \vdots & \ddots & \\ 0 & 0 & \cdots & a_{nn} \end{vmatrix} + 0 + 0 + \cdots + 0$$

$$= a_{11}a_{22} \begin{vmatrix} a_{33} & & \\ & & \\ 0 & \cdots & a_{nn} \end{vmatrix} + 0 + 0 + \cdots + 0$$

$$= \ldots = a_{11}a_{22}a_{33} \ldots a_{nn}.$$

■

Theorems 1, 2, 3, and 4 suggest that an effective strategy for evaluating the determinant of a matrix is to row reduce the matrix to upper triangular form. For $n > 3$, it can be shown that in general, this strategy will require fewer arithmetic operations than straight cofactor expansion. The following example illustrates this strategy.

EXAMPLE 12

$$\begin{vmatrix} 1 & 3 & 1 & 5 \\ 1 & 3 & -3 & -3 \\ 0 & 3 & 1 & 0 \\ 1 & 6 & 2 & 11 \end{vmatrix} = \begin{vmatrix} 1 & 3 & 1 & 5 \\ 0 & 0 & -4 & -8 \\ 0 & 3 & 1 & 0 \\ 0 & 3 & 1 & 6 \end{vmatrix} \quad \text{(by 2 row operations and Theorem 3)}$$

$$= (-1) \begin{vmatrix} 1 & 3 & 1 & 5 \\ 0 & 3 & 1 & 0 \\ 0 & 0 & -4 & -8 \\ 0 & 3 & 1 & 6 \end{vmatrix} \quad \text{(interchanging rows — Theorem 2)}$$

$$= (-1)(-4) \begin{vmatrix} 1 & 3 & 1 & 5 \\ 0 & 3 & 1 & 0 \\ 0 & 0 & 1 & 2 \\ 0 & 3 & 1 & 6 \end{vmatrix} \quad \text{(common factor, third row — Theorem 1)}$$

$$= 4 \begin{vmatrix} 1 & 3 & 1 & 5 \\ 0 & 3 & 1 & 0 \\ 0 & 0 & 1 & 2 \\ 0 & 0 & 0 & 6 \end{vmatrix} \quad \text{(by Theorem 3)}$$

$$= 4(1)(3)(1)(6) = 72 \quad \text{(by Theorem 4)}$$

In some cases, it may be appropriate to use some combination of row operations and cofactor expansion. The following example illustrates such a combination.

EXAMPLE 13

$$\begin{vmatrix} 1 & 5 & 6 & 7 \\ 1 & 8 & 7 & 9 \\ 1 & 5 & 6 & 10 \\ 0 & 1 & 4 & -2 \end{vmatrix} = \begin{vmatrix} 1 & 5 & 6 & 7 \\ 0 & 3 & 1 & 2 \\ 0 & 0 & 0 & 3 \\ 0 & 1 & 4 & -2 \end{vmatrix} \quad \text{(by 2 elementary row operations)}$$

$$= 1\begin{vmatrix} 3 & 1 & 2 \\ 0 & 0 & 3 \\ 1 & 4 & -2 \end{vmatrix} + 0 \quad \text{(by expansion along the first column)}$$

$$= (-3)\begin{vmatrix} 3 & 1 \\ 1 & 4 \end{vmatrix} \quad \text{(by expansion along the second row)}$$

$$= (-3)(12 - 1) = -33$$

The Determinant and Invertibility of Matrices

It follows from Theorems 1, 2, 3, and 4 that there is a connection between the determinant of a square matrix, its rank, and its invertibility.

THEOREM 5. Suppose that A is an n by n matrix. Then: (i) $\det A \neq 0 \Leftrightarrow$ (ii) the rank of A is $n \Leftrightarrow$ (iii) A is invertible.

Proof. It was proved in Section 3–5 that (ii) if and only if (iii), so it is only necessary to prove that (i) if and only if (ii).

Notice that Theorems 1, 2, and 3 indicate that if $\det A \neq 0$, then matrices obtained from A by elementary row operations will also have non-zero determinant. Every matrix is row equivalent to a matrix in row echelon form; this row echelon form has leading 1's in every entry on the main diagonal if and only if the rank of the matrix is n. Hence, a given matrix A is of rank n if and only if $\det A \neq 0$.

■

We shall see how to use the determinant in calculating the inverse in the next section. It is worth noting that Theorem 5 implies that "almost all" square matrices are invertible, because a square matrix fails to be invertible only if it satisfies the special condition $\det A = 0$. (The idea of "almost all" can be made precise in calculus.)

The Determinant and the Inverse of Non-linear Differentiable Functions

The determinant also plays an important role in calculus in establishing whether a *non-linear* differentiable mapping $\mathbf{f}: \mathbb{R}^n \to \mathbb{R}^n$ has an inverse. Such a mapping will have n component functions $\mathbf{f} = (f_1, f_2, \ldots, f_n)$, so it is possible to construct at each point an n by n matrix of partial derivatives. For example, consider the mapping $\mathbf{f}(u, v) = (f_1(u, v), f_2(u, v))$, then the matrix is $\begin{bmatrix} \dfrac{\partial f_1}{\partial u} & \dfrac{\partial f_1}{\partial v} \\ \dfrac{\partial f_2}{\partial u} & \dfrac{\partial f_2}{\partial v} \end{bmatrix}$. The theory of differentiation says that this matrix (sometimes called the "derivative matrix" of $\mathbf{f}$ at (u, v)) is the matrix of the "best linear approximation" to $\mathbf{f}$ at (u, v). The *determinant* of the derivative matrix is called the **Jacobian** of $\mathbf{f}$ at (u, v). The Inverse Function Theorem (in calculus) says that if the Jacobian of $\mathbf{f}$ at (u, v) is non-zero, then $\mathbf{f}$ has an inverse defined near the point $\mathbf{f}(u, v)$.

EXERCISES 5–2

A1 Use row operations and triangular form to compute the determinant of each of the following matrices. Show your work clearly. Decide whether each matrix is invertible.

(a) $\begin{bmatrix} 1 & 2 & 4 \\ 3 & 1 & 0 \\ -1 & 3 & 2 \end{bmatrix}$ (b) $\begin{bmatrix} 3 & 2 & 2 \\ 2 & 2 & 1 \\ 1 & 1 & 1 \end{bmatrix}$

(c) $\begin{bmatrix} 5 & 2 & -1 & 1 \\ 1 & 2 & -1 & 1 \\ 3 & 2 & 1 & 4 \\ -2 & 0 & 3 & 5 \end{bmatrix}$ (d) $\begin{bmatrix} 5 & 10 & 5 & -5 \\ 1 & 3 & 5 & 7 \\ 1 & 2 & 6 & 3 \\ -1 & 7 & 1 & 1 \end{bmatrix}$

A2 Use a combination of row operations and cofactor expansion (as in Example 13) to evaluate the determinant of each of these matrices.

(a) $\begin{bmatrix} 1 & -1 & 2 \\ 1 & 1 & -2 \\ 1 & 2 & 3 \end{bmatrix}$ (b) $\begin{bmatrix} 1 & 2 & 1 & 2 \\ 2 & 4 & 1 & 5 \\ 3 & 6 & 5 & 9 \\ 1 & 3 & 4 & 3 \end{bmatrix}$

A3 Use row operations to compute the determinant of each of the following matrices. In each case determine all values of p such that the matrix is invertible.

(a) $\begin{bmatrix} 2 & 3 & 1 \\ 1 & 1 & -1 \\ 4 & p & -2 \end{bmatrix}$ (b) $\begin{bmatrix} 2 & 3 & 1 & p \\ 0 & 1 & 2 & 1 \\ 0 & 1 & 7 & 6 \\ 1 & 0 & 1 & 0 \end{bmatrix}$ (c) $\begin{bmatrix} 1 & 1 & 1 & 1 \\ 1 & 2 & 3 & 4 \\ 1 & 4 & 9 & 16 \\ 1 & 8 & 27 & p \end{bmatrix}$

A4 Suppose that B is a 4 by 4 matrix, with j^{th} row $b_{j\rightarrow}$, and $\det B = 7$.

(a) If D is obtained from B by interchanging rows 1 and 3, what is $\det D$?

(b) If E is obtained from B by replacing the second row by $2b_{2\rightarrow} + 3b_{3\rightarrow}$, what is $\det E$?

(c) If F is obtained from B by replacing the second row by $2b_{1\rightarrow} + 3b_{3\rightarrow}$, what is $\det F$?

(d) What is $\det(2B)$?

B1 Use row operations and triangular form to compute the determinant of each of the following matrices. Show your work clearly. Decide whether each matrix is invertible.

(a) $\begin{bmatrix} 2 & 3 & 2 \\ 2 & -1 & 1 \\ 1 & 1 & 4 \end{bmatrix}$ (b) $\begin{bmatrix} 1 & 2 & 3 \\ 2 & -4 & 1 \\ 3 & 5 & -6 \end{bmatrix}$

(c) $\begin{bmatrix} 7 & 1 & -1 & 1 \\ 3 & 3 & -4 & 5 \\ 3 & 2 & 1 & 4 \\ 1 & 1 & -1 & 1 \end{bmatrix}$ (d) $\begin{bmatrix} 1 & 10 & 7 & -9 \\ 7 & -7 & 7 & 7 \\ 2 & -2 & 6 & 2 \\ -3 & -3 & 4 & 1 \end{bmatrix}$

B2 Use a combination of row operations and cofactor expansion (as in Example 13) to evaluate the determinant of each of these matrices.

(a) $\begin{bmatrix} 1 & 1 & -2 & 1 \\ 1 & 3 & -1 & -1 \\ 2 & 2 & -2 & 7 \\ 1 & 1 & 0 & 2 \end{bmatrix}$ (b) $\begin{bmatrix} 2 & 3 & 2 & -2 \\ 3 & 1 & 4 & 1 \\ 1 & 2 & 4 & 4 \\ 2 & -1 & 5 & 6 \end{bmatrix}$

B3 Use row operations to compute the determinant of each of the following matrices. In each case determine all values of p such that the matrix is invertible.

(a) $\begin{bmatrix} 1 & 0 & -1 \\ 2 & 1 & 2 \\ p & 1 & -2 \end{bmatrix}$ (b) $\begin{bmatrix} 1 & 1 & 1 \\ 2 & p & p \\ 2 & 2 & 1 \end{bmatrix}$

(c) $\begin{bmatrix} 2 & 5 & 2 & 4 \\ 0 & 1 & -1 & 1 \\ 0 & 1 & 4 & 2 \\ 1 & 0 & 1 & p \end{bmatrix}$ (d) $\begin{bmatrix} 1 & 1 & 1 & 1 \\ 1 & 2 & 4 & 8 \\ 1 & 3 & 9 & 27 \\ 1 & 4 & 16 & p \end{bmatrix}$

B4 Suppose that B is a 4 by 4 matrix, with j^{th} row $b_{j\rightarrow}$. Suppose that $\det B = 5$.

(a) If D is obtained from B by interchanging rows 2 and 4, what is $\det D$?

(b) If E is obtained from B by replacing the third row by $3b_{2\rightarrow} - 4b_{3\rightarrow}$, what is $\det E$?

(c) If F is obtained from B by replacing the fourth row by $2b_{1\rightarrow} + 3b_{3\rightarrow}$, what is $\det F$?

(d) What is $\det(3B)$?

CONCEPTUAL EXERCISES

D1 A square matrix A is **skew-symmetric** if $A^T = -A$. If A is an n by n skew-symmetric matrix with n odd, prove that $\det A = 0$.

D2 (a) Prove that

$$\det\begin{bmatrix} a+p & b+q & c+r \\ d & e & f \\ g & h & i \end{bmatrix} = \det\begin{bmatrix} a & b & c \\ d & e & f \\ g & h & i \end{bmatrix} + \det\begin{bmatrix} p & q & r \\ d & e & f \\ g & h & i \end{bmatrix}.$$

(b) Use part (a) to express $\det\begin{bmatrix} a+p & b+q & c+r \\ d+x & e+y & f+z \\ g & h & i \end{bmatrix}$ as the sum of four determinants of matrices whose entries are *not* sums.

(c) Let A and B be 3 by 3 matrices. Express $\det(A+B)$ as a sum of eight determinants of the form $\det\begin{bmatrix} a_{1\to} \\ b_{2\to} \\ a_{3\to} \end{bmatrix}$, and so on.

***D3** Prove that

$$\det\begin{bmatrix} a+b & p+q & u+v \\ b+c & q+r & v+w \\ c+a & r+p & w+u \end{bmatrix} = 2\det\begin{bmatrix} a & p & u \\ b & q & v \\ c & r & w \end{bmatrix}.$$

D4 Prove that $\det\begin{bmatrix} 1 & 1 & 1 \\ 1 & 1+a & 1+2a \\ 1 & (1+a)^2 & (1+2a)^2 \end{bmatrix} = 2a^3$.

***D5** (If you know calculus)
Suppose that A is a 4 by 4 matrix, and that the third row of A is replaced by $[t \quad 2t \quad 3t \quad 4t]$; call the resulting matrix $B(t)$. Obtain a simple expression for $\frac{d}{dt}(\det B(t))$ in terms of cofactors of A.

SECTION

5–3 The Inverse by the Cofactor Method; Cramer's Rule

The determinant provides an alternative method of calculating the inverse of a square matrix. Because determinant calculations are generally much longer than row reduction, this method of finding the inverse is not as efficient as the earlier method based on row reduction. However, it is useful in some theoretical applications because it provides a **formula** for A^{-1} in terms of the entries of A.

This method is based on a simple calculation, which makes use of the fact that $a_{j\to} \cdot C_{j\to} = \det A$, while if $j \neq k$, then $a_{k\to} \cdot C_{j\to} = 0$ by the False Expansion Theorem of Section 5–2. The transpose of the matrix of cofactors plays a major role; the cofactors of the jth row of A form the jth **row** of Cof(A), so they form the jth **column** of $(\text{Cof}(A))^T$. The calculation is given for the 3 by 3 case, but it is easy to see that it generalizes to the n by n case.

$$A(\text{Cof}(A))^T = \begin{bmatrix} a_{1\rightarrow} \\ a_{2\rightarrow} \\ a_{3\rightarrow} \end{bmatrix} \begin{bmatrix} C_{11} & C_{21} & C_{31} \\ C_{12} & C_{22} & C_{32} \\ C_{13} & C_{23} & C_{33} \end{bmatrix}$$

(Since $C_{1\rightarrow}$ is the first row of Cof(A), it is the first column of $(\text{Cof}(A))^T$.)

$$= \begin{bmatrix} a_{1\rightarrow}\cdot C_{1\rightarrow} & a_{1\rightarrow}\cdot C_{2\rightarrow} & a_{1\rightarrow}\cdot C_{3\rightarrow} \\ a_{2\rightarrow}\cdot C_{1\rightarrow} & a_{2\rightarrow}\cdot C_{2\rightarrow} & a_{2\rightarrow}\cdot C_{3\rightarrow} \\ a_{3\rightarrow}\cdot C_{1\rightarrow} & a_{3\rightarrow}\cdot C_{2\rightarrow} & a_{3\rightarrow}\cdot C_{3\rightarrow} \end{bmatrix}$$

$$= \begin{bmatrix} \det A & 0 & 0 \\ 0 & \det A & 0 \\ 0 & 0 & \det A \end{bmatrix} = (\det A)I,$$

where I is the identity matrix. (The off-diagonal entries are zero because of the False Expansion Theorem.)

If $\det A \neq 0$, it follows that $A\left(\frac{1}{\det A}\right)(\text{Cof}(A))^T = I$, and therefore,

$$A^{-1} = \left(\frac{1}{\det A}\right)(\text{Cof}(A))^T.$$

If $\det A = 0$, then A is not invertible by Theorem 5 of Section 5–2. We shall refer to this method of finding the inverse as the "cofactor method". (Some people refer to the transpose of the matrix of cofactors as the "adjoint matrix", and therefore call this the "adjoint method". The word "adjoint" is used to mean other things in mathematics, so we avoid its use here.)

EXAMPLE 14

Find the inverse of the matrix A by the cofactor method, where

$$A = \begin{bmatrix} 2 & 4 & -1 \\ 0 & 3 & 1 \\ 6 & -2 & 5 \end{bmatrix}.$$

SOLUTION:

$$\text{Cof}(A) = \begin{bmatrix} 17 & 6 & -18 \\ -18 & 16 & 28 \\ 7 & -2 & 6 \end{bmatrix}, \text{ so}$$

$$(\text{Cof}(A))^T = \begin{bmatrix} 17 & -18 & 7 \\ 6 & 16 & -2 \\ -18 & 28 & 6 \end{bmatrix}.$$

Note that there is a check available: $A\,(\mathrm{Cof}(A))^T$ is supposed to equal $(\det A)I$. Here,

$$\begin{bmatrix} 2 & 4 & -1 \\ 0 & 3 & 1 \\ 6 & -2 & 5 \end{bmatrix} \begin{bmatrix} 17 & -18 & 7 \\ 6 & 16 & -2 \\ -18 & 28 & 6 \end{bmatrix} = \begin{bmatrix} 76 & 0 & 0 \\ 0 & 76 & 0 \\ 0 & 0 & 76 \end{bmatrix},$$

so $\det A = 76$, and the matrix $(\mathrm{Cof}(A))^T$ is correct. It follows that

$$A^{-1} = \frac{1}{76} \begin{bmatrix} 17 & -18 & 7 \\ 6 & 16 & 2 \\ -18 & -28 & 6 \end{bmatrix}.$$

For 3 by 3 matrices, this method requires the evaluation of nine 2 by 2 determinants, which is manageable, but more work than would be required by the method of row reduction. To find the inverse of a 4 by 4 matrix by this method would require the evaluation of sixteen 3 by 3 determinants, and this method then becomes extremely unattractive.

Cramer's Rule

Consider the system of n linear equations in n variables, $A\mathbf{x} = \mathbf{b}$. If $\det A \neq 0$ so that A is invertible, the solution may be written in the form:

$$[\mathbf{x}] = A^{-1}[\mathbf{b}] = (\frac{1}{\det A})(\mathrm{Cof}(A))^T[\mathbf{b}]$$

$$= (\frac{1}{\det A}) \begin{bmatrix} C_{11} & C_{21} & C_{31} & \dots & C_{n1} \\ C_{12} & C_{22} & C_{32} & \dots & C_{n2} \\ C_{13} & C_{23} & C_{33} & \dots & C_{n3} \\ \vdots & \vdots & \vdots & \ddots & \vdots \\ C_{1n} & C_{2n} & C_{3n} & \dots & C_{nn} \end{bmatrix} \begin{bmatrix} b_1 \\ b_2 \\ b_3 \\ \vdots \\ b_n \end{bmatrix}.$$

Consider the value of the first component of $\mathbf{x}$:

$$x_1 = (1/\det A)(b_1C_{11} + b_2C_{21} + b_3C_{31} + \cdots + b_nC_{n1}).$$

This is $(1/\det A)$ multiplied by the dot product of the vector $\mathbf{b}$ with the first **row** of $(\mathrm{Cof}(A))^T$. But the first **row** of $(\mathrm{Cof}(A))^T$ is the first **column** of the original matrix of cofactors $\mathrm{Cof}(A)$, so

$$x_1 = (1/\det A)(\mathbf{b} \cdot C_{\downarrow 1}).$$

The Substitution Lemma of Section 5–2 states what happens when we substitute $\mathbf{b}$ for a row of the matrix A. Since $\det(A^T) = \det A$, there is a corresponding column version of this Lemma. By this column version, it follows that $x_1 = \left(\dfrac{\mathbf{b} \cdot C_{\downarrow 1}}{\det A}\right) = \left(\dfrac{\text{Num}_1}{\det A}\right)$, where this first numerator Num_1 is the determinant of the matrix obtained from A by substituting $[\mathbf{b}]$ for the first column:

$$\text{Num}_1 = \mathbf{b} \cdot C_{\downarrow 1} = \det \begin{bmatrix} b_1 & a_{12} & a_{13} & \cdots & a_{1n} \\ b_2 & a_{22} & a_{23} & \cdots & a_{2n} \\ b_3 & a_{32} & a_{33} & \cdots & a_{3n} \\ \vdots & \vdots & \vdots & \ddots & \vdots \\ b_n & a_{n2} & a_{n3} & \cdots & a_{nn} \end{bmatrix}.$$

The other components of x can be expressed in similar fashion, so the solution is $\mathbf{x} = \left(\dfrac{\text{Num}_1}{\det A}, \dfrac{\text{Num}_2}{\det A}, \ldots, \dfrac{\text{Num}_n}{\det A}\right)$, where Num_j is the determinant of the matrix obtained from A by substituting $[\mathbf{b}]$ for the jth column. This is called Cramer's Rule (or Method).

EXAMPLE 15

Solve the system of equations by Cramer's Rule.

$$\begin{aligned} x + y - z &= U \\ 2x + 4y + 5z &= V \\ x + y + 2z &= W \end{aligned}$$

SOLUTION: The coefficient matrix is $A = \begin{bmatrix} 1 & 1 & -1 \\ 2 & 4 & 5 \\ 1 & 1 & 2 \end{bmatrix}$, so

$$\det A = 1(8-5) + 1(-1)(4-5) + (-1)(2-4) = 6.$$

Hence,

$$(x, y, z) = \left(\frac{\det \begin{bmatrix} U & 1 & -1 \\ V & 4 & 5 \\ W & 1 & 2 \end{bmatrix}}{6}, \frac{\det \begin{bmatrix} 1 & U & -1 \\ 2 & V & 5 \\ 1 & W & 2 \end{bmatrix}}{6}, \frac{\det \begin{bmatrix} 1 & 1 & U \\ 2 & 4 & V \\ 1 & 1 & W \end{bmatrix}}{6} \right)$$

$$= \left(\frac{3U - 3V + 9W}{6}, \frac{U + 3V - 7W}{6}, \frac{-2U + 2W}{6} \right).$$

To solve a system of n equations in n variables by Cramer's Method would require the evaluation of the determinant of $n+1$ matrices, each of which is n by n. The most efficient way to evaluate these determinants is to row reduce each matrix to upper triangular form. Solving the system by Cramer's Method thus requires far more calculation than elimination with back-substitution, so Cramer's Method is not considered a computationally useful solution method. However, Cramer's Method is sometimes used to write a "formula" for the solution of a problem.

EXERCISES 5–3

A1 Determine the inverse of each of the following matrices by the cofactor method. Verify by multiplication.

(a) $A = \begin{bmatrix} 3 & -5 \\ 2 & -1 \end{bmatrix}$ (b) $B = \begin{bmatrix} 4 & 1 & 7 \\ 2 & -3 & 1 \\ -2 & 6 & 0 \end{bmatrix}$

A2 Let $A = \begin{bmatrix} 2 & -3 & k \\ -2 & 0 & 1 \\ 3 & 1 & 1 \end{bmatrix}$.

(a) Determine the matrix of cofactors, $\text{Cof}(A)$.

(b) Calculate $A(\text{Cof}(A))^T$, and determine $\det(A)$ and A^{-1}.

A3 Use Cramer's Rule to solve the following systems.

(a)
$$\begin{aligned} 2x_1 - 3x_2 &= 6 \\ 3x_1 + 5x_2 &= 7 \end{aligned}$$

(b)
$$\begin{aligned} 2x_1 + 3x_2 - 5x_3 &= 2 \\ 3x_1 - x_2 + 2x_3 &= -1 \\ 5x_1 + 4x_2 - 6x_3 &= 3 \end{aligned}$$

B1 Determine the inverse of each of the following matrices by the cofactor method. Verify by multiplication.

(a) $A = \begin{bmatrix} -3 & 4 \\ 7 & 2 \end{bmatrix}$ (b) $B = \begin{bmatrix} 0 & 2 & -3 \\ -3 & 3 & 0 \\ 2 & 5 & 4 \end{bmatrix}$

(c) $C = \begin{bmatrix} 5 & 4 \\ -2 & 3 \end{bmatrix}$ (d) $D = \begin{bmatrix} 1 & 0 & 2 \\ -2 & 1 & 0 \\ 3 & -1 & 1 \end{bmatrix}$

B2 Let $A = \begin{bmatrix} 2 & 1 & 3 \\ 3 & a & -1 \\ -2 & 1 & 4 \end{bmatrix}$.

(a) Determine the matrix of cofactors, $\text{Cof}(A)$.

(b) Calculate $A(\text{Cof}(A))^T$, and determine $\det(A)$ and A^{-1}.

B3 Use Cramer's Rule to solve the following systems.

(a)
$$\begin{aligned} 2x_1 - 7x_2 &= 3 \\ 5x_1 + 4x_2 &= -17 \end{aligned}$$

(b)
$$\begin{aligned} x_1 - 5x_2 - 2x_3 &= -2 \\ 2x_1 \qquad + 3x_3 &= 3 \\ 4x_1 + 1x_2 - 1x_3 &= 1 \end{aligned}$$

(c)
$$\begin{aligned} 3x_1 + 5x_2 &= -2 \\ x_1 + 3x_2 &= -3 \end{aligned}$$

(d)
$$\begin{aligned} x_1 \qquad + 2x_3 &= -2 \\ 3x_1 - x_2 + 3x_3 &= 5 \\ -2x_1 + x_2 \qquad &= 0 \end{aligned}$$

CONCEPTUAL EXERCISES

D1 Suppose that A is an invertible n by n matrix.

(a) Verify by Cramer's Method that the system of equations $A[\mathbf{x}] = [a_{\downarrow j}]$ has the unique solution $\mathbf{x} = \mathbf{e}_j$ (the jth standard basis vector).

(b) Explain the result of part (a) in terms of linear transformations and/or block multiplication.

***D2** Let $A = \begin{bmatrix} 2 & -1 & 0 & 1 \\ 0 & -1 & 3 & 2 \\ 0 & 1 & 0 & 0 \\ 0 & 2 & 0 & 3 \end{bmatrix}$.

Use the cofactor method to calculate $(A^{-1})_{23}$ and $(A^{-1})_{42}$. (If you calculate more than these two entries of A^{-1}, you have missed the point.)

SECTION

5–4 The Determinant of a Product, and Volume

Often it is necessary to calculate the determinant of the product of two matrices, $\det(AB)$. When you remember that each entry of AB is the dot product of a row from A and a column from B, and that the rule for calculating determinants is quite complicated, you might expect a very complicated rule. The following theorem should be a welcome surprise.

THEOREM 1. If A and B are n by n matrices, then $\det(AB) = (\det A)(\det B)$.

Proof. To prove this theorem by straightforward computation is obviously not sensible (except maybe for the 2 by 2 case). The usual proof depends on some facts about elementary matrices introduced in Section 3–6; it is deferred until the end of this section.

■

EXAMPLE 16

Verify Theorem 1 for the case

$$A = \begin{bmatrix} 3 & 0 & 1 \\ 2 & -1 & 4 \\ 5 & 2 & 0 \end{bmatrix} \quad \text{and} \quad B = \begin{bmatrix} -1 & 2 & 4 \\ 7 & 1 & 0 \\ 1 & -2 & 3 \end{bmatrix}.$$

SOLUTION:

$\det A = -15$, $\det B = -105$, so $(\det A)(\det B) = 1575$. On the other hand, $AB = \begin{bmatrix} -2 & 4 & 15 \\ -5 & -5 & 20 \\ 9 & 12 & 20 \end{bmatrix}$, so $\det AB = 1575$, and therefore $\det AB = (\det A)(\det B)$.

COROLLARY to Theorem 1 If $\det A \neq 0$, $\det(A^{-1}) = 1/\det A$.

Proof. Since $AA^{-1} = I$, $(\det A)(\det A^{-1}) = \det(AA^{-1}) = \det I = 1$, and the result follows by division by $\det A$.

■

Although the rule for the determinant of a product does not have a simple algebraic proof, there is a connection between determinants and volumes that gives a satisfying interpretation of the rule for products.

The Determinant and Volume

Recall from Section 1–4 that the volume of the parallelepiped determined by vectors **a**, **b**, **c** in $\mathbb{R}^3$ is given by

$$\text{the absolute value of det} \begin{bmatrix} c_1 & c_2 & c_3 \\ a_1 & a_2 & a_3 \\ b_1 & b_2 & b_3 \end{bmatrix}.$$

By interchanging rows twice, we may rewrite this as

$$\text{the absolute value of det} \begin{bmatrix} a_1 & a_2 & a_3 \\ b_1 & b_2 & b_3 \\ c_1 & c_2 & c_3 \end{bmatrix}.$$

Since the determinant of a matrix is equal to the determinant of its transpose, this may be rewritten:

$$\text{volume}(\mathbf{a}, \mathbf{b}, \mathbf{c}) = \left|\det\left[[\mathbf{a}] \ [\mathbf{b}] \ [\mathbf{c}]\right]\right|,$$

where as usual, [**a**] denotes a **column** matrix.

Now suppose that the 3 by 3 matrix A is the (standard) matrix of a linear transformation $L : \mathbb{R}^3 \to \mathbb{R}^3$. Then the images of **a**, **b**, and **c** are A**a**, A**b**, and A**c** respectively, and the volume of the **image parallelepiped** is

$$\text{volume}(A\mathbf{a}, A\mathbf{b}, A\mathbf{c}) = \left|\det\left[A[\mathbf{a}] \;\; A[\mathbf{b}] \;\; A[\mathbf{c}]\right]\right|.$$

By block multiplication

$$\left[A[\mathbf{a}] \;\; A[\mathbf{b}] \;\; A[\mathbf{c}]\right] = A\left[[\mathbf{a}] \;\; [\mathbf{b}] \;\; [\mathbf{c}]\right],$$

and therefore,

$$\text{volume}(A\mathbf{a}, A\mathbf{b}, A\mathbf{c}) = \left|\det A\right|\left|\det\left[[\mathbf{a}] \;\; [\mathbf{b}] \;\; [\mathbf{c}]\right]\right| = \left|\det A\right| \text{volume}(\mathbf{a}, \mathbf{b}, \mathbf{c}).$$

In words, this result can be described as follows: the absolute value of the determinant of A is the factor by which volume is changed under the linear transformation with matrix A. The result is illustrated in Figure 5–4–1, and restated in the following theorem.

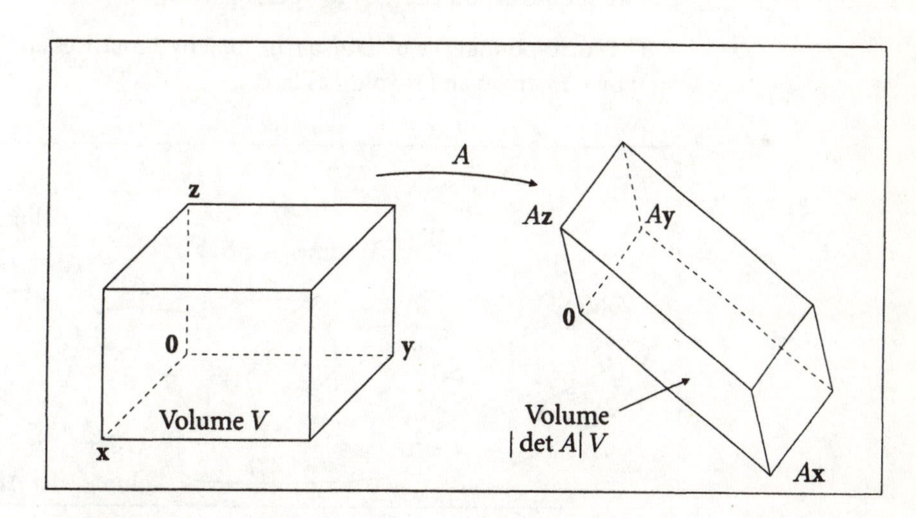

FIGURE 5-4-1. *Under a linear transformation with matrix A, the volume of a figure is changed by factor | det A|.*

THEOREM 2. Suppose that the 3 by 3 matrix A is the matrix of a linear transformation, and that the vectors **a**, **b**, and **c** in $\mathbb{R}^3$ determine a parallelepiped of non-zero volume. Then

$$|\det A| = \frac{\text{volume of the image parallelepiped under } A}{\text{volume of the original parallelipiped}}.$$

Proof. Calculation above. ∎

Notice that the calculation leading to this result has used the fact that the determinant of a product is the product of determinants. However, Theorem 2 also leads to the following geometrical interpretation of Theorem 1. (See Figure 5–4–2.)

Suppose that A is the matrix of a linear transformation $L_A : \mathbb{R}^3 \to \mathbb{R}^3$, and that B is the matrix of another linear transformation $L_B : \mathbb{R}^3 \to \mathbb{R}^3$. Then the composite map $L_A \circ L_B$ is a linear transformation with matrix AB, and under this transformation, volume is multiplied by the factor $|\det AB|$. By geometrical arguments, we expect that this should give the same result as first transforming by L_B (thus multiplying volume by factor $|\det B|$), and then transforming by L_A (thus multiplying volume by factor $|\det A|$). It follows that we should expect $|\det AB| = |\det A||\det B|$.

The corresponding result is also true for 2 by 2 matrices and areas in $\mathbb{R}^2$, or for n by n matrices and n-volumes in $\mathbb{R}^n$.

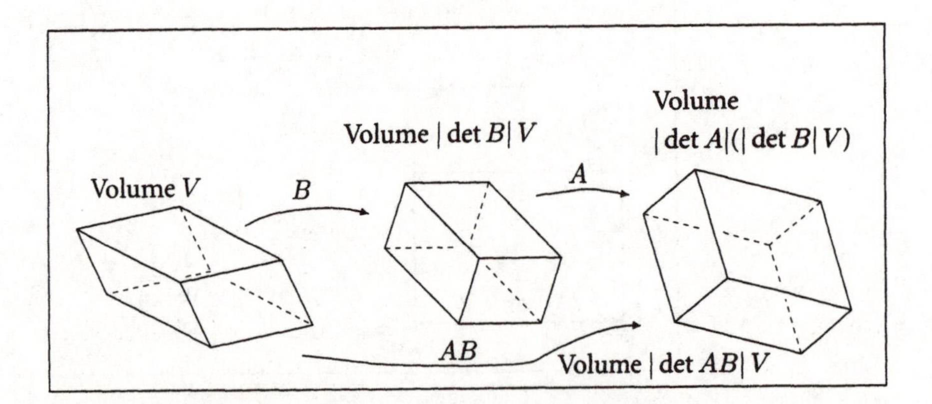

FIGURE 5-4-2. *Illustration of the fact that $|\det AB| = |\det A||\det B|$.*

EXAMPLE 17

Verify Theorem 2 for the case $\mathbf{a} = (1, -1, 0)$, $\mathbf{b} = (0, 1, 2)$, $\mathbf{c} = (-1, 5, 1)$, and

$$A = \begin{bmatrix} 4 & 1 & -1 \\ 2 & 4 & 1 \\ 1 & 1 & 4 \end{bmatrix}.$$

SOLUTION: The volume determined by **a**, **b**, and **c** is

$$\left|\det\begin{bmatrix} [\mathbf{a}] & [\mathbf{b}] & [\mathbf{c}] \end{bmatrix}\right| = \left|\det\begin{bmatrix} 1 & 0 & -1 \\ -1 & 1 & 5 \\ 0 & 2 & 1 \end{bmatrix}\right| = |-7| = 7;$$

the volume determined by the images $A\mathbf{a}$, $A\mathbf{b}$, and $A\mathbf{c}$ is

$$\left|\det\begin{bmatrix} A[\mathbf{a}] & A[\mathbf{b}] & A[\mathbf{c}] \end{bmatrix}\right| = \left|\det\begin{bmatrix} 3 & -1 & 0 \\ -2 & 6 & 19 \\ 0 & 9 & 8 \end{bmatrix}\right| = |-385| = 385.$$

Hence, the ratio $\dfrac{\text{volume of image}}{\text{original volume}} = \dfrac{385}{7} = 55.$

By direct computation, $|\det A| = |4(16-1) - 1(8-1) - 1(2-4)| = 55$ and the verification is complete.

The absolute value of $\det A$ now has an interpretation; what about the sign of $\det A$? Recall that $\det\begin{bmatrix} [\mathbf{a}] & [\mathbf{b}] & [\mathbf{c}] \end{bmatrix} = \mathbf{a} \cdot (\mathbf{b} \times \mathbf{c})$. Suppose that $\mathbf{a} \cdot \mathbf{b} \times \mathbf{c} > 0$, so that (by Section 1–4) the vectors **a**, **b**, and **c**, taken in this order, form a right-handed triple of vectors. Since

$$\det\begin{bmatrix} [A\mathbf{a}] & [A\mathbf{b}] & [A\mathbf{c}] \end{bmatrix} = (\det A)\det\begin{bmatrix} [\mathbf{a}] & [\mathbf{b}] & [\mathbf{c}] \end{bmatrix},$$

the ordered triple of image vectors $\{A\mathbf{a}, A\mathbf{b}, A\mathbf{c}\}$ is right-handed if $\det A$ is positive and left-handed if $\det A$ is negative. We conclude that the sign of $\det A$ determines whether the linear transformation is "orientation preserving" or "orientation reversing". (These ideas actually provide the definition of "orientation" in higher dimensions.)

Proof of Theorem 1. (This proof depends on facts from Section 3–6.)

The case where $\det A = 0$.

If $\det A = 0$, the rank of A is less than n, and it follows that the rank of AB is less than n. (This is a consequence of Theorem 3 of Section 3–5, or the matrix version of Exercise D2 of Section 4–5.) Since the rank of AB is less than n, (AB) is not invertible, and $\det AB = 0$. Thus in this case, $(\det A)(\det B) = 0 = \det AB$ (whatever the value of $\det B$).

The case where $\det A \neq 0$.

Since $\det A \neq 0$, A is invertible and thus has rank n. In Section 3–6, elementary matrices were introduced and it was shown (Theorem 2) that if A has rank n then A is the product of elementary matrices $A = F_1F_2\ldots F_k$. Consider $\det(AB) = \det(F_1F_2\ldots F_kB) = \det(F_1(F_2\ldots F_kB))$. If it can be shown that $\det FC = (\det F)(\det C)$ for **any** elementary matrix F, and **any** matrix C, then by repeated applications of this fact, it will follow that

$$\begin{aligned}\det A &= \det(F_1F_2\ldots F_k) = (\det F_1)(\det F_2\ldots F_k) = \ldots \\ &= (\det F_1)(\det F_2)\ldots(\det F_k)\end{aligned}$$

and then,

$$\begin{aligned}\det AB &= \det(F_1F_2\ldots F_kB) = (\det F_1)(\det F_2\ldots F_kB) = \ldots \\ &= (\det F_1)(\det F_2)\ldots(\det F_k)(\det B) \\ &= (\det A)(\det B).\end{aligned}$$

Thus the theorem is a consequence of the following lemma.

■

LEMMA. If F is an n by n elementary matrix and C is any n by n matrix, then $\det FC = (\det F)(\det C)$.

Proof. The three different types of elementary matrices must be considered. Suppose that F is the matrix that corresponds to multiplying the jth row by a constant k. Then F is obtained from I by this row operation, and by Theorem 1 in Section 5–2, $\det F = k \det I = k$. On the other hand, by the same theorem, since FC is the matrix obtained from C by multiplying its jth row by k, $\det FC = k \det C$. It follows that $\det FC = (\det F)(\det C)$.

It is left to the reader to show that the lemma is true in the cases of the other two types of elementary matrix.

■

Determinant and Change of Variable of Integration in Calculus

At the end of Section 5–2, the Jacobian (determinant) of a differentiable mapping was introduced. The Jacobian also plays a key role in the "Change of Variables of Integration Theorem" for functions of several variables. A typical non-linear change of variables in $\mathbb{R}^3$ can be described by a differentiable mapping $\mathbf{f} : \mathbb{R}^3 \to \mathbb{R}^3$. Because the definite integral of a function of three variables is (roughly speaking) the limit of a sum of contributions from typical volumes ΔV, and because these volumes are multiplied by the determinant of the mapping (in the case of **f**, by the Jacobian determinant of the best linear approximation to **f**), the Jacobian is an essential part of the formula for change of variables.

EXERCISES 5-4

A1 Verify Theorem 1 if

(a) $A = \begin{bmatrix} 2 & -1 \\ 3 & 5 \end{bmatrix}$, $B = \begin{bmatrix} 3 & 2 \\ -1 & 4 \end{bmatrix}$;

(b) $A = \begin{bmatrix} 2 & 4 & 1 \\ -1 & 2 & -1 \\ 3 & 0 & 2 \end{bmatrix}$, $B = \begin{bmatrix} 1 & 3 & 2 \\ -1 & 0 & 5 \\ 4 & 1 & 1 \end{bmatrix}$.

A2 (a) Calculate (by means of a 2 by 2 determinant) the area of the parallelogram determined by the vectors $(3, 5)$ and $(2, 7)$ in $\mathbb{R}^2$.

(b) Determine the image of these vectors under the linear mapping with matrix $\begin{bmatrix} -4 & 6 \\ 3 & 2 \end{bmatrix}$.

(c) Compute the determinant of this matrix.

(d) Compute the area of the parallelogram determined by the image vectors in two ways.

A3 (a) Compute the volume of the parallelepiped with edge vectors $(2, 3, 4)$, $(2, -1, -5)$, and $(1, 5, 2)$.

(b) Compute the determinant of $\begin{bmatrix} 1 & -1 & 3 \\ 4 & 0 & 1 \\ 0 & 2 & 5 \end{bmatrix}$.

(c) What is the volume of the image of the parallelepiped of part (a) under the linear mapping with the matrix of part (b)?

A4 Repeat Exercise A3 with vectors $(0, 2, 3)$, $(1, -5, 5)$, $(2, 1, 6)$ and matrix $\begin{bmatrix} 3 & -1 & 2 \\ 3 & 2 & -1 \\ 1 & 4 & 5 \end{bmatrix}$.

A5 (a) Calculate the 4-volume of the 4-dimensional parallelepiped determined by the vectors $(1, 0, 2, 0)$, $(0, 1, 1, 3)$, $(0, 2, 3, 0)$, $(1, 0, 2, 5)$.

(b) Calculate the 4-volume of the image of this parallelepiped under the linear mapping with matrix

$$\begin{bmatrix} 2 & 3 & 1 & 1 \\ 5 & 4 & 3 & 0 \\ 0 & 0 & 7 & 3 \\ 0 & 0 & 0 & 1 \end{bmatrix}.$$

B1 Verify Theorem 1 if

(a) $A = \begin{bmatrix} 4 & -1 \\ -3 & 2 \end{bmatrix}$, $B = \begin{bmatrix} -5 & 1 \\ 1 & 2 \end{bmatrix}$;

(b) $A = \begin{bmatrix} 3 & 1 & -1 \\ 1 & -3 & 4 \\ 0 & 1 & 1 \end{bmatrix}$, $B = \begin{bmatrix} 2 & 1 & 3 \\ 3 & 1 & 0 \\ 1 & 1 & 4 \end{bmatrix}$.

B2 (a) Calculate (by means of a 2 by 2 determinant) the area of the parallelogram determined by the vectors $(3,7)$ and $(2,4)$ in $\mathbb{R}^2$.

(b) Determine the image of these vectors under the linear mapping with matrix $\begin{bmatrix} 2 & 8 \\ -1 & 5 \end{bmatrix}$.

(c) Compute the determinant of this matrix.

(d) Compute the area of the parallelogram determined by the image vectors in two ways.

B3 (a) Compute the volume of the parallelepiped with edge vectors $(2,-1,0)$, $(1,-3,-1)$, and $(0,5,3)$.

(b) Compute the determinant of $\begin{bmatrix} 2 & -1 & 3 \\ 4 & 5 & 0 \\ -1 & 0 & 0 \end{bmatrix}$.

(c) What is the volume of the image of the parallelepiped of part (a) under the linear mapping with the matrix of part (b)?

B4 Repeat Exercise B3 with vectors $(1,1,-4)$, $(2,-1,3)$, $(-1,3,5)$ and matrix $\begin{bmatrix} 4 & -2 & 1 \\ 0 & 1 & 3 \\ -1 & 4 & 2 \end{bmatrix}$.

B5 (a) Calculate the 4-volume of the 4-dimensional parallelepiped determined by the vectors $(1,1,2,1)$, $(1,1,1,3)$, $(1,2,3,0)$, $(1,0,5,7)$.

(b) Calculate the 4-volume of the image of this parallelepiped under the linear mapping with matrix

$$\begin{bmatrix} 2 & 3 & 1 & 1 \\ 2 & 3 & 1 & 0 \\ -1 & 3 & 7 & 0 \\ 0 & 2 & 1 & 1 \end{bmatrix}.$$

CONCEPTUAL EXERCISES

***D1** Suppose that A is a 3 by 3 matrix and that $\det A = 0$. What can you say about the rows of A? Argue geometrically that the range of the linear mapping with matrix A cannot be all of $\mathbb{R}^3$, and that its nullspace cannot be trivial.

D2 Suppose that all the entries in the n by n matrix A are integers, and that $\det A = \pm 1$. Prove that the entries in A^{-1} are all integers.

***D3** (a) The linear mapping from $\mathbb{R}^2$ to $\mathbb{R}^2$ with matrix $A = \begin{bmatrix} k & 0 \\ 0 & 1/k \end{bmatrix}$ is said to be **area-preserving** because $|\det A| = 1$. Make a list of some other area-preserving linear maps of $\mathbb{R}^2$. For each such map, give the matrix, and describe the map geometrically (in words).

(b) Make a list of some volume-preserving linear maps of $\mathbb{R}^3$. For each such map, give a matrix, and a geometrical description in words.

D4 If $\det AB = 0$, is it necessarily true that $\det A = 0$ or $\det B = 0$? Prove it, or give a counter-example. (Assume that A and B are square.)

D5 (a) Let I denote the 3 by 3 identity matrix. What linear mapping from $\mathbb{R}^3$ to $\mathbb{R}^3$ has standard matrix kI, where k is a positive scalar?

(b) Explain in terms of volumes and part (a) why we should expect $\det(kI) = k^3$.

(c) Use (a) and (b) to explain why $\det(kA) = k^3 \det A$ for any 3 by 3 matrix A.

Chapter Review

Suggestions for Student Review

1 Define "cofactor", and explain cofactor expansion. Be especially careful about signs. (Section 5–1)

2 State as many facts as you can that simplify the evaluation of determinants; for each fact, explain why it is true. (Almost all of the theorems of Sections 5–1 and 5–2 are relevant; so is Theorem 1 of Section 5–4.)

3 Explain and justify the cofactor method for finding a matrix inverse. Write down a 3 by 3 matrix and calculate $A(\text{cof}(A))^T$.

4 How and why are determinants connected to volumes?

Chapter Quiz

1 By cofactor expansion along some column, evaluate $\det \begin{bmatrix} 2 & -4 & 0 & 0 \\ 1 & -2 & 2 & 9 \\ -3 & 6 & 0 & 3 \\ 1 & -1 & 0 & 0 \end{bmatrix}$.

2 By row-reducing to upper triangular form, evaluate

$$\det \begin{bmatrix} 3 & 2 & 7 & -8 \\ -6 & -1 & -9 & 20 \\ 3 & 8 & 21 & -17 \\ 3 & 5 & 12 & 1 \end{bmatrix}.$$

3 Evaluate $\det \begin{bmatrix} 0 & 2 & 0 & 0 & 0 \\ 0 & 0 & 0 & 3 & 0 \\ 0 & 0 & 0 & 0 & 1 \\ 0 & 0 & 4 & 0 & 0 \\ 5 & 0 & 0 & 0 & 6 \end{bmatrix}$.

4 Determine all values of k such that the matrix $\begin{bmatrix} k & 2 & 1 \\ 0 & 3 & k \\ 2 & -4 & 1 \end{bmatrix}$ is invertible.

5 Suppose that A is a 5 by 5 matrix; $\det A = 7$.
(a) If B is obtained from A by multiplying the fourth row of A by 3, what is $\det B$?
(b) If C is obtained from A by moving the first row to the bottom, and moving all other rows up, what is $\det C$? (Make it clear how you decided.)
(c) What is $\det(2A)$?
(d) What is $\det(A^{-1})$?
(e) What is $\det(A^T A)$?

6 Let $A = \begin{bmatrix} 2 & 3 & 1 \\ 1 & 1 & 1 \\ -2 & 0 & 2 \end{bmatrix}$. Determine $(A^{-1})_{31}$ (the entry in the third row, first column of A^{-1}) by the cofactor method.

7 Determine x_2 by Cramer's Method if

$$\begin{aligned} 2x_1 + 3x_2 + x_3 &= 1 \\ x_1 + x_2 - x_3 &= -1 \\ -2x_1 + 2x_3 &= 1. \end{aligned}$$

8 (a) What is the volume of the parallelepiped determined by the vectors $\mathbf{a} = (1, 1, -2)$, $\mathbf{b} = (2, -1, 3)$, $\mathbf{c} = (0, 3, 4)$?
(b) If A is the matrix $\begin{bmatrix} 2 & 1 & 5 \\ 0 & 3 & -2 \\ 0 & 0 & -4 \end{bmatrix}$, what is the volume of the parallelepiped determined by the vectors $A\mathbf{a}$, $A\mathbf{b}$, $A\mathbf{c}$?

Further Exercises

F1 Suppose that A is an n by n matrix with all row sums equal to zero. (That is, $\sum_{j=1}^{n} a_{ij} = 0$ for $i = 1, 2, \ldots, n$.) Prove that $\det A = 0$.

F2 Suppose that A and A^{-1} both have all integer entries. Prove that $\det A = \pm 1$. (Compare Exercise 5–4–D2.)

F3 Consider a triangle in the plane with vertices A, B, and C; we also denote the angles at these vertices by A, B, and C. Let the sides opposite to A, B, and C have lengths a, b, and c respectively. By trigonometry, show that

$$c = b \cos A + a \cos B.$$

Write similar equations for the other two sides, and use Cramer's Rule to show that

$$\cos A = \frac{b^2 + c^2 - a^2}{2bc}.$$

F4 (a) Let $V_3(a,b,c) = \det \begin{bmatrix} 1 & a & a^2 \\ 1 & b & b^2 \\ 1 & c & c^2 \end{bmatrix}$. Without expanding, argue that $(a-b)$, $(b-c)$, and $(c-a)$ are all factors of $V_3(a,b,c)$. By considering the cofactor of c^2, argue that

$$V_3(a,b,c) = (c-a)(c-b)(b-a).$$

(b) Let $V_4(a,b,c,d) = \det \begin{bmatrix} 1 & a & a^2 & a^3 \\ 1 & b & b^2 & b^3 \\ 1 & c & c^2 & c^3 \\ 1 & d & d^2 & d^3 \end{bmatrix}$. By arguments similar to those in part (a) (and without expanding the determinant), argue that

$$V_4(a,b,c,d) = (d-a)(d-b)(d-c)V_3(a,b,c)$$

(V_3 and V_4 are called Vandermonde determinants).

F5 Suppose that A is a 4 by 4 matrix partitioned into 2 by 2 blocks:

$$A = \left[\begin{array}{c|c} A_1 & A_2 \\ \hline A_3 & A_4 \end{array}\right].$$

(a) If $A_3 = 0$, show that $\det A = \det A_1 \det A_4$.

(b) Give an example to show that in general,

$$\det A \neq \det A_1 \det A_4 - \det A_2 \det A_3.$$

F6 Suppose that A is 3 by 3, B is 2 by 2, and $0_{m,n}$ denotes the m by n zero matrix.

(a) Show that $\det \left[\begin{array}{c|c} A & 0_{3,2} \\ \hline 0_{2,3} & B \end{array}\right] = \det A \det B$.

(b) What is $\det \left[\begin{array}{c|c} 0_{2,3} & B \\ \hline A & 0_{3,2} \end{array}\right]$?

F7 Suppose that $\mathbf{x}$ is determined as the solution of the system of equations $A\mathbf{x} = \mathbf{b}$, where A is n by n and invertible. How reliable are the values $x_1, x_2, \ldots, x_n$ if there are "small" errors in the components of $\mathbf{b}$, or the entries of A? To answer this question, use Cramer's Rule to express x_j as a function of A and $\mathbf{b}$, and determine the following derivatives in terms of A, $\mathbf{b}$, $\det A$, cofactors of A, etc.

(a) $\dfrac{\partial x_2}{\partial b_1}$

(b) $\dfrac{\partial x_2}{\partial b_2}$

(c) $\dfrac{\partial x_2}{\partial a_{12}}$

(d) $\dfrac{\partial x_2}{\partial a_{34}}$

CHAPTER

6

Eigenvectors and Diagonalization

The information contained in the matrix of a linear transformation is most clearly revealed if a basis can be found such that the matrix is in diagonal form. Such a basis must consist of eigenvectors of the linear transformation.

SECTION

6–1 Diagonalization

Recall from Section 4–6 that if $L : \mathbb{R}^n \to \mathbb{R}^n$ is a linear transformation, its matrix with respect to basis $\mathcal{B}$ is determined from its standard matrix by the equation

$$[L]_{\mathcal{B}} = P^{-1}[L]_S P,$$

where $P = \begin{bmatrix}[\mathbf{v}_1] & [\mathbf{v}_2] & \dots & [\mathbf{v}_n]\end{bmatrix}$ is the change of basis matrix; the columns of P are the standard coordinate vectors of the vectors of the basis $\mathcal{B}$. Examples 36, 37, and 39 in Section 4–6 show that we can more easily give a geometrical interpretation of a linear mapping L if there is a basis $\mathcal{B}$ such that the $\mathcal{B}$-matrix $[L]_{\mathcal{B}}$ is in diagonal form. Before exploring when such diagonalization is possible, we introduce some vocabulary and sketch some applications of diagonalization.

In Section 4–6, the emphasis was on the linear transformation L: how is L represented by a matrix relative to the basis $\mathcal{B}$? However, the actual calculations are done in terms of matrices, so we describe the problem purely in terms of matrices.

Let A be an n by n matrix. An n by n matrix B is **similar** to the matrix A if there exists an invertible matrix P such that

$$B = P^{-1}AP.$$

Notice that if B is similar to A, then by putting $Q = P^{-1}$, we have $A = Q^{-1}BQ$, so that A is also similar to B; we may say that "A and B are similar matrices". Note that "similar" is now a technical word when applied to matrices.

DEFINITION

An n by n matrix D is said to be **diagonal** or **in diagonal form** if it is of the form

$$D = \begin{bmatrix} d_1 & 0 & 0 & \dots & 0 \\ 0 & d_2 & 0 & \dots & 0 \\ 0 & 0 & d_3 & \dots & 0 \\ \vdots & \vdots & \vdots & \ddots & \vdots \\ 0 & 0 & 0 & \dots & d_n \end{bmatrix},$$

that is, if all off-diagonal entries are zero.

The identity matrix is obviously diagonal. So is the zero matrix, since the diagonal entries of a diagonal matrix can take any real values including zero.

The Big Question stated near the end of Section 4–6 can now be restated: is a given matrix A similar to a diagonal matrix? That is, given an n by n matrix A, is there an invertible matrix P such that $P^{-1}AP = D$, where D is in diagonal form? If there is such a matrix P, we shall say that P **diagonalizes** A, and that A is **diagonable.** (Some people prefer "diagonalizable", but it is awkward to say, so "diagonable" is quite common.)

Some Applications of Diagonalization

Geometrical mappings such as projections, reflections, and stretches (discussed in Section 3–3) are all most easily recognized if the basis $\mathcal{B}$ is chosen so that the $\mathcal{B}$-matrix of the mapping is diagonal. This was discussed in Section 4–6, and will be reviewed in Examples 1 and 2 below.

Another geometrical application of diagonalization occurs when we try to picture the graph of a quadratic equation in two variables, such as

$$ax^2 + 2bxy + cy^2 = d.$$

It turns out that we should consider the associated matrix $\begin{bmatrix} a & b \\ b & c \end{bmatrix}$; by diagonalizing this matrix, we can recognize the graph as an ellipse or a hyperbola, or perhaps as some degenerate case. This problem, together with a three-dimensional version, will be discussed in Section 8–3.

A physical application related to these geometrical applications is the analysis of the deformation of a solid. Imagine, for example, a small steel block that experiences a small deformation when some forces are applied. The change of shape in the block can be described in terms of a 3 by 3 "strain matrix". This matrix can always be diagonalized, so it turns out that we can identify the change of shape as the composition of three stretches, along mutually orthogonal directions. This application is discussed in Section 8–5.

Diagonalization is also an important tool for studying systems of linear difference equations, which arise in many settings. Consider, for example, a population that is divided into two groups; we count these two groups at regular intervals (say, once a month) so that at every time n, we have a vector $(p_1(n), p_2(n))$ that tells us how many are in each group. For some situations, the change from month to month can be described by saying that the vector $\mathbf{p}$ changes according to the rule

$$\mathbf{p}(n+1) = A\mathbf{p}(n),$$

where A is some known 2 by 2 matrix. It follows that $\mathbf{p}(n) = A^n\mathbf{p}(0)$. We are often interested in understanding what happens to the population "in the long run"; this requires us to calculate A^n for n large. This problem is easy to deal with if we can diagonalize A. Particular examples of this kind are Fibonacci sequences, discussed in Exercise D5 of this section, and Markov processes, discussed in Section 6–5.

One very important application of diagonalization, and the related idea of **eigenvectors** is the solution of systems of linear differential equations. If you know about derivatives and exponential functions, you can read about such systems in Section 6–4. The ideas introduced there have already been used in discussion of coupled springs in the Essay on Linearity and Superposition.

When Is A Similar to a Diagonal Matrix?

The following chain of argument leads to an important necessary and sufficient condition for diagonalization.

$P^{-1}AP = D$ if and only if $AP = PD$, or

$$A\left[[p_{\downarrow 1}]\ [p_{\downarrow 2}]\ \dots\ [p_{\downarrow n}]\right] = \left[[p_{\downarrow 1}]\ [p_{\downarrow 2}]\ \dots\ [p_{\downarrow n}]\right] \begin{bmatrix} d_1 & 0 & 0 & \dots & 0 \\ 0 & d_2 & 0 & \dots & 0 \\ 0 & 0 & d_3 & \dots & 0 \\ \vdots & \vdots & \vdots & \ddots & \vdots \\ 0 & 0 & 0 & \dots & d_n \end{bmatrix}.$$

The left-hand side of this equation is dealt with by a familiar form of block multiplication. The product on the right-hand side (RHS) can also be dealt with by block multiplication: the first column of RHS is obtained with the columns of P as blocks, and with the entries in the first column of D:

$$d_1[p_{\downarrow 1}] + 0[p_{\downarrow 2}] + \cdots + 0[p_{\downarrow n}] = d_1[p_{\downarrow 1}].$$

By a similar argument, the second column of RHS is $d_2[p_{\downarrow 2}]$, and so on, with the last column of RHS being $d_n[p_{\downarrow n}]$. Thus $AP = PD$ if and only if

$$\left[A[p_{\downarrow 1}]\quad A[p_{\downarrow 2}]\quad \dots\quad A[p_{\downarrow n}]\right] = \left[d_1[p_{\downarrow 1}]\quad d_2[p_{\downarrow 2}]\quad \dots\quad d_n[p_{\downarrow n}]\right].$$

Since two matrices are equal if and only if the corresponding columns are equal, the argument so far says that

$$P^{-1}AP = D \quad \text{if and only if} \quad A[p_{\downarrow j}] = d_j[p_{\downarrow j}] \quad \text{for } j = 1, 2, \dots, n.$$

Thus, P diagonalizes A if and only if the columns of P have the very special property that A multiplied by the jth column of P gives a scalar multiple of the jth column itself. In terms of the linear transformation with matrix A, this means that the image of $[p_{\downarrow j}]$ under the linear transformation is a scalar multiple of $[p_{\downarrow j}]$. A non-zero vector with this property is called an **eigenvector** of the linear transformation, and the corresponding scalar multiplier (d_j in this case) is called the **corresponding eigenvalue.** Thus, $[p_{\downarrow j}]$ is an eigenvector of A with eigenvalue d_j because

$$A[p_{\downarrow j}] = d_j[p_{\downarrow j}].$$

The definition and properties of eigenvectors will be examined more thoroughly in the next section. For now, note that the story about diagonalization can be restated as follows: *P diagonalizes A if and only if the columns of P are eigenvectors of A; moreover, the diagonal entries in D will be the eigenvalues corresponding to these eigenvectors.*

One final remark leads to the theorem on diagonalization. Since the matrix P is invertible, its columns must be linearly independent, and hence, by Corollary 1 to Theorem 2 (Section 4–4), the columns of P form a basis for $\mathbb{R}^n$. Thus any matrix P that diagonalizes A has columns that form a basis of $\mathbb{R}^n$, so P may be interpreted as a change of basis matrix. (This is not surprising since the idea of change of basis lies behind the definition of similarity of matrices; however, the definition of similarity did not explicitly require P to be a change of basis matrix.)

We have proved the following theorem.

THEOREM 1. **(Diagonalization Theorem)** An n by n matrix A can be diagonalized if and only if a basis of $\mathbb{R}^n$ can be found consisting of eigenvectors of A. If such a basis exists, the matrix P with these eigenvectors as columns diagonalizes A. If the resulting diagonal matrix is called D, the diagonal entries in $D = P^{-1}AP$ are the eigenvalues of A. The order of the eigenvalues on the diagonal of D is determined by the order of the eigenvectors as columns of P.

To use this theorem, we need to know more about eigenvectors and eigenvalues, so these ideas are explored in Section 6–2, and the discussion of diagonalization will be resumed in Section 6–3. Two examples from Section 4–6 are worth reviewing in the light of the Diagonalization Theorem.

EXAMPLE 1

In Example 36 of Section 4–6, we saw that the matrix $A = [\mathbf{proj}_{(3,4)}]_{\mathcal{S}} = \begin{bmatrix} 9/25 & 12/25 \\ 12/25 & 16/25 \end{bmatrix}$ was **diagonalized** to the diagonal matrix $\begin{bmatrix} 1 & 0 \\ 0 & 0 \end{bmatrix}$ by changing from the standard basis to the basis $\mathcal{B} = \{(3,4), (-4,3)\}$. The change of basis matrix has these basis vectors as its columns, so

$$P = \begin{bmatrix} 3 & -4 \\ 4 & 3 \end{bmatrix}, \qquad P^{-1} = \begin{bmatrix} 3/25 & 4/25 \\ -4/25 & 3/25 \end{bmatrix},$$

and

$$P^{-1}AP = \begin{bmatrix} 1 & 0 \\ 0 & 0 \end{bmatrix}.$$

The Diagonalization Theorem says that the columns of P should form a basis for $\mathbb{R}^2$, consisting of eigenvectors of A. We already know that $\mathcal{B}$ is a basis;

let us check that the columns of P are eigenvectors.

$$\begin{bmatrix} 9/25 & 12/25 \\ 12/25 & 16/25 \end{bmatrix} \begin{bmatrix} 3 \\ 4 \end{bmatrix} = \begin{bmatrix} 75/25 \\ 100/25 \end{bmatrix} = \begin{bmatrix} 3 \\ 4 \end{bmatrix} = 1 \begin{bmatrix} 3 \\ 4 \end{bmatrix},$$

so $(3,4)$ is an eigenvector with eigenvalue 1;

$$\begin{bmatrix} 9/25 & 12/25 \\ 12/25 & 16/25 \end{bmatrix} \begin{bmatrix} -4 \\ 3 \end{bmatrix} = \begin{bmatrix} 0 \\ 0 \end{bmatrix} = 0 \begin{bmatrix} -4 \\ 3 \end{bmatrix},$$

so $(-4,3)$ is an eigenvector with eigenvalue 0. Thus, the columns of P are eigenvectors of A.

The diagonal entries in the diagonal matrix are 1 and 0 in that order; 1 appears first because its eigenvector $(3,4)$ is the first column of P. The diagonal matrix $\begin{bmatrix} 1 & 0 \\ 0 & 0 \end{bmatrix}$ tells us that $\mathbf{proj}_{(3,4)}$ maps the vector with $\mathcal{B}$-coordinates $(\tilde{x}_1, \tilde{x}_2)$ to the vector with $\mathcal{B}$-coordinates $(\tilde{x}_1, 0)$. This is a particularly simple description of projection onto the $\tilde{x}_1$-axis, where the $\tilde{x}_1$-axis is in the direction of the vector with standard coordinates $(3,4)$.

EXAMPLE 2

In Example 39 in Section 4–6, we saw that

$$A = \begin{bmatrix} -3 & 5 & -5 \\ -7 & 9 & -5 \\ -7 & 7 & -3 \end{bmatrix}$$

was diagonalized by changing to the basis $\mathcal{B} = \{(1,1,0), (1,1,1), (0,1,1)\}$. Let us check that these new basis vectors are eigenvectors of A as the Diagonalization Theorem claims.

$$\begin{bmatrix} -3 & 5 & -5 \\ -7 & 9 & -5 \\ -7 & 7 & -3 \end{bmatrix} \begin{bmatrix} 1 \\ 1 \\ 0 \end{bmatrix} = \begin{bmatrix} 2 \\ 2 \\ 0 \end{bmatrix} = 2 \begin{bmatrix} 1 \\ 1 \\ 0 \end{bmatrix},$$

$$\begin{bmatrix} -3 & 5 & -5 \\ -7 & 9 & -5 \\ -7 & 7 & -3 \end{bmatrix} \begin{bmatrix} 1 \\ 1 \\ 1 \end{bmatrix} = \begin{bmatrix} -3 \\ -3 \\ -3 \end{bmatrix} = -3 \begin{bmatrix} 1 \\ 1 \\ 1 \end{bmatrix}, \qquad \text{and}$$

$$\begin{bmatrix} -3 & 5 & -5 \\ -7 & 9 & -5 \\ -7 & 7 & -3 \end{bmatrix} \begin{bmatrix} 0 \\ 1 \\ 1 \end{bmatrix} = \begin{bmatrix} 0 \\ 4 \\ 4 \end{bmatrix} = 4 \begin{bmatrix} 0 \\ 1 \\ 1 \end{bmatrix}.$$

Thus, $(1,1,0)$ is an eigenvector with eigenvalue 2, $(1,1,1)$ is an eigenvector with eigenvalue -3, and $(0,1,1)$ is an eigenvector with eigenvalue 4. The change of basis with these three vectors as the new basis vectors diagonalizes A. The diagonal entries are $2, -3, 4$ in that order, because the eigenvalue 2 corresponds to the first column of P, -3 corresponds to the second column, and 4 corresponds to the third column.

Note that

$$\begin{bmatrix} 2 & 0 & 0 \\ 0 & -3 & 0 \\ 0 & 0 & 4 \end{bmatrix} = \begin{bmatrix} 2 & 0 & 0 \\ 0 & 1 & 0 \\ 0 & 0 & 1 \end{bmatrix} \begin{bmatrix} 1 & 0 & 0 \\ 0 & -3 & 0 \\ 0 & 0 & 1 \end{bmatrix} \begin{bmatrix} 1 & 0 & 0 \\ 0 & 1 & 0 \\ 0 & 0 & 4 \end{bmatrix}.$$

Thus, the diagonal form of A shows us that A is the matrix of a linear transformation that is the composition of three mappings: a stretch by factor 4 in the direction of $(0, 1, 1)$; a stretch by factor 3, and a reflection in the direction of $(1, 1, 1)$; and a stretch by factor 2 in the direction of $(0, 1, 1)$.

The following facts about similar matrices are frequently useful. The proofs are left as Exercises D1, D2, D3, and D4.

THEOREM 2. If A is similar to B and B is similar to C, then A is similar to C.

THEOREM 3. If A is similar to B, then A^2 is similar to B^2, and for all positive integers n, A^n is similar to B^n.

THEOREM 4. If B is similar to A and A is invertible, then B is invertible.

THEOREM 5. If A is similar to B, then $\det A = \det B$.

EXERCISES 6–1

Note that Sections 6–2 and 6–3 contain further computational exercises on eigenvectors and diagonalization.

A1 By checking whether columns of P are eigenvectors of A, determine whether P diagonalizes A. If so, determine P^{-1}, and check that $P^{-1}AP$ is diagonal.

(a) $A = \begin{bmatrix} 11 & 6 \\ 9 & -4 \end{bmatrix}$, $\quad P = \begin{bmatrix} 2 & -1 \\ 1 & 3 \end{bmatrix}$

(b) $A = \begin{bmatrix} 6 & 5 \\ 3 & -7 \end{bmatrix}$, $\quad P = \begin{bmatrix} 1 & 2 \\ 1 & 1 \end{bmatrix}$

(c) $A = \begin{bmatrix} 5 & -8 \\ 4 & -7 \end{bmatrix}$, $P = \begin{bmatrix} 2 & 1 \\ 1 & 1 \end{bmatrix}$

A2 Verify that each column of P is an eigenvector of A, and determine the corresponding eigenvalues. Check that the rank of P is 3, so the columns are linearly independent. Hence (without further calculation), determine $P^{-1}AP$.

(a) $A = \begin{bmatrix} -1 & -6 & -12 \\ -1 & -3 & -7 \\ 1 & 2 & 6 \end{bmatrix}$, $P = \begin{bmatrix} 3 & 3 & -2 \\ 1 & 2 & -1 \\ -1 & -1 & 1 \end{bmatrix}$

(b) $A = \begin{bmatrix} 5 & 3 & -6 \\ -4 & -2 & 4 \\ 3 & 3 & -4 \end{bmatrix}$, $P = \begin{bmatrix} 1 & -1 & 3 \\ 0 & 1 & -1 \\ 1 & 0 & 3 \end{bmatrix}$

B1 By checking whether columns of P are eigenvectors of A, determine whether P diagonalizes A. If so, determine P^{-1}, and check that $P^{-1}AP$ is diagonal.

(a) $A = \begin{bmatrix} 4 & 2 \\ -5 & 3 \end{bmatrix}$, $P = \begin{bmatrix} 1 & 3 \\ -1 & 1 \end{bmatrix}$

(b) $A = \begin{bmatrix} 1 & 3 \\ 3 & 1 \end{bmatrix}$, $P = \begin{bmatrix} 1 & 1 \\ 1 & -1 \end{bmatrix}$

(c) $A = \begin{bmatrix} 1 & 2 \\ 3 & 2 \end{bmatrix}$, $P = \begin{bmatrix} 2 & 1 \\ 3 & -1 \end{bmatrix}$

B2 Verify that each column of P is an eigenvector of A, and determine the corresponding eigenvalues. Check that the rank of P is 3, so the columns are linearly independent. Hence (without further calculation), determine $P^{-1}AP$.

(a) $A = \begin{bmatrix} -7 & 2 & -4 \\ 8 & -1 & 4 \\ 18 & -6 & 11 \end{bmatrix}$, $P = \begin{bmatrix} -1 & -1 & 1 \\ 1 & 2 & -1 \\ 3 & 3 & -2 \end{bmatrix}$

(b) $A = \begin{bmatrix} 16 & -6 & 48 \\ -7 & 2 & -21 \\ -7 & 2 & -21 \end{bmatrix}$, $P = \begin{bmatrix} 3 & 3 & -2 \\ -1 & 0 & 1 \\ -1 & -1 & 1 \end{bmatrix}$

CONCEPTUAL EXERCISES

***D1** Prove Theorem 2.

D2 Prove Theorem 3.

***D3** Prove Theorem 4.

D4 Prove Theorem 5.

D5 **(How Diagonalization Helps Us Understand Fibonacci Sequences)**

(Partial answers are given at the end of this exercise.)

An infinite sequence of real numbers $\underline{\alpha} = (\alpha_0, \alpha_1, \alpha_2, \ldots, \alpha_n, \ldots)$ can be thought of as an "∞-tuple". We can add such sequences, and multiply them by real numbers in obvious ways, so that the set of sequences forms a vector space.

We define a **Fibonacci sequence** to be a sequence $\underline{\alpha}$ such that $\alpha_{n+1} = \alpha_n + \alpha_{n-1}$ for every $n = 1, 2, 3, \ldots$.

It follows that a Fibonacci sequence is completely determined by its first two terms α_0 and α_1. The best known Fibonacci sequence is

$$\underline{\alpha} = (0, 1, 1, 2, 3, 5, 8, \ldots).$$

Let $\underline{\alpha}$ be a Fibonacci sequence; the aim of this exercise is to find a formula for α_n in terms of α_0 and α_1.

(a) Set $\mathbf{a}_n = \begin{bmatrix} \alpha_{n+1} \\ \alpha_n \end{bmatrix}$, for $n = 0, 1, 2, 3, \ldots$. Find a matrix F such that $\mathbf{a}_{n+1} = F\mathbf{a}_n$ for all n. Show that $\mathbf{a}_n = F^n\mathbf{a}_0$. Our problem now is to find F^n.

(b) Let T denote the linear mapping whose standard matrix is F as in part (a).
(i) Let $\lambda = (1 + \sqrt{5})/2$ and $\mu = (1 - \sqrt{5})/2$. Verify that $\mathbf{v} = \begin{bmatrix} \lambda \\ 1 \end{bmatrix}$ is an eigenvector of T with eigenvalue λ, and that $\mathbf{w} = \begin{bmatrix} \mu \\ 1 \end{bmatrix}$ is an eigenvalue of T with eigenvalue μ. (Here and elsewhere, it is useful to note that $\lambda\mu = -1$, $\lambda + \mu = 1$, $\lambda + (1/\lambda) = \lambda - \mu = \sqrt{5}$.)
(ii) Show that $\mathcal{B} = \{\mathbf{v}, \mathbf{w}\}$ is a basis for $\mathbb{R}^2$.
(iii) By Theorem 1, $D = [T]_{\mathcal{B}}$ (the matrix of T with respect to $\mathcal{B}$) is diagonal: what is D?

(c) Let $P = \begin{bmatrix} [\mathbf{v}] & [\mathbf{w}] \end{bmatrix}$. By Theorem 1, $D = P^{-1}FP$. Write F in terms of D and P, and use this expression to determine F^n in terms of D^n. (See Theorem 3, Exercise D2.)

(d) Use the result of part (c) to determine a **formula** for α_n, the nth term of the Fibonacci sequence.

(e) Check that for the sequence $(0, 1, 1, 2, 3, 5, \ldots)$, where $\mathbf{a}_0 = \begin{bmatrix} 1 \\ 0 \end{bmatrix}$,

$$\alpha_n = \frac{1}{\sqrt{5}}\left\{\left((1 + \sqrt{5})/2\right)^n - \left((1 - \sqrt{5})/2\right)^n\right\} \quad \text{(really!)}.$$

Verify that this is correct for $n = 6$ ($\alpha_6 = 8$).
Recall that $(a, ar, ar^2, \ldots, ar^n, \ldots)$ *is a* **geometric** *sequence with* **ratio** r. *The Fibonacci sequence of part (e) is obviously not geometric, and in fact most Fibonacci sequences are not geometric. (Most examples of Fibonacci sequences have integer entries, but we allow real numbers as entries.)*

(f) A Fibonacci sequence is determined once $\mathbf{a}_0 = \begin{bmatrix} \alpha_1 \\ \alpha_0 \end{bmatrix}$ is known; we call $\mathbf{a}_0$ the *initial condition* for the sequence. Prove that a geometric sequence with ratio r is a Fibonacci sequence if and only if $r = \lambda$ with "initial condition" $\mathbf{a}_0$ a multiple of $\mathbf{v} = \begin{bmatrix} \lambda \\ 1 \end{bmatrix}$, or $r = \mu$ with $\mathbf{a}_0$ a multiple of $\mathbf{w} = \begin{bmatrix} \mu \\ 1 \end{bmatrix}$.

Let $\underline{\Lambda}_1 = (1, \lambda, \lambda^2, \lambda^3, \ldots)$ *and* $\underline{\Lambda}_2 = (1, \mu, \mu^2, \mu^3, \ldots)$, *the Fibonacci sequences with initial conditions* $\mathbf{v}$ *and* $\mathbf{w}$ *respectively.*

(g) Make sense of and prove the following theorem.

THEOREM The set of all Fibonacci sequences is a vector space with basis $\underline{\Lambda}_1$ and $\underline{\Lambda}_2$. Moreover, if we write the initial condition $\mathbf{a}_0 = \begin{bmatrix} \alpha_1 \\ \alpha_0 \end{bmatrix}$ of $\underline{\alpha}$ as $\mathbf{a}_0 = c_1\mathbf{v} + c_2\mathbf{w}$, then $\alpha_n = c_1\lambda^n + c_2\mu^n$.

(h) As an example of (g), take the initial condition $\mathbf{a}_0 = \begin{bmatrix} 1 \\ 0 \end{bmatrix}$ and determine c_1 and c_2 such that $\mathbf{a}_0 = c_1\mathbf{v} + c_2\mathbf{w}$, and use this and the conclusion of (g) to recover the formula for α_n in part (e).

Partial Answers to Exercise D5

(a) $F = \begin{bmatrix} 1 & 1 \\ 1 & 0 \end{bmatrix}$; (b) $D = \begin{bmatrix} \lambda & 0 \\ 0 & \mu \end{bmatrix}$;

(c) $F^n = \dfrac{1}{\sqrt{5}} \begin{bmatrix} \lambda^{n+1} - \mu^{n+1} & \lambda^n - \mu^n \\ \lambda^n - \mu^n & \lambda^{n-1} - \mu^{n-1} \end{bmatrix}$;

(d) $\alpha_n = \dfrac{1}{\sqrt{5}}(\alpha_1(\lambda^n - \mu^n) + \alpha_0(\lambda^{n-1} - \mu^{n-1}))$; (h) $c_1 = -c_2 = 1/\sqrt{5}$.

SECTION

6–2 Eigenvectors and Eigenvalues

In the discussion of diagonalization in the previous section, **eigenvectors** were introduced. An eigenvector is a special or preferred vector that has the property that it is mapped by a linear transformation to a multiple of itself. Such preferred vectors play an important role in many applications. For example, they identify the "normal modes of oscillation" that are crucial in the analysis and design of some mechanical structures or electrical systems (see the last example in the Essay on Linearity and Superposition in Physics). As we shall see in examples following the definition, they occur in natural ways for many geometrical transformations.

DEFINITION

Suppose that $L : \mathbb{R}^n \to \mathbb{R}^n$ is a linear transformation. A non-zero vector $\mathbf{v}$ such that $L(\mathbf{v}) = \lambda\mathbf{v}$ for some real number λ is called an **eigenvector** of L; the scalar λ is called the **eigenvalue corresponding to v.** If λ is given first, then $\mathbf{v}$ is called an eigenvector corresponding to the eigenvalue λ. (Use of the Greek letter λ [lambda] is fairly traditional for eigenvalues, but you may use other letters if you prefer.)

The restriction that $\mathbf{v}$ be non-zero is natural and important. It is natural because $L(\mathbf{0}) = \mathbf{0}$ for every linear transformation, so it is completely uninteresting to consider $\mathbf{0}$ as an eigenvector. It is important because many of the things we want to do with eigenvectors only make sense for non-zero vectors. In particular, the Diagonalization Theorem directs us to look for a basis of eigenvectors, and $\mathbf{0}$ is useless for this purpose.

Some books (particularly older British books) call these **characteristic vectors and values.** (A few call them "proper" values and vectors; "proper" is one translation of the German adjective "eigen". Other meanings of "eigen" emphasize "belonging" [for example, "mein eigenes Auto" means "my own car"] or the property of being "special to". Thus an eigenvector of a linear transformation is a vector belonging to, or special to the linear transformation.)

EXAMPLE 3

(Eigenvectors and eigenvalues of projections and reflections in $\mathbb{R}^3$)

Since $\mathbf{proj_n}(\mathbf{n}) = 1\mathbf{n}$, $\mathbf{n}$ is an eigenvector of $\mathbf{proj_n}$ with corresponding eigenvalue 1. If $\mathbf{v}$ is orthogonal to $\mathbf{n}$, $\mathbf{proj_n}(\mathbf{v}) = \mathbf{0} = 0\mathbf{v}$, so $\mathbf{v}$ is an eigenvector of $\mathbf{proj_n}$ with corresponding eigenvalue 0; notice in this example that there is a whole plane of eigenvectors corresponding to the eigenvalue 0. Notice that for an arbitrary vector $\mathbf{u}$ in $\mathbb{R}^3$, $\mathbf{proj_n}(\mathbf{u})$ is a multiple of $\mathbf{n}$, so that $\mathbf{u}$ is definitely **not** an eigenvector of $\mathbf{proj_n}$ unless it is a multiple of $\mathbf{n}$ or orthogonal to $\mathbf{n}$.

On the other hand $\mathbf{perp_n}(\mathbf{n}) = \mathbf{0} = 0\mathbf{n}$, so $\mathbf{n}$ is an eigenvector of $\mathbf{perp_n}$ with eigenvalue 0. Since $\mathbf{perp_n}(\mathbf{v}) = 1\mathbf{v}$ for any $\mathbf{v}$ orthogonal to $\mathbf{n}$, such a $\mathbf{v}$ is an eigenvector of $\mathbf{perp_n}$ with eigenvalue 1.

Since $\mathbf{refl_n}(\mathbf{n}) = -1\mathbf{n}$, $\mathbf{n}$ is an eigenvector of $\mathbf{refl_n}$ with eigenvalue (-1). For $\mathbf{v}$ orthogonal to $\mathbf{n}$, $\mathbf{refl_n}(\mathbf{v}) = 1\mathbf{v}$, so such a $\mathbf{v}$ is an eigenvector of $\mathbf{refl_n}$ with eigenvalue 1.

EXAMPLE 4

(Eigenvectors and eigenvalues of rotations)

Consider the rotation $R_\theta : \mathbb{R}^2 \to \mathbb{R}^2$ with matrix $\begin{bmatrix} \cos\theta & -\sin\theta \\ \sin\theta & \cos\theta \end{bmatrix}$, where θ is not an integer multiple of π. By geometry, it is clear that there is no non-zero $\mathbf{x}$ in $\mathbb{R}^2$ such that $R_\theta(\mathbf{x}) = \lambda\mathbf{x}$ for some real number λ. This linear transformation has no eigenvectors, and no real eigenvalues.

Now let R_θ denote the rotation of $\mathbb{R}^3$ with matrix $\begin{bmatrix} \cos\theta & -\sin\theta & 0 \\ \sin\theta & \cos\theta & 0 \\ 0 & 0 & 1 \end{bmatrix}$.
The only eigenvectors of R_θ are multiples of $\mathbf{e}_3$, with eigenvalue 1.

EXAMPLE 5

Let $A = \begin{bmatrix} 17 & -15 \\ 20 & -18 \end{bmatrix}$, and let $L : \mathbb{R}^2 \to \mathbb{R}^2$ be the linear transformation defined by $L(\mathbf{x}) = A\mathbf{x}$. Determine which of the following vectors are eigenvectors of L, and give the corresponding eigenvalues: $(1, 1)$, $(-2, -2)$, $(1, 3)$, $(3, 4)$.

SOLUTION: To test whether $(1, 1)$ is an eigenvector, we calculate $L(1, 1)$.

$$[L(1,1)] = \begin{bmatrix} 17 & -15 \\ 20 & -18 \end{bmatrix} \begin{bmatrix} 1 \\ 1 \end{bmatrix} = \begin{bmatrix} 2 \\ 2 \end{bmatrix} = 2\begin{bmatrix} 1 \\ 1 \end{bmatrix},$$

so $(1, 1)$ is an eigenvector of L, with eigenvalue 2.

Since $(-2, -2) = -2(1, 1)$, and L is linear,

$$L(-2,-2) = (-2)L(1,1) = (-2)(2)(1,1) = 2(-2,-2),$$

so $(-2, -2)$ is also an eigenvector of L with eigenvalue 2. In fact, by a similar argument, any non-zero multiple of $(1, 1)$ is an eigenvector with eigenvalue 2.

$$[L(1,3)] = \begin{bmatrix} 17 & -15 \\ 20 & -18 \end{bmatrix} \begin{bmatrix} 1 \\ 3 \end{bmatrix} = \begin{bmatrix} -28 \\ -34 \end{bmatrix} \neq \lambda \begin{bmatrix} 1 \\ 3 \end{bmatrix}$$

for any real number λ, so $(1, 3)$ is **not** an eigenvector of L.

$$[L(3,4)] = \begin{bmatrix} 17 & -15 \\ 20 & -18 \end{bmatrix} \begin{bmatrix} 3 \\ 4 \end{bmatrix} = \begin{bmatrix} -9 \\ -12 \end{bmatrix} = -3\begin{bmatrix} 3 \\ 4 \end{bmatrix},$$

so $(3, 4)$ (or any non-zero multiple of it) is an eigenvector of L with eigenvalue (-3).

Can a Matrix Have Eigenvectors?

The geometric meaning of eigenvectors is much clearer when they are thought of as belonging to linear transformations. However, in many applications of these ideas, it is a matrix A that is given. In such situations, we shall speak of the eigenvectors of the matrix A, *with the understanding that A is the matrix of a linear transformation with respect to the standard basis.*

Thus in Example 3, we may say that $(1, 1)$ and $(3, 4)$ are eigenvectors of the matrix $\begin{bmatrix} 17 & -15 \\ 20 & -18 \end{bmatrix}$, with eigenvalues 2 and -3 respectively.

Finding Eigenvalues and Eigenvectors

If eigenvectors and eigenvalues are going to be of any use, we need a systematic method for finding them. Suppose that a square matrix A is given; then $\mathbf{x}$ is an eigenvector if and only if $A\mathbf{x} = \lambda\mathbf{x}$. This condition can be rewritten

$$A\mathbf{x} - \lambda\mathbf{x} = \mathbf{0}.$$

It is tempting to write this as $(A - \lambda)\mathbf{x} = 0$, but this is **wrong** because A is a matrix and λ is a number, so their difference is not defined. To get around this, use the fact that $\mathbf{x} = I\mathbf{x}$. Then the eigenvector condition can be rewritten

$$(A - \lambda I)\mathbf{x} = \mathbf{0}.$$

The eigenvector $\mathbf{x}$ is thus a solution of the homogeneous system of linear equations with coefficient matrix $(A - \lambda I)$. A homogeneous system of n equations in n variables has non-trivial solutions if and only if it has rank less than n, and this is true if and only if $\det(A - \lambda I) = 0$. This is the key result in the procedure for finding eigenvalues and eigenvectors, so it is worth summarizing as a theorem.

THEOREM 1. Suppose that A is an n by n matrix. A real number λ is an eigenvalue of A if and only if λ satisfies the equation

$$\det(A - \lambda I) = 0.$$

If λ is an eigenvalue, a corresponding eigenvector may be found as a non-zero solution of the homogeneous system

$$(A - \lambda I)\mathbf{x} = \mathbf{0}.$$

Proof. Discussion preceding theorem. ■

> *REMARK*
>
> The theory on homogeneous systems of linear equations guarantees that for an eigenvalue λ, there is at least one eigenvector (that is, one non-trivial solution of the homogeneous system of rank $< n$).

EXAMPLE 6

Let us apply Theorem 1 to find the eigenvalues and eigenvectors of the matrix of Example 5. Let $A = \begin{bmatrix} 17 & -15 \\ 20 & -18 \end{bmatrix}$. Then

$$A - \lambda I = \begin{bmatrix} 17 & -15 \\ 20 & -18 \end{bmatrix} - \lambda \begin{bmatrix} 1 & 0 \\ 0 & 1 \end{bmatrix} = \begin{bmatrix} 17 - \lambda & -15 \\ 20 & -18 - \lambda \end{bmatrix}.$$

(You should set up your calculation like this: you will need $A - \lambda I$ later when you find the eigenvectors.) Then

$$\begin{aligned} \det(A - \lambda I) &= -306 - \lambda(17 - 18) + \lambda^2 + 300 \\ &= \lambda^2 + \lambda - 6 = (\lambda + 3)(\lambda - 2), \end{aligned}$$

so

$$\det(A - \lambda I) = 0 \text{ when } \lambda = -3 \text{ or } \lambda = 2.$$

These are the eigenvalues of A.

Consider the case $\lambda = -3$. To find $\mathbf{x}$ satisfying $(A - \lambda I)\mathbf{x} = \mathbf{0}$, we write the matrix $A - \lambda I$ and row reduce.

$$A - \lambda I = \begin{bmatrix} 20 & -15 \\ 20 & -15 \end{bmatrix} \to \begin{bmatrix} 4 & -3 \\ 0 & 0 \end{bmatrix},$$

so that the general solution of $(A - \lambda I)\mathbf{x} = \mathbf{0}$ is $\mathbf{x} = t(3/4, 1)$. Thus, an eigenvector corresponding to the eigenvalue -3 is $(3/4, 1)$. Any non-zero multiple of $(3/4, 1)$ is also an eigenvector; since the diagonalization procedure uses the eigenvectors as columns of P, it is more convenient to have eigenvectors with integer coordinates, so take the eigenvector corresponding to $\lambda = -3$ to be $(3, 4)$.

Now consider $\lambda = 2$.

$$A - \lambda I = \begin{bmatrix} 15 & -15 \\ 20 & -20 \end{bmatrix} \to \begin{bmatrix} 1 & -1 \\ 0 & 0 \end{bmatrix};$$

the general solution of $(A - \lambda I)\mathbf{x} = \mathbf{0}$ is $\mathbf{x} = t(1, 1), t \in \mathbb{R}$, so that an eigenvector in this case is $(1, 1)$ (or any non-zero multiple of $(1, 1)$).

These eigenvalues and eigenvectors agree with those given in Example 5.

Since $\{(3, 4), (1, 1)\}$ is a basis for $\mathbb{R}^2$, consisting of eigenvectors of A, Theorem 1 of Section 6–1 says that A is diagonalized by the matrix $P = \begin{bmatrix} 3 & 1 \\ 4 & 1 \end{bmatrix}$; the resulting diagonal matrix is $\begin{bmatrix} -3 & 0 \\ 0 & 2 \end{bmatrix}$, where the diagonal entries are the eigenvalues. You should check that this is what you get if you calculate $P^{-1}AP$. (If you take P to be $\begin{bmatrix} 1 & 3 \\ 1 & 4 \end{bmatrix}$, the diagonal matrix is $\begin{bmatrix} 2 & 0 \\ 0 & -3 \end{bmatrix}$.)

DEFINITION

The equation $\det(A - \lambda I) = 0$ is called **the characteristic equation of the matrix** A, and $\det(A - \lambda I)$ is called **the characteristic polynomial.**

For an n by n matrix A, the characteristic equation is a polynomial equation of degree n. The easiest way to see this is by recalling that the determinant is a signed sum of products with one factor from each row and one from each column; thus there will be one product that has a term of degree n in λ, and no products with terms of any higher degree. Note that the term of highest degree (λ^n) has coefficient $(-1)^n$; some other books prefer to work with the polynomial $\det(\lambda I - A)$ so that the coefficient of λ^n is always 1. In our notation the constant term in the characteristic polynomial $\det(A - \lambda I)$ is $\det A$—see Exercise D6(b).

It is relevant here to recall some facts about real solutions or "roots" of polynomial equations. Let us denote the characteristic polynomial by

$$C(\lambda) = (-1)^n\lambda^n + a_{n-1}\lambda^{n-1} + a_{n-2}\lambda^{n-2} + \cdots + a_1\lambda + a_0,$$

so that the characteristic equation is $C(\lambda) = 0$. Then: (1) λ_1 is a real root of the characteristic equation if and only if $(\lambda - \lambda_1)$ is a factor of $C(\lambda)$; (2) complex roots of the equation occur in "conjugate pairs", so the total number of complex roots must be even; (3) the total number of roots (real and complex, counting repetitions) is n; (4) if n is odd, there must be at least one real root; (5) if the entries in A are integers, since the leading coefficient of the characteristic polynomial is ± 1, any rational root must in fact be an integer.

EXAMPLE 7

Find the eigenvectors and eigenvalues of $A = \begin{bmatrix} -3 & 5 & -5 \\ -7 & 9 & -5 \\ -7 & 7 & -3 \end{bmatrix}$. (This is the matrix of Example 39 in Section 4–6 and Example 2 in Section 6–1.)

SOLUTION: To apply Theorem 1, we first write

$$A - \lambda I = \begin{bmatrix} -3-\lambda & 5 & -5 \\ -7 & 9-\lambda & -5 \\ -7 & 7 & -3-\lambda \end{bmatrix}.$$

Then, by expansion along the first row, the characteristic equation is

$$\begin{aligned} \det(A - \lambda I) &= (-3-\lambda)(\lambda^2 - 6\lambda + 8) - 5(7\lambda - 14) - 5(-7\lambda + 14) \\ &= -\lambda^3 + 3\lambda^2 + 10\lambda - 24 = 0. \end{aligned}$$

This is a little more convenient in the form $\lambda^3 - 3\lambda^2 - 10\lambda + 24 = 0$. Try $\lambda = 1$? Not a root. $\lambda = -1$? Not a root. $\lambda = 2$? Yes, this satisfies the characteristic equation, so 2 is an eigenvalue.

Divide $\lambda^3 - 3\lambda^2 - 10\lambda + 24$ by $(\lambda - 2)$, and find that

$$\lambda^3 - 3\lambda^2 - 10\lambda + 24 = (\lambda - 2)(\lambda^2 - \lambda - 12) = (\lambda - 2)(\lambda + 3)(\lambda - 4),$$

so that the eigenvalues are 2, -3, and 4.

For $\lambda = 2$:

$$A - \lambda I = \begin{bmatrix} -5 & 5 & -5 \\ -7 & 7 & -5 \\ -7 & 7 & -5 \end{bmatrix} \rightarrow \begin{bmatrix} 1 & -1 & 1 \\ 0 & 0 & 1 \\ 0 & 0 & 0 \end{bmatrix} \rightarrow \begin{bmatrix} 1 & -1 & 0 \\ 0 & 0 & 1 \\ 0 & 0 & 0 \end{bmatrix};$$

the general solution for the homogeneous system with this coefficient matrix is $x_2(1, 1, 0)$, where x_2 is an arbitrary real number, so an eigenvector corresponding to $\lambda = 2$ is $(1, 1, 0)$.

For $\lambda = -3$,

$$A - \lambda I = \begin{bmatrix} 0 & 5 & -5 \\ -7 & 12 & -5 \\ -7 & 7 & 0 \end{bmatrix} \to \begin{bmatrix} 1 & -1 & 0 \\ 0 & 1 & -1 \\ 0 & 0 & 0 \end{bmatrix} \to \begin{bmatrix} 1 & 0 & -1 \\ 0 & 1 & -1 \\ 0 & 0 & 0 \end{bmatrix}.$$

Since the general solution of $(A - \lambda I)\mathbf{x} = \mathbf{0}$ is $x_3(1, 1, 1), x_3 \in \mathbb{R}$, we see that an eigenvector corresponding to the eigenvalue -3 is $(1, 1, 1)$.

For $\lambda = 4$,

$$A - \lambda I = \begin{bmatrix} -7 & 5 & -5 \\ -7 & 5 & -5 \\ -7 & 7 & -5 \end{bmatrix} \to \begin{bmatrix} 7 & -5 & 5 \\ 0 & 0 & 0 \\ 0 & 2 & -2 \end{bmatrix} \to \begin{bmatrix} 1 & 0 & 0 \\ 0 & 1 & -1 \\ 0 & 0 & 0 \end{bmatrix}.$$

Hence an eigenvector for $\lambda = 4$ is $(0, 1, 1)$.

Thus, A is diagonalized by the matrix $P = \begin{bmatrix} 1 & 1 & 0 \\ 1 & 1 & 1 \\ 0 & 1 & 1 \end{bmatrix}$, and the resulting diagonal matrix has 2, -3, and 4 as its diagonal entries. This agrees with what was found in Example 34 in Section 4–6 and Example 2 in Section 6–1.

Examples 6 and 7 illustrate the best possible cases: in each case we found enough eigenvectors to form a basis for the appropriate $\mathbb{R}^n$, so the original matrix can be diagonalized. They illustrate an attractive and very useful theorem.

THEOREM 2. Suppose that A is an n by n matrix, and that $\lambda_1, \lambda_2, \ldots, \lambda_k$ are distinct ($\lambda_i \neq \lambda_j$) eigenvalues with corresponding eigenvectors $\mathbf{v}_1, \mathbf{v}_2, \ldots, \mathbf{v}_k$ respectively. Then $\{\mathbf{v}_1, \mathbf{v}_2, \ldots, \mathbf{v}_k\}$ is linearly independent. In particular, if there are n distinct eigenvalues, then the corresponding eigenvectors form a basis of $\mathbb{R}^n$ and the matrix A is diagonable.

Proof. To illustrate the idea of the proof, we first consider the case where there are two distinct eigenvalues. To test the corresponding eigenvectors for linear independence, consider the equation

$$c_1\mathbf{v}_1 + c_2\mathbf{v}_2 = \mathbf{0}. \qquad (*)$$

To use the fact that $\mathbf{v}_1$ and $\mathbf{v}_2$ are eigenvectors, we multiply equation $(*)$ by A.

$$A(c_1\mathbf{v}_1 + c_2\mathbf{v}_2) = c_1\lambda_1\mathbf{v}_1 + c_2\lambda_2\mathbf{v}_2 = \mathbf{0}.$$

Now multiply equation $(*)$ by $\lambda_1 I$:

$$\lambda_1 I(c_1\mathbf{v}_1 + c_2\mathbf{v}_2) = c_1\lambda_1\mathbf{v}_1 + c_2\lambda_1\mathbf{v}_2 = \mathbf{0}.$$

Subtract the last equation from the previous equation; the terms in $\mathbf{v}_1$ cancel, so we have

$$c_2(\lambda_2 - \lambda_1)\mathbf{v}_2 = \mathbf{0}.$$

Since $\lambda_2 \neq \lambda_1$ and $\mathbf{v}_2 \neq \mathbf{0}$, it follows that $c_2 = 0$.

We need to do similar calculations to complete the proof, so it is convenient to describe what we have done more compactly. We multiply equation $(*)$ by $(A - \lambda_1 I)$. We note that

$$A\mathbf{v}_1 = \lambda_1\mathbf{v}_1 \quad \text{so} \quad (A - \lambda_1 I)\mathbf{v}_1 = \mathbf{0},$$

and

$$A\mathbf{v}_2 = \lambda_2\mathbf{v}_2 \quad \text{so} \quad (A - \lambda_1 I)\mathbf{v}_2 = (\lambda_2 - \lambda_1)\mathbf{v}_2.$$

Thus,

$$(A - \lambda_1 I)(c_1\mathbf{v}_1 + c_2\mathbf{v}_2) = c_2(\lambda_2 - \lambda_1)\mathbf{v}_2 = \mathbf{0},$$

so $c_2 = 0$, as before. Now we are ready to move on.

Consider what happens when we look at λ_2 instead of λ_1. By arguments similar to those above,

$$(A - \lambda_2 I)\mathbf{v}_1 = (\lambda_1 - \lambda_2)\mathbf{v}_1 \quad \text{and} \quad (A - \lambda_2 I)\mathbf{v}_2 = \mathbf{0}.$$

Thus, from equation $(*)$ it follows that

$$(A - \lambda_2 I)(c_1\mathbf{v}_1 + c_2\mathbf{v}_2) = c_1(\lambda_1 - \lambda_2)\mathbf{v}_1 = \mathbf{0},$$

and since $\lambda_1 \neq \lambda_2$ and $\mathbf{v}_2 \neq \mathbf{0}$, it follows that $c_1 = 0$.

We have shown that in the case of two distinct eigenvalues, equation $(*)$ is true only if $c_1 = c_2 = 0$, so $\{\mathbf{v}_1, \mathbf{v}_2\}$ is a linearly independent set.

Now let us turn to the general case, with k distinct eigenvalues. Note that for $i = 1, 2, \ldots, k$,

$$(A - \lambda_i I)\mathbf{v}_i = \mathbf{0}.$$

We say that $(A - \lambda_i I)$ *annihilates* $\mathbf{v}_i$. On the other hand, if $i \neq j$,

$$(A - \lambda_i I)\mathbf{v}_j = (\lambda_j - \lambda_i)\mathbf{v}_j, \qquad \text{for } j = 1, \ldots, k, (j \neq i).$$

We need to test whether $\{\mathbf{v}_1, \ldots, \mathbf{v}_k\}$ is linearly independent, so we consider the equation

$$c_1\mathbf{v}_1 + c_2\mathbf{v}_2 + \cdots + c_k\mathbf{v}_k = \mathbf{0}. \qquad (**)$$

We multiply this equation by $(A - \lambda_1 I)$ and obtain

$$\mathbf{0} + c_2(\lambda_2 - \lambda_1)\mathbf{v}_2 + \ldots + c_k(\lambda_k - \lambda_1)\mathbf{v}_k = \mathbf{0}.$$

Then we multiply this equation by

$$(A - \lambda_{k-1}I)(A - \lambda_{k-2}I)\ldots(A - \lambda_2 I),$$

the product of all the *other* annihilating factors *except* $(A - \lambda_k I)$. The only surviving term is

$$c_k(\lambda_k - \lambda_1)(\lambda_k - \lambda_2)\ldots(\lambda_k - \lambda_{k-1})\mathbf{v}_k = \mathbf{0}.$$

Since the λ's are distinct and $\mathbf{v}_k \neq \mathbf{0}$, it follows that $c_k = 0$.

Similarly, by multiplying equation $(**)$ by the product of all the annihilating factors *except* the jth factor $(A - \lambda_j I)$, we obtain the result

$$c_j(\lambda_j - \lambda_1)(\lambda_j - \lambda_2)\ldots(\lambda_j - \lambda_k)\mathbf{v}_j = \mathbf{0},$$

where there is no factor $(\lambda_j - \lambda_j)$. It follows that $c_j = 0$ for every $j = 1, 2, \ldots, k$.

Since equation $(**)$ is satisfied only if

$$c_1 = c_2 = \ldots = c_k = 0,$$

the set $\{\mathbf{v}_1, \mathbf{v}_2, \ldots, \mathbf{v}_k\}$ is linearly independent.

If there are n distinct eigenvalues, there are n linearly independent eigenvectors. Since A is n by n, these eigenvectors are vectors in $\mathbb{R}^n$, and by the Corollary to Theorem 2 in Section 4–4, n linearly independent eigenvectors in $\mathbb{R}^n$ form a basis of $\mathbb{R}^n$. In this case, by the Diagonalization Theorem, A is diagonable.

■

EXAMPLE 8

Let $A = \begin{bmatrix} 1 & 0 & -1 \\ 11 & -4 & -7 \\ -7 & 3 & 4 \end{bmatrix}$. Determine whether A is diagonable, and if so, determine the diagonal matrix similar to A.

SOLUTION: We begin by finding the eigenvalues of A.

$$A - \lambda I = \begin{bmatrix} 1-\lambda & 0 & -1 \\ 11 & -4-\lambda & -7 \\ -7 & 3 & 4-\lambda \end{bmatrix},$$

so

$$\begin{aligned}\det(A - \lambda I) &= (1-\lambda)(\lambda^2+5) + 0 - 1(33 - 28 - 7\lambda) \\ &= -\lambda^3 + \lambda^2 + 2\lambda = -\lambda(\lambda-2)(\lambda+1).\end{aligned}$$

It follows that the eigenvalues are $0, 2, -1$. Since the three eigenvalues are distinct, by Theorem 2 the corresponding eigenvectors are necessarily linearly independent, and A is diagonable, with diagonal form $\begin{bmatrix} 0 & 0 & 0 \\ 0 & 2 & 0 \\ 0 & 0 & -1 \end{bmatrix}$. Notice that it is not required to find the diagonalizing matrix P, although it is often wise to do so as a check on your calculations.

REMARK

In this book, most eigenvalues turn out to be integers. This is somewhat unrealistic; in real applications, eigenvalues are often not rational numbers. Effective computer methods for finding eigenvalues depend on the theory of eigenvectors and eigenvalues and similarity.

EXERCISES 6–2

Note that Exercises 6–3 will give further practice in determining eigenvectors and eigenvalues, and on diagonalization.

A1 Let $A = \begin{bmatrix} -10 & 9 & 5 \\ -10 & 9 & 5 \\ 2 & 3 & -1 \end{bmatrix}$. Determine whether the following vectors are eigenvectors of A, and if so, determine the corresponding eigenvalues. You should answer without calculating the characteristic equation.
$(1,0,1)$, $(1,0,2)$, $(1,1,-1)$, $(1,-1,1)$, $(1,1,1)$.

A2 Find the eigenvectors of the matrix A of Example 8, and hence determine a matrix P that diagonalizes A. Find P^{-1} and check that $P^{-1}AP$ is diagonal.

A3 For each of the following matrices, determine the eigenvalues and eigenvectors; explain whether you can determine whether the matrix is diagonable with the information you have. If so, write the diagonalizing matrix P and the resulting diagonal matrix D. Finally, check that $P^{-1}AP = D$.

(a) $A = \begin{bmatrix} 0 & 1 \\ -6 & 5 \end{bmatrix}$ (b) $A = \begin{bmatrix} 1 & 3 \\ 0 & 1 \end{bmatrix}$

(c) $A = \begin{bmatrix} -26 & 10 \\ -75 & 29 \end{bmatrix}$ (d) $A = \begin{bmatrix} 1 & 3 \\ 4 & 2 \end{bmatrix}$

A4 Determine the eigenvalues and eigenvectors of the matrix $\begin{bmatrix} 0 & -5 & 3 \\ -2 & -6 & 6 \\ -2 & -7 & 7 \end{bmatrix}$. State whether the matrix is diagonable, and if so, give a matrix P that diagonalizes it and the corresponding diagonal form.

B1 (a) Let $A = \begin{bmatrix} -3 & 1 & -1 \\ 8 & -3 & 8 \\ 8 & -1 & 6 \end{bmatrix}$. Determine whether the following vectors are eigenvectors of A, and if so, determine the corresponding eigenvalues. You should answer without calculating the characteristic equation.
$(1,0,1)$, $(1,0,-1)$, $(1,1,0)$, $(1,-1,-1)$, $(1,1,2)$.

(b) Let $A = \begin{bmatrix} -3 & 5 & 5 \\ 1 & -3 & -1 \\ -1 & 5 & 3 \end{bmatrix}$. Determine whether the following vectors are eigenvectors of A, and if so, determine the corresponding eigenvalues. You should answer without calculating the characteristic equation.
$(1,0,1)$, $(1,0,-1)$, $(1,1,0)$, $(0,1,-1)$, $(1,-1,1)$.

B2 For each of the following matrices, determine the eigenvalues and eigenvectors; explain whether you can determine whether the matrix is diagonable with the information you have. If so, write the diagonalizing matrix P and the resulting diagonal matrix D. Finally, check that $P^{-1}AP = D$.

(a) $\begin{bmatrix} 4 & -1 \\ -2 & 5 \end{bmatrix}$ (b) $A = \begin{bmatrix} 2 & 1 \\ -1 & 4 \end{bmatrix}$

(c) $\begin{bmatrix} -2 & 2 \\ -3 & 5 \end{bmatrix}$ (d) $A = \begin{bmatrix} 2 & 2 \\ -3 & -5 \end{bmatrix}$

B3 Determine the eigenvalues and eigenvectors of the matrix $\begin{bmatrix} -4 & 6 & 6 \\ -2 & 2 & 4 \\ -1 & 3 & 1 \end{bmatrix}$. State whether the matrix is diagonable, and if so, give a matrix P that diagonalizes it and the corresponding diagonal form.

B4 Determine the eigenvalues and eigenvectors of the matrix $\begin{bmatrix} -4 & 18 & 4 \\ -3 & 11 & 2 \\ 5 & -14 & -1 \end{bmatrix}$. State whether the matrix is diagonable, and if so, give a matrix P that diagonalizes it and the corresponding diagonal form.

COMPUTER RELATED EXERCISES

C1 Some computer software packages have routines for testing whether a given number is an eigenvalue of a given matrix. If you have such a routine, test whether $0, \pm 1, \pm 2$ are eigenvalues of the 3 by 3 matrices of the exercises in the next section (Section 6–3).

C2 Let $A = \begin{bmatrix} 2.89316 & -1.28185 & 2.42918 \\ -0.70562 & 0.76414 & -0.67401 \\ 1.67682 & -0.83198 & 2.34270 \end{bmatrix}$. Use a computer to verify that $(1.21, -0.34, 0.87)$, $(1.31, 2.15, -0.21)$, and $(-1.85, 0.67, 2.10)$ are (approximately) eigenvectors of A with corresponding eigenvalues 5, 0.4, and 0.6 respectively.

C3 Does your computer software calculate the characteristic polynomial of a given matrix? If so, find out how to use it; apply it to the matrices of A and B Exercises in Sections 6–2 and 6–3.

C4 Does your computer software determine eigenvalues for a given matrix? Are there restrictions or conditions on its use? If applicable, apply it to some of the matrices in Exercises 6–3, and to Exercise 6–2 C2.

CONCEPTUAL EXERCISES

***D1** Suppose that $\mathbf{v}$ is an eigenvector both of the matrix A and of the matrix B, with corresponding eigenvalue λ for A, and corresponding eigenvalue μ for B. Show that $\mathbf{v}$ is an eigenvector of $(A + B)$ and of AB, and determine the corresponding eigenvalues.

D2 (a) Show that if λ is an eigenvalue of a matrix A, then λ^n is an eigenvalue of A^n. How are the corresponding eigenvectors related?
(b) Give an example of a 2 by 2 matrix A such that A has no real eigenvalues, but A^3 does have real eigenvalues. [Hint: Exercise 3–3 D6]

D3 Show that if A is invertible and $\mathbf{v}$ is an eigenvector of A, then $\mathbf{v}$ is also an eigenvector of A^{-1}. How are the corresponding eigenvalues related?

D4 Suppose that Q is an invertible matrix. Show that the eigenvalues of $Q^{-1}AQ$ are the same as the eigenvalues of A. Do this two ways: first by considering the characteristic equations, second by showing that any eigenvector of the matrix A determines a corresponding eigenvector of the matrix $Q^{-1}AQ$. (Discussion: In fact, if we think of eigenvectors in terms of linear transformations, A is the matrix of a linear transformation L with respect to the standard basis, and $Q^{-1}AQ$ is the matrix of L with respect to some other basis, say $\mathcal{C}$. Then the "eigenvector $\mathbf{v}$ of A" is really the eigenvector $\mathbf{v}$ of L, so $L(\mathbf{v}) = \lambda\mathbf{v}$ and $\mathbf{v}$ is an eigenvector of L no matter what basis we use. However, the $\mathcal{C}$-coordinate vector of $\mathbf{v}$ is an eigenvector of the $\mathcal{C}$-matrix of L.)

***D5** Suppose that A is an n by n matrix such that the sum of the entries in each row is the same. That is, $\sum_{k=1}^{n} a_{jk} = C$, for all $j = 1, 2, \dots, n$. Show that $(1, 1, \dots, 1)$ is an eigenvector of A. (Such matrices arise in probability theory.)

D6 (a) Suppose that A is diagonable. Prove that $\det A$ is equal to the product of the eigenvalues of A (including repeated eigenvalues) by considering $P^{-1}AP$.
(b) Show that the constant term in the characteristic polynomial is $\det A$. [Hint: How do you find the constant term in any polynomial $p(\lambda)$?]
(c) Without assuming that A is diagonable, show that $\det A$ is equal to the product of the roots of the characteristic equation of A (including any repeated roots and complex roots). [Hint: Consider the constant term in the characteristic equation, and the factored version of that equation.]

D7 Suppose that a square matrix A has no eigenvalues equal to 0. Prove that A is invertible. [Hint: D6]

SECTION

6–3 Diagonalization (Continued)

In the preceding section, Theorem 2 guarantees that if an n by n matrix has n distinct real eigenvalues it is diagonable. What happens when there are fewer than n distinct real eigenvalues? We consider some examples with complex characteristic roots and repeated real roots, and then summarize the relevant facts about diagonalization of such matrices.

The Case of Complex Roots

EXAMPLE 9

The matrix $\begin{bmatrix} 0 & -1 \\ 1 & 0 \end{bmatrix}$ has characteristic equation $\lambda^2 + 1 = 0$. There are obviously no real roots, so the matrix has no real eigenvalues or eigenvectors and is not diagonable. If you recognize that this is the matrix of a rotation in the plane through angle $\pi/2$, you will not be surprised that it is not diagonable.

The complex roots of the characteristic equation are $\lambda_1 = i$ and $\lambda_2 = -i$ (where $i^2 = -1$). Note that these form a complex conjugate pair.

EXAMPLE 10

The matrix $\begin{bmatrix} 1 & 2 & -2 \\ 2 & 1 & 2 \\ 2 & -2 & -1 \end{bmatrix}$ has characteristic polynomial

$$\begin{aligned} \det \begin{bmatrix} 1-\lambda & 2 & -2 \\ 2 & 1-\lambda & 2 \\ 2 & -2 & -1-\lambda \end{bmatrix} &= (1-\lambda)(\lambda^2+3) - 2(-2\lambda-6) - 2(2\lambda-6) \\ &= -(\lambda^3 - \lambda^2 + 3\lambda - 27) \\ &= -(\lambda-3)(\lambda^2+2\lambda+9). \end{aligned}$$

There is only one real eigenvalue, $\lambda = 3$; the other roots of the characteristic equation are complex.

We look for any eigenvectors corresponding to $\lambda = 3$.

$$A - 3I = \begin{bmatrix} -2 & 2 & -2 \\ 2 & -2 & 2 \\ 2 & -2 & -4 \end{bmatrix} \to \begin{bmatrix} 1 & -1 & 0 \\ 0 & 0 & 1 \\ 0 & 0 & 0 \end{bmatrix}.$$

The general solution of $(A - \lambda I)\mathbf{x} = \mathbf{0}$ is $\mathbf{x} = t(1, 1, 0), t \in \mathbb{R}$, so an eigenvector is $(1, 1, 0)$. Since the rank of the reduced row echelon form is 2, there are no other eigenvectors corresponding to $\lambda = 3$. There are no other real eigenvalues, so there are no other eigenvectors. Since there is not a basis for $\mathbb{R}^3$ consisting of eigenvectors of this matrix, the matrix is not diagonable.

The complex characteristic roots are $-1 + i2\sqrt{2}$ and $-1 - i2\sqrt{2}$; again, the second root is the complex conjugate of the first.

The case of complex roots is discussed further in Section 6–6.

The Case of a Repeated Root

EXAMPLE 11

Consider the matrix $A = \begin{bmatrix} 2 & 0 \\ 0 & 2 \end{bmatrix}$. The characteristic equation is $(\lambda - 2)^2 = 0$. The matrix has 2 as a repeated characteristic root, and any vector in the plane is an eigenvector, so it is easy to give a basis of eigenvectors; the matrix is already diagonal so it is diagonable in a trivial way.

EXAMPLE 12

Consider the matrix $B = \begin{bmatrix} 2 & 1 \\ 0 & 2 \end{bmatrix}$. The characteristic equation is $(\lambda - 2)^2 = 0$, so the root 2 is repeated. When $\lambda = 2$, $B - \lambda I = \begin{bmatrix} 0 & 1 \\ 0 & 0 \end{bmatrix}$, so the only eigenvector is $(1, 0)$ or multiples of it. In this example there is no basis of $\mathbb{R}^2$ consisting of eigenvectors of B, so **the matrix B is not diagonable.**

An elaboration of this example is the matrix $C = \begin{bmatrix} 2 & 1 & 0 \\ 0 & 2 & 0 \\ 0 & 0 & 2 \end{bmatrix}$. This time 2 is a three-fold root of the characteristic equation $(\lambda - 2)^3 = 0$. Since $C - \lambda I = \begin{bmatrix} 0 & 1 & 0 \\ 0 & 0 & 0 \\ 0 & 0 & 0 \end{bmatrix}$, any vector with $x_2 = 0$ is an eigenvector. Thus, we can find a set of two linearly independent eigenvectors $\{\mathbf{e}_1, \mathbf{e}_3\}$, but not a set of three linearly independent eigenvectors. Since there is no basis of eigenvectors, the matrix is not diagonable.

REMARK

In examples such as Examples 11 and 12, avoid saying "repeated eigenvalue", because it suggests that for each repetition of the root we will find a new eigenvector. Instead, we speak of a "repeated root of the characteristic equation", or a "repeated characteristic root". Thus, in Examples 11 and 12, there is only one eigenvalue (namely, 2), but 2 is a repeated characteristic root.

EXAMPLE 13

Can the matrix $A = \begin{bmatrix} 0 & 3 & -2 \\ -2 & 5 & -2 \\ -2 & 3 & 0 \end{bmatrix}$ be diagonalized?

SOLUTION:

$$A - \lambda I = \begin{bmatrix} 0 - \lambda & 3 & -2 \\ -2 & 5 - \lambda & -2 \\ -2 & 3 & 0 - \lambda \end{bmatrix},$$

so the characteristic equation is

$$0 = (0 - \lambda)(\lambda^2 - 5\lambda + 6) - 3(2\lambda - 4) - 2(-2\lambda + 4) = -(\lambda^3 - 5\lambda^2 + 8\lambda - 4).$$

By trial, $\lambda = 1$ is one solution, so divide by $(\lambda - 1)$ to get

$$0 = -(\lambda - 1)(\lambda^2 - 4\lambda + 4) = -(\lambda - 1)(\lambda - 2)^2.$$

The roots are $1, 2, 2$. Since the roots are not distinct, we cannot use Theorem 2 of Section 6–2 to decide whether A is diagonable. We must determine whether there is a basis consisting of eigenvectors.

Let us look for eigenvectors for the repeated root $\lambda = 2$. Then

$$A - \lambda I = \begin{bmatrix} -2 & 3 & -2 \\ -2 & 3 & -2 \\ -2 & 3 & -2 \end{bmatrix} \rightarrow \begin{bmatrix} 2 & -3 & 2 \\ 0 & 0 & 0 \\ 0 & 0 & 0 \end{bmatrix}.$$

Solutions to $(A - \lambda I)\mathbf{x} = \mathbf{0}$ may be found by taking x_2 and x_3 as any real numbers and solving for x_1: the general solution in the usual form is $\mathbf{x} = x_2(3/2, 1, 0) + x_3(-1, 0, 1)$, and we see that two linearly independent eigenvectors corresponding to the eigenvalue 2 are $(3, 2, 0)$ and $(-1, 0, 1)$. (Other choices are possible, but this is the most obvious choice.)

Now Theorem 2 tells us that the eigenvector corresponding to $\lambda = 1$ and any eigenvector corresponding to $\lambda = 2$ will be linearly independent. Hence we may conclude at this point that A is diagonable, and that the diagonal form is $\begin{bmatrix} 2 & 0 & 0 \\ 0 & 2 & 0 \\ 0 & 0 & 1 \end{bmatrix}$.

To write down a diagonalizing matrix P, we must solve for an eigenvector corresponding to eigenvalue 1: it is easy to determine that such an eigenvector is $(1, 1, 1)$. Therefore the matrix A is diagonalized by $P = \begin{bmatrix} 3 & -1 & 1 \\ 2 & 0 & 1 \\ 0 & 1 & 1 \end{bmatrix}$.

EXAMPLE 14

Is the matrix $A = \begin{bmatrix} -1 & 7 & -5 \\ -4 & 11 & -6 \\ -4 & 8 & -3 \end{bmatrix}$ diagonable?

SOLUTION:

$$A - \lambda I = \begin{bmatrix} -1 - \lambda & 7 & -5 \\ -4 & 11 - \lambda & -6 \\ -4 & 8 & -3 - \lambda \end{bmatrix},$$

so the characteristic equation is

$$\begin{aligned} 0 &= (-1-\lambda)(\lambda^2 - 8\lambda + 15) - 7(4\lambda - 12) - 5(-4\lambda + 12) \\ &= -(\lambda^3 - 7\lambda^2 + 15\lambda - 9) \\ &= -(\lambda - 1)(\lambda - 3)^2. \end{aligned}$$

The roots are 1, 3, 3.

For $\lambda = 3$, $A - \lambda I = \begin{bmatrix} -4 & 7 & -5 \\ -4 & 8 & -6 \\ -4 & 8 & -6 \end{bmatrix} \rightarrow \begin{bmatrix} 4 & 0 & -2 \\ 0 & 1 & -1 \\ 0 & 0 & 0 \end{bmatrix}$. Since the matrix has rank 2, the solution space is one-dimensional. There cannot be two linearly independent eigenvectors for $\lambda = 3$, there can be only one.

For $\lambda = 1$, $A - \lambda I = \begin{bmatrix} -2 & 7 & -5 \\ -4 & 10 & -6 \\ -4 & 8 & -4 \end{bmatrix} \rightarrow \begin{bmatrix} 1 & 0 & -1 \\ 0 & 1 & -1 \\ 0 & 0 & 0 \end{bmatrix}$.

The only corresponding eigenvector is $(1, 1, 1)$ (or any non-zero multiple).

In this example, there are only two linearly independent eigenvectors. Thus there is no basis for $\mathbb{R}^3$ consisting of eigenvectors of this matrix, so the matrix is not diagonable.

In the following discussion, it is explained that a simple (not repeated) characteristic root always provides exactly one eigenvector, as is the case in this example for $\lambda = 1$.

Eigenvalues and Diagonalization: Some Useful Facts

The following facts are useful in working with diagonalization problems. Since a complete discussion of repeated characteristic roots is beyond the scope of this book, we cannot give complete proofs of these facts. In the following statements, A is an n by n matrix.

(1) If $\mathbf{v}$ is an eigenvector of A corresponding to the eigenvalue λ, $\mathbf{v}$ cannot also be an eigenvector corresponding to a different eigenvalue μ: if $\lambda \neq \mu$, we cannot have $A\mathbf{v} = \lambda\mathbf{v}$ and $A\mathbf{v} = \mu\mathbf{v}$.

(2) A characteristic root is **simple** if it is not a repeated root. Suppose that λ is a simple real eigenvalue of A. Then the set of all eigenvectors corresponding to λ is the set of scalar multiples of a single eigenvector $\mathbf{v}$. We speak of a one-dimensional **eigenspace** corresponding to λ. In particular, if λ is a simple real eigenvalue of A, there cannot be two linearly independent eigenvectors corresponding to λ.

If A has n distinct real eigenvalues, the fact that each eigenvalue corresponds to a one-dimensional eigenspace follows from Theorem 2 of Section 6–2, together with Fact 1 in this list. In more general situations, this fact depends on results about repeated roots and complex roots, described below.

It is worth looking at Example 10, in the light of the facts just described. We can now conclude that the simple eigenvalue $\lambda = 3$ can provide only one eigenvector of the matrix, without even considering the system of equations for the corresponding eigenvector. Similarly in Example 14, as soon as we see that the repeated root $\lambda = 3$ gives only one eigenvector, we can conclude that there cannot be a basis of eigenvectors, because $\lambda = 1$ provides only one eigenvector.

(3) Suppose that λ is a simple *complex* characteristic root for A. Since the characteristic equation for A has real coefficients, we know that the complex conjugate $\overline{\lambda}$ is also a characteristic root. It will be shown in Section 6–6 that corresponding to the complex conjugate pair $\lambda, \overline{\lambda}$, there is a two-dimensional subspace of $\mathbb{R}^n$ that is **invariant** under A. A subspace $\mathbb{S}$ is invariant under A if $A\mathbf{v}$ is in $\mathbb{S}$ for every $\mathbf{v}$ in $\mathbb{S}$. It can be shown that the invariant subspace of $\mathbb{R}^n$ corresponding to λ and $\overline{\lambda}$ contains no real eigenvectors of A. In this case, we must therefore be lacking at least two of the eigenvectors needed to make up a basis for $\mathbb{R}^n$.

It follows that a real matrix A with a complex characteristic root is not diagonable in terms of real numbers. Thus, in Example 10, as soon as we realize that there are complex roots, we can conclude that the matrix is not diagonable.

(4) To discuss cases of repeated roots, it is helpful to introduce the concepts of **algebraic multiplicity** of an eigenvalue (the number of times λ is repeated as a root of the characteristic equation) and **geometric multiplicity** (the maximum number of linearly independent eigenvectors corresponding to λ). In Example 11, $\lambda = 2$ has both algebraic and geometric multiplicity 2; in Example 12, B has eigenvalue 2 with algebraic multiplicity 2 and geometric multiplicity 1, while C has eigenvalue 2 with algebraic multiplicity 3 and geometric multiplicity 2.

An important fact is that if λ is a real eigenvalue of algebraic multiplicity k, then there is a k-dimensional subspace of $\mathbb{R}^n$ that is invariant under A; this subspace can contain no eigenvectors for any eigenvalue other than λ. However, the geometric multiplicity may be less than k, and in this case it will certainly be impossible to find a basis for $\mathbb{R}^n$ consisting of eigenvectors of A. As an illustration of such an invariant subspace, consider the matrix C in Example 12: the subspace spanned by $\mathbf{e}_1 = (1, 0, 0)$ and $\mathbf{e}_2 = (0, 1, 0)$ is invariant under C but contains only one eigenvector $\mathbf{e}_1$. A complete discussion of these ideas is beyond the scope of this book.

(5) Ideas from both (3) and (4) above are used to deal with repeated complex characteristic roots.

(6) **Summary.** A square matrix is diagonable over the reals if and only if its characteristic roots are all real, and for each k-fold repeated real characteristic root there are k linearly independent eigenvectors. Examples 13 and 8 illustrate this fact; Example 14 illustrates the case where geometric multiplicity is less than algebraic multiplicity for some λ and diagonalization is impossible.

REMARK

Complex characteristic roots are discussed further in Section 6–6. For a more advanced discussion of repeated characteristic roots, see *Linear Algebra* by K. Hoffman and R. Kunze, Prentice-Hall, 1971.

A Note on Jordan Normal Form

Exploration of the cases where geometric multiplicity differs from algebraic multiplicity leads to **Jordan normal forms.** In Example 12, the matrices B and C are in Jordan normal form, which is as close as you can come to diagonal for a matrix that is not diagonable. Notice that the eigenvalues appear as the diagonal entries, but there are some "superdiagonal" 1's, so that the matrix is not diagonal. In fact, the matrix B appears as a 2 by 2 "Jordan block" in the matrix C.

To find out more about Jordan normal form, consult the book cited in the Remark above.

EXERCISES 6–3

A1 For the following matrices, determine the eigenvalues and any corresponding eigenvectors, and determine whether each matrix is diagonable. If it is diagonable, give a matrix P that diagonalizes it. (Hint: One has complex roots; the others have repeated roots; one is diagonable.)

$$A = \begin{bmatrix} 6 & -9 & -5 \\ -4 & 9 & 4 \\ 9 & -17 & -8 \end{bmatrix} \quad B = \begin{bmatrix} -2 & 7 & 3 \\ -1 & 2 & 1 \\ 0 & 2 & 1 \end{bmatrix} \quad C = \begin{bmatrix} -1 & 6 & 3 \\ 3 & -4 & -3 \\ -6 & 12 & 8 \end{bmatrix}$$

A2 Same instructions and hint as for Exercise A1.

$$E = \begin{bmatrix} 0 & 6 & -8 \\ -2 & 4 & -4 \\ -2 & 2 & -2 \end{bmatrix} \quad F = \begin{bmatrix} 2 & 0 & 0 \\ -1 & 0 & 1 \\ -1 & -2 & 3 \end{bmatrix} \quad G = \begin{bmatrix} -3 & -3 & 5 \\ 13 & 10 & -13 \\ 3 & 2 & -1 \end{bmatrix}$$

B1 Same instructions as for Exercise A1.

$$A = \begin{bmatrix} 1 & 6 & 3 \\ 0 & -2 & 0 \\ 3 & 6 & 1 \end{bmatrix} \quad B = \begin{bmatrix} -3 & 2 & 1 \\ 4 & -2 & -4 \\ -9 & 2 & 7 \end{bmatrix} \quad C = \begin{bmatrix} 2 & -1 & -2 \\ -3 & 2 & 3 \\ 4 & -3 & -4 \end{bmatrix}$$

B2 Same instructions as for Exercise A1.

$$E = \begin{bmatrix} -7 & 2 & 12 \\ -3 & 0 & 6 \\ -3 & 1 & 5 \end{bmatrix} \quad F = \begin{bmatrix} 3 & 0 & -4 \\ 1 & 1 & -2 \\ 1 & 0 & -1 \end{bmatrix} \quad G = \begin{bmatrix} -1 & -2 & 2 \\ 10 & 5 & -10 \\ 2 & 0 & -1 \end{bmatrix}$$

CONCEPTUAL EXERCISES

***D1** (a) Suppose that P diagonalizes the square matrix A, and that the diagonal form is D. Show that $A = PDP^{-1}$.
(b) Use the result of part (a) and properties of eigenvectors to calculate a matrix that has eigenvalues 2 and 3 with corresponding eigenvectors $(1, 2)$ and $(1, 3)$ respectively.
(c) Determine a matrix that has eigenvalues $2, -2, 3$ with corresponding eigenvectors $(1, 0, 1)$, $(1, 1, -1)$, $(1, -1, 2)$ respectively.

D2 (a) Suppose that P diagonalizes the square matrix A, and that the diagonal form is D. Show that $A^n = PD^nP^{-1}$.
(b) Use the result of part (a) to calculate C^5, where C is the matrix of Exercise A1. (The appropriate D and P are given in the Answers.)

SECTION
6–4 Diagonalization and Differential Equations

This section requires knowledge of the exponential function and its derivative. The ideas are not used elsewhere in this book. We begin with a simple example.

Consider two tanks Y and Z, each containing 1000 litres of a salt solution. At the initial time, $t = 0$ (in hours), the concentration of salt in tank Y is different from the concentration in tank Z. In each tank the solution is well stirred, so that the concentration is constant throughout the tank. The two tanks are joined by two pipes; through one pipe, solution is pumped from Y to Z at the rate of 20 L/h; through the other, solution is pumped from Z to Y at the same rate. The problem is to determine the amount of salt in each tank at time t.

Let $y(t)$ be the amount of salt (in kilograms) in tank Y at time t, and let $z(t)$ be the amount in tank Z. Then the concentration in Y at time t is $(y/1000)$ kg/L, and similarly $(z/1000)$ kg/L is the concentration in Z. Then for tank Y, salt is flowing out through one pipe at the rate of $(20)(y/1000)$ kg/h and in through the other pipe at the rate of $(20)(z/1000)$ kg/h. Since the rate of change is measured by the derivative, we have $\frac{dy}{dt} = -0.02y + 0.02z$. By consideration of Z, we get a second differential equation, so y and z are the solutions of the **system of linear ordinary differential equations:**

$$\frac{dy}{dt} = -0.02y + 0.02z$$
$$\frac{dz}{dt} = 0.02y - 0.02z.$$

It is convenient to rewrite this system in the form

$$\frac{d}{dt}\begin{bmatrix} y \\ z \end{bmatrix} = \begin{bmatrix} -0.02 & 0.02 \\ 0.02 & -0.02 \end{bmatrix}\begin{bmatrix} y \\ z \end{bmatrix}.$$

How can we solve this system? Well, it might be easier if we could change variables so that the 2 by 2 matrix was diagonalized. By standard methods, one eigenvalue of $\begin{bmatrix} -0.02 & 0.02 \\ 0.02 & -0.02 \end{bmatrix}$ is $\lambda_1 = 0$, with corresponding eigenvector $(1, 1)$, and the other eigenvalue is $\lambda_2 = -0.04$, with corresponding eigenvector $(-1, 1)$. Hence, the matrix is diagonalized by $P = \begin{bmatrix} 1 & -1 \\ 1 & 1 \end{bmatrix}$, with $P^{-1} = \begin{bmatrix} 1/2 & 1/2 \\ -1/2 & 1/2 \end{bmatrix}$.

Introduce new coordinates $(\tilde{y}, \tilde{z})$ by the change of coordinates equation as in Section 4–6: $\begin{bmatrix} y \\ z \end{bmatrix} = P\begin{bmatrix} \tilde{y} \\ \tilde{z} \end{bmatrix}$. Substitute this for $\begin{bmatrix} y \\ z \end{bmatrix}$ on both sides of the system to obtain

$$\frac{d}{dt}P\begin{bmatrix} \tilde{y} \\ \tilde{z} \end{bmatrix} = \begin{bmatrix} -0.02 & 0.02 \\ 0.02 & -0.02 \end{bmatrix} P\begin{bmatrix} \tilde{y} \\ \tilde{z} \end{bmatrix}.$$

Since the entries in P are constants, it is easy to check that

$$\frac{d}{dt}P\begin{bmatrix}\tilde{y}\\ \tilde{z}\end{bmatrix} = P\frac{d}{dt}\begin{bmatrix}\tilde{y}\\ \tilde{z}\end{bmatrix}.$$

Multiply both sides of the system of equations (on the left) by P^{-1}. Since P diagonalizes the 2 by 2 coefficient matrix, we get

$$\frac{d}{dt}\begin{bmatrix}\tilde{y}\\ \tilde{z}\end{bmatrix} = P^{-1}\begin{bmatrix}-0.02 & 0.02\\ 0.02 & -0.02\end{bmatrix}P\begin{bmatrix}\tilde{y}\\ \tilde{z}\end{bmatrix} = \begin{bmatrix}0 & 0\\ 0 & -0.04\end{bmatrix}\begin{bmatrix}\tilde{y}\\ \tilde{z}\end{bmatrix}.$$

Now write the pair of equations:

$$\frac{d\tilde{y}}{dt} = 0, \text{ and } \frac{d\tilde{z}}{dt} = -0.04\tilde{z}.$$

These equations are "decoupled" and we can easily solve each of them by simple one-variable calculus.

The only functions satisfying $\frac{d\tilde{y}}{dt} = 0$ are constants: we write $\tilde{y}(t) = a$. The only functions satisfying an equation of the form $\frac{dx}{dt} = kx$ are exponentials of the form $x(t) = (\text{constant})e^{kt}$, so from $\frac{d\tilde{z}}{dt} = -0.04\tilde{z}$, we obtain $\tilde{z}(t) = be^{-0.04t}$, where b is a constant.

Now we need to express the solution in terms of the original variables y and z:

$$\begin{bmatrix}y\\ z\end{bmatrix} = P\begin{bmatrix}\tilde{y}\\ \tilde{z}\end{bmatrix} = \begin{bmatrix}1 & -1\\ 1 & 1\end{bmatrix}\begin{bmatrix}\tilde{y}\\ \tilde{z}\end{bmatrix} = \begin{bmatrix}\tilde{y}-\tilde{z}\\ \tilde{y}+\tilde{z}\end{bmatrix}$$
$$= \begin{bmatrix}a - be^{-0.04t}\\ a + be^{-0.04t}\end{bmatrix}.$$

For later use, it is helpful to rewrite this as

$$\begin{bmatrix}y\\ z\end{bmatrix} = a\begin{bmatrix}1\\ 1\end{bmatrix} + be^{-0.04t}\begin{bmatrix}-1\\ 1\end{bmatrix}.$$

This is the general solution of the problem. To determine the constants a and b, we would need to know the amounts $y(0)$ and $z(0)$ at the initial time $t = 0$, and then y and z would be known for all t. Note that as $t \to \infty$, y and z tend to a common value (a), as one might expect.

A Practical Solution Procedure

The usual solution procedure takes advantage of the understanding obtained from this diagonalization argument, but it takes a major shortcut. Now that the expected form of solution is known, we simply look for a solution of the form $\begin{bmatrix} y \\ z \end{bmatrix} = ce^{\lambda t}\begin{bmatrix} a \\ b \end{bmatrix}$. Substitute this into the original system and use the fact that $\frac{d}{dt}ce^{\lambda t}\begin{bmatrix} a \\ b \end{bmatrix} = \lambda ce^{\lambda t}\begin{bmatrix} a \\ b \end{bmatrix}$:

$$\lambda ce^{\lambda t}\begin{bmatrix} a \\ b \end{bmatrix} = \begin{bmatrix} -0.02 & 0.02 \\ 0.02 & -0.02 \end{bmatrix} ce^{\lambda t}\begin{bmatrix} a \\ b \end{bmatrix}.$$

After the common factor $ce^{\lambda t}$ is cancelled, this tells us that $\begin{bmatrix} a \\ b \end{bmatrix}$ is an eigenvector of $\begin{bmatrix} -0.02 & 0.02 \\ 0.02 & -0.02 \end{bmatrix}$, with eigenvalue λ. We find the two eigenvalues and corresponding eigenvectors as above, and observe that since our problem is a linear homogeneous problem, the general solution will be an arbitrary linear combination of the two solutions $e^{0t}\begin{bmatrix} 1 \\ 1 \end{bmatrix}$ and $e^{-0.04t}\begin{bmatrix} -1 \\ 1 \end{bmatrix}$; this is exactly the general solution we found above.

General Discussion

There are many other problems that give rise to systems of linear homogeneous ordinary differential equations. For example, in the Essay on Linearity and Superposition, a mechanical system consisting of three springs was described (and solved by the shortcut method just explained). Electrical circuits also give rise to such systems. Many of these systems are much larger (that is, involve more variables) than the examples we have considered. Methods for solving these systems make extensive use of eigenvectors and eigenvalues, and require methods for dealing with cases where the characteristic equation has complex or repeated roots.

EXERCISES 6–4

A1 Find the general solution of each of the following systems of linear differential equations.

(a) $\frac{d}{dt}\begin{bmatrix} y \\ z \end{bmatrix} = \begin{bmatrix} 3 & 2 \\ 4 & -4 \end{bmatrix}\begin{bmatrix} y \\ z \end{bmatrix}$

(b) $\frac{d}{dt}\begin{bmatrix} y \\ z \end{bmatrix} = \begin{bmatrix} 0.2 & 0.7 \\ 0.1 & -0.4 \end{bmatrix}\begin{bmatrix} y \\ z \end{bmatrix}$

B1 Find the general solution of each of the following systems of linear differential equations.

(a) $\frac{d}{dt}\begin{bmatrix} y \\ z \end{bmatrix} = \begin{bmatrix} -0.5 & 0.3 \\ 0.1 & -0.7 \end{bmatrix}\begin{bmatrix} y \\ z \end{bmatrix}$

(b) $\frac{d}{dt}\begin{bmatrix} y \\ z \end{bmatrix} = \begin{bmatrix} -1 & 4 \\ 8 & -5 \end{bmatrix}\begin{bmatrix} y \\ z \end{bmatrix}$

(c) $\frac{d}{dt}\begin{bmatrix} x \\ y \\ z \end{bmatrix} = \begin{bmatrix} -1 & -1 & 0 \\ -13 & 3 & 8 \\ 11 & -5 & -8 \end{bmatrix}\begin{bmatrix} x \\ y \\ z \end{bmatrix}$

SECTION

6–5 Powers of Matrices; Markov Processes

There are some applications of linear algebra where it is necessary to calculate powers of a matrix. If the matrix A is diagonalized by P to diagonal matrix D, it follows from $D = P^{-1}AP$ that $A = PDP^{-1}$, and then

$$A^m = PD^mP^{-1}.$$

Thus, knowledge of the eigenvalues of A and the theory of diagonalization should be valuable tools in these applications. One such application, the study of Fibonacci sequences, appeared in Exercise D5 of Section 6–1. A more important application of powers of a matrix is the study of Markov processes. After discussing Markov processes, we turn the question around and show how the "power method" uses powers of a matrix A to determine an eigenvalue of A. We begin with an example of a Markov process.

EXAMPLE 15

Smith and Jones are the only competing suppliers of communication services in their community. At the present time, they each have a 50% share of the market. However, Smith has recently upgraded his service, and a survey indicates that from one month to the next, 90% of Smith's customers remain loyal while 10% switch to Jones. On the other hand, 70% of Jones's customers remain loyal and 30% switch to Smith. If this goes on for six months, how large are their market shares? If this goes on for a long time, how big will Smith's share become?

SOLUTION:

Let S_m be Smith's market share (as a decimal) at the end of the mth month, and let J_m be Jones's share. Then $S_m + J_m = 1$, since between them they have 100% of the market. At the end of the $(m+1)^{\text{th}}$ month, Smith has 90% of his previous customers and 30% of Jones's previous customers, so

$$S_{m+1} = 0.9S_m + 0.3J_m.$$

Similarly,

$$J_{m+1} = 0.1S_m + 0.7J_m.$$

We may rewrite these equations in matrix-vector form

$$\begin{bmatrix} S_{m+1} \\ J_{m+1} \end{bmatrix} = \begin{bmatrix} 0.9 & 0.3 \\ 0.1 & 0.7 \end{bmatrix} \begin{bmatrix} S_m \\ J_m \end{bmatrix}.$$

The matrix $\begin{bmatrix} 0.9 & 0.3 \\ 0.1 & 0.7 \end{bmatrix}$ is called the **transition matrix** for this problem: it describes the transition (change) from the **state** (S_m, J_m) at time m to the state at time $(m+1)$. We denote the matrix by T. Then we have answers to the questions if we can determine $T^6\begin{bmatrix} 0.5 \\ 0.5 \end{bmatrix}$, and $T^m\begin{bmatrix} 0.5 \\ 0.5 \end{bmatrix}$ for m large.

To answer the first question, we might compute T^6 directly, but for the second question this approach is not reasonable. Instead we diagonalize. By the methods of Section 6–2, we find that $\lambda_1 = 1$ is an eigenvalue, with eigenvector $\mathbf{v}_1 = (3, 1)$; and $\lambda_2 = 0.6$ is the other eigenvalue, with eigenvector $\mathbf{v}_2 = (1, -1)$. Thus

$$P = \begin{bmatrix} 3 & 1 \\ 1 & -1 \end{bmatrix} \quad \text{and} \quad P^{-1} = \frac{1}{4}\begin{bmatrix} 1 & 1 \\ 1 & -3 \end{bmatrix}.$$

It follows that

$$T^m = \begin{bmatrix} 3 & 1 \\ 1 & -1 \end{bmatrix} \begin{bmatrix} 1^m & 0 \\ 0 & (0.6)^m \end{bmatrix} \frac{1}{4} \begin{bmatrix} 1 & 1 \\ 1 & -3 \end{bmatrix}.$$

We could now answer our question directly, but we will get a simpler calculation if we observe that the eigenvectors form a basis so we can write

$$\begin{bmatrix} S_0 \\ J_0 \end{bmatrix} = c_1 \begin{bmatrix} 3 \\ 1 \end{bmatrix} + c_2 \begin{bmatrix} 1 \\ -1 \end{bmatrix} = P \begin{bmatrix} c_1 \\ c_2 \end{bmatrix}.$$

Then

$$\begin{bmatrix} c_1 \\ c_2 \end{bmatrix} = \begin{bmatrix} 3 & 1 \\ 1 & -1 \end{bmatrix}^{-1} \begin{bmatrix} S_0 \\ J_0 \end{bmatrix} = \frac{1}{4} \begin{bmatrix} S_0 + J_0 \\ S_0 - 3J_0 \end{bmatrix}.$$

Then by linearity,

$$T^m \begin{bmatrix} S_0 \\ J_0 \end{bmatrix} = c_1 T^m \mathbf{v}_1 + c_2 T^m \mathbf{v}_2$$
$$= c_1 \lambda_1^m \mathbf{v}_1 + c_2 \lambda_2^m \mathbf{v}_2$$
$$= \frac{1}{4}(S_0 + J_0) \begin{bmatrix} 3 \\ 1 \end{bmatrix} + \frac{1}{4}(S_0 - 3J_0)(0.6)^m \begin{bmatrix} 1 \\ -1 \end{bmatrix}.$$

Now $S_0 = J_0 = 0.5$; when $m = 6$,

$$\begin{bmatrix} S_6 \\ J_6 \end{bmatrix} = \frac{1}{4} \begin{bmatrix} 3 \\ 1 \end{bmatrix} - \frac{1}{4}(0.6)^6 \begin{bmatrix} 1 \\ -1 \end{bmatrix}$$
$$\approx \frac{1}{4} \begin{bmatrix} 3 - 0.0117 \\ 1 + 0.0117 \end{bmatrix} \approx \begin{bmatrix} 0.747 \\ 0.253 \end{bmatrix}.$$

Thus, after six months, Smith has 74.7% of the market.

When m is very large, $(0.6)^m$ is nearly zero, so for m large enough ($m \to \infty$) we have $S_\infty = 0.75$, $J_\infty = 0.25$.

Thus, in this problem, Smith's share approaches 75% as m gets large, but never gets larger than 75%. Now look carefully: we get the same answer "in the long run", no matter what the initial values of S_0 and J_0 are because $(0.6)^m \to 0$, and $S_0 + J_0 = 1$.

By emphasizing some features of this example, we will be led to an important definition and several general properties.

(1) Each column of T has sum 1. This means that all of Smith's customers show up a month later as customers of Smith or Jones, and similarly for Jones's customers. No customers are lost from the system and none are added after the process begins.

(2) It is natural to interpret the entries t_{jk} as **probabilities**. For example, $t_{11} = 0.9$ is the probability that a Smith customer remains a Smith customer, while $t_{21} = 0.1$ is the probability that a Smith customer becomes a Jones customer. If we consider "Smith customer" as a "state 1" and "Jones customer" as a "state 2", then t_{jk} is the probability of **transition** from state k to state j between time m and time $m + 1$.

(3) The "initial state vector" is $\begin{bmatrix} S_0 \\ J_0 \end{bmatrix}$; $T^m \begin{bmatrix} S_0 \\ J_0 \end{bmatrix}$ is the state vector at time m.

(4) Note that

$$\begin{bmatrix} S_1 \\ J_1 \end{bmatrix} = T \begin{bmatrix} S_0 \\ J_0 \end{bmatrix} = S_0 \begin{bmatrix} t_{11} \\ t_{21} \end{bmatrix} + J_0 \begin{bmatrix} t_{12} \\ t_{22} \end{bmatrix}.$$

Since $t_{11} + t_{21} = 1$ and $t_{12} + t_{22} = 1$, it follows that

$$S_1 + J_1 = S_0 + J_0.$$

Thus, it follows from (1) that each state vector has the same column sum. In our example, S_0 and J_0 were decimal fractions, so $S_0 + J_0 = 1$, but one could consider a process whose states had some other constant column sum.

(5) Note that 1 is an eigenvalue of T with eigenvector $(3, 1)$. To get a state vector with appropriate sum, we take the eigenvector to be $(\frac{3}{4}, \frac{1}{4})$. Thus,

$$T \begin{bmatrix} \frac{3}{4} \\ \frac{1}{4} \end{bmatrix} = \begin{bmatrix} \frac{3}{4} \\ \frac{1}{4} \end{bmatrix},$$

and the state vector $(\frac{3}{4}, \frac{1}{4})$ is **fixed** or **invariant** under the transformation with matrix T. Moreover, it is this fixed vector that is the limiting state approached by $T^m \begin{bmatrix} S_0 \\ J_0 \end{bmatrix}$ for any $\begin{bmatrix} S_0 \\ J_0 \end{bmatrix}$.

The following definition captures the essential properties of this example.

An n by n matrix T is the **transition matrix** of an n-state **Markov process** (or T is a "Markov matrix") if
(1) $t_{jk} \geq 0$, for each j and k, and
(2) each column sum is one: $\sum_{j=1}^{n} t_{jk} = 1$, for each k.

We take possible states of the process to be vectors $\mathbf{s} = (s_1, s_2, \ldots, s_n)$ such that $s_j \geq 0$ for each j, and $s_1 + s_2 + \cdots + s_n = 1$. (With minor changes, we could develop the theory with $s_1 + s_2 + \cdots + s_n =$ constant.)

The problem with the Markov process is to establish the behavior of a sequence of states $\mathbf{s}, T\mathbf{s}, T^2\mathbf{s}, \ldots, T^p\mathbf{s}$, and if possible, say something about the limit of $T^p\mathbf{s}$ as $p \to \infty$. As we saw in the example, diagonalization of T is a key to solving the problem. It is beyond the scope of this book to establish all the properties of Markov processes, but some of the properties are easy to prove, and others are easy to illustrate if we make extra assumptions.

PROPERTY 1. One eigenvalue of a Markov matrix is $\lambda_1 = 1$.

Proof. Since each column of T has sum 1, each column of $(T - 1I)$ has sum 0. Hence the sum of the rows of $(T - 1I)$ is $(0, 0, \ldots, 0)$. Thus the rows are linearly dependent, and $(T - 1I)$ has rank $< n$. Hence there is a non-trivial solution to $(T - 1I)\mathbf{x} = \mathbf{0}$ and 1 is an eigenvalue.

■

PROPERTY 2. The eigenvector $\mathbf{s}^*$ for $\lambda_1 = 1$ has $s_j^* \geq 0$ for each $j = 1, 2, \ldots, n$.

Discussion. The proof is omitted. This property is important because it means that the eigenvector $\mathbf{s}^*$ is a real state of the process. In fact, it is a **fixed** or **invariant** state:

$$T\mathbf{s}^* = \lambda_1\mathbf{s}^* = \mathbf{s}^* \qquad (\text{since } \lambda_1 = 1).$$

PROPERTY 3. All other eigenvectors satisfy $|\lambda_i| \leq 1$.

Discussion. To see why we expect this, let us assume that T is diagonable, with distinct eigenvalues $1, \lambda_2, \ldots, \lambda_n$ and corresponding eigenvectors $\mathbf{s}^*, \mathbf{s}_2, \ldots, \mathbf{s}_n$. Then any initial state $\mathbf{s}$ can be written

$$\mathbf{s} = c_1\mathbf{s}^* + c_2\mathbf{s}_2 + \cdots + c_n\mathbf{s}_n.$$

It follows that

$$T^m\mathbf{s} = c_1 1^m\mathbf{s}^* + c_2\lambda_1^m\mathbf{s}_2 + \cdots + c_n\lambda_n^m\mathbf{s}_n.$$

If any $|\lambda_k| > 1$, the term $|\lambda_k^m|$ would become much larger than the other terms when m is large, and also larger than 1; it would follow that $T^m\mathbf{s}$ has some coordinates with magnitude greater than 1. This is impossible because state coordinates satisfy $0 \leq s_k \leq 1$, so we must have $|\lambda_k| \leq 1$.

PROPERTY 4. Suppose that for some m all the entries in T^m are not zero. Then all the eigenvalues of T except for $\lambda_1 = 1$ satisfy $|\lambda_k| < 1$. In this case, for any initial state $\mathbf{s}$, $T^m\mathbf{s} \to \mathbf{s}^*$ as $m \to \infty$: all states tend to the invariant state $\mathbf{s}^*$ under the process.

Discussion. The proof is omitted. Notice that in the diagonable case, the fact that $T^m\mathbf{s} \to \mathbf{s}^*$ follows from the expression for $T^m\mathbf{s}$ given under Property 3. The Markov matrix $\begin{bmatrix} 0 & 1 \\ 1 & 0 \end{bmatrix}$ has eigenvalues 1 and -1; it does not satisfy the conclusions of Property 4. However, it does not satisfy the extra assumption of Property 4 either. It is simple and worthwhile to explore this "bad" case for a few initial states. What is the eigenvector for $\lambda = 1$?

Markov processes are discussed in many books; one such book is *Markov Chains with Stationary Probabilities* by K. L. Chung, Spring-Verlag, 1960. Properties of Markov matrices are also discussed in *Nonnegative Matrices* by H. Minc, Wiley, 1988 (in this book, these matrices are called "column stochastic").

Systems of Linear Difference Equations

If A is an n by n matrix, the matrix vector equation

$$\mathbf{s}(m+1) = A\mathbf{s}(m)$$

may be regarded as a system of n linear first-order difference equations, describing the coordinates $s_1, s_2, \ldots, s_n$ at time $m+1$ in terms of those at time m. They are "first-order difference" equations because they involve only one time difference from m to $m+1$; the Fibonacci equation $s(m+1) = s(m) + s(m-1)$ is a second-order difference equation.

Markov processes form a special class of this larger class of systems of linear difference equations, but there are applications that do not fit the Markov assumptions. For example, in **population models**, we might wish to consider deaths (so that some column sums of A would be less than 1) or births, even multiple births (so that some entries in A would be greater than 1). Similar considerations apply to some **economic** models, which are represented by matrix models. A proper discussion of such models requires more theory than is developed in this book.

The Power Method for Determining Eigenvalues

Practical applications of eigenvalues often involve larger matrices with noninteger entries. Such problems often require efficient computer methods for determining eigenvalues. A thorough discussion of such methods is beyond the scope of this book, but we can indicate how powers of matrices provide one tool for finding eigenvalues.

Let A be an n by n matrix. To simplify the discussion, we suppose that A has n distinct real eigenvalues $\lambda_1, \lambda_2, \ldots, \lambda_n$, with corresponding eigenvectors $\mathbf{v}_1, \mathbf{v}_2, \ldots, \mathbf{v}_n$. We suppose that $|\lambda_1| > |\lambda_j|$ for $j = 2, 3, \ldots, n$. λ_1 is called the

dominant eigenvalue. Any vector $\mathbf{x}$ in $\mathbb{R}^n$ can be written

$$\mathbf{x} = c_1\mathbf{v}_1 + c_2\mathbf{v}_2 + \cdots + c_n\mathbf{v}_n.$$

Then

$$A\mathbf{x} = c_1\lambda_1\mathbf{v}_1 + c_2\lambda_2\mathbf{v}_2 + \cdots + c_n\lambda_n\mathbf{v}_n$$

and

$$A^m\mathbf{x} = c_1\lambda_1^m\mathbf{v}_1 + c_2\lambda_2^m\mathbf{v}_2 + \cdots + c_n\lambda_n^m\mathbf{v}_n.$$

For m large, $|\lambda_1^m|$ is much greater than all other terms. If we divide by $c_1\lambda_1^m$, then all the terms on the right-hand side will be negligibly small except for $\mathbf{v}_1$, so we will be able to identify $\mathbf{v}_1$. By calculating $A\mathbf{v}_1$ we then determine λ_1.

To make this into an effective procedure, we must control the size of the vectors: if $\lambda_1 > 1$, then $\lambda_1^m \to \infty$ as m gets large, and the procedure would break down. Similarly, if all eigenvalues are between 0 and 1, then $A^m\mathbf{x} \to 0$, and the procedure would fail. To avoid these problems, we normalize the vector at each step (that is, convert it to a vector of length one).

The procedure is as follows.

Guess $\mathbf{x}_0$; normalize: $\mathbf{y}_0 = \mathbf{x}_0/\|\mathbf{x}_0\|$;

$\mathbf{x}_1 = A\mathbf{y}_0$; normalize: $\mathbf{y}_1 = \mathbf{x}_1/\|\mathbf{x}_1\|$;

$\mathbf{x}_2 = A\mathbf{y}_1$; normalize: $\mathbf{y}_2 = \mathbf{x}_2/\|\mathbf{x}_2\|$;

and so on.

We seek convergence of y_m to some limiting vector; if such a vector exists, it must be $\mathbf{v}_1$, the eigenvector for the largest eigenvalue λ_1, so $A\mathbf{v}_1 = \lambda_1\mathbf{v}_1$.

This procedure is illustrated in the following example, which is simple enough that you can check the calculations with a calculator.

EXAMPLE 16

Determine the eigenvalue of largest absolute value for the matrix $A = \begin{bmatrix} 13 & 6 \\ -12 & -5 \end{bmatrix}$ by the power method.

SOLUTION: Choose any starting vector: let $\mathbf{x}_0 = (1,1)$. Then:

$$\mathbf{y}_0 = \frac{1}{\sqrt{2}}(1,1) \approx (.707, .707);$$

$$\mathbf{x}_1 = A\mathbf{y}_0 \approx \begin{bmatrix} 13 & 6 \\ -12 & -5 \end{bmatrix}\begin{bmatrix} 0.707 \\ 0.707 \end{bmatrix} \approx \begin{bmatrix} 13.44 \\ -12.02 \end{bmatrix},$$

$$\mathbf{y}_1 = (\mathbf{x}_1/\|\mathbf{x}_1\|) \approx \begin{bmatrix} 0.745 \\ -0.667 \end{bmatrix};$$

$$\mathbf{x}_2 = A[\mathbf{y}_1] \approx \begin{bmatrix} 5.683 \\ -5.605 \end{bmatrix}, \quad \mathbf{y}_2 \approx \begin{bmatrix} 0.712 \\ -0.702 \end{bmatrix};$$

$$\mathbf{x}_3 = A[\mathbf{y}_2] \approx \begin{bmatrix} 5.044 \\ -5.034 \end{bmatrix}, \quad \mathbf{y}_3 \approx \begin{bmatrix} 0.7078 \\ -0.7063 \end{bmatrix};$$

$$\mathbf{x}_4 = A[\mathbf{y}_3] \approx \begin{bmatrix} 4.9636 \\ -4.9621 \end{bmatrix}, \quad \mathbf{y}_4 \approx \begin{bmatrix} 0.7072 \\ -0.7070 \end{bmatrix}.$$

At this point, we judge that $\mathbf{y}_m \to \begin{bmatrix} 0.707 \\ -0.707 \end{bmatrix}$, so $\mathbf{v}_1 = \begin{bmatrix} 0.707 \\ -0.707 \end{bmatrix}$ is an eigenvector of A, and the corresponding dominant eigenvalue is $\lambda_1 = 7$. (The answer is easy to check by standard methods.)

There are many questions that arise with the power method. What if we make a bad choice of initial vector? If we choose $\mathbf{x}_0$ in the subspace spanned by all eigenvectors of A *except* $\mathbf{v}_1$, the method would fail to give $\mathbf{v}_1$. How do we decide when to stop repeating the steps of the procedure? For a computer version of the algorithm, it would be important to have tests to decide that the procedure has converged—or that it will never converge. (It will not converge if the maximum $|\lambda|$ occurs for a complex characteristic root.)

Once we have determined the dominant eigenvalue of A, how can we determine other eigenvalues? If A is invertible, the dominant eigenvalue of A^{-1} would give the reciprocal of the eigenvalue of A with smallest absolute value. Another approach is to observe that if one eigenvalue λ_1 is known, then eigenvalues of $A - \lambda_1 I$ will give us information about eigenvalues of A. (See Exercise D3.) For a discussion of numerical determination of eigenvalues see *Numerical Analysis: A Practical Approach*, Third Edition by M. J. Maron and R. J. Lopez, Wadsworth, 1991.

EXERCISES 6–5

A1 (a) Determine which of the following matrices are Markov matrices.

(b) For each Markov matrix, determine the invariant or fixed state (corresponding to the eigenvalue $\lambda = 1$).

$$A = \begin{bmatrix} 0.2 & 0.6 \\ 0.8 & 0.3 \end{bmatrix} \qquad B = \begin{bmatrix} 0.3 & 0.6 \\ 0.7 & 0.4 \end{bmatrix}$$

$$C = \begin{bmatrix} 0.7 & 0.3 & 0.0 \\ 0.1 & 0.6 & 0.1 \\ 0.2 & 0.2 & 0.9 \end{bmatrix} \qquad D = \begin{bmatrix} 0.9 & 0.1 & 0.0 \\ 0.0 & 0.9 & 0.1 \\ 0.1 & 0.0 & 0.9 \end{bmatrix}$$

A2 Suppose that census data show that every decade, 15% of people dwelling in rural areas move into towns and cities while 5% of urban dwellers move into rural areas.

(a) What would be the eventual steady state population distribution?

(b) If the population were 50% urban, 50% rural at some census, what would be the distribution after 50 years?

A3 A car rental company serving one city has three locations: the airport, the train station, and the city centre. Of the cars rented at the airport, 8/10 are returned to the airport, 1/10 are left at the train station, and 1/10 are left at the city centre. Of cars rented at the train station, 3/10 are left at the airport, 6/10 are returned to the train station, and 1/10 are left at the city centre. Of cars rented at the city centre, 3/10 go to the airport, 1/10 go to the train station, and 6/10 are returned to the city centre. Model this as a Markov process, and determine the steady state (fixed) distribution for the cars.

A4 To see how the power method works, use it to determine the largest eigenvalue of the given matrix, starting with the given initial vector. (You will need a calculator or computer.)

(a) $\begin{bmatrix} 5 & 0 \\ 0 & -2 \end{bmatrix}$, $\mathbf{x}_0 = (1, 1)$ (b) $\begin{bmatrix} 27 & 84 \\ -7 & -22 \end{bmatrix}$, $\mathbf{x}_0 = (1, 0)$

B1 (a) Determine which of the following matrices are Markov matrices.

(b) For each Markov matrix, determine the invariant or fixed state (corresponding to the eigenvalue $\lambda = 1$).

$$A = \begin{bmatrix} 0.4 & 0.7 \\ 0.5 & 0.3 \end{bmatrix} \qquad B = \begin{bmatrix} 0.5 & 0.6 \\ 0.5 & 0.4 \end{bmatrix}$$

$$C = \begin{bmatrix} 0.8 & 0.3 & 0.2 \\ 0.0 & 0.6 & 0.2 \\ 0.2 & 0.1 & 0.6 \end{bmatrix} \qquad D = \begin{bmatrix} 0.8 & 0.1 & 0.2 \\ 0.1 & 0.9 & 0.6 \\ 0.1 & 0.1 & 0.2 \end{bmatrix}$$

B2 The town of Markov Centre has only two suppliers of widgets; all inhabitants buy their supply on the first day of each month. Neither supplier is very successful at keeping customers. 70% of the customers who deal with Johnson decide they will "try the other guy" next time. Thomson does even worse:

only 20% of his customers come back the next month, while the rest go to Johnson.

(a) Model this as a Markov process, and determine the steady state (fixed) distribution of customers.

(b) Determine a general expression for Johnson and Thomson's shares of the customers, given an initial state where Johnson has 25% and Thomson has 75%.

B3 A Student Society at a large university campus decides to create a pool of bicycles, which can be used by the members of the society. Bicycles can be borrowed or returned at the Residence, the Library, or the Athletic Centre. The first day, 200 marked bicycles are left at each location. At the end of the day, at the Residence, there are 160 bicycles that started at the Residence, 40 that started at the Library, and 60 that started at the Athletic Centre. At the Library, there are 20 that started at the Residence, 140 that started at the Library, and 40 that started at the Athletic Centre. At the Athletic Centre, there are 20 that started at the Residence, 20 that started at the Library, and 100 that started at the Athletic Centre. If this pattern is repeated every day, what is the steady state distribution of bicycles?

B4 Use the power method with initial vector $(1, 0)$ to determine the dominant eigenvalue of

$$\begin{bmatrix} 3.5 & 4.5 \\ 4.5 & 3.5 \end{bmatrix}.$$

Show your calculations clearly.

COMPUTER RELATED EXERCISES

C1 Use the power method with initial vector $(1, 0, 0)$ to determine the dominant eigenvalue of the matrix of Exercise C2 in Section 6–2. You may do this by using software that includes matrix operations. Or you may do this by writing a program to carry out the procedure; in this case, be sure to build in tests for exiting.

C2 Explore whether your computer software has built-in routines for numerical determination of eigenvalues.

CONCEPTUAL EXERCISES

D1 (a) Let T be the transition matrix for a 2-state Markov process. Show that the eigenvalue that is not 1 is $\lambda_2 = t_{11} + t_{22} - 1$.

(b) For a 2-state Markov process with $t_{21} = a$ and $t_{12} = b$, show that the fixed state (eigenvector for $\lambda = 1$) is $(\frac{b}{a+b}, \frac{a}{a+b})$.

*D2 Suppose that T is a Markov matrix.

(a) Show that for any state $\mathbf{x}$,

$$\sum_{k=1}^{n}(T\mathbf{x})_k = \sum_{k=1}^{n} x_k.$$

(b) Show that if $\mathbf{v}$ is an eigenvector of T with $\lambda \neq 1$, then $\sum_{k=1}^{n} v_k = 0$.

D3 Suppose that A is diagonalized by matrix P, and that the eigenvalues of A are $\lambda_1, \lambda_2, \ldots, \lambda_n$. Show that the eigenvalues of $(A - \lambda_1 I)$ are $0, \lambda_2 - \lambda_1, \lambda_3 - \lambda_1, \ldots, \lambda_n - \lambda_1$. [Hint: $A - \lambda_1 I$ is diagonalized by P.]

SECTION

6–6 Eigenvectors in Complex Vector Spaces with Application to Complex Characteristic Roots of Real Matrices

In the earlier sections of this chapter, we have essentially given up on diagonalization when we encountered complex roots of the characteristic equation. We would like to be able to say more about these cases. In order to do this, we first consider the larger problem of discussing eigenvectors and diagonalization in the case of linear mappings $L : \mathbb{C}^n \to \mathbb{C}^n$, or equivalently, in the case of n by n matrices with complex entries.

Eigenvalues and eigenvectors are defined in the same way as before, except that the scalars and the coordinates of the vectors are complex numbers.

DEFINITION

A complex number λ is an eigenvalue of the linear mapping $L : \mathbb{C}^n \to \mathbb{C}^n$, with corresponding eigenvector $\mathbf{z}$ in $\mathbb{C}^n$ if $L(\mathbf{z}) = \lambda\mathbf{z}$. A complex number λ is an eigenvalue of the n by n matrix A with complex entries, with corresponding eigenvector $\mathbf{z}$, if $A\mathbf{z} = \lambda\mathbf{z}$.

Since the theory of solving systems of equations and inverting matrices and finding coordinates with respect to a basis is exactly the same for complex vector spaces with complex scalars as the theory for real vector spaces, the basic theorem on diagonalization is unchanged except that the vector space is now $\mathbb{C}^n$. A complex n by n matrix A is diagonalized by a matrix P if and only if the columns of P form a basis for $\mathbb{C}^n$ consisting of eigenvectors of A; if such a basis exists, then $P^{-1}AP$ is diagonal, and the kth diagonal entry is the eigenvalue corresponding to the eigenvector that appears as the kth column of P.

We do not often have to carry out the diagonalization procedure for matrices with complex entries. A simple example is given to illustrate how the theory works.

EXAMPLE 17

Determine whether the matrix $A = \begin{bmatrix} 3-8i & -11+7i \\ -1-4i & -2+6i \end{bmatrix}$ is diagonable. If so, determine the diagonalizing matrix P, and verify that $P^{-1}AP$ is diagonal.

SOLUTION: Consider

$$A - \lambda I = \begin{bmatrix} 3-8i-\lambda & -11+7i \\ -1-4i & -2+6i-\lambda \end{bmatrix}.$$

Then the characteristic equation is

$$\det(A - \lambda I) = \lambda^2 - (1-2i)\lambda + (3-3i) = 0.$$

The roots of this equation are $\lambda_1 = 1+i$ and $\lambda_2 = -3i$.

For $\lambda_1 = 1+i$, a corresponding eigenvector is $(1+i, 1)$, and for $\lambda_2 = -3i$, a corresponding eigenvector is $(2+i, 1)$. (You should check that these satisfy $A\mathbf{z} = \lambda\mathbf{z}$. Also, you should realize that eigenvectors are determined only up to a scalar multiple: we could have chosen our first eigenvector to be $-i(1+i, i) = (1-i, 1)$.)

We now take

$$P = \begin{bmatrix} 1+i & 2+i \\ 1 & 1 \end{bmatrix} \text{ and find that } P^{-1} = \begin{bmatrix} -1 & 2+i \\ 1 & -1-i \end{bmatrix}.$$

Then, as the theory tells us to expect, we can verify that

$$P^{-1}AP = \begin{bmatrix} 1+i & 0 \\ 0 & -3i \end{bmatrix}.$$

Complex Characteristic Roots of a Real Matrix and a Real Canonical Form

Does diagonalizing a complex matrix tell us anything useful about diagonalizing a real matrix? First, note that since the real numbers form a subset of the complex numbers, we may regard a matrix A with real entries as being a matrix with complex entries—all of the entries just happen to have imaginary part equal to zero. In this context, if the real matrix A has a complex characteristic root λ, we speak of λ as a complex eigenvalue of A, with a corresponding complex eigenvector. We can then proceed to diagonalize this matrix A, as above. We speak of *diagonalizing A over* $\mathbb{C}$.

EXAMPLE 18

Let $A = \begin{bmatrix} 5 & -6 \\ 3 & -1 \end{bmatrix}$. Find its eigenvectors and eigenvalues, and diagonalize, over $\mathbb{C}$.

SOLUTION:

$$A - \lambda I = \begin{bmatrix} 5-\lambda & -6 \\ 3 & -1-\lambda \end{bmatrix},$$

so $\det(A - \lambda I) = \lambda^2 - 4\lambda + 13$, and the roots of the characteristic equation are $\lambda_1 = 2 + i3$ and $\lambda_2 = 2 - i3 = \overline{\lambda}_1$.

For $\lambda_1 = 2 + i3$,

$$A - \lambda I = \begin{bmatrix} 5-(2+i3) & -6 \\ 3 & -1-(2+i3) \end{bmatrix} = \begin{bmatrix} 3-i3 & -6 \\ 3l & -3-i3 \end{bmatrix},$$

and

$$\begin{bmatrix} 3-i3 & -6 \\ 3 & -3-i3 \end{bmatrix} \overset{(3+i3)R_1}{\longrightarrow} \begin{bmatrix} 18 & -18-i18 \\ 6 & -6-6i \end{bmatrix} \rightarrow \begin{bmatrix} 1 & -(1+i) \\ 0 & 0 \end{bmatrix}.$$

Hence, a complex eigenvector corresponding to $\lambda_1 = 2 + i3$ is $\mathbf{z}_1 = (1 + i, 1)$.

For $\lambda_2 = 2 - i3$, check that

$$A - \lambda I = \begin{bmatrix} 3+i3 & -6 \\ 3 & -3+i3 \end{bmatrix} \rightarrow \begin{bmatrix} 1 & -(1-i) \\ 0 & 0 \end{bmatrix},$$

so that an eigenvector corresponding to $\lambda_2 = 2 - i3$ is $\mathbf{z}_2 = (1 - i, 1)$.

It follows that A is diagonalized to $\begin{bmatrix} 2+i3 & 0 \\ 0 & 2-i3 \end{bmatrix}$ by

$$P = \begin{bmatrix} 1+i & 1-i \\ 1 & 1 \end{bmatrix}.$$

Check that $P^{-1} = \begin{bmatrix} -i & 1+i \\ i & 1-i \end{bmatrix}$, and that $P^{-1}AP$ is diagonal, as claimed.

The solution to Example 18 is not completely satisfying. Given a square matrix A with *real* entries, we would like to determine a similar matrix $P^{-1}AP$ that also has *real* entries and that reveals information about the eigenvalues of A, even if they are complex. The rest of this section is concerned with the problem of finding such a real matrix similar to A.

We begin by observing that complex eigenvalues of a real matrix have special properties. Notice in Example 18 that the two eigenvalues form a complex conjugate pair ($\lambda_2 = \overline{\lambda}_1$), and so do the two eigenvectors ($\mathbf{z}_2 = \overline{\mathbf{z}}_1$). The example illustrates the following theorem.

THEOREM 1. Suppose that A is an n by n matrix with real entries, and that λ is a complex eigenvalue of A with corresponding eigenvector $\mathbf{z}$. Then $\overline{\lambda}$ is also an eigenvalue, with corresponding eigenvector $\overline{\mathbf{z}}$.

Proof. Suppose that $A\mathbf{z} = \lambda\mathbf{z}$. Recall that for complex numbers,

$$\overline{a_1 z_1 + a_2 z_2} = \overline{a}_1 \overline{z}_1 + \overline{a}_2 \overline{z}_2.$$

If the coefficients a_1, a_2 are real, $\overline{a}_i = a_i$, so

$$\overline{a_1 z_1 + a_2 z_2} = a_1 \overline{z}_1 + a_2 \overline{z}_2.$$

We apply this to the product $A\mathbf{z}$, and use the fact that the entries in A are real to obtain

$$A\overline{\mathbf{z}} = \overline{A\mathbf{z}} = \overline{\lambda\mathbf{z}} = \overline{\lambda}\,\overline{\mathbf{z}}.$$

But this says that if λ is a complex eigenvalue of the matrix A with corresponding eigenvector $\mathbf{z}$, then $\overline{\lambda}$ is also a complex eigenvalue of A with corresponding eigenvector $\overline{\mathbf{z}}$.

■

We remark that Theorem 1 applies even if λ is a pure real (that is, $\overline{\lambda} = \lambda$), but it is uninteresting in this case. Accordingly, for the rest of this section we make the following assumption.

ASSUMPTION The real n by n matrix A has an eigenvalue $\lambda = \alpha + i\beta$, where α and β are real, and $\beta \neq 0$.

It is natural also to display the eigenvector $\mathbf{z}$ corresponding to α in terms of real and imaginary parts:

$$\begin{aligned} \mathbf{z} &= (z_1, z_2, \ldots, z_n) \\ &= (x_1, x_2, \ldots, x_n) + i(y_1, y_2, \ldots, y_n) \\ &= \mathbf{x} + i\mathbf{y}. \end{aligned}$$

Since $\beta \neq 0$, it follows that $\mathbf{y} \neq \mathbf{0}$; the proof is left as Exercise D2.

Now consider the equation $A\mathbf{z} = \lambda\mathbf{z}$, or

$$A(\mathbf{x} + i\mathbf{y}) = (\alpha + i\beta)(\mathbf{x} + i\mathbf{y}) = (\alpha\mathbf{x} - \beta\mathbf{y}) + i(\beta\mathbf{x} + \alpha\mathbf{y}).$$

It is a standard trick to take the real and complex parts of such equations, and we get

$$A\mathbf{x} = (\alpha\mathbf{x} - \beta\mathbf{y}) \qquad \text{and} \qquad A\mathbf{y} = (\beta\mathbf{x} + \alpha\mathbf{y}). \tag{\#}$$

This pair of equations tells us that **x** and **y** and (by linearity) any linear combination of **x** and **y** are mapped by A into the subspace of $\mathbb{R}^n$ spanned by **x** and **y**. We say that $\text{Sp}(\{\mathbf{x}, \mathbf{y}\})$ is an **invariant** subspace under the linear mapping with matrix A.

Note that **x** and **y** must be linearly independent, for if $\mathbf{x} = k\mathbf{y}$, equations (#) would say that they are real eigenvectors of A with corresponding real eigenvalues, and this is impossible with our assumption that $\beta \neq 0$. (See Exercise D3). Moreover, it can be shown that no vector in $\text{Sp}(\{\mathbf{x}, \mathbf{y}\})$ is a real eigenvector of A (Exercise D4). This discussion together with Exercises D2, D3, D4, is summarized as follows:

THEOREM 2. Suppose that λ is a complex eigenvalue with non-zero imaginary part of the real n by n matrix A, with corresponding complex eigenvector $\mathbf{z} = \mathbf{x} + i\mathbf{y}$. Then $\text{Sp}(\{\mathbf{x}, \mathbf{y}\})$ is a two-dimensional subspace of $\mathbb{R}^n$ that is invariant under A and contains no real eigenvector of A.

Let us simplify our problem by supposing that A is a 2 by 2 matrix. Then it follows from Theorem 2 that $\{\mathbf{x}, \mathbf{y}\}$ is a basis for $\mathbb{R}^2$; call this basis $\mathcal{B}$. Recall from Section 4–6 how the $\mathcal{B}$-matrix of a linear transformation is defined. The pair of equations (#) tells us that if we consider A as the standard matrix of a linear mapping from $\mathbb{R}^2$ to $\mathbb{R}^2$, then the $\mathcal{B}$-matrix of this mapping is $\begin{bmatrix} \alpha & \beta \\ -\beta & \alpha \end{bmatrix}$. That is, if $P = \big[[\mathbf{x}]\,[\mathbf{y}]\big]$, then

$$P^{-1}AP = \begin{bmatrix} \alpha & \beta \\ -\beta & \alpha \end{bmatrix}.$$

The matrix $\begin{bmatrix} \alpha & \beta \\ -\beta & \alpha \end{bmatrix}$ is called a **real canonical form** for A. It is "real" because its entries are real. "Canonical form" means something like "best possible standard form" — and since the matrix $\begin{bmatrix} \alpha & \beta \\ -\beta & \alpha \end{bmatrix}$ displays explicitly the real and imaginary parts of the eigenvalue λ, this is as much as we can hope for in a real matrix *similar* to A if A has a complex eigenvalue.

EXAMPLE 19

Find a real canonical form of the matrix $A = \begin{bmatrix} 5 & -6 \\ 3 & -1 \end{bmatrix}$ of Example 18.

SOLUTION: In Example 18, we saw that A had eigenvalues $\lambda = 2 + i3$, $\overline{\lambda} = 2 - i3$, so the theory tells us that a real canonical form for A is $\begin{bmatrix} 2 & 3 \\ -3 & 2 \end{bmatrix}$.

Let us check this by determining the change of basis matrix P. We had eigenvector $\mathbf{z} = (1+i, 1)$ for $\lambda = 2+i3$, so the real part of $\mathbf{z}$ is $\mathbf{x} = (1, 1)$, and the imaginary part is $\mathbf{y} = (1, 0)$. Thus, $P = \begin{bmatrix} 1 & 1 \\ 1 & 0 \end{bmatrix}$ and $P^{-1} = \begin{bmatrix} 0 & 1 \\ 1 & -1 \end{bmatrix}$. Check that $P^{-1}AP = \begin{bmatrix} 2 & 3 \\ -3 & 2 \end{bmatrix}$.

REMARKS

(1) A matrix of the form $\begin{bmatrix} \alpha & \beta \\ -\beta & \alpha \end{bmatrix}$ can be rewritten as

$$\sqrt{\alpha^2+\beta^2}\begin{bmatrix} \cos\theta & -\sin\theta \\ \sin\theta & \cos\theta \end{bmatrix},$$

where $\cos\theta = \alpha/\sqrt{\alpha^2+\beta^2}$, and $\sin\theta = -\beta/\sqrt{\alpha^2+\beta^2}$. This suggests that the linear transformation can be considered as a rotation followed by a dilatation or contraction. However, the new basis vectors $\mathbf{x}$ and $\mathbf{y}$ in the invariant subspace $\text{Sp}(\{\mathbf{x},\mathbf{y}\})$ are not necessarily orthogonal to each other, so the matrix $\begin{bmatrix} \cos\theta & -\sin\theta \\ \sin\theta & \cos\theta \end{bmatrix}$ does not represent a true rotation of $\text{Sp}(\{\mathbf{x},\mathbf{y}\})$.

(2) If we had taken $2-i3$ as our starting eigenvalue, we would have chosen the basis $\{(1,1),(-1,0)\}$ as our basis for the invariant subspace, with change of basis matrix $Q = \begin{bmatrix} 1 & -1 \\ 1 & 0 \end{bmatrix}$, and $Q^{-1}AQ = \begin{bmatrix} 2 & -3 \\ 3 & 2 \end{bmatrix}$.

(3) The complex eigenvector $\mathbf{z}$ is determined only up to multiplication by an *arbitrary* non-zero complex number. This means that the vectors $\mathbf{x}$ and $\mathbf{y}$ are not uniquely determined. For example, we could have taken $\mathbf{z}$ to be $(1-i)(1+i,1) = (2,1-i) = (2,1)+i(0,-1)$. Check that the choice $\mathbf{x} = (2,1)$, $\mathbf{y} = (0,-1)$ still leads to $\begin{bmatrix} 2 & 3 \\ -3 & 2 \end{bmatrix}$.

The Case of a 3 by 3 Matrix

Suppose that A is a 3 by 3 matrix with one real eigenvalue μ with corresponding eigenvector $\mathbf{v}$, and complex eigenvalues $\lambda = \alpha + i\beta$ and $\overline{\lambda} = \alpha - i\beta$, with corresponding eigenvectors $\mathbf{z} = \mathbf{x} + i\mathbf{y}$ and $\overline{\mathbf{z}} = \mathbf{x} - i\mathbf{y}$ respectively. Then we have the equations

$$A\mathbf{v} = \mu\mathbf{v}, \qquad A\mathbf{x} = (\alpha\mathbf{x} - \beta\mathbf{y}), \qquad A\mathbf{y} = (\beta\mathbf{x} + \alpha\mathbf{y}).$$

By the definition of the matrix of a linear transformation with respect to a basis $\mathcal{B}$ (Section 4–6), the matrix of A with respect to the basis $\{\mathbf{v},\mathbf{x},\mathbf{y}\}$ is $\begin{bmatrix} \mu & 0 & 0 \\ 0 & \alpha & \beta \\ 0 & -\beta & \alpha \end{bmatrix}$. The matrix A can be brought into this **real canonical form** by a change of basis with change of basis matrix $P = \begin{bmatrix} [\mathbf{v}] & [\mathbf{x}] & [\mathbf{y}] \end{bmatrix}$.

Concluding Remarks

For a real n by n matrix A with complex eigenvalue $\lambda = \alpha + i\beta$ and corresponding eigenvector $\mathbf{z} = \mathbf{x} + i\mathbf{y}$, Sp($\{\mathbf{x}, \mathbf{y}\}$) is still an invariant two-dimensional subspace of $\mathbb{R}^n$. If we use $\mathbf{x}$ and $\mathbf{y}$ as consecutive vectors in a basis for $\mathbb{R}^n$, with the other vectors being eigenvectors for other eigenvalues, then in the matrix similar to A there is a block $\begin{bmatrix} \alpha & \beta \\ -\beta & \alpha \end{bmatrix}$, with the α's occurring on the diagonal in positions determined by the position of $\mathbf{x}$ and $\mathbf{y}$ in the basis.

For repeated complex eigenvalues, the situation is the same as for repeated real eigenvalues. In some cases, it is possible to find a basis of eigenvectors and diagonalize the matrix over C. In other cases, further theory is required, leading to the Jordan normal form.

EXERCISES 6–6

A1 For each of the following matrices, determine a diagonal matrix similar to the given matrix over C. Also determine a real canonical form and give a change of basis matrix P that brings the matrix into this form.

(a) $\begin{bmatrix} 0 & 4 \\ -1 & 0 \end{bmatrix}$ (b) $\begin{bmatrix} -1 & 2 \\ -1 & -3 \end{bmatrix}$

(c) $\begin{bmatrix} 2 & 2 & -1 \\ -4 & 1 & 2 \\ 2 & 2 & -1 \end{bmatrix}$ (d) $\begin{bmatrix} 2 & 1 & -1 \\ 2 & 1 & 0 \\ 3 & -1 & 2 \end{bmatrix}$

B1 For each of the following matrices, determine a diagonal matrix similar to the given matrix over C. Also determine a real canonical form and give a change of basis matrix P that brings the matrix into this form.

(a) $\begin{bmatrix} 1 & -5 \\ 1 & -3 \end{bmatrix}$ (b) $\begin{bmatrix} 1 & -5 \\ 1 & 3 \end{bmatrix}$

B2 Same instructions as for Exercise B1.

(a) $\begin{bmatrix} 0 & -2 & 1 \\ 2 & 2 & -1 \\ 0 & -2 & 2 \end{bmatrix}$ (b) $\begin{bmatrix} -1 & 2 & -2 \\ -2 & -1 & -1 \\ 4 & -2 & 5 \end{bmatrix}$

B3 Same instructions as for Exercise B1.

(a) $\begin{bmatrix} 6 & 0 & -4 \\ 0 & 1 & 1 \\ 8 & -1 & -5 \end{bmatrix}$ (b) $\begin{bmatrix} 2 & 2 & 2 \\ 1 & 1 & -2 \\ -2 & -1 & 2 \end{bmatrix}$

CONCEPTUAL EXERCISES

D1 Verify that if $\mathbf{z}$ is an eigenvector of a matrix A with complex entries, then $\overline{\mathbf{z}}$ is an eigenvector of $\overline{A}$.

***D2** Suppose that A is an n by n matrix with real entries, and that $\lambda = \alpha + i\beta$ is a complex eigenvalue of A with $\beta \neq 0$. Let the corresponding eigenvector be $\mathbf{x} + i\mathbf{y}$. Prove that $\mathbf{y} \neq \mathbf{0}$ and $\mathbf{x} \neq \mathbf{0}$.

D3 Suppose that A is a real matrix, α and β are real numbers with $\beta \neq 0$, and $\mathbf{x}$ and $\mathbf{y}$ are vectors such that $A\mathbf{x} = (\alpha\mathbf{x} - \beta\mathbf{y})$, $A\mathbf{y} = (\beta\mathbf{x} + \alpha\mathbf{y})$ (equations (#) of this section). Show that $\mathbf{x} \neq k\mathbf{y}$ for any real number k. [Hint: Suppose $\mathbf{x} = k\mathbf{y}$ for some k, and show that this requires $\beta = 0$.]

D4 With the same assumptions as in Exercise D2, prove that $\text{Sp}(\{\mathbf{x}, \mathbf{y}\})$ contains no eigenvector of A corresponding to a real eigenvector of A.

Chapter Review

Suggestions for Student Review

1 What does it mean to say that matrices A and B are *similar*? Give one or two reasons why this is an important question. (Section 6–1)

2 Define *eigenvectors* and *eigenvalues* of a matrix A. Explain the connection between the statement that λ is an eigenvalue of A with eigenvector $\mathbf{x}$ and the condition $\det(A - \lambda I) = 0$. (Section 6–2)

3 Suppose you are told that the n by n matrix A has characteristic roots $\lambda_1, \lambda_2, \ldots, \lambda_n$, possibly including repeated roots.

(a) What conditions on these roots guarantees that A is diagonable over the real numbers?

(b) Is there any case where you can tell from the roots that A is not diagonable? (Section 6–3)

4 Use the idea suggested in Exercise D1 of Section 6–3 to create matrices for your classmates to diagonalize.

5 Suppose that $P^{-1}AP = D$ where D is a diagonal matrix with distinct diagonal entries $\lambda_1, \lambda_2, \ldots, \lambda_n$. How can we use this information to solve the system of linear differential equations

$$\frac{d}{dt}\mathbf{x} = A\mathbf{x}?$$

(Section 6–4)

6 Explain how diagonalization of matrices over $\mathbb{C}$ differs from diagonalization over $\mathbb{R}$. (Section 6–6)

Chapter Quiz

1 Let $A = \begin{bmatrix} 5 & -16 & -4 \\ 2 & -7 & -2 \\ -2 & 8 & 3 \end{bmatrix}$.
Determine whether the following vectors are eigenvectors of A, and if so, state the corresponding eigenvalue.
(a) $(3, 1, 0)$; (b) $(1, 0, 1)$; (c) $(4, 1, 0)$; (d) $(2, 1, -1)$

2 Determine whether the matrix $\begin{bmatrix} -3 & 1 & 0 \\ 13 & -7 & -8 \\ -11 & 5 & 4 \end{bmatrix}$ is diagonable (in terms of real numbers). If so, give a diagonalizing matrix P and the corresponding diagonal form.

3 If λ is an eigenvalue of the invertible matrix A, prove that λ^{-1} is an eigenvalue of A^{-1}.

4 Suppose that A is a 3 by 3 matrix such that

$$\det A = 0, \det(A + 2I) = 0, \det(A - 3I) = 0.$$

Answer the following questions, giving a brief explanation in each case.
(a) What is the dimension of the solution space of the system $A\mathbf{x} = \mathbf{0}$?
(b) What is the dimension of the solution space of $(A - 2I)\mathbf{x} = \mathbf{0}$?
(c) What is the rank of A?

5 (Section 6–4) Find the general solution of the system of differential equations

$$\frac{d}{dt}\begin{bmatrix} y \\ z \end{bmatrix} = \begin{bmatrix} 0.1 & 0.2 \\ 0.3 & 0.2 \end{bmatrix}\begin{bmatrix} y \\ z \end{bmatrix}.$$

6 (Section 6–5) Let $A = \begin{bmatrix} 0.9 & 0.1 & 0.0 \\ 0.0 & 0.8 & 0.1 \\ 0.1 & 0.1 & 0.9 \end{bmatrix}$. Verify that A is a Markov matrix, and determine its invariant state $\mathbf{x}$ such that $\sum_{j=1}^{3} x_j = 1$.

7 Let $A = \begin{bmatrix} 0 & 13 \\ -1 & 4 \end{bmatrix}$. Determine a diagonal matrix similar to A over the complex numbers, and give a diagonalizing matrix P.

Further Exercises

F1 (a) Suppose that A and B are square matrices such that $AB = BA$. Suppose that the eigenvalues of A are distinct. Prove that any eigenvector of A is also an eigenvector of B.
(b) Give an example to illustrate that the result of part (a) may not be true if A has repeated characteristic roots.

F2 If $\det B \neq 0$, prove that AB and BA have the same characteristic equation, and hence the same eigenvalues.

F3 Suppose that A is an n by n matrix with n distinct real eigenvalues $\lambda_1, \lambda_2, \ldots, \lambda_n$ with corresponding eigenvectors $\mathbf{v}_1, \mathbf{v}_2, \ldots, \mathbf{v}_n$ respectively. By representing $\mathbf{x}$ with respect to the basis of eigenvectors, show that $(A - \lambda_1 I)(A - \lambda_2 I)\ldots(A - \lambda_n I)\mathbf{x} = \mathbf{0}$ for every $\mathbf{x}$ in $\mathbb{R}^n$, and hence conclude that "A satisfies its characteristic equation": if the characteristic equation is

$$(-1)^n\lambda^{n-1} + a_{n-1}\lambda^{n-1} + \ldots + a_1\lambda + a_0 = 0,$$

then

$$(-1)^n A^n + a_{n-1}A^{n-1} + \ldots + a_1 A + a_0 I = 0.$$

(Hint: Write the characteristic polynomial in factored form.) This result, called the Cayley-Hamilton Theorem, is true for any square matrix, even if there are repeated and/or complex roots, but more theory is required to prove it.

F4 For an invertible n by n matrix, use the Cayley-Hamilton Theorem (Exercise F3) to show that A^{-1} can be written as a polynomial of degree less than or equal to $n - 1$ in A (that is, a linear combination of $I, A, A^2, \ldots, A^{n-1}$). (Section 3–5, Exercise D2 may be helpful.)

CHAPTER

7

Orthonormal Bases

The problems of determining coordinates and changing basis have special simplifying features when we use orthonormal bases. For general vector spaces, the concept of an inner product generalizes the dot product in $\mathbb{R}^n$ and makes possible the discussion of orthogonality.

SECTION

7–1 Orthonormal Bases and Orthogonal Matrices

Most of our intuition about coordinate geometry is based on experience with the standard basis for $\mathbb{R}^n$. It is therefore a little uncomfortable for many beginners to deal with the arbitrary bases that arise in Chapters 4 and 6. Fortunately, for many problems, it is possible to work with bases that have the most essential properties of the standard basis: the basis vectors are **mutually orthogonal** (that is, orthogonal to each other), and each basis vector is of **unit length.**

A set of vectors $\{\mathbf{v}_1, \mathbf{v}_2, \ldots, \mathbf{v}_k\}$ in $\mathbb{R}^n$ is **orthogonal** if $\mathbf{v}_h \cdot \mathbf{v}_j = 0$ whenever $h \neq j$. (We also say that the vectors $\mathbf{v}_1, \mathbf{v}_2, \ldots, \mathbf{v}_k$ are mutually orthogonal in this case.)

EXAMPLE 1

The set $\{(1,1,1,1), (1,-1,1,-1), (-1,0,1,0)\}$ is an orthogonal set of vectors in $\mathbb{R}^4$. (Check the dot products yourself.) The set $\{(1,1,1,1), (1,-1,-1,1), (0,0,0,0)\}$ is also an orthogonal set.

If the zero vector is excluded, orthogonal sets automatically have one very nice property.

THEOREM 1. If $\{\mathbf{v}_1, \mathbf{v}_2, \ldots, \mathbf{v}_k\}$ is an orthogonal set of non-zero vectors in $\mathbb{R}^n$, it is linearly independent.

Proof. Consider the equation $c_1\mathbf{v}_1 + c_2\mathbf{v}_2 + \cdots + c_k\mathbf{v}_k = 0$. Take the dot product of $\mathbf{v}_j$ with each side: since $\mathbf{v}_h \cdot \mathbf{v}_j = 0$ unless $h = j$, the result is

$$0 + 0 + \cdots + 0 + c_j\mathbf{v}_j \cdot \mathbf{v}_j + 0 + \cdots + 0 = 0.$$

Since $\mathbf{v}_j$ was assumed to be non-zero, it follows that $c_j = 0$. Since this is true for every j, it follows that $\{\mathbf{v}_1, \mathbf{v}_2, \ldots, \mathbf{v}_k\}$ is linearly independent.

■

REMARK

The trick used in this proof of taking the dot product of each side with one of the vectors $\mathbf{v}$ is an amazingly useful trick. Many of the things we do with orthogonal sets depend on it.

In addition to being mutually orthogonal, the vectors should have unit length.

DEFINITION

A set $\{\mathbf{v}_1, \mathbf{v}_2, \ldots, \mathbf{v}_k\}$ of vectors in $\mathbb{R}^n$ is **orthonormal** if (1) it is orthogonal, and (2) each vector $\mathbf{v}_j$ is of unit length (that is, "normalized").

Notice that the vectors in an orthonormal set are necessarily non-zero, since they have length one. It follows from Theorem 1 that orthonormal sets are necessarily linearly independent.

EXAMPLE 2

Any subset of the standard basis vectors in $\mathbb{R}^n$ is an orthonormal set; for example, in $\mathbb{R}^6$, $\{\mathbf{e}_1, \mathbf{e}_2, \mathbf{e}_5, \mathbf{e}_6\}$ is an orthonormal set of four vectors (where, as usual, $\mathbf{e}_j$ is the jth standard basis vector).

EXAMPLE 3

$\{\frac{1}{2}(1,1,1,1), \frac{1}{2}(1,-1,1,-1), \frac{1}{\sqrt{2}}(-1,0,1,0)\}$ is an orthonormal set in $\mathbb{R}^4$. (The vectors are multiples of the vectors in Example 1, so they are certainly mutually orthogonal; they have been normalized so that each has length one.)

Many arguments based on orthonormal sets could be given for orthogonal sets of non-zero vectors. The general arguments are slightly simpler for the orthonormal case since $\|\mathbf{v}_j\| = 1$ in this case. In specific examples, it may be simpler to use orthogonal sets and postpone until the end the normalization that often introduces square roots (compare Examples 1 and 3). The arguments here will usually be given for orthonormal sets.

Coordinates with Respect to an Orthonormal Basis

An orthonormal set of n vectors in $\mathbb{R}^n$ is necessarily a basis for $\mathbb{R}^n$ since it is automatically linearly independent, and by the Corollary to Theorem 2 in Section 4–4, a linearly independent set of n vectors in $\mathbb{R}^n$ is a basis for $\mathbb{R}^n$. It is attractive to use an orthonormal basis partly because we feel that we understand geometry better with such a basis than with an arbitrary basis. There are also several important technical advantages to orthonormal bases.

The first of these advantages is that it is very easy to find **coordinates with respect to an orthonormal basis.** Suppose that $\mathcal{B} = \{\mathbf{v}_1, \mathbf{v}_2, \ldots, \mathbf{v}_n\}$ is an orthonormal basis for $\mathbb{R}^n$, and that $\mathbf{x}$ is any vector in $\mathbb{R}^n$. To find the $\mathcal{B}$-coordinates of $\mathbf{x}$, we must find $(\tilde{x}_1, \tilde{x}_2, \ldots, \tilde{x}_n)$ such that

$$\mathbf{x} = \tilde{x}_1\mathbf{v}_1 + \tilde{x}_2\mathbf{v}_2 + \cdots + \tilde{x}_n\mathbf{v}_n.$$

If $\mathcal{B}$ were an arbitrary basis, the procedure (from Chapter 4) would be to write the vectors in column form and solve the resulting system of n equations in n variables. Since $\mathcal{B}$ is an orthonormal basis, we can use our amazingly useful trick: take the dot product of $\mathbf{x} = \tilde{x}_1\mathbf{v}_1 + \tilde{x}_2\mathbf{v}_2 + \cdots + \tilde{x}_n\mathbf{v}_n$ with $\mathbf{v}_j$, and find that

$$\mathbf{x} \cdot \mathbf{v}_j = \tilde{x}_j,$$

because $\mathbf{v}_j \cdot \mathbf{v}_j = 1$, and $\mathbf{v}_h \cdot \mathbf{v}_j = 0$ for all $j \neq h$. The result of this argument is important enough to summarize as a theorem.

THEOREM 2. If $\{\mathbf{v}_1, \mathbf{v}_2, \ldots, \mathbf{v}_n\}$ is an orthonormal basis for $\mathbb{R}^n$, then the jth coordinate of a vector $\mathbf{x}$ with respect to this basis is

$$\tilde{x}_j = \mathbf{x} \cdot \mathbf{v}_j.$$

It follows that the vector $\mathbf{x}$ can be written as

$$\mathbf{x} = (\mathbf{x} \cdot \mathbf{v}_1)\mathbf{v}_1 + (\mathbf{x} \cdot \mathbf{v}_2)\mathbf{v}_2 + \cdots + (\mathbf{x} \cdot \mathbf{v}_n)\mathbf{v}_n.$$

The proof is described in the preceding paragraph.

EXAMPLE 4

Let $\mathcal{B}$ denote the set $\{\frac{1}{2}(1,1,1,1), \frac{1}{2}(1,-1,1,-1), \frac{1}{\sqrt{2}}(-1,0,1,0), \frac{1}{\sqrt{2}}(0,1,0,-1)\}$. Verify that $\mathcal{B}$ is an orthonormal basis for $\mathbb{R}^4$, and find the coordinates of $(1,2,3,4)$ with respect to $\mathcal{B}$.

SOLUTION: Call the vectors $\mathbf{v}_1, \mathbf{v}_2, \mathbf{v}_3, \mathbf{v}_4$ respectively. Then

$$\mathbf{v}_1 \cdot \mathbf{v}_1 = \frac{1}{4}(1+1+1+1) = 1, \quad \mathbf{v}_1 \cdot \mathbf{v}_2 = \frac{1}{4}(1-1+1-1) = 0,$$
$$\mathbf{v}_1 \cdot \mathbf{v}_3 = \frac{1}{2\sqrt{2}}(-1+0+1+0) = 0, \quad \mathbf{v}_1 \cdot \mathbf{v}_4 = \frac{1}{2\sqrt{2}}(0+1+0-1) = 0;$$

similarly, $\mathbf{v}_2 \cdot \mathbf{v}_2 = 1$, $\mathbf{v}_2 \cdot \mathbf{v}_3 = 0 = \mathbf{v}_2 \cdot \mathbf{v}_4$, $\mathbf{v}_3 \cdot \mathbf{v}_3 = 1$, $\mathbf{v}_3 \cdot \mathbf{v}_4 = 0$, $\mathbf{v}_4 \cdot \mathbf{v}_4 = 1$. Notice that since the dot product is symmetric ($\mathbf{a} \cdot \mathbf{b} = \mathbf{b} \cdot \mathbf{a}$), it is not necessary to check the remaining six dot products ($\mathbf{v}_2 \cdot \mathbf{v}_1$, etc.). Since the vectors are mutually orthogonal and of unit length, they are orthonormal. An orthonormal set is linearly independent, and a linearly independent set of four vectors in $\mathbb{R}^4$ is a basis of $\mathbb{R}^4$.

Now to find the coordinates of $(1,2,3,4)$, suppose that $(1,2,3,4) = c_1\mathbf{v}_1 + c_2\mathbf{v}_2 + c_3\mathbf{v}_3 + c_4\mathbf{v}_4$. Then

$$c_1 = \mathbf{v}_1 \cdot (1,2,3,4) = \frac{1}{2}(1+2+3+4) = 5,$$
$$c_2 = \mathbf{v}_2 \cdot (1,2,3,4) = \frac{1}{2}(1-2+3-4) = -1,$$

$$c_3 = \mathbf{v}_3 \cdot (1,2,3,4) = \frac{1}{\sqrt{2}}(-1+0+3+0) = 2/\sqrt{2} = \sqrt{2},$$
$$c_4 = \mathbf{v}_4 \cdot (1,2,3,4) = \frac{1}{\sqrt{2}}(0+2+0-4) = -\sqrt{2}.$$

Thus the $\mathcal{B}$-coordinate vector of $(1,2,3,4)$ is $(1,2,3,4)_{\mathcal{B}} = (5,-1,\sqrt{2},-\sqrt{2})$.

Another technical advantage of using orthonormal bases is related to the first one. Often it is necessary to calculate lengths and dot products of vectors whose coordinates are given with respect to some basis other than the standard basis. If the basis is not orthonormal, the calculations are a little ugly, but they are extremely simple when the basis is orthonormal: since $\mathbf{v}_j \cdot \mathbf{v}_k = 0$ if $j \neq k$, and $\mathbf{v}_j \cdot \mathbf{v}_j = 1$,

$$\begin{aligned}\mathbf{x} \cdot \mathbf{y} &= \left(\tilde{x}_1\mathbf{v}_1 + \tilde{x}_2\mathbf{v}_2 + \cdots + \tilde{x}_n\mathbf{v}_n\right) \cdot \left(\tilde{y}_1\mathbf{v}_1 + \tilde{y}_2\mathbf{v}_2 + \cdots + \tilde{y}_n\mathbf{v}_n\right) \\ &= \tilde{x}_1\tilde{y}_1 + \tilde{x}_2\tilde{y}_2 + \cdots + \tilde{x}_n\tilde{y}_n,\end{aligned}$$

and

$$\|\mathbf{x}\|^2 = \mathbf{x} \cdot \mathbf{x} = \tilde{x}_1^2 + \tilde{x}_2^2 + \cdots + \tilde{x}_n^2.$$

Thus the formulas in the new coordinates look exactly like the formulas in standard coordinates. This fact will be used in Section 7–2.

Dot products and orthonormal bases in $\mathbb{R}^n$ have important generalizations to "inner products" and orthonormal bases in more general vector spaces. One particularly important case is Fourier series. Some of this generalization is sketched in Sections 7–4 and 7–5.

Change of Basis and Orthogonal Matrices

A third technical advantage of using orthonormal bases is that it is very easy to invert the change of basis matrix from the standard basis to the new orthonormal basis. To keep the writing short we give the argument in $\mathbb{R}^3$, but the corresponding argument works in any dimension.

Let $\mathcal{B} = \{\mathbf{v}_1, \mathbf{v}_2, \mathbf{v}_3\}$ be an orthonormal basis for $\mathbb{R}^3$. From Section 4–6, the standard coordinates of $\mathbf{v}_1, \mathbf{v}_2, \mathbf{v}_3$ are the columns of the change of basis matrix P, so $P = \begin{bmatrix} [\mathbf{v}_1] & [\mathbf{v}_2] & [\mathbf{v}_3] \end{bmatrix}$. Consider the product of the transpose of P with P: the vectors $\mathbf{v}_1, \mathbf{v}_2, \mathbf{v}_3$ are the rows of P^T so

$$P^TP = \begin{bmatrix} \mathbf{v}_1 \\ \mathbf{v}_2 \\ \mathbf{v}_3 \end{bmatrix} \begin{bmatrix} [\mathbf{v}_1] & [\mathbf{v}_2] & [\mathbf{v}_3] \end{bmatrix} = \begin{bmatrix} \mathbf{v}_1 \cdot \mathbf{v}_1 & \mathbf{v}_1 \cdot \mathbf{v}_2 & \mathbf{v}_1 \cdot \mathbf{v}_3 \\ \mathbf{v}_2 \cdot \mathbf{v}_1 & \mathbf{v}_2 \cdot \mathbf{v}_2 & \mathbf{v}_2 \cdot \mathbf{v}_3 \\ \mathbf{v}_3 \cdot \mathbf{v}_1 & \mathbf{v}_3 \cdot \mathbf{v}_2 & \mathbf{v}_3 \cdot \mathbf{v}_3 \end{bmatrix} = \begin{bmatrix} 1 & 0 & 0 \\ 0 & 1 & 0 \\ 0 & 0 & 1 \end{bmatrix}.$$

It follows that if P is a matrix whose columns are orthonormal, then the inverse of P is simply the transpose of P. Matrices with this property are given a special name.

DEFINITION

An n by n matrix R such that $R^T R = I$ is called an **orthogonal matrix.** It follows that $R^{-1} = R^T$, and that $RR^T = I$.

(It might be a little easier to remember how things fit together if these were called orthonormal matrices, but the name "orthogonal matrix" is the name that everybody uses.)

The result of the calculation above is important enough to summarize as a theorem.

THEOREM 3. The change of basis matrix P from the standard basis to an orthonormal basis is an orthogonal matrix.

It is also important to observe that the definition of orthogonal matrix is equivalent to the orthonormality of either the columns or the rows of the matrix.

THEOREM 4. R is orthogonal

if and only if the columns of R form an orthonormal set

if and only if the rows of R form an orthonormal set.

Proof. This really just repeats the calculation above. By the usual rule for matrix multiplication, and because the hth row of R^T is the hth column of R, the hj^{th} entry of $R^T R$ is $r_{\downarrow h} \cdot r_{\downarrow j}$. Thus $R^T R = I$ if and only if the columns form an orthonormal set. The result for the rows of R follows from consideration of the product $RR^T = I$.

■

EXAMPLE 5

For any θ, the set $\{(\cos\theta, \sin\theta), (-\sin\theta, \cos\theta)\}$ is orthonormal (check), so the matrix $P = \begin{bmatrix} \cos\theta & -\sin\theta \\ \sin\theta & \cos\theta \end{bmatrix}$ is an orthogonal matrix, with $P^{-1} = P^T = \begin{bmatrix} \cos\theta & \sin\theta \\ -\sin\theta & \cos\theta \end{bmatrix}$. This matrix is the matrix of a change of basis (the basis vectors are "rotated" to new basis vectors); earlier this matrix appeared in a different role as the matrix of a rotation **transformation** of $\mathbb{R}^2$. The distinction is discussed at greater length in a note at the end of this section.

EXAMPLE 6

The set $\{(1,1,0),(1,-1,-1),(1,-1,2)\}$ is orthogonal. (Check!) If the vectors are normalized (that is, divided by their lengths), the resulting set is orthonormal, so the following matrix R is orthogonal:

$$R = \begin{bmatrix} 1/\sqrt{2} & 1/\sqrt{3} & 1/\sqrt{6} \\ 1/\sqrt{2} & -1/\sqrt{3} & -1/\sqrt{6} \\ 0 & -1/\sqrt{3} & 2/\sqrt{6} \end{bmatrix};$$

$$R^{-1} = R^T = \begin{bmatrix} 1/\sqrt{2} & 1/\sqrt{2} & 0 \\ 1/\sqrt{3} & -1/\sqrt{3} & -1/\sqrt{3} \\ 1/\sqrt{6} & -1/\sqrt{6} & 2/\sqrt{6} \end{bmatrix}.$$

Check that $RR^T = I$.

EXAMPLE 7

The vectors of Example 4 are orthonormal, so the following matrix is orthogonal:

$$\begin{bmatrix} 1/2 & 1/2 & -1/\sqrt{2} & 0 \\ 1/2 & -1/2 & 0 & 1/\sqrt{2} \\ 1/2 & 1/2 & 1/\sqrt{2} & 0 \\ 1/2 & -1/2 & 0 & -1/\sqrt{2} \end{bmatrix}.$$

The most important application of orthogonal matrices considered in this book is in the diagonalization of symmetric matrices in Chapter 8, but there are many other geometrical applications as well. In the next example, an orthogonal change of basis is used to find the standard matrix of a rotation transformation about an axis that is not a coordinate axis. (This was one question we could not answer in Chapter 3.)

EXAMPLE 8

Find the standard matrix of the linear transformation L of $\mathbb{R}^3$ that rotates vectors about the axis defined by the vector $(1,1,1)$ counterclockwise through an angle $\pi/3$.

SOLUTION: If the rotation were about the standard x_3-axis (that is, the axis defined by $\mathbf{e}_3$), the matrix of the rotation would be

$$R_1 = \begin{bmatrix} \cos\pi/3 & -\sin\pi/3 & 0 \\ \sin\pi/3 & \cos\pi/3 & 0 \\ 0 & 0 & 1 \end{bmatrix} = \begin{bmatrix} 1/2 & -\sqrt{3}/2 & 0 \\ \sqrt{3}/2 & 1/2 & 0 \\ 0 & 0 & 1 \end{bmatrix}.$$

This will also be the $\mathcal{B}$-matrix of the rotation in this problem if there exists a basis $\mathcal{B} = \{\mathbf{f}_1, \mathbf{f}_2, \mathbf{f}_3\}$ such that: (1) $\mathbf{f}_3$ is a unit vector in the direction of the axis $(1,1,1)$; (2) $\mathcal{B}$ is orthonormal; and (3) $\mathcal{B}$ is right-handed (so that we can correctly include the counterclockwise sense of rotation with respect to a right-handed basis). Let us find such a basis.

To start, let $\mathbf{f}_3 = (1/\sqrt{3})(1,1,1)$. We must find two vectors that are orthogonal to $\mathbf{f}_3$ and to each other. By solving the equation $(1,1,1)\cdot\mathbf{x} = 0$ by inspection, and then requiring that two solutions be orthogonal to each other, we find the vectors $(1,-1,0)$ and $(1,1,-2)$. (There are infinitely many other choices for this pair of vectors; this is just one simple choice.) These three vectors will be normalized to form an orthonormal basis.

For the basis to be right-handed, the vectors must be taken in an order such that the matrix with the basis vectors as columns has positive **determinant** (see Sections 1–4 and 5–4). The conclusion of this test is that

$$\mathbf{f}_1 = \frac{1}{\sqrt{2}}(1,-1,0) \quad \text{and} \quad \mathbf{f}_2 = \frac{1}{\sqrt{6}}(1,1,-2).$$

The required right-handed orthonormal basis is thus

$$\mathcal{B} = \{\frac{1}{\sqrt{2}}(1,-1,0), \frac{1}{\sqrt{6}}(1,1,-2), \frac{1}{\sqrt{3}}(1,1,1)\},$$

and the orthogonal change of basis matrix from the standard basis to this basis is

$$P = \begin{bmatrix} 1/\sqrt{2} & 1/\sqrt{6} & 1/\sqrt{3} \\ -1/\sqrt{2} & 1/\sqrt{6} & 1/\sqrt{3} \\ 0 & -2/\sqrt{6} & 1/\sqrt{3} \end{bmatrix}.$$

Since $[L]_{\mathcal{B}} = R_1$ (above), the standard matrix is given by

$$[L]_{\mathcal{S}} = P[L]_{\mathcal{B}}P^{-1} = \begin{bmatrix} 2/3 & -1/3 & 2/3 \\ 2/3 & 2/3 & -1/3 \\ -1/3 & 2/3 & 2/3 \end{bmatrix}. \qquad \textit{(Check.)}$$

It is easy to check that $(1,1,1)$ is an eigenvector of this matrix with eigenvalue 1, which it should be since it defines the axis of the rotation represented by the matrix. Notice also that the matrix $[L]_S$ is itself an orthogonal matrix. Since a rotation transformation always maps the standard basis to a new orthonormal basis, its matrix can always be taken as a change of basis matrix and it must be orthogonal.

A Note on Rotation Transformations and Rotation of Axes as a Change of Basis in $\mathbb{R}^2$

In Example 5, the matrix $\begin{bmatrix} \cos\theta & -\sin\theta \\ \sin\theta & \cos\theta \end{bmatrix}$ appeared in the role of a change of basis matrix for the change from the standard basis to the basis $\mathcal{B} = \{(\cos\theta, \sin\theta), (-\sin\theta, \cos\theta)\}$ of $\mathbb{R}^2$. This change of basis is often described as a "rotation of **axes** through angle θ" because each of the basis vectors in $\mathcal{B}$ is obtained from the corresponding standard basis vector by a rotation through angle θ. Such rotations of axes are sometimes discussed in high school.

Treatments of rotation of axes often emphasize the **change of coordinates equation.** Recall from Section 4–6 that if P is the change of basis matrix, the change of coordinates equation can be written in the form $[\mathbf{x}]_{\mathcal{B}} = P^{-1}[\mathbf{x}]_{\mathcal{S}}$. If the change of basis is a rotation of axes, then P is an orthogonal matrix, so $P^{-1} = P^T$, and thus the change of coordinates equation for this rotation of axes is $[\mathbf{x}]_{\mathcal{B}} = P^T[\mathbf{x}]_{\mathcal{S}}$. Since $P^T = \begin{bmatrix} \cos\theta & \sin\theta \\ -\sin\theta & \cos\theta \end{bmatrix}$, the change of coordinates equation can be written

$$\begin{bmatrix} \tilde{x}_1 \\ \tilde{x}_2 \end{bmatrix} = \begin{bmatrix} \cos\theta & \sin\theta \\ -\sin\theta & \cos\theta \end{bmatrix} \begin{bmatrix} x_1 \\ x_2 \end{bmatrix}$$

or as two equations for the "new" coordinates in terms of the "old",

$$\begin{aligned} \tilde{x}_1 &= x_1\cos\theta + x_2\sin\theta \\ \tilde{x}_2 &= -x_1\sin\theta + x_2\cos\theta. \end{aligned}$$

This last pair of equations can be derived by fairly simple trigonometry from a standard picture. Books that discuss only rotation of axes in the plane often start with these equations.

The matrix $\begin{bmatrix} \cos\theta & -\sin\theta \\ \sin\theta & \cos\theta \end{bmatrix}$ appeared in Section 3–3 as the standard matrix $[L]$ of the linear transformation of $\mathbb{R}^2$ that rotates vectors counterclockwise through angle θ. Conceptually, this is quite different from a rotation of axes.

It sometimes worries beginners that the matrix for rotation through θ as a linear transformation is the transpose of the matrix that appears in the change of coordinates equation for a rotation of axes through θ. In fact, what seems even worse, if you replace θ by $(-\theta)$, one matrix turns into the other (because $\cos(-\theta) = \cos\theta$ and $\sin(-\theta) = -\sin\theta$). One way to try to understand this is to imagine what happens to the vector $\mathbf{e}_1 = (1, 0)$ under two different scenarios. First consider L to be the transformation that rotates each vector by angle θ, with $0 < \theta < \pi/2$; then $L(\mathbf{e}_1)$ is a vector in the first quadrant that makes an angle $+\theta$ with the x_1-axis. Next consider a rotation of axes through $(-\theta)$; $\mathbf{e}_1$ has not moved but the axes have, and with respect to the **new** axes, $\mathbf{e}_1$ is in the **new** first quadrant and makes an angle of $(+\theta)$ with respect to the (new) $\tilde{x}_1$-axis. Thus the new coordinates of $\mathbf{e}_1$ relative to the rotated axes should be exactly the same as the **standard** coordinates of $L(\mathbf{e}_1)$. See Figures 7–1–1 and 7–1–2.

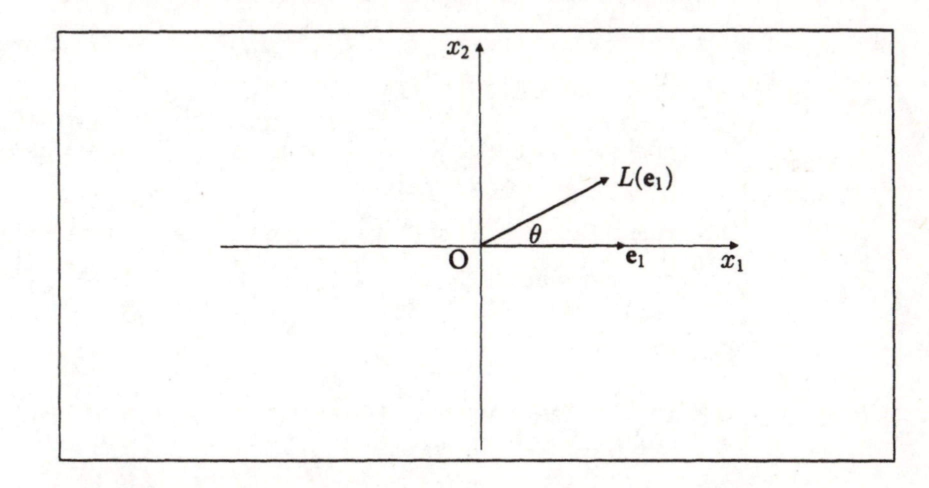

FIGURE 7-1-1. *The transformation $L : \mathbb{R}^2 \to \mathbb{R}^2$ rotates vectors by angle θ.*

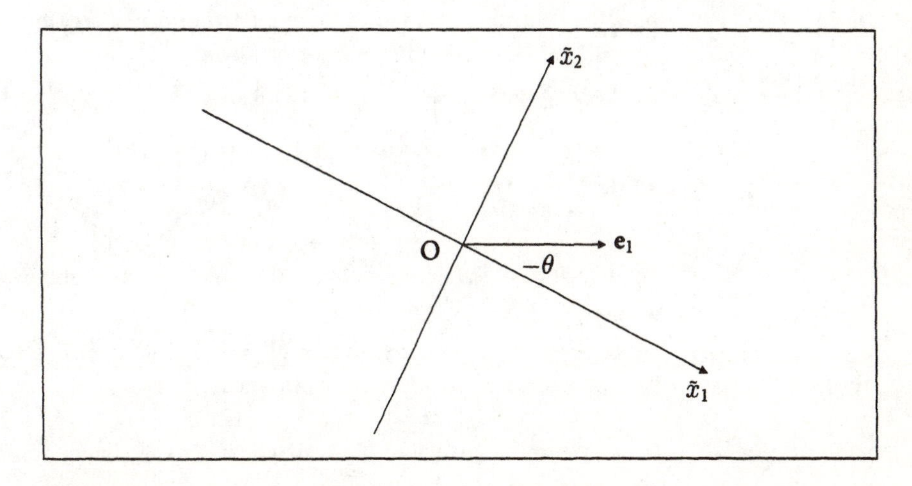

FIGURE 7-1-2. *The standard basis vector $\mathbf{e}_1$ is shown relative to axes obtained from the standard axes by rotation through $(-\theta)$.*

EXERCISES 7–1

A1 Each of the following is a set of vectors in some $\mathbb{R}^n$; for each, (i) state n, (ii) determine if the set is orthogonal, (iii) determine if the set is orthonormal, (iv) if the set is orthogonal but not orthonormal, produce the corresponding orthonormal set, (v) if the set (by (iii) or (iv)) provides an orthonormal basis of $\mathbb{R}^n$, write the corresponding orthogonal change of basis matrix, and if not, say why not.

(a) $\mathcal{T} = \{(1,2),(2,-1)\}$
(b) $\mathcal{U} = \{(1,-1,1),(1,2,1),(2,-1,1)\}$
(c) $\mathcal{V} = \{(1,1,3),(3,0,-1),(1,-10,3)\}$
(d) $\mathcal{W} = \{(1,0,1,1),(2,1,-1,-1),(1,-3,-1,0)\}$

A2 Find the coordinates of $\mathbf{x} = (4,3,5)$ and $\mathbf{y} = (3,-7,2)$ relative to the orthonormal basis $\mathcal{B} = \{\frac{1}{3}(1,-2,2), \frac{1}{3}(2,2,1), \frac{1}{3}(-2,1,2)\}$.

A3 Find the coordinates of $\mathbf{w} = (2,-4,6)$ and $\mathbf{z} = (6,6,3)$ with respect to the basis of Exercise A2.

A4 Find the coordinates of $(2,4,-3,5)$ and $(-4,1,3,-5)$ with respect to the orthonormal basis $\{\frac{1}{2}(1,1,1,1), \frac{1}{2}(1,-1,1,-1), \frac{1}{\sqrt{2}}(-1,0,1,0), \frac{1}{\sqrt{2}}(0,1,0,-1)\}$.

A5 Find the coordinates of $(3,1,0,1)$ and $(-1,3,2,-3)$ with respect to the basis of Exercise A4.

A6 For each of the following matrices, denote the matrix by A, (i) calculate A^TA, (ii) decide whether A is orthogonal, and (iii) if it is *not* orthogonal, indicate how the columns of A fail to form an orthonormal set (for example, "the second and third columns are not orthogonal").

(a) $\begin{bmatrix} 5/13 & 12/13 \\ 12/13 & -5/13 \end{bmatrix}$ (b) $\begin{bmatrix} 3/5 & 4/5 \\ -4/5 & -3/5 \end{bmatrix}$ (c) $\begin{bmatrix} 2/5 & -1/5 \\ 1/5 & 2/5 \end{bmatrix}$

(d) $\begin{bmatrix} 1/3 & 2/3 & -2/3 \\ 2/3 & -2/3 & 1/3 \\ 2/3 & 1/3 & 2/3 \end{bmatrix}$ (e) $\begin{bmatrix} 1/3 & 2/3 & 2/3 \\ 2/3 & -2/3 & 1/3 \\ 2/3 & 1/3 & -2/3 \end{bmatrix}$

A7 Let $L : \mathbb{R}^3 \to \mathbb{R}^3$ be the rotation through angle $(\frac{\pi}{4})$ about the axis determined by $(2,2,-1)$.

(a) Let $\mathbf{g}_3 = (2,2,-1)$. Verify that $\mathbf{g}_1 = (-1,2,2)$ is orthogonal to $\mathbf{g}_3$. Define $\mathbf{g}_2 = \mathbf{g}_3 \times \mathbf{g}_1$, and calculate the components of $\mathbf{g}_2$.

(b) Let $\mathbf{f}_i = \mathbf{g}_i/\|\mathbf{g}_i\|$, for $i = 1,2,3$, so that $\mathcal{B} = \{\mathbf{f}_1,\mathbf{f}_2,\mathbf{f}_3\}$ is an orthonormal basis. Write the change of basis matrix R for the change from the standard basis to this basis in the form $R = \frac{1}{3}\left[\ \ddots\ \right]$.

(c) Write the $\mathcal{B}$-matrix of L, $[L]_{\mathcal{B}}$. For later calculation it is probably easiest to write this in the form $[L]_{\mathcal{B}} = \frac{1}{\sqrt{2}}\begin{bmatrix} \cdot & \\ & \cdot \end{bmatrix}$.

(d) Determine the standard matrix of $[L]_{\mathcal{S}}$. (See Example 8. With care, you can carry out the calculation by hand. It might be more natural to compute the matrix product in decimal form with the help of a computer.)

B1 Each of the following is a set of vectors in some $\mathbb{R}^n$; for each, (i) state n, (ii) determine if the set is orthogonal, (iii) determine if the set is orthonormal, (iv) if the set is orthogonal but not orthonormal, produce the corresponding orthonormal set, (v) if the set (by (iii) or (iv)) provides an orthonormal basis of $\mathbb{R}^n$, write the corresponding orthogonal change of basis matrix, and if not, say why not.

(a) $\mathcal{T} = \{(1,3),(-3,1)\}$

(b) $\mathcal{U} = \{(2,-1,1),(1,0,-2),(2,5,1)\}$

(c) $\mathcal{V} = \{(1,1,1),(-1,0,-1),(1,-1,0)\}$

(d) $\mathcal{W} = \{(2,0,1,1),(2,1,-2,0),(1,-2,0,-1)\}$

B2 Find the coordinates of $\mathbf{x} = (6,2,1)$ and $\mathbf{y} = (-4,2,3)$ relative to the orthonormal basis $\mathcal{B} = \{\frac{1}{\sqrt{2}}(1,0,1), \frac{1}{\sqrt{3}}(-1,1,1), \frac{1}{\sqrt{6}}(1,2,-1)\}$.

B3 Find the coordinates of $\mathbf{w} = (3,3,-5)$ and $\mathbf{z} = (-1,4,2)$ with respect to the orthonormal basis $\mathcal{B}$ of Exercise B2.

B4 (a) Let $\mathbf{w}_1 = (1,1,2)$ and $\mathbf{w}_2 = (-2,0,1)$. Determine a third vector $\mathbf{w}_3$ such that $\mathbf{w}_1, \mathbf{w}_2, \mathbf{w}_3$ form a right-handed orthogonal set.

(b) Let $\mathbf{v}_i = \mathbf{w}_i/\|\mathbf{w}_i\|$, so that $\mathcal{B} = \{\mathbf{v}_1, \mathbf{v}_2, \mathbf{v}_3\}$ is an orthonormal basis. Find $(2,3,-4)_{\mathcal{B}}$ and $(-3,1,3)_{\mathcal{B}}$.

B5 Find the coordinates of $(3,-2,6,1)$ and $(2,-4,0,4)$ with respect to the orthonormal basis $\{\frac{1}{2}(1,1,1,1), \frac{1}{2}(1,-1,1,-1), \frac{1}{\sqrt{2}}(-1,0,1,0), \frac{1}{\sqrt{2}}(0,1,0,-1)\}$.

B6 Find the coordinates of $(5,0,-2,2)$ and $(4,2,-2,3)$ with respect to the orthonormal basis $\{\frac{1}{2}(1,1,1,1), \frac{1}{2}(1,-1,1,-1), \frac{1}{\sqrt{2}}(-1,0,1,0), \frac{1}{\sqrt{2}}(0,1,0,-1)\}$.

B7 For each of the following matrices, denote the matrix by A, (i) calculate $A^T A$, (ii) decide whether A is orthogonal, and (iii) if it is *not* orthogonal, indicate how the columns of A fail to form an orthonormal set (for example, "the second and third columns are not orthogonal").

(a) $\begin{bmatrix} 2/\sqrt{5} & -1/\sqrt{5} \\ 1/\sqrt{5} & -2/\sqrt{5} \end{bmatrix}$ (b) $\begin{bmatrix} 2/\sqrt{5} & -1/\sqrt{5} \\ -1/\sqrt{5} & -2/\sqrt{5} \end{bmatrix}$ (c) $\begin{bmatrix} 1/2 & 1/2 \\ -1/2 & 1/2 \end{bmatrix}$

(d) $\begin{bmatrix} 1/\sqrt{3} & 1/\sqrt{2} & -1/\sqrt{6} \\ -1/\sqrt{3} & 1/\sqrt{2} & 1/\sqrt{6} \\ 1/\sqrt{3} & 0 & 2/\sqrt{6} \end{bmatrix}$ (e) $\begin{bmatrix} 1/\sqrt{3} & 1/\sqrt{6} & 1/\sqrt{2} \\ 1/\sqrt{3} & 1/\sqrt{6} & 1/\sqrt{2} \\ 1/\sqrt{3} & -2/\sqrt{6} & 0 \end{bmatrix}$

CONCEPTUAL EXERCISES

D1 Verify that the product of two orthogonal matrices is an orthogonal matrix.

***D2** (a) Prove that if R is an orthogonal matrix, then $\det R = \pm 1$.
(b) Give an example of a 2 by 2 matrix A that has $\det A = 1$, but is not orthogonal.

D3 (a) Observe that the dot product of two vectors $\mathbf{x}$ and $\mathbf{y}$ can be written as

$$\mathbf{x} \cdot \mathbf{y} = \begin{bmatrix} x_1 & x_2 & \dots & x_n \end{bmatrix} \begin{bmatrix} y_1 \\ y_2 \\ \vdots \\ y_n \end{bmatrix} = [\mathbf{x}]^T [\mathbf{y}].$$

Use this fact to show that if an n by n matrix R is orthogonal, then $\|R\mathbf{x}\| = \|\mathbf{x}\|$ for every $\mathbf{x}$ in $\mathbb{R}^n$.
(b) Show that any real eigenvalue of an orthogonal matrix must be either 1 or -1.

SECTION

7–2 General Projections and the Gram-Schmidt Procedure

Projections onto and Perpendicular to a Subspace

The projection of one vector **b** onto another vector **a** was defined in Chapter 1; it could be viewed as the projection of **b** onto the subspace spanned by **a**. It is natural and useful to define the projection of vectors onto more general subspaces.

DEFINITION

Suppose that S is a k-dimensional subspace of $\mathbb{R}^n$; let $\mathcal{B} = \{\mathbf{v}_1, \mathbf{v}_2, \dots, \mathbf{v}_k\}$ be an **orthonormal** basis of S. If $\mathbf{x}$ is any vector in $\mathbb{R}^n$, the **projection of x onto S** is defined to be

$$\mathbf{proj}_S(\mathbf{x}) = (\mathbf{x} \cdot \mathbf{v}_1)\mathbf{v}_1 + (\mathbf{x} \cdot \mathbf{v}_2)\mathbf{v}_2 + \cdots + (\mathbf{x} \cdot \mathbf{v}_k)\mathbf{v}_k.$$

The **projection of x perpendicular to S** is defined to be

$$\mathbf{perp}_S(\mathbf{x}) = \mathbf{x} - \mathbf{proj}_S(\mathbf{x}).$$

Note that $\mathbf{proj}_S(\mathbf{x})$ is a vector in S. It follows easily from the definition that $\mathbf{perp}_S(\mathbf{x})$ is orthogonal to every vector in S because

$$\begin{aligned}\mathbf{v}_j \cdot \mathbf{perp}_S(\mathbf{x}) &= \mathbf{v}_j \cdot \mathbf{x} - \mathbf{v}_j \cdot \big((\mathbf{x} \cdot \mathbf{v}_1)\mathbf{v}_1 + (\mathbf{x} \cdot \mathbf{v}_2)\mathbf{v}_2 + \cdots + (\mathbf{x} \cdot \mathbf{v}_k)\mathbf{v}_k\big)\\ &= \mathbf{v}_j \cdot \mathbf{x} - \big(0 + 0 + \cdots + (\mathbf{x} \cdot \mathbf{v}_j)(\mathbf{v}_j \cdot \mathbf{v}_j) + \cdots + 0\big)\\ &= 0 \qquad \text{for } j = 1, 2, \dots, k.\end{aligned}$$

Note also that $\mathbf{x} = \mathbf{proj}_S(\mathbf{x}) + \mathbf{perp}_S(\mathbf{x})$.

EXAMPLE 9

Let S be the subspace of $\mathbb{R}^4$ spanned by $\{(1, 1, 1, 1), (1, -1, 1, -1)\}$. Determine $\mathbf{proj}_S(2, 5, -7, 3)$ and $\mathbf{perp}_S(2, 5, -7, 3)$.

SOLUTION: To apply the definition of projection we require an orthonormal basis for S. The two spanning vectors are already mutually orthogonal (check), so it is only necessary to normalize them. An orthonormal basis for S is

$$\mathcal{B} = \{\mathbf{v}_1, \mathbf{v}_2\} = \{\tfrac{1}{2}(1, 1, 1, 1), \tfrac{1}{2}(1, -1, 1, -1)\}.$$

Since

$$\begin{aligned}\mathbf{proj}_S(\mathbf{x}) &= (\mathbf{x} \cdot \mathbf{v}_1)\mathbf{v}_1 + (\mathbf{x} \cdot \mathbf{v}_2)\mathbf{v}_2,\\ \mathbf{proj}_S(2, 5, -7, 3) &= \Big((2, 5, -7, 3) \cdot \frac{1}{2}(1, 1, 1, 1)\Big)\frac{1}{2}(1, 1, 1, 1)\\ &\quad + \Big((2, 5, -7, 3) \cdot \frac{1}{2}(1, -1, 1, -1)\Big)\frac{1}{2}(1, -1, 1, -1)\end{aligned}$$

$$= \frac{1}{4}(3)(1,1,1,1) + \frac{1}{4}(-13)(1,-1,1,-1)$$

$$= \frac{1}{4}\left(-10,16,-10,16\right) = \left(\frac{-5}{2},4,\frac{-5}{2},4\right).$$

It follows from the definition of $\mathbf{perp}_S$ that

$$\mathbf{perp}_S\,(2,5,-7,3) = \left(2,5,-7,3\right) - \left(\frac{-5}{2},4,\frac{-5}{2},4\right) = \left(\frac{9}{2},1,-\frac{9}{2},-1\right).$$

REMARKS

(1) Notice that the definition of $\mathbf{proj}_S\,(\mathbf{x})$ is a generalization of the definition in Section 1–3 of $\mathbf{proj}_{\mathbf{n}}(\mathbf{x})$ for the case where $\mathbf{n}$ was a unit vector: $\mathbf{proj}_{\mathbf{n}}(\mathbf{x}) = (\mathbf{x} \cdot \mathbf{n})\mathbf{n}$. For this one-dimensional case, the vector $\mathbf{n}$ is used to identify the one-dimensional subspace S that it spans.

(2) Notice also that the definition of $\mathbf{proj}_S\,(\mathbf{x})$ is a truncated version of the expression given in Section 7–1 for $\mathbf{x}$ in terms of an orthonormal basis $\{\mathbf{v}_1, \mathbf{v}_2, \ldots, \mathbf{v}_n\}$ for $\mathbb{R}^n$:

$$\mathbf{x} = (\mathbf{x} \cdot \mathbf{v}_1)\mathbf{v}_1 + (\mathbf{x} \cdot \mathbf{v}_2)\mathbf{v}_2 + \cdots + (\mathbf{x} \cdot \mathbf{v}_n)\mathbf{v}_n.$$

Projection onto the subspace spanned by the first k of these basis vectors simply takes the sum of the first k terms.

(3) The definition of $\mathbf{perp}_S$ is exactly analogous to the definition of $\mathbf{perp}_{\mathbf{n}}$ in Section 1–3.

DEFINITION

Let S be a subspace of $\mathbb{R}^n$. We shall say that a vector $\mathbf{x}$ is **orthogonal to** S if $\mathbf{x}$ is orthogonal to every vector in S. The set of all vectors orthogonal to S is again a subspace of $\mathbb{R}^n$ (Exercise 3–4–D2). This subspace is called the **orthogonal complement** of S and denoted $S^{\perp}$ (so that $S^{\perp} = \{\mathbf{x} \in \mathbb{R}^n \mid \mathbf{x} \cdot \mathbf{s} = 0 \text{ for every } \mathbf{s} \text{ in } S\}$).

For example, if S is a plane in $\mathbb{R}^3$ with normal vector $\mathbf{n}$, we say that $\mathbf{n}$ is orthogonal to the plane, and the plane is the orthogonal complement of $\mathrm{Sp}(\{\mathbf{n}\})$.

The following facts about S and $S^{\perp}$ are intuitively plausible, and easy to prove. *(See what they mean in the cases where* S *is a one- or two-dimensional subspace of* $\mathbb{R}^3$.*)* You are asked to prove these facts as Exercises D1, D2, D3, and D4.

(i) $S \cap S^{\perp} = \{\mathbf{0}\}$;

(ii) if S is k-dimensional, then $S^{\perp}$ is $(n-k)$-dimensional;

(iii) if $\{\mathbf{v}_1, \mathbf{v}_2, \ldots, \mathbf{v}_k\}$ is an orthonormal basis for S, and $\{\mathbf{v}_{k+1}, \mathbf{v}_{k+2}, \ldots, \mathbf{v}_n\}$ is an orthonormal basis for $S^{\perp}$, then $\{\mathbf{v}_1, \ldots, \mathbf{v}_k, \mathbf{v}_{k+1}, \ldots, \mathbf{v}_n\}$ is an orthonormal basis for $\mathbb{R}^n$;

(iv) $\mathbf{perp}_S = \mathbf{proj}_{S^{\perp}}$.

Because of the last fact, $\mathbf{perp}_S$ is usually called the **projection onto the orthogonal complement of** S; we shall continue to denote it by $\mathbf{perp}_S$.

We have defined projection onto a subspace in terms of an orthonormal basis for the subspace. You might worry that if we used a different basis we would get a different projection onto the subspace; if that were true, we would not have a well-defined concept of projection. The following theorem indicates that projection has properties that are independent of the choice of basis.

THEOREM 1. Suppose that S is a k-dimensional subspace of $\mathbb{R}^n$.

(1) If $\mathbf{x}$ is any vector in $\mathbb{R}^n$, the unique vector $\mathbf{s}$ in S that minimizes the distance $\|\mathbf{x}-\mathbf{s}\|$ is $\mathbf{s} = \mathbf{proj}_S(\mathbf{x})$, and the distance from $\mathbf{x}$ to $\mathbf{proj}_S(\mathbf{x})$ is $\|\mathbf{perp}_S(\mathbf{x})\|$.

(2) The only way to write $\mathbf{x}$ as a sum of a vector in S and a vector in $S^{\perp}$ is $\mathbf{x} = \mathbf{proj}_S(\mathbf{x}) + \mathbf{perp}_S(\mathbf{x})$.

Proof. (1) Let $\{\mathbf{v}_1, \mathbf{v}_2, \ldots, \mathbf{v}_k\}$ be an orthonormal basis for S, and let $\{\mathbf{v}_{k+1}, \mathbf{v}_{k+2}, \ldots, \mathbf{v}_n\}$ be an orthonormal basis for $S^{\perp}$ (the orthogonal complement of S). Taken together, the two bases form an orthonormal basis of $\mathbb{R}^n$, so that for any vector $\mathbf{x}$ in $\mathbb{R}^n$ we may write

$$\mathbf{x} = \tilde{x}_1\mathbf{v}_1 + \cdots + \tilde{x}_k\mathbf{v}_k + \tilde{x}_{k+1}\mathbf{v}_{k+1} + \tilde{x}_n\mathbf{v}_n.$$

Any vector $\mathbf{s}$ in S can be expressed as $\mathbf{s} = \tilde{s}_1\mathbf{v}_1 + \tilde{s}_2\mathbf{v}_2 + \cdots + \tilde{s}_k\mathbf{v}_k$, so that (from Section 7–1) the square of the distance from $\mathbf{x}$ to $\mathbf{s}$ is given by

$$\|\mathbf{x}-\mathbf{s}\|^2 = \left((\tilde{x}_1 - \tilde{s}_1)^2 + (\tilde{x}_2 - \tilde{s}_2)^2 + \cdots + (\tilde{x}_k - \tilde{s}_k)^2\right) + \left(\tilde{x}_{k+1}^2 + \cdots + \tilde{x}_n^2\right).$$

Now how should $\mathbf{s}$ be chosen in S to minimize this distance? Obviously, we must choose $\tilde{s}_j = \tilde{x}_j$, for $j = 1, 2, \ldots, k$. But this means that

$$\mathbf{s} = \tilde{x}_1\mathbf{v}_1 + \cdots + \tilde{x}_k\mathbf{v}_k = \mathbf{proj}_S(\mathbf{x}),$$

and this is what was to be proved. Moreover, the minimum distance is

$$\left(\tilde{x}_{k+1}^2 + \cdots + \tilde{x}_n^2\right)^{1/2} = \|\mathbf{x} - \mathbf{proj}_S(\mathbf{x})\| = \|\mathbf{perp}_S(\mathbf{x})\|.$$

The calculation shows that if $\tilde{s}_j \neq \tilde{x}_j$, the distance $\|\mathbf{x} - \mathbf{s}\|$ is strictly greater than $\|\mathbf{x} - \mathbf{proj}_S(\mathbf{x})\|$, so $\mathbf{proj}_S(\mathbf{x})$ is the unique vector in S that minimizes this distance.

(2) Certainly $\mathbf{x} = \mathbf{proj}_S(\mathbf{x}) + \mathbf{perp}_S(\mathbf{x})$; suppose also that $\mathbf{x} = \mathbf{s}_1 + \mathbf{p}_1$, where $\mathbf{s}_1$ is in S and $\mathbf{p}_1$ is in $S^\perp$. Then

$$\mathbf{proj}_S(\mathbf{x}) + \mathbf{perp}_S(\mathbf{x}) = \mathbf{s}_1 + \mathbf{p}_1,$$

so

$$\mathbf{proj}_S(\mathbf{x}) - \mathbf{s}_1 = \mathbf{p}_1 - \mathbf{perp}_S(\mathbf{x}).$$

The vector on the left of this equation is in S (because the subspace S is closed under subtraction). Similarly, the vector on the right is in $S^\perp$. Therefore, their common value is in $S \cap S^\perp = \{\mathbf{0}\}$. It follows that

$$\mathbf{proj}_S(\mathbf{x}) = \mathbf{s}_1$$

and

$$\mathbf{p}_1 = \mathbf{perp}_S(\mathbf{x}),$$

so we have a unique decomposition, as claimed. ■

The properties of minimizing distance or of giving a unique orthogonal decomposition are independent of the choice of basis, so that the idea of projection does not depend on which basis for S we use in the definition. If we knew enough about the existence of minimum values, we could have begun by defining projection in terms of this property. Minimizing distance is important as a method of dealing with problems of "best fit" or "best approximation". Sections 7–3 and 7–4 give some important applications; the following example gives a simple illustration.

EXAMPLE 10

Let S be the subspace spanned by $\{(1,1,1,1),(1,-1,1,-1)\}$. Find the point in S that is closest to $\mathbf{x} = (2,-3,3,-4)$, and find the distance from $\mathbf{x}$ to S.

SOLUTION: This is the same subspace as in Example 9; the closest point is

$$\begin{aligned}\mathbf{proj}_S(\mathbf{x}) &= \left((2,-3,3,-4)\cdot\frac{1}{2}(1,1,1,1)\right)\frac{1}{2}(1,1,1,1)\\ &\quad+\left((2,-3,3,-4)\cdot\frac{1}{2}(1,-1,1,-1)\right)\frac{1}{2}(1,-1,1,-1)\\ &= \frac{-2}{4}(1,1,1,1)+\frac{12}{4}(1,-1,1,-1)=(\frac{5}{2},-\frac{7}{2},\frac{5}{2},-\frac{7}{2}).\end{aligned}$$

Since $\mathbf{perp}_S(\mathbf{x}) = (2,-3,3,-4)-(\frac{5}{2},-\frac{7}{2},\frac{5}{2},-\frac{7}{2}) = (-\frac{1}{2},\frac{1}{2},\frac{1}{2},-\frac{1}{2})$, the distance from $\mathbf{x}$ to S is

$$\|\mathbf{perp}_S(\mathbf{x})\| = \frac{1}{2}\sqrt{1+1+1+1}=1.$$

The Gram-Schmidt Procedure for Producing an Orthonormal Basis

For many of the calculations of this chapter, we need an orthonormal basis for a subspace S of $\mathbb{R}^n$. If S is a k-dimensional subspace of some $\mathbb{R}^n$ (possibly with $k = n$), it is certainly possible by the methods of Chapter 4 to produce some basis $\{\mathbf{z}_1, \mathbf{z}_2, \ldots, \mathbf{z}_k\}$ for S. The Gram-Schmidt procedure converts this to an orthonormal basis for S. To simplify the description, we first produce an **orthogonal** basis; note that if $\{\mathbf{w}_1, \mathbf{w}_2, \ldots, \mathbf{w}_k\}$ is an **orthogonal** basis for S, then

$$\mathbf{proj}_S(\mathbf{x}) = \frac{1}{\|\mathbf{w}_1\|^2}(\mathbf{x}\cdot\mathbf{w}_1)\mathbf{w}_1 + \frac{1}{\|\mathbf{w}_2\|^2}(\mathbf{x}\cdot\mathbf{w}_2)\mathbf{w}_2 + \cdots + \frac{1}{\|\mathbf{w}_k\|^2}(\mathbf{x}\cdot\mathbf{w}_k)\mathbf{w}_k.$$

When we have produced the orthogonal basis, it is easy to convert it to an orthonormal basis by normalizing each vector.

The construction is inductive. We know from Chapter 1 how to project onto and perpendicular to a one-dimensional subspace; we use that knowledge to construct an orthogonal basis for a two-dimensional subspace. Next, we use this basis to project onto and perpendicular to the two-dimensional subspace, and thus to construct an orthogonal basis for a three-dimensional subspace. In general, given an orthogonal basis for a $(j-1)$-dimensional subspace, the procedure tells us how to construct an orthogonal basis for the j-dimensional subspace.

Restating the problem: given an arbitrary basis $\{\mathbf{z}_1, \mathbf{z}_2, \ldots, \mathbf{z}_k\}$ for the subspace S, produce an orthonormal basis $\{\mathbf{v}_1, \mathbf{v}_2, \ldots, \mathbf{v}_k\}$.

First Step: Let $\mathbf{w}_1 = \mathbf{z}_1$. Then the one-dimensional subspace spanned by $\mathbf{w}_1$ is obviously the same as the subspace spanned by $\mathbf{z}_1$. In the notation of Section 3–4, $\mathrm{Sp}(\{\mathbf{w}_1\}) = \mathrm{Sp}(\{\mathbf{z}_1\})$. For later reference it is convenient to denote this subspace by $\mathbb{S}_1$.

Second Step: Take the perpendicular part of the second vector. We need a second vector in a basis for $\mathbb{S}$ that is orthogonal to the first vector $\mathbf{w}_1$. Let $\mathbf{w}_2 = \mathbf{perp}_{\mathbb{S}_1}(\mathbf{z}_2)$, so that $\mathbf{w}_2$ is orthogonal to $\mathbf{w}_1$. Then $\mathbf{w}_2$ is not the zero vector because $\{\mathbf{z}_1, \mathbf{z}_2, \dots, \mathbf{z}_k\}$ is linearly independent. Since $\mathbf{w}_1$ is a multiple of $\mathbf{z}_1$ and

$$\mathbf{w}_2 = \mathbf{z}_2 - \frac{1}{\|\mathbf{w}_1\|^2}(\mathbf{z}_2 \cdot \mathbf{w}_1)\mathbf{w}_1,$$

it follows that

$$\mathrm{Sp}(\{\mathbf{w}_1, \mathbf{w}_2\}) = \mathrm{Sp}(\{\mathbf{z}_1, \mathbf{z}_2\})$$

(see Exercise D5). Denote this two-dimensional subspace by $\mathbb{S}_2$.

j^{th} **Step:** Now suppose that $j - 1$ steps have been carried out so that $\{\mathbf{w}_1, \mathbf{w}_2, \dots, \mathbf{w}_{j-1}\}$ is orthogonal by construction, and

$$\mathbb{S}_{j-1} = \mathrm{Sp}(\{\mathbf{z}_1, \mathbf{z}_2, \dots, \mathbf{z}_{j-1}\}) = \mathrm{Sp}(\{\mathbf{w}_1, \mathbf{w}_2, \dots, \mathbf{w}_{j-1}\}).$$

Take the perpendicular part of the next vector $\mathbf{z}_j$. To produce a vector in $\mathbb{S}$ that is orthogonal to $\{\mathbf{w}_1, \mathbf{w}_2, \dots, \mathbf{w}_{j-1}\}$, let

$$\begin{aligned}\mathbf{w}_j &= \mathbf{perp}_{\mathbb{S}_{j-1}}(\mathbf{z}_j)\\ &= \mathbf{z}_j - \frac{1}{\|\mathbf{w}_1\|^2}(\mathbf{z}_j \cdot \mathbf{w}_1)\mathbf{w}_1 - \cdots - \frac{1}{\|\mathbf{w}_{j-1}\|^2}(\mathbf{z}_j \cdot \mathbf{w}_{j-1})\mathbf{w}_{j-1}.\end{aligned}$$

As before, $\mathbf{w}_j$ cannot be the zero vector because this would mean that $\mathbf{z}_j$ was a linear combination of the vectors $\{\mathbf{z}_1, \mathbf{z}_2, \dots, \mathbf{z}_{j-1}\}$, and we know that the set $\{\mathbf{z}_1, \mathbf{z}_2, \dots, \mathbf{z}_k\}$ is linearly independent. As above, let

$$\mathbb{S}_j = \mathrm{Sp}(\{\mathbf{w}_1, \mathbf{w}_2, \dots, \mathbf{w}_j\}) = \mathrm{Sp}(\{\mathbf{z}_1, \mathbf{z}_2, \dots, \mathbf{z}_j\}).$$

Thus $\{\mathbf{w}_1, \mathbf{w}_2, \dots, \mathbf{w}_j\}$ is an orthogonal basis for $\mathbb{S}_j$.

Now continue the procedure until $j = k$, so that

$$\mathbb{S}_k = \mathrm{Sp}(\{\mathbf{w}_1, \mathbf{w}_2, \dots, \mathbf{w}_k\}) = \mathrm{Sp}(\{\mathbf{z}_1, \mathbf{z}_2, \dots, \mathbf{z}_k\}) = \mathbb{S},$$

and an orthogonal basis has been produced for the original subspace $\mathbb{S}$. To get an orthonormal basis, simply let $\mathbf{v}_j = \frac{1}{\|\mathbf{w}_j\|}\mathbf{w}_j$, for each j.

Summary and Remarks

(1) Note that at the jth step, we take the part of the jth basis vector that is orthogonal to the first $(j-1)$ vectors in the basis. This is the one essential step in the Gram-Schmidt procedure.

(2) It is an important feature of the construction that each $\mathbb{S}_{j-1}$ is a subspace of the next $\mathbb{S}_j$.

(3) We can modify the procedure to normalize each $\mathbf{w}_j$ immediately after we obtain it, if it seems more convenient. Alternatively, since it is really only the direction of $\mathbf{w}_j$ that is important in this procedure, we may temporarily re-scale each $\mathbf{w}_j$ in any convenient fashion to simplify the calculation, and then normalize all vectors at the end of the procedure.

EXAMPLE 11

Use the Gram-Schmidt procedure to find an orthonormal basis for the subspace $\mathbb{S} = \mathrm{Sp}(\{(1,1,0,1,1), (-1,2,1,0,1), (0,1,1,1,2)\})$ of $\mathbb{R}^5$.

SOLUTION: Call the vectors of the basis $\mathbf{z}_1$, $\mathbf{z}_2$, $\mathbf{z}_3$ respectively.

First Step: Let $\mathbf{w}_1 = \mathbf{z}_1 = (1,1,0,1,1)$.

Second Step: Take the part of $\mathbf{z}_2$ perpendicular to $\mathbf{w}_2$: let

$$\begin{aligned}\mathbf{w}_2 = \mathbf{perp}_{\mathbf{w}_1}(\mathbf{z}_2) &= \mathbf{z}_2 - \mathbf{proj}_{\mathbf{w}_1}(\mathbf{z}_2) = \mathbf{z}_2 - \frac{1}{\|\mathbf{w}_1\|^2}(\mathbf{z}_2 \cdot \mathbf{w}_1)\mathbf{w}_1 \\ &= (-1,2,1,0,1) - \frac{1}{4}\big((-1,2,1,0,1)\cdot(1,1,0,1,1)\big)(1,1,0,1,1) \\ &= (-1,2,1,0,1) - \frac{2}{4}(1,1,0,1,1) = \left(\frac{-3}{2}, \frac{3}{2}, 1, \frac{-1}{2}, \frac{1}{2}\right).\end{aligned}$$

(It is wise to check your arithmetic by verifying that $\mathbf{w}_1 \cdot \mathbf{w}_2 = 0$.*)*

Since it is really only the direction of $\mathbf{w}_2$ that is essential in finding an orthogonal basis, to simplify calculations, we redefine $\mathbf{w}_2 = (-3,3,2,-1,1)$. This has the direction determined by our earlier version of $\mathbf{w}_2$.

Third Step: Take the perpendicular part of $\mathbf{z}_3$: with $S_2 = \mathrm{Sp}(\{\mathbf{w}_1, \mathbf{w}_2\})$, let

$$\begin{aligned}
\mathbf{w}_3 = \mathbf{perp}_{S_2}(\mathbf{z}_3) &= \mathbf{z}_3 - \frac{1}{\|\mathbf{w}_1\|^2}(\mathbf{z}_3 \cdot \mathbf{w}_1)\mathbf{w}_1 - \frac{1}{\|\mathbf{w}_2\|^2}(\mathbf{z}_3 \cdot \mathbf{w}_2)\mathbf{w}_2 \\
&= (0,1,1,1,2) - \frac{1}{4}\big((0,1,1,1,2) \cdot (1,1,0,1,1)\big)(1,1,0,1,1) \\
&\quad - \frac{1}{24}\big((0,1,1,1,2) \cdot (-3,3,2,-1,1)\big)(-3,3,2,-1,1) \\
&= (0,1,1,1,2) - \frac{4}{4}(1,1,0,1,1) - \frac{6}{24}(-3,3,2,-1,1) \\
&= (-1,0,1,0,1) - \left(\frac{-3}{4}, \frac{3}{4}, \frac{2}{4}, \frac{-1}{4}, \frac{1}{4}\right) = \left(\frac{-1}{4}, \frac{-3}{4}, \frac{2}{4}, \frac{1}{4}, \frac{3}{4}\right).
\end{aligned}$$

(Again, it is wise to check that $\mathbf{w}_3$ is orthogonal to $\mathbf{w}_1$ and $\mathbf{w}_2$.)

$\{\mathbf{w}_1, \mathbf{w}_2, \mathbf{w}_3\}$ is an orthogonal basis for $S = \mathrm{Sp}(\{\mathbf{z}_1, \mathbf{z}_2, \mathbf{z}_3\})$. To obtain an orthonormal basis for this subspace, we simply divide each vector in this orthogonal basis by its length.

Conclusion: $\left\{\frac{1}{2}(1,1,0,1,1), \frac{1}{2\sqrt{6}}(-3,3,2,-1,1), \frac{1}{2\sqrt{6}}(-1,-3,2,1,3)\right\}$ is the required orthonormal basis.

REMARKS

(1) Notice that the order of the vectors in the original arbitrary basis has an important role in the calculation, because each step takes **perp** of the **next** vector. If the original vectors were given in a different order, the procedure might produce a different orthonormal basis for S.

(2) For simple cases, it may be easier to produce an orthonormal basis for a given subspace S by guessing and trial and error.

(3) The procedure actually produces the required orthonormal basis if you start with any finite spanning set for S (even if the spanning set is not linearly independent). If you start with a linearly **dependent** spanning set, at some step when you take the perpendicular part of the next vector, you will get the zero vector; go on to the next vector in the spanning set.

EXERCISES 7–2

A1 Each of the following sets is orthogonal but not orthonormal. Determine the projection of $\mathbf{x} = (2,3,5,6)$ onto the subspace spanned by each set.
(a) $\mathcal{A} = \{(1,-1,-1,1),(1,2,1,2)\}$
(b) $\mathcal{B} = \{(1,0,1,0),(0,1,0,1),(1,0,-1,0)\}$

A2 Use the Gram-Schmidt procedure to produce an orthonormal basis from the given basis in each of the following cases.
(a) $\{(1,0,0),(1,1,0),(1,1,1)\}$
(b) $\{(1,1,1),(1,1,0),(1,0,0)\}$

A3 Use the Gram-Schmidt procedure to produce an orthonormal basis from the given basis in each of the following cases.
(a) $\{(1,1,0,1),(0,1,1,1),(1,-1,-1,-1)\}$
(b) $\{(1,0,1,0,1),(1,0,-1,1,0),(1,1,1,1,1)\}$

A4 Let $\mathbb{S}$ be the solution space of the homogeneous system of equations

$$x_1 + x_2 - x_3 - x_4 = 0 \quad \text{and} \quad x_2 - x_3 + x_4 = 0.$$

Determine the point in $\mathbb{S}$ that is closest to $(1,4,0,1)$, and determine the distance from $(1,4,0,1)$ to $\mathbb{S}$.

B1 Each of the following sets is orthogonal but not orthonormal. Determine the projection of $\mathbf{x} = (4,3,-2,5)$ onto the subspace spanned by each set.
(a) $\mathcal{A} = \{(1,0,-1,1),(1,1,-1,-2)\}$
(b) $\mathcal{B} = \{(2,1,0,1),(-1,1,1,1)\}$
(c) $\mathcal{C} = \{(1,0,1,1),(0,1,-1,1),(1,1,0,-1)\}$
(d) $\mathcal{D} = \{(1,1,1,0),(1,0,-1,1),(0,1,-1,-1)\}$

B2 Use the Gram-Schmidt procedure to produce an orthonormal basis from the given basis in each of the following cases.
(a) $\{(1,0,-1,1),(1,1,1,1),(2,0,1,1)\}$
(b) $\{(1,0,0,1),(1,1,0,2),(0,-1,1,1)\}$

B3 Use the Gram-Schmidt procedure to produce an orthonormal basis from the given basis in each of the following cases.
(a) $\{(1,1,0,1),(0,-1,1,1),(3,0,1,1)\}$
(b) $\{(1,1,0,1,0),(-1,0,1,1,1),(1,1,0,2,1)\}$

B4 Let $\mathbb{S}$ be the solution space of the homogeneous system of equations

$$x_1 + x_3 + x_4 = 0 \quad \text{and} \quad x_2 + 2x_3 - x_4 = 0.$$

Determine the point in $\mathbb{S}$ that is closest to $(2,3,0,3)$, and determine the distance from $(2,3,0,3)$ to $\mathbb{S}$.

B5 Let S be the solution space of the homogeneous system of equations

$$2x_1 + x_2 + x_3 = 0 \quad \text{and} \quad x_1 - 3x_3 + x_4 = 0.$$

Determine the point in S that is closest to $(1, 0, 2, -2)$, and determine the distance from $(1, 0, 2, -2)$ to S.

CONCEPTUAL EXERCISES

***D1** Prove that if S is a k-dimensional subspace of $\mathbb{R}^n$, then $S \cap S^\perp = \{\mathbf{0}\}$.

D2 Prove that if S is a k-dimensional subspace of $\mathbb{R}^n$, then $S^\perp$ is an $(n-k)$-dimensional subspace. [Hint: Introduce a basis for S, and consider $S^\perp$ as a solution space.]

D3 Prove that if $\{\mathbf{v}_1, \mathbf{v}_2, \ldots, \mathbf{v}_k\}$ is an orthonormal basis for S, and $\{\mathbf{v}_{k+1}, \mathbf{v}_{k+2}, \ldots, \mathbf{v}_n\}$ is an orthonormal basis for $S^\perp$, then $\{\mathbf{v}_1, \ldots, \mathbf{v}_k, \mathbf{v}_{k+1}, \ldots, \mathbf{v}_n\}$ is an orthonormal basis for $\mathbb{R}^n$.

D4 Prove that $\mathbf{perp}_S = \mathbf{proj}_{S^\perp}$ for any subspace S of $\mathbb{R}^n$.

D5 Suppose that $\mathbf{v}_1$ and $\mathbf{v}_2$ are vectors in $\mathbb{R}^n$. Show that $\text{Sp}(\{\mathbf{v}_1, \mathbf{v}_2\})$ is equal to $\text{Sp}(\{k\mathbf{v}_1, a\mathbf{v}_1 + b\mathbf{v}_2\})$ for any scalars k, a, b such that k and b are not zero. (This result and its generalization underly the Gram-Schmidt procedure. A more general version for rows of a matrix was proved in Section 3–4 in discussing the rowspace.)

***D6** If $\{\mathbf{v}_1, \mathbf{v}_2, \ldots, \mathbf{v}_k\}$ is an orthonormal basis for a subspace S, verify that the standard matrix of $\mathbf{proj}_S$ can be written (generalizing Exercise 3–2–D1)

$$[\mathbf{proj}_S] = [\mathbf{v}_1][\mathbf{v}_1]^T + [\mathbf{v}_2][\mathbf{v}_2]^T + \cdots + [\mathbf{v}_k][\mathbf{v}_k]^T.$$

SECTION

7–3 Projections and the Method of Least Squares

Suppose that an experimenter measures a variable y at times $t_1, t_2, \ldots, t_n$, and obtains the values $y_1, y_2, \ldots, y_n$. For example, y might be the position of a particle at time t, or the temperature of some body of fluid. Suppose that the experimenter believes that the data fit (more or less) a curve of the form $y = a + bt + ct^2$. How should she choose a, b, and c to get the best fitting curve of this form?

Let us consider some particular $a, b,$ and c. If the data fit the curve $y = a + bt + ct^2$ perfectly, then for each j, $y_j = a + bt_j + ct_j^2$. However, for arbitrary a, b, c, we expect that at each t_j there will be an error, denoted by e_j and measured by the vertical distance

$$e_j = y_j - (a + bt_j + ct_j^2),$$

as shown in Figure 7–3–1.

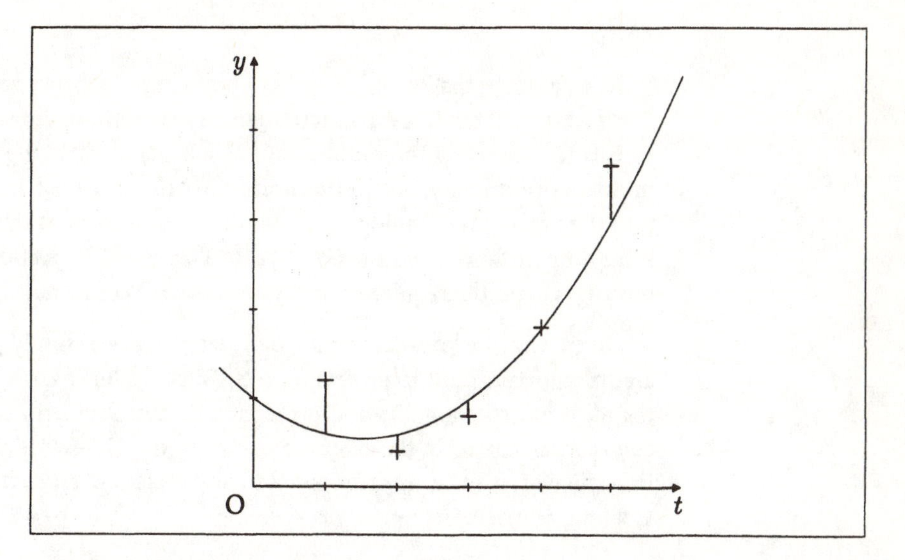

FIGURE 7-3-1. *Some data points and a curve $y = a + bt + ct^2$; vertical line segments measure the error in the fit at each t_j.*

One approach to finding the best fitting curve might be to try to minimize the *total* error $\sum_{j=1}^{n} e_j$. This would be unsatisfactory, however, because we might get a small total error by having large positive errors cancelled by large negative errors.

The standard test for finding the a, b, c that give the best fitting curve is to minimize the sum of the squares of the errors: $\sum_{j=1}^{n}(y_j - (a + bt_j + t_j^2))^2$. This is called the **method of least squares.**

To find the "parameters" a, b, c that minimize this expression for given vectors $(t_1, t_2, \ldots, t_n)$ and $(y_1, y_2, \ldots, y_n)$, one could proceed by calculus, but we proceed by using projection. This requires us to adopt the following point of view.

Let $\mathbf{y}$ denote the n-tuple $(y_1, y_2, \ldots, y_n)$, let $\mathbf{1} = (1, 1, \ldots, 1)$, let $\mathbf{t} = (t_1, t_2, \ldots, t_n)$, and let $\mathbf{t}^2 = (t_1^2, t_2^2, \ldots, t_n^2)$. Now consider the **distance** from $\mathbf{y}$ to $(a\mathbf{1} + b\mathbf{t} + c\mathbf{t}^2)$: **the square of this distance is exactly the sum of the squares of the errors**

$$\|\mathbf{y} - (a\mathbf{1} + b\mathbf{t} + c\mathbf{t}^2)\|^2 = \sum_{j=1}^{n} (y_j - (a + bt_j + t_j^2))^2.$$

Next, observe that $(a\mathbf{1} + b\mathbf{t} + c\mathbf{t}^2)$ is a point in the subspace of $\mathbb{R}^n$ spanned by $\mathcal{B} = \{\mathbf{1}, \mathbf{t}, \mathbf{t}^2\}$. If the t_j are distinct, these three vectors are linearly independent, so $\mathcal{B}$ is the basis of a three-dimensional subspace, which we call $\mathbb{S}$. Thus the problem of finding a, b, c to minimize the sum of the squares is transformed into the problem of finding a, b, c so that $a\mathbf{1} + b\mathbf{t} + c\mathbf{t}^2$ is the point in $\mathbb{S}$ with minimum distance from the point $\mathbf{y}$. By Theorem 1 of Section 7–2, this point is $\mathbf{proj}_{\mathbb{S}}(\mathbf{y})$ and the required a, b, and c are the $\mathcal{B}$-coordinates in $\mathbb{S}$ of $\mathbf{proj}_{\mathbb{S}}(\mathbf{y})$.

Given what we know so far, it might seem that we should proceed by using an orthonormal basis to project onto $\mathbb{S}$. There is, however, a better way to use the ideas of orthogonality and projection on this problem. If a, b, and c have been chosen correctly, the **error vector** $\mathbf{e} = \mathbf{y} - a\mathbf{1} - b\mathbf{t} - c\mathbf{t}^2$ is equal to $\mathbf{perp}_{\mathbb{S}}(\mathbf{y})$. In particular, it must be orthogonal to any vectors in $\mathbb{S}$, so it is orthogonal to $\mathbf{1}$, $\mathbf{t}$, and $\mathbf{t}^2$. Therefore

$$\begin{aligned} \mathbf{1} \cdot \mathbf{e} &= \mathbf{1} \cdot (\mathbf{y} - a\mathbf{1} - b\mathbf{t} - c\mathbf{t}^2) = 0, \\ \mathbf{t} \cdot \mathbf{e} &= \mathbf{t} \cdot (\mathbf{y} - a\mathbf{1} - b\mathbf{t} - c\mathbf{t}^2) = 0, \\ \mathbf{t}^2 \cdot \mathbf{e} &= \mathbf{t}^2 \cdot (\mathbf{y} - a\mathbf{1} - b\mathbf{t} - c\mathbf{t}^2) = 0. \end{aligned}$$

The required a, b, c are determined as the solutions of this system of three homogeneous linear equations in three variables.

It is helpful to rewrite these equations by introducing the matrix

$$X = \begin{bmatrix} [\mathbf{1}] & [\mathbf{t}] & [\mathbf{t}^2] \end{bmatrix},$$

and the vector $\mathbf{a} = (a, b, c)$ of parameters. Then the error vector can be written as $[\mathbf{e}] = [\mathbf{y}] - X[\mathbf{a}]$; since the three equations are obtained by taking dot products with the columns of X, the system of equations can be written in the form

$$X^T([\mathbf{y}] - X[\mathbf{a}]) = [\mathbf{0}].$$

The equations in this form are called the normal equations for the least squares fit. Since the columns of X are linearly independent, the matrix X^TX has rank three; since it is a 3 by 3 matrix, this means that it is invertible, and the system of equations has a solution. The matrix X depends on the model curve and on the way the data is collected. It is called the design matrix.

These ideas can easily be generalized to the least square fitting of data to other model curves.

EXAMPLE 12

Suppose that the experimenter's data are as follows:

t	1.0	2.1	3.1	4.0	4.9	6.0
y	6.1	12.6	21.1	30.2	40.9	55.5

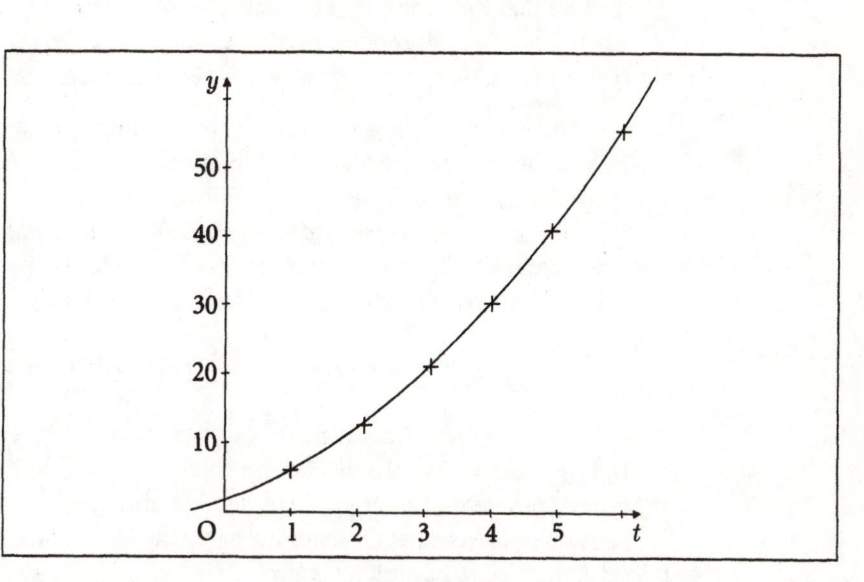

FIGURE 7-3-2. *The data points and the best fitting curve from Example 12.*

As in the discussion, she wishes to find the curve $y = a + bt + ct^2$ that fits the data best in the sense of minimizing the sum of the squares of the errors. Then the matrix X satisfies

$$X^T = \begin{bmatrix} 1 & 1 & 1 & 1 & 1 & 1 \\ 1.0 & 2.1 & 3.1 & 4.0 & 4.9 & 6.0 \\ 1.0 & 4.41 & 9.61 & 16.0 & 24.01 & 36.0 \end{bmatrix}$$

and $\mathbf{y} = (6.1, 12.6, 21.1, 30.2, 40.9, 55.5)$. Using a microcomputer program that does rational arithmetic, and converting the final answers to decimal form, we find that the solution for the system $X^T X[\mathbf{a}] = X^T[\mathbf{y}]$ is $(a, b, c) = (1.63175, 3.38382, 0.93608)$. The data does not justify retaining so many decimal places, so we take the best fitting quadratic curve to be $y = 1.63 + 3.38t + 0.94t^2$.

Overdetermined Systems

The problem of finding the best fitting curve can be viewed as a special case of the problem of "solving" an **overdetermined system.** Suppose that $A\mathbf{x} = \mathbf{b}$ is a system of p equations in q variables, where p is greater than q. With more equations than variables, we expect the system to be inconsistent unless $\mathbf{b}$ has some special properties. If the system is inconsistent, we say that the system is **overdetermined** — there are too many equations to be satisfied.

Note that the problem in our example of finding the best fitting quadratic curve was of this form: we needed to solve $X\mathbf{a} = \mathbf{y}$, for the three variables (parameters a, b, c), where there were n equations (one for each data point).

The argument in the general case is analogous to our argument for the special case. If there is no $\mathbf{x}$ such that $A\mathbf{x} = \mathbf{b}$, the next best "solution" is to find $\mathbf{x}$ that minimizes the "error" $\|A\mathbf{x} - \mathbf{b}\|$. However, $A\mathbf{x} = x_1[a_{\downarrow 1}] + \ldots + x_q[a_{\downarrow q}]$, and this is a vector in the columnspace of A. Our problem thus is to find the $\mathbf{x}$ such that $A\mathbf{x}$ is the point in the columspace closest to $\mathbf{b}$. By an argument analogous to that in the special case, we can show that $\mathbf{x}$ must satisfy the equation

$$A^T(A\mathbf{x} - \mathbf{b}) = \mathbf{0}, \text{ or } A^TA\mathbf{x} = A^T\mathbf{b}.$$

In the problem of finding the best fitting curve , the matrix X has special features because of the way that the columns $[\mathbf{1}], [\mathbf{t}], [\mathbf{t}^2]$ arise. For the more general problem of an overdetermined system, the columns of A may have no such special structure; the column space may have dimension less than q, and A^TA may fail to have an inverse.

EXERCISES 7–3

A1 Find a and b to obtain the best fitting equation of the form $y = a + bt$ for the given data. Note that X^TX is a 2 by 2 matrix in these cases. Make a graph showing the data and the best fitting line.

(a)

t	1	2	3	4	5
y	9	6	5	3	1

(b)

t	−2	−1	0	1	2
y	2	2	4	4	5

A2 Verify that the following systems are inconsistent, then determine for each system the **x** that minimizes $\|A\mathbf{x} - \mathbf{b}\|$ (where A is the coefficient matrix of the system, **b** is the right-hand side).

(a) $$\begin{aligned} x_1 + 2x_2 &= 5 \\ 2x_1 - 3x_2 &= 6 \\ x_1 - 12x_2 &= -4 \end{aligned}$$

(b) $$\begin{aligned} 2x_1 + 3x_2 &= -4 \\ 3x_1 - 2x_2 &= 4 \\ x_1 - 6x_2 &= 7 \end{aligned}$$

B1 Same instructions as in Exercise A1.

(a)

t	-2	-1	0	1	2
y	9	8	5	3	1

(b)

t	1	2	3	4	5
y	4	3	4	5	5

B2 The same instructions as for Exercise A2.

(a) $$\begin{aligned} x_1 - x_2 &= 4 \\ 3x_1 + 2x_2 &= 5 \\ x_1 - 6x_2 &= 10 \end{aligned}$$

(b) $$\begin{aligned} x_1 + x_2 &= 7 \\ x_1 - x_2 &= 4 \\ x_1 + 3x_2 &= 14 \end{aligned}$$

B3 Find a, b, and c to obtain the best fitting equation of the form $y = a + bt + ct^2$ for the following data, and make a graph showing the data and the curve.

t	-2	-1	0	1	2
y	3	2	0	2	8

COMPUTER RELATED EXERCISES

C1 Find a, b, and c to obtain the best fitting equation of the form $y = a + bt + ct^2$ for the following data, and make a graph showing the data and the curve.

t	0.0	1.1	1.9	3.0	4.1	5.2
y	4.0	3.6	4.1	5.6	7.9	11.8

CONCEPTUAL EXERCISES

D1 Show that if $X = \begin{bmatrix} [\mathbf{1}] & [\mathbf{t}] & [\mathbf{t}^2] \end{bmatrix}$, where $\mathbf{t} = (t_1, t_2, \ldots, n)$, then

$$X^T X = \begin{bmatrix} n & \sum_{j=1}^{n} t_j & \sum_{j=1}^{n} t_j^2 \\ \sum_{j=1}^{n} t_j & \sum_{j=1}^{n} t_j^2 & \sum_{j=1}^{n} t_j^3 \\ \sum_{j=1}^{n} t_j^2 & \sum_{j=1}^{n} t_j^3 & \sum_{j=1}^{n} t_j^4 \end{bmatrix}$$

SECTION

7–4 Inner Product Spaces and Fourier Series

Inner Products

The dot product plays an essential role in the discussion of lengths, distances, and projections in $\mathbb{R}^n$. In Chapter 4, we saw that the ideas of vector space and linear mapping apply to more general sets, including some function spaces. If ideas such as projection are going to be used in these more general spaces, it will be necessary to have a generalization of the dot product to general vector spaces. Consideration of the most essential properties of the dot product in Section 1–2 leads to the following definition.

DEFINITION

Let $\mathbb{V}$ be a vector space over the real numbers. An **inner product on** $\mathbb{V}$ is a function $\langle\ ,\ \rangle : \mathbb{V} \times \mathbb{V} \to \mathbb{R}$ such that

(1) $\langle \mathbf{v}, \mathbf{v} \rangle \geq 0$ for all $\mathbf{v}$ in $\mathbb{V}$, and $\langle \mathbf{v}, \mathbf{v} \rangle = 0$ only if $\mathbf{v} = \mathbf{0}$ ("$\langle\ ,\ \rangle$ is *positive definite*");

(2) $\langle \mathbf{v}, \mathbf{w} \rangle = \langle \mathbf{w}, \mathbf{v} \rangle$ for any $\mathbf{v}$ and $\mathbf{w}$ in $\mathbb{V}$ ("$\langle\ ,\ \rangle$ is *symmetric*");

(3) $\langle \mathbf{v}, a\mathbf{w} + b\mathbf{z} \rangle = a\langle \mathbf{v}, \mathbf{w} \rangle + b\langle \mathbf{v}, \mathbf{z} \rangle$, for any real numbers a and b, and for any $\mathbf{v}$, $\mathbf{w}$, and $\mathbf{z}$ in $\mathbb{V}$; (because of (1), a similar rule holds if there is a linear combination in the first slot in the inner product, so "$\langle\ ,\ \rangle$ is *bilinear*", which means linear in both slots).

Since these properties of inner products mimic the properties of the dot product, it makes sense to define the norm or length of a vector, and the distance between vectors, in terms of the inner product. (Recall that these can be defined in $\mathbb{R}^n$ in terms of the dot product.) Define the **norm** (or **length**) of $\mathbf{v}$ to be $\|\mathbf{v}\| = \left(\langle \mathbf{v}, \mathbf{v} \rangle\right)^{1/2}$, and the **distance** between two elements $\mathbf{v}$ and $\mathbf{w}$ of $\mathbb{V}$ to be $\|\mathbf{v} - \mathbf{w}\|$. As in Chapter 1, all the usual properties of lengths and distances (such as the Cauchy-Schwarz inequality and the triangle inequality) actually follow from the definition in terms of the dot product. In vector spaces over the complex numbers, some modifications are required for the definition of an inner product and in some related proofs; this is discussed in Section 7–5.

A vector space $\mathbb{V}$ with an inner product defined is called **an inner product space**. The most obvious example is $\mathbb{R}^n$ with the usual dot product. Other scalar products can be defined on $\mathbb{R}^n$, but they turn out not to be different in interesting ways, as discussed in Exercises D1 and 8–2 D2. However, there are simple inner products on spaces of polynomials, illustrated in Example 13, and on spaces of functions, considered after this example.

EXAMPLE 13

Let P_2 be the space of polynomials of degree less than or equal to 2 *in a real variable* x. Verify that $\langle p, q\rangle = p(0)q(0) + p(1)q(1) + p(2)q(2)$ defines an inner product on P_2, and determine $\langle 1 + x + x^2, 2 - 3x^2\rangle$ and $\|1 - 2x - x^2\|$ with respect to this inner product.

SOLUTION: We first verify that $\langle\ ,\ \rangle$ satisfies the three properties of an inner product.

(1) $\langle p, p\rangle = (p(0))^2 + (p(1))^2 + (p(2))^2 \geq 0$ for all p in P_2. Moreover $\langle p, p\rangle = 0$ if and only if $p(0) = p(1) = p(2) = 0$, and the only $p \in P_2$ that is zero for three values of x is the zero polynomial, $p(x) = 0$ for all x. Thus $\langle\ ,\ \rangle$ is positive definite.

(2) $\langle q, p\rangle = q(0)p(0) + q(1)p(1) + q(2)p(2) = \langle p, q\rangle$, so $\langle\ ,\ \rangle$ is symmetric.

(3) For real numbers a and b,

$$\begin{aligned}\langle p, aq + br\rangle &= p(0)(aq(0) + br(0)) + p(1)(aq(1) + br(1)) + p(2)(aq(2) + br(2)) \\ &= a\langle p, q\rangle + b\langle p, r\rangle,\end{aligned}$$

so $\langle\ ,\ \rangle$ is bilinear.

Since these three properties are satisfied, $\langle\ ,\ \rangle$ is an inner product on P_2.

With respect to this inner product,

$$\begin{aligned}\langle 1 + x + x^2, 2 - 3x^2\rangle &= (1 + 0 + 0)(2 - 0) + (1 + 1 + 1)(2 - 3) \\ &\quad + (1 + 2 + 4)(2 - 12) = 2 - 3 - 70 = -71; \\ \|1 - 2x - x^2\| &= \langle 1 - 2x - x^2, 1 - 2x - x^2\rangle^{1/2} \\ &= (1^2 + (1 - 2 - 1)^2 + (1 - 4 - 4)^2)^{1/2} = \sqrt{54}.\end{aligned}$$

The Inner Product $\int_a^b f(x)g(x)\,dx$

Let $C[a, b]$ be the space of functions from $\mathbb{R}$ to $\mathbb{R}$ that are continuous on the interval $[a, b]$. Thus, $f : \mathbb{R} \to \mathbb{R}$ is in $C[a, b]$ if f is continuous for all x in $[a, b]$. By a standard theorem of calculus, such a function is integrable on $[a, b]$ (that is, $\int_a^b f(x)\,dx$ exists for such a function). The product of continuous functions is continuous, so the product is also integrable, and it makes sense to define an inner product as follows.

The inner product $\langle\ ,\ \rangle$ is defined on $C[a, b]$ by

$$\langle f, g\rangle = \int_a^b f(x)g(x)\,dx.$$

The three properties of an inner product are satisfied because

(1) $\int_a^b f(x)f(x)\,dx \geq 0$ for all continuous f, and $\int_a^b f(x)f(x)\,dx = 0$ only if $f(x) = 0$ for all x in $[a, b]$;

(2) $\int_a^b f(x)g(x)\,dx = \int_a^b g(x)f(x)dx$;

(3) $\int_a^b f(x)(cg(x) + kh(x))\,dx = c\int_a^b f(x)g(x)\,dx + k\int_a^b f(x)h(x)\,dx$, for any constants c and k.

Since an integral is the limit of sums, this inner product defined as the **integral of the product of the values of f and g at each x** is a fairly natural generalization of the dot product in $\mathbb{R}^n$ defined as a **sum of the product of the jth components of x and y for each j.**

One interesting consequence is that the norm of a function f with respect to this inner product is $\|f\| = \left(\int_a^b f^2(x)\,dx\right)^{1/2}$; intuitively, this is quite satisfactory as a measure of how far the function f is from the zero function. Note also that the distance between two functions f and g is $\|f - g\| = \left(\int_a^b (f-g)^2(x)\,dx\right)^{1/2}$.

One of the most interesting and important applications of this inner product is to Fourier series.

Fourier Series

Let $CP_{2\pi}$ denote the space of continuous real-valued functions of a real variable that are periodic with period 2π: such functions satisfy $f(x + 2\pi) = f(x)$ for all x. Examples of such functions are $f(x) = c$ for any constant c, $\cos x$, $\sin x$, $\cos 2x$, $\sin 2x$, $\cos nx$, $\sin nx$. (Note that the function $\cos 2x$ is periodic with period 2π because $\cos 2(x+2\pi) = \cos 2x$; however, its "fundamental (smallest) period" is π.) In some electrical engineering applications, it is of interest to consider a signal described by functions such as that shown in Figure 7–4–1. f is defined by

$$f(x) = \begin{cases} -\pi - x & \text{if } -\pi \leq x \leq -\pi/2 \\ x & \text{if } -\pi/2 \leq x \leq \pi/2 \\ \pi - x & \text{if } \pi/2 \leq x \leq \pi \end{cases}$$

and $f(x + 2\pi) = f(x)$ for all x.

f is obviously continuous and periodic.

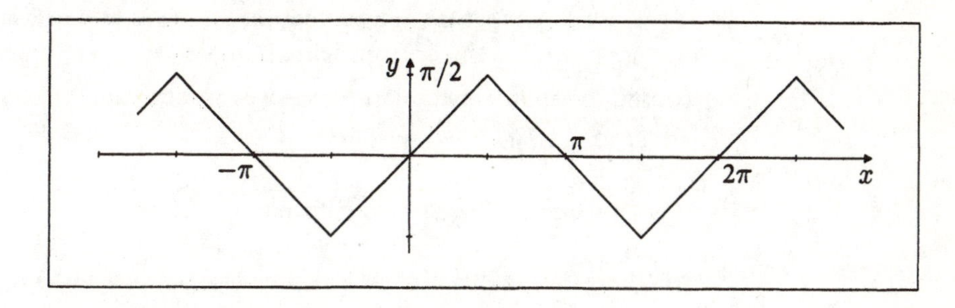

FIGURE 7-4-1. *A continuous periodic function.*

In the early nineteenth century, while studying the problem of the conduction of heat, Fourier had the brilliant idea of trying to represent an arbitrary function in $CP_{2\pi}$ as a linear combination of the set of functions $\{1, \cos x, \sin x, \cos 2x, \sin 2x, \ldots, \cos nx, \sin nx, \ldots\}$. The framework we now use to understand Fourier series was not yet developed in 1800, and some major mathematicians of that era were doubtful about the usefulness and correctness of Fourier's approach. Nevertheless Fourier used these ideas to solve important problems. His ideas developed into Fourier analysis, which is now one of the essential tools in quantum physics and communications engineering, as well as in many other areas.

With modern hindsight, we can formulate the questions and ideas as follows. (The proofs of the statements are discussed below.)

(i) For any n, the set of functions

$$\{1, \cos x, \sin x, \cos 2x, \sin 2x, \ldots, \cos nx, \sin nx\}$$

is an orthogonal set with respect to the inner product

$$\langle f, g \rangle = \int_{-\pi}^{\pi} f(x)g(x)\,dx;$$

the set is therefore an orthogonal basis for the subspace that it spans. This subspace will be denoted $CP_{2\pi,n}$.

(ii) Given an arbitrary function f in $CP_{2\pi}$, how well can it be approximated by a function in $CP_{2\pi,n}$? We expect from our experience with distance and subspaces that the closest approximation to f in $CP_{2\pi,n}$ is $\mathbf{proj}_{CP_{2\pi,n}}(f)$. The coefficients for Fourier's representation of f by a linear combination of $\{1, \cos x, \sin x, \cos 2x, \sin 2x, \ldots, \cos nx, \sin nx, \ldots\}$ are found by considering this projection.

(iii) We hope that the approximation improves as n gets larger. Since the distance from f to the nth approximation $\mathbf{proj}_{CP_{2\pi,n}}(f)$ is $\|\mathbf{perp}_{CP_{2\pi,n}}(f)\|$, to test if the approximation improves we must examine whether $\|\mathbf{perp}_{CP_{2\pi,n}}(f)\| \to 0$ as $n \to \infty$.

Let us consider these statements in more detail.

(i) **The orthogonality of constants, sines, and cosines with respect to the inner product $\langle f, g\rangle = \int_{-\pi}^{\pi} f(x)g(x)\,dx$.**

These results follow by standard trigonometric integrals and trigonometric identities.

Consider $\int_{-\pi}^{\pi} \sin nx\,dx = -\frac{1}{n}\cos nx\,|_{-\pi}^{\pi} = 0$.

Similarly, $\int_{-\pi}^{\pi} \cos nx\,dx = \frac{1}{n}\sin nx\,|_{-\pi}^{\pi} = 0$.

Also, $\int_{-\pi}^{\pi} \cos mx \sin nx\,dx = \int_{-\pi}^{\pi} \frac{1}{2}\big(\sin(m+n)x - \sin(m-n)x\big)\,dx = 0$;

and if $m \neq n$,

$\int_{-\pi}^{\pi} \cos mx \cos nx\,dx = \int_{-\pi}^{\pi} \frac{1}{2}\big(\cos(m+n)x + \cos(m-n)x\big)\,dx = 0$;

and if $m \neq n$,

$\int_{-\pi}^{\pi} \sin mx \sin nx\,dx = \int_{-\pi}^{\pi} \frac{1}{2}\big(\cos(m-n)x - \cos(m+n)x\big)\,dx = 0$.

Hence the set $\{1, \cos x, \sin x, \cos 2x, \sin 2x, \ldots, \cos nx, \sin nx\}$ is orthogonal. To use this as a basis for projection arguments, it is necessary to calculate $\|1\|^2$, $\|\cos mx\|^2$, and $\|\sin mx\|^2$.

$$\|1\|^2 = \int_{-\pi}^{\pi} 1\,dx = 2\pi;$$

$$\begin{aligned}\|\cos mx\|^2 &= \int_{-\pi}^{\pi} \cos mx \cos mx\,dx = \int_{-\pi}^{\pi} \cos^2 mx\,dx \\ &= \int_{-\pi}^{\pi} \frac{1}{2}(1 + \cos 2mx)\,dx = \pi;\end{aligned}$$

$$\|\sin mx\|^2 = \int_{-\pi}^{\pi} \sin mx \sin mx\, dx = \int_{-\pi}^{\pi} \sin^2 mx\, dx$$
$$= \int_{-\pi}^{\pi} \frac{1}{2}(1 - \cos 2mx)\, dx = \pi.$$

(ii) The Fourier coefficients of f as coordinates of a projection, with respect to the orthogonal basis of $CP_{2\pi,n}$.

To find the closest approximation $\mathbf{proj}_{CP_{2\pi,n}}(f)$ in $CP_{2\pi,n}$ to an arbitrary function f in $CP_{2\pi}$, the procedure is parallel to the procedure in Sections 7–1 and 7–2. $\mathbf{proj}_{CP_{2\pi,n}}(f)$ can be expressed as a linear combination of the members of the orthogonal basis of the subspace $CP_{2\pi,n}$. There is a standard way to label the coefficients of this linear combination.

$$\mathbf{proj}_{CP_{2\pi,n}} f = \frac{a_0}{2} 1 + a_1 \cos x + a_2 \cos 2x + \cdots + a_n \cos nx$$
$$+ b_1 \sin x + b_2 \sin 2x + \ldots + b_n \sin nx.$$

The factor $\frac{1}{2}$ in the coefficient of 1 is there because $\|1\|^2$ is equal to 2π, while the other basis vectors have norm squared equal to π.

We require the projection formula from Section 7–2 for the case of an orthogonal (not orthonormal) basis; recall that we need to divide each basis vector by its length, so for an orthogonal basis the general formula becomes

$$\mathbf{proj}_S(\mathbf{x}) = \left(\mathbf{x} \cdot \frac{\mathbf{v}_1}{\|\mathbf{v}_1\|}\right) \frac{\mathbf{v}_1}{\|\mathbf{v}_1\|} + \left(\mathbf{x} \cdot \frac{\mathbf{v}_2}{\|\mathbf{v}_2\|}\right) \frac{\mathbf{v}_2}{\|\mathbf{v}_2\|} + \cdots + \left(\mathbf{x} \cdot \frac{\mathbf{v}_k}{\|\mathbf{v}_k\|}\right) \frac{\mathbf{v}_k}{\|\mathbf{v}_k\|}$$
$$= \left(\frac{\mathbf{x} \cdot \mathbf{v}_1}{\|\mathbf{v}_1\|^2}\right) \mathbf{v}_1 + \left(\frac{\mathbf{x} \cdot \mathbf{v}_2}{\|\mathbf{v}_2\|^2}\right) \mathbf{v}_2 + \cdots + \left(\frac{\mathbf{x} \cdot \mathbf{v}_k}{\|\mathbf{v}_k\|^2}\right) \mathbf{v}_k.$$

For the present case, this means that

$$\frac{a_0}{2} = \langle f, \frac{1}{\|1\|^2} \rangle = \frac{1}{2\pi} \int_{-\pi}^{\pi} f(x)\, dx,$$

so that

$$a_0 = \frac{1}{\pi} \int_{-\pi}^{\pi} f(x)\, dx;$$

$$a_m = \langle f, \frac{\cos mx}{\|\cos mx\|^2} \rangle = \frac{1}{\pi} \int_{-\pi}^{\pi} f(x) \cos mx\, dx;$$

$$b_m = \langle f, \frac{\sin mx}{\|\sin mx\|^2} \rangle = \frac{1}{\pi} \int_{-\pi}^{\pi} f(x) \sin mx\, dx.$$

These are the standard **Fourier coefficients**; we have obtained them by arguments using orthogonality and projection.

(iii) **Is $\text{proj}_{CP_{2\pi,n}}(f)$ equal to f in the limit as $n \to \infty$?**

This is equivalent to the question: does $\|\text{perp}_{CP_{2\pi,n}}(f)\| \to 0$ as $n \to \infty$? As $n \to \infty$, the sum becomes an infinite series (the **Fourier series** for f), and the question being asked is a question about the convergence of series, and in fact, about series of functions. Such questions are questions in calculus (or in "analysis") and are beyond the scope of this book. (The short answer is "yes, the series converges provided that f is continuous and a bit more". The problem becomes more complicated if f is allowed to be piecewise continuous.) Questions about convergence are important in physical and engineering applications.

If you are interested in a further discussion of this topic, many books on differential equations include a discussion of Fourier series and applications.

EXERCISES 7–4

A1 For the inner product $\langle\ ,\ \rangle$ on P_2 defined in Example 13, calculate the following.

(a) $\langle x - 2x^2, 1 + 3x\rangle$
(b) $\langle 2 - x + 3x^2, 4 - 3x^2\rangle$
(c) $\|3 - 2x + x^2\|$
(d) $\|9 + 9x + 9x^2\|$

A2 In each of the following cases, determine whether $\langle\ ,\ \rangle$ defines an inner product on P_2.

(a) $\langle p, q\rangle = p(0)q(0) + p(1)q(1)$
(b) $\langle p, q\rangle = |p(0)q(0)| + |p(1)q(1)| + |p(2)q(2)|$
(c) $\langle p, q\rangle = p(-1)q(1) + 2p(0)q(0) + p(1)q(-1)$

A3 If you know how to do integration by parts and exploit symmetries of the integrand, calculate the Fourier coefficients of the function of Figure 7–4–1, and verify that $a_n = 0$ for all n, while $b_n = (4/\pi n^2)\sin(n\pi/2)$. Since $\sin(n\pi/2) = 0$ if n is even, and ± 1 if n is odd, this leads to the Fourier series representation

$$f(x) = (4/\pi)\left(\sin x - \frac{1}{9}\sin 3x + \frac{1}{25}\sin 5x - \cdots\right).$$

B1 For the inner product $\langle\ ,\ \rangle$ on P_2 defined in Example 13, calculate the following.

(a) $\langle 1 - 3x^2, 1 + x + 2x^2\rangle$
(b) $\langle 3 - x, -2 - 1 - x^2\rangle$
(c) $\|1 - 5x + 2x^2\|$
(d) $\|73x + 73x^2\|$

B2 In each of the following cases, determine whether $\langle\ ,\ \rangle$ defines an inner product on P_3.

(a) $\langle p, q\rangle = p(-1)q(-1) + p(0)q(0) + p(1)q(1)$

(b) $\langle p, q\rangle = p(0)q(0) + p(1)q(1) + p(3)q(3) + p(4)q(4)$

(c) $\langle p, q\rangle = p(-1)q(0) + p(1)q(1) + p(2)q(2) + p(3)q(3)$

COMPUTER RELATED EXERCISES

C1 Use a computer to obtain the graph of the $f(x)$ of Figure 7–4–1 and Exercise A3, and compare this graph with the graphs of

$$f_1(x) = (4/\pi)\sin x, \qquad f_3 = (4/\pi)\left(\sin x - \frac{1}{9}\sin 3x\right),$$

$$f_5 = (4/\pi)\left(\sin x - \frac{1}{9}\sin 3x + \frac{1}{25}\sin 5x\right), \quad f_7, \text{ and so on.}$$

CONCEPTUAL EXERCISES

D1 (a) Let $\{\mathbf{e}_1, \mathbf{e}_2\}$ be the standard basis for $\mathbb{R}^2$, and suppose that $\langle\ ,\ \rangle$ is an inner product on $\mathbb{R}^2$. Show that if $\mathbf{x}$ and $\mathbf{y}$ are vectors in $\mathbb{R}^2$,

$$\langle \mathbf{x}, \mathbf{y}\rangle = x_1y_1\langle \mathbf{e}_1, \mathbf{e}_1\rangle + x_1y_2\langle \mathbf{e}_1, \mathbf{e}_2\rangle + x_2y_1\langle \mathbf{e}_2, \mathbf{e}_1\rangle + x_2y_2\langle \mathbf{e}_2, \mathbf{e}_2\rangle.$$

(b) For the inner product of part (a), define a matrix G (*the standard matrix of the inner product* $\langle\ ,\ \rangle$) by $g_{ij} = \langle \mathbf{e}_i, \mathbf{e}_j\rangle$ for $i, j = 1, 2$. Show that G is symmetric and that

$$\langle \mathbf{x}, \mathbf{y}\rangle = \sum_{i,j=1}^{2} g_{ij}x_iy_j = [\mathbf{x}]^T G[\mathbf{y}].$$

(c) Apply the Gram-Schmidt procedure, using the inner product $\langle\ ,\ \rangle$ and the corresponding norm, to produce a basis $\mathcal{B} = \{\mathbf{v}_1, \mathbf{v}_2\}$ for $\mathbb{R}^2$, such that

$$\langle \mathbf{v}_1, \mathbf{v}_1\rangle = \langle \mathbf{v}_2, \mathbf{v}_2\rangle = 1,$$
$$\langle \mathbf{v}_1, \mathbf{v}_2\rangle = 0.$$

Thus $\mathcal{B}$ is an orthonormal basis with respect to the inner product $\langle\ ,\ \rangle$.

(d) Define the matrix $\tilde{G} = \begin{bmatrix} \tilde{g}_{11} & \tilde{g}_{12} \\ \tilde{g}_{21} & \tilde{g}_{22} \end{bmatrix}$ (the $\mathcal{B}$-matrix of the inner product $\langle\ ,\ \rangle$, by

$$\tilde{g}_{ij} = \langle \mathbf{v}_i, \mathbf{v}_j\rangle, \quad i, j = 1, 2.$$

Show that $\tilde{G} = I$, and that for $\mathbf{x} = \tilde{x}_1\mathbf{v}_1 + \tilde{x}_2\mathbf{v}_2$ and $\mathbf{y} = \tilde{y}_1\mathbf{v}_1 + \tilde{y}_2\mathbf{v}_2$,

$$\langle \mathbf{x}, \mathbf{y} \rangle = \tilde{x}_1\tilde{y}_1 + \tilde{x}_2\tilde{y}_2.$$

Conclusion. For an arbitrary inner product $\langle\ ,\ \rangle$ on $\mathbb{R}^2$, there exists a basis for $\mathbb{R}^2$ that is orthonormal with respect to $\langle\ ,\ \rangle$; when $\langle \mathbf{x}, \mathbf{y} \rangle$ is expressed in terms of this basis, it looks just like the standard dot product in $\mathbb{R}^2$. This argument generalizes in a straightforward way to $\mathbb{R}^n$; see Exercise 8–2 D2.

***D2** Consider P_2 with the inner product $\langle\ ,\ \rangle$ as defined in Example 13. Using $\langle\ ,\ \rangle$ to define projections, apply the Gram-Schmidt procedure to the basis $\{1, x, x^2\}$ for P_2 to obtain a basis for P_2 that is orthonormal with respect to $\langle\ ,\ \rangle$.

D3 (a) Suppose that $\{\mathbf{v}_1, \mathbf{v}_2, \mathbf{v}_3\}$ is a basis for a real vector space V of dimension three, with inner product. Define a 3 by 3 matrix G by

$$g_{ij} = \langle \mathbf{v}_i, \mathbf{v}_j \rangle.$$

Explain why G is symmetric, and show that if $\mathbf{x} = x_1\mathbf{v}_1 + x_2\mathbf{v}_2 + x_3\mathbf{v}_3$ and $\mathbf{y} = y_1\mathbf{v}_1 + y_2\mathbf{v}_2 + y_3\mathbf{v}_3$,

$$\langle \mathbf{x}, \mathbf{y} \rangle = [x_1 \quad x_2 \quad x_3] G \begin{bmatrix} y_1 \\ y_2 \\ y_3 \end{bmatrix}.$$

(b) Let P_2 be the space of polynomials of degree less than or equal to 2 in a real variable x. Verify that each of the following defines an inner product on P_2.

(i) If $p(x) = p_0 + p_1x + p_2x^2$ and $q(x) = q_0 + q_1x + q_2x^2$,

$$\langle p, q \rangle = p_0q_0 + p_1q_1 + p_2q_2.$$

(ii) For polynomials in P_2 defined only for $-1 \le x \le 1$,

$$\langle p, q \rangle = \int_{-1}^{1} p(x)q(x)dx.$$

(c) Determine the matrix G of the inner products of part (b) with respect to the basis $\{1, x, x^2\}$ for P_2. Also determine G for the inner product of Example 13.

SECTION

7–5 Inner Products in Complex Vector Spaces; Unitary Matrices

We would like to have an inner product defined in a vector space with complex scalars, because the concepts of length, orthogonality, and projection are powerful tools for solving some kinds of problems.

We could extend the dot product to $\mathbf{C}^n$:

$$\mathbf{w} \cdot \mathbf{z} = w_1 z_1 + w_2 z_2 + \cdots + w_n z_n.$$

However, while we may continue to use this dot product as a useful notation, **we cannot use the dot product as a rule for defining an inner product in $\mathbf{C}^n$.**

The reason is that we require an inner product to satisfy $\langle \mathbf{z}, \mathbf{z} \rangle \geq 0$ for all $\mathbf{z}$ in $\mathbf{C}^n$, so that we can define $\|\mathbf{z}\| = \langle \mathbf{z}, \mathbf{z} \rangle^{1/2}$. However

$$\begin{aligned} \mathbf{z} \cdot \mathbf{z} &= z_1^2 + z_2^2 + \cdots + z_n^2 \\ &= (x_1^2 + x_2^2 + \cdots + x_n^2 - y_1^2 - \cdots - y_n^2) + 2i(x_1 y_1 + x_2 y_2 + \cdots + x_n y_n) \end{aligned}$$

(where $z_k = x_k + iy_k$, $k = 1, 2, \ldots, n$).

We see that $\mathbf{z} \cdot \mathbf{z}$ is not even real, in general, and certainly cannot be expected to satisfy $\mathbf{z} \cdot \mathbf{z} \geq 0$.

The properties of the complex numbers $\mathbf{C}$ point us in the right direction for defining an inner product. Recall that for a complex number z, $\|z\|^2 = \bar{z}z$.

DEFINITION

In $\mathbf{C}^n$, the standard inner product $\langle\ ,\ \rangle$ is defined by

$$\begin{aligned} \langle \mathbf{w}, \mathbf{z} \rangle &= \overline{\mathbf{w}} \cdot \mathbf{z} \\ &= \overline{w_1} z_1 + \overline{w_2} z_2 + \cdots + \overline{w_n} z_n, \quad \text{for } \mathbf{z}, \mathbf{w} \in \mathbf{C}^n. \end{aligned}$$

EXAMPLE 14

Let $\mathbf{u} = (1+i, 2-i)$, $\mathbf{v} = (-2+i, 3+2i)$. Determine $\langle \mathbf{u}, \mathbf{v} \rangle$, $\langle \mathbf{v}, \mathbf{u} \rangle$, and $\langle (2-i)\mathbf{u}, \mathbf{v} \rangle$.

SOLUTION:

$$\begin{aligned}\langle \mathbf{u}, \mathbf{v}\rangle &= \overline{\mathbf{u}} \cdot \mathbf{v} = \overline{(1+i, 2-i)} \cdot (-2+i, 3+2i)\\ &= (1-i, 2+i) \cdot (-2+i, 3+2i)\\ &= (1-i)(-2+i) + (2+i)(3+2i)\\ &= -1+3i+4+7i = 3+10i\\ \langle \mathbf{v}, \mathbf{u}\rangle &= (-2-i, 3-2i) \cdot (1+i, 2-i)\\ &= -1-3i+4-7i = 3-10i\end{aligned}$$

Note that $\langle \mathbf{v}, \mathbf{u}\rangle = \overline{\langle \mathbf{u}, \mathbf{v}\rangle}$. We discuss this below.

Finally,

$$\begin{aligned}\langle (2-i)\mathbf{u}, \mathbf{v}\rangle &= \overline{(2-i)\mathbf{u}} \cdot \mathbf{v} = \left(\overline{(2-i)}\ \overline{\mathbf{u}}\right) \cdot \mathbf{v}\\ &= (2+i)(1-i, 2+i) \cdot (-2+i, 3+2i)\\ &= (2+i)(3+10i) = (2+i)\langle \mathbf{u}, \mathbf{v}\rangle\end{aligned}$$

As discussed below, this is not quite what we might have expected from our experience of the dot product in $\mathbb{R}^n$ or general inner products in real vector spaces.

Properties of Complex Inner Products

Example 14 warns us that for complex vector spaces, we must modify the requirements of symmetry and bilinearity stated for real inner products. You should check from the definition that the standard inner product in $\mathbb{C}^n$ has the following three properties.

Properties of the Standard Inner Product in $\mathbb{C}^n$

(1) $\langle \mathbf{z}, \mathbf{z}\rangle \geq 0$ for all $\mathbf{z}$ in $\mathbb{C}^n$, and $\langle \mathbf{z}, \mathbf{z}\rangle = 0$ if and only if $\mathbf{z} = \mathbf{0}$.

(2) $\langle \mathbf{w}, \mathbf{z}\rangle = \overline{\langle \mathbf{z}, \mathbf{w}\rangle}$ for all $\mathbf{w}, \mathbf{z}$ in $\mathbb{C}^n$.

(3) For all $\mathbf{u}, \mathbf{v}, \mathbf{w}, \mathbf{z}$ in $\mathbb{C}^n$, and α in $\mathbb{C}$,

(i) $\langle \mathbf{w}, \mathbf{u}+\mathbf{v}\rangle = \langle \mathbf{w}, \mathbf{u}\rangle + \langle \mathbf{w}, \mathbf{v}\rangle$;

(ii) $\langle \mathbf{u}+\mathbf{v}, \mathbf{w}\rangle = \langle \mathbf{u}, \mathbf{w}\rangle + \langle \mathbf{v}, \mathbf{w}\rangle$;

(iii) $\langle \mathbf{w}, \alpha\mathbf{z}\rangle = \alpha\langle \mathbf{w}, \mathbf{z}\rangle$, but

(iv) $\langle \alpha\mathbf{w}, \mathbf{z}\rangle = \overline{\alpha}\langle \mathbf{w}, \mathbf{z}\rangle$.

For general vector spaces over the complex numbers, we take these properties as "characterizing properties" of an inner product. That means that a rule $\langle\, ,\, \rangle$ that assigns a scalar to each pair of vectors in a vector space is an inner product if and only if it satisfies these properties. (As an application, Fourier transform theory is defined in terms of an inner product on a vector space of complex-valued functions.)

Note that property (1) allows us to define (standard) length by

$$\|\mathbf{z}\| = \langle \mathbf{z}, \mathbf{z} \rangle^{1/2}.$$

Property (2) is the **Hermitian** property of the inner product. Notice that if the scalars are all real, we can say that $\overline{\langle \mathbf{z}, \mathbf{w} \rangle} = \langle \mathbf{z}, \mathbf{w} \rangle$, so the Hermitian property simplifies to symmetry.

Property (3) says that the inner product is not quite bilinear. However, this property reduces to bilinearity when the scalars are real so that $\overline{\alpha} = \alpha$.

The Cauchy-Schwarz-Buniakowski and Triangle Inequalities

Hermitian inner products also satisfy these inequalities for all vectors in the appropriate space; however, new proofs are required.

(4) $|\langle \mathbf{w}, \mathbf{z} \rangle| \leq \|\mathbf{w}\| \|\mathbf{z}\|$;

(5) $\|\mathbf{w} + \mathbf{z}\| \leq \|\mathbf{w}\| + \|\mathbf{z}\|$.

Proof of (4). This may be regarded as a variation on the "rabbit-out-of-the-hat" proof in Section 1–2 for the Cauchy-Schwarz inequality.

Suppose that $\mathbf{z} \neq \mathbf{0}$, and let

$$\alpha = \frac{\langle \mathbf{z}, \mathbf{w} \rangle}{\langle \mathbf{z}, \mathbf{z} \rangle}.$$

The following calculation uses standard facts about complex numbers (for example, $\alpha\overline{\alpha} = |\alpha|^2$), and the Hermitian property of the inner product.

$$\begin{aligned}
0 &\leq \langle \mathbf{w} - \alpha \mathbf{z}, \mathbf{w} - \alpha \mathbf{z} \rangle \\
&= \langle \mathbf{w}, \mathbf{w} \rangle - \alpha \langle \mathbf{w}, \mathbf{z} \rangle - \overline{\alpha} \langle \mathbf{z}, \mathbf{w} \rangle + \overline{\alpha}\alpha \langle \mathbf{z}, \mathbf{z} \rangle \\
&= \langle \mathbf{w}, \mathbf{w} \rangle - \frac{|\langle \mathbf{w}, \mathbf{z} \rangle|^2}{\langle \mathbf{z}, \mathbf{z} \rangle} - \frac{|\langle \mathbf{w}, \mathbf{z} \rangle|^2}{\langle \mathbf{z}, \mathbf{z} \rangle} + \frac{|\langle \mathbf{w}, \mathbf{z} \rangle|^2}{\langle \mathbf{z}, \mathbf{z} \rangle^2} \langle \mathbf{z}, \mathbf{z} \rangle \\
&= \frac{1}{\langle \mathbf{z}, \mathbf{z} \rangle} \Big(\langle \mathbf{z}, \mathbf{z} \rangle \langle \mathbf{w}, \mathbf{w} \rangle - |\langle \mathbf{w}, \mathbf{z} \rangle|^2 \Big).
\end{aligned}$$

Hence the Cauchy-Schwarz inequality is established. ■

The proof of the triangle inequality requires only minor changes from the proof for the real case; you are asked to provide the details in Exercise D1.

Orthogonality in $\mathbb{C}^n$ and Unitary Matrices

With an inner product defined, we can proceed as in Sections 1–2, 1–3, 7–1, and 7–2 to introduce orthogonality, projections, and distance. No new approaches or ideas are required, so we omit a detailed discussion. Notice that since $\langle \mathbf{w}, \mathbf{z} \rangle$ may be complex, there is no obvious way to define the angle between vectors.

Recall that a change of basis in $\mathbb{R}^n$ from the standard basis to a new orthonormal basis was described by an orthogonal matrix, that is, a matrix whose columns form an orthonormal basis for $\mathbb{R}^n$. The corresponding matrices for $\mathbb{C}^n$ are called **unitary** matrices.

DEFINITION

An n by n matrix with complex entries is said to be **unitary** if its columns form an orthonormal basis for $\mathbb{C}^n$.

For an orthogonal matrix R, we saw that the defining property was equivalent to the matrix condition $R^T R = I$, or $R^{-1} = R^T$. However, the corresponding condition for a matrix to be unitary involves complex conjugation.

THEOREM 1. An n by n complex matrix U is unitary if and only if

$$\overline{U}^T U = I.$$

It follows that U^{-1} is the conjugate transpose of $U : U^{-1} = \overline{U}^T$.

Proof. By the standard rule for matrix multiplication,

$$\begin{aligned}\left(\overline{U}^T U\right)_{jk} &= \left(\overline{U}^T\right)_{j\rightarrow} \cdot U_{\downarrow k} = \overline{U}_{\downarrow j} \cdot U_{\downarrow k} \\ &= \langle U_{\downarrow j}, U_{\downarrow k} \rangle.\end{aligned}$$

Thus, if $\overline{U}^T U = I$, we have

$$\langle U_{\downarrow j}, U_{\downarrow k} \rangle = \begin{cases} 1 & \text{if } \ j = k \\ 0 & \text{if } \ j \neq k, \end{cases}$$

so the columns of U are orthonormal. Conversely, if the columns are orthonormal, $\overline{U}^T U = I$.

■

EXAMPLE 15

Are the matrices $U = \begin{bmatrix} 1 & -i \\ i & 1 \end{bmatrix}$ and $V = \begin{bmatrix} \frac{1}{\sqrt{3}}(1+i) & \frac{1}{\sqrt{6}}(1+i) \\ -\frac{1}{\sqrt{3}}i & \frac{2}{\sqrt{6}}i \end{bmatrix}$ unitary?

SOLUTION: $\overline{U} = \begin{bmatrix} 1 & i \\ -i & 1 \end{bmatrix}$, so $\overline{U}^T = \begin{bmatrix} 1 & -i \\ i & 1 \end{bmatrix}$. Thus

$$\overline{U}^T U = \begin{bmatrix} 1 & -i \\ i & 1 \end{bmatrix}\begin{bmatrix} 1 & -i \\ i & 1 \end{bmatrix} = \begin{bmatrix} 2 & -2i \\ 2i & 2 \end{bmatrix} \neq I,$$

so U is not unitary. (If you are not careful, you could easily be misled by the fact that the **dot** product of the columns of U is zero; the **inner** product, however, is not zero).

$$\overline{V}^T = \begin{bmatrix} \frac{1}{\sqrt{3}}(1-i) & \frac{i}{\sqrt{3}} \\ \frac{1}{\sqrt{6}}(1-i) & -\frac{2i}{\sqrt{6}} \end{bmatrix}$$

$$\begin{aligned} \overline{V}^T V &= \begin{bmatrix} \frac{1}{\sqrt{3}}(1-i) & \frac{i}{\sqrt{3}} \\ \frac{1}{\sqrt{6}}(1-i) & -\frac{2i}{\sqrt{6}} \end{bmatrix}\begin{bmatrix} \frac{1}{\sqrt{3}}(1+i) & \frac{1}{\sqrt{6}}(1+i) \\ -\frac{1}{\sqrt{3}}i & \frac{2}{\sqrt{6}}i \end{bmatrix} \\ &= \begin{bmatrix} \frac{1}{3}(2+1) & \frac{1}{3\sqrt{2}}(2-2) \\ \frac{1}{3\sqrt{2}}(2-2) & \frac{1}{6}(2+4) \end{bmatrix} \\ &= \begin{bmatrix} 1 & 0 \\ 0 & 1 \end{bmatrix}, \end{aligned}$$

so V is unitary.

EXERCISES 7–5

A1 Use the standard inner product in $\mathbb{C}^n$ to calculate $\langle \mathbf{u}, \mathbf{v} \rangle$, $\langle \mathbf{v}, \mathbf{u} \rangle$, $\|\mathbf{u}\|$, and $\|\mathbf{v}\|$ if

(a) $\mathbf{u} = (2+3i, -1-2i)$, $\mathbf{v} = (-2i, 2-5i)$;

(b) $\mathbf{u} = (-1+4i, 2-i)$, $\mathbf{v} = (3+i, 1+3i)$.

A2 Check whether the following matrices are unitary.

(a) $A = \frac{1}{\sqrt{3}}\begin{bmatrix} 1 & 1+i \\ 1+i & 1 \end{bmatrix}$ (b) $B = \begin{bmatrix} i & 0 \\ 0 & -1 \end{bmatrix}$

(c) $C = \begin{bmatrix} \frac{1}{\sqrt{3}}(-1+i) & \frac{1}{\sqrt{6}}(1-i) \\ \frac{1}{\sqrt{3}} & \frac{2}{\sqrt{6}} \end{bmatrix}$

A3 (a) Verify that $\mathbf{u}$ is orthogonal to $\mathbf{v}$ if $\mathbf{u} = (1, 0, i)$, $\mathbf{v} = (i, 1, 1)$.
(b) Determine the projection of $\mathbf{w} = (1+i, 2+i, 3-i)$ onto the subspace of $\mathbb{C}^3$ spanned by $\mathbf{u}$ and $\mathbf{v}$. (Remember to adapt the projection formula from Section 7–2 for the case of the Hermitian inner product in $\mathbb{C}^3$.)

B1 Use the standard inner product in $\mathbb{C}^n$ to calculate $\langle \mathbf{u}, \mathbf{v} \rangle$, $\langle \mathbf{v}, \mathbf{u} \rangle$, $\|\mathbf{u}\|$, and $\|\mathbf{v}\|$ if

(a) $\mathbf{u} = (-2-3i, 2+i)$, $\mathbf{v} = (4-i, 4+i)$;
(b) $\mathbf{u} = (3-i, 1+2i)$, $\mathbf{v} = (1+i, 2+i)$.

B2 Check whether the following matrices are unitary.

(a) $A = \begin{bmatrix} \frac{1}{\sqrt{7}}(1+i) & \frac{-5}{\sqrt{35}} \\ \frac{1}{\sqrt{7}}(1+2i) & \frac{1}{\sqrt{35}}(3+i) \end{bmatrix}$ (b) $B = \begin{bmatrix} \frac{1}{\sqrt{6}}(1+i) & \frac{1}{\sqrt{3}}(1+i) \\ \frac{2}{\sqrt{6}}i & \frac{1}{\sqrt{3}}i \end{bmatrix}$

B3 (a) Verify that $\mathbf{u}$ is orthogonal to $\mathbf{v}$ if $\mathbf{u} = (1+i, 1, 2)$, $\mathbf{v} = (1-i, 2i, 0)$.
(b) Determine the projection of $\mathbf{w} = (2+i, 2-i, 1-2i)$ onto the subspace of $\mathbb{C}^3$ spanned by $\mathbf{u}$ and $\mathbf{v}$. (Remember to adapt the projection formula from Section 7–2 for the case of the Hermitian inner product in $\mathbb{C}^3$.)

CONCEPTUAL EXERCISES

***D1** Prove the triangle inequality for the standard inner product in $\mathbb{C}^n$ by modifying the proof in Section 1–2. You may need to use the fact that

$$\langle \mathbf{w}, \mathbf{z} \rangle + \langle \mathbf{z}, \mathbf{w} \rangle = 2\text{Re}\langle \mathbf{w}, \mathbf{z} \rangle \leq 2|\langle \mathbf{w}, \mathbf{z} \rangle|.$$

D2 Prove that for any complex n by n matrix,

$$\det \overline{A} = \overline{\det A}.$$

(Begin by checking the 2 by 2 case.)

***D3** Prove that any unitary matrix satisfies

$$|\det U| = 1.$$

(However, it does not follow that $\det U = \pm 1$ because $\det U$ need not be real. See, for example, Exercise A2 (b).)

D4 In $\mathbb{C}^3$, define $\mathbf{w} \times \mathbf{z}$ by

$$\mathbf{w} \times \mathbf{z} = \begin{bmatrix} \mathbf{e}_1 & \mathbf{e}_2 & \mathbf{e}_3 \\ w_1 & w_2 & w_3 \\ z_1 & z_2 & z_3 \end{bmatrix}$$

where $\mathbf{e}_1 = (1, 0, 0)$, $\mathbf{e}_2 = (0, 1, 0)$, $\mathbf{e}_3 = (0, 0, 1)$ in $\mathbb{C}^3$. Does $\mathbf{w} \times \mathbf{z}$ satisfy $\langle \mathbf{w}, \mathbf{w} \times \mathbf{z} \rangle = 0$? What about $\overline{\mathbf{w} \times \mathbf{z}}$?

D5 (a) Show that if U is unitary, then $\|U\mathbf{z}\| = \|\mathbf{z}\|$ for all $\mathbf{z}$ in $\mathbb{C}^n$.

(b) Hence show that if U is unitary, all its eigenvalues satisfy $|\lambda| = 1$.

Chapter Review

Suggestions for Student Review

1 What is meant by "an orthogonal set of vectors" in $\mathbb{R}^n$? What is the difference between an orthogonal basis and an orthonormal basis?

2 Why is it easier to determine coordinates with respect to an orthonormal basis than with respect to an arbitrary basis? What are some special features of the change of basis matrix from the standard basis to an orthonormal basis? What is an orthogonal matrix?

3 Does every subspace of $\mathbb{R}^n$ have an orthonormal basis? What about the zero subspace? How do you find an orthonormal basis? Sketch the Gram-Schmidt procedure.

4 What are the essential geometrical properties of a projection onto a subspace of $\mathbb{R}^n$? (Section 7–2, Theorem 1) How do you calculate a projection onto a subspace?

5 Outline how to use ideas about orthogonality to find the best fitting *line* for a given set of data points $\{(t_i, y_i) \mid i = 1, \ldots, n\}$. (Section 7–3)

6 What are the essential properties of an inner product? Give an example of an inner product other than the standard dot product. (Section 7–4)

7 Discuss the standard inner product in $\mathbb{C}^n$; how are the essential properties of an inner product modified in generalizing from the real to the complex case? (Section 7–5)

Chapter Quiz

1 Determine whether the following sets are orthogonal, orthonormal, or neither. Show how you decide.

(a) $\{\mathbf{u}_1 = \frac{1}{3}(1, 0, 1, 1), \mathbf{u}_2 = \frac{1}{3}(0, 1, 1, -1), \mathbf{u}_3 = \frac{1}{2}(0, 1, -1, 0)\}$

(b) $\{\mathbf{v}_1 = \frac{1}{\sqrt{3}}(1, 0, 1, 1), \mathbf{v}_2 = \frac{1}{\sqrt{5}}(0, 0, 1, -2)\}$

(c) $\{\mathbf{v}_1 = \frac{1}{\sqrt{3}}(1,0,1,1), \mathbf{v}_2 = \frac{1}{\sqrt{3}}(1,1,-1,0), \mathbf{v}_3 = \frac{1}{\sqrt{3}}(0,-1,-1,1)\}$.

2 Consider the orthonormal set $\mathcal{B} = \{\frac{1}{2}(1,0,1,1,1), \frac{1}{\sqrt{3}}(1,1,0,0,-1), \frac{1}{\sqrt{3}}(-1,1,0,1,0)\}$. Let $\mathbb{S}$ be the subspace of $\mathbb{R}^5$ spanned by $\mathcal{B}$. Given that $(2,5,1,3,-2)$ is a vector in $\mathbb{S}$, use the orthonormality of $\mathcal{B}$ to determine the coordinates of $(2,5,1,3,-2)$ with respect to $\mathcal{B}$.

3 (a) Prove that if R is an orthogonal matrix, $\det R = \pm 1$.
(b) Prove that if R and S are orthogonal n by n matrices, then so is RS.

4 Let $\mathbb{S}$ be the subspace of $\mathbb{R}^4$ defined by

$$\mathbb{S} = \mathrm{Sp}(\{(1,0,1,0), (1,1,1,1), (1,3,3,1)\}).$$

(a) Apply the Gram-Schmidt procedure to the given spanning set to produce an orthonormal basis for $\mathbb{S}$.
(b) Determine the point in $\mathbb{S}$ closest to $(1,-2,-1,0)$.

5 (Section 7–4) Determine whether each of the following definitions $\langle\ ,\ \rangle$ actually defines an inner product on $\mathcal{M}(2,2)$, the space of 2 by 2 matrices. Explain how you decide in each case.
(a) $\langle A, B\rangle = \det(AB)$
(b) $\langle A, B\rangle = a_{11}b_{11} + 2a_{12}b_{12} + 2a_{21}b_{21} + a_{22}b_{22}$

6 (Section 7–5)
(a) In $\mathbb{C}^3$, determine $\langle(3-i, i, 2), (1, 3, 4-i)\rangle$.
(b) Prove that $\frac{1}{\sqrt{3}}\begin{bmatrix} 1-i & -i \\ 1 & -1+i \end{bmatrix}$ is a unitary matrix.

Further Exercises

F1 (ISOMETRIES OF $\mathbb{R}^3$)
(a) A linear mapping is an **isometry** of $\mathbb{R}^3$ if $\|L(\mathbf{x})\| = \|\mathbf{x}\|$ for every $\mathbf{x}$ in $\mathbb{R}^3$. Prove that an isometry preserves dot products and angles as well as lengths.
(b) Show that L is an isometry if and only if the standard matrix of L is orthogonal. [Hint: See Exercise F5 in Chapter 3 Review, and Exercise D3 in Section 7–1.]
(c) Explain why an isometry of $\mathbb{R}^3$ must have 1 or 3 real characteristic roots, counting multiplicity. By Exercise 7–1 D3 part (b), these must be 1 or -1.
(d) Let A be the standard matrix of L. Suppose that 1 is an eigenvalue of A with eigenvector $\mathbf{u}$, so that $A\mathbf{u} = \mathbf{u}$. Let $\mathbf{v}$ and $\mathbf{w}$ be vectors such that $\{\mathbf{u}, \mathbf{v}, \mathbf{w}\}$ is an orthonormal basis for $\mathbb{R}^3$, and let $R = [\mathbf{u}\ \mathbf{v}\ \mathbf{w}]$. Show that

$$R^T A R = \begin{bmatrix} 1 & O_{12} \\ O_{21} & A^* \end{bmatrix},$$

where the right-hand side is a partioned (block) matrix, with 0_{12} being a 1 by 2 zero matrix, 0_{21} being a 2 by 1 zero matrix, and A^* being a 2 by 2 orthogonal matrix. Moreover, show that the characteristic roots of A are 1 and the characteristic roots of A^*.

Remark Note that an analogous form can be obtained for $R^T AR$ in the case where one eigenvalue is -1.

(e) Use Exercise 5 from Chapter 3 Review to analyze the A^* of part (d), and explain why every isometry of $\mathbb{R}^3$ is the identity mapping, or a reflection, or a composition of reflections, or a rotation, or a composition of a reflection and a rotation.

F2 A linear mapping $L : \mathbb{R}^n \to \mathbb{R}^n$ is called an **involution** if $L \circ L =$ Identity. In terms of its standard matrix this means that $A^2 = I$. Prove that any two of the following imply the third.
(a) A is the matrix of an involution.
(b) A is symmetric.
(c) A is an isometry.

F3 If S is a subspace of a real finite-dimensional inner product space V, prove that $(\mathrm{S}^\perp)^\perp = \mathrm{S}$. (See Exercise 3–4–D2.) Discuss the relation of this fact with Theorem 1 and Exercises D1–4 of Section 7–2.

F4 The sum $\mathrm{S} + \mathrm{T}$ of subspaces of a finite-dimensional vector space V was defined in the paragraph preceding Exercise F2 of Chapter 4 Review. Prove that
(a) $(\mathrm{S} + \mathrm{T})^\perp = \mathrm{S}^\perp \cap \mathrm{T}^\perp$;
(b) $(\mathrm{S} \cap \mathrm{T})^\perp = \mathrm{S}^\perp + \mathrm{T}^\perp$.

F5 Often, a problem of finding a *sequence* of approximations to some vector (or function) $\mathbf{v}$ can be described by requiring the jth approximation to be the closest vector to $\mathbf{v}$ in some finite-dimensional subspace S_j of a "big" inner product space V, where the subspaces are required to satisfy

$$\mathrm{S}_1 \subset \mathrm{S}_2 \subset \ldots \subset \mathrm{S}_j \subset \mathrm{S}_{j+1} \subset \ldots \subset \mathrm{V}.$$

The jth approximation is then $\mathbf{proj}_{\mathrm{S}_j}(\mathbf{v})$.
(One example of this kind of approximation is Fourier series; approximations by polynomials of increasing degree may also fit this description.) Prove that if V is finite-dimensional, then the approximations improve as j increases in the sense that

$$\|\mathbf{v} - \mathbf{proj}_{\mathrm{S}_{j+1}}(\mathbf{v})\| \leq \|\mathbf{v} - \mathbf{proj}_{\mathrm{S}_j}(\mathbf{v})\|.$$

F6 The dual vector space V^* of a vector space V was defined in Exercise F8 in Chapter 4 Review. If V is a real finite-dimensional vector space, prove that the mapping $\mathcal{P} : \mathrm{V} \to \mathrm{V}^*$ defined by $(\mathcal{P}(\mathbf{v}))(\mathbf{w}) = \langle \mathbf{v}, \mathbf{w} \rangle$ for all $\mathbf{w}$ in V is an isomorphism.

F7 (a) Prove the following statement (used in Section 7–3). If X is an n by 3 matrix with linearly independent columns, then X^TX has rank 3. [Hint: Consider a vector in the nullspace of X^TX.]
(b) Generalize the result of part (a) to the case of $n \times p$ matrices, where $n \geq p$.
(c) Suppose that A is an n by 3 matrix, where $n \geq 3$, and the rank of A is 2. Prove that A^TA has rank 2. [Hint: Let B be obtained from A by interchanging two columns of A. How is B^TB related to A^TA?]

F8 **QR-factorization.** Suppose that A is an invertible n by n matrix. Prove that A can be written as the product of an orthogonal matrix Q and an upper triangular matrix R:

$$A = QR.$$

[Hint: Apply the Gram-Schmidt procedure to the columns of A, starting at the first column. If instead you start at the last column, the factor R is lower triangular.]

Remark This QR-factorization is important in a numerical procedure for determining eigenvalues of symmetric matrices.

CHAPTER

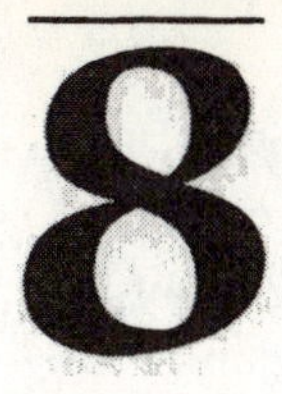

Symmetric Matrices and Quadratic Forms

Many applications of matrices involve symmetric matrices. Symmetric matrices can always be diagonalized by an orthogonal change of basis.

SECTION

8–1 Diagonalization of Symmetric Matrices

Introduction

DEFINITION

A matrix A is **symmetric** if $A^T = A$, or equivalently, if $a_{jk} = a_{kj}$ for all j and k. Obviously this is only possible if A is square.

Symmetric matrices arise naturally in many physical applications. For example, the strain matrix describing the deformation of a solid and the inertia tensor of a rotating body are symmetric (Section 8–5). In many of these physical applications of symmetric matrices, the diagonalization theorem is referred to as the Principal Axis Theorem.

Symmetric matrices also have important applications because of their connection with quadratic forms, discussed in Section 8–2. In $\mathbb{R}^2$, equations involving quadratic forms describe ellipses and hyperbolas; there are higher dimensional generalizations (Section 8–3). In calculus the second order terms of a Taylor approximation give a quadratic form, and diagonalizing this form helps us to decide whether a critical point is a local maximum, a local minimum, or a saddle point (Section 8–4).

Eigenvectors and Eigenvalues of a Symmetric Matrix

In Chapter 6, we saw that diagonalization of a square matrix might not be possible if some of the roots of its characteristic equation were complex, or if some of its real roots were repeated. *Symmetric matrices can always be diagonalized: all the roots of its characteristic equation are real, and a basis of eigenvectors can always be found. In fact, the basis of eigenvectors can be chosen to be orthonormal so the diagonalizing matrix is orthogonal.* Before considering why this works, we give two examples.

EXAMPLE 1

Determine the eigenvalues and corresponding eigenvectors of the symmetric matrix $A = \begin{bmatrix} 0 & 1 \\ 1 & -2 \end{bmatrix}$. What is the diagonal matrix corresponding to A, and what is the orthogonal matrix that diagonalizes A?

SOLUTION:

$A-\lambda I = \begin{bmatrix} 0-\lambda & 1 \\ 1 & -2-\lambda \end{bmatrix}$, so $\det(A-\lambda I) = \lambda^2+2\lambda-1$, and by the standard quadratic formula, the roots of the characteristic equation are $-1+\sqrt{2}$ and $-1-\sqrt{2}$. Thus the resulting diagonal matrix is

$$D = \begin{bmatrix} -1-\sqrt{2} & 0 \\ 0 & -1+\sqrt{2} \end{bmatrix}.$$

When $\lambda = -1+\sqrt{2}$,

$$A-\lambda I = \begin{bmatrix} 1-\sqrt{2} & 1 \\ 1 & -1-\sqrt{2} \end{bmatrix} \rightarrow \begin{bmatrix} 1 & -1-\sqrt{2} \\ 0 & 0 \end{bmatrix}$$

(where the fact that $(1+\sqrt{2})(1-\sqrt{2}) = 1-2 = -1$ has been used in the row reduction). It follows that a corresponding eigenvector is $\mathbf{v} = (1+\sqrt{2}, 1)$.

(Check that $A\mathbf{v} = \lambda\mathbf{v}$:

$$\begin{bmatrix} 0 & 1 \\ 1 & -2 \end{bmatrix} \begin{bmatrix} 1+\sqrt{2} \\ 1 \end{bmatrix} = \begin{bmatrix} 1 \\ -1+\sqrt{2} \end{bmatrix} = (-1+\sqrt{2}) \begin{bmatrix} 1+\sqrt{2} \\ 1 \end{bmatrix}.)$$

When $\lambda = -1 - \sqrt{2}$,

$$A - \lambda I = \begin{bmatrix} 1+\sqrt{2} & 1 \\ 1 & -1+\sqrt{2} \end{bmatrix} \to \begin{bmatrix} 1 & -1+\sqrt{2} \\ 0 & 0 \end{bmatrix},$$

so a corresponding eigenvector is $(1 - \sqrt{2}, 1)$. *(Check. Are the eigenvectors orthogonal to each other?)*

To find the orthogonal matrix R that diagonalizes A, it is necessary to use normalized eigenvectors as columns of R. Rounded to four decimal places, these eigenvectors give the matrix $R = \begin{bmatrix} 0.9239 & -0.3827 \\ 0.3827 & 0.9239 \end{bmatrix}$.

EXAMPLE 2

Diagonalize the symmetric matrix $A = \begin{bmatrix} 5 & -4 & -2 \\ -4 & 5 & -2 \\ -2 & -2 & 8 \end{bmatrix}$. Find an orthogonal matrix R that diagonalizes A.

SOLUTION:

$$A - \lambda I = \begin{bmatrix} 5-\lambda & -4 & -2 \\ -4 & 5-\lambda & -2 \\ -2 & -2 & 8-\lambda \end{bmatrix},$$

so $\det(A - \lambda I) = -\lambda^3 + 18\lambda^2 - 81\lambda = -\lambda(\lambda - 9)^2$. The eigenvalues are 0,9,9. Since A is symmetric, the theory developed in the following pages guarantees that A can be diagonalized to $\begin{bmatrix} 0 & 0 & 0 \\ 0 & 9 & 0 \\ 0 & 0 & 9 \end{bmatrix}$. To find the diagonalizing matrix R, it is necessary to find an orthonormal basis of eigenvectors.

When $\lambda = 0$,

$$A - \lambda I = \begin{bmatrix} 5 & -4 & -2 \\ -4 & 5 & -2 \\ -2 & -2 & 8 \end{bmatrix} \to \begin{bmatrix} 1 & 0 & -2 \\ 0 & 1 & -2 \\ 0 & 0 & 0 \end{bmatrix},$$

and a corresponding eigenvector is $(2, 2, 1)$. Call this $\mathbf{w}_1$.

When $\lambda = 9$,

$$A - \lambda I = \begin{bmatrix} -4 & -4 & -2 \\ -4 & -4 & -2 \\ -2 & -2 & -1 \end{bmatrix} \to \begin{bmatrix} 2 & 2 & 1 \\ 0 & 0 & 0 \\ 0 & 0 & 0 \end{bmatrix}.$$

The general solution of the homogeneous system is $\mathbf{x} = s(-1, 1, 0) + t(-1, 0, 2)$, so there is a two-dimensional "eigenspace" of eigenvectors corresponding to $\lambda = 9$. The pair $\{(-1,1,0),(-1,0,2)\}$ is linearly independent, but not orthogonal. We apply the Gram-Schmidt procedure to $\{(-1,1,0),(-1,0,2)\}$:

let $\mathbf{w}_2 = (-1,1,0)$,

let $\mathbf{w}_3 = \mathbf{perp}_{(-1,1,0)}(-1,0,2) = \frac{1}{2}(-1,-1,4)$.

Then $\{\mathbf{w}_2, \mathbf{w}_3\}$ is an orthogonal basis for the eigenspace corresponding to $\lambda = 9$, and $\{\mathbf{w}_1, \mathbf{w}_2, \mathbf{w}_3\}$ is an orthogonal basis for $\mathbb{R}^3$ consisting of eigenvectors of A.

(Alternatively, we could determine $\mathbf{w}_3$ by taking $\mathbf{w}_1 \times \mathbf{w}_2$ in this example, but this procedure would work only for 3 by 3 matrices. In other dimensions, the cross-product is not available.)

Normalize these vectors $\mathbf{w}_1, \mathbf{w}_2, \mathbf{w}_3$ to obtain the columns of the diagonalizing orthogonal matrix $R = \begin{bmatrix} 2/3 & -1/\sqrt{2} & -1/3\sqrt{2} \\ 2/3 & 1/\sqrt{2} & -1/3\sqrt{2} \\ 1/3 & 0 & 4/3\sqrt{2} \end{bmatrix}$. Note that the order of these columns corresponds to the chosen order of the eigenvalues in the diagonal matrix.

If A and B are square matrices and R is an orthogonal matrix such that

$$B = R^T A R,$$

we say that B is **orthogonally equivalent** to A. If A is orthogonally equivalent to a diagonal matrix, we say that A is **orthogonally diagonable**. Examples 1 and 2 suggest that every symmetric matrix is orthogonally diagonable, as we shall prove in the following pages. It is left as an exercise to prove the converse: any orthogonally diagonable matrix must be symmetric.

Notation Reviewed and Revised

To explain why symmetric matrices are orthogonally diagonable, it is useful to observe that dot products of vectors can be written in terms of matrix products. Careful use of notation will make this discussion easier to follow.

Recall that in Section 3–1 we used $[\mathbf{x}]$ to denote an n by 1 (column) matrix whose entries were the coordinates of the vector $\mathbf{x}$ in $\mathbb{R}^n$. In Section 3–4, we agreed to use $\mathbf{x}$ itself (without brackets) to denote the column matrix, particularly when we considered the equation $\mathbf{y} = A\mathbf{x}$, where both $\mathbf{x}$ and $\mathbf{y}$ must be understood as n by 1 matrices. (However, we continued to use the brackets when we wanted to emphasize that the vector must be interpreted as a column, or when we needed to indicate the basis with respect to which coordinates were calculated.)

Now if **x** denotes an n by 1 matrix $\begin{bmatrix} x_1 \\ x_2 \\ \vdots \\ x_n \end{bmatrix}$, then $\mathbf{x}^T$ denotes the transpose of **x**, and this is a 1 by n (row) matrix:

$$\mathbf{x}^T = [\, x_1 \quad x_2 \quad \ldots \quad x_n \,].$$

It is very useful to observe that

$$x_1y_1 + x_2y_2 + \cdots + x_ny_n = [\, x_1 \quad x_2 \quad \ldots \quad x_n \,] \begin{bmatrix} y_1 \\ y_2 \\ \vdots \\ y_n \end{bmatrix}.$$

It follows that *we may express the dot product of two vectors in terms of the matrix product:*

$$\mathbf{x} \cdot \mathbf{y} = \mathbf{x}^T\mathbf{y}.$$

Usually we think of the dot product as simply a real number, but this notation suggests that we may consider $\mathbf{x}^T\mathbf{y}$ as a 1 by 1 matrix. Such a matrix must be symmetric, so

$$\mathbf{x}^T\mathbf{y} = (\mathbf{x}^T\mathbf{y})^T = \mathbf{y}^T(\mathbf{x}^T)^T = \mathbf{y}^T\mathbf{x}.$$

This simply expresses the known fact that the dot product is symmetric: $\mathbf{x} \cdot \mathbf{y} = \mathbf{y} \cdot \mathbf{x}$. This kind of manipulation will be useful in the proof of the following lemma, and throughout this chapter.

LEMMA. An n by n matrix A is symmetric if and only if $\mathbf{x} \cdot (A\mathbf{y}) = (A\mathbf{x}) \cdot \mathbf{y}$, for all **x** and **y** in $\mathbb{R}^n$.

Proof. Suppose that A is symmetric. For any **x** and **y** in $\mathbb{R}^n$,

$$\mathbf{x} \cdot (A\mathbf{y}) = \mathbf{x}^T A\mathbf{y}.$$

The 1 by 1 matrix on the right-hand side is equal to its transpose, so

$$\begin{aligned} \mathbf{x} \cdot (A\mathbf{y}) &= (\mathbf{x}^T A\mathbf{y})^T \\ &= \mathbf{y}^T A^T\mathbf{x} \qquad \text{(because } (CD)^T = D^TC^T\text{)} \\ &= \mathbf{y}^T A\mathbf{x} \qquad \text{(because } A \text{ is symmetric)} \\ &= \mathbf{y} \cdot A\mathbf{x} = A\mathbf{x} \cdot \mathbf{y}. \end{aligned}$$

Conversely, if $\mathbf{x} \cdot A\mathbf{y} = \mathbf{y} \cdot A\mathbf{x}$ for all $\mathbf{x}, \mathbf{y}$ in $\mathbb{R}^n$,

$$\begin{aligned} \mathbf{x}^T A\mathbf{y} &= \mathbf{y}^T A\mathbf{x} = (\mathbf{y}^T A\mathbf{x})^T \\ &= \mathbf{x}^T A^T \mathbf{y}, \quad \text{for all } \mathbf{x}, \mathbf{y} \text{ in } \mathbb{R}^n. \end{aligned}$$

It follows that

$$\mathbf{x}^T (A - A^T)\mathbf{y} = 0, \quad \text{for all } \mathbf{x}, \mathbf{y} \text{ in } \mathbb{R}^n.$$

By a double application of Theorem 2 of Section 3–1, it follows that $A - A^T = 0$, or $A = A^T$, so A is symmetric.

■

The Theory of Symmetric Diagonalization

We begin by discussing the orthogonality of the eigenvectors for a symmetric matrix.

THEOREM 1. If $\mathbf{v}_1$ and $\mathbf{v}_2$ are eigenvectors of a symmetric matrix A corresponding to distinct eigenvalues λ_1 and λ_2, then $\mathbf{v}_1$ is orthogonal to $\mathbf{v}_2$.

Proof. Consider

$$\mathbf{v}_1 \cdot A\mathbf{v}_2 = \mathbf{v}_1 \cdot \lambda_2\mathbf{v}_2 = \lambda_2(\mathbf{v}_1 \cdot \mathbf{v}_2).$$

Since A is symmetric, by the preceding lemma,

$$\mathbf{v}_1 \cdot A\mathbf{v}_2 = \mathbf{v}_2 \cdot A\mathbf{v}_1 = \mathbf{v}_2 \cdot \lambda_1\mathbf{v}_1 = \lambda_1(\mathbf{v}_2 \cdot \mathbf{v}_1).$$

It follows that $\lambda_1(\mathbf{v}_2 \cdot \mathbf{v}_1) = \lambda_2(\mathbf{v}_1 \cdot \mathbf{v}_2)$, or

$$(\lambda_1 - \lambda_2)(\mathbf{v}_1 \cdot \mathbf{v}_2) = 0.$$

It was assumed that $\lambda_1 \neq \lambda_2$, so $\mathbf{v}_1 \cdot \mathbf{v}_2$ must be zero, and the eigenvectors corresponding to distinct eigenvalues are mutually orthogonal, as claimed.

■

At the beginning of this section, we claimed that a symmetric matrix has all real eigenvalues. The first step is to prove the following theorem.

THEOREM 2. If A is a symmetric square matrix, then it has at least one real eigenvalue.

The proof of Theorem 2 is postponed until the end of this section.

THEOREM 3. Suppose that λ_1 is a real eigenvalue of the symmetric n by n matrix A, with corresponding unit eigenvector $\mathbf{v}_1$. Then there is an orthonormal basis $\mathcal{B}$ for $\mathbb{R}^n$ with $\mathbf{v}_1$ as its first vector. Let R be the corresponding orthogonal change of basis matrix. Then

$$R^T AR = \left[\begin{array}{c|c} \lambda_1 & 0_{1,n-1} \\ \hline 0_{n-1,1} & A_1^* \end{array}\right],$$

where A_1^* is a symmetric $(n-1)$ by $(n-1)$ matrix, and the upper right block and the lower left block are zero matrices with $n-1$ entries.

Proof. By the methods of Chapter 4 and the Gram-Schmidt procedure, we can produce an orthonormal basis $\mathcal{B} = \{\mathbf{v}_1, \mathbf{w}_2, \mathbf{w}_3, \ldots, \mathbf{w}_n\}$ with $\mathbf{v}_1$ as its first vector. The corresponding orthogonal matrix is

$$R = \begin{bmatrix} \mathbf{v}_1 & \mathbf{w}_2 & \mathbf{w}_3 & \ldots & \mathbf{w}_n \end{bmatrix}.$$

Then

$$R^T AR = \begin{bmatrix} \mathbf{v}_1^T \\ \mathbf{w}_2^T \\ \mathbf{w}_3^T \\ \vdots \\ \mathbf{w}_n^T \end{bmatrix} A \begin{bmatrix} \mathbf{v}_1 & \mathbf{w}_2 & \mathbf{w}_3 & \ldots & \mathbf{w}_n \end{bmatrix}$$

$$= \begin{bmatrix} \mathbf{v}_1 \cdot A\mathbf{v}_1 & \mathbf{v}_1 \cdot A\mathbf{w}_2 & \ldots & \mathbf{v}_1 \cdot A\mathbf{w}_n \\ \mathbf{w}_2 \cdot A\mathbf{v}_1 & \mathbf{w}_2 \cdot A\mathbf{w}_2 & \ldots & \mathbf{w}_2 \cdot A\mathbf{w}_n \\ \vdots & \vdots & \ddots & \vdots \\ \mathbf{w}_n \cdot A\mathbf{v}_1 & \mathbf{w}_n \cdot A\mathbf{w}_2 & \ldots & \mathbf{w}_n \cdot A\mathbf{w}_n \end{bmatrix}.$$

$R^T AR$ is symmetric because

$$(R^T AR)^T = R^T A^T R^{TT} = R^T AR.$$

$\mathbf{v}_1$ is an eigenvector, so $A\mathbf{v}_1 = \lambda_1 \mathbf{v}_1$. Since $\mathbf{v}_1 \cdot \mathbf{v}_1 = 1$, the first entry in the first column is λ_1; since $\mathbf{w}_j \cdot \mathbf{v}_1 = 0$, all the other entries in the first column are zeros. By symmetry, all the entries in the first row except for the first must also be zeros. Also by the symmetry of $R^T AR$, the $(n-1)$ by $(n-1)$ block in the lower right corner must be symmetric; denote this block by A_1^* and Theorem 3 is proved.

■

REMARK

The result of Theorem 3 (the reduction of the problem from the n by n matrix A to the $(n-1)$ by $(n-1)$ matrix A_1^*) is called **deflation.** By repeating this step $n-1$ times, we can prove the following theorem.

THEOREM 4. **(The Symmetric Diagonalization Theorem, or the Principal Axis Theorem;** also known as the Spectral Theorem.) Suppose that A is a symmetric n by n matrix. Then there is an orthonormal basis of eigenvectors of A, with corresponding orthogonal matrix R, such that $R^T AR$ is a diagonal matrix with diagonal entries equal to the real eigenvalues of A. The order in which the eigenvalues appear on the diagonal is determined by the order of the eigenvectors as columns of R. (The coordinate axes determined by the eigenvectors are called the **principal axes** for the symmetric matrix.)

Proof. The proof is by induction on n. *Students may omit the details without loss of continuity, but the Remarks following the proof are important.*

When $n = 2$, Theorems 2 and 3 tell us that there is an orthonormal basis $\{\mathbf{v}_1, \mathbf{w}_2\}$, with corresponding orthogonal matrix $R = [\,\mathbf{v}_1 \quad \mathbf{w}_2\,]$ such that

$$R^T AR = \left[\begin{array}{c|c} \lambda_1 & 0 \\ \hline 0 & A_1^* \end{array}\right].$$

In this case A_1^* is 1 by 1, so it is a real number λ_2. Then $R^T AR$ is diagonal, λ_2 is an eigenvalue, $\{\mathbf{v}_1, \mathbf{w}_2\}$ is an orthonormal basis of eigenvectors, so Theorem 4 is true if $n = 2$. It is worth remarking that λ_2 may be equal to λ_1; repeated characteristic roots do not cause problems or require special treatment in this argument.

Now suppose that the result is true for $(n-1)$ by $(n-1)$ symmetric matrices, and consider the n by n symmetric matrix A. By Theorem 2, A has at least one real eigenvalue λ_1; there is at least one corresponding eigenvector, and we may normalize to obtain a unit eigenvector $\mathbf{v}_1$. Then by Theorem 3, there is an orthonormal basis $\mathcal{B} = \{\mathbf{v}_1, \mathbf{w}_2, \ldots, \mathbf{w}_n\}$, with orthogonal matrix $R = [\,\mathbf{v}_1 \quad \mathbf{w}_2 \quad \mathbf{w}_3 \quad \ldots \quad \mathbf{w}_n\,]$, such that $R^T AR = \left[\begin{array}{c|c} \lambda_1 & 0_{1,n-1} \\ \hline 0_{n-1,1} & A_1^* \end{array}\right]$.

Now consider A_1^*; since this is $(n-1)$ by $(n-1)$, by our hypothesis there is an orthonormal basis for $\mathbb{R}^{n-1}$ consisting of eigenvectors of A_1^*. Denote these eigenvectors by $\mathbf{v}_2^*, \mathbf{v}_3^*, \ldots, \mathbf{v}_n^*$, and the corresponding eigenvalues by

$\lambda_2, \lambda_3, \ldots, \lambda_n$, so that

$$A_1^* \mathbf{v}_j^* = \lambda_j \mathbf{v}_j^*, \quad \text{for } j = 2, 3, \ldots, n.$$

(Notice again that it is not required that the λ's be distinct.)

These eigenvectors, together with $\mathbf{v}_1$, essentially provide us with the n orthonormal eigenvectors required to diagonalize A, but there are two important technical issues to tidy up.

First, $\mathbf{v}_j^*$ is in $\mathbb{R}^{n-1}$, and eigenvectors of A must be in $\mathbb{R}^n$. To deal with this, for each vector $\mathbf{x}$ in $\mathbb{R}^{n-1}$, we define a corresponding vector in $\mathbb{R}^n$, which we denote by $(0, \mathbf{x})$:

$$\text{if } \mathbf{x} = (x_1, x_2, \ldots, x_{n-1}) \in \mathbb{R}^{n-1}, (0, \mathbf{x}) = (0, x_1, x_2, \ldots, x_{n-1}) \in \mathbb{R}^n.$$

Then the set of vectors $\{(0, \mathbf{v}_j^*) \mid j = 2, 3, \ldots, n\}$ is a set of vectors in $\mathbb{R}^n$. Moreover, because $\{\mathbf{v}_j^* \mid j = 2, 3, \ldots, n\}$ is an orthonormal set in $\mathbb{R}^{n-1}$, it can be shown that $\{(0, \mathbf{v}_j^*) \mid j = 2, 3, \ldots, n\}$ is an orthonormal set in $\mathbb{R}^n$.

The second issue is that the matrix $\left[\begin{array}{c|c} \lambda_1 & 0_{1,n-1} \\ \hline 0_{n-1,1} & A_1^* \end{array}\right]$ was determined with respect to the basis $\mathcal{B}$. Note that with respect to this basis, we have $(\mathbf{v}_1)_\mathcal{B} = (1, 0, 0, \ldots, 0)$.

It follows that $(\mathbf{v}_1)_\mathcal{B} \cdot (0, \mathbf{v}_j^*) = 0$, for $j = 2, 3, \ldots, n$. For this to make sense, we must interpret $(0, \mathbf{v}_j^*)$ as the *$\mathcal{B}$-coordinate vector* of a vector $\mathbf{v}_j$ in $\mathbb{R}^n$:

$$(\mathbf{v}_j)_\mathcal{B} = (0, \mathbf{v}_j^*), \quad j = 2, 3, \ldots, n.$$

Then $\{\mathbf{v}_1, \mathbf{v}_2, \mathbf{v}_3, \ldots, \mathbf{v}_n\}$ is an orthonormal basis for $\mathbb{R}^n$.

To conclude, we must show that each $\mathbf{v}_j$ is an eigenvector of A, for $j = 2, 3, \ldots, n$. We do this in terms of the $\mathcal{B}$-matrix of A and the $\mathcal{B}$-coordinates of $\mathbf{v}_j$: by block multiplication,

$$\left[\begin{array}{c|c} \lambda_1 & 0_{1,n-1} \\ \hline 0_{n-1,1} & A_1^* \end{array}\right] \left[\begin{array}{c} 0 \\ \hline \mathbf{v}_j^* \end{array}\right] = \lambda_j \left[\begin{array}{c} 0 \\ \hline \mathbf{v}_j^* \end{array}\right], \quad j = 2, 3, \ldots, n.$$

If we rewrite this information with respect to the standard basis, it says that

$$A\mathbf{v}_j = \lambda_j \mathbf{v}_j, \quad j = 2, 3, \ldots, n.$$

We have shown that $\{\mathbf{v}_1, \mathbf{v}_2, \ldots, \mathbf{v}_n\}$ is an orthonormal basis of $\mathbb{R}^n$ consisting of eigenvectors of A, so that the orthogonal matrix $R = \begin{bmatrix} \mathbf{v}_1 & \mathbf{v}_2 & \ldots & \mathbf{v}_n \end{bmatrix}$ diagonalizes A.

■

REMARKS

(1) Theorem 2 tells us that a symmetric matrix A has at least one real eigenvalue. By the Principal Axis Theorem, we know that A is diagonalized by an orthogonal change of basis. The resulting diagonal entries are real eigenvalues of A corresponding to the n orthonormal eigenvectors that appear as columns of the diagonalizing orthogonal matrix. Since we already have a basis of eigenvectors, there cannot be any other eigenvalues or eigenvectors (real or complex), so *all* eigenvalues of the symmetric matrix A are real.

(2) Theorem 4 asserts that for a symmetric matrix A, there is always an orthonormal basis of eigenvectors. From this fact, and from the discussion at the end of Section 6–3, it follows that for a k-times repeated characteristic root λ of A, there must be k orthonormal eigenvectors corresponding to λ. (The "geometric multiplicity" of λ is always equal to its "algebraic multiplicity" for a symmetric matrix A.)

(3) If you have read Section 4–7, you may find it interesting to restate some of the technical details of the proof of Theorem 4 in the language of isomorphisms. The correspondence in the proof between $\mathbf{x}$ in $\mathbb{R}^{n-1}$ and $(0, \mathbf{x})$ in $\mathbb{R}^n$ can be described as a linear mapping $inj : \mathbb{R}^{n-1} \to \mathbb{R}^n$ such that $inj(\mathbf{x}) = (0, \mathbf{x})$. Then it is straightforward to show that inj is a one-to-one mapping, and that it is an isomorphism from $\mathbb{R}^{n-1}$ onto the $(n-1)$-dimensional subspace $\text{Sp}(\{\mathbf{w}_2, \mathbf{w}_3, \ldots, \mathbf{w}_n\})$ of $\mathbb{R}^n$. Moreover, this mapping preserves inner products ($inj(\mathbf{x}) \cdot inj(\mathbf{y}) = \mathbf{x} \cdot \mathbf{y}$) — it is an "isometric" mapping — so the orthonormal eigenvectors of A_1^* are mapped to orthonormal eigenvectors of the original matrix.

Finally, we return to Theorem 2.

Proof of Theorem 2. *(Uses complex arithmetic.)*

If we allow complex roots (as in Section 6–5), then the characteristic equation of A certainly has at least one root λ by the Fundamental Theorem of Algebra. Then, with complex arithmetic, the matrix $(A - \lambda I)$ has rank less than n, and there exists at least one non-trivial solution $\mathbf{v}$. If λ is complex, then $\mathbf{v}$ is complex, and we may write it as $\mathbf{v} = \mathbf{u} + i\mathbf{w}$, where $\mathbf{u}$ and $\mathbf{w}$ are vectors in $\mathbb{R}^n$, and $i^2 = -1$. Let $\overline{\lambda}$ denote the complex conjugate of λ; by the standard rules of complex arithmetic, since A has real entries, $A\overline{\mathbf{v}} = \overline{\lambda}\overline{\mathbf{v}}$. Note that

$$\overline{\mathbf{v}} \cdot \mathbf{v} = (\mathbf{u} - i\mathbf{w}) \cdot (\mathbf{u} + i\mathbf{w}) = (\|\mathbf{u}\|^2 + \|\mathbf{w}\|^2),$$

so that $\overline{\mathbf{v}} \cdot \mathbf{v}$ is real.

Now consider

$$\overline{\mathbf{v}} \cdot A\mathbf{v} = \overline{\mathbf{v}} \cdot \lambda\mathbf{v} = \lambda(\overline{\mathbf{v}} \cdot \mathbf{v}) = \lambda(\|\mathbf{u}\|^2 + \|\mathbf{w}\|^2),$$

while

$$\mathbf{v} \cdot A\overline{\mathbf{v}} = \mathbf{v} \cdot \overline{\lambda}\overline{\mathbf{v}} = \overline{\lambda}(\mathbf{v} \cdot \overline{\mathbf{v}}) = \overline{\lambda}(\|\mathbf{u}\|^2 + \|\mathbf{w}\|^2).$$

Since A is symmetric, $\overline{\mathbf{v}} \cdot A\mathbf{v} = \mathbf{v} \cdot A\overline{\mathbf{v}}$, so $\lambda = \overline{\lambda}$, and this means that the eigenvalue λ must be real. A thus has at least one real eigenvalue, and the proof is complete.

It is also possible to give a proof based on calculus ideas. The quadratic form $Q(\mathbf{x}) = \mathbf{x}^T A\mathbf{x}$ is a continuous function of $\mathbf{x}$, and so when $\mathbf{x}$ is restricted to the unit hypersphere $\|\mathbf{x}\|^2 = \mathbf{x} \cdot \mathbf{x} = 1$, Q must have a maximum value, say λ, at some point, say $\mathbf{v}$, on the hypersphere. (This depends on the fact that a continuous function defined on a closed bounded set attains its maximum value.) Then by considering $Q(\mathbf{v} + \epsilon\mathbf{w}) \leq Q(\mathbf{v})$ for arbitrary $\mathbf{w}$ and for all positive ϵ, or by using partial derivatives and the Lagrange Multiplier Theorem, one can show that the vector $\mathbf{v}$ must be an eigenvector of A, with eigenvalue λ. We omit the details.

■

EXERCISES 8–1

A1 For each of the following symmetric matrices, (i) find a corresponding diagonal matrix, and (ii) find an orthogonal matrix R that diagonalizes the given matrix.

(a) $A = \begin{bmatrix} 4 & 2 \\ 2 & 1 \end{bmatrix}$ (b) $B = \begin{bmatrix} 5 & 3 \\ 3 & -3 \end{bmatrix}$ (c) $C = \begin{bmatrix} 0 & 1 & 1 \\ 1 & 0 & 1 \\ 1 & 1 & 0 \end{bmatrix}$

(d) $D = \begin{bmatrix} 1 & 0 & -2 \\ 0 & -1 & -2 \\ -2 & -2 & 0 \end{bmatrix}$ (e) $E = \begin{bmatrix} 1 & 8 & 4 \\ 8 & 1 & -4 \\ 4 & -4 & 7 \end{bmatrix}$

B1 For each of the following symmetric matrices, (i) find a corresponding diagonal matrix, and (ii) find an orthogonal matrix R that diagonalizes the given matrix.

(a) $A = \begin{bmatrix} 5 & 2 \\ 2 & 2 \end{bmatrix}$ (b) $B = \begin{bmatrix} 7 & 3 \\ 3 & -1 \end{bmatrix}$ (c) $C = \begin{bmatrix} 0 & 1 & -1 \\ 1 & 0 & 1 \\ -1 & 1 & 0 \end{bmatrix}$

(d) $D = \begin{bmatrix} 1 & 0 & -1 \\ 0 & 1 & 1 \\ -1 & 1 & 2 \end{bmatrix}$ (e) $E = \begin{bmatrix} 1 & 2 & -4 \\ 2 & -2 & -2 \\ -4 & -2 & 1 \end{bmatrix}$

COMPUTER RELATED EXERCISES

C1 Use a computer to determine the eigenvalues of the following matrices.

(a) The matrix of A1(e).

(b) $\begin{bmatrix} 4.1 & 1.9 & 0.5 \\ 1.9 & 1.2 & 0.6 \\ 0.5 & 0.6 & -2.1 \end{bmatrix}$ (c) $\begin{bmatrix} 0.15 & 0.05 & 0.95 & 0.25 \\ 0.05 & 0.15 & 0.25 & 0.95 \\ 0.95 & 0.25 & 0.15 & 0.05 \\ 0.25 & 0.95 & 0.05 & 0.15 \end{bmatrix}$

C2 Let $S(t)$ denote the symmetric matrix

$$S(t) = \begin{bmatrix} 2+t & -2t & t \\ -2t & 2-t & -t \\ t & -t & 1+t \end{bmatrix}.$$

By calculating the eigenvalues of $S(-0.1)$, $S(-0.05)$, $S(0)$, $S(0.05)$, $S(0.1)$, explore how the eigenvalues of $S(t)$ change as t varies.

CONCEPTUAL EXERCISES

***D1** Show that if A and B are symmetric matrices of the same size, then $A+B$ is also symmetric. Is AB symmetric? A^2?

D2 Show that if A is orthogonally diagonable (that is, if there is a diagonal matrix D and an orthogonal matrix R such that $R^T AR = D$), then A is symmetric.

SECTION

8–2 Quadratic Forms

Expressions such as $2x^2 - 4xy - y^2$ or $x^2 + 8xy + 11y^2$ are called **quadratic forms** on $\mathbb{R}^2$ (or in the variables (x, y)). Quadratic forms appear in geometry through equations such as $2x^2 - 4xy - y^2 = 1$ (see Section 8–3). They also appear in calculus as the second degree terms in Taylor expansions (see Section 8–4). A formal definition of quadratic forms on $\mathbb{R}^n$ will be given below, but first we discuss the connection between quadratic forms and matrices.

Consider the symmetric matrix $A = \begin{bmatrix} a & b \\ b & c \end{bmatrix}$. Using the notation of Section 8–1, we observe that if $\mathbf{x} = \begin{bmatrix} x \\ y \end{bmatrix}$, then

$$\begin{aligned} \mathbf{x}^T A\mathbf{x} &= \begin{bmatrix} x & y \end{bmatrix} \begin{bmatrix} a & b \\ b & c \end{bmatrix} \begin{bmatrix} x \\ y \end{bmatrix} \\ &= \begin{bmatrix} x & y \end{bmatrix} \begin{bmatrix} ax + by \\ bx + cy \end{bmatrix} \\ &= (ax^2 + bxy) + (bxy + cy^2) = ax^2 + 2bxy + cy^2. \end{aligned}$$

Thus, corresponding to the symmetric matrix A, there is a corresponding quadratic form

$$Q(\mathbf{x}) = ax^2 + 2bxy + cy^2.$$

Given the quadratic form $Q(\mathbf{x}) = ax^2 + 2bxy + cy^2$, we reconstruct the symmetric matrix $A = \begin{bmatrix} a & b \\ b & c \end{bmatrix}$ by placing the coefficient of x^2 as the top left entry, the coefficient of y^2 as the bottom right entry, and *half* the coefficient of xy as the entry in the first row, second column, and *half* in the second row, first column. We deal with the coefficient of xy this way to ensure that A is symmetric.

EXAMPLE 3

Determine the symmetric matrix corresponding to the quadratic form

$$Q(\mathbf{x}) = 2x^2 - 4xy - y^2.$$

SOLUTION: The matrix is $\begin{bmatrix} 2 & -2 \\ -2 & -1 \end{bmatrix}$. Notice that half of the coefficient of

xy appears as the entry in the first row, second column of this matrix, and half appears as the entry in the second row, first column.

Notice that we could have written $ax^2+2bxy+cy^2$ in terms of other *unsymmetric* matrices. For example,

$$ax^2 + 2bxy + cy^2 = \mathbf{x}^T \begin{bmatrix} a & 2b \\ 0 & c \end{bmatrix} \mathbf{x} = \mathbf{x}^T \begin{bmatrix} a & 3b \\ -b & c \end{bmatrix} \mathbf{x}.$$

Many choices are possible. However, we agree always to choose the *symmetric* matrix, as in Example 3, for two reasons. First, it gives us a unique (symmetric) matrix corresponding to a given quadratic form. Second, the choice of the symmetric matrix A allows us to apply the special theory available for symmetric matrices. (A third reason is a connection to general inner products discussed in Exercise D2.)

DEFINITION

A **quadratic form** on $\mathbb{R}^n$, with corresponding **symmetric** matrix A, is a real-valued function $Q : \mathbb{R}^n \to \mathbb{R}$ defined by

$$\begin{aligned} Q(\mathbf{x}) = \sum_{j,k=1}^{n} a_{jk}x_jx_k &= a_{11}x_1^2 + a_{12}x_1x_2 + \cdots + a_{1n}x_1x_n \\ &\quad + a_{21}x_2x_1 + a_{22}x_2^2 + \cdots + a_{2n}x_2x_n + \cdots \\ &\quad + a_{n1}x_nx_1 + a_{n2}x_nx_2 + \cdots + a_{nn}x_n^2 \\ &= \sum_{j=1}^{n} a_{jj}x_j^2 + \sum_{\substack{j,k=1 \\ j<k}}^{n} 2a_{jk}x_jx_k. \end{aligned}$$

It is straightforward to check that this can be written in terms of the dot product:

$$\begin{aligned} Q(\mathbf{x}) &= \mathbf{x} \cdot A\mathbf{x} = \mathbf{x}^T A\mathbf{x} \\ &= [\,x_1 \quad x_2 \quad \ldots \quad x_n\,] \begin{bmatrix} a_{11}x_1 + a_{12}x_2 + \cdots + a_{1n}x_n \\ a_{21}x_1 + a_{22}x_2 + \cdots + a_{2n}x_n \\ \cdots \\ \cdots \\ a_{n1}x_1 + a_{n2}x_2 + \cdots + a_{nn}x_n \end{bmatrix}. \end{aligned}$$

A quadratic form $Q(\mathbf{x})$ is in **diagonal form** if all the coefficients a_{jk} with $j \neq k$ are equal to zero. Equivalently, $Q(\mathbf{x})$ is in diagonal form if it is of the

form

$$Q(\mathbf{x}) = a_{11}x_1^2 + a_{22}x_2^2 + \cdots + a_{nn}x_n^2.$$

This means that $A = \begin{bmatrix} a_{11} & 0 & 0 & \dots & 0 \\ 0 & a_{22} & 0 & \dots & 0 \\ 0 & 0 & a_{33} & \dots & 0 \\ \vdots & \vdots & \vdots & \ddots & \vdots \\ 0 & \dots & \dots & \dots & a_{nn} \end{bmatrix}$. In this case, we see immediately that $a_{11}, a_{22}, \dots, a_{nn}$ are the eigenvalues of A. (Some of these eigenvalues may be zero.)

THEOREM 1. Let $Q(\mathbf{x}) = \mathbf{x}^T A\mathbf{x}$ be a quadratic form on $\mathbb{R}^n$. Then there is an orthonormal basis $\mathcal{B}$ such that when $Q(\mathbf{x})$ is expressed in terms of $\mathcal{B}$-coordinates, $Q(\mathbf{x})$ is in diagonal form. Moreover, the orthogonal matrix R whose columns are the basis vectors in $\mathcal{B}$ is the matrix that diagonalizes the symmetric matrix A, and the coefficients in the diagonal version of $Q(\mathbf{x})$ are the eigenvalues of A.

Proof. For simplicity, we give the proof in the case $n = 3$. Since we need to indicate the basis with respect to which coordinates are calculated, we use the bracket notation for column vectors.

Since A is symmetric, it has three real eigenvalues; suppose that they are $\lambda_1, \lambda_2, \lambda_3$ (with repetition permitted so that possibly $\lambda_j = \lambda_{j+1}$). Suppose that the orthogonal matrix that diagonalizes A is R, so that $R^T AR$ is diagonal with $\lambda_1, \lambda_2, \lambda_3$ as the entries on the diagonal. As usual, denote the coordinates with respect to the new basis of eigenvectors by $(\mathbf{x})_{\mathcal{B}} = (\tilde{x}_1, \tilde{x}_2, \tilde{x}_3)$, and recall from Section 4–6 that the change of coordinates equation is $[\mathbf{x}]_{\mathcal{S}} = R[\mathbf{x}]_{\mathcal{B}}$. Then

$$\begin{aligned} Q(\mathbf{x}) = [\mathbf{x}]_{\mathcal{S}}^T A[\mathbf{x}]_{\mathcal{S}} = \left[R[\mathbf{x}]_{\mathcal{B}}\right]^T AR[\mathbf{x}]_{\mathcal{B}} &= [\mathbf{x}]_{\mathcal{B}}^T R^T AR[\mathbf{x}]_{\mathcal{B}} \\ &= [\,\tilde{x}_1 \quad \tilde{x}_2 \quad \tilde{x}_3\,] \begin{bmatrix} \lambda_1 & 0 & 0 \\ 0 & \lambda_2 & 0 \\ 0 & 0 & \lambda_3 \end{bmatrix} \begin{bmatrix} \tilde{x}_1 \\ \tilde{x}_2 \\ \tilde{x}_3 \end{bmatrix} \\ &= \lambda_1\tilde{x}_1^2 + \lambda_2\tilde{x}_2^2 + \lambda_3\tilde{x}_3^2. \end{aligned}$$

Thus we see that the same change of basis that diagonalizes A also diagonalizes the corresponding quadratic form $Q(\mathbf{x})$.

■

EXAMPLE 4

Diagonalize the quadratic form from Example 3,

$$Q(\mathbf{x}) = 2x^2 - 4xy - y^2.$$

SOLUTION: The symmetric matrix of the quadratic form is

$$A = \begin{bmatrix} 2 & -2 \\ -2 & -1 \end{bmatrix}.$$

Then $A - \lambda I = \begin{bmatrix} 2-\lambda & -2 \\ -2 & -1-\lambda \end{bmatrix}$, so the characteristic equation is

$$\lambda^2 - \lambda - 6 = (\lambda - 3)(\lambda + 2) = 0,$$

and the eigenvalues are 3 and -2. Hence $Q(\mathbf{x}) = 3\tilde{x}^2 - 2\tilde{y}^2$ (or, if we choose to put the eigenvalues in the other order, $Q(\mathbf{x}) = -2\tilde{x}^2 + 3\tilde{y}^2$). To find the diagonalizing matrix R (or, equivalently, to see how the new coordinate axes are related to the standard axes), we must find the eigenvectors of A.

When $\lambda = 3$,

$$A - \lambda I = \begin{bmatrix} -1 & -2 \\ -2 & -4 \end{bmatrix} \rightarrow \begin{bmatrix} 1 & 2 \\ 0 & 0 \end{bmatrix},$$

so a corresponding eigenvector is $(2, -1)$.

When $\lambda = -2$,

$$A - \lambda I = \begin{bmatrix} 4 & -2 \\ -2 & 1 \end{bmatrix} \rightarrow \begin{bmatrix} 2 & -1 \\ 0 & 0 \end{bmatrix},$$

so a corresponding eigenvector is $(1, 2)$. Note that the two eigenvectors are mutually orthogonal as expected.

An orthogonal matrix that diagonalizes A is thus $R = \frac{1}{\sqrt{5}}\begin{bmatrix} 2 & 1 \\ -1 & 2 \end{bmatrix}$.

You may find it instructive to substitute the corresponding change of coordinate equations

$$x = \frac{1}{\sqrt{5}}(2\tilde{x} + \tilde{y}), \quad y = \frac{1}{\sqrt{5}}(-\tilde{x} + 2\tilde{y})$$

into the expression for $Q(\mathbf{x})$ to verify that

$$2x^2 - 4xy - y^2 = 3\tilde{x}^2 - 2\tilde{y}^2.$$

DEFINITION

A quadratic form $Q(\mathbf{x})$ on $\mathbb{R}^n$ is **positive definite** if $Q(\mathbf{x}) > 0$ for all non-zero $\mathbf{x}$ in $\mathbb{R}^n$. It is negative definite if $Q(\mathbf{x}) < 0$ for all non-zero $\mathbf{x}$.

These concepts are useful in applications. For example, we shall see in Section 8–3 that the graph of $Q(\mathbf{x}) = 1$ in $\mathbb{R}^2$ is an ellipse if and only if $Q(\mathbf{x})$ is positive definite, and in Section 8–4 that a critical point of a function $f : \mathbb{R}^n \to \mathbb{R}$ gives a local minimum value if the quadratic terms in the Taylor expansion form a positive definite quadratic form.

THEOREM 2. $Q(\mathbf{x}) = \mathbf{x}^T A\mathbf{x}$ is positive definite if and only if all eigenvalues of A are positive.

Proof. This follows immediately from Theorem 1. ■

REMARKS

(1) Many quadratic forms are neither positive definite nor negative definite. For example, on $\mathbb{R}^2$, $x^2 = x^2 + 0y^2$, and $x^2 - 2y^2$ are neither positive definite nor negative definite.

(2) Because of Theorem 2, we say that a symmetric matrix is positive definite if all its eigenvalues are positive.

EXERCISES 8–2

A1 For each quadratic form $Q(\mathbf{x})$, determine the corresponding symmetric matrix A. By diagonalizing A, express $Q(\mathbf{x})$ in diagonal form. Give an orthogonal matrix R that diagonalizes A. Determine whether each quadratic form is positive definite.

(a) $Q(\mathbf{x}) = x^2 - 3xy + y^2$　　(b) $Q(\mathbf{x}) = 5x^2 - 4xy + 2y^2$

(c) $Q(\mathbf{x}) = x^2 + y^2 - 3z^2 - 2xy + 6xz + 6yz$

B1 Same instructions as in A1.

(a) $Q(\mathbf{x}) = 7x^2 + 4xy + 4y^2$ (b) $Q(\mathbf{x}) = 2x^2 + 6xy + 2y^2$

(c) $Q(\mathbf{x}) = 3x^2 + 5y^2 + 3z^2 - 2xy - 2xz + 2yz$

(d) $Q(\mathbf{x}) = 2x^2 + 2y^2 - 3z^2 - 4xy + 6xz + 6yz$

COMPUTER RELATED EXERCISES

C1 Decide whether the following quadratic forms are positive definite, negative definite, or neither, by determining eigenvalues for the corresponding matrix with the help of a computer.

(a) $Q(\mathbf{x}) = -0.1x_1^2+2.1x_2^2+1.3x_3^2+1.1x_4^2-0.8x_1x_2+1.2x_1x_4+1.6x_2x_3+4.2x_3x_4$

(b) $Q(\mathbf{x}) = 0.85(x_1^2 + x_2^2 + x_3^2 + x_4^2) - 0.1x_1x_2 + 0.6x_1x_3 + 0.2x_1x_4 + 0.2x_2x_3$
$+0.6x_2x_4 - 0.1x_3x_4$

CONCEPTUAL EXERCISES

***D1** Given a square matrix A, define the **symmetric part** of A to be

$$A^+ = \frac{1}{2}(A + A^T),$$

and the **skew-symmetric** part of A to be

$$A^- = \frac{1}{2}(A - A^T).$$

(A^- is sometimes called the anti-symmetric part.)

(a) Verify that A^+ is symmetric, A^- is skew-symmetric (which means $(A^-)^T = -A^-$) and $A = A^+ + A^-$.

(b) Determine expressions for the typical entries $(A^+)_{jk}$ and $(A^-)_{jk}$ in terms of entries in A. Note that the diagonal entries of A^- must be 0.

(c) Prove that for every $\mathbf{x}$ in $\mathbb{R}^n$,

$$\mathbf{x}^T A\mathbf{x} = \mathbf{x}^T A^+\mathbf{x}.$$

Do this in two ways. (i) For the case where A is 2 by 2 and 3 by 3, calculate explicit expressions in terms of the coordinates of $\mathbf{x}$. (ii) Use the fact that $A = A^+ + A^-$, and prove that $\mathbf{x}^T A^-\mathbf{x} = 0$. (Note that the result of part (c) is a further explanation of why we choose the *symmetric* matrix corresponding to a quadratic form.)

D2 In Section 7–4, we said that general inner products on $\mathbb{R}^n$ were not different in interesting ways from the standard inner product. This exercise explains that statement.

Let $\langle\ ,\ \rangle$ be an inner product on $\mathbb{R}^n$, and let $S = \{\mathbf{e}_1, \mathbf{e}_2, \ldots, \mathbf{e}_n\}$ be the standard basis.

(a) Verify that for any $\mathbf{x}$ and $\mathbf{y}$ in $\mathbb{R}^n$,

$$\langle \mathbf{x}, \mathbf{y} \rangle = \sum_{j,k=1}^{n} x_j y_k \langle \mathbf{e}_j, \mathbf{e}_k \rangle.$$

(b) Let G be the n by n matrix defined by $g_{jk} = \langle \mathbf{e}_j, \mathbf{e}_k \rangle$ and verify that

$$\langle \mathbf{x}, \mathbf{y} \rangle = \mathbf{x}^T G \mathbf{y}.$$

(c) Use the properties of an inner product to verify that G is symmetric and positive definite. (A symmetric matrix is positive definite if the corresponding quadratic form is positive definite.)

(d) By adapting the proof of Theorem 1, show that there is a basis $\mathcal{B} = \{\mathbf{v}_1, \mathbf{v}_2, \ldots, \mathbf{v}_n\}$ such that in $\mathcal{B}$-coordinates,

$$\langle \mathbf{x}, \mathbf{y} \rangle = \lambda_1 \tilde{x}_1 \tilde{y}_1 + \lambda_2 \tilde{x}_2 \tilde{y}_2 + \cdots + \lambda_n \tilde{x}_n \tilde{y}_n,$$

where $\lambda_1, \lambda_2, \ldots, \lambda_n$ are the eigenvalues of G. In particular,

$$\langle \mathbf{x}, \mathbf{x} \rangle = \|\mathbf{x}\|^2 = \sum_{j=1}^{n} \lambda_j \tilde{x}_j^2.$$

> *REMARK*
> In a formula such as this for the square of the norm, the numbers $\lambda_1, \lambda_2, \ldots, \lambda_n$ are sometimes called *weights*, and the sum is described as a "weighted sum of squares."

(e) Introduce a new basis $\mathcal{C} = \{\mathbf{w}_1, \mathbf{w}_2, \ldots, \mathbf{w}_n\}$ by defining $\mathbf{w}_j = \mathbf{v}_j / \sqrt{\lambda_j}$. Use an asterisk to denote $\mathcal{C}$-coordinates, so that $\mathbf{x} = x_1^* \mathbf{w}_1 + x_2^* \mathbf{w}_2 + \cdots + x_n^* \mathbf{w}_n$. Verify that

$$\langle \mathbf{w}_j, \mathbf{w}_k \rangle = \begin{cases} 1 & \text{if } j = k \\ 0 & \text{if } j \neq k \end{cases}$$

and that

$$\langle \mathbf{x}, \mathbf{y} \rangle = x_1^* y_1^* + x_2^* y_2^* + \cdots + x_n^* y_n^*.$$

Thus, with respect to the inner product $\langle\ ,\ \rangle$, $\mathcal{C}$ is an orthonormal basis, and in $\mathcal{C}$-coordinates, the inner product of two vectors looks just like the standard dot product.

The Graphs of Quadratic Equations

In $\mathbb{R}^2$, it is often of interest to know the graph of an equation of the form $Q(\mathbf{x}) = k$, where $Q(\mathbf{x})$ is a quadratic form and k is a constant. If we were only interested in one or two particular graphs, it might be sensible to simply use a computer to produce the graphs. However, by applying diagonalization to the problem of determining these graphs, we see a very clear interpretation of eigenvectors; we also consider a concrete useful application of change of coordinates. Moreover, this approach to these graphs leads to a *classification* of the various possibilities; all of the graphs of the form $Q(\mathbf{x}) = k$ in $\mathbb{R}^2$ can be divided into a few standard cases. Classification is a useful process because it allows us to say: "I really only need to understand these few standard cases." A classification of these graphs is given after the next two examples.

EXAMPLE 5

Sketch the graph of the equation $3x^2 + 4xy = 16$.

SOLUTION: The quadratic form $Q(\mathbf{x}) = 3x^2 + 4xy$ corresponds to the symmetric matrix $A = \begin{bmatrix} 3 & 2 \\ 2 & 0 \end{bmatrix}$. Then $A - \lambda I = \begin{bmatrix} 3-\lambda & 2 \\ 2 & 0-\lambda \end{bmatrix}$, so the characteristic equation is $\lambda^2 - 3\lambda - 4 = 0$, and the eigenvalues are 4 and -1. Thus, by an orthogonal change of basis, the equation can be brought into the form

$$4\tilde{x}^2 - \tilde{y}^2 = 16, \quad \text{or} \quad \frac{\tilde{x}^2}{4} - \frac{\tilde{y}^2}{16} = 1.$$

This is the equation of a hyperbola, and we can sketch the graph with respect to $\tilde{x}$- and $\tilde{y}$-axes. We observe that there are $\tilde{x}$-intercepts at $(2,0)$ and $(-2,0)$, and that there are no intercepts on the $\tilde{y}$-axis. The asymptotes of the hyperbola are determined by the equation $4\tilde{x}^2 - \tilde{y}^2 = 0$. By factoring, we determine that the asymptotes are the lines with equations $2\tilde{x} - \tilde{y} = 0$ and $2\tilde{x} + \tilde{y} = 0$. With this information, we obtain the graph in Figure 8–3–1.

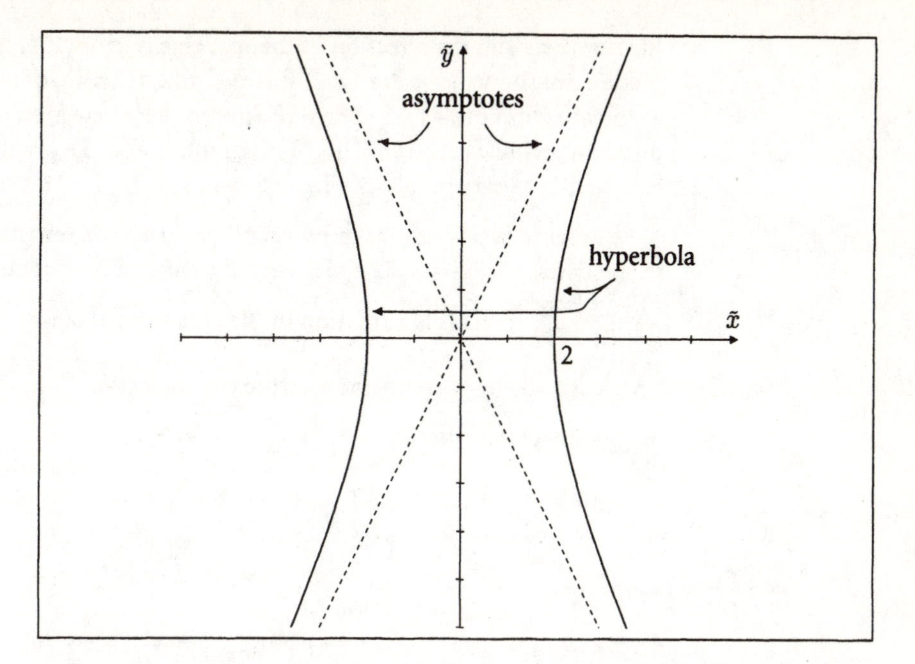

FIGURE 8-3-1. *The graph of* $4\tilde{x}^2 - \tilde{y}^2 = 16$, *shown with horizontal* $\tilde{x}$*-axis, and asymptotes.*

However, we want a picture of the graph of $3x^2 + 4xy = 16$ relative to the original x- and y-axes. We need to know how the new axes are related to these original axes, so we must find the eigenvectors of A.

When $\lambda = 4$,

$$A - \lambda I = \begin{bmatrix} -1 & 2 \\ 2 & -4 \end{bmatrix} \to \begin{bmatrix} 1 & -2 \\ 0 & 0 \end{bmatrix},$$

so a corresponding eigenvector is $(2, 1)$. These are the standard coordinates of the vector determining the new $\tilde{x}$-axis (since the eigenvalue 4 is associated with this eigenvector).

When $\lambda = -1$,

$$A - \lambda I = \begin{bmatrix} 4 & 2 \\ 2 & 1 \end{bmatrix} \to \begin{bmatrix} 2 & 1 \\ 0 & 0 \end{bmatrix},$$

so a corresponding eigenvector is $(-1, 2)$. Note that this eigenvector is orthogonal to the one for $\lambda = 4$; this was expected because A is symmetric. The new $\tilde{y}$-axis is determined by this vector. (We could have chosen $(1, -2)$, but $(-1, 2)$ is better because the basis $\{(2, 1), (-1, 2)\}$ is right-handed.)

Now we sketch the graph of $3x^2 + 4xy = 16$. In the standard xy-plane, draw the new $\tilde{x}$-axis in the direction of the first eigenvector $(2, 1)$; for clarity in the picture, we show the vector $(4, 2)$ instead of $(2, 1)$. Also, draw the new $\tilde{y}$-axis in the direction of $(-1, 2)$. Then, relative to these new axes, sketch the graph of the hyperbola $4\tilde{x}^2 - \tilde{y}^2 = 16$. This graph is also the graph of the original equation $3x^2 + 4xy = 16$. See Figure 8–3–2.

In order to include the asymptotes in the sketch, we rewrite their equations in standard coordinates. The orthogonal change of basis matrix in this case is $\frac{1}{\sqrt{5}}\begin{bmatrix} 2 & -1 \\ 1 & 2 \end{bmatrix}$. (This is a rotation of the axes through angle $\theta = 0.46$ radians, since $\cos\theta = \frac{2}{\sqrt{5}}$ — see the note at the end of Section 7–2.) Thus the change of coordinates equation is

$$\begin{bmatrix} \tilde{x} \\ \tilde{y} \end{bmatrix} = \frac{1}{\sqrt{5}}\begin{bmatrix} 2 & 1 \\ -1 & 2 \end{bmatrix}\begin{bmatrix} x \\ y \end{bmatrix},$$

or

$$\tilde{x} = \frac{1}{\sqrt{5}}(2x + y) \quad \text{and} \quad \tilde{y} = \frac{1}{\sqrt{5}}(-x + 2y).$$

Then one asymptote is

$$0 = 2\tilde{x} + \tilde{y} = \frac{2}{\sqrt{5}}(2x + y) + \frac{1}{\sqrt{5}}(-x + 2y) = \frac{1}{\sqrt{5}}(3x + 4y),$$

and the other is

$$0 = 2\tilde{x} - \tilde{y} = \frac{2}{\sqrt{5}}(2x + y) - \frac{1}{\sqrt{5}}(-x + 2y) = \frac{1}{\sqrt{5}}(5x).$$

Thus, in standard coordinates, the asymptotes are $3x + 4y = 0$ and $x = 0$. The original y-axis turns out to be an asymptote of this graph; with hindsight, one can see that this must be the case by considering points on the graph with $x = 0$.

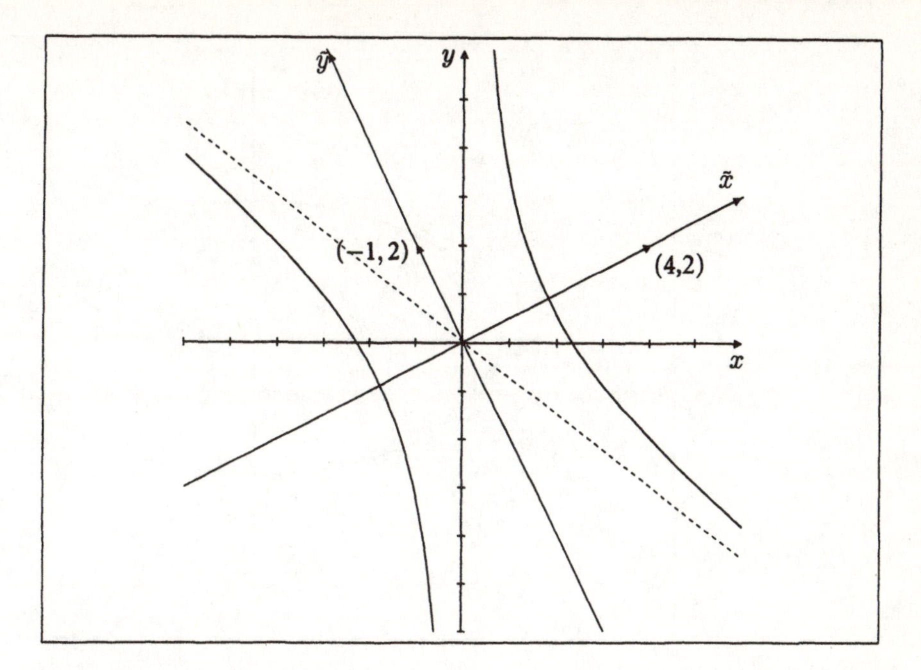

FIGURE 8-3-2. *The graph of* $3x^2 + 4xy = 16 \Leftrightarrow 4\bar{x}^2 - \bar{y} = 16$. *Note that the* y*-axis is an asymptote.*

EXAMPLE 6

Sketch the graph of the equation $6x^2 + 4xy + 3y^2 = 14$.

SOLUTION: The corresponding symmetric matrix is $\begin{bmatrix} 6 & 2 \\ 2 & 3 \end{bmatrix}$. The eigenvalues are 2 and 7; an eigenvector corresponding to $\lambda = 2$ is $(1, -2)$, and an eigenvector corresponding to $\lambda = 7$ is $(2, 1)$. If $(1, -2)$ is taken to define the $\bar{x}$-axis, and $(2, 1)$ to define the $\bar{y}$-axis, then the original equation is equivalent to

$$2\bar{x}^2 + 7\bar{y}^2 = 14, \quad \text{or} \quad \frac{\bar{x}^2}{7} + \frac{\bar{y}^2}{2} = 1.$$

This is the equation of an ellipse with $\bar{x}$-intercepts $(\sqrt{7}, 0)$ and $(-\sqrt{7}, 0)$, and $\bar{y}$-intercepts $(0, \sqrt{2})$ and $(0, -\sqrt{2})$.

In Figure 8–3–3, the ellipse is shown relative to the $\bar{x}$- and $\bar{y}$-axes. In Figure 8–3–4, the new $\bar{x}$- and $\bar{y}$-axes determined by the eigenvectors are shown relative to the standard axes, and the ellipse from Figure 8–3–3 is rotated into place. The resulting ellipse is the graph of $6x^2 + 4xy + 3y^2 = 14$.

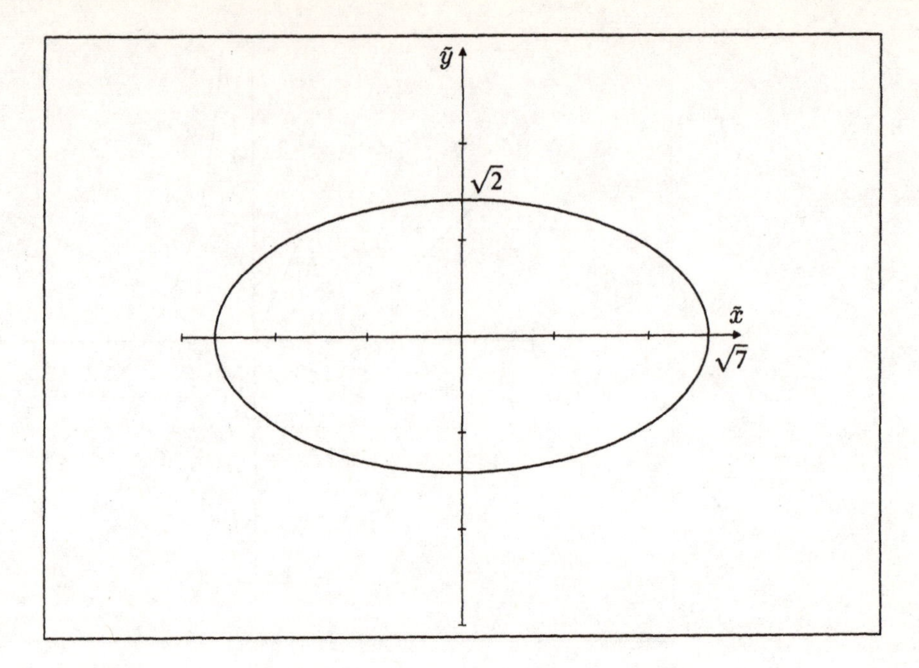

FIGURE 8-3-3. *The graph of* $2\tilde{x}^2 + 7\tilde{y}^2 = 14$, *shown with horizontal* $\tilde{x}$*-axis.*

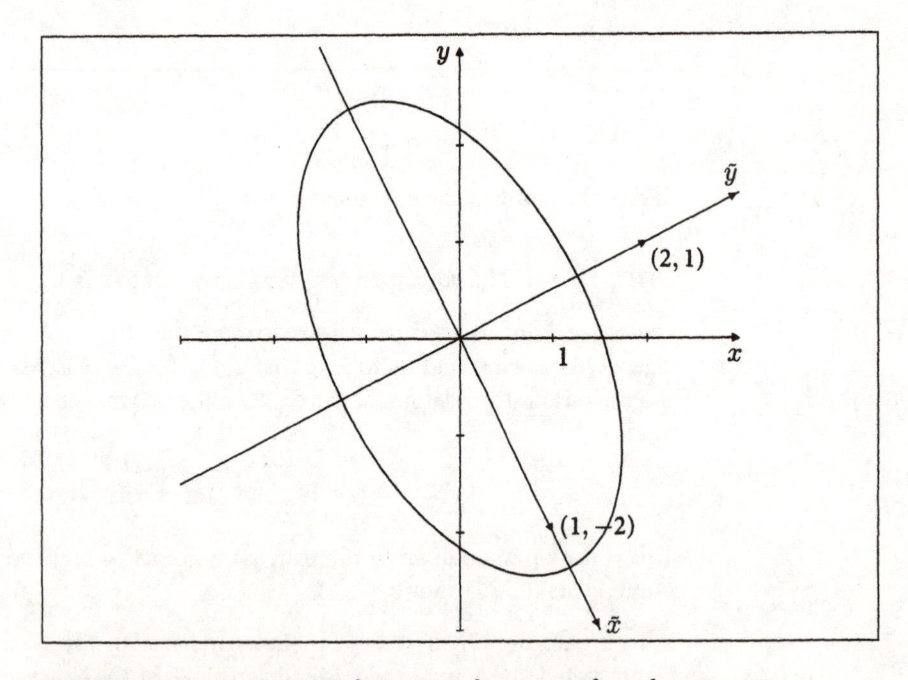

FIGURE 8-3-4. *The graph of* $6x^2 + 4xy + 3y^2 = 14 \Leftrightarrow 2\tilde{x}^2 + 7\tilde{y}^2 = 14$.

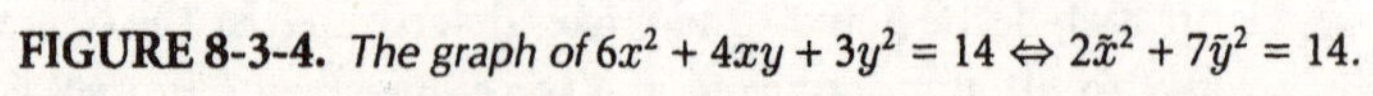

To classify all the graphs of equations of the form $Q(\mathbf{x}) = k$, we diagonalize, and rewrite the equation in the form $\lambda_1 x^2 + \lambda_2 y^2 = k$. Here, λ_1 and λ_2 are the eigenvalues associated with the quadratic form $Q(\mathbf{x})$, and the tildes over the new coordinates x and y are omitted for simplicity. It is sufficient to consider the cases $k = 1, 0$, or -1, because if $k > 0$, the graph of $\lambda_1 x^2 + \lambda_2 y^2 = k$ is just the graph of $\lambda_1 x^2 + \lambda_2 y^2 = 1$ enlarged by the dilatation with factor $\sqrt{k}$. The distinct possibilities are displayed in the following table.

TABLE I. *Graphs of* $\lambda_1 x^2 + \lambda_2 y^2 = k$.

	$k = 1$	$k = 0$	$k = -1$
$\lambda_1 > 0,\ \lambda_2 > 0$	ellipse	point $(0,0)$	empty set
$\lambda_1 > 0,\ \lambda_2 = 0$	parallel lines $x = \pm 1/\sqrt{\lambda_1}$	line $x = 0$	empty set
$\lambda_1 > 0,\ \lambda_2 < 0$	hyperbola $\lambda_1 x^2 - \lvert\lambda_2\rvert y^2 = 1$	intersecting lines $\sqrt{\lambda_1}x = \sqrt{\lvert\lambda_2\rvert}y$	hyperbola $\lambda_1 x^2 - \lvert\lambda_2\rvert y^2 = -1$
$\lambda_1 = 0,\ \lambda_2 < 0$	empty set	line $y = 0$	parallel lines $y = \pm 1/\sqrt{\lvert\lambda_2\rvert}$
$\lambda_1 < 0,\ \lambda_2 < 0$	empty set	point $(0,0)$	ellipse

The cases where $k = 0$ or one eigenvalue is zero may be regarded as *degenerate* (not general) cases. The *nondegenerate* cases are the ellipses and hyperbolas, which are among the **conic sections** — curves obtained in $\mathbb{R}^3$ as the intersection of a cone and a plane. Notice that the cases of a single point, a single line, or intersecting lines can also be obtained as the intersection of a cone and a plane passing through the vertex of the cone. However, the cases of parallel lines (in the table) are not obtained as the intersection of a cone and a plane.

It is also important to realize that one class of conic sections, the parabolas, does not appear in our table. In $\mathbb{R}^2$, the equation of a parabola is a quadratic equation, but it contains first-degree terms (for example, $y^2 - 4x = 0$). Since the quadratic form $Q(\mathbf{x})$ contains only second-degree terms, an equation of the form $Q(\mathbf{x}) = k$ cannot be the equation of a parabola.

The classification provided by the table suggests that it might be interesting to consider how degenerate cases arise as limiting cases of nondegenerate cases. For example, Figure 8–3–5 shows that the case of parallel lines ($y = \pm$constant) arises from the family of ellipses $\lambda x^2 + y^2 = 1$ as λ becomes 0.

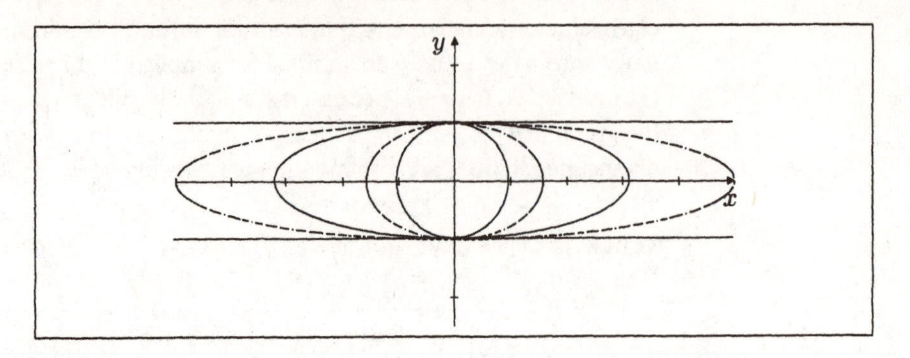

FIGURE 8-3-5. *A family of ellipses $\lambda x^2 + y^2 = 1$. The circle occurs for $\lambda = 1$; as λ decreases, the ellipse gets "fatter"; for $\lambda = 0$, the graph is a pair of lines.*

Table I could have been constructed using only the cases $k = 1$ and $k = 0$. The graphs obtained for $k = -1$ are all obtained also for $k = 1$, although they may be oriented differently. For example, the graph of $x^2 - y^2 = -1$ is the same as the graph of $-x^2 + y^2 = 1$, and this hyperbola may be obtained from the hyperbola $x^2 - y^2 = 1$ by the reflection $\bar{x} = y$, $\bar{y} = x$. However, for purposes of illustration, it is convenient to include both $k > 0$ and $k < 0$. Figure 8–3–6

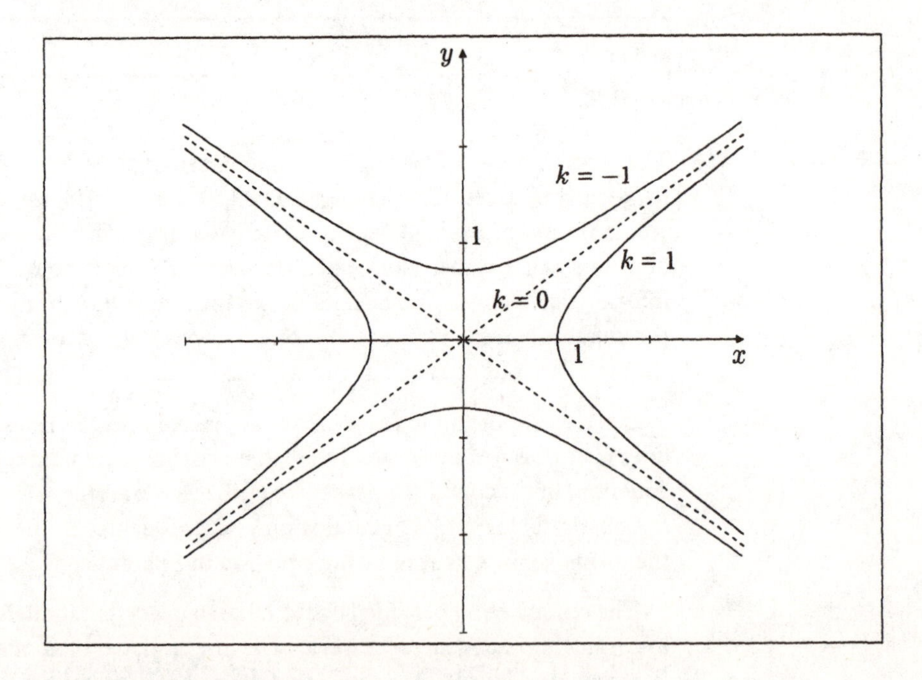

FIGURE 8-3-6. *The graphs of $x^2 - 2y^2 = k$ for $k = 1, 0, -1$.*

shows that the case of intersecting lines ($k = 0$) separates the hyperbolas with intercepts on the x-axis ($x^2 - 2y^2 = k > 0$) from the hyperbolas with intercepts on the y-axis ($x^2 - 2y^2 = k < 0$).

Graphs of $Q(\mathbf{x}) = k$ in $\mathbb{R}^3$

For a quadratic equation of the form $Q(\mathbf{x}) = k$ in $\mathbb{R}^3$, there are similar results — but because there are three variables instead of two, there are more possibilities. The nondegenerate cases give *ellipsoids, hyperboloids of one sheet, and hyperboloids of two sheets.* The graphs are called *quadric surfaces.*

The usual standard form for an ellipsoid is $\frac{x^2}{a^2} + \frac{y^2}{b^2} + \frac{z^2}{c^2} = 1$. This is the case obtained by diagonalizing $Q(\mathbf{x}) = k$ if the eigenvalues and k all have the same sign, and if we write

$$\lambda_1/k = 1/a^2, \quad \lambda_2/k = 1/b^2, \quad \lambda_3/k = 1/c^2.$$

An ellipsoid is shown in Figure 8–3–7; it looks like a sphere that has been stretched by different factors along three orthogonal axes, so that cross-sections of the ellipsoid parallel to any of the coordinate planes are ellipses. The positive intercepts on the coordinate axes are $(a, 0, 0)$, $(0, b, 0)$, $(0, 0, c)$.

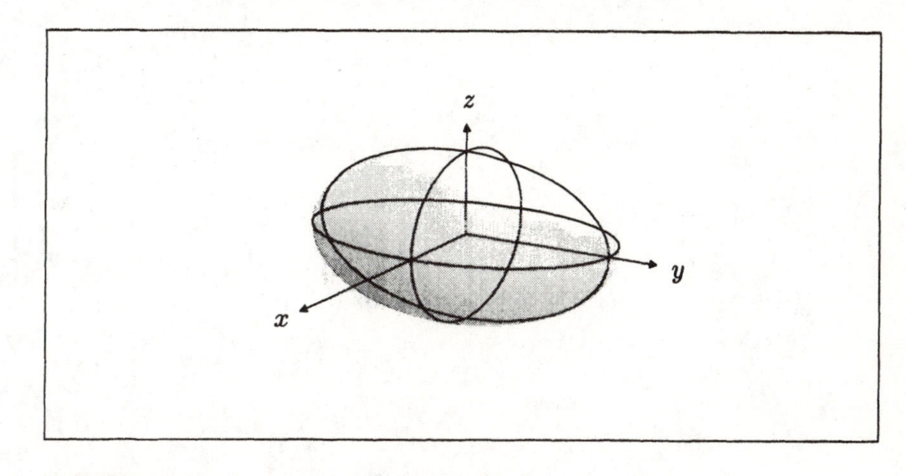

FIGURE 8-3-7. *An ellipsoid in standard position.*

The standard form of the equation for a hyperboloid of one sheet is $\frac{x^2}{a^2} + \frac{y^2}{b^2} - \frac{z^2}{c^2} = 1$. This form is obtained when k and two eigenvalues of the matrix of Q are positive and the third eigenvalue is negative (or when k and two eigenvalues are negative and the other eigenvalue is positive). Notice that if this is rewritten $\frac{x^2}{a^2} + \frac{y^2}{b^2} = 1 + \frac{z^2}{c^2}$, it is clear that for every z there are values of x and y that satisfy the equation, so that the surface is all "one piece" (or one sheet). It is also clear that by taking z constant, we get elliptical cross-sections, while x or y constant gives hyperbolic cross-sections. A hyperboloid of one sheet is shown in Figure 8–3–8(a).

The standard form of the equation for a hyperboloid of two sheets is $\frac{x^2}{a^2} + \frac{y^2}{b^2} - \frac{z^2}{c^2} = -1$. (Two positive eigenvalues, one negative eigenvalue, k negative; or two negative eigenvalues, one positive eigenvalue, k positive.) If this is rewritten $\frac{x^2}{a^2} + \frac{y^2}{b^2} = -1 + \frac{z^2}{c^2}$, it is clear that for $|z| < c$ there are no values of x and y that satisfy the equation. Therefore, the graph consists of two "pieces" (or two sheets), one with $z \geq c$ and one with $z \leq -c$. A hyperboloid of two sheets is shown in Figure 8–3–8(c).

It is interesting to consider the family of surfaces obtained by varying k in the equation $\frac{x^2}{a^2} + \frac{y^2}{b^2} - \frac{z^2}{c^2} = k$, as in Figure 8–3–8. When $k = 1$, the surface is a hyperboloid of one sheet; as k decreases toward 0, the "waist" of the hyperboloid shrinks until at $k = 0$ it has "pinched in" to a single point and the hyperboloid becomes a cone; as k decreases further towards -1, the waist has disappeared and the graph is now a hyperboloid of two sheets.

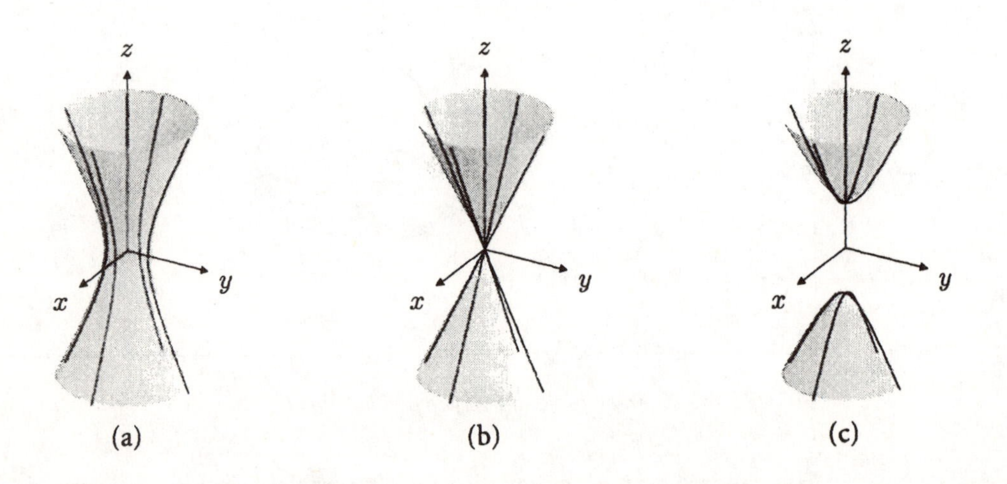

FIGURE 8–3–8. *Graphs of $4x^2 + 4y^2 - z^2 = k$. (a) $k = 1$. A hyperboloid of one sheet. (b) $k = 0$. A cone. (c) $k = -1$. A hyperboloid of two sheets.*

Table II displays the possible cases for $Q(\mathbf{x}) = k$ in $\mathbb{R}^3$. If $k \neq 0$, we may divide both sides by k (possibly changing signs in the quadratic form), so it is sufficient to consider $k = 1$ and $k = 0$. The nondegenerate cases are the ellipsoids and hyperboloids. Note that the hyperboloid of two sheets appears in the form $\dfrac{x^2}{a^2} - \dfrac{y^2}{b^2} - \dfrac{z^2}{c^2} = 1$ in this table.

TABLE II. *Graphs of $\lambda_1 x^2 + \lambda_2 y^2 + \lambda_3 z^2 = k$.*

	$k = 1$	$k = 0$
$\lambda_1, \lambda_2, \lambda_3 > 0$	ellipsoid	point $(0, 0, 0)$
$\lambda_1, \lambda_2 > 0;\ \lambda_3 = 0$	elliptic cylinder	z-axis
$\lambda_1, \lambda_2 > 0;\ \lambda_3 < 0$	hyperboloid of one sheet	cone
$\lambda_1 > 0;\ \lambda_2 = 0;\ \lambda_3 < 0$	hyperbolic cylinder	intersecting planes
$\lambda_1 > 0;\ \lambda_2, \lambda_3 < 0$	hyperboloid of two sheets	cone
$\lambda_1 = 0;\ \lambda_2, \lambda_3 < 0$	empty set	x-axis
$\lambda_1, \lambda_2, \lambda_3 < 0$	empty set	point $(0, 0, 0)$

Figure 8–3–9 shows some degenerate quadric surfaces. Note that paraboloidal surfaces do not appear as graphs of the form $Q(\mathbf{x}) = k$ in $\mathbb{R}^3$ for the same reason that parabolas do not appear in Table I for $\mathbb{R}^2$: their equations contain first-degree terms.

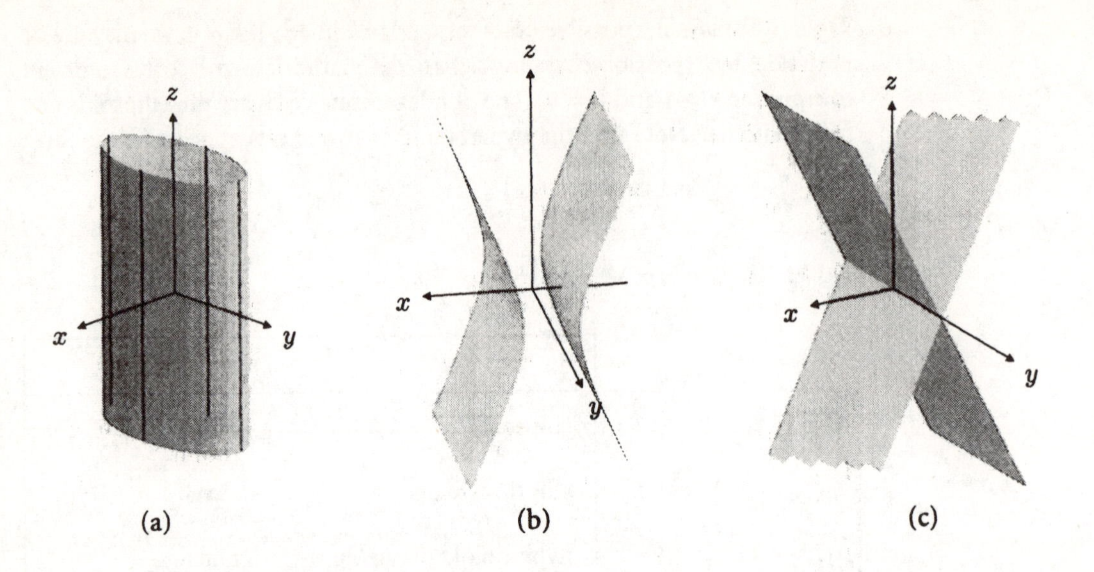

FIGURE 8–3–9. *Some degenerate quadric surfaces. (a) An elliptic cylinder $x^2 + \lambda y^2 = 1$, parallel to the z-axis. (b) A hyperbolic cylinder $\lambda x^2 - z^2 = 1$, parallel to the y-axis. (c) Intersecting planes $\lambda x^2 - z^2 = 0$.*

EXERCISES 8–3

A1 Sketch the graph of the equation $2x^2 + 4xy - y^2 = 6$. Show both the original axes and the new axes.

A2 Sketch the graph of the equation $2x^2+6xy+10y^2 = 11$. Show both the original axes and the new axes.

A3 Suppose that the matrices of Exercise 8–1–A1 are the matrices of quadratic forms. Identify the sets (as ellipses or hyperboloids, etc.).

(a) $\mathbf{x} \cdot A\mathbf{x} = 1$ and $\mathbf{x} \cdot A\mathbf{x} = -1$.

(b) $\mathbf{x} \cdot B\mathbf{x} = 1$ and $\mathbf{x} \cdot B\mathbf{x} = -1$.

(c) $\mathbf{x} \cdot C\mathbf{x} = 1$ and $\mathbf{x} \cdot C\mathbf{x} = -1$.

(d) $\mathbf{x} \cdot D\mathbf{x} = 1$ and $\mathbf{x} \cdot D\mathbf{x} = -1$.

(e) $\mathbf{x} \cdot E\mathbf{x} = 1$ and $\mathbf{x} \cdot E\mathbf{x} = -1$.

B1 Sketch the graph of the equation $6x^2+6xy-2y^2 = 21$. Show both the original axes and the new axes.

B2 Sketch the graph of the equation $7x^2 + 12xy + 12y^2 = 48$. Show both the original axes and the new axes.

B3 Sketch the graph of the equation $9x^2 + 4xy + 6y^2 = 90$. Show both the original axes and the new axes.

B4 Sketch the graph of the equation $x^2 + 6xy - 7y^2 = 32$. Show both the original axes and the new axes.

B5 In each of the following cases, diagonalize the quadratic form. Then determine the nature of the surface $Q(\mathbf{x}) = k$, for $k = 1, 0, -1$. Two of the quadratic forms are degenerate.

(a) $x^2 + y^2 + 5z^2 + 6xy + 2xz + 2yz = k$

(b) $x^2 + 5y^2 + 5z^2 + 4xy + 4xz + 6yz = k$

(c) $-x^2 + y^2 - z^2 + 2xy - 6xz - 2yz = k$

(d) $4x^2 + 5y^2 + 4z^2 + 2xy - 2yz = k$

COMPUTER RELATED EXERCISES

C1 Some computer software can be used to determine eigenvalues of symmetric matrices. Alternatively, you can determine approximate values of the eigenvalues by graphically determining the zeros of the characteristic polynomial. Use such methods to identify the following surfaces as ellipsoids, hyperboloids of one sheet, etc. (The graphical method is not sufficiently accurate for serious applications, but it will suffice for these problems where we need only the signs of the eigenvalues.)

(a) $x^2 - y^2 - 14xy + 6xz + 8yz = 10$

(b) $3x^2 - 6z^2 + 10xy + 4xz + 16yz = 37$

SECTION

8-4 Quadratic Forms and Maximum and Minimum Values

In the calculus of functions of several variables, one important problem is determining the points where maximum and minimum values occur. Recall that $\mathbf{a}$ is a local minimum point of the function $f : \mathbb{R}^n \to \mathbb{R}$ if $f(\mathbf{x}) \geq f(\mathbf{a})$ for all $\mathbf{x}$ in a neighbourhood of $\mathbf{a}$. For a local maximum point, $f(\mathbf{x}) \leq f(\mathbf{a})$ for all $\mathbf{x}$ in a neighbourhood of $\mathbf{a}$. It is a standard simple extension of results from one-variable calculus to show that if f is differentiable at the local minimum or maximum point $\mathbf{a}$, then $\dfrac{\partial f}{\partial x_j}(\mathbf{a}) = 0$ for $j = 1, 2, \dots, n$. This test identifies candidates: to determine whether $\mathbf{a}$ is actually a local minimum or a local maximum—or perhaps a saddle point—it is necessary to look at the second derivatives of f.

The tool for using second derivatives is Taylor's Theorem; the version needed here is that if f has continuous third partial derivatives in a neighbourhood of **a**, then for **x** near **a**,

$$f(\mathbf{x}) - f(\mathbf{a}) = \sum_{j=1}^{n} \frac{\partial f}{\partial x_j}(\mathbf{a})(x_j - a_j) + \frac{1}{2}\sum_{j,k=1}^{n} \frac{\partial^2 f}{\partial x_j \partial x_k}(\mathbf{a})(x_j - a_j)(x_k - a_k) + R,$$

where the remainder R satisfies $\lim_{\mathbf{x}\to\mathbf{a}} \frac{R}{\|\mathbf{x}-\mathbf{a}\|^2} = 0$. This condition on the remainder guarantees that for **x** close to **a**, under most (but not all) circumstances, the remainder term can be ignored in comparison to terms in $(x_j - a_j)(x_k - a_k)$.

Suppose that **a** is a "critical point", where $\frac{\partial f}{\partial x_j}(\mathbf{a}) = 0$ for all j. Then the sign of $f(\mathbf{x}) - f(\mathbf{a})$ is what determines whether **a** is a local maximum or local minimum (or neither), and since the first partial derivatives vanish, the sign of $f(\mathbf{x}) - f(\mathbf{a})$ is determined by the terms involving the second derivatives — unless we are in one of the exceptional cases where the remainder R is important; exceptional cases are illustrated in Examples 7 and 8.

To simplify the writing, let $z_j = (x_j - a_j)$, and let $h_{jk} = \frac{\partial^2 f}{\partial x_j \partial x_k}(\mathbf{a})$. The matrix H with entries h_{jk} is called the **Hessian matrix of f at a.** With this auxiliary notation, Taylor's formula at a critical point **a** can be written

$$f(\mathbf{x}) - f(\mathbf{a}) = \frac{1}{2}\sum_{j,k=1}^{n} h_{jk} z_j z_k + R = \frac{1}{2}\mathbf{z}^T H \mathbf{z} + R.$$

It is now clear that the sign of $f(\mathbf{x}) - f(\mathbf{a})$ is determined by the **quadratic form** $Q(\mathbf{z}) = \frac{1}{2}\sum_{j,k=1}^{n} h_{jk} z_j z_k = \frac{1}{2}\mathbf{z}^T H \mathbf{z}$ (except in cases where R becomes important). Note that H is automatically symmetric, provided that the second derivatives are continuous (by Clairaut's Theorem in calculus).

Recall from Section 8–2 that $Q(\mathbf{z})$ is positive definite if $Q(\mathbf{z}) > 0$ for all non-zero **z**, and (by Theorem 2 of that section) this is true if and only if all the eigenvalues of the corresponding matrix H are positive. This is exactly what we require to have $f(\mathbf{x}) - f(\mathbf{a}) > 0$ for all **x** near **a** — **a** is a local minimum in this case. Similarly $Q(\mathbf{z})$ is a negative definite ($Q(\mathbf{z}) < 0$ for all non-zero **z**) if and only if all the eigenvalues of H are negative. If the Hessian form $\frac{1}{2}\mathbf{z}^T H \mathbf{z}$ is negative definite, then $f(\mathbf{x}) - f(\mathbf{a}) < 0$ for all **x** near **a**, and **a** is a local maximum. In contrast, suppose that at least one eigenvalue of the Hessian matrix H is positive, and at least one is negative: say $\lambda_1 > 0$, $\lambda_2 < 0$. We can always diagonalize, so that

$$Q(\mathbf{z}) = \lambda_1 \tilde{z}_1^2 + \lambda_2 \tilde{z}_2^2 + \cdots + \lambda_n \tilde{z}_n^2.$$

Then if $(\tilde{z}_1, \tilde{z}_2, \ldots, \tilde{z}_n) = (1, 0, \ldots, 0)$, it follows that $Q(\mathbf{z}) > 0$, while if $(\tilde{z}_1, \tilde{z}_2, \ldots, \tilde{z}_n) = (0, 1, 0, \ldots, 0)$, $Q(\mathbf{z}) < 0$. Thus, if $Q(\mathbf{z})$ is the Hessian form for $f(\mathbf{x}) - f(\mathbf{a})$, it follows in this case that $f(\mathbf{x}) - f(\mathbf{a})$ is sometimes positive, sometimes negative, so $\mathbf{a}$ is a saddle point for f.

We summarize this discussion. Since $f(\mathbf{x}) - f(\mathbf{a}) = Q(\mathbf{z}) + R$, where R can usually be neglected, we have the following conclusions.

(1) Suppose that all the eigenvalues of H are positive; then $\mathbf{a}$ is a local minimum point.

(2) Suppose that all the eigenvalues of H are negative; then $\mathbf{a}$ is a local maximum point.

(3) If some of the eigenvalues are positive and some are negative, then $\mathbf{a}$ is a **saddle point** for f.

We must consider two further possible cases.

(4) Suppose that (i) some of the eigenvalues are positive and the rest are zero; or (ii) some of the eigenvalues are negative and the rest are zero. In these cases, the remainder R becomes important; it actually determines the sign of $f(\mathbf{x}) - f(\mathbf{a})$ for some points. (These points lie on lines through $\mathbf{a}$ whose direction is an eigenvector corresponding to eigenvalue 0 of the Hessian H of f at $\mathbf{a}$.) Hence, in these cases, the "second derivative test" for maximum, minimum, or saddle points fails.

Proper proofs of these statements require careful attention to limit arguments from calculus, so they will not be presented here. We give some examples; first, three examples where the test fails (case (4) in the list above).

EXAMPLE 7

Let $f(x, y) = x^2 + y^3$. Then $\dfrac{\partial f}{\partial x} = 2x$, and $\dfrac{\partial f}{\partial y} = 3y^2$, so the only critical point is $(0, 0)$. Since $\dfrac{\partial^2 f}{\partial x^2} = 2$, $\dfrac{\partial^2 f}{\partial x \partial y} = 0$, $\dfrac{\partial^2 f}{\partial y^2} = 6y$, it follows that H at $(0, 0) = \begin{bmatrix} 2 & 0 \\ 0 & 0 \end{bmatrix}$. Clearly the eigenvalues are 2 and 0, so that this is a case that is not determined by the standard test. However, note that $f(0, 1) = 1 > f(0, 0)$, while $f(0, -1) = -1 < f(0, 0)$, so that $(0, 0)$ is neither a local maximum point nor a local minimum point for f.

EXAMPLE 8

Let $f(x, y) = x^2 + y^4$. It is easy to see that $(0, 0)$ is again the only critical point, and that the eigenvalues of the Hessian H at $(0, 0)$ are again 2 and 0. However, in this case, $f(x, y) > 0 = f(0, 0)$ for all $(x, y) \neq (0, 0)$, so that $(0, 0)$ is a (strict) local minimum.

EXAMPLE 9

Let $f(x, y) = x^2 - y^4$. The first and second derivatives at $(0, 0)$ are as in Example 7. However, $f(x, 0) > 0$ if $x \neq 0$, while $f(0, y) < 0$ if $y \neq 0$, so that $(0, 0)$ is a saddle point for this function.

EXAMPLE 10

Find and classify the critical points of

$$f(x, y) = x^2 - 4xy + y^3 + 4y.$$

Here $\dfrac{\partial f}{\partial x} = 2x - 4y$, and $\dfrac{\partial f}{\partial y} = -4x + 3y^2 + 4$. At a critical point, these are both zero, so $x = 2y$; substitute this into the second equation to get

$$3y^2 - 8y + 4 = (3y - 2)(y - 2) = 0.$$

It follows that the critical points are $(4, 2)$ and $(4/3, 2/3)$.

For general (x, y), the Hessian matrix (of second partial derivatives) is $\begin{bmatrix} 2 & -4 \\ -4 & 6y \end{bmatrix}$.

It follows that at $(4, 2)$, $H = \begin{bmatrix} 2 & -4 \\ -4 & 12 \end{bmatrix}$. The characteristic equation is $\lambda^2 - 14\lambda + 8 = 0$, so that $\lambda = 7 \pm (\sqrt{164})/2 \approx 7 \pm 6.4 \approx 0.6$ or 13.4. Since both eigenvalues of the Hessian matrix at the critical point $(4, 2)$ are positive, $(4, 2)$ is a local minimum point of f.

At the other critical point $(4/3, 2/3)$, $H = \begin{bmatrix} 2 & -4 \\ -4 & 4 \end{bmatrix}$, so the characteristic equation is $\lambda^2 - 6\lambda - 8 = 0$. The eigenvalues are $3 \pm (\sqrt{68})/2 = 3 \pm 4.2 \approx 7.2$ or -1.2. Since the eigenvalues have opposite signs, the critical point $(4/3, 2/3)$

is a saddle point of f.

An Alternative Test

It is more usual in introductory calculus texts to present a different "second derivative test" for maximum, minimum, and saddle points. The usual version for functions of two variables is as follows:

Suppose that $\mathbf{a}$ is a critical point of $f : \mathbb{R}^2 \to \mathbb{R}$, and that H is the Hessian matrix at $\mathbf{a}$.

(a) If $\det H > 0$, and $\dfrac{\partial^2 f}{\partial x^2}(\mathbf{a}) > 0$, then $\mathbf{a}$ is a local minimum point.

(b) If $\det H > 0$, and $\dfrac{\partial^2 f}{\partial x^2}(\mathbf{a}) < 0$, then $\mathbf{a}$ is a local maximum point.

(c) If $\det H < 0$, then $\mathbf{a}$ is a saddle point.

(d) If $\det H = 0$, then this test gives no conclusion.

You can connect this with the test given above in terms of eigenvalues if you use the fact that for any diagonable matrix A, the determinant of A is equal to the product of the eigenvalues of A. (The last statement is true because if P diagonalizes A to the diagonal matrix D, then $P^{-1}AP = D$, and

$$\det(D) = \det(P^{-1}AP) = \det(P^{-1})\det(A)\det(P) = \det(A);$$

$\det(D)$ is the product of its diagonal entries, which are the eigenvalues. Exercise 6-2-D6.)

There is a version of this test for a function of n variables involving the "principal minors" of the Hessian matrix, but we omit it. Either test involves a considerable amount of calculation if there are 3 or more variables.

EXERCISES 8-4

A1 Find any critical points of the following functions. Then determine the eigenvalues of the Hessian matrix at each critical point to determine whether the critical point is a local maximum, local minimum, or a saddle point.

(a) $f(x, y) = 2xy + y^2 + 2x - 2y + 7$ (b) $f(x, y) = 3 - x^2 - 2y^2 - 2xy - 6y + 3$

B1 Same instructions as in A1.

(a) $f(x, y) = xy + \dfrac{8}{x} + \dfrac{8}{y}$ (b) $f(x, y) = xy - \dfrac{8}{x} + \dfrac{8}{y}$

COMPUTER RELATED EXERCISES

C1 Suppose that each matrix given below is the Hessian matrix of a function f at a critical point $\mathbf{a}$ of f. By determining the signs of the eigenvalues, determine whether the critical point is a local maximum, a local minimum, or a saddle point of f, or perhaps not determined by the second derivative test. (See the note in Exercise 8–3–C1.)

(a) $\begin{bmatrix} 1 & 7 \\ 7 & 4 \end{bmatrix}$ (b) $\begin{bmatrix} -2 & -3 & 4 \\ -3 & 0 & -5 \\ 4 & -5 & -4 \end{bmatrix}$ (c) $\begin{bmatrix} 4 & -2 & -5 \\ -2 & -1 & 3 \\ -5 & 3 & -5 \end{bmatrix}$

Two Physical Applications of Symmetric Matrices

Some may think of mathematics as only a set of rules for doing calculations. A theorem such as the Principal Axis Theorem (Section 8.1) is often important because it provides a simple way of thinking about complicated situations. The Principal Axis Theorem plays an important role in the two applications described here.

Small Deformations

Summary. The small deformation of a solid body may be understood as the composition of three **stretches** along the **principal axes** of a symmetric matrix together with a rigid rotation of the body.

Outline

Consider a body of material that can be deformed when it is subjected to some external forces. This might be, for example, a piece of steel under some load. Fix an origin of coordinates $\mathbf{0}$ in the body; to simplify the story, suppose that this origin is left unchanged by the deformation. Suppose that a material point in the body, which is at $\mathbf{x}$ before the forces are applied, is moved by the forces to the point $\mathbf{f}(\mathbf{x}) = (f_1(x), f_2(x), f_3(x))$; we have assumed that $\mathbf{f}(\mathbf{0}) = \mathbf{0}$. The problem is to understand this deformation $\mathbf{f}$ so that it can be related to the properties of the body. (Note that $\mathbf{f}$ represents the displacement of the point initially at $\mathbf{x}$, not the force at $\mathbf{x}$.)

For many materials under reasonable forces, the deformation is small; this means that the point $\mathbf{f}(\mathbf{x})$ is not far from $\mathbf{x}$. It is convenient to introduce a parameter β to describe how small the deformation is, and a function $\mathbf{h}(\mathbf{x})$, and write $\mathbf{f}(\mathbf{x}) = \mathbf{x} + \beta\mathbf{h}(\mathbf{x})$. (This equation is really the definition of $\mathbf{h}(\mathbf{x})$ in terms of the point $\mathbf{x}$, the given function $\mathbf{f}(\mathbf{x})$, and the parameter β.)

For many materials, an arbitrary small deformation is well approximated by its "best linear approximation", the derivative. In this case, the map $\mathbf{f} : \mathbb{R}^3 \to \mathbb{R}^3$ is approximated near the origin by the linear transformation with matrix $\left[\frac{\partial f_j}{\partial x_k}(\mathbf{0})\right]$, so that in this approximation, a point originally at $\mathbf{v}$ is moved (approximately) to $\left[\frac{\partial f_j}{\partial x_k}(\mathbf{0})\right][\mathbf{v}]$. (This is a standard calculus approximation, mentioned in Sections 5–2 and 5–4.)

In terms of the parameter β and the function $\mathbf{h}$, this matrix may be written $\left[\frac{\partial f_j}{\partial x_k}(\mathbf{0})\right] = I + \beta G$, where $G = \left[\frac{\partial h_j}{\partial x_k}(\mathbf{0})\right]$. It is convenient to represent G as the sum of its symmetric part $E = \frac{1}{2}(G + G^T)$, and its skew-symmetric part $W = \frac{1}{2}(G - G^T)$, so that $G = E + W$. (The symmetric part and the skew-symmetric part were introduced in Exercise 8–2 D1.)

The next step is to observe that we can write

$$I + \beta G = I + \beta(E + W) = (I + \beta E)(I + \beta W) - \beta^2 EW;$$

since β is assumed to be small, β^2 is very small and may be ignored. (Such treatment of terms like β^2 can be justified by careful discussion of the limit as $\beta \to 0$.)

The small deformation we started with is now described as the composition of two linear transformations, one with matrix $I + \beta E$, the other with matrix $I + \beta W$. It can be shown that $I + \beta W$ describes a small rigid rotation of the body; a rigid rotation does not alter the distance between any two points in the body. (The matrix βW is called an infinitesimal rotation.)

Finally, we have the linear transformation with matrix $I + \beta E$. This matrix is symmetric, so there exist principal axes such that the symmetric matrix is diagonalized to $\begin{bmatrix} 1+\epsilon_1 & 0 & 0 \\ 0 & 1+\epsilon_2 & 0 \\ 0 & 0 & 1+\epsilon_3 \end{bmatrix}$. (It is equivalent to diagonalize βE and add the result to I, because I is transformed to itself under any orthonormal change of basis.) Since β is small, it follows that the numbers ϵ_j are small in magnitude, and therefore $1 + \epsilon_j > 0$. This diagonalized matrix can be written as the product of three matrices:

$$\begin{bmatrix} 1+\epsilon_1 & 0 & 0 \\ 0 & 1+\epsilon_2 & 0 \\ 0 & 0 & 1+\epsilon_3 \end{bmatrix} = \begin{bmatrix} 1+\epsilon_1 & 0 & 0 \\ 0 & 1 & 0 \\ 0 & 0 & 1 \end{bmatrix} \begin{bmatrix} 1 & 0 & 0 \\ 0 & 1+\epsilon_2 & 0 \\ 0 & 0 & 1 \end{bmatrix} \begin{bmatrix} 1 & 0 & 0 \\ 0 & 1 & 0 \\ 0 & 0 & 1+\epsilon_3 \end{bmatrix},$$

and it is now apparent that, excluding rotation, the small deformation can be represented as the composition of three stretches along the principal axes of the matrix βE. The quantities $\epsilon_1, \epsilon_2, \epsilon_3$ are related to external and internal forces in the material by elastic properties of the material. (βE is called the infinitesimal strain; this notation is not quite the standard notation, in case you choose to read further about this topic in a book on "continuum mechanics".)

The Inertia Tensor

Summary. For the purpose of discussing the rotating motion of a rigid body, information about the mass distribution within the body is summarized in a symmetric matrix N called the **inertia tensor.** (1) This tensor is easiest to understand if principal axes are used, so that the matrix is diagonal; in this case, the diagonal entries are simply the moments of inertia about the principal axes, and the moment of inertia about any other axis can be calculated in terms of these **principal moments of inertia.** (2) In general, the **angular momentum** vector $\mathbf{J}$ of the rotating body is equal to $N\boldsymbol{\omega}$, where $\boldsymbol{\omega}$ is the **instantaneous angular velocity** vector. $\mathbf{J}$ is a scalar multiple of $\boldsymbol{\omega}$ if and only if $\boldsymbol{\omega}$ is an eigenvector of N, that is if and only if the axis of rotation is one of the principal axes of the body. This is a beginning to an explanation of how the body wobbles during rotation ($\boldsymbol{\omega}$ need not be constant) even though $\mathbf{J}$ is a conserved quantity (that is, $\mathbf{J}$ is constant if no external force is applied).

Outline

Suppose that a rigid body is rotating about some point in the body that remains fixed in space throughout the rotation. Make this fixed point the origin $\mathbf{0}$. Suppose that there are coordinate axes fixed in space and also three reference axes that are fixed in the body (so that they rotate with the body). At any time t these body axes make certain angles with respect to the space axes; at a later time $t+\Delta t$ the body axes have moved to a new position. Since $\mathbf{0}$ is fixed and the body is rigid, the body axes have moved only by a rotation, and it is a fact that any rotation in $\mathbb{R}^3$ is determined by its axis and an angle. Call the unit vector along this axis $\mathbf{u}(t+\Delta t)$, and denote the angle by $\Delta\theta$. Now let $\Delta t \to 0$; the unit vector $\mathbf{u}(t+\Delta t)$ must tend to a limit $\mathbf{u}(t)$, and this determines the **instantaneous**

axis of rotation at time t. Also as $\Delta t \to 0$, $\dfrac{\Delta\theta}{\Delta t} \to \dfrac{d\theta}{dt}$, the **instantaneous rate of rotation about the axis.** The **instantaneous angular velocity** is defined to be the vector $\boldsymbol{\omega} = \left(\dfrac{d\theta}{dt}\right)\mathbf{u}(t)$.

(It is a standard exercise to show that the instantaneous linear velocity $\mathbf{v}(t)$ at some point in the body whose space coordinates are given by $\mathbf{x}(t)$ is determined by $\mathbf{v} = \boldsymbol{\omega} \times \mathbf{x}$.)

To use concepts such as energy and momentum in the discussion of rotating motion, it is necessary to introduce moments of inertia.

For a single mass m at the point $\mathbf{x} = (x_1, x_2, x_3)$ the **moment of inertia about the x_3-axis** is defined to be $m(x_1^2 + x_2^2)$; this will be denoted by n_{33}. The factor $(x_1^2+x_2^2)$ is simply the square of the distance of the mass from the x_3-axis. There are similar definitions of the moments of inertia about the x_1-axis (denoted by n_{11}) and about the x_2-axis (denoted by n_{22}).

For a general axis ℓ through the origin with unit direction vector $\mathbf{u}$, the moment of inertia of the mass about ℓ is defined to be m multiplied by the square of the distance of m from ℓ. Thus the moment of inertia in this case is

$$m\|\mathbf{perp}_{\mathbf{u}}(\mathbf{x})\|^2 = m\left[\mathbf{x} - (\mathbf{x}\cdot\mathbf{u})\mathbf{u}\right]^T\left[\mathbf{x} - (\mathbf{x}\cdot\mathbf{u})\mathbf{u}\right] = m\left(\|\mathbf{x}\|^2 - (\mathbf{x}\cdot\mathbf{u})^2\right).$$

With some manipulation, using $[\mathbf{u}]^T[\mathbf{u}] = 1$ and $(\mathbf{x}\cdot\mathbf{u}) = [\mathbf{x}]^T[\mathbf{u}]$, we can verify that this is equal to the expression

$$[\mathbf{u}]^T m\left(\|\mathbf{x}\|^2 I - [\mathbf{x}][\mathbf{x}]^T\right)[\mathbf{u}].$$

Because of this, for the single point mass m at $\mathbf{x}$, we define the **inertia tensor** N to be the 3 by 3 matrix

$$N = m\left(\|\mathbf{x}\|^2 I - [\mathbf{x}][\mathbf{x}]^T\right).$$

(Vectors and matrices are special kinds of "tensors"; for our present purposes we simply treat N as a matrix.) With this definition, the moment of inertia about an axis with unit direction vector $\mathbf{u}$ is

$$\mathbf{u}^T N\mathbf{u}.$$

It is easy to check that N is the matrix with components n_{11}, n_{22}, n_{33} as given above, and for $j \neq k$, $n_{jk} = -mx_jx_k$. It is clear that this matrix N is symmetric because $\mathbf{x}\mathbf{x}^T$ is a symmetric 3 by 3 matrix. (The term mx_jx_k is called a "product of inertia". This name has no special meaning; the term is simply a product that appears as an entry in the inertia tensor.)

It is easy to extend the definition of moments of inertia and the inertia tensor to bodies that are more complicated than a single point mass. Consider a rigid body that can be thought of as k masses joined to each other by weightless rigid rods. The moment of inertia of the body about the x_3-axis is determined by taking the moment of inertia about the x_3-axis of each mass and simply **adding** these moments; the moments about the x_1- and x_2-axes and the products of inertia are defined similarly. The inertia tensor of this body is just the **sum** of the inertia tensors of the k masses; since it is the sum of symmetric matrices, it is also symmetric. If the mass is distributed continuously, the various moments and products of inertia are determined by definite integrals. In any case, the inertia tensor N is still defined, and is still a symmetric matrix.

Since N is a symmetric matrix, it can be brought into diagonal form by the Principal Axis Theorem. The diagonal entries are then the moments of inertia with respect to the principal axes, and these are called the **principal moments of inertia.** Denote these by N_1, N_2, N_3. Let $\mathcal{P}$ denote the orthonormal basis consisting of unit eigenvectors of the matrix N (which means that these vectors are unit vectors along the principal axes). Suppose an arbitrary axis ℓ is determined by the unit vector $\mathbf{u}$ such that $[\mathbf{u}]_{\mathcal{P}} = (\tilde{u}_1, \tilde{u}_2, \tilde{u}_3)$. Then, from the discussion of quadratic forms in Section 8–2, the moment of inertia about this axis ℓ is simply

$$\mathbf{u}^T N\mathbf{u} = [\,\tilde{u}_1 \quad \tilde{u}_2 \quad \tilde{u}_3\,] \begin{bmatrix} N_1 & 0 & \\ 0 & N_2 & 0 \\ 0 & 0 & N_3 \end{bmatrix} \begin{bmatrix} \tilde{u}_1 \\ \tilde{u}_2 \\ \tilde{u}_3 \end{bmatrix} = N_1\tilde{u}_1^2 + N_2\tilde{u}_2^2 + N_3\tilde{u}_3^2.$$

This formula is greatly simplified because of the use of principal axes.

It is important to get equations for rotating motion that correspond to Newton's equation:

the rate of change of momentum equals the applied force.

The appropriate equation is:

the rate of change of angular momentum is the applied torque.

It turns out that the right way to define the angular momentum vector $\mathbf{J}$ for a general body is

$$\mathbf{J}(t) = N(t)\boldsymbol{\omega}(t).$$

Note that in general, N is a function of t, since it depends on the positions at time t of each of the masses making up the solid body. Understanding the possible motions of a rotating body depends on determining $\boldsymbol{\omega}(t)$, or at least saying something about it. In general, this is a very difficult problem, but there will often be important simplifications if N is diagonalized by the Principal Axis

Theorem. Note that $\mathbf{J}(t)$ is parallel to $\boldsymbol{\omega}(t)$ if and only if $\boldsymbol{\omega}(t)$ is an eigenvector of $N(t)$.

EXERCISES 8–5

CONCEPTUAL EXERCISES

***D1** Show that if R is an orthogonal matrix that diagonalizes the symmetric matrix βE to a matrix with diagonal entries $\epsilon_1, \epsilon_2, \epsilon_3$, then R also diagonalizes $(I + \beta E)$ to a matrix with diagonal entries $1 + \epsilon_1$, $1 + \epsilon_2$, $1 + \epsilon_3$.

SECTION

8–6 Unitary Diagonalization of Hermitian Matrices

In Section 8–1, we saw that the problem of diagonalizing a matrix with real entries has attractive special features when the matrix is symmetric. Is there a comparable simplification in the case of matrices with complex entries? What is the "right" special property for complex matrices to take the place of symmetry for real matrices?

There are two approaches to identifying the desired special property. First, observe that in Section 8–1 the essential use made of symmetry of the matrix A was to show that

$$\begin{aligned}\mathbf{x} \cdot A\mathbf{y} &= \mathbf{x}^T A\mathbf{y} = (\mathbf{x}^T A\mathbf{y})^T \\ &= \mathbf{y}^T A^T \mathbf{x} = \mathbf{y}^T A\mathbf{x} = (A\mathbf{x}) \cdot \mathbf{y},\end{aligned}$$

for all $\mathbf{x}, \mathbf{y}$ in $\mathbb{R}^n$, because $A = A^T$. If we expect to apply this with vectors in $\mathbb{C}^n$, we should use the inner product in $\mathbb{C}^n$ (Section 7–5). We want the *complex* matrix A to satisfy

$$\langle \mathbf{x}, A\mathbf{y} \rangle = \langle A\mathbf{x}, \mathbf{y} \rangle, \quad \text{for all } \mathbf{x}, \mathbf{y} \text{ in } \mathbb{C}^n.$$

Since $\langle \mathbf{x}, A\mathbf{y} \rangle = \overline{\mathbf{x}} \cdot A\mathbf{y}$, we write $\langle \mathbf{x}, A\mathbf{y} \rangle = \overline{\mathbf{x}}^T \cdot A\mathbf{y}$. Therefore, the condition $\langle \mathbf{x}, A\mathbf{y} \rangle = \langle A\mathbf{x}, \mathbf{y} \rangle$ is equivalent to

$$\overline{\mathbf{x}}^T A\mathbf{y} = (\overline{A\mathbf{x}})^T \mathbf{y} = \overline{\mathbf{x}}^T \overline{A}^T \mathbf{y}, \quad \text{for all } \mathbf{x}, \mathbf{y} \text{ in } \mathbb{C}^n,$$

and by a double application of Theorem 2, Section 3–1, this is true if and only if $A = \overline{A}^T$.

An alternative approach is to recall that a real symmetric matrix is diagonalized by an orthogonal matrix. For the complex case, we would expect to diagonalize by a unitary matrix. (Recall from Section 7–5 that U is unitary if $\overline{U}^T U = I$.) If a unitary matrix U diagonalizes A to a diagonal matrix D with *real* entries, we must have

$$\overline{U}^T AU = D \quad \text{or} \quad A = UD\overline{U}^T.$$

Then

$$A^T = \overline{U}D^T U^T = \overline{U}DU^T$$

so

$$\overline{A}^T = U\overline{D}\,\overline{U}^T = UD\overline{U}^T = A.$$

Both approaches identify the same property, so we make the following definition.

DEFINITION

An n by n matrix A with complex entries is **Hermitian** if $\overline{A} = A^T$, or equivalently, if $\langle \mathbf{x}, A\mathbf{y}\rangle = \langle A\mathbf{x}, \mathbf{y}\rangle$ for all $\mathbf{x}, \mathbf{y}$ in $\mathbb{C}^n$.

EXAMPLE 11

Which of the following matrices are Hermitian?

$$A = \begin{bmatrix} 2 & 3-i \\ 3+i & 4 \end{bmatrix} \quad B = \begin{bmatrix} 1 & 2i \\ -2i & 3-i \end{bmatrix} \quad C = \begin{bmatrix} 0 & i & i \\ -i & 0 & i \\ -i & i & 0 \end{bmatrix}$$

SOLUTION: $\overline{A} = \begin{bmatrix} 2 & 3+i \\ 3-i & 4 \end{bmatrix}$, so $\overline{A} = A^T$ and A is Hermitian.

$\overline{B} = \begin{bmatrix} 1 & -2i \\ 2i & 3+i \end{bmatrix}$, but $B^T = \begin{bmatrix} 1 & -2i \\ 2i & 3-i \end{bmatrix}$, so B is not Hermitian.

$\overline{C} = \begin{bmatrix} 0 & -i & -i \\ i & 0 & -i \\ i & -i & 0 \end{bmatrix}$. Since $\overline{c}_{32} \neq c_{23}$, C is not Hermitian.

REMARKS

(1) Notice that the Hermitian property is a generalization of symmetry: if A happens to be real, then $\overline{A}^T = A^T$, so a Hermitian matrix with real entries is in fact symmetric.

(2) A linear mapping is Hermitian if $\langle \mathbf{x}, L(\mathbf{y}) \rangle = \langle L(\mathbf{x}), \mathbf{y} \rangle$, for all $\mathbf{x}$, $\mathbf{y}$ in the vector space. Hermitian linear mappings in infinite-dimensional spaces play a central role in quantum mechanics.

THEOREM 1. Suppose that A is an n by n Hermitian matrix. Then:
(1) All eigenvalues of A are real.
(2) Eigenvectors corresponding to distinct eigenvalues are orthogonal to each other.
(3) The matrix A can be diagonalized by a unitary change of basis matrix.

Discussion. The proof is essentially the same as for the real case in Section 8–1, with appropriate changes to allow for complex numbers. To demonstrate this, we give proofs for (1) and (2).

Proof of (1). Suppose that λ is an eigenvalue, with corresponding unit eigenvector $\mathbf{v}$.

Then $\langle \mathbf{v}, A\mathbf{v} \rangle = \langle \mathbf{v}, \lambda\mathbf{v} \rangle = \lambda$, while $\langle A\mathbf{v}, \mathbf{v} \rangle = \langle \lambda\mathbf{v}, \mathbf{v} \rangle = \overline{\lambda}$.

But $\langle \mathbf{v}, A\mathbf{v} \rangle = \langle A\mathbf{v}, \mathbf{v} \rangle$, so $\lambda = \overline{\lambda}$, and λ is real.

Proof of (2). Suppose that λ_1 and λ_2 are distinct eigenvalues with corresponding eigenvectors $\mathbf{v}_1$ and $\mathbf{v}_2$. Then

$$\langle \mathbf{v}_1, A\mathbf{v}_2 \rangle = \lambda_2 \langle \mathbf{v}_1, \mathbf{v}_2 \rangle,$$

and

$$\langle A\mathbf{v}_1, \mathbf{v}_2 \rangle = \overline{\lambda_1} \langle \mathbf{v}_1, \mathbf{v}_2 \rangle = \lambda_1 \langle \mathbf{v}_1, \mathbf{v}_2 \rangle.$$

Since A is Hermitian, $\langle \mathbf{v}_1, A\mathbf{v}_2 \rangle = \langle A\mathbf{v}_1, \mathbf{v}_2 \rangle$, so

$$(\lambda_2 - \lambda_1) \langle \mathbf{v}_1, \mathbf{v}_2 \rangle = 0.$$

Since $\lambda_1 \neq \lambda_2$, $\langle \mathbf{v}_1, \mathbf{v}_2 \rangle = 0$, as claimed.

REMARKS

(3) If A and B are matrices such that for some unitary matrix U, $B = \overline{U}^T AU$, we say that B is **unitarily equivalent** to A. The theorem can thus be summarized by saying that *a Hermitian matrix is unitarily equivalent to a diagonal matrix with real entries.*

(4) Note that a complex matrix that is not Hermitian may be diagonable with *complex* eigenvalues. An example is $\begin{bmatrix} i & 0 \\ 0 & -1 \end{bmatrix}$. If the eigenvalues are not real, the diagonalizing matrix need not be unitary.

EXAMPLE 12

Let $A = \begin{bmatrix} 2 & 1+i \\ 1-i & 3 \end{bmatrix}$. Check that A is Hermitian, determine its eigenvalues and corresponding eigenvectors, and carry out the diagonalization.

SOLUTION: We leave it to the reader to check that A is Hermitian. Consider $A - \lambda I = \begin{bmatrix} 2-\lambda & 1+i \\ 1-i & 3-\lambda \end{bmatrix}$.

The characteristic equation is $\lambda^2 - 5\lambda + 6 - 2 = 0$, so $\lambda = 4$, or $\lambda = 1$.

If $\lambda = 4$,

$$A - \lambda I = \begin{bmatrix} -2 & 1+i \\ 1-i & -1 \end{bmatrix} \begin{matrix} \\ (1+i)R_2 \end{matrix} \to \begin{bmatrix} -2 & 1+i \\ 2 & -(1+i) \end{bmatrix} \to \begin{bmatrix} -2 & 1+i \\ 0 & 0 \end{bmatrix}.$$

A corresponding eigenvector is $(1+i, 2)$.

If $\lambda = 1$,

$$A - \lambda I = \begin{bmatrix} 1 & 1+i \\ 1-i & 2 \end{bmatrix} \begin{matrix} \\ (1+i)R_2 \end{matrix} \to \begin{bmatrix} 1 & 1+i \\ 2 & 2(1+i) \end{bmatrix} \to \begin{bmatrix} 1 & 1+i \\ 0 & 0 \end{bmatrix}.$$

An eigenvector is $(1+i, -1)$.

A unitary diagonalizing matrix is

$$U = \begin{bmatrix} \frac{1+i}{\sqrt{6}} & \frac{1+i}{\sqrt{3}} \\ \frac{2}{\sqrt{6}} & -\frac{1}{\sqrt{3}} \end{bmatrix}.$$

(Remember that the columns must have length one.)

$$\overline{U}^T = \begin{bmatrix} \dfrac{1-i}{\sqrt{6}} & \dfrac{2}{\sqrt{6}} \\ \dfrac{1-i}{\sqrt{6}} & -\dfrac{1}{\sqrt{3}} \end{bmatrix}.$$

Then

$$D = \begin{bmatrix} \dfrac{1-i}{\sqrt{6}} & \dfrac{2}{\sqrt{6}} \\ \dfrac{1-i}{\sqrt{3}} & -\dfrac{1}{\sqrt{3}} \end{bmatrix} \begin{bmatrix} 2 & 1+i \\ 1-i & 3 \end{bmatrix} \begin{bmatrix} \dfrac{1+i}{\sqrt{6}} & \dfrac{1+i}{\sqrt{3}} \\ \dfrac{2}{\sqrt{6}} & -\dfrac{1}{\sqrt{3}} \end{bmatrix}$$

$$= \begin{bmatrix} 4 & 0 \\ 0 & 1 \end{bmatrix}. \qquad \text{(Check!)}$$

EXERCISES 8–6

A1 For each of the following matrices:

(a) Determine whether it is Hermitian.

(b) If it is Hermitian, determine its eigenvalues and corresponding eigenvectors; check that the eigenvectors are mutually orthogonal.

$$C = \begin{bmatrix} 4 & \sqrt{2}+i \\ \sqrt{2}-i & 2 \end{bmatrix} \qquad D = \begin{bmatrix} 5 & \sqrt{2}-i \\ \sqrt{2}+i & \sqrt{3}+i \end{bmatrix}$$

$$E = \begin{bmatrix} 6 & \sqrt{3}-i \\ \sqrt{3}+i & 3 \end{bmatrix} \qquad F = \begin{bmatrix} 1 & 1+i & 0 \\ 1-i & 0 & 1-i \\ 0 & 1+i & -1 \end{bmatrix}$$

B1 For each of the following matrices:

(a) Determine whether it is Hermitian.

(b) If it is Hermitian, determine its eigenvalues and corresponding eigenvectors; check that the eigenvectors are mutually orthogonal.

$$G = \begin{bmatrix} 2 & \sqrt{2}+i \\ \sqrt{2}+i & \sqrt{3} \end{bmatrix} \qquad H = \begin{bmatrix} 5 & \sqrt{2}-i \\ \sqrt{2}+i & 3 \end{bmatrix}$$

$$J = \begin{bmatrix} 5 & \sqrt{3}+i \\ \sqrt{3}-i & 2 \end{bmatrix} \qquad K = \begin{bmatrix} 1 & i & -i \\ -i & -1 & i \\ i & -i & 0 \end{bmatrix}$$

CONCEPTUAL EXERCISES

***D1** Suppose that A and B are Hermitian n by n matrices, and that A is invertible. Determine which of the following are Hermitian.
(a) AB (b) A^2 (c) A^{-1}

D2 Prove (without appealing to diagonalization) that if A is Hermitian, then $\det A$ is real.

D3 A general 2 by 2 Hermitian matrix can be written $\begin{bmatrix} a & b+ci \\ b-ci & d \end{bmatrix}$, $a, b, c, d \in \mathbb{R}$.
(a) What can you say about a, b, c, d if A is unitary as well as Hermitian?
(b) What can you say about a, b, c, d if A is Hermitian, unitary, and diagonal?
(c) What can you say about the form of a 3 by 3 matrix that is Hermitian, unitary, and diagonal?

Chapter Review

Suggestions for Student Review

1 How does the theory of diagonalization of symmetric matrices differ from the theory for general square matrices? (Section 8–1)

2 Explain the connection between quadratic forms and symmetric matrices. How do you find the symmetric matrix corresponding to a quadratic form? How does diagonalization of the symmetric matrix enable us to diagonalize the quadratic form? (Section 8–2)

3 What role do eigenvectors play in helping us to understand the graphs of equations $Q(\mathbf{x}) = k$, where $Q(\mathbf{x})$ is a quadratic form? (Section 8–3)

4 Outline how quadratic forms arise in the theory of finding maximum and minimum values of functions of two or three variables. How do the concepts of "positive definite" and "negative definite" help in this theory? (Section 8–4)

5 What is a Hermitian matrix? State what you can about diagonalizing a Hermitian matrix. (Section 8–6)

Chapter Quiz

1 Let $A = \begin{bmatrix} 2 & -3 & 2 \\ -3 & 3 & 3 \\ 2 & 3 & 2 \end{bmatrix}$. Determine the diagonal matrix that is orthogonally equivalent to A, and the diagonalizing orthogonal matrix.

2 By diagonalizing the quadratic form, make a sketch of the graph of

$$5x^2 - 2xy + 5y^2 = 12,$$

in the xy-plane. Show the new coordinate axes (the axes of the conic section).

3 In each of the following cases, state whether the matrix A is diagonable, and explain briefly.

(a) A is a 3 by 3 matrix with characteristic equation $\lambda(\lambda - 1)(\lambda + 1) = 0$.

(b) A is a 3 by 3 matrix with characteristic equation $(\lambda - 1)^2(\lambda + 1) = 0$.

(c) A is a 3 by 3 matrix with characteristic equation $\lambda^3 - 1 = 0$.

(d) A is a symmetric 3 by 3 matrix with characteristic equation $(\lambda - 1)^2(\lambda + 1) = 0$.

4 Prove that if A is a symmetric 4 by 4 matrix with characteristic equation $(\lambda - 3)^4 = 0$, then $A = 3I$.

5 Let $A = \begin{bmatrix} 0 & 3 + ki \\ 3 + i & 3 \end{bmatrix}$.

(a) Determine k such that A is Hermitian.

(b) With the value of k as determined in part (a), find the eigenvalues of A, and corresponding eigenvectors, and verify that the eigenvectors are orthogonal.

Further Exercises

F1 In Exercise F8 at the end of Chapter 7, we saw the QR-factorization: an invertible n by n matrix A can be expressed in the form $A = QR$, where Q is orthogonal and R is upper triangular. Let $A_1 = RQ$, and prove that A_1 is orthogonally equivalent to A and hence has the same eigenvalues as A. (By repeating this process: $A = Q_1R_1$, $A_1 = R_1Q_1$, $A_1 = Q_2R_2$, $A_2 = R_2Q_2, \ldots$, one obtains an effective numerical procedure for determining eigenvalues of a symmetric matrix. See *Numerical Analysis: A Practical Approach, Third Edition* by M.J. Maron and R.J. Lopez, Wadsworth, 1991.)

F2 (a) Suppose that A is a positive definite symmetric n by n matrix. Prove that A "has a square root": that is, there is a symmetric positive definite matrix B such that $B^2 = A$. [Hint: Suppose that Q diagonalizes A to D, so that $Q^TAQ = D$; define C to be a positive square root for D, and let $B = QCQ^T$.]

(b) Prove that the result of part (a) is still true if it is assumed that A is non-negative definite (that is, all eigenvalues satisfy $\lambda \geq 0$).

F3 (a) If A is any n by n real matrix, prove that A^TA is symmetric and non-negative definite (that is, all eigenvalues of A^TA are greater than or equal to zero). [Hint: Consider $A\mathbf{x} \cdot A\mathbf{x}$.]

(b) If A is invertible, prove that A^TA is positive definite.

F4 (a) Suppose that A is an invertible n by n matrix. Prove that A can be expressed as a product of an orthogonal matrix Q and a positive definite symmetric matrix U, $A = QU$. This is known as a **polar decomposition** of A. [Hint: Use Exercises F2 and F3, and let U be the square root of A^TA, and

$Q = AU^{-1}$.]

(b) Let $V = QUQ^T$. Show that V is symmetric and that $A = VQ$. Moreover, show that $V^2 = AA^T$, so that V is a symmetric positive definite square root of AA^T.

(c) Suppose that the 3 by 3 matrix A is the matrix of an orientation-preserving linear mapping L. Show that L is the composition of a rotation following three stretches along mutually orthogonal axes. [This follows from part (a), facts about isometries of $\mathbb{R}^3$, and ideas in Section 8–5. In fact, this is a finite version of the result for infinitesimal strain in Section 8–5.]

F5 Generalize the results of Exercises F2, F3, and F4(a) for the case of complex matrices. You will have to replace "symmetric" by "Hermitian", and "orthogonal" by "unitary", and you will have to use $\overline{A}^T$ in place of A^T.

Complex Numbers

Introduction

The first numbers we encounter as children are the *natural numbers* $1, 2, 3, \ldots$ and so on. In school, we soon find that in order to perform certain subtractions we must extend our concept of number to the *integers*, including negative integers such as -1, -3, and -10. Then, so that division can always be carried out, the concept of number is extended to the *rational numbers*, which include the integers and quotients of integers such as $\frac{1}{2}, \frac{2}{3}, \frac{3}{7}$. Next we have to extend our understanding of number to the *real numbers*, which include all the rationals, but also include irrational numbers such as $\sqrt{2}$ and π.

To solve the equation $x^2 + 1 = 0$ we have to extend our concept of number one more time. We *introduce* a number i, which is a solution of this equation, so that $i^2 = -1$. The system of numbers consisting of numbers of the form $a + bi$, where a and b are real numbers, is called the *complex numbers.* Note that the real numbers are included as those complex numbers with $b = 0$. As is the case with all the previous extensions of our understanding of number, some people are initially uncertain about the "meaning" of the "new" numbers. However, the complex numbers have a consistent set of rules of arithmetic, and the extension to complex numbers is justified by the fact that they allow us to solve important mathematical and physical problems that we could not solve using only the real numbers.

The Arithmetic of Complex Numbers

DEFINITION

A complex number is a number of the form $z = x + yi$, where x and y are real numbers, and i is an element such that $i^2 = -1$. The set of all complex numbers is denoted by $\mathbb{C}$.

Addition of complex numbers $z_1 = x_1 + y_1 i$ and $z_2 = x_2 + y_2 i$ is defined by

$$z_1 + z_2 = x_1 + y_1 i + x_2 + y_2 i = (x_1 + x_2) + (y_1 + y_2)i.$$

Multiplication of complex numbers z_1 and z_2 is defined by

$$\begin{aligned} z_1 z_2 &= (x_1 + y_1 i)(x_2 + y_2 i) = x_1 x_2 + x_1 y_2 i + x_2 y_1 i + y_1 y_2 i^2 \\ &= (x_1 x_2 - y_1 y_2) + (x_1 y_2 + x_2 y_1) i. \end{aligned}$$

EXAMPLE 1

$$(2 + 3i) + (5 - 4i) = 7 - i$$

$$(2 + 3i) - (5 - 4i) = -3 + 7i$$

$$(3 - 2i)(-2 + 5i) = \big(3(-2) - (-2)(5)\big) + \big(3(5) + (-2)(-2)\big)i = 4 + 19i$$

REMARKS

(1) Notice that $z = 0$ if and only if $x = 0$ and $y = 0$.

(2) If $z = x + y\,i$, we say that the **real part** of z is x and write $\mathrm{Re}(z) = x$. The **imaginary part** of z is y (*not* $y\,i$), and we write $\mathrm{Im}(z) = y$. It is important to remember that the imaginary part of z is a *real* number.

(3) If $x = 0$, $z = y\,i$ is said to be "pure imaginary". If $y = 0$, $z = x$ is "pure real". Notice that the usual real numbers are included in C as the subset of pure real complex numbers.

(4) The use of the words "real" and "imaginary" in this way is unfortunate because it may suggest an unintended distinction between "genuine" and "fake". Imaginary numbers are perfectly genuine mathematical tools. It is best to think of "real" and "imaginary" as *technical* words, which are used in this way for historical reasons.

(5) It is best not to use $\sqrt{-1}$ as a notation for the number whose square is -1; this can lead to confusion in some cases. In physics and engineering, where i is often used to denote electrical current, it is common to use j for the number such that $j^2 = -1$.

(6) It is sometimes convenient to write $x + iy$ instead of $x + y\,i$. This is particularly common with the polar form for complex numbers, which is discussed below.

The Complex Conjugate and Division

We have not yet discussed division of complex numbers. From the multiplication in Example 1, we can say that

$$\frac{4+19i}{3-2i} = -2+5i.$$

In order to give a systematic method for expressing the quotient of two complex numbers as a complex number in standard form, it is useful to introduce the complex conjugate.

DEFINITION

The **complex conjugate** of the complex number $z = x + yi$ is $x - yi$. The complex conjugate of z is denoted $\overline{z}$.

EXAMPLE 2

$$\overline{x+yi} = x - yi$$

$$\overline{2+5i} = 2-5i; \quad \overline{-3-2i} = -3+2i$$

Properties of the Complex Conjugate

(1) $\overline{(\overline{z})} = z$

(2) z is pure real if and only if $\overline{z} = z$.

(3) z is pure imaginary if and only if $\overline{z} = -z$.

(4) $\overline{(z_1 + z_2)} = \overline{z_1} + \overline{z_2}$

(5) $\overline{z_1 z_2} = \overline{z_1}\ \overline{z_2}$

(6) $\overline{z^n} = (\overline{z})^n$

(7) $z + \overline{z} = 2\,\mathrm{Re}(z)$ (pure real)

(8) $z - \overline{z} = i\,2\,\mathrm{Im}(z)$ (pure imaginary)

(9) $z\overline{z} = x^2 + y^2$ (non-negative real)

The proofs of these properties all follow directly from the definition. The best way for you to become comfortable with complex arithmetic is to prove these for yourself. (Exercise D1)

The **quotient** of two complex numbers can now be displayed as a complex number in standard form by multiplying both the numerator and the denominator by the complex conjugate of the denominator, and simplifying: if $w = a + bi$ and $z = c + di$, then

$$\begin{aligned}\frac{w}{z} = \frac{w\overline{z}}{z\overline{z}} &= \frac{(a+bi)(c-di)}{(c+di)(c-di)} \\ &= \frac{(ac+bd)+(bc-ad)i}{c^2+d^2} \\ &= \frac{ac+bd}{c^2+d^2} + \frac{bc-ad}{c^2+d^2}\,i.\end{aligned}$$

Notice that the quotient is defined for every pair of complex numbers w, z provided that the denominator is not zero.

EXAMPLE 3

$$\begin{aligned}\frac{2+5i}{3-4i} = \frac{(2+5i)(3+4i)}{(3-4i)(3+4i)} &= \frac{(6-20)+(8+15)i}{9+16} \\ &= -\frac{14}{25} + \frac{23}{25}\,i\end{aligned}$$

Roots of Polynomial Equations—Conjugate Pairs

The numbers $x + y\,i$ and $x - y\,i$ are called a **conjugate pair** of complex numbers. Conjugate pairs occur naturally as solutions or *roots* of quadratic equations with real coefficients. For example, i and $-i$ are the roots of $x^2 + 1 = 0$, and $(-1+i)$ and $(-1-i)$ are the roots of $x^2 + 2x + 2 = 0$.

A **polynomial** equation with real coefficients is an equation of the form

$$a_n x^n + a_{n-1} x^{n-1} + \cdots + a_1 x + a_0 = 0,$$

where the coefficients $a_n, a_{n-1}, \ldots, a_0$ are real numbers.

THEOREM Complex roots of a polynomial equation with real coefficients occur in conjugate pairs. That is, if z is a root, so is $\overline{z}$. (Note that a real root r satisfies $\overline{r} = r$, so we do not have a pair in this case.)

Proof. Suppose that z is a root, so that

$$a_n z^n + a_{n-1} z^{n-1} + \cdots + a_1 z + a_0 = 0.$$

Then

$$\overline{a_n z^n + a_{n-1} z^{n-1} + \cdots + a_1 z + a_0} = 0.$$

Use the properties of conjugation with respect to sums and products, and note that $\overline{a_k} = a_k$ because the coefficients are real. It follows that

$$a_n(\overline{z})^n + a_{n-1}(\overline{z})^{n-1} + \cdots + a_1(\overline{z}) + a_0 = 0,$$

so $\overline{z}$ is a root, as claimed.

■

EXAMPLE 4

Find the roots of $x^3 + 1 = 0$.

SOLUTION: By trial, $x = -1$ is one root, and $(x + 1)$ is a factor of $x^3 + 1 = 0$. Then

$$x^3 + 1 = (x + 1)(x^2 - x + 1).$$

We find the other roots by using the formula for quadratic equations: remember that the solutions of $ax^2 + bx + c = 0$ are

$$\frac{-b + \sqrt{b^2 - 4ac}}{2a} \quad \text{and} \quad \frac{-b - \sqrt{b^2 - 4ac}}{2a}.$$

Applying this formula to the equation $x^2 - x + 1 = 0$ gives the other roots as

$$\frac{1 + \sqrt{3}\,i}{2} \quad \text{and} \quad \frac{1 - \sqrt{3}\,i}{2},$$

and this is a conjugate pair.

Geometrical Representation—the Complex Plane

For some purposes, it is convenient to represent complex numbers as ordered pairs of real numbers: instead of $z = x + y\,i$, we write $z = (x, y)$. Then addition and multiplication appear as follows:

$$z_1 + z_2 = (x_1, y_1) + (x_2, y_2) = (x_1 + x_2, y_1 + y_2),$$
$$z_1 z_2 = (x_1, y_1)(x_2, y_2) = (x_1 x_2 - y_1 y_2, x_1 y_2 + x_2 y_1).$$

In terms of this ordered pair notation, it is natural to represent complex numbers as points in the plane, with the real part of z being the x-coordinate and the imaginary part being the y-coordinate. We will speak of the "real axis" and the "imaginary axis" in the "complex plane". See Figure A–1. A picture of this kind is sometimes called an Argand diagram.

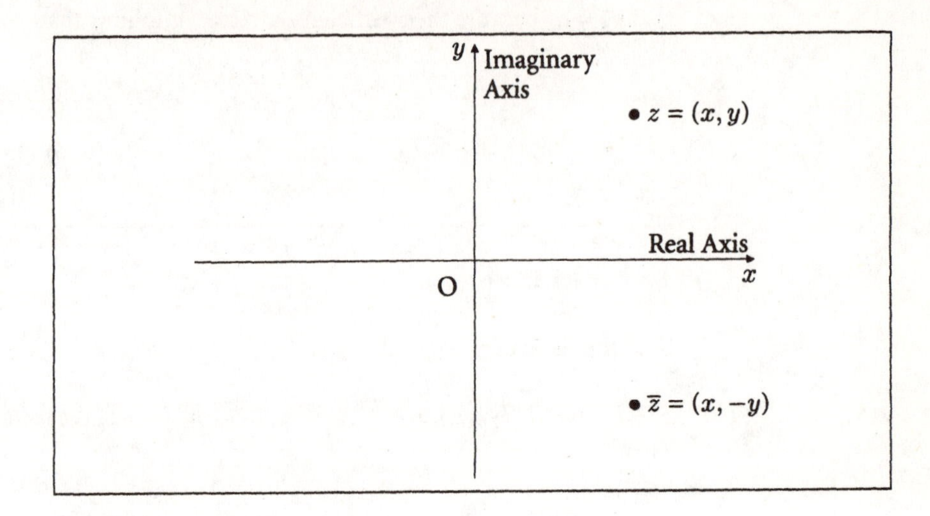

FIGURE A–1. *The complex plane.*

If k is a pure real number, then

$$kz = k(x, y) = (kx, ky).$$

Thus, with respect to addition and multiplication by *real* scalars, the complex plane is just like the usual plane $\mathbb{R}^2$. However, in the complex plane we can also multiply by complex scalars. This has a natural geometrical interpretation, which we shall describe in terms of the polar representation below. The next example describes one special case of multiplication.

EXAMPLE 5

(a) What is the point in the Argand diagram corresponding to i? (b) Give a geometrical interpretation of multiplication of complex numbers by i.

SOLUTION: (a) $i = (0, 1)$.

(b) $iz = (0, 1)(x, y) = (-y, x)$.

$(-y, x)$ is the point in the plane obtained from (x, y) by a rotation of the plane through angle $\pi/2$ radians counterclockwise about the origin.

Polar Form

DEFINITION

Given a complex number $z = x + yi$, let

$$|z| = r = \sqrt{x^2 + y^2},$$

and if $|z| \neq 0$, let θ be the angle measured counterclockwise from the positive x-axis such that

$$x = r\cos\theta \quad \text{and} \quad y = r\sin\theta.$$

$|z|$ is the **modulus** of z, and θ is the **argument** of z. The **polar form** of z is

$$z = r(\cos\theta + i\sin\theta).$$

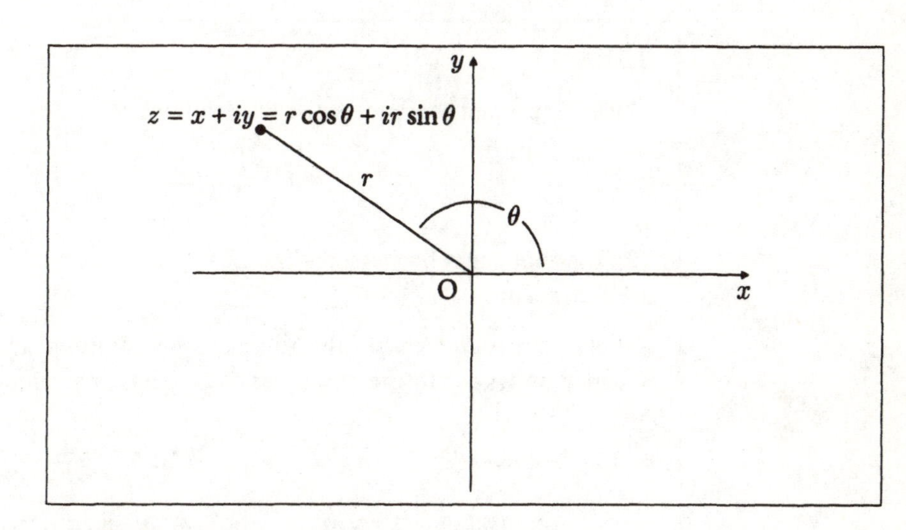

FIGURE A–2. *Polar representation of z.*

EXAMPLE 6

Determine the modulus, argument, and polar form of
(a) $2 - 2i$ and (b) $-1 + \sqrt{3}i$.

SOLUTION:

(a) $|2 - 2i| = \sqrt{4 + 4} = 2\sqrt{2}$.

θ satisfies

$$2 = 2\sqrt{2}\cos\theta, \; -2 = 2\sqrt{2}\sin\theta,$$

so $\cos\theta = \dfrac{1}{\sqrt{2}}$, $\sin\theta = -\dfrac{1}{\sqrt{2}}$, and $\theta = -\frac{\pi}{4}$.

$$2 - 2i = 2\sqrt{2}\left(\cos\left(-\frac{\pi}{4}\right) + i\sin\left(-\frac{\pi}{4}\right)\right).$$

(b) $|-1 + \sqrt{3}i| = \sqrt{1+3} = 2.$

Since $-1 = 2\cos\theta$ and $\sqrt{3} = 2\sin\theta$, $\theta = \dfrac{2\pi}{3}$.

$$z = 2\left(\cos\frac{2\pi}{3} + i\sin\frac{2\pi}{3}\right).$$

However, see Remark 3 below.

REMARKS

(1) An important consequence of the definition is

$$|z|^2 = x^2 + y^2 = z\bar{z}.$$

(2) The angles may be measured in radians or degrees. In this Appendix we use radians.

(3) Notice that for a given z the argument θ is *not* uniquely determined. We may always add integer multiples of 2π. Thus, in Example 6,

$$\text{argument of } (2 - 2i) = -\frac{\pi}{4} + 2k\pi, \quad k = 0, \pm 1, \pm 2, \ldots$$

$$\text{argument of } (-1 + \sqrt{3}i) = \frac{2\pi}{3} + 2k\pi, \quad k = 0, \pm 1, \pm 2, \ldots.$$

(4) It is tempting, but *very misleading*, to write $\theta = \arctan(y/x)$. Remember that you need *two* trigonometric functions to locate the correct quadrant for z. Also note that y/x is not defined if $x = 0$.

(5) If $z = r(\cos\theta + i\sin\theta)$, then the modulus of $\bar{z}$ is also r, and the argument of $\bar{z}$ is $(-\theta)$.

(6) The plural of modulus is **moduli**. We will need this in the statements of rules for products and quotients below.

Polar form is particularly convenient for multiplication and division, because of the trigonometric identities

$$\cos(\theta_1 + \theta_2) = \cos\theta_1 \cos\theta_2 - \sin\theta_1 \sin\theta_2$$
$$\sin(\theta_1 + \theta_2) = \sin\theta_1 \cos\theta_2 + \cos\theta_1 \sin\theta_2.$$

It follows that

$$\begin{aligned} z_1 z_2 &= r_1(\cos\theta_1 + i\sin\theta_1) r_2(\cos\theta_2 + i\sin\theta_2) \\ &= r_1 r_2 \Big(\cos\theta_1 \cos\theta_2 - \sin\theta_1 \sin\theta_2 + i(\sin\theta_1 \cos\theta_2 + \cos\theta_1 \sin\theta_2) \Big) \\ &= r_1 r_2 \Big(\cos(\theta_1 + \theta_2) + i \sin(\theta_1 + \theta_2) \Big). \end{aligned}$$

In words, the modulus of a product is the **product** of the moduli of the factors, while the argument of a product is the **sum** of the arguments.

Similarly, you should check that the identities lead to

$$\frac{z_1}{z_2} = \frac{r_1}{r_2}\Big(\cos(\theta_1 - \theta_2) + i\sin(\theta_1 - \theta_2)\Big),$$

so the modulus of a quotient is the **quotient** of the moduli of the factors, while the argument of the quotient is the **difference** of the arguments.

EXAMPLE 7

We give examples of multiplication and division in polar form.

(a) $$\begin{aligned} (1 - i)(-\sqrt{3} + i) &= \sqrt{2}\left(\cos\left(\frac{-\pi}{4}\right) + i\sin\left(\frac{-\pi}{4}\right)\right) 2\left(\cos\frac{5\pi}{6} + i\sin\frac{5\pi}{6}\right) \\ &= 2\sqrt{2}\left(\cos\frac{7\pi}{12} + i\sin\frac{7\pi}{12}\right) \\ &\approx -0.732 + i(2.732) \end{aligned}$$

(b) $$\begin{aligned} \frac{2 + 2i}{1 + \sqrt{3}i} &= \frac{2\sqrt{2}\left(\cos\frac{\pi}{4} + i\sin\frac{\pi}{4}\right)}{2\left(\cos\frac{\pi}{3} + i\sin\frac{\pi}{3}\right)} \\ &= \sqrt{2}\left(\cos\left(\frac{-\pi}{12}\right) + i\sin\left(\frac{-\pi}{12}\right)\right) \\ &\approx 1.366 - i(0.366) \end{aligned}$$

The polar representation allows us to give a geometrical interpretation to multiplication of complex numbers by some fixed complex number.

EXAMPLE 8

Interpret multiplication by $(-1+\sqrt{3}i)$ as a geometric mapping of the complex plane.

SOLUTION: Let $z = r(\cos\theta + i\sin\theta)$. Then

$$\begin{aligned}(-1+\sqrt{3}i)z &= 2\left(\cos\frac{2\pi}{3} + i\sin\frac{2\pi}{3}\right) r(\cos\theta + i\sin\theta)\\ &= 2r\left(\cos\left(\theta + \frac{2\pi}{3}\right) + i\sin\left(\theta + \frac{2\pi}{3}\right)\right).\end{aligned}$$

Thus to get geometrically from z to $(-1+\sqrt{3}i)z$, we multiply the modulus by 2 and increase the argument by $\frac{2\pi}{3}$. We have a *dilatation* of the plane by factor 2 followed by a *rotation* counterclockwise through $\frac{2\pi}{3}$ radians.

Powers and the Complex Exponential

From the rule for products, we find that

$$z^2 = r^2(\cos 2\theta + i\sin 2\theta).$$

Then

$$\begin{aligned}z^3 = z^2 z &= r^2 r(\cos(2\theta+\theta) + i\sin(2\theta+\theta))\\ &= r^3(\cos 3\theta + i\sin 3\theta).\end{aligned}$$

By induction, it can be shown that

$$z^n = r^n(\cos n\theta + i\sin n\theta).$$

This is called **de Moivre's formula.**

EXAMPLE 9

$$(2+2i)^3 = \left(2\sqrt{2}(\cos\frac{\pi}{4} + i\sin\frac{\pi}{4})\right)^3 = 16\sqrt{2}\left(\cos\frac{3\pi}{4} + i\sin\frac{3\pi}{4}\right)$$
$$= 16\sqrt{2}\left(-\frac{1}{\sqrt{2}} + i\frac{1}{\sqrt{2}}\right) = -16 + 16i$$

In the case where $r = 1$, de Moivre's formula reduces to

$$(\cos\theta + i\sin\theta)^n = (\cos n\theta + i\sin n\theta).$$

This is formally just like one of the exponential laws, $(\exp(\theta))^n = \exp(n\theta)$. Since we have not previously dealt with exponentials with complex variables, we make the following definition.

DEFINITION

$e^{i\theta} = \cos\theta + i\sin\theta$, where e is the usual preferred base for exponentials in calculus ($e \approx 2.71828\ldots$).

The formula is called Euler's formula. To see why this is a sensible definition, we would really need to think about what we mean by a function of a complex variable. We can, however, do some things to convince ourselves that Euler's formula is reasonable.

First, we can check that certain properties hold for both sides of Euler's formula. For example,

$$e^{i0} = e^0 = 1 = (\cos 0 + i\sin 0).$$

Also

$$e^{i\theta_1}e^{i\theta_2} = e^{i(\theta_1+\theta_2)} \qquad \text{(by the exponential law)}$$

and this corresponds to the product rule

$$(\cos\theta_1 + i\sin\theta_1)(\cos\theta_2 + i\sin\theta_2) = \cos(\theta_1+\theta_2) + i\sin(\theta_1+\theta_2).$$

Alternatively, if one knows the Taylor (Maclaurin) series for e^x, $\cos x$, and

$\sin x$, one has (formally)

$$\begin{aligned} e^{i\theta} &= 1 + i\theta + \frac{i^2\theta^2}{2} + \frac{i^3\theta^3}{3} + \cdots \\ &= (1 - \frac{\theta^2}{2} + \frac{\theta^4}{4!} + \cdots) + i(\theta - \frac{\theta^3}{3!} + \cdots) \\ &= \cos\theta + i\sin\theta. \end{aligned}$$

We often write $e^{i\theta}$ in place of $(\cos\theta + i\sin\theta)$. One interesting consequence of Euler's formula is the following:

$$e^{i\pi} = -1, \qquad \text{or} \qquad e^{i\pi} + 1 = 0.$$

In one formula, we have five of the most important numbers in mathematics: $0, 1, e, i, \pi$.

One area where Euler's formula has important applications is ordinary differential equations. There, one often uses the fact that

$$e^{(a+bi)t} = e^{at}e^{ibt} = e^{at}(\cos bt + i\sin bt).$$

EXAMPLE 10

(a) $(2i)^3 = \left(2e^{\frac{i\pi}{2}}\right)^3 = 8e^{i\frac{3\pi}{2}} = -8i$

(b) $$\begin{aligned} (\sqrt{3} + i)^5 &= \left(2e^{i\frac{\pi}{6}}\right)^5 = 32e^{i\frac{5\pi}{6}} \\ &= 32\left(\cos\frac{5\pi}{6} + i\sin\frac{5\pi}{6}\right) \\ &= 32\left(\frac{-\sqrt{3}}{2} + \frac{1}{2}i\right) = -16\sqrt{3} + 16i \end{aligned}$$

---

n^{th} Roots

The formula of de Moivre for n^{th} powers is the key to finding n^{th} roots. Suppose that we need to find the nth root of the non-zero complex number $z = re^{i\theta}$. That is, we need a number w such that $w^n = z$. Suppose that $w = Re^{i\phi}$. Then $w^n = z$ implies that

$$R^n e^{in\phi} = re^{i\theta}.$$

Then R is the real n^{th} root of the positive real number r. However, because arguments of complex numbers are determined only up to the addition of $2k\pi$, all we can say about ϕ is that

$$n\phi = \theta + 2k\pi, \quad k = 0, \pm 1, \pm 2, \ldots$$

or

$$\phi = \frac{\theta + 2k\pi}{n}, \quad k = 0, \pm 1, \pm 2, \ldots.$$

EXAMPLE 11

Find all the cube roots of 8.

SOLUTION: $8 = 8e^{i(0+2k\pi)}$, $k = 0, \pm 1, \pm 2, \ldots$. Thus $8^{1/3} = 2e^{i\frac{2k\pi}{3}}$, $k = 0, \pm 1, \pm 2, \ldots$.

If $k = 0$, we have the root $w_0 = 2e^0 = 2$.

If $k = 1$, we have the root $w_1 = 2e^{\frac{i2\pi}{3}} = -1 + \sqrt{3}i$.

If $k = 2$, we have the root $w_2 = 2e^{\frac{i4\pi}{3}} = -1 - \sqrt{3}i$.

If $k = 3$, we have the root $2e^{i2\pi} = 2 = w_0$.

By increasing k further, we simply repeat the roots we have already found, and similarly, consideration of negative k gives us no further roots. The number 8 has three third roots w_0, w_1, w_2. We should have expected this since we are really solving the equation $w^3 - 8 = 0$.

This same pattern will always occur: *a non-zero complex number $z = re^{i\theta}$ has n distinct n^{th} roots,*

$$r^{1/n}e^{i\frac{(\theta+2k\pi)}{n}}, \quad n = 0, 1, 2, \ldots, n-1.$$

EXAMPLE 12

Find the 4th roots of (-81).

SOLUTION: $-81 = 81e^{i(\pi+2k\pi)}$.

Thus the fourth roots are $3e^{i\frac{\pi}{4}}, 3e^{i\frac{3\pi}{4}}, 3e^{i\frac{5\pi}{4}}, 3e^{i\frac{7\pi}{4}}$, or

$$\frac{3\sqrt{2}}{2}(1+i), \frac{3\sqrt{2}}{2}(-1+i), \frac{3\sqrt{2}}{2}(-1-i), \frac{3\sqrt{2}}{2}(1-i).$$

In the previous two examples, we took roots of numbers that were pure real: we were really solving $z^n - a = 0$, where a was real. By our earlier theorem, when the coefficients of the equations are real, the roots that are not real occur in complex conjugate pairs. As a contrast, let us consider roots of a number that is not real.

EXAMPLE 13

Find the cube roots of $5i$ and illustrate in an Argand diagram.

SOLUTION: $5i = 5e^{i(\frac{\pi}{2}+2k\pi)}$, so the cube roots are

$$w_0 = 5^{1/3}e^{i\frac{\pi}{6}} = 5^{1/3}\left(\frac{\sqrt{3}}{2} + i\frac{1}{2}\right), \quad \text{for } k = 0;$$

$$w_1 = 5^{1/3}e^{i\frac{5\pi}{6}} = 5^{1/3}\left(-\frac{\sqrt{3}}{2} + i\frac{1}{2}\right), \quad \text{for } k = 1;$$

and $w_2 = 5^{1/3}e^{i\frac{9\pi}{6}} = 5^{1/3}(-i)$, for $k = 2$.

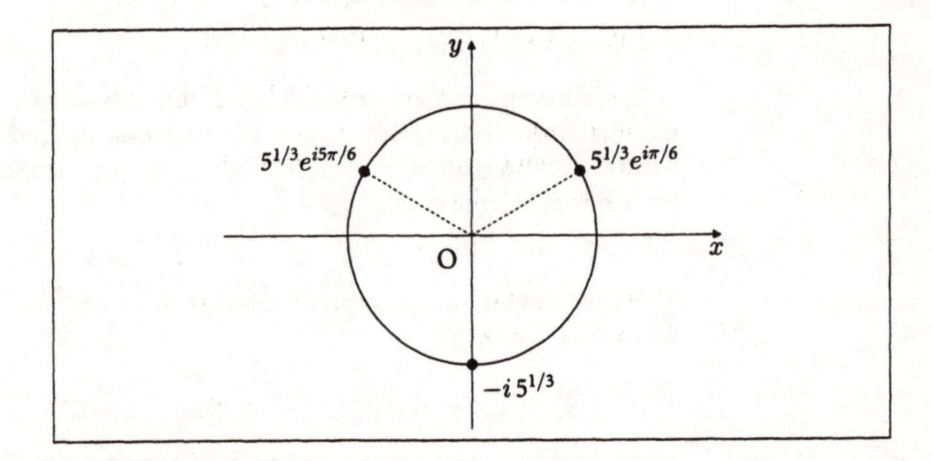

FIGURE A–3. *The three cube roots of $5i$.*

Notice that all three roots are points on the circle of radius $5^{1/3}$ centred at the origin, and that they are separated by equal angles $\frac{2\pi}{3}$.

Examples 11, 12 and 13 all illustrate a general rule: The nth roots of a complex number $z = re^{i\theta}$ all lie on the circle of radius $r^{1/n}$ and they are separated by equal angles $2\pi/n$.

EXERCISES

A1 Determine the following sums or differences.

(a) $(2+5i)+(3+2i)$ (b) $(2-7i)+(-5+3i)$

(c) $(-3+5i)-(4+3i)$ (d) $(-5-6i)-(-9-11i)$

A2 Express the following products in standard form.

(a) $(1+3i)(3-2i)$ (b) $(-2-4i)(3-i)$

(c) $(1-6i)(-4+i)$ (d) $(-1-i)(1-i)$

A3 Determine the complex conjugates of the following numbers.

(a) $(3-5i)$ (b) $(2+7i)$

A4 Determine the real and imaginary parts of

(a) $3-6i$; (b) $(2+5i)(1-3i)$; (c) $\dfrac{4}{6-i}$.

A5 Express the following quotients in standard form.

(a) $\dfrac{1}{2+3i}$ (b) $\dfrac{3}{2-7i}$

(c) $\dfrac{2-5i}{3+2i}$ (d) $\dfrac{1+6i}{4-i}$

A6 Use polar form to determine wz and $\dfrac{w}{z}$ if

(a) $w=1+i,\ z=1+\sqrt{3}i$; (b) $w=-\sqrt{3}-i,\ z=1-i$;

(c) $w=1+2i,\ z=-2-3i$; (d) $w=-3+i,\ z=6-i$.

A7 Use polar form to determine

(a) $(3-3i)^3$; (b) $(-1-\sqrt{3}i)^4$; (c) $(-2\sqrt{3}+2i)^5$.

A8 Use polar form to determine all the indicated roots.

(a) $(-1)^{1/5}$ (b) $(-16i)^{1/4}$

(c) $(-\sqrt{3}-i)^{1/3}$ (d) $(1+4i)^{1/3}$

B1 Determine the following sums or differences.

(a) $(3+4i)+(1+5i)$ (b) $(3-2i)+(-7+6i)$

(c) $(-5+7i)-(2+6i)$ (c) $(-7-2i)-(-8-9i)$

B2 Express the following products in standard form.

(a) $(2+i)(5-3i)$ (b) $(-3-2i)(5-2i)$

(c) $(3-5i)(-+6i)$ (d) $(-3-i)(3-i)$

B3 Determine the complex conjugates of the following numbers.

(a) $(4-8i)$ (b) $(5+11i)$

B4 Determine the real and imaginary parts of

(a) $4-7i$; (b) $(3+2i)(2-3i)$; (c) $\dfrac{5}{4-i}$.

B5 Express the following quotients in standard form.

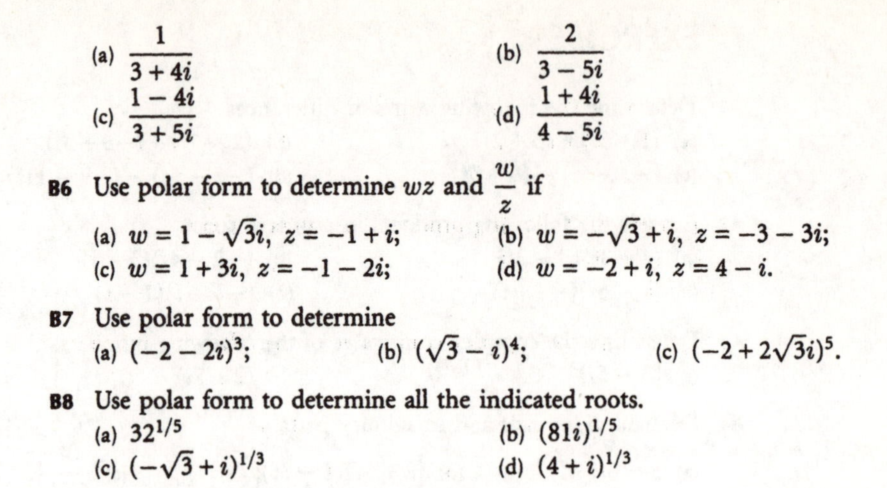

(a) $\dfrac{1}{3+4i}$ (b) $\dfrac{2}{3-5i}$

(c) $\dfrac{1-4i}{3+5i}$ (d) $\dfrac{1+4i}{4-5i}$

B6 Use polar form to determine wz and $\dfrac{w}{z}$ if

(a) $w = 1 - \sqrt{3}i,\ z = -1 + i$; (b) $w = -\sqrt{3} + i,\ z = -3 - 3i$;

(c) $w = 1 + 3i,\ z = -1 - 2i$; (d) $w = -2 + i,\ z = 4 - i$.

B7 Use polar form to determine

(a) $(-2-2i)^3$; (b) $(\sqrt{3}-i)^4$; (c) $(-2+2\sqrt{3}i)^5$.

B8 Use polar form to determine all the indicated roots.

(a) $32^{1/5}$ (b) $(81i)^{1/5}$

(c) $(-\sqrt{3}+i)^{1/3}$ (d) $(4+i)^{1/3}$

CONCEPTUAL EXERCISES

D1 Give a detailed verification in the general case of properties (1)–(9) of complex conjugation.

D2 Give a geometrical interpretation of complex conjugation as a mapping of the complex plane. (See Figure A–1.)

D3 If $z = r(\cos\theta + i\sin\theta)$, what is $|\bar{z}|$? What is the argument of $\bar{z}$?

D4 Write out the details of the proof that

$$\frac{z_1}{z_2} = \frac{r_1}{r_2}\big(\cos(\theta_1 - \theta_2) + i\sin(\theta_1 - \theta_2)\big).$$

D5 Use Euler's formula to show that

(a) $\overline{e^{i\theta}} = e^{-i\theta}$; (b) $\cos t = \dfrac{e^{it} + e^{-it}}{2}$; (c) $\sin t = \dfrac{e^{it} - e^{-it}}{2i}$.

ANSWERS TO A EXERCISES

CHAPTER 1

SECTION 1-1

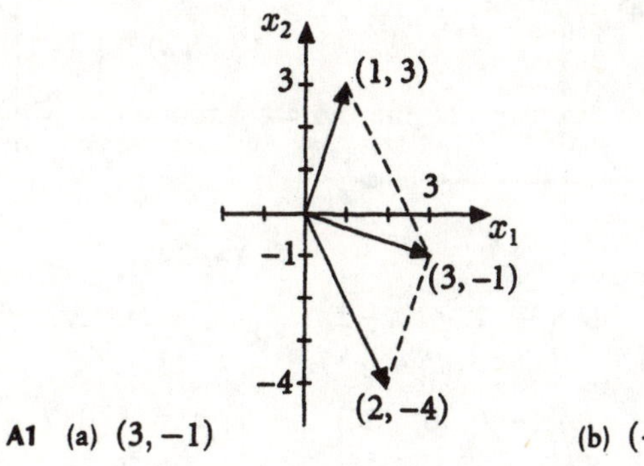

A1 (a) $(3, -1)$

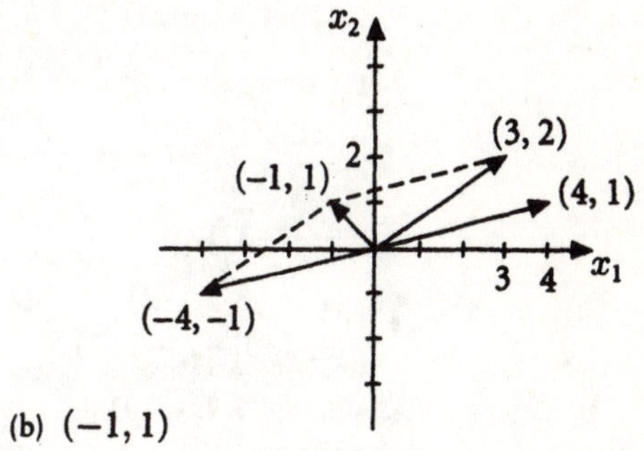

(b) $(-1, 1)$

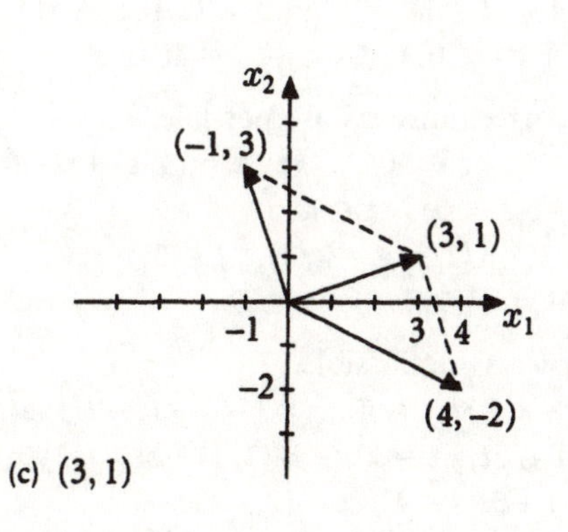

(c) $(3, 1)$

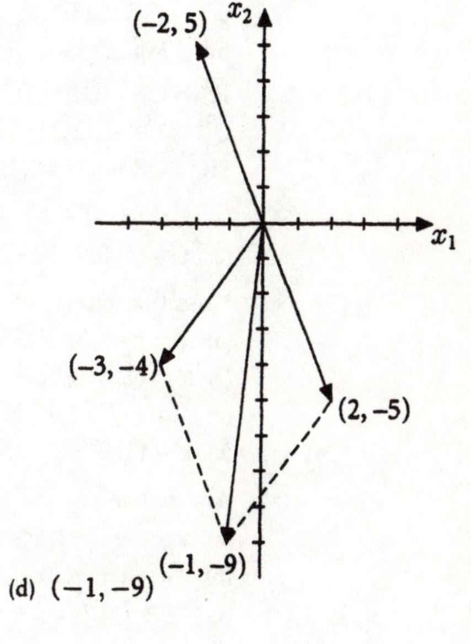

(d) $(-1, -9)$

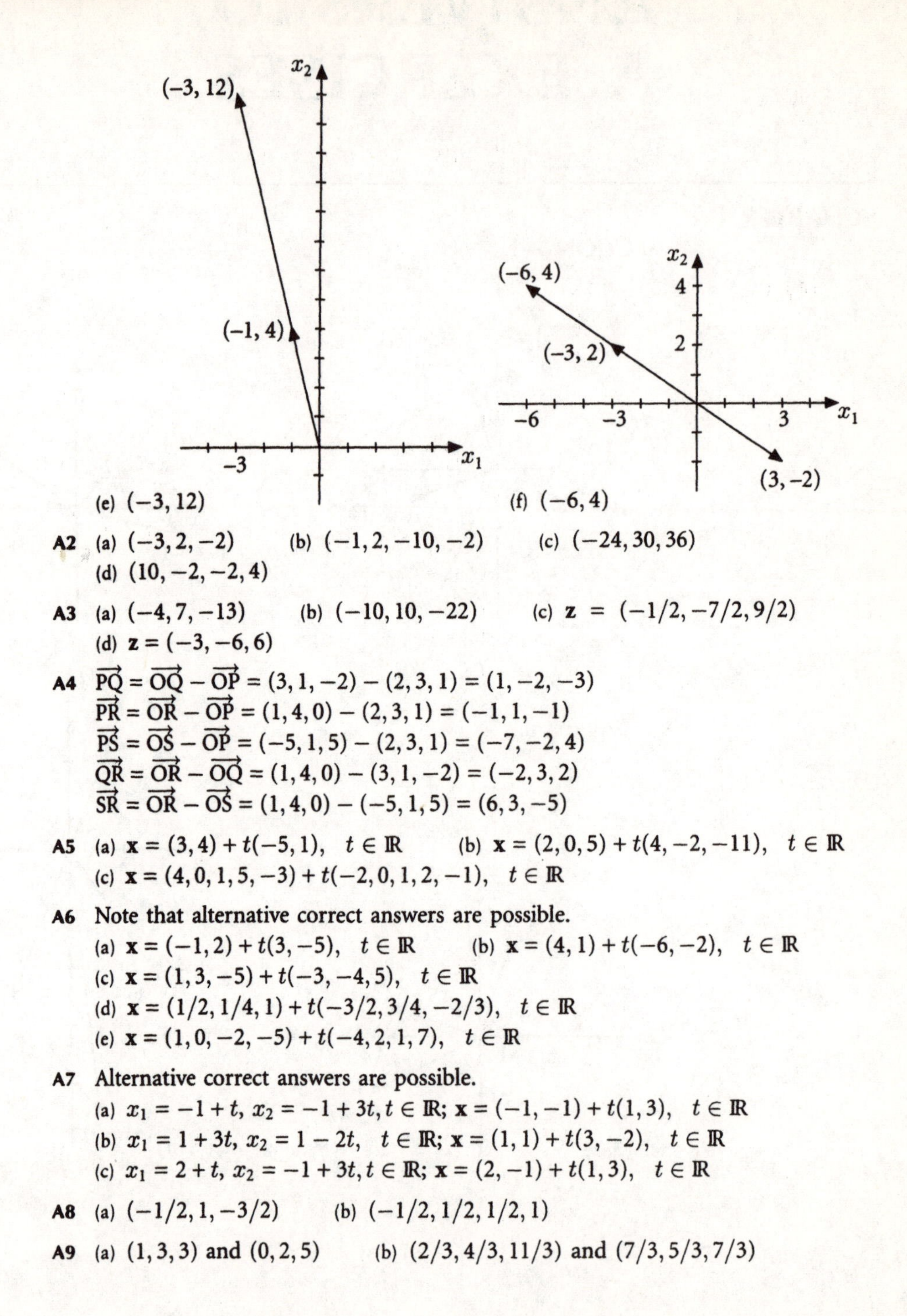

(e) $(-3, 12)$

(f) $(-6, 4)$

A2 (a) $(-3, 2, -2)$ (b) $(-1, 2, -10, -2)$ (c) $(-24, 30, 36)$
(d) $(10, -2, -2, 4)$

A3 (a) $(-4, 7, -13)$ (b) $(-10, 10, -22)$ (c) $\mathbf{z} = (-1/2, -7/2, 9/2)$
(d) $\mathbf{z} = (-3, -6, 6)$

A4 $\overrightarrow{PQ} = \overrightarrow{OQ} - \overrightarrow{OP} = (3, 1, -2) - (2, 3, 1) = (1, -2, -3)$
$\overrightarrow{PR} = \overrightarrow{OR} - \overrightarrow{OP} = (1, 4, 0) - (2, 3, 1) = (-1, 1, -1)$
$\overrightarrow{PS} = \overrightarrow{OS} - \overrightarrow{OP} = (-5, 1, 5) - (2, 3, 1) = (-7, -2, 4)$
$\overrightarrow{QR} = \overrightarrow{OR} - \overrightarrow{OQ} = (1, 4, 0) - (3, 1, -2) = (-2, 3, 2)$
$\overrightarrow{SR} = \overrightarrow{OR} - \overrightarrow{OS} = (1, 4, 0) - (-5, 1, 5) = (6, 3, -5)$

A5 (a) $\mathbf{x} = (3, 4) + t(-5, 1), \quad t \in \mathbb{R}$ (b) $\mathbf{x} = (2, 0, 5) + t(4, -2, -11), \quad t \in \mathbb{R}$
(c) $\mathbf{x} = (4, 0, 1, 5, -3) + t(-2, 0, 1, 2, -1), \quad t \in \mathbb{R}$

A6 Note that alternative correct answers are possible.
(a) $\mathbf{x} = (-1, 2) + t(3, -5), \quad t \in \mathbb{R}$ (b) $\mathbf{x} = (4, 1) + t(-6, -2), \quad t \in \mathbb{R}$
(c) $\mathbf{x} = (1, 3, -5) + t(-3, -4, 5), \quad t \in \mathbb{R}$
(d) $\mathbf{x} = (1/2, 1/4, 1) + t(-3/2, 3/4, -2/3), \quad t \in \mathbb{R}$
(e) $\mathbf{x} = (1, 0, -2, -5) + t(-4, 2, 1, 7), \quad t \in \mathbb{R}$

A7 Alternative correct answers are possible.
(a) $x_1 = -1 + t,\ x_2 = -1 + 3t, t \in \mathbb{R}$; $\mathbf{x} = (-1, -1) + t(1, 3), \quad t \in \mathbb{R}$
(b) $x_1 = 1 + 3t,\ x_2 = 1 - 2t, \quad t \in \mathbb{R}$; $\mathbf{x} = (1, 1) + t(3, -2), \quad t \in \mathbb{R}$
(c) $x_1 = 2 + t,\ x_2 = -1 + 3t, t \in \mathbb{R}$; $\mathbf{x} = (2, -1) + t(1, 3), \quad t \in \mathbb{R}$

A8 (a) $(-1/2, 1, -3/2)$ (b) $(-1/2, 1/2, 1/2, 1)$

A9 (a) $(1, 3, 3)$ and $(0, 2, 5)$ (b) $(2/3, 4/3, 11/3)$ and $(7/3, 5/3, 7/3)$

A10 (a) $(0, 13/4, -11/4)$

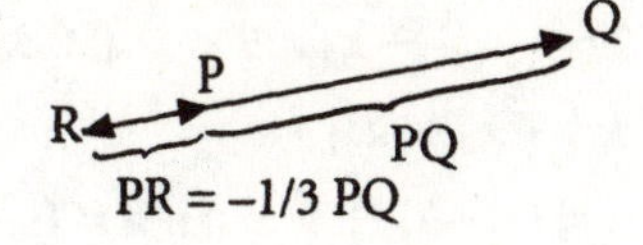

(b) $(0, -1, -2/3, 8)$

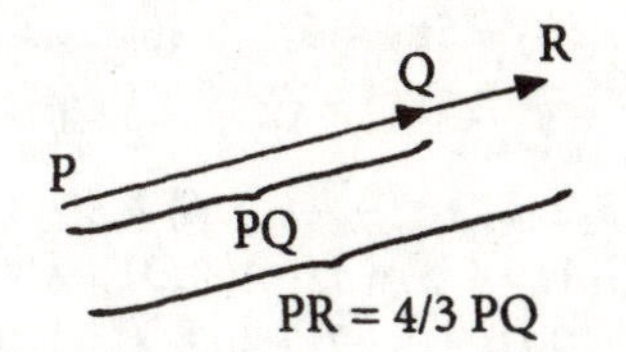

(c) $(-14/3, 1, 6)$

A11 (a) $(-25/17, -36/17)$ (b) $(0, 1, 2)$ (c) no point of intersection
(d) $(7, -2, 5)$

A12 (a) $\vec{AB} = r\vec{AC}$, for some $r \in \mathbb{R}$

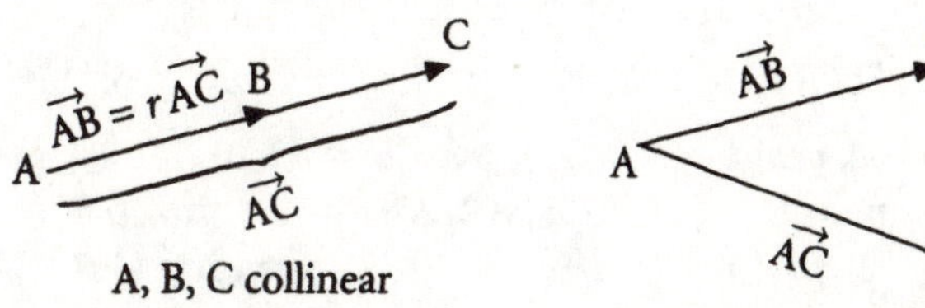

A, B, C non-collinear

(b) Since $-2\vec{PQ} = (-6, 2, -4, -2) = \vec{PR}$, the points P, Q and R must be collinear. (c) The points S, T and U are not collinear because $\vec{SU} \neq r\vec{ST}$ for any r.

SECTION 1–2

A1 (a) $\sqrt{29}$ (b) $\sqrt{17}$ (c) $\sqrt{251}/5$ (d) $\sqrt{6}$

A2 (a) $2\sqrt{10}$ (b) 5 (c) $\sqrt{170}$ (d) $3\sqrt{6}$

A3 (a) $\|\mathbf{x}\| = \sqrt{26}$; $\|\mathbf{y}\| = \sqrt{30}$; $\|\mathbf{x} + \mathbf{y}\| = 2\sqrt{22}$; $|\mathbf{x} \cdot \mathbf{y}| = 16$; The triangle inequality: $2\sqrt{22} \approx 9.38 \leq \sqrt{26} + \sqrt{30} \approx 10.58$; The Cauchy-Schwarz inequality: $16 \leq \sqrt{26(30)} \approx 27.93$;
(b) $\|\mathbf{x}\| = \sqrt{6}$; $\|\mathbf{y}\| = \sqrt{29}$; $\|\mathbf{x} + \mathbf{y}\| = \sqrt{41}$; $|\mathbf{x} \cdot \mathbf{y}| = 3$; The triangle inequality: $\sqrt{41} \approx 6.40 \leq \sqrt{6} + \sqrt{29} \approx 7.83$; The Cauchy-Schwarz inequality: $3 \leq \sqrt{6(29)} \approx 13.19$

A4 (a) $\cos\theta \approx 0.47619$, $\theta \approx 1.074$ radians. (b) $\cos\theta \approx 0.12910$, $\theta \approx 1.441$ radians. (c) $\cos\theta \approx 0.38100$, $\theta \approx 1.180$ radians.

A5 (a) $(1,3,2)\cdot(2,-2,2)=0$; these vectors are orthogonal.
(b) $(-3,1,7)\cdot(2,-1,1)=0$; these vectors are orthogonal.
(c) $(2,1,1)\cdot(-1,4,2)\neq 0$; these vectors are not orthogonal.
(d) $(4,1,0,-2)\cdot(-1,4,3,0)=0$; these vectors are orthogonal.

A6 (a) $k=6$ (b) $k=0$ or $k=3$ (c) $k=-3$ (d) any $k\in\mathbb{R}$

A7 (a) $2x_1+4x_2-x_3=9$ (b) $3x_1+5x_3=26$ (c) $3x_1-4x_2+x_3=8$

A8 (a) $3x_1+x_2+4x_3+x_4=-2$ (b) $x_2+3x_3+3x_4=1$

A9 (a) $\mathbf{n}=(3,-2,1)$ (b) $\mathbf{n}=(-4,3,-5)$ (c) $\mathbf{n}=(1,-1,2,-3)$

A10 (a) $2x_1-3x_2+5x_3=6$ (b) $x_2=-2$

A11 (a) $(20/13,51/13,37/13)$ (b) $(7/3,-1/3,-2/3)$

A12 (a) $\mathbf{x}=(3,2,1)+r(-7,-1,6)+s(-1,-2,-1)\quad r,s\in\mathbb{R}$.
(b) $\mathbf{x}=(-1,-4,3)+r(-1,8,3)+s(4,5,-7)\quad r,s\in\mathbb{R}$.
(c) $\mathbf{x}=(1,0,0)+r(-1,1,0)+s(-1,0,1)\quad r,s\in\mathbb{R}$.

A13 (a) The line is parallel to the plane.
(b) The line is orthogonal to the plane.
(c) The line is neither parallel nor orthogonal to the plane, $\theta\approx 0.702$ radians.
(d) The line is orthogonal to the plane.
(e) The line is parallel to the plane.

SECTION 1–3

A1 (a) $\mathbf{proj_a b}=(0,-5)$, $\mathbf{perp_a b}=(3,0)$ (b) $\mathbf{proj_a b}=(36/25,48/25)$, $\mathbf{perp_a b}=(-136/25,102/25)$ (c) $\mathbf{proj_a b}=(0,5,0)$, $\mathbf{perp_a b}=(-3,0,2)$
(d) $\mathbf{proj_a b}=(-4/9,8/9,-8/9)$, $\mathbf{perp_a b}=(40/9,1/9,-19/9)$

A2 (a) $\hat{\mathbf{u}}=(2/7,6/7,3/7)$ (b) $110/7$ (c) $(220/49,660/49,330/49)$
(d) $(270/49,222/49,-624/49)$

A3 (a) $(3/\sqrt{14},1/\sqrt{14},-2/\sqrt{14})$ (b) $16/\sqrt{14}$ (c) $(24/7,8/7,-16/7)$
(d) $(-3/7,69/7,30/7)$

A4 (a) $(-2/17,-3/17,2/17)$, $(70/17,-14/17,49/17)$
(b) $(3/2,3/2,-3)$, $(5/2,-1/2,1)$
(c) $(14/3,-7/3,7/3)$, $(1/3,4/3,2/3)$
(d) $(1/3,-2/3,-1/3,1)$, $(5/3,-1/3,7/3,0)$

A5 (a) $(5/2,5/2)$, $5/\sqrt{2}$ (b) $(58/17,91/17)$, $6/\sqrt{17}$
(c) $(17/6,1/3,-1/6)$, $\sqrt{29/6}$ (d) $(5/3,11/3,-1/3)$, $\sqrt{6}$

A6 (a) Let $A=(0,-5,0)$ be a point in the plane, $B=(2,3,1)$,
$d=\|\mathbf{proj_n}\vec{AB}\|=\dfrac{2}{\sqrt{26}}$ (b) $\dfrac{13}{\sqrt{38}}$ (c) $\dfrac{4}{\sqrt{5}}$ (d) $\sqrt{6}$

A7 (a) Let $A=(0,0,0,0)$ be a point in the hyperplane, $B=(2,4,3,4)$,
$\mathbf{perp_n}\vec{AB}=\frac{1}{3}(0,14,1,10)$ (b) $\frac{1}{7}(-12,11,9,-9)$

SECTION 1–4

A1 (a) $(-27,-9,-9)$ (b) $(-31,-34,8)$ (c) $(-4,5,-4)$

A2 (a) $(0,0,0)$ (b) $\mathbf{p}\times\mathbf{q} = (-6,5,-13)$ (c) $\mathbf{p}\times 3\mathbf{r} = (6,9,-15)$
(d) $\mathbf{p}\times(\mathbf{q}+\mathbf{r}) = (-4,8,-18)$ (e) $\mathbf{p}\times(\mathbf{q}\times\mathbf{r}) = (-46,-19,15)$,
$(\mathbf{p}\times\mathbf{q})\times\mathbf{r} = (-44,-32,8)$

A3 (a) $\sqrt{35}$ (b) $\sqrt{11}$ (c) 9 (d) 13

A4 (a) Point of intersection: $\mathbf{x}_0 = (-3,1,3)$, $-x_1+5x_2+3x_3 = 17$
(b) No point of intersection, $\dfrac{9}{\sqrt{35}}$

A5 (a) No point of intersection, $\dfrac{11}{\sqrt{10}}$ (b) point of intersection, $\mathbf{x}_0 = (1,0,7)$,
$3x_1 - x_2 - x_3 = -4$,

A6 (a) $x_1 - 4x_2 - 10x_3 = -85$ (b) $2x_1 - 2x_2 + 3x_3 = -5$
(c) $-5x_1 - 2x_2 + 6x_3 = 15$

A7 (a) $39x_1 + 12x_2 + 10x_3 = 140$ (b) $11x_1 - 21x_2 - 17x_3 = -56$
(c) $-12x_1 + 3x_2 - 19x_3 = -14$

A8 (a) Method 1: $\mathbf{x} = (46/11, 3/11, 0) + t(-2,-3,-11)$, $t\in\mathbb{R}$
Method 2: $\mathbf{x} = (46/11, 3/11, 0) + s(2/11, 3/11, 1)$, $s\in\mathbb{R}$
(b) Method 1: $\mathbf{x} = (7/2, 4, 0) + t(3,-4,2)$, $t\in\mathbb{R}$
Method 2: $\mathbf{x} = (7/2, 4, 0) + s(3/2, -2, 1)$, $s\in\mathbb{R}$

A9 (a) 126 (b) 5

Chapter 1 Quiz

1 $\mathbf{x} = (-2,1,-4) + t(7,-3,5)$, $t\in\mathbb{R}$

2 $8x_1 - x_2 + 7x_3 = 9$.

3 (a) The line is neither parallel nor orthogonal to the plane.
(b) The line is orthogonal to the plane.
(c) The line is parallel to the plane.

4 $\cos\alpha = \dfrac{2}{\sqrt{14}}$, $\cos\beta = \dfrac{-3}{\sqrt{14}}$, $\cos\gamma = \dfrac{1}{\sqrt{14}}$

5 $(18/11, -12/11, 18/11)$

6 $(5/2, -5/2, -1/2, 3/2)$, 1

7 $(2,-1,7)$

8 Hence, $\overrightarrow{PQ}$ and $\overrightarrow{SR}$ are parallel, and $\overrightarrow{QR}$ and $\overrightarrow{PS}$ are parallel; so $PQRS$ forms a parallelogram. The area is $\sqrt{2}$.

9 $|(\mathbf{a}+k\mathbf{b})\cdot(\mathbf{b}\times\mathbf{c})| = |\mathbf{a}\cdot(\mathbf{b}\times\mathbf{c}) + 0| = |\mathbf{a}\cdot(\mathbf{b}\times\mathbf{c})|$

10 (a) False; many examples. (b) True. $\mathbf{d}_1 = r\mathbf{d}_2$ $\mathbf{d}_2 = s\mathbf{d}_3$ Hence $\mathbf{d}_1 = rs\mathbf{d}_3$
(c) False. The three points $(0,0,0)$, $(1,4,-3)$ and $(-3,-12,9)$ are collinear,

and thus lie in infinitely many planes. (d) True, unless the vector is the zero vector. $\mathbf{x} \cdot \mathbf{x} = x_1^2 + x_2^2 + x_3^2$ (e) False. Any example where $\mathbf{a}$ is not parallel or orthogonal to $\mathbf{b}$. (f) True. $\|\mathbf{a} \times (\mathbf{b} + 3\mathbf{a})\| = \|(\mathbf{a} \times \mathbf{b}) + (\mathbf{a} \times 3\mathbf{a}\| = \|\mathbf{a} \times \mathbf{b}\|$

CHAPTER 2

SECTION 2-1

A1 (a) $(17, 4)$ (b) $(13, 0, 6) + t(-2, 1, 0),\ t \in \mathbb{R}$ (c) $(32, -8, 2)$
(d) $(-1, -3, 2, 0) + t(-1, 1, -1, 1),\ t \in \mathbb{R}$
(e) $(3, -1, 0, 1, 0) + s(5, -1, 1, 0, 0) + t(-36, 15, 0, 4, 1),\ s, t \in \mathbb{R}$

A2 (a) $\begin{bmatrix} 1 & 2 & 5 & -2 \\ 0 & 1 & 3 & -1 \\ 0 & 0 & 3 & 7 \end{bmatrix}$ (b) $\begin{bmatrix} 7 & 3 & 5 & 4 \\ 0 & 1 & 18 & 34 \\ 0 & 0 & -24 & -56 \end{bmatrix}$

(c) $\begin{bmatrix} 1 & 13 & -9 & 8 \\ 0 & -12 & 8 & -6 \\ 0 & -64 & 48 & -37 \\ 0 & 70 & -51 & 53 \end{bmatrix}$ (d) $\begin{bmatrix} 1 & 1 & -1 & 2 \\ 0 & 12 & -8 & 6 \\ 0 & -4 & 8 & -7 \\ 0 & -2 & -3 & 17 \end{bmatrix}$

A3 (a) Matrix A is not in row echelon form because the first non-zero entry of row 3 is not a 1. Matrix C is not in row echelon form because the leading 1 of the third row is not further to the right than the leading 1 of the second row. Matrix D is not in row echelon form because the first non-zero entry of row 2 is not a 1. Matrix B is in row echelon form.

(b) $\begin{bmatrix} 1 & 2 & 3 \\ 0 & 1 & -2 \\ 0 & 0 & 0 \end{bmatrix}$ is in row echelon form. The system is inconsistent.

$\begin{bmatrix} 0 & 1 & 2 \\ 0 & 0 & 1 \\ 0 & 0 & 0 \end{bmatrix}$ is in row echelon form. The system is consistent. $\begin{bmatrix} 1 & -1 & -2 \\ 0 & 1 & 2 \\ 0 & 1 & 0 \end{bmatrix}$

is not in row echelon form. The system is consistent. $\begin{bmatrix} 1 & 0 & 2 \\ 0 & 2 & 1 \\ 0 & 0 & 1 \end{bmatrix}$ is not in row echelon form. The system is consistent.

A4 (a) $\begin{bmatrix} 1 & 1/4 & 1/4 \\ 0 & 1 & -7/13 \end{bmatrix}$ (b) $\begin{bmatrix} 1 & -1 & 5/2 & 4 \\ 0 & 0 & 1 & 2 \\ 0 & 0 & 0 & 1 \end{bmatrix}$ (c) $\begin{bmatrix} 1 & -1 & -1 \\ 0 & 1 & 0 \\ 0 & 0 & 1 \\ 0 & 0 & 0 \end{bmatrix}$

(d) $\begin{bmatrix} 1 & 0 & 1 & 0 \\ 0 & 1 & 1 & 2 \\ 0 & 0 & 1 & 2 \\ 0 & 0 & 0 & 0 \end{bmatrix}$ (e) $\begin{bmatrix} 1 & 2 & 1 & 1 \\ 0 & 1 & 2 & 1 \\ 0 & 0 & 1 & 5/7 \\ 0 & 0 & 0 & 1 \end{bmatrix}$

(f) $\begin{bmatrix} 1 & 0 & 3 & 0 & 1 \\ 0 & 1 & -1 & 2 & 1 \\ 0 & 0 & 1 & 1/24 & 11/24 \\ 0 & 0 & 0 & 0 & 1 \end{bmatrix}$

A5 (a) Inconsistent.

(b) Consistent. The solution is not unique.
$\mathbf{x} = (1, -1, 0, 3) + t(-1, -1, 1, 0), \quad t \in \mathbb{R}.$
(c) Consistent. The solution is not unique.
$\mathbf{x} = \left(\frac{19}{2}, 0, \frac{5}{2}, -2\right) + t(-1, 1, 0, 0), \quad t \in \mathbb{R}.$

A6 (a) $\begin{bmatrix} 1 & -5/3 & 2/3 \\ 0 & 1 & 10/11 \end{bmatrix}$ Consistent. The solution is unique. $\left(\frac{24}{11}, \frac{10}{11}\right)$

(b) $\begin{bmatrix} 1 & 2 & 1 & 5 \\ 0 & 1 & 0 & 4/7 \end{bmatrix}$ Consistent. The solution is not unique.
$\mathbf{x} = \left(\frac{27}{7}, \frac{4}{7}, 0\right) + t(-1, 0, 1), \quad t \in \mathbb{R}$, a line in $\mathbb{R}^3$.

(c) $\begin{bmatrix} 1 & 2 & -3 & 8 \\ 0 & 1 & -2 & 3 \\ 0 & 0 & 0 & 0 \end{bmatrix}$ Consistent. The solution is not unique.
$\mathbf{x} = (2, 3, 0) + t(-1, 2, 1), \quad t \in \mathbb{R}$, a line in $\mathbb{R}^3$.

(d) $\begin{bmatrix} 1 & -2 & -16/3 & -12 \\ 0 & 1 & 8/3 & 7 \\ 0 & 0 & 1 & 3 \end{bmatrix}$ Consistent. The solution is unique.
$\mathbf{x} = (2, -1, 3)$ A point in $\mathbb{R}^3$.

(e) $\begin{bmatrix} 1 & 2 & -1 & 4 \\ 0 & 1 & 3 & 2 \\ 0 & 0 & 0 & 1 \end{bmatrix}$ Inconsistent.

(f) $\begin{bmatrix} 1 & 2 & -3 & 0 & -5 \\ 0 & 1 & 1 & 4 & 9 \\ 0 & 0 & 0 & 1 & 2 \end{bmatrix}$ Consistent. The solution is not unique.
$\mathbf{x} = (-7, 1, 0, 2) + t(5, -1, 1, 0), \quad t \in \mathbb{R}$, a line in $\mathbb{R}^4$.

(g) $\begin{bmatrix} 1 & 2 & -3 & 1 & 4 & 1 \\ 0 & 1 & -1 & 0 & 1 & 1 \\ 0 & 0 & 1 & 1 & 0 & 1 \\ 0 & 0 & 0 & 1 & 1 & 2 \end{bmatrix}$ Consistent. The solution is not unique.
$\mathbf{x} = (-4, 0, -1, 2, 0) + t(0, 0, 1, -1, 1), \quad t \in \mathbb{R}$, a line in $\mathbb{R}^5$.

A7 (a) If $a \neq 0, b \neq 0$, this system is consistent, the solution is unique. If $a = 0, b \neq 0$, this system is consistent but the solution is not unique. If $a \neq 0, b = 0$, the system is inconsistent. If $a = 0, b = 0$, this system is consistent but the solution is not unique.
(b) If $c \neq 0, d \neq 0$, the system is consistent, the solution is unique. If $d = 0$, the system is consistent only if $c = 0$. If $c = 0$, the system is consistent for all values of d, but the solution is not unique.

A8 $a + b + c = 1500,$
$120a + 140b + 160c = 208000,$
$25a + 20b + 30c = 38000;$
600 apples, 400 bananas, and 500 oranges.

A9 $2A + 3C + 5P = 840,$
$A + C + P = 249,$

$A + C = 165$;
75% in algebra, 90% in calculus, and 84% in physics.

SECTION 2–2

A1 (a) $\begin{bmatrix} 1 & 0 & 0 \\ 0 & 1 & 0 \\ 0 & 0 & 1 \end{bmatrix}$ the rank of A is 3. (b) $\begin{bmatrix} 1 & 0 & 0 \\ 0 & 1 & 0 \\ 0 & 0 & 1 \end{bmatrix}$ the rank of B is 3.

(c) $\begin{bmatrix} 1 & 1 & 0 & 0 \\ 0 & 0 & 1 & 0 \\ 0 & 0 & 0 & 1 \end{bmatrix}$ the rank of C is 3. (d) $\begin{bmatrix} 1 & 0 & 0 & 0 \\ 0 & 1 & 0 & -2 \\ 0 & 0 & 1 & 3 \end{bmatrix}$ the rank of D is 3. (e) $\begin{bmatrix} 1 & 0 & 0 & -1/2 \\ 0 & 1 & 0 & 3/2 \\ 0 & 0 & 1 & 1/2 \\ 0 & 0 & 0 & 0 \end{bmatrix}$ the rank of E is 3.

(f) $\begin{bmatrix} 1 & 0 & 0 & 0 & -56 \\ 0 & 1 & 0 & 0 & 17 \\ 0 & 0 & 1 & 0 & 23 \\ 0 & 0 & 0 & 1 & -6 \end{bmatrix}$ the rank of F is 4.

A2 (a) (i) The system is inconsistent. (ii) The system is necessarily consistent. There are infinitely many solutions. 1 parameter. $\mathbf{x} = t(-2, 1, 1, 0)$, $t \in \mathbb{R}$
(b) (i) The system is consistent. There are infinitely many solutions. 1 parameter. $\mathbf{x} = (0, -2, 0, 1) + t(-2, 1, 1, 0)$, $t \in \mathbb{R}$ (ii) The system is necessarily consistent. There are infinitely many solutions. 2 parameters. $\mathbf{x} = s(-2, 1, 1, 0, 0) + t(0, 2, 0, -1, 1)$ $t \in \mathbb{R}$
(c) (i) The system is inconsistent. (ii) The system is necessarily consistent. There are infinitely many solutions. 1 parameter. $\mathbf{x} = t(0, -1, 1, 0, 0)$, $t \in \mathbb{R}$

A3 (a) $\begin{bmatrix} 0 & 2 & -5 \\ 1 & 2 & 3 \\ 1 & 4 & -3 \end{bmatrix} \rightarrow \begin{bmatrix} 1 & 0 & 0 \\ 0 & 1 & 0 \\ 0 & 0 & 1 \end{bmatrix}$ rank is 3. 0 parameters. Trivial solution.

(b) $\begin{bmatrix} 3 & 1 & -9 \\ 1 & 1 & -5 \\ 2 & 1 & -7 \end{bmatrix} \rightarrow \begin{bmatrix} 1 & 0 & -2 \\ 0 & 1 & -3 \\ 0 & 0 & 0 \end{bmatrix}$ rank is 2. 1 parameter. $\mathbf{x} = t(2, 3, 1)$, $t \in \mathbb{R}$

(c) $\begin{bmatrix} 1 & -1 & 2 & -3 \\ 3 & -3 & 8 & -5 \\ 2 & -2 & 5 & -4 \\ 3 & -3 & 7 & -7 \end{bmatrix} \rightarrow \begin{bmatrix} 1 & -1 & 0 & -7 \\ 0 & 0 & 1 & 2 \\ 0 & 0 & 0 & 0 \\ 0 & 0 & 0 & 0 \end{bmatrix}$ rank is 2. 2 parameters.
$\mathbf{x} = s(1, 1, 0, 0) + t(7, 0, -2, 1)$, $t \in \mathbb{R}$

(d) $\begin{bmatrix} 0 & 1 & 2 & 2 & 0 \\ 1 & 2 & 5 & 3 & -1 \\ 2 & 1 & 5 & 1 & -3 \\ 1 & 1 & 4 & 2 & -2 \end{bmatrix} \rightarrow \begin{bmatrix} 1 & 0 & 0 & -2 & 0 \\ 0 & 1 & 0 & 0 & 2 \\ 0 & 0 & 1 & 1 & -1 \\ 0 & 0 & 0 & 0 & 0 \end{bmatrix}$ rank is 3. 2 parameters.
$\mathbf{x} = s(2, 0, -1, 1, 0) + t(0, -2, 1, 0, 1)$, $s, t \in \mathbb{R}$

SECTION 2–3

A1
$$\begin{bmatrix} R_1+R_2 & -R_2 & 0 & 0 & 0 & E_1 \\ -R_2 & R_2+R_3 & -R_3 & 0 & 0 & 0 \\ 0 & -R_3 & R_3+R_4+R_8 & 0 & -R_8 & 0 \\ 0 & 0 & 0 & R_5+R_6 & -R_6 & 0 \\ 0 & 0 & -R_8 & -R_6 & R_6+R_7+R_8 & E_2 \end{bmatrix}$$

A2 To simplify writing, let $\alpha = \dfrac{1}{\sqrt{2}}$.
Total horizontal force: $R_1 + R_2 = 0$.
Total vertical force: $R_V - F_V = 0$.
Total moment about A: $R_1 s + F_V(2s) = 0$.
The horizontal and vertical equations at the joints A,B,C,D,E are
$\alpha N_2 + R_2 = 0$ and $N_1 + \alpha N_2 + R_V = 0$;
$N_3 + \alpha N_4 + R_1 = 0$ and $-N_1 + \alpha N_4 = 0$;
$-N_3 + \alpha N_6 = 0$ and $-\alpha N_2 + N_5 + \alpha N_6 = 0$;
$-\alpha N_4 + N_7 = 0$ and $-\alpha N_4 - N_5 = 0$;
$-N_7 - \alpha N_6 = 0$ and $-\alpha N_6 - F_V = 0$.

A3 140 (occuring at the vertex $(100, 40)$)

SECTION 2–4

A1 (a) Consistent. $(1 - i, \frac{4}{5} + \frac{2}{5}i, -\frac{1}{5} - \frac{3}{5}i)$
(b) Consistent. $(5 + i, -2 + 2i, -i, 0) + t(-1 + 2i, 0, -i, 1)$

Chapter 2 Quiz

1 $\begin{bmatrix} 1 & -1 & 1 & 0 & 2 \\ 0 & 1 & 0 & 0 & 7 \\ 0 & 0 & -2 & 1 & -5 \\ 0 & 0 & 0 & 0 & 1 \end{bmatrix}$. Inconsistent.

2 $\begin{bmatrix} 1 & 0 & 0 & 0 & 1 \\ 0 & 1 & 0 & 0 & 0 \\ 0 & 0 & 1 & 0 & -1/3 \\ 0 & 0 & 0 & 1 & 1/3 \end{bmatrix}$

3 (a) The system is inconsistent for all (a, b, c) of the form $(a, b, 1)$ or $(a, -2, c)$, and is consistent for all (a, b, c) where $b \neq -2$ and $c \neq 1$.
(b) unique solution if and only if $b \neq -2$ and $c^2 \neq 1$.

4 (a) $s(-2, -1, 1, 0, 0) + t\left(\frac{11}{4}, -\frac{11}{2}, 0, -\frac{5}{4}, 1\right)$, $s, t \in \mathbb{R}$.
(b) $\mathbf{v} \cdot \mathbf{a} = 0$, $\mathbf{v} \cdot \mathbf{b} = 0$, and $\mathbf{v} \cdot \mathbf{c} = 0$, yields a homogeneous system of three linear equations with five variables. Since the number of variables is greater than the number of equations, the system will always have non-trivial solutions.

5 (a) False. (b) False. (c) True.

CHAPTER 3

SECTION 3–1

A1 (a) $\begin{bmatrix} -1 & -6 & 4 \\ 6 & -4 & 2 \end{bmatrix}$ (b) $\begin{bmatrix} -3 & 6 \\ -6 & 3 \\ -12 & 6 \end{bmatrix}$ (c) $\begin{bmatrix} -1 & 9 \\ -11 & -17 \end{bmatrix}$

A2 (a) $\begin{bmatrix} -11 & -1 & 12 \\ 8 & 27 & -1 \end{bmatrix}$ (b) $\begin{bmatrix} 12 & 11 \\ 9 & 4 \\ 3 & 15 \end{bmatrix}$ (c) $\begin{bmatrix} -4 & 6 & 13 & -4 \\ 0 & -3 & -5 & 1 \\ 8 & 15 & 19 & -1 \end{bmatrix}$

(d) The product is not defined since the number of entries in each row of the first matrix does not equal the number of entries in each column of the second matrix.

A3 (a) $A(B+C) = AB + AC = \begin{bmatrix} -13 & 10 \\ 14 & 7 \end{bmatrix}$ $A(3B) = \begin{bmatrix} 9 & 12 \\ 30 & -18 \end{bmatrix} = 3(AB)$

(b) $A(B+C) = \begin{bmatrix} 6 & -16 \\ 4 & 14 \end{bmatrix} = AB + AC$ $A(3B) = \begin{bmatrix} 3 & -18 \\ -27 & 51 \end{bmatrix} = 3(AB)$

A4 (a) $A+B$ is defined. AB is not defined. $(A+B)^T = \begin{bmatrix} -3 & 2 & 1 \\ -1 & 2 & 3 \end{bmatrix}$

(b) $A+B$ is not defined. AB is defined. $(AB)^T = \begin{bmatrix} -21 & -10 \\ 15 & -27 \end{bmatrix}$

A5 (a) $AB = \begin{bmatrix} 13 & 31 & 2 \\ 10 & 12 & 10 \end{bmatrix}$. (b), (c), (d), (e) do not exist because the matrices are not of the correct size for these products to be defined.

(f) $\begin{bmatrix} 11 & 7 & 3 & 15 \\ 7 & 9 & 11 & 1 \end{bmatrix}$ (g) $\begin{bmatrix} 52 & 139 \\ 62 & 46 \end{bmatrix}$ (h) $\begin{bmatrix} 13 & 10 \\ 31 & 12 \\ 2 & 10 \end{bmatrix}$

(i) $D^TC = (C^TD)^T = \begin{bmatrix} 11 & 7 \\ 7 & 9 \\ 3 & 11 \\ 15 & 1 \end{bmatrix}$

A6 (a) $\begin{bmatrix} 12 \\ 17 \\ 3 \end{bmatrix}, \begin{bmatrix} 8 \\ 4 \\ -4 \end{bmatrix}, \begin{bmatrix} -2 \\ 5 \\ 1 \end{bmatrix}$ (b) $\begin{bmatrix} 12 & 8 & -2 \\ 17 & 4 & 5 \\ 3 & -4 & 1 \end{bmatrix}$

A7 (a) $(1,0,0)$ (b) $(2,0,0)$ (c) $(0,0,3)$

A8 (a) $\begin{bmatrix} -6 & -18 \\ 2 & 6 \end{bmatrix}$ (b) 0 (c) $\begin{bmatrix} 10 & 8 & -6 \\ -5 & -4 & 3 \\ 15 & 12 & -9 \end{bmatrix}$ (d) -3

A9 $\begin{bmatrix} -13 & 16 \\ -27 & 0 \end{bmatrix}$

SECTION 3–2

A1 (a) Domain $\mathbb{R}^2$, codomain $\mathbb{R}^4$

(b) $f_A(2,-5)=(-19,6,-23,38)$, $f_A(-3,4)=(18,-9,17,-36)$.

(c) $(-2,3,1,4)$ $(3,0,5,-6)$

(d) $(-2x_1+3x_2, 3x_1, x_1+5x_2, 4x_1-6x_2)$

A2 (a) Domain $\mathbb{R}^4$, codomain $\mathbb{R}^3$

(b) $f_A(2,-2,3,1)=(-11,9,7)$ $f_A(-3,1,4,2)=(-13,-1,3)$

(c) $(1,2,1)$, $(2,-1,0)$, $(-3,0,2)$, $(0,3,-1)$

(d) $(x_1+2x_2-3x_3,\ 2x_1-x_2+3x_4,\ x_1+2x_3-x_4)$

A3 (a) f is not linear. (b) g is linear. (c) h is not linear.
(d) k is linear. (e) l is not linear.

A4 (a) Domain $\mathbb{R}^3$, codomain $\mathbb{R}^2$ $\begin{bmatrix} 2 & -3 & 1 \\ 0 & 1 & -5 \end{bmatrix}$

(b) Domain $\mathbb{R}^4$, codomain $\mathbb{R}^2$ $\begin{bmatrix} 5 & 0 & 3 & -1 \\ 0 & 1 & -7 & 3 \end{bmatrix}$

(c) Domain $\mathbb{R}^4$, codomain $\mathbb{R}^4$ $\begin{bmatrix} 1 & 0 & -1 & 1 \\ 1 & 2 & -1 & -3 \\ 0 & 1 & 1 & 0 \\ 1 & -1 & 1 & -1 \end{bmatrix}$

A5 $\dfrac{1}{5}\begin{bmatrix} 4 & -2 \\ -2 & 1 \end{bmatrix}$

A6 $\dfrac{1}{17}\begin{bmatrix} 16 & -4 \\ -4 & 1 \end{bmatrix}$

A7 $\dfrac{1}{9}\begin{bmatrix} 4 & 4 & -2 \\ 4 & 4 & -2 \\ -2 & -2 & 1 \end{bmatrix}$

SECTION 3–3

A1 (a) The domain of S is $\mathbb{R}^3$ and the codomain of S is $\mathbb{R}^2$. The domain of T is $\mathbb{R}^2$ and the codomain of T is $\mathbb{R}^3$.

(b) $[S\circ T]=\begin{bmatrix} 7 & 15 \\ 3 & 4 \end{bmatrix}$, $[T\circ S]=\begin{bmatrix} 0 & 1 & 7 \\ -4 & -1 & 1 \\ 1 & 2 & 12 \end{bmatrix}$

A2 (a) The domain of S is $\mathbb{R}^4$ and the codomain of S is $\mathbb{R}^2$. The domain of T is $\mathbb{R}^2$ and the codomain of T is $\mathbb{R}^4$.

(b) $[S\circ T]=\begin{bmatrix} 6 & -19 \\ 10 & -10 \end{bmatrix}$, $[T\circ S]=\begin{bmatrix} -3 & 5 & 16 & 9 \\ 6 & 8 & 4 & 0 \\ -6 & -8 & -4 & 0 \\ -9 & -17 & -16 & -5 \end{bmatrix}$

A3 (a) $\begin{bmatrix} 0 & -1 \\ 1 & 0 \end{bmatrix}$ (b) $\begin{bmatrix} -1 & 0 \\ 0 & -1 \end{bmatrix}$ (c) $\begin{bmatrix} \frac{\sqrt{2}}{2} & \frac{\sqrt{2}}{2} \\ -\frac{\sqrt{2}}{2} & \frac{\sqrt{2}}{2} \end{bmatrix}$

A4 a $\begin{bmatrix} 1 & 0 \\ 0 & 5 \end{bmatrix}$ (b) $\begin{bmatrix} \cos\theta & -5\sin\theta \\ \sin\theta & 5\cos\theta \end{bmatrix}$ (c) $\begin{bmatrix} \cos\theta & -\sin\theta \\ 5\sin\theta & 5\cos\theta \end{bmatrix}$

A5 (a) $\begin{bmatrix} 4/5 & -3/5 \\ -3/5 & -4/5 \end{bmatrix}$ (b) $\begin{bmatrix} -3/5 & 4/5 \\ 4/5 & 3/5 \end{bmatrix}$

A6 (a) $\frac{1}{3}\begin{bmatrix} 1 & -2 & -2 \\ -2 & 1 & -2 \\ -2 & -2 & 1 \end{bmatrix}$ (b) $\frac{1}{9}\begin{bmatrix} 1 & 8 & 4 \\ 8 & 1 & -4 \\ 4 & -4 & 7 \end{bmatrix}$

A7 (a) Both the domain and codomain of $L \circ M$ are $\mathbb{R}^3$.
(b) Both the domain and codomain of $M \circ L$ are $\mathbb{R}^2$.
(c) not defined (d) not defined (e) not defined
(f) The domain of $N \circ M$ is $\mathbb{R}^3$ and the codomain of $N \circ M$ is $\mathbb{R}^4$.

A8 (a) $\begin{bmatrix} 5 & 0 & 0 \\ 0 & 5 & 0 \\ 0 & 0 & 0 \\ 0 & 0 & 5 \end{bmatrix}$ (b) $\begin{bmatrix} 0 & 1 & 0 \\ 2 & 0 & 1 \end{bmatrix}$

(c) There is no shear T in $\mathbb{R}^2$ such that $T \circ P = P \circ S$. (d) $\begin{bmatrix} 1 & 0 & 0 \\ 0 & 1 & 0 \end{bmatrix}$

SECTION 3–4

A1 (a) (3, 1, 6, 1) is not in the range of L. (b) $L(1, 1, -2) = (3, -5, 1, 5)$

A2 ((a)) (1, 0, 0) and (0, 1, 0,) are in the set but their sum is not, so this is not a subspace.
(b) This is a subspace, in fact it is all of $\mathbb{R}^5$
(c) This is a subspace spanned by (1, 1, 1) and (1, 2, 3) (write $\mathbf{x} = (s+1)(1,1,1) + (t+1)(1,2,3)$)
(d) Not closed under addition, not a subspace.
(e) Since the set is non-empty and is closed under addition and multiplication by scalars, the set forms a subspace of $\mathbb{R}^3$.

A3 (a) S is a subspace of $\mathbb{R}^4$.
(b) T is not a subspace of $\mathbb{R}^4$
(c) U is not closed under multiplication by scalars and is therefore not a subspace of $\mathbb{R}^4$.
(d) V is a subspace of $\mathbb{R}^4$.
(e) W is not closed under multiplication by scalars, so W cannot be a subspace of $\mathbb{R}^4$.

A4 $\left[\begin{array}{ccc|c|c} 1 & 2 & -1 & -3 & 5 \\ 0 & 1 & 1 & 2 & 4 \\ 1 & 0 & 2 & 8 & 6 \\ 1 & 1 & 1 & 4 & 7 \end{array}\right] \to \left[\begin{array}{ccc|c|c} 1 & 0 & 0 & 2 & 12/5 \\ 0 & 1 & 0 & -1 & 11/5 \\ 0 & 0 & 1 & 3 & 9/5 \\ 0 & 0 & 0 & 0 & 3/5 \end{array}\right]$, so
(a) $(-3, 2, 8, 4) = 2(1, 0, 1, 1) - 1(2, 1, 0, 1) + 3(-1, 1, 2, 1)$;
(b) $(5, 4, 6, 7)$ is not in Sp(V).

A5 (a) $(3,2,-1,-1)$ is not in $\text{Sp}(V)$.
(b) $(-7,3,0,8) = -2(1,-1,1,0) + 3(-1,1,0,2) - 2(1,1,-1,-1)$.

A6 (a) $\{(2,-3,-1,1,0),(-5,-4,-2,0,1)\}$ spans the nullspace of L.
(b) $\{(-2,1,0,0,0,0),(2,0,1,5,1,0)\}$ spans the nullspace of L.

A7 The plane with equation $x_1 + x_2 - x_3 = 0$ is the subspace of $\mathbb{R}^3$ spanned by $\{(1,1,2),(1,0,1)\}$.

A8 $\text{Sp}(\{(1,1,2),(3,-1,0)\})$ is the solution space of the plane with equation $x_1 + 3x_2 - 2x_3 = 0$.

A9 The plane is not a subspace because it does not contain 0, so it does not have a spanning set.

A10 The matrix of L is any multiple of $\begin{bmatrix} 1 & -1 \\ 2 & -2 \\ 3 & -3 \end{bmatrix}$.

A11 The matrix of L is any multiple of $\begin{bmatrix} 2 & 1 \\ 2 & 1 \\ 2 & 1 \end{bmatrix}$.

SECTION 3–5

A1 $A^{-1} = \dfrac{1}{23}\begin{bmatrix} 5 & 4 \\ -2 & 3 \end{bmatrix}$, $B^{-1} = \begin{bmatrix} 2 & 0 & -1 \\ -1 & 1 & -1 \\ -1 & 0 & 1 \end{bmatrix}$.

C is not invertible. $D^{-1} = \begin{bmatrix} 0 & -1 & 1 \\ -1 & 1 & 0 \\ 1 & 0 & 0 \end{bmatrix}$.

$E^{-1} = \begin{bmatrix} 6 & 10 & -5/2 & -7/2 \\ 1 & 2 & -1/2 & -1/2 \\ -2 & -3 & 1 & 1 \\ 0 & -3 & 0 & 1 \end{bmatrix}$, $F^{-1} = \begin{bmatrix} 1 & 0 & -1 & 1 & -2 \\ 0 & 1 & 0 & -1 & 2 \\ 0 & 0 & 1 & -1 & 1 \\ 0 & 0 & 0 & 1 & -2 \\ 0 & 0 & 0 & 0 & 1 \end{bmatrix}$

A2 (a) $\begin{bmatrix} 1 \\ -1 \\ 0 \end{bmatrix}$ (b) $\begin{bmatrix} -3 \\ 0 \\ 2 \end{bmatrix}$ (c) $\begin{bmatrix} -2 \\ -1 \\ 2 \end{bmatrix}$

A3 (a) $A^{-1} = \begin{bmatrix} 2 & -1 \\ -3 & 2 \end{bmatrix}$, $B^{-1} = \begin{bmatrix} -5 & 2 \\ 3 & -1 \end{bmatrix}$ (b) $(AB)^{-1} = \begin{bmatrix} -16 & 9 \\ 9 & -5 \end{bmatrix}$
(c) $(3A)^{-1} = \begin{bmatrix} 2/3 & -1/3 \\ -1 & 2/3 \end{bmatrix}$ (d) $(A^T)^{-1} = \begin{bmatrix} 2 & -3 \\ -1 & 2 \end{bmatrix}$

A4 (a) $[R_{\pi/6}]^{-1} = [R_{-\pi/6}] = \begin{bmatrix} \frac{\sqrt{3}}{2} & \frac{1}{2} \\ -\frac{1}{2} & \frac{\sqrt{3}}{2} \end{bmatrix}$ (b) $\begin{bmatrix} 1 & 3 \\ 0 & 1 \end{bmatrix}$
(c) $\begin{bmatrix} 1/5 & 0 \\ 0 & 1/5 \end{bmatrix}$ (d) $\begin{bmatrix} 1 & 0 \\ 0 & -1 \end{bmatrix}$

A5 (a) $[S] = \begin{bmatrix} 1 & 0 \\ 2 & 1 \end{bmatrix}$, $[S^{-1}] = \begin{bmatrix} 1 & 0 \\ -2 & 1 \end{bmatrix}$

(b) $[R] = \begin{bmatrix} 0 & 1 \\ 1 & 0 \end{bmatrix}$, $[R^{-1}] = \begin{bmatrix} 0 & 1 \\ 1 & 0 \end{bmatrix}$

(c) $[(R \circ S)^{-1}] = \begin{bmatrix} 0 & 1 \\ 1 & -2 \end{bmatrix}$, $[(S \circ R)^{-1}] = \begin{bmatrix} -2 & 1 \\ 1 & 0 \end{bmatrix}$

SECTION 3–6

A1 (a) $\begin{bmatrix} 1 & -5 & 0 \\ 0 & 1 & 0 \\ 0 & 0 & 1 \end{bmatrix}$ (b) $\begin{bmatrix} 1 & 0 & 0 \\ 0 & 0 & 1 \\ 0 & 1 & 0 \end{bmatrix}$ (c) $\begin{bmatrix} 1 & 0 & 0 \\ 0 & 1 & 0 \\ 0 & 0 & -1 \end{bmatrix}$

(d) $\begin{bmatrix} 1 & 0 & 0 \\ 0 & 6 & 0 \\ 0 & 0 & 1 \end{bmatrix}$ (e) $\begin{bmatrix} 1 & 0 & 0 \\ 0 & 1 & 0 \\ 4 & 0 & 1 \end{bmatrix}$

A2 (a) $\begin{bmatrix} 1 & 0 & 0 & 0 \\ 0 & 1 & 0 & 0 \\ 0 & 0 & 1 & 0 \\ 0 & 0 & -3 & 1 \end{bmatrix}$ (b) $\begin{bmatrix} 1 & 0 & 0 & 0 \\ 0 & 0 & 0 & 1 \\ 0 & 0 & 1 & 0 \\ 0 & 1 & 0 & 0 \end{bmatrix}$ (c) $\begin{bmatrix} 1 & 0 & 0 & 0 \\ 0 & 1 & 0 & 0 \\ 0 & 0 & -3 & 0 \\ 0 & 0 & 0 & 1 \end{bmatrix}$

A3 (a) A is elementary, add (-4) times row 2 to row 3.

(b) B is not elementary, both row 1 and row 3 have been multiplied by (-1).

(c) C is not elementary, row 1 has been multiplied by 3 and row 3 has been added to row 1.

(d) D is elementary, interchange rows 1 and 3.

(e) E is not elementary, two row interchanges are needed to obtain E from I.

A4 (a) $R_2 \updownarrow R_3$; $(\frac{1}{2})R_3$; $R_1 - 4R_3$; $R_1 - 3R_2$

(b) $E_1 = \begin{bmatrix} 1 & 0 & 0 \\ 0 & 0 & 1 \\ 0 & 1 & 0 \end{bmatrix}$ $E_2 = \begin{bmatrix} 1 & 0 & 0 \\ 0 & 1 & 0 \\ 0 & 0 & 1/2 \end{bmatrix}$ $E_3 = \begin{bmatrix} 1 & 0 & -4 \\ 0 & 1 & 0 \\ 0 & 0 & 1 \end{bmatrix}$

$E_4 = \begin{bmatrix} 1 & -3 & 0 \\ 0 & 1 & 0 \\ 0 & 0 & 1 \end{bmatrix}$

(c) $A^{-1} = E_4E_3E_2E_1 = \begin{bmatrix} 1 & -2 & -3 \\ 0 & 0 & 1 \\ 0 & 1/2 & 0 \end{bmatrix}$

(d) $A = \begin{bmatrix} 1 & 0 & 0 \\ 0 & 0 & 1 \\ 0 & 1 & 0 \end{bmatrix} \begin{bmatrix} 1 & 0 & 0 \\ 0 & 1 & 0 \\ 0 & 0 & 2 \end{bmatrix} \begin{bmatrix} 1 & 0 & 4 \\ 0 & 1 & 0 \\ 0 & 0 & 1 \end{bmatrix} \begin{bmatrix} 1 & 3 & 0 \\ 0 & 1 & 0 \\ 0 & 0 & 1 \end{bmatrix}$

A5 (a) $R_3 - 2R_1$, $R_2 - 3R_3$, $R_1 - 2R_3$, $R_1 - 2R_2$,

(b) $E_1 = \begin{bmatrix} 1 & 0 & 0 \\ 0 & 1 & 0 \\ -2 & 0 & 1 \end{bmatrix}$, $E_2 = \begin{bmatrix} 1 & 0 & 0 \\ 0 & 1 & -3 \\ 0 & 0 & 1 \end{bmatrix}$, $E_3 = \begin{bmatrix} 1 & 0 & -2 \\ 0 & 1 & 0 \\ 0 & 0 & 1 \end{bmatrix}$,

$$E_4 = \begin{bmatrix} 1 & -2 & 0 \\ 0 & 1 & 0 \\ 0 & 0 & 1 \end{bmatrix}$$

(c) $A^{-1} = \begin{bmatrix} -7 & -2 & 4 \\ 6 & 1 & -3 \\ -2 & 0 & 1 \end{bmatrix}$

(d) $A = \begin{bmatrix} 1 & 0 & 0 \\ 0 & 1 & 0 \\ 2 & 0 & 1 \end{bmatrix} \begin{bmatrix} 1 & 0 & 0 \\ 0 & 1 & 3 \\ 0 & 0 & 1 \end{bmatrix} \begin{bmatrix} 1 & 0 & 2 \\ 0 & 1 & 0 \\ 0 & 1 & 1 \end{bmatrix} \begin{bmatrix} 1 & 2 & 0 \\ 0 & 1 & 0 \\ 0 & 0 & 1 \end{bmatrix}$

Chapter 3 Quiz

1 (a) $\begin{bmatrix} -14 & 1 & -17 \\ -1 & 10 & -39 \end{bmatrix}$

(b) The product is not defined since the number of entries in each row of B does not equal the number of entries in each column of A.

(c) $\begin{bmatrix} -3 & -38 \\ 0 & -23 \\ -8 & -42 \end{bmatrix}$

2 (a) $f_A(\mathbf{c}) = (-11, 0)$, $f_A(\mathbf{d}) = (-16, 17)$ (b) $\begin{bmatrix} -16 & -11 \\ 17 & 0 \end{bmatrix}$

3 (a) $[L] = [R_{\pi/3}] = \begin{bmatrix} \cos\frac{\pi}{3} & -\sin\frac{\pi}{3} & 0 \\ \sin\frac{\pi}{3} & \cos\frac{\pi}{3} & 0 \\ 0 & 0 & 1 \end{bmatrix} = \begin{bmatrix} 1/2 & -\sqrt{3}/2 & 0 \\ \sqrt{3}/2 & 1/2 & 0 \\ 0 & 0 & 1 \end{bmatrix}$

(b) $[M] = [\mathbf{refl}_{(-1,-1,2)}] = \begin{bmatrix} 2/3 & -1/3 & 2/3 \\ -1/3 & 2/3 & 2/3 \\ 2/3 & 2/3 & -1/3 \end{bmatrix}$

(c) $[L \circ M] = \begin{bmatrix} 1/3 + \sqrt{3}/6 & -1/6 - \sqrt{3}/3 & 1/3 - \sqrt{3}/3 \\ \sqrt{3}/3 - 1/6 & -\sqrt{3}/6 + 1/3 & \sqrt{3}/3 + 1/3 \\ 2/3 & 2/3 & -1/3 \end{bmatrix}$

4 The solution space of $A\mathbf{x} = \mathbf{0}$ is all the vectors of the form $\mathbf{x} = (-2s - t, s, s, t, 0) = s(-2, 1, 1, 0, 0) + t(-1, 0, 0, 1, 0)$, $s, t, \in \mathbb{R}$. The solution set of $A\mathbf{x} = \mathbf{b}$ is $\mathbf{x} = (5 - 2s - t, 6s, s, t, 7) = (5, 6, 0, 0, 7) + s(-2, 1, 1, 0, 0) + t(-1, 0, 0, 1, 0)$, $s, t, \in \mathbb{R}$. The solution set is obtained from the solution space of $A\mathbf{x} = \mathbf{0}$ by translating by the fixed vector $(5, 6, 0, 0, 7)$.

5 (a) $\mathbf{c}$ is not in the columnspace of B. $\mathbf{d}$ is in the columnspace of B.
(b) $\mathbf{x} = (-1, -2, -3)$ (c) $\mathbf{y} = (0, 1, 0)$

6 (a) The set of all vectors orthogonal to $\mathbf{a}$ and $\mathbf{b}$ is the solution space of the system $A\mathbf{x} = \mathbf{0}$. By theorem 5 of Section 3–4, the solution space of a system of homogeneous equations is a subspace of $\mathbb{R}^n$.
(b) The set of all vectors $\mathbf{x} = (1, 1, 0, 0) + s(0, 0, 1, 2) + t(0, 0, 0, -3)$ does not contain the zero vector so it does not form a subspace of $\mathbb{R}^4$.
(c) By theorem 7 of Section 3–4, this set forms a subspace of $\mathbb{R}^5$.

7 $A^{-1} = \begin{bmatrix} 2/3 & 0 & 0 & 1/3 \\ 1/6 & 0 & 1/2 & -1/6 \\ 0 & 1 & 0 & 0 \\ -1/3 & 0 & 0 & 1/3 \end{bmatrix}$

8 The matrix is invertible only for $p \neq 1$.
The inverse is
$\frac{1}{1-p}\begin{bmatrix} -1 & p & -p \\ -1 & 1-2p & p \\ -1 & -1 & 1 \end{bmatrix}.$

9 The range of L is closed under addition and multiplication by scalars, so it is a subspace.

10 (a) $K = I_3$.
(b) There is no matrix K.
(c) There is no such transformation.
(d) The matrix of L is $\begin{bmatrix} 1 & -2/3 \\ 1 & -2/3 \\ 2 & -4/3 \end{bmatrix}$, or any multiple of it.
(e) There is no such transformation L.
(f) There is no such matrix.

11 (a) $(\frac{1}{2})R_2$; $(\frac{1}{4})R_3$; $R_1 + 2R_3$; $R_2 + \frac{3}{2}R_3$; $E_1 = \begin{bmatrix} 1 & 0 & 0 \\ 0 & 1/2 & 0 \\ 0 & 0 & 1 \end{bmatrix}$;
$E_2 = \begin{bmatrix} 1 & 0 & 0 \\ 0 & 1 & 0 \\ 0 & 0 & 1/4 \end{bmatrix}$; $E_3 = \begin{bmatrix} 1 & 0 & 2 \\ 0 & 1 & 0 \\ 0 & 0 & 1 \end{bmatrix}$; $E_4 = \begin{bmatrix} 1 & 0 & 0 \\ 0 & 1 & 3/2 \\ 0 & 0 & 1 \end{bmatrix}$
(b) $A = \begin{bmatrix} 1 & 0 & 0 \\ 0 & 2 & 0 \\ 0 & 0 & 1 \end{bmatrix}\begin{bmatrix} 1 & 0 & 0 \\ 0 & 1 & 0 \\ 0 & 0 & 4 \end{bmatrix}\begin{bmatrix} 1 & 0 & -2 \\ 0 & 1 & 0 \\ 0 & 0 & 1 \end{bmatrix}\begin{bmatrix} 1 & 0 & 0 \\ 0 & 1 & -3/2 \\ 0 & 0 & 1 \end{bmatrix}$

CHAPTER 4

SECTION 4–1

A1 (a) subspace
(b) If $n = 3$, then a matrix in row echelon form is $A = \begin{bmatrix} 1 & 0 & 2 \\ 0 & 1 & 0 \\ 0 & 0 & 1 \end{bmatrix}$. However, $2A = \begin{bmatrix} 2 & 0 & 4 \\ 0 & 2 & 0 \\ 0 & 0 & 2 \end{bmatrix}$ is not in row echelon form because the first coefficient of each row is not a 1. Hence, the subset is not closed under multiplication by scalars and is therefore not a subspace of $M(n,n)$. (The subset also fails to be closed under addition.)
(c) subspace
(d) does not contain the n by n zero matrix; not a subspace
(e) subspace

A2 (a) subspace (b) subspace (c) subspace (d) subspace

A3 (a) subspace
(b) not closed under multiplication by scalars, not a subspace of F (The subset also fails to be closed under addition.)
(c) subspace
(d) not closed under multiplication by scalars, not a subspace

SECTION 4–2

A1 (a) $\begin{bmatrix} 1 & 0 & 1 \\ 1 & 1 & 2 \\ 1 & 2 & 4 \end{bmatrix} \to \begin{bmatrix} 1 & 0 & 0 \\ 0 & 1 & 0 \\ 0 & 0 & 1 \end{bmatrix}$; the solution is unique.
(b) $\{\mathbf{v}_1, \mathbf{v}_2, \mathbf{v}_3\}$ has the unique representation property.
(c) linearly independent
(d) $\begin{bmatrix} 1 & 0 & 1 \\ 1 & 1 & 1 \\ -1 & -2 & -3 \end{bmatrix} \to \begin{bmatrix} 1 & 0 & 0 \\ 0 & 1 & 0 \\ 0 & 0 & 1 \end{bmatrix}$; the solution is unique. $\{\mathbf{w}_1, \mathbf{w}_2, \mathbf{w}_3\}$has the unique representation property, and is linearly independent.

A2 (a) $\begin{bmatrix} 1 & 1 & 2 \\ 0 & 1 & -1 \\ -1 & 1 & -4 \\ 1 & 2 & 1 \end{bmatrix} \to \begin{bmatrix} 1 & 0 & 3 \\ 0 & 1 & -1 \\ 0 & 0 & 0 \\ 0 & 0 & 0 \end{bmatrix}$ the solution is not unique.
(b) $\{\mathbf{v}_1, \mathbf{v}_2, \mathbf{v}_3\}$ does not have the unique representation property.
(c) linearly dependent.
(d) $\begin{bmatrix} 1 & 1 & 2 \\ 1 & 2 & 1 \\ -1 & 1 & 4 \\ 0 & 1 & 1 \end{bmatrix} \to \begin{bmatrix} 1 & 0 & 0 \\ 0 & 1 & 0 \\ 0 & 0 & 1 \\ 0 & 0 & 0 \end{bmatrix}$ the solution is unique. $\{\mathbf{w}_1, \mathbf{w}_2, \mathbf{w}_3\}$has the unique representation property. Linearly independent.

A3 (a) linearly independent
(b) linearly dependent
$-3t(1,0,1,0) - 2t(0,1,1,1) - t(0,0,1,1) + t(3,2,6,3) = \mathbf{0}$
(c) linearly dependent: $2t(1,1,0,1,1) - t(2,3,1,3,3) + t(0,1,1,1,1) = \mathbf{0}$
(d) linearly dependent: $t(\cos^2 x - \sin^2 x - \cos 2x) = 0$

A4 (a) linearly independent for all $k \neq -3$
(b) linearly independent for all $k \neq -5/2$

A5 (a) $\begin{bmatrix} 1 & 1 & 2 \\ 1 & -1 & 1 \\ 2 & -1 & 1 \end{bmatrix} \to I_3$, so as in Example 19, the given set spans $\mathbb{R}^3$ and is linearly independent, so it is a basis for $\mathbb{R}^3$
(b) Not a basis. 4 vectors in $\mathbb{R}^3$ cannot be linearly independent.
(c) $\begin{bmatrix} 1 & 1 & 3 \\ -1 & 2 & 0 \\ 1 & -1 & 1 \end{bmatrix} \to \begin{bmatrix} 1 & 0 & 2 \\ 0 & 1 & 1 \\ 0 & 0 & 0 \end{bmatrix}$, so the given vectors are linearly dependent and do not form a basis for $\mathbb{R}^3$.

A6 We consider $A\mathbf{x} = \mathbf{b}$, where $A = \begin{bmatrix} 1 & 2 & 3 & 2 \\ 1 & 3 & 3 & 3 \\ 1 & 1 & 4 & 1 \\ 2 & 3 & 11 & 4 \end{bmatrix}$; row reduce: $A \rightarrow I$. Since $A\mathbf{x} = \mathbf{b}$ has a solution for every $\mathbf{b}$, the given vectors (the columns of A) form a spanning set for $\mathbb{R}^4$; since $A\mathbf{x} = \mathbf{0}$ has only the trivial solution, the vectors are linearly independent, and they form a basis for $\mathbb{R}^4$.

A7 (a) $\mathbf{v}_1 = (-2, 1, 1)$ and two vectors orthogonal to it, say $\mathbf{v}_2 = (1, 0, 2)$, $\mathbf{v}_3 = (0, 1, -1)$.
(b) $\mathbf{v}_1 = (1, -3, -5)$, which is the normal vector to the plane of the reflection, and two vectors $\mathbf{v}_2$ and $\mathbf{v}_3$ which lie in the plane $\mathbf{v}_2 = (5, 0, 1)$ and $\mathbf{v}_3 = (3, 1, 0)$.

A8 (a) linearly dependent (b) linearly independent

SECTION 4–3

A1 (a) $(\mathbf{x})_{\mathcal{B}} = (3, -2)$, $(\mathbf{y})_{\mathcal{B}} = (2, 3)$ (b) $(\mathbf{x})_{\mathcal{B}} = (-2, 3, 1)$, $(\mathbf{y})_{\mathcal{B}} = (-2, 3, -1)$
(c) $(\mathbf{x})_{\mathcal{B}} = (5, -3, 4)$ $\mathbf{y}$ is not in the subspace spanned by $\mathcal{B}$

A2 (a) $\mathcal{B}$ is obviously linearly independent.
$\begin{bmatrix} 1 & 0 & b_1 \\ 2 & -2 & b_2 \\ 0 & 1 & b_3 \end{bmatrix} \rightarrow \begin{bmatrix} 1 & 0 & b_1 \\ 0 & 1 & b_3 \\ 0 & 0 & 2b_1 - b_2 - 2b_3 \end{bmatrix}$, so $\mathbf{b}$ is in Sp($\mathcal{B}$) if and only if $2b_1 - b_2 - 2b_3 = 0$, so $\mathcal{B}$ is a basis for the plane. (b) $(3, 2, 1)$ and $(5, 2, 3)$ are not in the plane; $(3, 2, 2)_{\mathcal{B}} = (3, 2)$

A3 (a) $\begin{bmatrix} 1 & 1 & -1 \\ 0 & -1 & -1 \\ 1 & 2 & 1 \end{bmatrix} \rightarrow I_3$, so $\mathcal{B}$ is a basis for $\mathbb{R}^3$.
(b) $(1, 0, 0)_{\mathcal{B}} = (-1, 1, -1)$; $(4, -2, 7)_{\mathcal{B}} = (4, 1, 1)$; $(-2, -2, 3)_{\mathcal{B}} = (2, -1, 3)$
(c) $(2, -4, 10)_{\mathcal{B}} = (6, 0, 4)$

A4 (a) (i) $\begin{bmatrix} 1 & 2 & -2 & -1 \\ 2 & 1 & 2 & 1 \\ 1 & -1 & 4 & 2 \\ 3 & 2 & 10 & 7 \end{bmatrix} \rightarrow \begin{bmatrix} 1 & 0 & 0 & -1/2 \\ 0 & 1 & 0 & 1/2 \\ 0 & 0 & 1 & 3/4 \\ 0 & 0 & 0 & 0 \end{bmatrix}$, $\mathbf{w}$ is in Sp($\mathcal{V}$)
(ii) the homogeneous system has $c_1 = c_2 = c_3 = 0$ as its only solution, vectors in $\mathcal{V}$ do form a basis for Sp($\mathcal{V}$) (iii) $\mathbf{w}_{\mathcal{V}} = (-1/2, 1/2, 3/4)$
(b) (i) $\begin{bmatrix} 1 & 1 & 0 & 4 \\ 3 & -2 & 1 & -1 \\ 2 & 1 & -1 & 9 \\ 3 & 2 & 1 & 3 \end{bmatrix} \rightarrow \begin{bmatrix} 1 & 0 & 0 & 2 \\ 0 & 1 & 0 & 2 \\ 0 & 0 & 1 & -3 \\ 0 & 0 & 0 & -4 \end{bmatrix}$, $\mathbf{w}$ is not in Sp($\mathcal{V}$) (ii) homogeneous system has $c_1 = c_2 = c_3 = 0$ as its only solution, so the vectors in $\mathcal{V}$ do form a basis for Sp($\mathcal{V}$).

SECTION 4–4

A1 (a) Denote the vectors in S by $\mathbf{v}_1, \mathbf{v}_2, \mathbf{v}_3, \mathbf{v}_4$. $\{\mathbf{v}_1, \mathbf{v}_2\}$ is a basis for Sp(S). The dimension of Sp(S) is 2.

(b) Denote the vectors in S by $\mathbf{v}_1, \mathbf{v}_2, \mathbf{v}_3, \mathbf{v}_4, \mathbf{v}_5$. $\{\mathbf{v}_1, \mathbf{v}_3, \mathbf{v}_5\}$ is a basis for Sp(S). The dimension of Sp(S) is 3.

A2 (a) Denote the vectors in S by $\mathbf{v}_1, \mathbf{v}_2, \mathbf{v}_3, \mathbf{v}_4$. $\{\mathbf{v}_1, \mathbf{v}_2, \mathbf{v}_3\}$ is a basis for Sp(S).
(b) $\{\mathbf{v}_1, \mathbf{v}_2, \mathbf{v}_3, \mathbf{v}_5\}$ is a basis for Sp(S).
(c) $\{\mathbf{v}_1, \mathbf{v}_2, \mathbf{v}_3, \mathbf{v}_4\}$ is a basis for Sp(S).

A3 (a) The dimension of Sp(S) is 3. (b) The dimension of Sp(S) is 3.

A4 (a) Vectors in the plane $\mathbf{v}_1 = (1, 2, 0)$, $\mathbf{v}_2 = (1, 0, 2)$. $\{\mathbf{v}_1, \mathbf{v}_2\}$ is linearly independent, a basis for the plane
(b) Let $\mathbf{w}_3 = \mathbf{e}_2 = (0, 1, 0)$. (Another natural choice for $\mathbf{w}_3$ is the normal vector $(2, -1, -1)$.)

A5 (a) Three vectors in the hyperplane are $\mathbf{v}_1 = (1, 0, 0, 1)$, $\mathbf{v}_2 = (1, 0, -1, 0)$, $\mathbf{v}_3 = (1, 1, 0, 0)$. $\{\mathbf{v}_1, \mathbf{v}_2, \mathbf{v}_3\}$ is linearly independent and a basis for the hyperplane.
(b) $\mathbf{w}_4 = \mathbf{e}_1 = (1, 0, 0, 0)$. (Another natural choice for $\mathbf{w}_4$ is the normal vector $(1, -1, 1, -1)$.)

SECTION 4–5

A1 (a) Number of variables is 4. Rank of A is 2. Dimension of the solution space is 2.
(b) Number of variables in the system is 5. Rank of B is 3. Dimension of the solution space is 2.
(c) Number of variables in the system is 5. Rank of C is 2. Dimension of the solution space is 3.
(d) Number of variables in the system is 6. Rank of D is 3. Dimension of the solution space is 3.

A2 (a) Reduced row echelon form is I. A basis for the rowspace is $\{(1, 0, 0), (0, 1, 0), (0, 0, 1)\}$. A basis for the columnspace is $\{(1, 1, 1), (2, 1, 0), (8, 5, -2)\}$. The nullspace is trivial $\{\mathbf{0}\}$, so there is no basis.

(b) Reduced row echelon form $\begin{bmatrix} 1 & 0 & -1 & 0 \\ 0 & 1 & -2 & 0 \\ 0 & 0 & 0 & 1 \end{bmatrix}$. A basis for the rowspace is $\{(1, 0, -1, 0), (0, 1, -2, 0), (0, 0, 0, 1)\}$; a basis for the columnspace is $\{(1, 2, 0), (1, 3, 1), (1, 4, 3)\}$. A basis for the nullspace is $\{(1, 2, 1, 0)\}$.

(c) RREF $\begin{bmatrix} 1 & 2 & 0 & 3 & 0 \\ 0 & 0 & 1 & 4 & 0 \\ 0 & 0 & 0 & 0 & 1 \\ 0 & 0 & 0 & 0 & 0 \end{bmatrix}$. A basis for the rowspace is $\{(1, 2, 0, 3, 0), (0, 0, 1, 4, 0), (0, 0, 0, 0, 1\}$; for the columnspace $\{(1, 1, 2, 3), (0, 1, 0, 1), (0, 1, 1, 2)\}$; for the nullspace $\{(-2, 1, 0, 0, 0), (-3, 0, -4, 1, 0)\}$.

A3 (a) A basis for the nullspace is $\{(2,1,0),(2,0,-1)\}$ (or any pair of linearly independent vectors orthogonal to (1,-2,2)); a basis for the range is $\{(1,-2,2)\}$; the rank is 1.
(b) For the nullspace a basis is $\{(3,1,2)\}$; for the range $\{(1,-3,0),(0,-2,1)\}$; the rank is 2.
(c) The nullspace is trivial, no basis. The range is $\mathbb{R}^3$ so take any basis of $\mathbb{R}^3$ (for example, the standard basis). The rank is 3.

A4 (a) $n = 5$
(b) $\{(1,0,2,0,3),(0,1,-1,0,1),(0,0,0,1,1)\}$ is a basis for the rowspace, since these vectors form a basis for the rowspace of R which is equal to the rowspace of A.
(c) $m = 4$
(d) A basis for the columnspace of A is $\{(1, 2, 1, 1), (1, 3, 1, 2), 1, 2, 3, -1)\}$; these columns are linearly independent because the first, second, fourth columns of R are linearly independent (Theorem 2, compare Example 31), and no larger set of columns is independent. In R, fifth column is equal to 3 times first plus 1 times second plus 1 times fourth, so the same is true in A, and $(5, 11, 7, 4)_{\mathcal{B}} = (3, 1, 1)$.
(e) $\mathbf{x} = s(-2, 1, 1, 0, 0) + t(-3, -1, 0, -1, 1)$, so a basis is $\{(-2, 1, 1, 0, 0), (-3, -1, 0, -1, 1)\}$
(f, g) rank of A is 3. Therefore, a basis for the solution space has two elements.
(h) $a_{i\rightarrow} \cdot \mathbf{x} = 0$, $i = 1,2,3,4$. Intuitively (Exercise 4–2 D5), orthogonality implies that solutions are linearly independent of rows, so the two bases together form a basis for $\mathbb{R}^5$; a more precise answer requires ideas from Section 7–1.

SECTION 4–6

A1 (a) $[L]_{\mathcal{B}} = \begin{bmatrix} 0 & 2 \\ 1 & -1 \end{bmatrix}$; $(L(\mathbf{x}))_{\mathcal{B}} = (6, 1)$

(b) $[L]_{\mathcal{B}} = \begin{bmatrix} 2 & 2 & 0 \\ 0 & 0 & 4 \\ -1 & -1 & 5 \end{bmatrix}$; $(L(\mathbf{x}))_{\mathcal{B}} = (12, -4, -11)$

A2 (a) $\begin{bmatrix} -3 & 0 \\ 0 & 4 \end{bmatrix}$ (b) $\begin{bmatrix} 0 & 2 \\ 1 & 0 \end{bmatrix}$

A3 (a) $\begin{bmatrix} 0 & 1 & 5 \\ 1 & 0 & 0 \\ 0 & 0 & 0 \end{bmatrix}$ (b) $\begin{bmatrix} 0 & 1 & 0 \\ 0 & 0 & 1 \\ 1 & 0 & 0 \end{bmatrix}$

A4 In each case, the first vector in $\mathcal{B}$ is the vector used in defining the mapping, and the other vector(s) is (are) orthogonal to it.
(a) $\mathcal{B} = \{(1,-2),(2,1)\}$, $[\mathbf{refl}_{(1,-2)}]_{\mathcal{B}} = \begin{bmatrix} -1 & 0 \\ 0 & 1 \end{bmatrix}$
(b) $\mathcal{B} = \{(2,1,-1),(1,0,2),(0,1,1)\}$ (other choices possible for second and

third vectors), $[\mathbf{proj}_{(2,1,-1)}]_{\mathcal{B}} = \begin{bmatrix} 1 & 0 & 0 \\ 0 & 0 & 0 \\ 0 & 0 & 0 \end{bmatrix}$

(c) $\mathcal{B} = \{(-1,-1,1),(1,1,0),(0,1,1)\}$ (other choices possible for second and third vectors, $[\mathbf{refl}_{(2,1,-1)}]_{\mathcal{B}} = \begin{bmatrix} -1 & 0 & 0 \\ 0 & 1 & 0 \\ 0 & 0 & 1 \end{bmatrix}$

A5 (a) $(1,2,4)_{\mathcal{B}} = (2,-1,1)$ (b) $[L]_{\mathcal{B}} = \begin{bmatrix} 2 & 0 & 0 \\ -1 & 0 & 2 \\ 1 & 1 & 0 \end{bmatrix}$

(c) $[L(1,2,4)]_{\mathcal{B}} = \begin{bmatrix} 2 & 0 & 0 \\ -1 & 0 & 2 \\ 1 & 1 & 0 \end{bmatrix} \begin{bmatrix} 2 \\ -1 \\ 1 \end{bmatrix} = \begin{bmatrix} 4 \\ 0 \\ 1 \end{bmatrix}$, so that

$[L(1,2,4)]_{\mathcal{S}} = \begin{bmatrix} 1 & 1 & 0 \\ 0 & -1 & 1 \\ 1 & 0 & 2 \end{bmatrix} \begin{bmatrix} 4 \\ 0 \\ 1 \end{bmatrix} = \begin{bmatrix} 4 \\ 1 \\ 6 \end{bmatrix}$

A6 (a) $(5,3,-5)_{\mathcal{B}} = (2,3,-3)$ (b) $[L]_{\mathcal{B}} = \begin{bmatrix} 0 & -2 & 2 \\ 0 & 0 & 3 \\ 1 & 0 & -3 \end{bmatrix}$

(c) $[L(5,3,-5)]_{\mathcal{B}} = \begin{bmatrix} 0 & -2 & 2 \\ 0 & 0 & 3 \\ 1 & 0 & -3 \end{bmatrix} \begin{bmatrix} 2 \\ 3 \\ -3 \end{bmatrix} = \begin{bmatrix} -12 \\ -9 \\ 11 \end{bmatrix}$, so,

$[L(5,3,-5)]_{\mathcal{S}} = \begin{bmatrix} 1 & 1 & 0 \\ 0 & 2 & 1 \\ -1 & 0 & 1 \end{bmatrix} \begin{bmatrix} -12 \\ -9 \\ 11 \end{bmatrix} = \begin{bmatrix} -21 \\ -7 \\ 23 \end{bmatrix}$

A7 (a) $\begin{bmatrix} 11 & 16 \\ -4 & -3 \end{bmatrix}$ (b) $\begin{bmatrix} 5 & 0 \\ 0 & -5 \end{bmatrix}$ (c) $\begin{bmatrix} -40 & -118 \\ 18 & 52 \end{bmatrix}$

(d) $\begin{bmatrix} 4 & 0 \\ 0 & 6 \end{bmatrix}$ (e) $\begin{bmatrix} 4 & 2 & 0 \\ 0 & 4 & 2 \\ 0 & 0 & 4 \end{bmatrix}$ (f) $\begin{bmatrix} 2 & 0 & 0 \\ 0 & -2 & 0 \\ 0 & 0 & 3 \end{bmatrix}$

SECTION 4–7

A1 (a) Define $L : P_3 \to \mathbb{R}^4$ by $L(a_0 + a_1x + a_2x^2 + a_3x^3) = (a_0, a_1, a_2, a_3)$

(b) Define $L : \mathcal{M}(2,2) \to \mathbb{R}^4$ by $L\left(\begin{bmatrix} a_1 & a_2 \\ a_3 & a_4 \end{bmatrix}\right) = (a_1, a_2, a_3, a_4)$

(c) Define $L : P_3 \to \mathcal{M}(2,2)$ by $L(a_0 + a_1x + a_2x^2 + a_3x^3) = \begin{bmatrix} a_0 & a_1 \\ a_2 & a_3 \end{bmatrix}$

SECTION 4–8

A1 (a) $-5 - 3i$ (b) $(5 - 3i, 7 + 8i, -1 - 9i)$ (c) $(-10 + 4i, 4 + 6i)$
(d) $(-4 - 3i, -1 - 7i, -12 + i)$

A2 (a) $[L] = \begin{bmatrix} 1 + 2i & 3 + i \\ 1 & 1 - i \end{bmatrix}$ (b) $(3 - 4i, -1 - 2i)$

(c) the range of L is $\{t(1 + 2i, 1) | t \in \mathbb{R}\}$. The nullspace of L is

$\{t(-1+i, 1) \mid t \in \mathbb{R}\}$.

Chapter 4 Quiz

1 (a) vector space (b) vector space (c) not a vector space

2 (a) linearly dependent (b) linearly dependent
(c) linearly independent

3 (a, b) A basis for $\mathbb{S}$. $\mathcal{B} = \{(1,0,1,1,3),(1,1,0,1,1),(3,3,1,0,2)\}$. The dimension of $\mathbb{S}$ is 3.
(c) $(0,2,-1,-3,-5)_{\mathcal{B}} = (-2,-1,1)$

4 A basis for the columnspace of A, and therefore the range of the linear mapping, is $\{(1,2,0,3),(0,1,2,3),(1,2,1,4)\}$. A basis for the nullspace is $\{(-1,1,1,0,0),(1,-3,0,-2,1)\}$.

5 $[L]_{\mathcal{B}} = \begin{bmatrix} 0 & 2/3 & 11/3 \\ -1 & 2/3 & -10/3 \\ 0 & 1/3 & 1/3 \end{bmatrix}$

6 $[L]_{\mathcal{B}} = \begin{bmatrix} -1 & 0 & 0 \\ 0 & 1 & 0 \\ 0 & 0 & 1 \end{bmatrix}$, $[L]_{S} = \begin{bmatrix} 0 & 0 & 1 \\ 0 & 1 & 0 \\ 1 & 0 & 0 \end{bmatrix}$.

8 (a) False. (b) True. (c) True. (d) False. (e) True.

CHAPTER 5

SECTION 5–1

A1 (a) The first row equals the third row.
(b) The third column consists of all zeros.
(c) The second column equals the fourth column.

A2 (a) 18 (b) -46 (c) 98 (d) 0 (e) -420

A3 (a) $\mathrm{Cof}(A) = \begin{bmatrix} 5 & -7 \\ 4 & 2 \end{bmatrix}$, $A(\mathrm{Cof}(A))^T = 38\begin{bmatrix} 1 & 0 \\ 0 & 1 \end{bmatrix}$

(b) $\mathrm{Cof}(A) = \begin{bmatrix} -13 & 7 & 5 \\ -13 & 7 & 5 \\ 13 & -7 & -5 \end{bmatrix}$, $A(\mathrm{Cof}(A))^T = 0\,I$

(c) $\mathrm{Cof}(A) = \begin{bmatrix} 34 & 14 & 25 \\ -33 & 6 & -21 \\ -28 & -5 & -1 \end{bmatrix}$, $A(\mathrm{Cof}(A))^T = -111I$

SECTION 5–2

A1 (a) 30 (b) 1 (c) 8 (d) -1120

A2 (a) 14 (b) -5

A3 (a) $3p - 14$, invertible for all $p \neq \dfrac{14}{3}$

(b) $\det = -5p - 20$; invertible for all $p \neq -4$
(c) $\det = 2p - 116$; invertible for all $p \neq 58$

A4 (a) -7 (b) 14 (c) 0 (d) 112

SECTION 5–3

A1 (a) $\begin{bmatrix} -1/7 & 5/7 \\ -2/7 & 3/7 \end{bmatrix}$ (b) $\begin{bmatrix} -3/8 & 21/8 & 11/8 \\ -1/8 & 7/8 & 5/8 \\ 3/8 & -13/8 & -7/8 \end{bmatrix}$

A2 (a) $\mathrm{Cof}(A) = \begin{bmatrix} -1 & 5 & -2 \\ 3+k & 2-3k & -11 \\ -3 & -2-2k & -6 \end{bmatrix}$

(b) $A(\mathrm{Cof}(A))^T = \begin{bmatrix} -2k-17 & 0 & 0 \\ 0 & -2k-17 & 0 \\ 0 & 0 & -2k-17 \end{bmatrix}, = (-2k-17)I$, so
$\det A = -2k - 17$, and provided that $-2k - 17 \neq 0$,

$$A^{-1} = \frac{1}{-2k-17}\begin{bmatrix} -1 & 3+k & -3 \\ 5 & 2-3k & -2-2k \\ -2 & -11 & -6 \end{bmatrix}$$

A3 (a) $(51/19, -4/19)$ (b) $(-1/5, 14/5, 6/5)$

SECTION 5–4

A1 (a) $\det A = 13$, $\det B = 14$, $\det \begin{bmatrix} 7 & 0 \\ 4 & 26 \end{bmatrix} = 182$

(b) $\det A = -2$, $\det B = 56$, $\det \begin{bmatrix} 2 & 7 & 25 \\ -7 & -4 & 7 \\ 11 & 11 & 8 \end{bmatrix} = -112$

A2 (a) 11 (b) $(18, 19)$ and $(34, 20)$ (c) -26 (d) 286

A3 (a) 63 (b) 42 (c) 2646

A4 (a) 41 (b) 78 (c) 3198

A5 (a) 5 (b) 245

Chapter 5 Quiz

1 12

2 180

3 120

4 invertible for all $k \neq -\frac{7}{8} \pm \frac{\sqrt{145}}{8}$

5 (a) 21 (b) 7 (c) 224 (d) $\frac{1}{7}$ (e) 49

6 $(A^{-1})_{31} = \frac{1}{\det A}\left((\mathrm{Cof}(A))^T\right)_{31} = -\frac{1}{3}$

7 $\left(\frac{2}{3}, -\frac{1}{2}, \frac{7}{6}\right)$

8 (a) 33 (b) 792

CHAPTER 6

SECTION 6–1

A1 (a) $(2, 1)$ is an eigenvector of A with eigenvalue 14, and $(-1, 3)$ is an eigenvector of A with eigenvalue -7. $P^{-1} = \frac{1}{7}\begin{bmatrix} 3 & 1 \\ -1 & 2 \end{bmatrix}$, $P^{-1}AP = \begin{bmatrix} 14 & 0 \\ 0 & -7 \end{bmatrix}$

$P^{-1}AP$ is diagonal.

(b) P does not diagonalize A.

(c) $(2, 1)$ is an eigenvector of A with eigenvalue 1, and $(1, 1)$ is an eigenvector of A with eigenvalue -3. $P^{-1}AP = \begin{bmatrix} 1 & 0 \\ 0 & -3 \end{bmatrix}$. $P^{-1}AP$ is diagonal.

A2 (a) $(3, 1, -1), (3, 2, -1)$, and $(-2, -1, 1)$ are eigenvectors of A with eigenvalues $1, -1$, and 2 respectively. $P^{-1}AP = \begin{bmatrix} 1 & 0 & 0 \\ 0 & -1 & 0 \\ 0 & 0 & 2 \end{bmatrix}$

(b) $(1, 0, 1), (-1, 1, 0)$, and $(3, -1, 3)$ are eigenvectors of A with eigenvalues $-1, 2$, and -2 respectively. $P^{-1}AP = \begin{bmatrix} -1 & 0 & 0 \\ 0 & 2 & 0 \\ 0 & 0 & -2 \end{bmatrix}$

SECTION 6–2

A1 Calculate $A\mathbf{v}$; $(1, 0, 1)$ and $(1, -1, 1)$ are not eigenvectors, the others are; for $(1, 0, 2)$, $\lambda = 0$; for $(1, 1, -1)$, $\lambda = -6$; for $(1, 1, 1)$, $\lambda = 4$

A2 The diagonalizing matrix for Example 8 is $P = \begin{bmatrix} 1 & 1 & 1 \\ 1 & 3 & -1 \\ 1 & -1 & 2 \end{bmatrix}$. (Your matrix may differ by a constant factor in each column.)

A3 (a) $\lambda = 2$, eigenvector $(1, 2)$; $\lambda = 3$, eigenvector $(1, 3)$. $P = \begin{bmatrix} 1 & 1 \\ 2 & 3 \end{bmatrix}$, diagonal matrix $D = \begin{bmatrix} 2 & 0 \\ 0 & 3 \end{bmatrix}$

(b) $\lambda = 1$, all eigenvectors of A are scalar multiples of $(1, 0)$, the matrix A is not diagonable.

(c) $\lambda = -1$, eigenvector $(2, 5)$; $\lambda = 4$, eigenvector $(1, 3)$; $P = \begin{bmatrix} 2 & 1 \\ 5 & 3 \end{bmatrix}$; diagonal matrix $\begin{bmatrix} -1 & 0 \\ 0 & 4 \end{bmatrix}$

(d) $\lambda = 5$, eigenvector, $(3, 4)$, $\lambda = -2$, eigenvector $(-1, 1)$, $P = \begin{bmatrix} 3 & -1 \\ 4 & 1 \end{bmatrix}$, diagonal matrix $D = \begin{bmatrix} 5 & 0 \\ 0 & -2 \end{bmatrix}$

A4 Eigenvalues are 1, 2, -2; eigenvectors are columns of $P = \begin{bmatrix} 3 & 1 & 1 \\ 0 & -1 & 1 \\ 1 & -1 & 1 \end{bmatrix}$.

Matrix is diagonable; similar to $\begin{bmatrix} 1 & 0 & 0 \\ 0 & 2 & 0 \\ 0 & 0 & -2 \end{bmatrix}$.

SECTION 6–3

A1 The characteristic equation of A has roots 1,1,5. There is only one eigendirection for $\lambda = 1$, so A is not diagonable. For B, 1 is a simple root of the characteristic equation, and the only real eigenvalue. B is not diagonable. C has eigenvalues 2, 2, -1, and is diagonalized by $P = \begin{bmatrix} 1 & 1 & 1 \\ 0 & 1 & -1 \\ 1 & -1 & 2 \end{bmatrix}$. Other choices of P are possible.

A2 The characteristic equation of E has roots 2, 2, -2. There is only one eigendirection for $\lambda = 2$, so E is not diagonable. F has eigenvalues 1, 2, 2, and is diagonalized by $P = \begin{bmatrix} 0 & 2 & 3 \\ 1 & -1 & -1 \\ 1 & 0 & 1 \end{bmatrix}$. Other choices of P are possible. G has only one real eigenvalue, 2; the other roots are complex, and so G is not diagonable.

SECTION 6–4

A1 (a) $ae^{-5t}\begin{bmatrix} -1 \\ 4 \end{bmatrix} + be^{4t}\begin{bmatrix} 2 \\ 1 \end{bmatrix}, \quad a, b \in \mathbb{R}$

(b) $ae^{-0.5t}\begin{bmatrix} -1 \\ 1 \end{bmatrix} + be^{0.3t}\begin{bmatrix} 7 \\ 1 \end{bmatrix}, \quad a, b \in \mathbb{R}$

SECTION 6–5

A1 (a) A and C are not Markov matrices. B and D are Markov matrices.
(b) For B, the invariant state is $(6/13, 7/13)$. For D, the invariant state is $(1/3, 1/3, 1/3)$.

A2 (a) In the long run, 25% of the population will be rural dwellers and 75% will be urban dwellers.
(b) After 5 decades, approximately 33% of the population will be rural dwellers and 67% will be urban dwellers.

A3 $T = \frac{1}{10}\begin{bmatrix} 8 & 3 & 3 \\ 1 & 6 & 1 \\ 1 & 1 & 6 \end{bmatrix}$. In the long run, 60% of the cars will be at the airport, 20% of the cars will be at the train station, and 20% of the cars will be at the city centre.

A4 (a) dominant eigenvalue is $\lambda = 5$. (b) dominant eigenvalue is $\lambda = 6$.

SECTION 6–6

A1 (a) diagonal matrix $D = \begin{bmatrix} 2i & 0 \\ 0 & -2i \end{bmatrix}$, real canonical form: $\begin{bmatrix} 0 & 2 \\ -2 & 0 \end{bmatrix}$, $P = \begin{bmatrix} 0 & -2 \\ 1 & 0 \end{bmatrix}$

(b) diagonal matrix $D = \begin{bmatrix} -2+i & 0 \\ 0 & -2-i \end{bmatrix}$, real canonical form: $\begin{bmatrix} -2 & 1 \\ -1 & -2 \end{bmatrix}$, $P = \begin{bmatrix} -1 & -1 \\ 1 & 0 \end{bmatrix}$

(c) diagonal matrix $D = \begin{bmatrix} 0 & 0 & 0 \\ 0 & 1+2i & 0 \\ 0 & 0 & 1-2i \end{bmatrix}$, real canonical form: $\begin{bmatrix} 0 & 0 & 0 \\ 0 & 1 & 2 \\ 0 & -2 & 1 \end{bmatrix}$, $P = \begin{bmatrix} 1 & 1 & 0 \\ 0 & 0 & 1 \\ 2 & 1 & 0 \end{bmatrix}$

(d) diagonal matrix $D = \begin{bmatrix} 1 & 0 & 0 \\ 0 & 2+i & 0 \\ 0 & 0 & 2-i \end{bmatrix}$, real canonical form: $\begin{bmatrix} 1 & 0 & 0 \\ 0 & 2 & 1 \\ 0 & -1 & 2 \end{bmatrix}$, $P = \begin{bmatrix} 0 & 1 & 2 \\ 1 & 3 & 1 \\ 1 & 5 & 0 \end{bmatrix}$

Chapter 6 Quiz

1 $(3, 1, 0)$ is not an eigenvector of A. $(1, 0, 1)$, $(4, 1, 0)$, and $(2, 1, -1)$ are eigenvectors of A with corresponding eigenvalues $1, 1$, and -1 respectively.

2 The matrix is diagonable. $P = \begin{bmatrix} 1 & 1 & 1 \\ 3 & -1 & 1 \\ -1 & 2 & 1 \end{bmatrix}$, diagonal matrix $D = \begin{bmatrix} 0 & 0 & 0 \\ 0 & -4 & 0 \\ 0 & 0 & -2 \end{bmatrix}$

3 $A^{-1}\mathbf{x} = \dfrac{1}{\lambda}\mathbf{x} = \lambda^{-1}\mathbf{x}$

4 (a) one-dimensional (b) zero-dimensional (c) rank of A is two

5 $ae^{-0.1t}\begin{bmatrix} -1 \\ 1 \end{bmatrix} + be^{0.4t}\begin{bmatrix} 2 \\ 3 \end{bmatrix}$

6 $(\frac{1}{4}, \frac{1}{4}, \frac{1}{2})$

7 $D = \begin{bmatrix} 2+3i & 0 \\ 0 & 2-3i \end{bmatrix}$, $P = \begin{bmatrix} 2+3i & 2-3i \\ 1 & 1 \end{bmatrix}$

CHAPTER 7

SECTION 7–1

A1 (a) $n = 2$, orthogonal but not orthonormal, orthonormal set is an orthonormal basis $\{(1/\sqrt{5})(1,2), (1/\sqrt{5})(2,-1)\}$, orthogonal matrix $(1/\sqrt{5})\begin{bmatrix} 1 & 2 \\ 2 & -1 \end{bmatrix}$

(b) $n = 3$, not orthogonal therefore cannot be orthonormal, therefore cannot provide an orthogonal matrix

(c) $n = 3$, orthogonal but not orthonormal, orthonormal set is an orthonormal basis $\{(1/\sqrt{11})(1,1,3), (1/\sqrt{10})(3,0,-1), (1/\sqrt{110})(1,-10,3)\}$; orthogonal matrix is $\begin{bmatrix} 1/\sqrt{11} & 3/\sqrt{10} & 1/\sqrt{110} \\ 1/\sqrt{11} & 0 & -10\sqrt{110} \\ 3/\sqrt{11} & -1/\sqrt{10} & 3/\sqrt{110} \end{bmatrix}$

(d) $n = 4$, orthogonal, not orthonormal; orthonormal set, $\{(1/\sqrt{3})(1,0,1,1), (1/\sqrt{7})(2,1,-1,1), (1/\sqrt{11})(1,-3,-1,0)\}$ is not a basis for $\mathbb{R}^4$ (not enough vectors), so there is no corresponding orthogonal matrix

A2 $(\mathbf{x})_B = (8/3, 19/3, 5/3)$; $(\mathbf{y})_B = (7,-2,-3)$

A3 $(\mathbf{w})_B = (22/3, 2/3, 4/3)$; $(\mathbf{z})_B = (0,9,0)$

A4 $(2,4,-3,5)_B = (4,-5,-5/\sqrt{2},-1/\sqrt{2})$;
$(-4,1,3,-5)_B = (-5/2, 3/2, 7/\sqrt{2}, 6/\sqrt{2})$

A5 $(3,1,0,1)_B = (5/2, 1/2, -3/\sqrt{2}, 0)$ $(-1,3,2,-3)_B = (1/2, 1/2, 3/\sqrt{2}, 6/\sqrt{2})$

A6 (a) $A^TA = \begin{bmatrix} 1 & 0 \\ 0 & 1 \end{bmatrix}$. A is orthogonal.

(b) $A^TA = \begin{bmatrix} 1 & 24/25 \\ 24/25 & 1 \end{bmatrix}$; not orthogonal, the two columns are not orthogonal

(c) $A^TA = \begin{bmatrix} 1/5 & 0 \\ 0 & 1/5 \end{bmatrix}$; not orthogonal, none of columns or the rows are of unit length

(d) $A^TA = \begin{bmatrix} 1 & 0 & 4/9 \\ 0 & 1 & -4/9 \\ 4/9 & -4/9 & 1 \end{bmatrix}$; not orthogonal; columns are not mutually orthogonal

(e) $A^TA = \begin{bmatrix} 1 & 0 & 0 \\ 0 & 1 & 0 \\ 0 & 0 & 1 \end{bmatrix}$; A is orthogonal.

A7 (a) $\mathbf{g}_2 = (6,-3,6)$.

(b) $R = \frac{1}{3}\begin{bmatrix} -1 & 2 & 2 \\ 2 & -1 & 2 \\ 2 & 2 & -1 \end{bmatrix}$

(c) $[L]_B = \frac{1}{\sqrt{2}}\begin{bmatrix} 1 & -1 & 0 \\ 1 & 1 & 0 \\ 0 & 0 & \sqrt{2} \end{bmatrix}$

(d) $[L]_S = \dfrac{1}{9\sqrt{2}} \begin{bmatrix} 5+4\sqrt{2} & -1+4\sqrt{2} & 8-2\sqrt{2} \\ -7+4\sqrt{2} & 5+4\sqrt{2} & -4-2\sqrt{2} \\ -4-2\sqrt{2} & 8-2\sqrt{2} & 8+\sqrt{2} \end{bmatrix}$

SECTION 7–2

A1 (a) $(5/2, 5, 5/2, 5)$ (b) $(2, 9/2, 5, 9/2)$

A2 (a) $\{(1,0,0), (0,1,0), (0,0,1)\}$

(b) $\left\{ \dfrac{1}{\sqrt{3}}(1,1,1), \dfrac{1}{\sqrt{6}}(1,1,-2), \dfrac{1}{\sqrt{2}}(1,-1,0) \right\}$

A3 (a) $\{(1/\sqrt{3})(1,1,0,1), (1/\sqrt{15})(-2,1,3,1), (1/\sqrt{10})(2,-1,2,-1)\}$;

(b) $\{(1/\sqrt{3})(1,0,1,0,1), (1/\sqrt{3})(1,0,-1,1,0), (1/\sqrt{15})(-1,3,1,2,0)\}$

A4 A basis for S is $\{(0,1,1,0), (2,-1,0,1)\}$; an orthonormal basis is $\{(1/\sqrt{2})(0,1,1,0), (1/\sqrt{22})(4,-1,1,2)\}$; the closest point is $\mathbf{proj}_S(1,4,0,1) = (1/11)(4,21,23,2)$; the distance is $\|\mathbf{perp}_S(1,4,0,1)\| = 6\sqrt{33}/11$

SECTION 7–3

A1 (a) $y = 10.5 - 1.9t$

(b) $y = 3.4 + 0.8t$

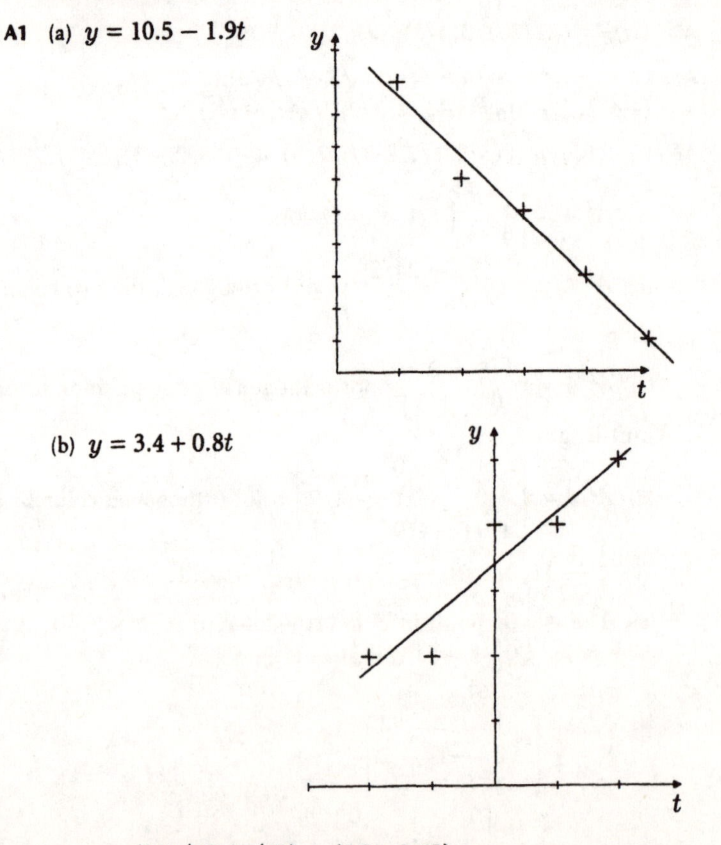

A2 (a) $\mathbf{x} = (383/98, 32/49) \approx (3.91, 0.65)$

(b) $\mathbf{x} = (167/650, -401/325) \approx (0.26, 1.23)$

SECTION 7–4

A1 (a) -46 (b) -84 (c) $\sqrt{22}$ (d) $9\sqrt{59}$

A2 (a) not an inner product (b) not an inner product (c) not an inner product

A3 $a_0 = 0$ because $f(x)$ is odd. $a_n = 0$ because $f(x)\cos nx$ is odd.

$b_n = \frac{2}{\pi}\left[\int_0^{\pi/2} x \sin nx\, dx + \int_{\pi/2}^{\pi} (\pi - x)\sin nx\, dx\right]$ Integrating by parts

$$b_n = \frac{2}{\pi}\left[\frac{\sin\frac{n\pi}{2}}{n^2} + \frac{\sin\frac{n\pi}{2}}{n^2}\right] = \left(\frac{4}{\pi n^2}\right)\left(\sin\frac{n\pi}{2}\right)$$

SECTION 7–5

A1 (a) $\langle \mathbf{u}, \mathbf{v}\rangle = 2 + 5i$, $\langle \mathbf{v}, \mathbf{u}\rangle = 2 - 5i$, $\|\mathbf{u}\| = \sqrt{18}$, $\|\mathbf{v}\| = \sqrt{33}$
(b) $\langle \mathbf{u}, \mathbf{v}\rangle = -6i$, $\langle \mathbf{v}, \mathbf{u}\rangle = 6i$, $\|\mathbf{u}\| = \sqrt{22}$, $\|\mathbf{v}\| = \sqrt{20}$

A2 (a) $\overline{A}^T A = \frac{1}{3}\begin{bmatrix} 3 & 2 \\ 2 & 3 \end{bmatrix} \neq I$, A is not unitary.
(b) $\overline{B}^T B = I$, B is unitary. (c) $\overline{C}^T C = I$, C is unitary.

A3 (a) $\langle \mathbf{u}, \mathbf{v}\rangle = \overline{(1, 0, i)} \cdot (i, 1, 1) = 0$
(b) $\mathbf{proj}_S(\mathbf{w}) = \left(-\frac{1}{3} + 3i, 2 + \frac{1}{3}i, 1 + \frac{1}{3}i\right)$

Chapter 7 Quiz

1 (a) $\mathbf{u}_1$ and $\mathbf{u}_3$ are not orthogonal.
(b) Since $\mathbf{v}_1$ and $\mathbf{v}_2$ are not orthogonal, the set is not orthogonal and therefore cannot be orthonormal.
(c) The set is orthonormal.

2 $(2, 5, 1, 3, -2)_{\mathcal{B}} = (2, \frac{9}{\sqrt{3}}, \frac{6}{\sqrt{3}})$

3 (a) $1 = \det I = \det(RR^T) = (\det R)(\det R^T) = (\det R)^2$
(b) $(RS)^T(RS) = S^T R^T RS = S^T IS = S^T S = I$

4 (a) $\left\{\frac{1}{\sqrt{2}}(1, 0, 1, 0), \frac{1}{\sqrt{2}}(0, 1, 0, 1), \frac{1}{2}(-1, 1, 1, -1)\right\}$
(b) The closest point in S to $(1, -2, -1, 0)$ is $(1, -2, -1, 0)$. The point $(1, -2, -1, 0)$ is already in S.

5 (a) not an inner product on $M(2, 2)$
(b) defines an inner product on $M(2, 2)$

6 (a) $11 - 4i$ (b) $\overline{A}^T A = \begin{bmatrix} 1 & 0 \\ 0 & 1 \end{bmatrix}$, A is therefore unitary.

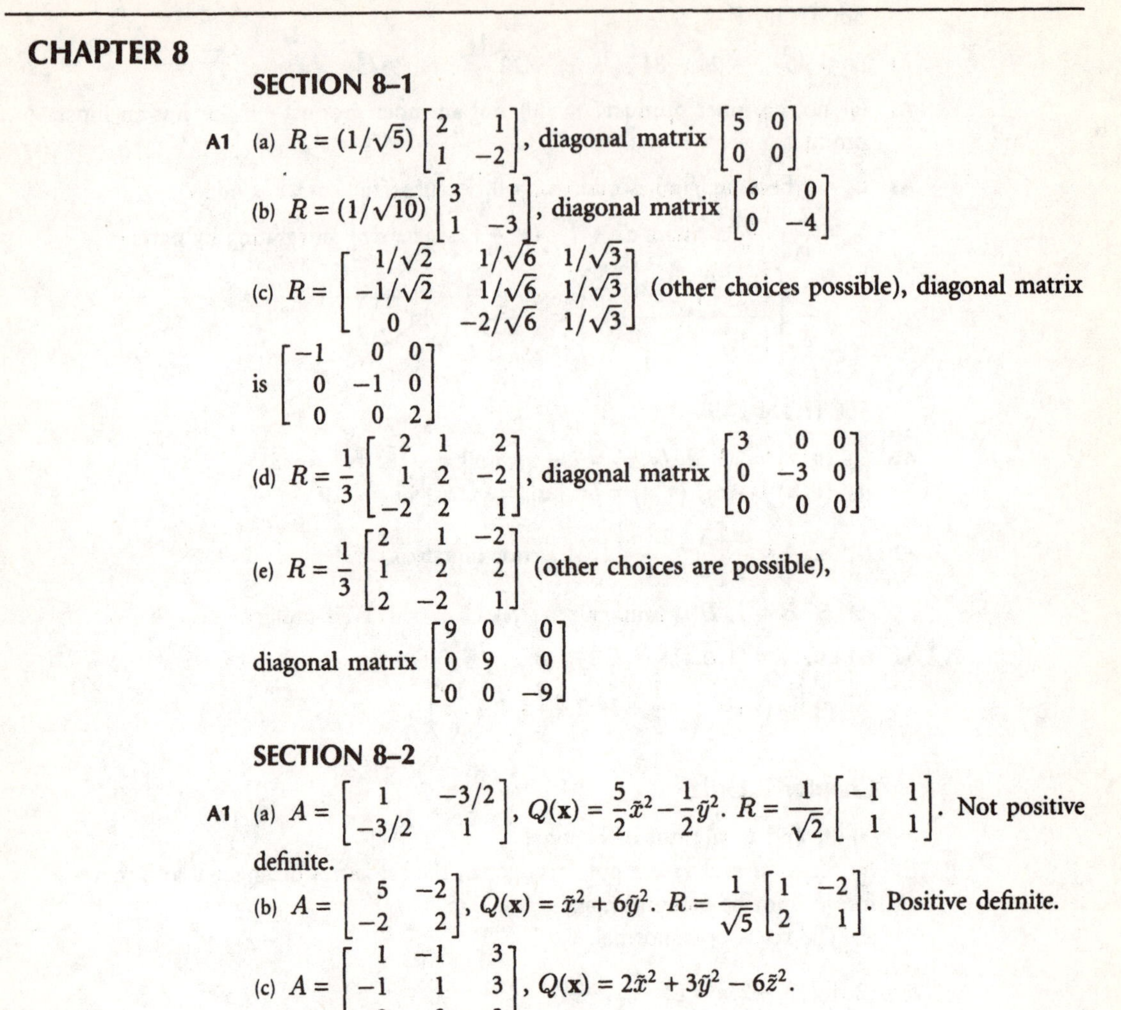

CHAPTER 8

SECTION 8–1

A1 (a) $R = (1/\sqrt{5})\begin{bmatrix} 2 & 1 \\ 1 & -2 \end{bmatrix}$, diagonal matrix $\begin{bmatrix} 5 & 0 \\ 0 & 0 \end{bmatrix}$

(b) $R = (1/\sqrt{10})\begin{bmatrix} 3 & 1 \\ 1 & -3 \end{bmatrix}$, diagonal matrix $\begin{bmatrix} 6 & 0 \\ 0 & -4 \end{bmatrix}$

(c) $R = \begin{bmatrix} 1/\sqrt{2} & 1/\sqrt{6} & 1/\sqrt{3} \\ -1/\sqrt{2} & 1/\sqrt{6} & 1/\sqrt{3} \\ 0 & -2/\sqrt{6} & 1/\sqrt{3} \end{bmatrix}$ (other choices possible), diagonal matrix is $\begin{bmatrix} -1 & 0 & 0 \\ 0 & -1 & 0 \\ 0 & 0 & 2 \end{bmatrix}$

(d) $R = \frac{1}{3}\begin{bmatrix} 2 & 1 & 2 \\ 1 & 2 & -2 \\ -2 & 2 & 1 \end{bmatrix}$, diagonal matrix $\begin{bmatrix} 3 & 0 & 0 \\ 0 & -3 & 0 \\ 0 & 0 & 0 \end{bmatrix}$

(e) $R = \frac{1}{3}\begin{bmatrix} 2 & 1 & -2 \\ 1 & 2 & 2 \\ 2 & -2 & 1 \end{bmatrix}$ (other choices are possible), diagonal matrix $\begin{bmatrix} 9 & 0 & 0 \\ 0 & 9 & 0 \\ 0 & 0 & -9 \end{bmatrix}$

SECTION 8–2

A1 (a) $A = \begin{bmatrix} 1 & -3/2 \\ -3/2 & 1 \end{bmatrix}$, $Q(\mathbf{x}) = \frac{5}{2}\tilde{x}^2 - \frac{1}{2}\tilde{y}^2$. $R = \frac{1}{\sqrt{2}}\begin{bmatrix} -1 & 1 \\ 1 & 1 \end{bmatrix}$. Not positive definite.

(b) $A = \begin{bmatrix} 5 & -2 \\ -2 & 2 \end{bmatrix}$, $Q(\mathbf{x}) = \tilde{x}^2 + 6\tilde{y}^2$. $R = \frac{1}{\sqrt{5}}\begin{bmatrix} 1 & -2 \\ 2 & 1 \end{bmatrix}$. Positive definite.

(c) $A = \begin{bmatrix} 1 & -1 & 3 \\ -1 & 1 & 3 \\ 3 & 3 & -3 \end{bmatrix}$, $Q(\mathbf{x}) = 2\tilde{x}^2 + 3\tilde{y}^2 - 6\tilde{z}^2$.

$R = \begin{bmatrix} -1/\sqrt{2} & 1/\sqrt{3} & 1/\sqrt{6} \\ 1/\sqrt{2} & 1/\sqrt{3} & 1/\sqrt{6} \\ 0 & 1/\sqrt{3} & -2/\sqrt{6} \end{bmatrix}$. Not positive definite.

SECTION 8–3

A1 Hyperbola $3\tilde{x}^2 - 2\tilde{y}^2 = 6$, with $\tilde{x}$-axis determined by (2,1), and $\tilde{y}$-axis determined by $(-1, 2)$

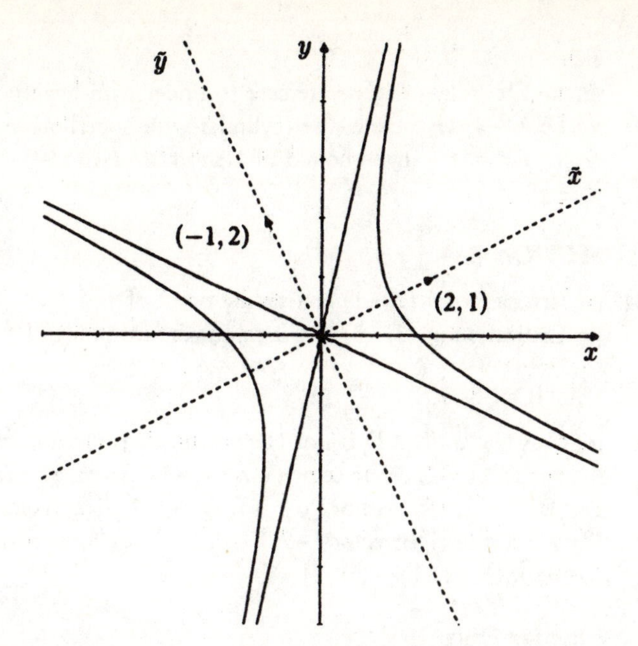

A2 Ellipse $11\tilde{x}^2+\tilde{y}^2 = 11$, with $\tilde{x}$-axis determined by $(1,3)$, and $\tilde{y}$-axis determined by $(-3,1)$

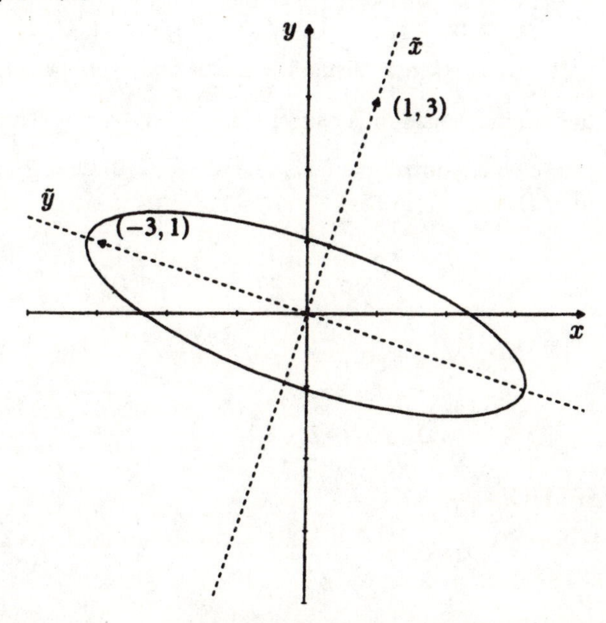

A3 (a) $\mathbf{x} \cdot A\mathbf{x} = 1$ – pair of parallel lines; $\mathbf{x} \cdot A\mathbf{x} = -1$ – empty set

(b) $\mathbf{x} \cdot B\mathbf{x} = 1$ – hyperbola, semi-transverse axis $1/\sqrt{6}$, semi-conjugate axis $1/2$ $\mathbf{x} \cdot B\mathbf{x} = -1$ – hyperbola, semi-transverse axis $1/2$, semi-conjugate axis $1/\sqrt{6}$

(c) $\mathbf{x} \cdot C\mathbf{x} = 1$ – hyperboloid of 2 sheets; $\mathbf{x} \cdot C\mathbf{x} = -1$ – hyperboloid of 1 sheet

(d) $\mathbf{x} \cdot D\mathbf{x} = 1$ – degenerate case (cylinder with hyperbola as cross-section)
$\mathbf{x} \cdot D\mathbf{x} = -1$ degenerate case (cylinder with hyperbola as cross-section)

(e) $\mathbf{x} \cdot E\mathbf{x} = 1$ – hyperboloid of 1 sheet; $\mathbf{x} \cdot E\mathbf{x} = -1$ – hyperboloid of 2 sheets

SECTION 8–4

A1 (a) Critical point $(2, -1)$ is a saddle point of f.
(b) Critical point $(3, -3)$ is a local maximum point of f.

SECTION 8–6

A1 (a) C is Hermitian. D is not Hermitian. E is Hermitian. F is Hermitian.
(b) For C: $\lambda = 1$, eigenvector $(\sqrt{2} + i, -3)$. $\lambda = 5$, eigenvector $(\sqrt{2} + i, 1)$.
For E: $\lambda = 7$, eigenvector $(\sqrt{3} - i, 1)$. $\lambda = 2$, eigenvector $(\sqrt{3} - i, -4)$.
For F: $\lambda = 0$, eigenvector $(-2, 1 - i, 2)$. $\lambda = \sqrt{5}$, eigenvector $\left(3 + \sqrt{5}, (1 - i)(1 + \sqrt{5}), 2\right)$. $\lambda = -\sqrt{5}$

Chapter Quiz

1 $D = \begin{bmatrix} 6 & 0 & 0 \\ 0 & -3 & 0 \\ 0 & 0 & 4 \end{bmatrix}$, $R = \begin{bmatrix} 1/\sqrt{6} & 1/\sqrt{3} & 1/\sqrt{2} \\ -2/\sqrt{6} & 1/\sqrt{3} & 0 \\ -1/\sqrt{6} & -1/\sqrt{3} & 1/\sqrt{2} \end{bmatrix}$

2 If $(1, -1)$ is taken to define the $\tilde{x}$-axis and $(1, 1)$ to define the $\tilde{y}$-axis, then the original equation is equivalent to $\dfrac{\tilde{x}^2}{2} + \dfrac{\tilde{y}^2}{3} = 1$. This is the equation of an ellipse with $\tilde{x}$-intercepts $(\sqrt{2}, 0)$ and $(-\sqrt{2}, 0)$, and $\tilde{y}$-intercepts $(0, \sqrt{3})$ and $(0, \sqrt{3})$.

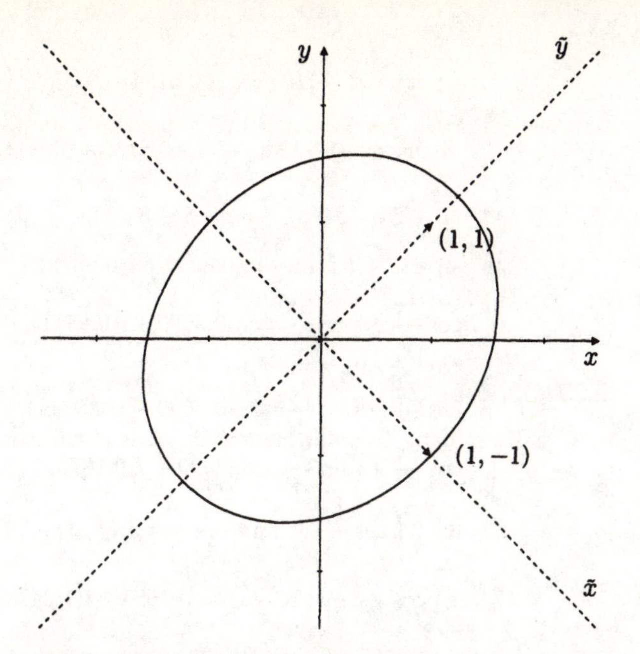

3 (a) distinct eigenvalues, so corresponding eigenvectors form a linearly independent set; diagonable
(b) may or may not be diagonable
(c) not diagonable; complex characteristic roots
(d) Any 3 by 3 *symmetric* matrix is diagonable.

4 $R^T AR = 3I$. so $A = R3IR^T = 3RR^T = 3I$.

5 (a) Hermitian for $k = -1$. (b) -2 and 5 are the eigenvalues of A. For $\lambda = -2$, eigenvector is $(3 - i, -2)$. For $\lambda = 5$, eigenvector is $(3 - i, 5)$.

APPENDIX

A1 (a) $5 + 7i$ (b) $-3 - 4i$ (c) $-7 + 2i$ (d) $4 + 5i$

A2 (a) $9 + 7i$ (b) $-10 - 10i$ (c) $2 + 25i$ (d) -2

A3 (a) $3 + 5i$ (b) $2 - 7i$

A4 (a) $\text{Re}(z) = 3$, $\text{Im}(z) = -6$. (b) $\text{Re}(z) = 17$, $\text{Im}(z) = -1$.
(c) $\text{Re}(z) = \frac{24}{37}$, $\text{Im}(z) = \frac{4}{37}$.

A5 (a) $\frac{2}{13} - \frac{3}{13}i$ (b) $\frac{6}{53} + \frac{21}{53}i$ (c) $\frac{-4}{13} - \frac{19}{13}i$ (d) $\frac{-2}{17} + \frac{25}{17}i$

A6 (a) $wz \approx -0.7321 + i(2.7321)$, $\frac{w}{z} \approx 0.6830 - i(0.1830)$
(b) $wz \approx -2.7321 + i(0.7321)$, $\frac{w}{z} \approx -0.3660 - i(1.3660)$

(c) $wz = 4 - 7i$, $\dfrac{w}{z} \approx -0.6154 - 0.0769i \left(= -\dfrac{8}{13} - \dfrac{1}{13}i \right)$

(d) $wz = -17 + 9i$; $\dfrac{w}{z} \approx -0.5135 + 0.0811\,i \left(= -\dfrac{19}{37} + \dfrac{3}{37}i \right)$

A7 (a) $-54 - 54i$ (b) $-8 - 8\sqrt{3}\,i$ (c) $512(\sqrt{3} + i)$

A8 (a) $\cos\dfrac{\pi}{5} + i\sin\dfrac{\pi}{5} \approx 0.8090 + i(0.5878)$,

$\cos\dfrac{3\pi}{5} + i\sin\dfrac{3\pi}{5} \approx -0.3090 + i(0.9511)$,

$\cos\pi + i\sin\pi = -1$,

$\cos\dfrac{7\pi}{5} + i\sin\dfrac{7\pi}{5} \approx -0.3090 - i(0.9511)$,

$\cos\dfrac{9\pi}{5} + i\sin\dfrac{9\pi}{5} \approx 0.8090 - i(0.5878)$

(b) $2\left(\cos\dfrac{-\pi}{8} + i\sin\dfrac{-\pi}{8}\right) \approx 1.8478 - i(0.7654)$,

$2\left(\cos\dfrac{3\pi}{8} + i\sin\dfrac{3\pi}{8}\right) \approx 0.7654 + i(1.8478)$,

$2\left(\cos\dfrac{7\pi}{8} + i\sin\dfrac{7\pi}{8}\right) \approx -1.8478 + i(0.7654)$,

$2\left(\cos\dfrac{11\pi}{8} + i\sin\dfrac{11\pi}{8}\right) \approx -0.7654 - i(1.8478)$

(c) $2^{1/3}\left(\cos\dfrac{-5\pi}{18} + i\sin\dfrac{-5\pi}{18}\right) \approx 0.8099 - i(0.9652)$,

$2^{1/3}\left(\cos\dfrac{7\pi}{18} + i\sin\dfrac{7\pi}{18}\right) \approx 0.4309 + i(1.1839)$,

$2^{1/3}\left(\cos\dfrac{19\pi}{18} + i\sin\dfrac{19\pi}{18}\right) \approx -1.2408 - i(0.2188)$

(d) $17^{1/6}\left(\cos\left(\dfrac{\theta}{3}\right) + i\sin\left(\dfrac{\theta}{3}\right)\right) \approx 1.4495 + i(0.6858)$,

$17^{1/6}\left(\cos\left(\dfrac{\theta + 2\pi}{3}\right) + i\sin\left(\dfrac{\theta + 2\pi}{3}\right)\right) \approx -1.3187 + i(0.9124)$,

$17^{1/6}\left(\cos\left(\dfrac{\theta + 4\pi}{3}\right) + i\sin\left(\dfrac{\theta + 4\pi}{3}\right)\right) \approx -0.1308 - i(1.5981)$

INDEX